BIOCHEMICAL JOURNAL

REVIEWS 1993

BIOCHEMICAL JOURNAL
REVIEWS 1993

edited by W.H. Evans and A.E. Pegg

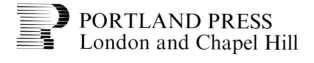

PORTLAND PRESS
London and Chapel Hill

The *Biochemical Journal* is published by Portland Press
on behalf of the Biochemical Society

Portland Press Ltd.
59 Portland Place
London W1N 3AJ
U.K.

Portland Press Inc.
P.O. Box 2191
Chapel Hill
NC 27515-2191
U.S.A.

ISBN 1 85578 040 2

British Library Cataloguing-in-Publication Data
A catalogue record for this book
is available from the British Library

Cover illustration: Putative mode of operation of the glucose
transporters based on molecular dynamics simulations of GLUT 1.
Top, outside glucose binding site open and inner binding site closed;
bottom, closed site outside and open site inside. For more details see
the review by Gould and Holman in this compendium. Illustration
reproduced by kind permission of Dr. G.D. Holman.

Printed in Great Britain by the University Press, Cambridge

CONTENTS

Biochem. J. (1993) **290**, 297–308 (Printed in Great Britain)

REVIEW ARTICLE

The inducible transcription factor NF-κB: structure–function relationship of its protein subunits

Stefan GRIMM and Patrick A. BAEUERLE*

Laboratory for Molecular Biology of the Ludwig-Maximilians University Munich, Gene Center, Am Klopferspitz 18a, D-8033 Martinsried, Federal Republic of Germany

INTRODUCTION

At the level of DNA, gene regulation is governed by *cis* regulatory elements (for a recent review, see Roeder, 1991). Most important for the control of transcription initiation are promoter elements that serve to provide an oriented entry site for DNA-dependent RNA polymerases. In contrast to prokaryotes, eukaryotic polymerases require several additional polypeptides binding in close proximity to the promoter in order to allow transcription initiation. Polymerase II (Pol II), which transcribes most genes in eukaryotes, needs promoter-bound TATA (TF-IID) or initiator binding proteins. A multitude of accessory factors, including TF-IID-associated factors (TAFs), TF-IIA, TF-IIB, TF-IIE and TF-IIF, is required for assembly of a functional transcription initiation complex containing Pol II. Finally, additional proteins binding upstream from TATA and initiator elements can improve the efficacy of a given promoter. Intensive research is going on in order to characterize functionally and structurally the transcription factors required in eukaryotic cells for the initiation of basic transcription.

Despite their complexity, core promoter elements usually provide very little specific regulatory information. Specific regulatory programs are conferred to genes by additional *cis*-regulatory elements, called enhancers. These are frequently found upstream from promoter elements but also in introns or downstream of genes. Enhancers can dramatically enhance the activity of promoters. The relative positional flexibility of enhancers with respect to the invariable position of a promoter might have its basis in the flexibility of DNA segments looping out between physically interacting enhancer- and promoter-binding proteins.

A great diversity of specific DNA-binding proteins are responsible for the specific regulatory potential of upstream promoter and enhancer elements. They can be grouped into transcription factors with activating or repressing potential. Some factors can display both activities. In addition, there are transcription factors that seem to serve an accessory role in sustaining and thereby controlling the effects of activators and repressors. The gene regulatory potential is further augmented by combination of multiple factor binding sites within enhancers and upstream promoters.

Gene regulatory programs are governed by the activity of transcription regulatory proteins. Among the various strategies that have evolved to control transcription factor activity, a common one is the *de novo* synthesis of a transcription factor. But this strategy calls for yet another factor(s) to turn on the gene. This leads into gene networks and hierarchies of transcription factors, a theme frequently exploited during differentiation and determination processes. For rapid gene induction in response to environmental signals, many post-translational modes of transcription factor activation have

evolved. The control of DNA binding via association of transcription factors with small diffusible ligands is a common theme in prokaryotes and, in eukaryotes, is the basis of how the many members of the steroid hormone receptor superfamily are activated. Another widespread mechanism is covalent modification of factors, for instance, by addition or removal of phosphoryl groups (reviewed by Hunter and Karin, 1992). Recently, the role of accessory proteins in gene regulation has received great attention (reviewed by Shaw, 1990). A paradigm is serum response factor (SRF), a transcription regulatory protein conferring to genes responsiveness to serum stimulation of cells. SRF is dependent in this property on a second polypeptide, called ternary complex factor, that associates with SRF and a DNA sequence adjacent to the SRF-binding motif. While this is a nuclear event, other transcription factors can comprise the cytoplasmic compartment for their process of activation (reviewed in Schmitz et al., 1991). This allows the factor to participate actively in cytoplasmic/nuclear signalling. Examples are the glucocorticoid receptor (Muller and Renkawitz, 1991), IGSF-3 (Levy et al., 1989) and NF-AT (Crabtree, 1989). A particularly well-studied system in which the activation of a specific transcription factor requires derepression of DNA binding and inducible nuclear uptake is NF-κB. The activity of this factor is controlled by at least three functionally distinct protein subunits. Previous reviews have covered the physiology of NF-κB and its relationship to structurally homologous proteins (Baeuerle and Baltimore, 1991; Baeuerle, 1991; Nolan and Baltimore, 1992; Blank et al., 1992; Grilli et al., 1992). A particular focus of this Review are functional and structural aspects of NF-κB subunits.

NF-κB is a transcription factor that is activated in many different cell types following a challenge with primary (viruses, bacteria, stress factors) or secondary pathogenic stimuli (inflammatory cytokines). The active factor then leads to a rapid induction of genes encoding defence and signalling proteins, suggesting that NF-κB has specialized during evolution as an immediate early mediator of immune and inflammatory responses. There is now increasing evidence that NF-κB and related proteins are also involved in growth control (Gilmore, 1991; Ohno et al., 1991; Neri et al., 1991; Narayanan et al., 1992).

THE DNA-BINDING SUBUNITS: FUNCTION AND STRUCTURE

DNA binding and dimerization

NF-κB was first described as an activity specifically retarding in electrophoretic mobility shift assays (EMSAs) DNA fragments containing the decameric DNA sequence motif 5'-GGGACTTTCC-3' (Sen and Baltimore, 1986a). This NF-κB binding site, called the B motif, was identified as a B-cell-specific

---

Abbreviations used: NF, nuclear factor; TF, transcription factor; TAF, TF-IID-associated factor; SRF, serum response factor; EMSA, electrophoretic mobility shift assay; NRD, NF-κB/Rel/dorsal domain; NLS, nuclear location signal.
*To whom correspondence and reprint requests should be addressed.

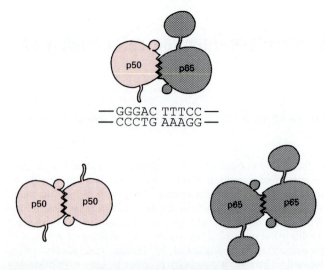

Figure 1   Homo- and hetero-dimeric NF-κB complexes

The p50 subunit is shown with an appendix corresponding to a glycine-rich linker sequence (see Figure 3), a blob corresponding to a nuclear location signal, and a zig-zag line corresponding to a dimerization motif. The additional blob on p65 represents the *trans*-activating C-terminal sequences. The DNA sequence under the p50–p65 heterodimer is found in the enhancers of the immunoglobulin κ light chain gene and in two copies in the HIV-1 LTR. The p50–p65 heterodimer is shown in its preferred orientation over its binding site.

element in the intronic κ light chain enhancer (Lenardo et al., 1988). Soon, it became evident that the element is also functional in pre-B and other cell types, however, not as constitutive but as phorbol ester- and lipopolysaccharide-inducible enhancer element (Sen and Baltimore, 1986b; Nabel and Baltimore, 1987; Pierce et al., 1988). NF-κB is now recognized as ubiquitous factor that occurs, with the exception of a few cell types, in an inducible form requiring certain stimuli in order to appear in nuclei in a DNA-binding form. The activation of NF-κB is independent of protein synthesis (Sen and Baltimore, 1986b). A treatment of cytoplasmic fractions with detergents resulted in a cell-free activation of NF-κB (Baeuerle and Baltimore, 1988a). These findings indicated that the activation of NF-κB involves post-translational mobilization of a sequestered cytoplasmic form.

DNA affinity purification of NF-κB from human cell lines (Kawakami et al., 1988; Baeuerle and Baltimore, 1989), human placenta (Zabel et al., 1991) or rabbit lung (Ghosh et al., 1990) using double-stranded multimers of the recognition sequence 5′-GGGACTTTCC-3′ yielded in each case two polypeptides with apparent molecular sizes of 50 and 65 kDa, referred to as p50 and p65. Reconstitution experiments showed that NF-κB forms a multisubunit complex containing p50 as well as p65 (Baeuerle and Baltimore, 1989). Initial u.v.-crosslinking studies and renaturation experiments using SDS-gel-purified subunits indicated that only p50 has κB-specific DNA binding activity (Kawakami et al., 1988; Baeuerle and Baltimore, 1989). An improved renaturation method and the use of DNA probes for u.v.-crosslinking, in which only one half-site of the decameric motif was photoreactive and radioactively labelled, allowed direct demonstration that the purified p65 subunit has κB-specific DNA binding activity on its own, and that, in NF-κB, both subunits contact the DNA (Urban et al., 1991). Also the p65 protein produced by *in vitro* translation (Ruben et al., 1991) or by baculovirus-infected insect cells (Fujita et al., 1992) could bind to DNA. In another study, DNA binding activity of p65

was only detected after C-terminal truncation of the protein (Nolan et al., 1991).

U.v.-crosslinking experiments showed that in NF-κB p50 and p65 contact DNA as a heterodimer, and glycerol gradient centrifugation analysis suggested that p50 and p65 form a heterodimer in solution (Urban et al., 1991) (Figure 1). A heterotypic dimerization of subunits is also observed with many other transcription factors binding to palindromic sequence motifs (reviewed in Lamb and McKnight, 1992). In the κB sequence 5′-GGGAAATTCC-3′ from the β-interferon enhancer, the p50 subunit in NF-κB preferred binding to the first half-site containing the three GC pairs (Urban et al., 1991). p65, on the other hand, showed a preference for the second half-site, which is usually more degenerate when κB motifs from known target genes are accordingly aligned and compared (Zabel et al., 1991; Baeuerle, 1991). The differential half-site recognition by p50 and p65 was also evident from DNA-binding assays using duplicated half-sites as competitor oligonucleotides (Urban et al., 1991). Upon gel filtration, NF-κB eluted with a size larger than that of immunoglobulin G (Baeuerle and Baltimore, 1989). It is therefore possible that NF-κB can form higher order complexes (for instance, a tetramer), but this awaits further analysis.

p50 and p65 subunits of NF-κB can also form homodimers (Figure 1). This is evident from the following observations. (i) The sedimentation coefficient of p50–p65 NF-κB was intermediate to that of isolated p50 and p65 (Urban et al., 1991). (ii) The mobility of a p50–p65–DNA complex in EMSAs was intermediate to that of a faster-migrating complex containing only p50 and a slower-migrating complex containing only p65. (iii) Isolated p50 and p65 subunits were u.v.-crosslinked equally well to both half-sites of the motif 5′-GGGAAATTCC-3′.

p50 homodimers can occur as constitutive factor in nuclei of certain cell types (Kieran et al., 1990; Kang et al., 1992). The DNA-binding activity of p50 was discovered in parallel to that of NF-κB and is referred to in the literature as KBF-1 (Israel et al., 1987; Kieran et al., 1990), EBP1 (Clark et al., 1990) or H2TF1 (Baldwin and Sharp, 1988). H2TF1 is now recognized to be a distinct factor (A. Baldwin, personal communication). It was noted that the p50 homodimer has a higher affinity for a palindromic 11-bp motif from the enhancer of MHC class I gene (5′-GGGGATTCCCC-3′) than it has for the less symmetric decameric κB motif 5′-GGGACTTTCC-3′ (Kieran et al., 1990; Urban and Baeuerle, 1991). This finding supports the notion that the slight asymmetry of most κB motifs might have evolved to bind preferentially p50–p65 heterodimers. Isolation of ideal binding motifs from a pool of random oligonucleotides using recombinant p50 homodimer showed that p50 prefers binding to highly symmetric GC-rich 11-bp motifs with the consensus 5′-GGGGPuNT/GPyCCC-3′ (Kunsch et al., 1992).

Oligonucleotides selected by p65 homodimers showed a consensus sequence markedly different from that of p50: 5′-(G)GGPuNTTTCC-3′ (Kunsch et al., 1992). There is no apparent requirement for an eleventh base pair and even a decreased requirement for a conserved tenth base pair. The GC content is much lower and the half-site sequence 5′-TTTCC-3′ extremely conserved. NF-κB is a rare example of a dimeric transcription factor in which the DNA-binding subunits have distinguishable DNA-binding specificity. This allows the evolution of *cis*-acting elements preferentially recognized by heterodimers but not by the respective homodimers. It is not yet clear to what extent homodimer-specific binding sites are used for gene regulation.

p65 homodimers were not yet identified in nuclear extracts, even when artificial motifs were used in EMSAs that preferentially bind to purified p65 (Urban and Baeuerle, 1990). When equal amounts of p65 and p50 homodimers are mixed, it takes less than

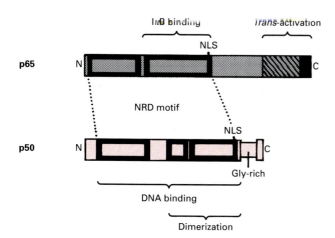

Figure 2   Structure and function of p65 and p50 NF-κB subunits

Bold boxes indicate the highly homologous parts of the NRD motif which are differently interrupted by unique sequences in p65 and p50. Dotted lines connect the N- and C-terminal boundaries of the NRD motif. The filled box in the *trans*-activating C-terminus of p65 indicates TA₁. TA₂ is contained within the shaded area. Both TA regions are separately active. N, N-terminus; C, C-terminus.

10 min at 37 °C for subunit exchange and complete conversion of the homodimers into the heterodimer (M. Urban and P. Baeuerle, unpublished work). This shows that p50 and p65 have a much higher tendency to form heterodimers than homodimers, and that the half life of homodimeric complexes is below 10 min. It is thus unlikely that appreciable amounts of p50 and p65 homodimers can coexist within the cell, unless they are stabilized by additional proteins or DNA.

The various subunit combinations of NF-κB bind to DNA with an extremely high affinity. Dissociation constants (K_D) of 0.4×10^{-12} and 0.9×10^{-12} M, respectively, were determined for p50–p65 and p50–p50 complexes formed with SDS-gel-purified NF-κB subunits from human placenta (Urban and Baeuerle, 1991). Using the same κB motif (5′-GGGACTTTCC-3′), proteins expressed in insect cells by the baculovirus system yielded dissociation constants of 5.7×10^{-12} (p50–p65) and 6.7×10^{-12} M (p50–p50) (Fujita et al., 1992). Depending on the position of the C-terminus (see below), bacterially expressed p50–p50 gave K_D values between 2.6×10^{-12} (443 amino acids) and 8.3×10^{-12} M (503 amino acids) (Kretzschmar et al., 1992). With p50–p65 NF-κB purified from the human cell line HeLa, the same workers found a K_D of 1.3×10^{-12} M. Given the systemic differences and potential sources for errors, the various numbers are in reasonable agreement and document an extremely high affinity of NF-κB for its cognate motif. In one case, a dissociation constant was determined for a p65 homodimer and amounted to 32.2×10^{-12} M (Fujita et al., 1992). This comparatively low affinity might explain why much less protein–DNA complex is usually obtained in EMSAs with p65 compared to similar amounts of p50 or NF-κB (see Urban et al., 1991; Schmitz and Baeuerle, 1991; Nolan et al., 1991). Very similar affinity constants for p50, p65 and NF-κB as above were obtained when three other κB motifs were tested (Fujita et al., 1992). The very low abundance of NF-κB in cells (Lenardo et al., 1989; Henkel et al., 1992) might be one reason why NF-κB requires such an extreme affinity for its cognate motifs.

It was noted that the position of the binding site 5′-GGGACTTTCC-3′ within circularly permutated DNA fragments strongly influenced the mobility of NF-κB–DNA complexes while the mobility of the uncomplexed DNA was not affected (Schreck et al., 1990). Such mobility changes are indicative for an induced alteration of the DNA structure at the site of protein binding. The relative temperature-independence of the mobility effect suggested that NF-κB caused DNA bending rather than an increased flexibility of the DNA. The estimated bend angles of DNA were more than 100 ° for complexes containing p50–65 and p65–65, and about 57 ° for p50–p50 DNA complexes. Polycations or a site-specific cleavage of the DNA backbone close to the κB motif strongly facilitated DNA binding of NF-κB. Presumably, this is due to a lowered energy requirement for bending which, otherwise, needs to be overcome upon binding of the protein.

The primary structures of p50 and p65

The similar DNA-binding specificity and the homo- as well as hetero-dimerization properties of p50 and p65 have their molecular basis in a 300-amino-acid-long region of sequence similarity (Kieran et al., 1990; Ghosh et al., 1990; Ruben et al., 1991; Nolan et al., 1991; Meyer et al., 1992) (Figure 2). The homology region was found much earlier to be shared by the v-*rel* oncogene product from the avian retrovirus REV-T and the morphogenic protein dorsal from the fruit fly *Drosophila melanogaster* (reviewed in Govind and Steward, 1991). However, the function of this region in the viral and fly proteins remained unknown until it was discovered and functionally analysed in the NF-κB subunits. Now, this region of sequence similarity, which we refer to as NF-κB/Rel/dorsal (NRD) domain, is recognized as the minimal requirement for DNA binding and dimerization of the proteins and defines a novel family of transcription factors (reviewed in Blank et al., 1992; Nolan and Baltimore, 1992). The other members of this family and their relationship to NF-κB will be briefly discussed below.

Compared to dimerization and DNA-binding motifs of other transcription factor families, sharing, for instance, basic leucine zipper, helix–loop–helix or homeo domains, the NRD domain is unusual in that it requires a fairly long and intact stretch of approximately 300 amino acids. As demonstrated by coimmunoprecipitation experiments with deleted forms of p50, the C-terminal half of the NRD domain in p50 is sufficient for dimerization (Logeat et al., 1991) (Figure 2). This points to a role of the N-terminal half in contacting the DNA. Preliminary results indeed suggest that the fine differences of DNA-binding specificity between p50 and p65 can be exchanged between the subunits by swopping less than 30 N-terminal amino acids of the NRD domains (S. Grimm and P. Baeuerle, unpublished work). Within this sequence, a highly conserved cysteine residue (position 62) was shown to interfere with DNA binding of p50 upon mutation or oxidation *in vitro* (Matthews et al., 1992). This might explain the earlier observation by Toledano and Leonard (1991) that the DNA binding of NF-κB can be reversibly controlled *in vitro* by oxidation/reduction. Similar observations were made with the homologous cysteine residue of the related c-Rel protein (Kumar et al., 1992). The physiological significance of these observations is not known.

Both p50 and p65 have at the C-terminal end of their NRD domains a cluster of positively charged amino acid residues: Arg-Lys-Arg-Gln-Lys and Lys-Arg-Lys-Arg, respectively. Similar sequences in nuclear proteins were reported to serve as signals for receptor-mediated nuclear uptake and are referred to as nuclear location signals (NLS) (for review see Garcia-Bustos et al., 1990) (Figure 2). Upon separate overexpression of p50 and p65, both proteins were detected in nuclei (Blank et al., 1991; Henkel et al., 1992; Beg et al., 1992; Zabel et al., 1993). While p50 was exclusively nuclear (Blank et al., 1991; Henkel et al.,

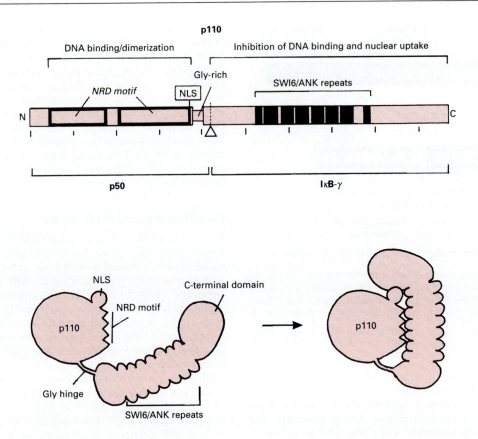

Figure 3   Structure and function of p110, the precursor for p50

The upper part of the panel shows the primary structure of p110. The bars indicate distances of 100 amino acids. The N-terminal (N) portion of p110 encompasses p50 while the C-terminal (C) portion contains IκB-γ. The NRD motif is shown as bold boxes and the eight SWI6/ANK repeats as filled boxes. Repeats 1 and 8 are more degenerate than the other six repeats. The NLS of p50 is boxed and a putative processing site indicated by an arrowhead. The lower left part of the panel shows a three-dimensional model of p110 highlighting the flexible linkage of p50 and the C-terminal domain such that it can properly mask the NLS upon intramolecular folding (lower right panel). Note that for interference with dimerization (NRD motif) fewer repeats are required, consistent with data from deletion experiments.

1992), p65 was at low concentrations cytoplasmic, but nuclei could be 'filled up' with p65 when higher amounts of p65 were expressed (Zabel et al., 1993). It seemed that p65 was retained by an endogenous activity in the cytoplasm (IκB?) which could be titrated out by overexpression of p65. Mutational alterations of basic amino acids in the putative NLS sequences of p50 and p65 into uncharged residues resulted in the accumulation of p50 and p65 in the cytoplasm, even when the proteins were expressed to high levels. This shows that the cluster of positively charged amino acids, which is conserved among all members of the NRD family of proteins, is indeed part of a NLS.

Isolation of the cDNA coding for p50 led to another surprise: p50 is not synthesized as active DNA-binding protein, but is contained in the N-terminal half of a non-DNA-binding precursor of 110 kDa, referred to as p110 (human p110 has 969 amino acids) (Kieran et al., 1990; Ghosh et al., 1990; Bours et al., 1990; Meyer et al., 1991) (Figure 3). The C-terminal half of p110 contains yet another sequence motif discovered earlier in several proteins involved in cell cycle control and cell architecture (reviewed in Blank et al., 1992). This sequence motif with a length of 30–33 amino acids, called SWI6/ankyrin (ANK) repeat, is reiterated eight times in p110. The precise cleavage site in p110 has not yet been identified. Experiments using recombinant p50 must therefore rely on a p50 product with an estimated C-terminus. The protease responsible for p110 processing is thought to be related to a ubiquitin-dependent enzyme (Fan and Maniatis,

1991). The natural cleavage site of p110 is close to a sequence highly enriched in glycine residues that could potentially serve as a 'hinge' region between the two halves of p110 (Figure 3; see below).

Are there functional differences between p50 and p65? The p50 molecule has, apart from the NRD domain, very little extra sequence, whereas p65 has a C-terminal extension of unique protein sequence (Figure 2). In addition to DNA-binding properties, transcription factors are known to require *trans*-activation domains, which are thought to interact with components of the basic transcription machinery. We therefore tested p50 and p65 for their *trans*-activating potential. This was perfomed in two ways. Firstly, *trans*-activation of a κB-controlled reporter gene construct by overexpressed p50 and p65 homodimers was tested. Secondly, p50 and p65 sequences were fused to the DNA-binding domain of the yeast GAL4 protein and the fusion proteins examined for *trans*-activation of a GAL4-controlled reporter gene construct.

Upon separate overexpression of p50 and p65 in COS cells, the respective p50–p50 and p65–p65 complexes were detected by EMSA (Schmitz and Baeuerle, 1991). Although the cell line contained endogenous NF-κB, this was not active and the endogenous NF-κB subunits could not detectably form heterodimers with the overexpressed proteins. In transient transfection assays using a chloramphenicol acetyltransferase (CAT) reporter plasmid with two κB sites, *trans*-activation was exclusively observed with p65. This was also reported by other laboratories (Ruben et al., 1992; Perkins et al., 1992; Ballard et al., 1992). Only when the *trans*-activation domain of the herpes virus protein vp16 was fused to the C-terminus of p50 could the protein induce κB-dependent reporter gene expression (Perkins et al., 1992; S. Grimm and P. Baeuerle, unpublished work). Also, when sequences of p50 and p65 were linked in full length or as fragments to the GAL4 domain, only sequences derived from p65 showed induction of a GAL4-controlled CAT reporter gene (Schmitz and Baeuerle, 1991; Ruben et al., 1992). The NRD domains of p50 and p65 were inactive, whereas C-terminal sequences from p65 could strongly stimulate transcription.

A fine mapping identified the most C-terminal 30 amino acids of p65 as a strong and independent *trans*-activation domain, called TA_1. The N-terminally adjacent 100 amino acids also showed activity and were referred to as TA_2. TA_2 contains a sequence motif homologous to TA_1 (TA_1') which is, however, more dependent on flanking sequences for independent activity than is TA_1 (M. L. Schmitz and P. Baeuerle, unpublished work). The negatively charged TA_1 and TA_2 have a high probability to be present in an α-helical conformation. In the case of TA_1 this structure prediction was supported by c.d. analysis of a synthetic peptide (M. L. Schmitz, M. dos Santos Silva and P. Baeuerle, unpublished work). Breakage of the α-helix by introduction of a proline residue or changing its surface by addition or deletion of one alanine residue in its middle strongly impaired the activity of TA_1, suggesting that the α-helix is an important element for recognition of p65 by adaptor proteins. The 'squelching' effect observed upon expression of vp16 indicated that TA_1 and TA_1' belong, like vp16, to the class of acidic activators.

When p50 is expressed in COS cells to higher levels than p65, a strong repression of the κB-dependent *trans*-activation is observed (Schmitz and Baeuerle, 1991). Because transcriptionally inactive p50 homodimers bind to the same sites as the *trans*-activating p50–p65 and p65–p65 complexes, this effect might come from p50 homodimers occupying limited binding sites. This idea found strong support in the observation that a GAL4–p65 chimaera, which displays a dual DNA-binding specificity, was only affected by p50 overexpression in its κB-dependent but not GAL4-dependent *trans*-activating activity (Schmitz and Baeuerle, 1991). The negative regulatory effect of p50 homodimers on IL-2 promoter activity was recently proposed to play a physiological role during T cell activation (Kang et al., 1992).

An intriguing finding is that p50 homodimers are transcriptionally active in cell-free transcription systems (Kretzschmar et al., 1992; Fujita et al., 1992). The addition of recombinant p50 produced in *Escherichia coli* or insect cells to *in vitro* assays strongly stimulated κB-dependent initiation of mRNA synthesis. The stimulation was strongest with the palindromic motif from the MHC class I enhancer and almost undetectable with that from the β-interferon enhancer (Fujita et al., 1992). It was proposed that the binding sites influenced the conformation of bound p50 such that a *trans*-activation domain is either exposed or sequestered. A conformational alteration of p50 dimers was indeed evident from the distinct protease susceptibility of p50 complexes formed on different cognate sequences. The reason for the opposite activities of p50 in assays *in vitro* and transient transfection assays using intact cells remains unclear.

In conclusion, the p65 subunit in NF-κB serves for strong transcriptional activation of genes, whereas a major function of the p50 subunit is to associate with p65 in order to form a heterodimer that binds with increased affinity to DNA. Thus, p50 can be considered as a 'helper' subunit imposing a limited

regulation on the *trans*-activating p65 subunit by increasing its affinity for DNA. This strategy is not without precedent. c-Myc, a *trans*-activator and proto-oncogene product, binds poorly to DNA (Blackwood et al., 1992). For high-affinity DNA binding it requires Max, a transcriptionally inactive heterodimerization partner. In the following section, additional subunits of NF-κB will be described that impose a very tight negative control on the activity of the DNA-binding subunits.

THE INHIBITORY IκB SUBUNITS: STRUCTURE AND FUNCTION

Purification and specificity

In non-stimulated cells, NF-κB DNA binding activity is not detectable in nuclear, cytosolic or membrane fractions. However, if cytosolic fractions are treated with the ionic detergent sodium deoxycholate, followed by a chase with the non-ionic detergent Nonidet P-40, the DNA-binding activity of a κB-specific factor can be generated. This *in vitro*-activated cytoplasmic factor is identical to the NF-κB found in nuclei of activated cells, as shown by purification, DNA-binding analyses and partial protein sequencing of p50 and p65 subunits (Baeuerle and Baltimore, 1989; Kieran et al., 1990; Ghosh et al., 1990; Zabel et al., 1991; Nolan et al., 1992). The treatment with deoxycholate released an activity from NF-κB that, upon re-addition, reversibly inhibited the DNA-binding activity of NF-κB (Baeuerle and Baltimore, 1988b). The inhibiting factor, termed inhibitor of NF-κB (IκB), could not interfere with the DNA-binding of any other nuclear factor tested. Treatment with deoxycholate (Baeuerle and Baltimore, 1989) or low pH (Zabel and Baeuerle, 1991) allowed dissociation of NF-κB and IκB and subsequent purification of NF-κB from cytosol by DNA affinity chromatography, as well as purification of IκB by conventional column chromatography methods (Zabel and Baeuerle, 1990; Ghosh et al., 1990; Ghosh and Baltimore, 1990; Link et al., 1992).

In our laboratory, two chromatographically distinct IκB variants were isolated from human placenta (Zabel and Baeuerle, 1990). IκB-α had an apparent molecular size of 37 kDa, very similar to the IκB variant isolated from rabbit lung (Ghosh and Baltimore, 1990), while IκB-β had a size of 43 kDa (Link et al., 1992). Both isoforms had isoelectric points between 4.8 and 5. A c-Rel-associated IκB was immunoisolated from chicken cells. The 40 kDa phosphoprotein, termed pp40, was immunologically related to human IκB-β (Kerr et al, 1991). There are two substantial differences between IκB-α and -β, suggesting that the proteins come from different genes. (i) While IκB-α was specific for NF-κB, IκB-β could, in addition, inhibit the DNA binding of the related c-Rel protein (Kerr et al., 1991). (ii) While *in vitro* treatments with protein kinases A and C abolished the inhibitory activity of both variants, a phosphatase treatment interfered only with the inhibiting activity of purified IκB-β (Kerr et al., 1991; Link et al., 1992). The physiological relevance of the *in vitro* phosphorylation data is presently unknown. The various studies have recently been reviewed in detail (Schmitz et al., 1991).

In NF-κB, both IκB variants bind preferentially to the p65 rather than the p50 subunit. This is evident from the following observations. (i) An excess of p65, but not p50, can prevent inhibition of NF-κB by purified IκB proteins (Urban and Baeuerle, 1990). Likewise, addition of p65 to an inactive complex of NF-κB and IκB leads to the release of active NF-κB. (ii) IκB proteins can only inhibit the DNA binding of NF-κB and p65 homodimers but not that of p50 homodimers (Baeuerle and Baltimore, 1989; Urban et al., 1991). (iii) p65 but not p50 can bind stoichiometric amounts of the IκB protein (see below), as tested by coimmunoprecipitation (Zabel et al., 1993). These results could mean that IκB proteins bind to sequences unique for p65 and not to the homologous NRD domain. However, Nolan et al. (1991) showed that a portion of p65 encompassing the NRD motif is susceptible to IκB inhibition. This suggests that the NRD domain, in addition to binding DNA and a second DNA-binding subunit, also interacts with IκB. The binding sequences for IκB were mapped to the C-terminal half of the NRD domain (Beg et al., 1992). This explains why IκB-β can inactivate both p65 and c-Rel, which share sequence similarity only within the NRD domain. Interestingly, antibodies to p50 could coimmunoprecipitate small amounts of MAD-3. Furthermore, a high excess of recombinant IκB (MAD-3) interfered with DNA binding (Liou et al., 1992) and nuclear uptake of p50 homodimers (Beg et al., 1993; Zabel et al., 1993). These observations suggest that the NRD domain of p50 also has a weak affinity for IκB proteins which, at physiological concentrations of the proteins, might not be relevant. Bcl-3 and IκB-γ are IκB proteins specifically binding to the NRD domain of p50 (see below).

Molecular cloning of the IκB proteins

As will be described in a separate section, the first cloned IκB protein was the p50 precursor which contains in its C-terminal half an IκB protein called IκB-γ. IκB-γ can arise from alternative splicing and has specificity for p50. When macrophages adhere to their substratum, they newly express a protein called MAD-3 (Haskill et al., 1991). Molecular cloning and *in vitro* translation revealed that human MAD-3 had an apparent molecular size of 35 kDa and contains five ankyrin repeats. The size of MAD-3, and its sequence similarity to the C-terminal portion of the p50 precursor, prompted the investigators to test MAD-3 for IκB-like activity. The protein could indeed inhibit specifically the DNA-binding activity of NF-κB and c-Rel *in vitro*. Tewari et al. (1992) isolated a MAD-3-encoding cDNA clone from rat liver as one induced upon hepatectomy, and called the protein RL/IF-1. In parallel, a cDNA clone encoding the chicken pp40 protein was isolated by immunoscreening of an expression library (Davis et al., 1991). pp40 was highly homologous to human MAD-3 and showed the same inhibiting specificity, suggesting that pp40 is the chicken homologue of MAD-3. Partial amino acid sequence indicated that the IκB protein purified earlier from rabbit lung (Ghosh and Baltimore, 1990) was highly related if not identical to MAD-3.

Another protein tested for IκB-like activity because of its SWI6/ANK repeats is the proto-oncogene product Bcl-3. Bcl-3 was discovered as being encoded adjacent to a translocation breakpoint on human chromosome 19 associated with chronic lymphocytic leukaemia (Ohno et al., 1990). The protein has an apparent molecular size of 47 kDa and contains seven SWI6/ANK repeats. Also Bcl-3 showed IκB-like activity (Hatada et al., 1992; Wulczyn et al., 1992). It was, however, specific for the p50 homodimer and could not significantly inhibit formation of complexes of NF-κB or c-Rel with DNA. Phosphopeptide mapping with *in vitro* $^{32}PO_4$-labelled Bcl-3, MAD-3 and placental IκB-α showed that IκB-α was unrelated to MAD-3 but shared phosphopeptides with Bcl-3 (Kerr et al. 1993). In strong support for the idea that IκB-α is part of Bcl-3, N-terminal truncation of Bcl-3 resulted in alteration of its inhibiting specificity. A shortened form of Bcl-3 was a specific inhibitor of p65 but not of c-Rel or p50 homodimers, as was reported earlier for IκB-α (Kerr et al., 1991). A truncation seems to occur in intact cells because Bcl-3-specific antibodies immunoprecipitated a 37 kDa protein apart from the 50 kDa Bcl-3. Future studies have to explore by what mechanism (proteolysis or alternative splicing) Bcl-3 is processed to yield IκB-α-like activity. Recently we observed that

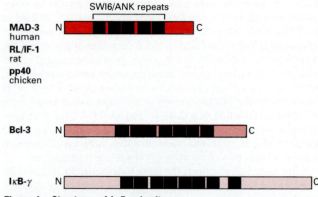

Figure 4   Structures of IκB subunits

The primary structures of IκB proteins are shown. The SWI6/ANK repeats are shown as filled boxes with their real distances. The C-terminal portion of p100 might yet contain another IκB protein, IκB-δ, not listed here. N, N-terminus; C, C-terminus.

monospecific antibodies to MAD-3 immunodeplete IκB-α but not IκB-β activity (Zabel et al., 1993). This suggests that IκB-α is highly related to MAD-3 and that IκB-β might not yet be cloned. Further studies are required to define the relationship between IκB-α, MAD-3 and Bcl-3. An interesting possibility is that Bcl-3 serves as a gene activator by relieving the negative regulatory effect of p50 homodimers occurring when p50 dimers occupy binding sites for transcriptionally active complexes. In transfection experiments using a reporter gene controlled by the HIV-1 enhancer, this activity of Bcl-3 was indeed demonstrated (Franzoso et al., 1992).

Structural features of IκB proteins

The primary structures of MAD-3 (pp40; RL/IF-1) and Bcl-3 have a few common features (Figure 4). The SWI6/ANK repeats are clustered in the middle of the molecules and are flanked by sequences rich in acidic, hydroxyl, proline and glycine residues. Nolan and Baltimore (1992) noted upon aligning the SWI6/ANK repeat domains from IκB-like proteins that the repeats show much greater sequence similarity when compared with respect to their position among the different proteins then when compared within one molecule. For instance, the first repeats in each protein were much more similar to each other than repeat 1 and 2 within the same protein. This might indicate that the repeats have individual functions, a question that can be addressed by swopping experiments.

The pp40 protein was subjected to a deletion and mutational analysis (Inoue et al., 1992a). The most highly conserved sequence within the SWI6/ANK repeats was mutated into a stretch of alanine residues. With the exception of repeat 3, this abolished in each case the capability of pp40 to inhibit DNA binding and to associate with c-Rel and p65. A portion of the molecule encompassing solely the SWI6/ANK repeats was inactive. Only when the C-terminal portion was present in addition to the repeats was pp40 active. Apparently the N-terminal portion was dispensable for the tested activities. These results show that the SWI6/ANK repeats are necessary but not sufficient for the activity of pp40. Very similar results were obtained with MAD-3 (T. Henkel and P. Baeuerle, unpublished work).

The various functions of IκB proteins

It is now well-established that IκB proteins inhibit the DNA-binding activity of proteins with NRD domains. The molecular mechanism of inhibition is, however, poorly understood. There is preliminary evidence that IκB proteins neither simply mask the DNA-binding domain nor interfere with dimerization of DNA-binding subunits. If IκB disrupts the association between p50 and p65, p50 would be released and become detectable as DNA-binding p50 homodimer. This is not the case. Moreover, sizing data suggested that the cytoplasmic NF-κB complex is a heterotrimer composed of p50, p65 and IκB molecules (Baeuerle and Baltimore, 1988b; Zabel and Baeuerle, 1990). It is, however, possible that IκB, by binding to p65, alters its association with p50 resulting in a reduced DNA-binding affinity or altered specificity. Future studies using recombinant proteins should therefore investigate whether the NF-κB–IκB complex retains some novel DNA-binding properties. The related c-Rel protein can apparently form a complex with DNA which contains IκB proteins, as was evident from the immunoreactivity of a c-Rel–DNA complex with antibodies to pp40 (Kerr et al., 1991).

We have observed that IκB proteins can reduce the half life of a NF-κB–DNA complex from 45 min to less than 7 min (Zabel and Baeuerle, 1990). This dissociation followed higher-order kinetics. In cell-free transcription assays, IκB could even disrupt a transcription initiation complex induced by NF-κB, thereby specifically terminating in vitro transcription (Kretzschmar et al., 1992). The findings indicate that IκB proteins do not simply cover the DNA-binding domain of NF-κB but rather exert an allosteric effect on the heterodimer. The IκB protein MAD-3 was found to be predominantly in the nucleus (Zabel et al., 1993). Taken together, these properties would allow IκB proteins to function in the nucleus as inhibitors of NF-κB-dependent transcription.

Apart from inhibiting DNA binding, another function of IκB proteins is to control the nuclear uptake of associated DNA-binding proteins. This was first evident from the findings that the complex of NF-κB with IκB was cytoplasmic upon subcellular fractionation and could not be removed by enucleation procedures from living cells (Baeuerle and Baltimore, 1988b). More direct proof was obtained by indirect immunofluorescence labelling of cells overexpressing MAD-3 and the DNA-binding subunits of NF-κB (Beg et al., 1992; Zabel et al., 1993). When MAD-3 was overexpressed on its own it was present in both cytoplasm and nucleus. It is possible that the protein passively entered the nucleus because its size of 37 kDa is below the cut-off of nuclear pores. When MAD-3 was coexpressed with p65, it completely prevented nuclear uptake of p65. Likewise, p65 interfered with nuclear appearance of MAD-3, suggesting that the two subunits mutually control their access to nuclei. Also, p50 and MAD-3 mutually affected their nuclear uptake; however, a high excess of MAD-3 was required, which is consistent with the much weaker affinity of MAD-3 for p50 (see above).

Because both p50 and p65 contain functional NLS signals, we tested, by the use of antibodies recognizing the NLS epitopes in p50 and p65, whether IκB proteins interfere with the accessibility of NLS sequences for NLS receptors involved in targeting proteins to nuclear pores. Recombinant MAD-3 or purified IκB-α could indeed prevent immunoprecipitation of p65 by anti-p65NLS IgG. Immunoprecipitation with another p65-specific antibody was not influenced by MAD-3. MAD-3 could also not block immunoprecipitation of p50 by anti-p50 NLS. The reactivity of anti-p50 NLS IgG was however affected by MAD-3 when p50 was in complex with p65. This suggests that in the p50–p65 heterodimer one IκB molecule can mask the NLS in both p65 and in p50, although it is bound to only one of the two subunits. Two observations by Beg et al. (1992) are consistent with NLS masking by IκB. These investigators showed that addition of a second NLS from SV40 large T antigen to the N-

terminus of p65 can over-ride the inhibiting effect of IκB on nuclear uptake. Moreover, mutation of all four basic residues in the NLS of p65 abolished binding of IκB. This indicates, but not necessarily proves, that IκB directly binds to the NLS of p65.

In conclusion, IκB proteins are proteins specialized in negatively controlling the DNA-binding subunits of NF-κB/Rel proteins. A particular advantage of such regulatory subunits is that they allow a post-translational induction of transcription factors via mechanisms simply releasing the inhibitors. This is much less time- and energy-consuming than de novo synthesis of transcription factors, as observed with c-Jun, c-Fos and c-Myc. Moreover, there is no requirement for primary transcription activators. Two well-studied functions of IκB proteins are inhibition of DNA binding and nuclear uptake of DNA-binding subunits. A third potential role of IκB proteins is downregulation of κB-dependent gene expression in the nucleus, but this possibility has to await further studies with intact cells.

THE PRECURSOR FOR p50: DNA BINDING AND INHIBITORY SUBUNIT IN ONE MOLECULE

While the IκB subunits for the trans-activating p65 subunit are produced by separate genes, one of the inhibitory subunits for p50 is produced in cis as the C-terminal part of the precursor molecule p110 (Figure 3). An obvious advantage of this strategy is that the inhibitor is always produced in a 1:1 ratio with its target p50. As a consequence, p50 cannot readily appear as active DNA-binding protein and, therefore, cannot operate after its synthesis as constitutive nuclear suppressor of transcription. This is of particular importance in view of the fact that the p50/p110 gene is transcriptionally upregulated by NF-κB (Ten et al., 1991), while the gene coding for the p65 subunit is apparently not (Ruben et al., 1991).

If the C-terminal portion of p110 (IκB-γ) is indeed functionally equivalent to separately encoded IκB proteins, one would expect that it could also interact with p50 in trans. This was demonstrated by coimmunoprecipitation of IκB-γ with p50, and by inhibition of the DNA-binding activity of p50 by bacterially expressed IκB-γ (Hatada et al., 1992; Inoue et al., 1992b; Henkel et al., 1992; Liou et al., 1992). An mRNA species encoding solely IκB-γ was detected in B cell lines (Inoue et al., 1992b; Liou et al., 1992). The alternative splice product allows overproduction of IκB-γ relative to p50 and could serve to control p50 homodimers that have escaped control by the coproduced IκB-γ. A second IκB protein specifically controlling p50 homodimers is Bcl-3 (see above). Future studies have to explore whether there are physiological stimuli that can release IκB-γ or Bcl-3 from p50 in order to allow formation of negative regulatory p50 homodimers.

The association of p50 and IκB-γ shows that there is no requirement for covalent linkage of the two parts of p110 in order to form an inactive complex. A proteolytic event generating p50 and IκB-γ from p110 is undoubtedly required for formation of p50 dimers (or p50–p65 heterodimers), but it appears insufficient. Therefore, protease(s) cleaving p110 must not necessarily be controlled or directly involved in the activation. Consistent with this idea, Fan and Maniatis (1991) presented evidence for the involvement of the constitutive, ubiquitin-dependent protease(s) in processing of p110. Moreover, we were so far unsuccessful in finding a treatment of cells that would enhance the slow conversion of p110 into p50 observed in cells overexpressing p110 (T. Henkel and P. Baeuerle, unpublished work). In such cells, an abundant cytoplasmic non-DNA-binding form of p50 is found, presumably a p50–IκB-γ complex (Henkel et al., 1992) or a p50–p110 complex (Rice et al., 1992). The step ultimately controlling formation of homo- and hetero-dimers containing p50 must therefore involve a mechanism dissociating IκB-γ or p110 from p50.

The protease encoded by HIV-1 was shown to cleave p110 at position 412 *in vitro* and upon HIV-1 infection of cells, which is N-terminal to the physiological site of cleavage (Rivière et al., 1991). Cells infected with HIV-1 showed however no increased amounts of p50 homodimers or nuclear NF-κB, suggesting that the cleaved precursor remained in an inactive form.

p110 sediments through a glycerol gradient with an *s* value of 5.2 S, indicating that it is present as a monomer (provided it has a globular shape). If this is indeed the case, association of p50 and IκB-γ portions would occur intramolecularly. A glycine-rich stretch of 30 amino acids, positioned precisely between the two functionally distinct portions of p110, could provide, as 'hinge', the molecular basis for a spatial approach and intramolecular association of the two functional domains. An alternative model would be that two p110 molecules dimerize via their NRD domains. The IκB-γ portions could then exert their function in *cis* or *trans*. Deletion of an acidic region between SWI6/ANK repeats 7 and 8 of p110 created a mutant protein with DNA-binding activity (Blank et al., 1991). Apparently, the mutation disturbed the presumed intramolecular association of the p50 and IκB-γ portions and caused an 'opening' of the molecule. This allowed dimerization of the NRD domains and DNA binding. Figure 3 shows a model of p110 in 'open' and 'closed' conformation.

Direct evidence for masking of domains within p110 came from an immunological study (Henkel et al., 1992). An antipeptide antibody raised against a C-terminal epitope of p50 was immunoreactive with p50, but not with p110. Only after denaturation or C-terminal truncation of p110 could the antibody recognize the p50 epitope in p110. The observation that the C-terminal 200 amino acids of p110 were sufficient to mask an epitope in the p50 portion that is separated by a linear distance of more than 400 amino acid residues (including a 'flexible' domain) argues strongly for an intramolecular association of large, independent domains in p110.

Deletion analysis of p110 showed that only one SWI6/ANK repeat has to remain with the IκB-γ portion in order to suppress DNA binding of the p50 portion (Kieran et al., 1990). This is surprising given the fact that IκB-γ produced in *trans* has to contain all of its ankyrin repeats in order to maintain its inhibiting activity (Hatada et al., 1992). Possibly, it is the covalent linkage of shortened IκB-γ sequences to p50 that can stabilize, due to the lack of diffusion control, their weak interaction with the p50 portion.

As shown by immunofluorescence studies, p110 is a cytoplasmic protein (Blank et al., 1991; Henkel et al., 1992). Its diffuse cytoplasmic distribution is identical to that of a SV40 large T antigen mutant protein impaired in nuclear transport, and there is no resemblance to the immunostainings observed with antibodies to cytoskeletal proteins. Upon subcellular fractionation, p110 partitions into a 100000 *g* supernatant, suggesting that it is a cytosolic protein. The exclusion of p110 from nuclei was unexpected since p110 contains the NLS sequence of p50. Studies with an antipeptide antibody specific for a sequence overlapping the NLS showed that in p110 this epitope was not accessible, unless the protein was treated with the ionic detergent SDS. The antibody is thought to mimic a physiological receptor(s) involved in recognizing NLS sequences and targetting proteins to nuclear pores. The observation that it reacts with p50, but not p110, suggests that the IκB-γ portion masks the NLS epitope. Immunoprecipitation of C-terminally truncated forms of p110 with the antibody showed that immunoreactivity was restored when only 200 amino acids were deleted from the C-

terminus, which removes only one SWI6/ANK repeat from IκB-γ. When this truncated form of p110 was expressed in cells, it was partially taken up into nuclei. A C-terminal domain of 200 amino acids is apparently sufficient to mask the NLS. The repeats seem not to be directly required for cytoplasmic retention, but might serve to properly position the C-terminal domain, as detailed in Figure 3.

p110 is found in complexes containing v-Rel and c-Rel (Capobianco et al., 1992; Kochel et al., 1991). It will be interesting to find out the stoichiometry and physiological significance of this interaction and what sequences of p110 are binding v-Rel. Can p110 unfold and use its dimerization domain to bind v-Rel, or does the IκB-γ portion have free valencies? Very recent studies provided evidence that the entire p110 molecule has an IκB-like function and controls DNA binding and nuclear uptake of c-Rel and p65 (Rice et al., 1992).

In conclusion, p110 appears as an unusual molecule. It is a very rare example of a non-viral cytoplasmic precursor protein requiring proteolysis for maturation. p110 combines two opposite functions in one molecule: a DNA-binding function with nuclear affinity and a specific inhibitor of the DNA-binding portion and its nuclear affinity. The sequences of p110 containing the two activities can apparently interact within the same molecule. The interaction of p110 with c-Rel and p65 is an intriguing finding; inducible cytoplasmic complexes could form in a single-step reaction.

BIOGENESIS OF NF-κB

An inducible transcription factor composed of several subunits that are encoded by different genes relies very much on a co-ordinate production and assembly of the subunits. In the case of NF-κB, overproduction of p50 would result in a constitutive DNA-binding protein with no or low *trans*-activating potential. On the other hand, overproduction of p65 would result in a constitutive activator bypassing the inducible control imposed by limited IκB. Overproduction of IκB might not be as deleterious as long as reactions inactivating NF-κB-bound IκB are not inhibited by an excess of free IκB. As discussed above, p50 brings along its own inhibitor within a precursor molecule. Therefore, it does not matter that the p110 gene is transcriptionally upregulated by NF-κB (Meyer et al., 1991; Ten et al., 1991); all p110 that is produced in excess over p65 would accumulate in an inactive form and remain as such in the cytoplasm, even after proteolytic processing (Figure 5). The gene encoding p65 seems to be expressed at a very low level (Nolan et al., 1991; Ruben et al., 1991), which is consistent with our finding that p65 protein is barely detectable in Western blots using total protein from various cell types. On the other hand, the IκB protein MAD-3 gives a much stronger signal in Western blots, suggesting that it is present in excess over p65. This imbalance would ensure a tight control of IκB over the *trans*-activating p65. There is good evidence that IκB proteins can inhibit DNA binding and nuclear uptake of p65 that is not yet complexed with p50 (Urban et al., 1991; Beg et al., 1992; Zabel et al., 1993).

The complexes ultimately used for assembly of the inducible heterotrimeric p50–p65–IκB complex in the cytoplasm would then be p50–IκB-γ and p65(homodimer)–IκB(-α or -β/MAD-3) (Figure 5). Future studies must explore whether there is a spontaneous exchange reaction releasing IκB-γ when p50–IκB-γ encounters p65–IκB, or whether additional proteins are required to control the process of assembly. In several studies, intact IκB-γ could not be detected in cells (Fan and Maniatis, 1992; Inoue et al., 1992b; Henkel et al., 1992), which could be explained by an extreme lability of the protein within living cells. By selective

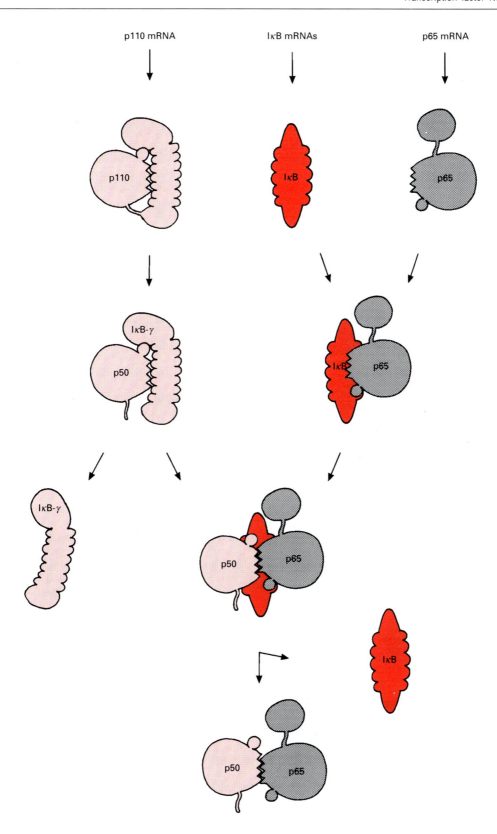

Figure 5   A model for the biogenesis of the inducible form of NF-κB

The inducible cytoplasmic form of NF-κB is assembled from the products of at least three different genes, encoding p110, IκB and p65. While the p110 and MAD-3 genes are inducible, the p65 gene appears to be constitutively expressed at low levels. For reasons of simplicity, the model shows a dimer of p65 and IκB, but a p65 homodimer with one or two bound IκB molecules is more likely to occur. p110 is shown as an intramolecularly folded molecule that, even after proteolysis, does not release p50. This assures that no inhibiting p50 homodimers are formed that could constitutively enter the nucleus. By an as yet unknown process, p50 is released from IκB-γ and incorporated into the complex with p65 and another IκB. Release of the IκB protein finally triggers gene activation by the p50–p65 heterodimer. An alternative pathway is direct association of p65 with p110. This complex must then undergo proteolysis prior to or during activation in order to yield p50.

degradation of IκB-γ, the reaction between p50–IκB-γ and p65–IκB could be shifted towards formation of the p50–p65–IκB trimer.

An alternative and much simpler assembly of an inducible cytoplasmic NF-κB complex could occur via p110. Newly synthesized p65 would in this model associate with p110 present in excess in the cytoplasm. The p110–p65 complex has then to undergo proteolysis in order to generate functional p50. It is presently unknown whether IκB-γ is released from a p50–p65–IκB-γ complex in response to extracellular stimuli.

CONTROL OF NF-κB ACTIVATION

In centre stage of the control of NF-κB transcription factor activity are inhibitory subunits rather than pretranslational regulatory steps. Activation of the factor appears to simply require disruption of the interaction between IκB and DNA-binding subunits (Figure 6). Both types of subunits could potentially serve as targets for dissociating reactions. Because *in vivo* activated nuclear p50–p65 is susceptible to inhibition by purified IκB, IκB rather than p65 seems to be the target (Baeuerle and Baltimore, 1988b). This is supported by the very recent finding that stimulation of pre-B cells with interleukin-1, tumour necrosis factor, phorbol ester and lipopolysaccharide all induce

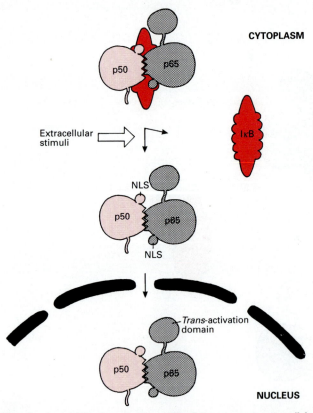

Figure 6   A model for the activation of NF-κB in response to extracellular stimuli

For details see the legends to Figures 1 and 5. The nuclear location signals of p50 and p65 are shown masked by a single IκB subunit. IκB is perhaps exclusively bound via the p65 subunit. Stimulation by a great variety of extracellular signals releases IκB, a reaction not only unmasking the NLS on both p50 and p65 but also restoring nuclear uptake and high-affinity DNA binding of the heterodimer. Finally, binding of NF-κB to enhancer elements in nuclear DNA initiates transcription, a process requiring the strong *trans*-activation domain in the p65 subunit.

a discrete mobility decrease of IκB followed by a depletion of the protein (T. Henkel and P. Baeuerle, unpublished work).

In cell-free systems, various reactions were reported to activate NF-κB through release of IκB, including treatments of NF-κB–IκB complexes with sodium deoxycholate, low pH, protein kinases and protein phosphatases (see above). While detergents and low pH might not be physiological activators, phosphorylation events are known to control protein activities in many biological systems. Pharmacological support for an involvement of protein kinases and phosphatases in NF-κB activation within intact cells comes from the inducing effect of the protein kinase C activator phorbol myristate acetate (Sen and Baltimore, 1986b) and the protein phosphatase inhibitor okadaic acid (Thévenin et al., 1991), and from the inhibiting effect of the tyrosine kinase inhibitor herbimycin A (Iwasaki et al., 1992). However, these findings do not prove a direct phosphoryl transfer onto IκB, as observed by *in vitro* kinase experiments (Ghosh and Baltimore, 1990). The only direct evidence that phosphorylation of IκB proteins influences IκB activity in intact cells comes from the observation that a phosphatase treatment of purified IκB-β and pp40 abolishes their inhibiting activity (Link et al., 1992; Kerr et al., 1991). The proteins were apparently purified in a phospho form and required the bound phosphate in order to bind and inhibit NF-κB. Whether this modification is involved in the process of activation, or rather has a modulatory role, is not known. $^{32}PO_4$-labelling studies with intact cells will allow a demonstration of whether there are changes in the state of phosphorylation of IκB proteins in response to physiological stimuli. Mutational analysis of sites must finally demonstrate the functional significance of any modification identified. Of particular interest will be the question whether the various IκB proteins respond differently to various inducing conditions.

One common intracellular reaction induced by many, if not all, NF-κB-activating stimuli is oxidative stress (reviewed in Schreck et al., 1993). There are now three lines of evidence suggesting that reactive oxygen intermediates, most probably peroxides, play a role in mobilization of NF-κB. (i) NF-κB is post-translationally activated by low concentrations of hydrogen peroxide (Schreck et al., 1991). (ii) Activation of NF-κB in response to all inducing agents tested so far is blocked by a variety of chemically distinct antioxidants (Schreck et al., 1991, 1992, 1993). (iii) Reports in the literature describe induction of oxidative stress by many agents activating NF-κB, for instance, tumour necrosis factor, interleukin-1, phorbol ester, lipopolysaccharide, anti-IgM and u.v. light (reviewed in Schreck et al., 1993). These observations suggested that NF-κB is an oxidative stress responsive transcription factor, and that reactive oxygen intermediates play a messenger function in the activation of the factor. How they can cause the release of IκB is not understood. Direct oxidative modification of IκB, as demonstrated for the prokaryotic factor oxyR (Storz et al., 1991), is one possibility; but the rather slow kinetics of NF-κB mobilization in response to H_2O_2 and the failure to activate NF-κB *in vitro* by treatment with reactive oxygen intermediates (Schreck et al., 1991) calls for other proteins sensing and transducing the signal to the cytoplasmic NF-κB complex. After all, (oxidative stress-responsive) protein kinases are very likely to be involved.

HETEROTYPIC DIMERIZATION AND THE RELATIONSHIP OF NF-κB TO OTHER MEMBERS OF THE FAMILY

There are currently five proteins in higher vertebrates known to contain the NRD domain (for reviews see Blank et al., 1992; Nolan and Baltimore, 1992). p50/p110 and p65 have been described in detail here. p49/p100 (also called p50B) is a protein highly related to p50/p110 (Schmid et al., 1991; Bours et al., 1992; Neri et al., 1992; Mercurio et al., 1992). It is therefore possible that p49/p50B fulfils the same 'helper' function as p50, and that p100 contains an inhibitory activity in its C-terminal half. A protein highly related to p65 is the proto-oncogene c-Rel (reviewed in Gilmore, 1990). Also c-Rel has strong *trans*-activating potential (Bull et al., 1990). Depending on the investigators, the fifth protein is called Rel-B (Ryseck et al., 1991) or I-Rel (Ruben et al., 1992). Rel-B was identified as immediate-early serum responsive factor and found to have *trans*-activating activity. A cDNA encoding I-Rel was isolated with the help of a PCR product amplified with degenerate DNA primers homologous to the NRD domain of p50. Although Rel-B and I-Rel seem to be identical, I-Rel was reported to have inhibitory activity. Future studies should address this controversy.

Apart from forming homodimers, it seems that most NRD family members can *in vitro* form heterodimers among each other, as tested with recombinant or *in vitro* translated proteins by immunoprecipitation or EMSA. Furthermore, all five NRD proteins (and various combinations thereof) can form complexes with the κB motif 5'-GGGACTTTCC-3' and positively or negatively affect *trans*-activation from κB-controlled reporter genes in transient transfection assays. This led to confusion about the composition of 'NF-κB-like complexes' detected in nuclear extracts by EMSA. A similar problem was encountered earlier with the factor AP-1 when it became clear that there is extensive heterotypic dimerization between members of the Jun/Fos/Fra/CREB family of basic/leucine zipper *trans*-activators (reviewed in Lamb and McKnight, 1992). Therefore, in future experiments, the following questions should be addressed. (1) Do all possible heterodimers (and homodimers) of NRD proteins exist in living cells? Or, perhaps, does the controlled assembly of subunits only allow certain subunit combinations? As detailed in this Review, there is very good evidence that the combination of p50 and p65 is of physiological relevance. Also, p65 and c-Rel can apparently form heterodimers in cells, but these seem to have a DNA-binding specificity markedly distinct from that of p50–p65 (Hansen et al., 1992). U.v.-crosslinking studies suggested that p50 and c-Rel can form complexes in activated T cells during a later stage of the activation process (Molitor et al., 1991). (2) Do the NRD proteins vary in their tissue- and cell type-specific expression? If so, only certain combinations would occur in a given cell type, the number of which might be further limited by a controlled assembly. (3) Do physiologically relevant complexes of p50 and p65 with other members of the NRD family have the same sequence specificity as NF-κB and do they bind to κB motifs with the same high affinity? The determination of affinity constants and assay methods to detect DNA target sequences will be required to address this question.

CONCLUSIONS

Future studies on the subunits of the NF-κB system will focus on the fine structure and sequence motifs of the subunits. X-ray crystallography and n.m.r. techniques are required to understand in detail how DNA-binding subunits of NF-κB homo- and hetero-dimerize and how they contact DNA. The techniques will also be helpful to solve the structure and function of SWI6/ANK repeats in the specific interaction of IκB proteins with DNA-binding subunits. Important questions that need to be addressed in the near future are the following. What combinations of NRD proteins do really exist in living cells and what is their specificity? By what mechanism is IκB released from NF-κB upon stimu-

lation? Finally, what is the role of NF-κB and IκB proteins in growth control?

This work was supported by the Bundesministerium für Forschung und Technologie and the Deutsche Forschungsgemeinschaft (SFB 217), and is a partial fulfilment of the doctoral thesis of S. G.

REFERENCES

Baeuerle, P. A. (1991) Biochim. Biophys. Acta **1072**, 68–80
Baeuerle, P. A. and Baltimore, D. (1988a) Cell **53**, 211–217
Baeuerle, P. A. and Baltimore, D. (1988b) Science **242**, 540–546
Baeuerle, P. A. and Baltimore, D. (1989) Genes Dev. **3**, 1689–1698
Baeuerle, P. A. and Baltimore, D. (1991) in Molecular Aspects of Cellular Regulation (Cohen, P. and Foulkes, J. G., eds.) , vol. 6, pp. 409–432, Elsevier/North Holland Biomedical Press
Baldwin, A. S. and Sharp, P. A. (1988) Proc. Natl. Acad. Sci. U.S.A. **85**, 723–727
Ballard, D. W., Dixon, E. P., Peffer, N. J., Bogerd, H., Doerre, S., Stein, B. and Greene, W. C. (1992) Proc. Natl. Acad. Sci. U.S.A. **89**, 1875–1879
Beg, A. A., Ruben, S. M., Scheinman, R. I., Haskill, S., Rosen, C. A. and Baldwin, A. S., Jr. (1992) Genes Dev. **6**, 1899–1913
Blackwood, E. M., Lüscher, B. and Eisenmann, B. (1992) Genes Dev. **6**, 71–80
Blank, V., Kourilsky, P. and Israel, A. (1991) EMBO J. **10**, 4159–4167
Blank, V., Kourilsky, P. and Israel, A. (1992) Trends Biochem. Sci. **17**, 135–140
Bours, V., Villalobos, J., Burd, P., Kelly, K. and Siebenlist, U. (1990) Nature (London) **348**, 76–80
Bours, V., Burd, P. R., Brown, K., Villalobos, J., Park, S., Ryseck, R. P., Bravo, R., Kelly, K. and Siebenlist, U. (1992) WHAT JOURNAL? **12**, 685–695
Bull, P., Morley, K. L., Hoekstra, M. F., Hunter, T. and Verma, I. (1990) Mol. Cell. Biol **10**, 5473–5485
Clark, L., Matthews, J. R. and Hay, R. T. (1990) J. Virol. **64**, 1335–1344
Crabtree, G. R. (1989) Science **243**, 355–361
Davis, N., Ghosh, S., Simmons, D. L., Tempst, P., Liou, H.-C., Baltimore, D. and Bose, H. R. (1991) Science **253**, 1268–1271
Fan, C.-M. and Maniatis, T. (1991) Nature (London) **354**, 395–398
Franzoso, G., Bours, V., Park, S., Tomita-Yamaguchi, M., Kelly, K. and Siebenlist, U. (1992) Nature (London) **359**, 339–342
Fujita, T., Nolan, G. P., Ghosh, S. and Baltimore, D. (1992) Genes Dev. **6**, 775–787
Garcia-Bustos, J., Heitmann, J. and Hall, M. (1990) Biochim. Biophys. Acta **1071**, 83–101
Ghosh, S. and Baltimore, D. (1990) Nature (London) **344**, 678–682
Ghosh, S., Gifford, A. M., Riviere, L. R., Tempst, P., Nolan, G. P. and Baltimore, D. (1990) Cell **62**, 1019–1029
Gilmore, T. D. (1991) Trends Genet. **7**, 318–322
Govind, S. and Steward, R. (1991) Trends Genet. **4**, 119–125
Grilli, M., Chiu, J. J.-S. and Lenardo, M. J. (1992) Annu. Rev. Immunol., in the press
Hansen, S. K., Nerlov, C., Zabel, U., Verde, P., Johnsen, M., Baeuerle, P. A. and Blasi, F. (1992) EMBO J. **11**, 205–213
Haskill, S., Beg, A. A., Tompkins, S. M., Morris, J. S., Yurochko, A. D., Sampson-Johannes, A., Mondal, K., Ralph, P. and Baldwin, A. S. (1991) Cell **65**, 1281–1289
Hatada, E. N., Nieters, A., Wulczyn, F. G., Naumann, M., Meyer, R., Nucifora, G., McKeithan, T. and Scheidereit, C. (1992) Proc. Natl. Acad. Sci. U.S.A. **89**, 2489–2493
Henkel, T., Zabel, U., van Zee, K., Müller, J. M., Fanning, E. and Baeuerle, P. A. (1992) Cell **68**, 1121–1133
Hunter, T. and Karin, M. (1992) Cell **70**, 375–387
Inoue, J.-i., Kerr, L., Rashid, D., Davis, N., Bose, H. R. and Verma, I. M. (1992a) Proc. Natl. Acad. Sci. U.S.A. **89**, 4333–4337
Inoue, J.-i., Kerr, L. D., Kakizuka, A. and Verma, I. M. (1992b) Cell **68**, 1109–1120
Israel, A., Kimura, A., Kieran, M., Yano, O., Kannelopoulos, J., Le Bail, O. and Kourilsky, P. (1987) Proc. Natl. Acad. Sci. U.S.A. **84**, 2653–2657
Iwasaki, T., Uehara, Y., Graves, L., Rachie, N. and Bomsztyk, K. (1992) FEBS Lett. **298**, 240–244
Kang, S.-M., Chen-Tran, A., Grilli, M. and Lenardo, M. J. (1992) Science **256**, 1452–1456
Kawakami, K., Scheidereit, C. and Roeder, R. G. (1988) Proc. Natl. Acad. Sci. U.S.A. **85**, 4700–4704
Kerr, L. D., Inoue, J.-i., Davis, N., Link, E., Baeuerle, P. A., Bose, H. R. and Verma, I. M. (1991) Genes Dev. **5**, 1464–1476
Kerr, L. D., Duckett, C. S., Wamsley, P., Zhang, Q., Chiao, P., Nabel, G., Baeuerle, P. A. and Verma, I. (1993) Genes Dev., in the press
Kieran, M., Blank, V., Logeat, F., Vanderckhove, J., Lottspeich, F., Le Bail, O., Urban, M. B., Kourilsky, P., Baeuerle, P. A. and Israel, A. (1990) Cell **62**, 1007–1018
Kochel, T., Mushinski, J. F. and Rice, N. R. (1991) Oncogene **6**, 615–626
Kretzschmar, M., Meisterernst, M., Scheidereit, C., Li, G. and Roeder, R. G. (1992) Genes. Dev. **6**, 761–774
Kumar, S., Rabson, A. B. and Gelinas, C. (1992) Mol. Cell. Biol. **12**, 3094–3106
Kunsch, C., Ruben, S. M. and Rosen, C. A. (1992) Mol. Cell. Biol. **12**, 4412–4421

Lamb, P. and McKnight, S. L. (1992) Trends Biochem. Sci. **16**, 417–422

Lenardo, M. J., Pierce, J. W. and Baltimore, D. (1987) Science **236**, 1573–1577

Levy, D. E., Kessler, D. S., Pine, R. and Darnell, J. E. (1989) Genes Dev. **3**, 1362–1371

Link, E., Kerr, L. D., Schreck, R., Zabel, U., Verma, I. M. and Baeuerle, P. A. (1992) J. Biol. Chem. **267**, 239–246

Liou, H.-C., Nolan, G. P., Ghosh, S., Fujita, T. and Baltimore, D. (1992) EMBO J. **11**, 3003–3009

Logeat, F., Israel, N., Ten., R. M., Blank, V., Le Bail, O., Kourilsky, P. and Israel, A. (1991) EMBO J. **10**, 1827–1832

Matthews, J. R., Wakasugi, N., Virelizier, J.-L., Yodoi, J. and Hay, R. T. (1992) Nucleic Acids Res. **20**, 3821–3830

Mercurio, F., Didonato, J., Rosette, C. and Karin, M. (1992) DNA Cell Biol. **11**, 523–537

Meyer, R., Hatada, E., Hohmann, H.-P., Haiker, M., Bartsch, C., Rothlisberger, U., Lahm, H.-W., Schlaeger, E. J., van Loon, A. P. G. M. and Scheidereit, C. (1991) Proc. Natl. Acad. Sci. U.S.A. **88**, 966–970

Molitor, J. A., Walker, W. H., Doerre, S., Ballard, D. W. and Greene, W. C. (1990) Proc. Natl. Acad. Sci. U.S.A. **87**, 10028–10032

Muller, M. and Renkawitz, R. (1991) Biochim. Biophys. Acta **1088**, 171–182

Nabel, G. and Baltimore, D. (1987) Nature (London) **335**, 683–689

Nolan, G. P. and Baltimore, D. (1992) Curr. Opin. Genet. Dev. **2**, 211–220

Nolan, G. P., Ghosh, S., Liou, H.-C., Tempst, P. and Baltimore, D. (1991) Cell **64**, 961–969

Perkins, N. L., Schmid, R. M., Duckett, C. S., Leung, K., Rice, N. R. and Nabel, G. J. (1992) Proc. Natl. Acad. Sci. U.S.A. **89**, 1529–1533

Pierce, J. W., Lenardo, M. and Baltimore, D. (1988) Proc. Natl. Acad. Sci. U.S.A. **85**,1482–1486

Rice, N. R., MacKichan, M. L. and Israel, A. (1992) Cell **71**, 243–253

Rivière, Y., Blank, V., Kourilsky, P. and Israel, A. (1991) Nature (London) **350**, 625–626

Roeder, R. G. (1991) Trends Biochem. Sci. **16**, 402–408

Ruben, S. M., Dillon, P. J., Schreck, R., Henkel, T., Chen, C. H., Maher, M., Baeuerle, P. A. and Rosen, C. A. (1991) Science **241**, 89–92

Ruben, S. M., Narayanan, R., Klement, J. F., Chen, C.-H. and Rosen, C. A. (1992a) Mol. Cell. Biol. **12**, 444–454

Ruben, S. M., Klement, J. F., Coleman, T. A., Maher, M., Chen, C.-H. and Rosen, C. A. (1992b) Genes Dev. **6**, 745–760

Ryseck, R.-P., Bull, M., Takamiya, V., Bours, U., Siebenlist, P., Dobrzanski, P. and Bravo, R. (1992) Mol. Cell. Biol. **12**, 674–684

Schmitz, M. L. and Baeuerle, P. A. (1991) EMBO J. **10**, 3805–3817

Schmitz, M. L., Henkel, T. and Baeuerle, P. A. (1991) Trends Cell Biol. **1**, 130–137

Schreck, R., Zorbas, H., Winnacker, E.-L. and Baeuerle, P. A. (1990) Nucleic Acids Res. **18**, 6497–6502

Schreck, R., Rieber, P. and Baeuerle, P. A. (1991) EMBO J. **10**, 2247–2258

Schreck, R., Meier, B., Männel, D., Dröge, W. and Baeuerle, P. A. (1992) J. Exp. Med. **175**, 1181–1194

Schreck, R., Albermann, K. and Baeuerle, P. A. (1993) Free Radical Res. Commun., in the press

Sen, R. and Baltimore, D. (1988a) Cell **46**, 705–716

Sen, R. and Baltimore, D. (1988b) Cell **47**, 921–928

Shaw, P. E. (1990) New Biologist **2**, 11–118

Storz, G., Tartaglia, L. A. and Ames, B. (1990) Science **248**, 189–194

Ten, R. M., Paya, C. V., Israel, N., LeBail, O., Mattei, M.-G., Virelizier, J.-L., Kourilsky, P. and Israel, A. (1992) EMBO J. **11**, 195–203

Tewari, M., Dobrzanski, P., Mohn, K. L., Cressman, P., Hsu, J.-C., Bravo, R. and Taub, R. (1992) Mol. Cell. Biol. **12**, 2898–2908

Thévenin, C., Kim, S.-J., Rieckmann, P., Fujiki, H., Norcross, M. A., Sporn, M. B., Fauci, A. S. and Kehrl, J. H. (1990) New Biologist **2**, 793–800

Toledano, M. B. and Leonard, W. J. (1991) Proc. Natl. Acad. Sci. U.S.A. **88**, 4328–4332

Urban, M. B. and Baeuerle, P. A. (1990) Genes Dev. **4**, 1975–1984

Urban, M. B. and Baeuerle, P. A. (1991) New. Biologist **3**, 279–288

Urban, M. B., Schreck, R. and Baeuerle, P. A. (1991) EMBO J. **10**, 1817–1825

Wulczyn, G., Naumann, M. and Scheidereit, C. (1992) Nature (London) **358**, 597–599

Zabel, U. and Baeuerle, P. A. (1990) Cell **61**, 255–265

Zabel, U., Schreck, R. and Baeuerle, P. A. (1991) J. Biol. Chem. **266**, 252–260

Zabel, U., Henkel, T., dos Santos Silva, M. and Baeuerle, P. A. (1993) EMBO J., in the press

Biochem. J. (1993) **296**, 521–541 (Printed in Great Britain)

REVIEW ARTICLE
Negative regulation of transcription in eukaryotes

Andrew R. CLARK* and Kevin DOCHERTY
Department of Medicine, University of Birmingham, Queen Elizabeth Hospital, Birmingham B15 2TH, U.K.

INTRODUCTION

The first stages of the study of eukaryotic gene expression revolved around the identification of positively acting transcription factors and the investigation of their function. It has since become clear that negative regulation of transcription is an equally vital process, in some way participating in the control of more or less all genes so far studied (for recent reviews, see [1–4]). Transcriptional repression is required for the establishment of the temporally and spatially complex patterns of gene expression which are characteristic of eukaryotes, and is also frequently involved in the modulation of gene expression in response to changes in the micro-environment of the cell. In this review we briefly discuss several mechanisms of negative regulation, before addressing the variety of cellular processes in which these mechanisms are involved.

MECHANISMS OF NEGATIVE REGULATION

The process of transcriptional regulation in eukaryotes is highly complex, allowing several stages at which negative regulatory events may intervene. For the purposes of discussion, some features of transcriptional regulation are illustrated schematically in Figure 1.

The activation of transcription may be controlled by cytoplasmic retention of transcription factors. In the best studied example, members of the rel family of activators are retained in the cytoplasm by interactions with IκB factors. Dissociation of these interactions, and the subsequent migration of rel factors to the nucleus, is controlled by phosphorylation of the IκB factors ([5–7]; also see later). In the absence of its ligand the glucocorticoid receptor (GR) is also retained in the cytoplasm, through interaction with a complex which includes the 90 kDa heat shock protein hsp90 [8]. This should not be regarded as an exclusively negative regulatory event, since hsp90 may assist in the interaction of the glucocorticoid receptor with its ligand [9].

In prokaryotes a common mechanism of negative regulation involves factors which bind near to or overlapping the site of binding of the polymerase complex, thus interfering with the formation of this complex (Figure 1, part 4; [10]). Unlike their prokaryotic counterparts, eukaryotic promoters must often in-

tegrate responses to several afferent signals which regulate transcription both positively and negatively. This kind of integrated response is not possible if the promoter is unoccupied, perhaps explaining why such a 'promoter occlusion' mechanism is infrequently employed by eukaryotes. Apparent examples include the autoregulation of the simian virus 40 (SV40) early promoter by the T antigen, an early-gene product [11], and several instances of negative regulation by nuclear hormone receptors [12–16].

A more commonly employed mechanism involves interference with the binding of transcriptional activators (binding competition; Figure 1, part 5). A negatively acting factor binds to a sequence adjacent to or overlapping the binding site for an activator, and prevents the binding of the activator by steric hindrance [17–23]. This is an essentially passive mechanism characterized by dependence upon the relative positioning of positive and negative sites. The negatively acting factor is required only to bind DNA, and regions other than its DNA-binding domain are dispensible [24,25] or replaceable [26].

Transcription factors belonging to the same family may show similar or identical DNA-binding specficity, and thus may compete for binding sites when present in the same cell (Figure 1, part 9). This is true for instance of the *Drosophila* homeodomain-containing proteins, which overlap extensively in specificity (see later). The cyclic AMP response element-binding factor (CREB) and the activator protein 1 (AP1) factors are related structurally and in DNA-binding specificity. CREB is able to recognize AP1 consensus sites, but not to stimulate transcription in this context, and thus can down-modulate AP1 activity by competition for access to DNA [27]. The antagonism between CREB and AP1 at a non-consensus binding site has been shown to be modulated by cyclic AMP-dependent phosphorylation of CREB [28].

As in the example above, the negatively acting displacement factor has often been previously described as a positive transcription factor, and it is then necessary to explain why the displacer does not activate transcription. A possibility deserving consideration is that the 'negative factor' may simply be the weaker of two activators which bind in a mutually exclusive fashion (see, for instance, ref. [29]). An alternative theory has arisen from studies of negative glucocorticoid response elements

---

Abbreviations used: AP1, activator protein 1; α1AT, α_1-antitrypsin; ATF, activating transcription factor; CArG, CC(A/T-rich)GG element; CEBP, CCAAT/enhancer-binding protein; CF1, common factor 1; COUP, chicken ovalbumin upstream promoter; CRE, cyclic AMP response element; CREB, CRE-binding factor; CREM, CRE modulator; DRTF1, developmentally regulated transcription factor 1; E2F, E2 element-binding factor; Elf-1, ets-like factor 1; emc, extramacrochaetae; eve, even-skipped; FAP, fos gene AP1-binding sequence; FGF, fibroblast growth factor; ftz, fushi tarazu; GR, glucocorticoid receptor; GRE, glucocorticoid response element; nGRE, negative GRE; GRM, general regulator of mating type; HLH, helix-loop-helix factor; bHLH, basic HLH; nHLH, negative HLH; HNF1α, hepatocyte nuclear factor 1α; HPFH, hereditary persistence of foetal haemoglobin; Id, inhibitor of differentiation; IκB, inhibitor of NFκB; ISRE, interferon-stimulated response element; IRF1, interferon-responsive transcription factor 1; ITF-1, immunoglobulin gene transcription factor 1; LCR, locus control region; LIP, liver-enriched inhibitory protein; mad, max dimerization partner; max, myc auxilliary factor; MRF4, muscle regulatory factor 4; myf5, myogenic factor 5; MyoD, myogenic factor D; mxi1, max-interactor 1; NFAT, nuclear factor of activated T cells; NFκB, nuclear factor of Igκ gene B site; NF-μNR, nuclear factor-μ enhancer negative regulator; NRD, negative regulatory domain; Oct, octamer binding factor; PEA2, polyomavirus enhancer-binding activity 2; PEPCK, phosphoenolpyruvate carboxykinase; Pit-1, pituitary-specific factor 1; POU, Pit-1–Oct–unc86 homology domain; PRD, positive regulatory domain; Rb, retinoblastoma gene product; SRE, serum response element; SRF, serum response factor; SV40, simian virus 40; TAT, tyrosine aminotransferase; T$_3$, thyroid hormone (tri-iodothyronine); T$_3$R, thyroid hormone receptor; TFE3, transcription factor of μE3 motif; TGF, transforming growth factor; TPA, 12-O-tetradecanoylphorbol 13-acetate (= PMA); TRE, TPA response element; TSE, tissue-specific extinguisher; WT1, Wilms tumour gene product; zen, zerknüllt; z2, zen-related factor.
* To whom correspondence should be addressed.

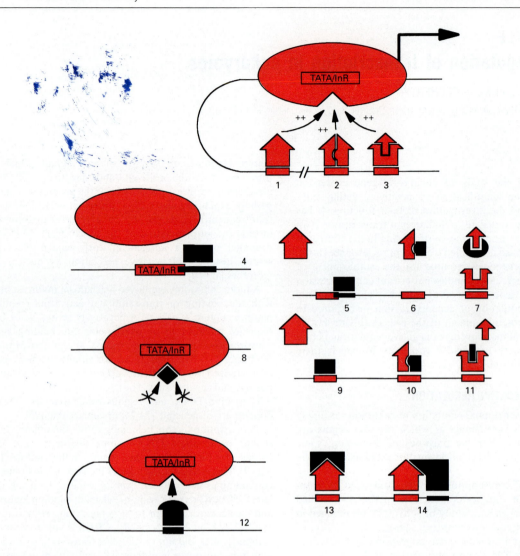

Figure 1   Features of transcriptional activation and negative regulation

In this and in other figures, the convention is adopted that positively acting sites and factors are shown in red, and negatively acting sites and factors are shown in black. The multicomponent polymerase complex is assembled at the TATA box (TATA) or at the initiator element (InR) in promoters lacking a canonical TATA box. The complex has low 'basal' activity which is stimulated by transcription factors bound in *cis* (that is, on the same molecule of DNA). Activators may operate from near to the polymerase complex, or from great distances away. In the latter case, contact with the polymerase complex is thought to be permitted by looping out of the intervening DNA (part 1). Other features are described in the text.

(GREs). These diverge from the GRE consensus sequence, leading to the suggestion that receptor–DNA interactions at such sites differ from interactions at canonical sites, so that the receptor conformation is altered and productive interactions with the transcription machinery are prevented [30–32]. Some evidence in favour of this hypothesis comes from the observation of a highly unusual trimeric glucocorticoid receptor complex at a negative response element of the pro-opiomelanocortin promoter [33]. Dimers of the rel-related p50 transcription factor show different *trans*-activating properties at alternative binding sites [34,35], and such differences correlate with changes in sensitivity to a proteinase, suggesting alternative conformations at activatory and non-activatory sites [35]. Ligand-activated thyroid hormone receptor (T_3R) is a transcriptional activator at the palindromic sequence AGGTCA.TGACCT. At the closely related oestrogen response element AGGTCA.NNN.TGACCT, T_3R is unable to activate transcription, but down-regulates the oestrogen response by competition for the binding site [36]. The position of the dyad symmetry axis may influence the regulatory

activity of the DNA-bound receptor dimer through effects upon its conformation. An oestrogen response element of the oxytocin promoter is similarly down-regulated by the retinoic acid receptor. A powerful activating domain can be fused to the retinoic acid receptor DNA-binding domain, generating a chimaera which binds strongly to the response element, but fails to activate transcription [37]. These observations challenge the view that simply tethering an activation domain to DNA is sufficient for stimulation of transcription, and suggest that sequence-directed features of the protein–DNA interaction may dictate regulatory properties. Attempts have also been made to relate differences in *trans*-activation to structural perturbations of the bound DNA, such as bending [38]. Bending of DNA has been ascribed to a growing number of transcription factors [39–41]; however, the functional significance of this effect is presently unclear.

Very many transcription factors bind to DNA as dimers (Figure 1, part 2), and possess specialized dimerization domains such as the 'leucine zipper' [42,43], the POU domain (named after the Pit-1, Oct and unc-86 proteins in which it was first described; [44]) and the helix-loop-helix motif [45]. There may be considerable (though not unlimited) dimerization promiscuity within a family of related proteins such as the activating transcription factor (ATF)/CREB, CCAAT/enhancer-binding protein (CEBP), T_3R and AP1 families. For instance, the prototypical AP1 factor is a heterodimer of c-Fos and c-Jun proteins which interact through leucine zipper motifs. There are at least four Fos-like and three Jun-like cellular proteins. Various heterodimeric combinations of Fos-type and Jun-type proteins are possible, in addition to homo- or hetero-dimers of Jun-type but not Fos-type proteins (reviewed in [46,47]). Such promiscuity is an enormous source of regulatory diversity, since different dimeric complexes may differ in DNA-binding and in *trans*-activating properties [48,48a].

Two types of negative regulatory interaction are observed. Firstly, a negatively acting factor may form a dimeric complex which fails to bind DNA, and thus represses transcription by sequestrating positive transcription factors in an inactive form (Figure 1, part 6). The helix-loop-helix antagonists Id and extramacrochaetae (emc) [49–51] are able to dimerize with other helix-loop-helix factors (HLHs), but lack essential DNA-binding surfaces and down-regulate the activating potential of their dimerization partners (see later). Similarly, members of the POU domain and CEBP families attenuate the DNA-binding activity of their respective dimerization partners [52,53]. Secondly, the negatively acting factor may form a dimeric complex which is able to bind DNA, but lacks domains required for transcriptional activation (Figure 1, part 10). In this case the negative factor not only may compete for occupation of *cis*-acting regulatory sites, but may also function by sequestering activators in transcriptionally inert heterodimeric complexes. The proto-oncogene product c-myc requires dimerization with its partner max (myc auxilliary factor) in order to bind DNA efficiently, to activate transcription and to achieve its oncogenic effect ([54,55], and references therein). Nevertheless, when in excess over myc, max is a negative regulator of transcription, forming homodimers which bind DNA but do not activate transcription [56,57]. This illustrates the potential for large changes in transcriptional activity through small changes in the ratios of dimerization partners. Myc-dependent transcriptional activation is also negatively regulated by alternative max partners, mad [58] and mxi1 [59], which form DNA-binding but non-activating complexes with max, thus competing for the essential cofactor of myc, and for occupancy of binding sites. Other examples of such behaviour have been described, including two members of the CREB family, CREM [60,61] and CREB-2 [62]; the CEBP-related factor LIP (liver-enriched inhibitory protein) [63]; a basic HLH leucine zipper factor mTFE3 [64]; and two members of the AP1 family, JunB [66,67] and FosB-S [68–70].

Interestingly, several negatively acting factors are the products of genes which also encode transcriptional activators, and are generated through alternative splicing of primary transcripts [61,64,68–71], or through alternative use of translation initiation codons [63,72]. The generation of positive and negative factors from the same transcript may be subject to developmental, tissue-specific or signal-responsive regulation, providing an additional level at which gene expression may be controlled [73].

As well as interactions between structurally related proteins, there may be unexpected negative regulatory interactions between factors belonging to entirely different classes, for instance between c-Jun and the glucocorticoid receptor [74,75] or between c-Jun and myogenic factor D (MyoD) [76]. It is interesting that some transcription factors have evolved not only for multiple interactions 'within-family', but also for cross-regulatory interactions with other families. The consequences of such cross-regulatory interactions are the subject of current debate (see later), and may include either interference with binding or interference with transcriptional activation.

Both DNA-binding and non-DNA-binding proteins may negatively regulate transcription by 'masking' the activation surfaces of target positive factors. GAL80, a regulator of galactose metabolism in yeast, does not itself bind DNA, but appears to mask the activation domain of the positive factor GAL4 (Figure 1, part 13; [77]). Similarly, the mammalian retinoblastoma gene product Rb does not bind directly to DNA, but modulates the regulatory activity of DNA-binding factors with which it interacts (Figure 1, part 11; [78]; see later). The c-*myc* gene is positively regulated by a binding site for the transcription factor CF1 (common factor 1) [79]. An adjacent sequence negatively regulates c-*myc* transcription, and because binding of proteins to the two sites is co-operative it is suggested that the negative regulatory factor acts by masking activation domains of CF1 (Figure 1, part 14; [80]). Genes expressed specifically in yeast **a**-type cells are activated by a non-cell-specific factor GRM (general regulator of mating type). Cell specificity of GRM activity appears to be regulated by an α cell-specific factor which binds co-operatively with GRM at sites flanking the GRM binding site, and may down-modulate transcription by masking the activation surface(s) of the positive factor [81] and/or by recruiting a general negative regulatory factor to this site [82].

There is strong evidence that transcription factors transmit their activating signals to the polymerase complex through the action of intermediary factors or cofactors (Figure 1, part 3; [83–85]). This is most clearly shown by the phenomenon of 'squelching', in which the presence of large excesses of transcription factors, or of their activation domains, interferes with transcription activation by sequestrating the putative cofactors [86]. Squelching events can be observed *in vivo* without artificially high concentrations of the negatively acting factor [87]; thus the cofactors may be limiting for some transcription activation processes, and squelching may be a physiologically significant mechanism of regulating gene expression.

The adenovirus E1a gene products not only activate transcription of viral genes, but are also able to repress *trans*-activation by a huge number of cellular transcription factors [88–97]. Binding to DNA is required for neither positive nor negative regulatory activity. One possible explanation of the breadth of E1a action is that it interferes with the transmission of activating signals by interaction with cofactors (Figure 1, part 7) or with the polymerase complex (Figure 1, part 8; [94]). An interaction of E1a with the TATA box-binding factor TFIID (transcription factor IID) has been demonstrated [98], and this may mediate both positive and negative effects upon transcription.

The mechanisms described above are essentially passive, involving interference with stages of transcriptional activation. In contrast, 'silencing' is an active process which may be regarded as a mirror image of enhancer activity (Figure 1, part 12). In other words, it involves direct inhibition of transcription initiation, possibly by causing a disruption of the polymerase complex at the promoter, or by otherwise impairing the catalytic activity of the complex. The first described silencer element [99] showed classical enhancer-like properties of acting upon *cis*-linked promoters at great distance and independently of orientation. Other silencers have demonstrated varying degrees of dependence upon position and orientation [100–105]. Several appear to resemble enhancers in that they are composite in structure [105–110], or show increased effects on transcription with increased copy number [110–113]. Finally, an important consequence of the enhancer analogy is that silencer factors are expected to possess protein domains, distinct from their DNA-binding domains, which will mediate negative regulation of transcription. The few distinct, transferrable silencing domains which have so far been identified have little in common [114–118], raising the possibility that silencing may be as functionally diverse as transcriptional activation.

Silencing and other negative regulatory processes may involve effects upon chromatin structure (reviewed in [4,119,120]). The most compelling evidence for this comes from studies in yeast. For instance, the archetypal yeast silencer, which switches off expression of the HMRa silent mating-type locus, contains a functional autonomous replication sequence [107,121]. Some observations suggest a role for replication in the establishment of the transcriptionally inert state of the silent mating-type loci [122,123]; however, this remains controversial [124]. It is not known how replication may influence transcriptional regulation. The establishment of domains of chromatin structure is emerging as an important stage in transcriptional regulation [125,126], and may also be involved in silencer function. A factor binding to the silencers of the yeast mating-type genes is required for nuclear scaffold attachment and for transcriptional repression [127]. Negatively acting sites which flank the mouse immunoglobulin heavy chain enhancer overlap with potential scaffold attachment sites, suggesting that repression of enhancer activity may be mediated through effects on chromatin loop formation ([128]; see later). In spite of these indications, the evidence for large-scale chromatin structural effects in silencer activity remains weak, at least as far as higher eukaryotes are concerned.

DNA methylation may theoretically play a role in the effect of replication upon transcription. Hypermethylation is known to be associated with a state of transcriptional silence (reviewed in [129]); however it has been difficult to establish a causal relationship (e.g. see [130]). CpG methylation has been shown to repress transcription *in vivo* and *in vitro*, either by directly influencing binding of transcription factors [131], or indirectly through the action of methyl-specific DNA-binding proteins which may block the access of transcription factors [132–135]. However, these factors bind to sequences unusually rich in CpG dinucleotides, and their relevance to transcription in general is unknown.

Finally, nucleosomes can be regarded as negative regulators of transcription, since they potentially interfere with DNA binding by transcription factors and the polymerase complex (reviewed in [136]). In yeast this negative effect is revealed by histone protein mutagenesis or dosage alteration experiments [137–139]; however, in higher eukaryotes there is as yet little evidence for gene-specific negative regulation of transcription by nucleosomes. Although condensation of DNA into heterochromatin is associated with transcriptional inactivity, it is not obviously a cause rather than an effect of transcriptional inactivation.

FUNCTIONS OF NEGATIVE REGULATION

Transcriptional repression is involved in a wide variety of tissue-specific, developmental, cell cycle and signal-responsive regulatory events. It will be clear to the reader that the boundaries between these classes of events are somewhat blurred, and we make such distinctions here simply for the sake of simplicity of discussion.

DEVELOPMENTAL AND TISSUE-SPECIFIC REGULATION

In higher eukaryotes, tissue-specific transcription is generally dependent upon activating factors which are themselves distributed in a tissue-specific manner. It is highly characteristic

that expression in inappropriate tissues is also prevented by negative mechanisms, as revealed by the ectopic activation of transcription when negatively acting sites are deleted or mutated [14,104,128,140–148]. Both promoters and enhancers are modular in structure [149], and contain binding sites for widely distributed as well as tissue-specific transcription factors. Dual positive–negative regulation of transcription appears to ensure fidelity of tissue-specific gene expression, in part through prevention of 'leaky' ectopic transcription driven by widely distributed transcription factors.

Where complex patterns of tissue-specific expression are observed, there may be correspondingly complex mechanisms of negative regulation to generate these patterns. This is true for instance of a major histocompatibility complex gene which is almost ubiquitously expressed, but at differing levels [22,150]. Transcriptional activity is regulated by an enhancer which overlaps with a silencer element. The enhancer binds uniformly distributed positive factors, whilst silencer-binding factors vary in concentration, and this variation is believed to dictate tissue-specific differences in steady-state transcriptional activity [22].

Developmental and tissue-specific regulation of gene expression may be regarded as temporal or spatial cross-sections of the same phenomenon. Central to both is the establishment of cellular identity, whether this specifies a static differentiated phenotype or a dynamic developmental fate. The study of developmentally regulated gene expression requires model systems in which developmental or differentiation events can be monitored *in vivo*, or recapitulated *in vitro*. Such systems form the subject of the following sections.

Extinction

The importance of negative regulation in developmental processes has been illuminated by the phenomenon of extinction, the inactivation of tissue-specific gene expression in somatic hybrids between differentiated and undifferentiated cells (reviewed in [151]). Extinction encompasses entire sets of tissue-specific genes [152], operates at the level of transcriptional regulation [153–160] and is thought to reflect mechanisms which block the expression of tissue-specific traits prior to differentiation and/or outside specific developmental lineages. It has been used as a starting point in attempts to identify diffusible, *trans*-dominant, negatively acting factors involved in such pre-differentiation blockage of tissue-specific gene expression.

Various cytogenetic and molecular biological techniques have been used to identify [161,162] and subsequently to clone [163,164] a tissue-specific extinguisher gene, *TSE1*, involved in the extinction of liver-specific genes such as those encoding tyrosine aminotransferase (TAT) and phosphoenolpyruvate carboxykinase (PEPCK) (reviewed in [165]). The target of TSE1 activity in the TAT gene maps to a far-upstream CRE, with TSE1-mediated repression being alleviated by activation of the protein kinase A pathway [166]. The TSE1 locus encodes an inhibitory subunit of protein kinase A [163,164]. Down-regulation of the TAT gene is thus explained by inhibition of the phosphorylation and activation of the CREB [167], and is a somewhat indirect, 'lack-of-activation' phenomenon. Other tissue-specific extinguisher loci certainly exist (see for instance [153,161,168,169]), and it remains to be seen whether these will similarly be found to function indirectly.

In several instances the loss of tissue-specific gene expression in intertypic hybrids has been correlated with the disappearance of a trans-acting positive regulatory factor thought to be essential for efficient transcription (Table 1). Frequently the loss of the *trans*-activator has been shown to occur at the level of tran-

Table 1 The disappearance of tissue-specific transcription factors implicated in the extinction of certain genes

Down-regulation of messenger RNA encoding the relevant transcription factor is indicated in the third column. Abbreviations: IEF1, insulin enhancer factor 1; TTF1, thyroglobulin transcription factor 1; LEF1, lymphoid enhancer-binding factor 1; ND, not determined.

Gene	Transcription factor	Regulated expression?	References
Insulin	IEF1	ND	170
Prolactin, growth hormone	Pit-1	Yes	156,171,172
Thyroglobulin	TTF1	ND	158
Immunoglobulin	Oct2	Yes	173,174
Albumin, α1AT, β-fibrinogen	HNF1α	Yes	155,175–177
T-cell receptor α	LEF1	Yes	178

scription of the corresponding gene (see Table 1). These findings suggest that the transcriptional silence of tissue-specific genes in somatic cell hybrids is again a lack-of-activation phenomenon. However, it remains to be established that the loss of a tissue-specific activator causes extinction, rather than being a consequence of an extinction event occurring at some higher regulatory level. Extinction of the immunoglobulin gene is frequently accompanied by, but does not require the loss of, the B-cell-specific Oct2 factor [173,179]. A transfected α_1-antitrypsin (α1AT) promoter is activated in fibroblasts by constitutive expression of the liver-specific factor HNF1α; however, such constitutive expression does not prevent extinction of the endogenous hepatocyte α1AT promoter in hybrids of hepatocytes and HNF1α-expressing fibroblasts [175]. This suggests that an absence of activators may not be sufficient to account for extinction. Extinction of the liver-specific TAT gene is associated with down-regulation of several transcriptional activators [180]. It may thus be necessary to shift attention to the complex mechanisms governing the expression of tissue-specific regulators, and their extinction in intertypic hybrids.

A few strands of evidence point to active repression of transcription rather than passive lack-of-activation as a mechanism of extinction, although the details remain to be elucidated. Single or multiple copies of an element from the immunoglobulin heavy chain enhancer render a heterologous promoter subject to extinction, an effect which is dominant over transcriptional activation by a strong promiscuous enhancer [179,181]. This can be explained only by an active repression mechanism. In other instances active negative regulatory mechanisms have been implicated in both the extinction [140,159] and the maintenance [144,156] of correct tissue-specific transcription. Observations of this kind suggest that the phenomenon of extinction may yet provide a handle upon factors responsible for the maintenance of tissue-specific gene expression.

Drosophila development

The dipteran fly *Drosophila* has proved an invaluable experimental organism for the study of developmental regulation of gene expression, revealing numerous mechanistic principles which are likely to apply to higher eukaryotes. During embryonic development, a system of coarse positional information dependent on maternally expressed morphogens is refined by the sequential and hierarchical action of gap, pair-rule and segment-polarity genes. At the end of this process the domains of expression of regulatory factors have been established at single-cell resolution (reviewed in [182–184]). Genetic analyses reveal multiple interactions between genes acting at the same or at different levels of the regulatory hierarchy, and suggest that

several interactions required for pattern formation are negative. Many of the genes of this network have been cloned and shown to encode nuclear-localized DNA-binding proteins, with DNA-binding domains belonging to the homeodomain [185,186], zinc-finger [187,188], rel-related [189], or basic leucine zipper [190] classes. The demonstration and analysis of transcriptional regulatory properties in a variety of systems permits the gap to be closed between genetically defined positive or negative function and detailed descriptions of molecular mechanisms of action. A few selected examples from this large field of research will serve to illustrate some of the general principles. We concentrate upon early events of anterior–posterior pattern formation. Similar negative regulatory mechanisms are involved in the action of homeotic selectors such as Antennapaedia and Ultrabithorax (see e.g. [191–195]), and in the dorsal–ventral pattern formation process (see e.g. [101,196–199]).

In the developing blastoderm, the expression of the gap gene *Krüppel* is activated by the maternal morphogen bicoid [200]. Restriction of *Krüppel* expression to a broad central domain is dependent upon negative regulatory interactions with other gap genes *giant*, *knirps* and *tail-less* [190,201–203]. In the Krüppel promoter, multiple binding sites for the activator bicoid overlap with tail-less, knirps or giant binding sites [190,203]. A small fragment containing overlapping bicoid and knirps binding sites activates transcription in response to bicoid, and this activation is blocked dose-dependently by knirps, which alone does not influence transcription [203]. This suggests a model for the generation of the Krüppel pattern in which overlapping gradients of activators and repressors compete for access to overlapping binding sites. This is a powerful mechanism for sensing and interpreting morphogen gradients, and appears to be frequently employed during *Drosophila* embryogenesis.

At the next level of the regulatory hierarchy, each of the pair-rule genes comes to be expressed in a series of seven narrow bands encircling the embryo. In the case of pair-rule genes such as *even-skipped* (*eve*) and *hairy*, individual domains of expression appear to be regulated by more or less distinct and independent *cis*-acting elements [204–207]. The sharp boundaries of the second stripe of eve expression are dictated by competition between the activators bicoid and hunchback, and repressors giant and Krüppel within the eve stripe two regulatory element ([108,208–210]; Figure 2).

It is suggested that repressors must act by passive, short-range mechanisms in order to permit the independent regulation of discrete stripe regulatory elements [208,210]. For instance, the eve stripe 3 regulatory element is activated in regions where the Krüppel concentration is high, and this cannot occur if Krüppel protein bound to the stripe 2 element exerts a dominant, silencer-like effect upon the eve promoter. Yet the regulatory properties

(a)

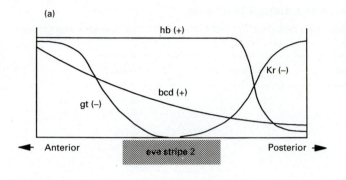

(b)

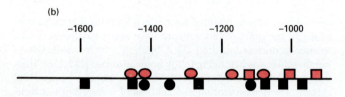

Figure 2   Negative regulation of the second stripe of *eve* expression in the *Drosophila* embryo

(**a**) Overlapping gradients of the positive regulators hunchback (hb) and bicoid (bcd) and the negative regulators Krüppel (Kr) and giant (gt) along the antero-posterior axis. (**b**) Overlapping binding sites for the positive regulators hb and bcd (red squares and circles respectively) and negative regulators Kr and gt (black squares and circles respectively) in the 5′ region of the *eve* gene. Co-ordinates are with respect to the major transcription start site.

of Krüppel appear more complex than predicted in this model, and direct silencer-like activity has been observed in cultured mammalian [115] and *Drosophila* [117] cells. In the latter study Krüppel also displayed concentration-dependent activation and repression of a synthetic promoter containing a single binding site, an effect seemingly dependent upon distinct regulatory properties of monomeric and dimeric Krüppel complexes [117a]. It is not yet clear how these complex properties of the Krüppel protein relate to its function in embryonic pattern formation.

The so-called secondary pair-rule gene *fushi tarazu* (*ftz*) is regulated somewhat differently from the *eve* and *hairy* genes. Rather than being activated within specific embryonic domains, it is activated in a broad region and specifically repressed within interstripe domains [211,212], at least in part through the action of other pair-rule genes such as *hairy* [213,214]. The ftz promoter contains several positive and negative sites [211,215]. Against a background of uniform expression driven by a non-cell-specific promoter, multiple copies of individual negative sites restore zebra striped expression [215].

In the final stage of embryonic pattern formation, parasegmental compartments are established through the action of segment polarity genes (for a definition of the term 'compartment', see [182]). Many of the regulatory proteins involved in the establishment of this pattern, and in subsequent developmental stages, contain a conserved DNA-binding structure known as the homeodomain. Conservation of the homeodomain results in substantial overlap of DNA-binding specificity, whilst greater divergence outside of the homeodomain may help to explain the very distinct regulatory properties ([186]; see also references in [216]). In a number of experimental systems, activation of transcription by homeodomain factors such as zerknüllt (zen) and zen-related (z2) can be blocked by eve or

engrailed proteins, which alone do not influence transcription [24,25,217]. This is a passive, binding competition effect which reflects similarities in sequence specificity of the activators and negative regulators. The deletion of a transcription-activating domain from z2 allows it to behave like eve and engrailed, blocking activation by competing with activators for access to DNA [25]. Although a silencer-like mode of action has also been described for eve [218], this may be related to an unusual binding competition mechanism, involving co-operative interactions between eve at a high-affinity distal site and at a low-affinity promoter-proximal site, where its binding interferes with the binding of transcription activators [219].

Both positive and negative properties have been demonstrated for a number of *Drosophila* transcription factors [117,191,196], underlining the context-dependence of regulatory activity. Bi-directional regulation may be dictated by variations in the sequences bound, or by different protein–protein interactions at different *cis*-acting elements [220,221], and clearly increases the regulatory repertoire possible with a finite number of *trans*-acting factors.

Differentiation *in vitro*

Certain cell lines can be induced to undergo differentiation *in vitro* by treatment with appropriate agents, and thus constitute useful systems for investigating molecular events in differentiation. For example, the retinoic acid-induced differentiation of embryonal carcinoma cell lines models early events in embryogenesis [222]. The response to retinoic acid includes the transcriptional activation of several cellular genes, and a switch from non-permissive to permissive behaviour for viral replication ([223,224]; reviewed in [225]). Retinoic acid influences the programme of gene expression both directly, by activating transcription of target genes (e.g. [226]), and indirectly, by altering expression of genes encoding transcription factors. For instance, the gene encoding the embryo-specific Oct3 factor is negatively regulated by retinoic acid [227]. Since Oct3 itself may negatively regulate some target promoters [228], its disappearance might lead to transcriptional derepression. The c-*jun* gene is derepressed by retinoic acid [229], and its product participates in differentiation-linked activation of several cellular genes [230]. Following the induction of differentiation the c-Jun-containing AP1 complex also activates the polyomavirus enhancer, displacing a labile repressor, PEA2, from its DNA-binding site [231,232]. Largely using the changes in viral regulation as an experimental system, other investigators have provided evidence for limiting amounts of labile negative regulators of transcription in the undifferentiated embryonic stem cell [233–235]. These putative repressors remain to be characterized.

The proliferating 3T3-L1 preadipocyte cell line can be induced to differentiate into mature adipocytes, and this is accompanied by the transcriptional activation of several adipocyte-specific genes (reviewed in [236]). De-repression of the stearoyl-CoA desaturase 2 gene is accompanied by changes in protein binding within a region which shows silencer activity in preadipocyte and HeLa cells [237]. The activation of the 442(aP2) promoter appears to involve displacement of a negative regulatory complex by an AP1 complex [23,238–240]. Finally, several adipocyte-specific genes are activated by the transcription factor CEBPα [240–242]. CEBPα is expressed in the course of terminal differentiation [242–244], and the expression of a conditionally active form in transfected cells causes growth arrest [245]. Thus CEBPα may constitute part of a proliferation–differentiation switch. This type of switching behaviour is better understood in the context of myogenesis.

Myogenesis is governed by basic helix-loop-helix factors (bHLHs) typefied by MyoD, myogenin and MRF4 (muscle regulatory factor 4) (reviewed in [246–248]). These muscle-specific proteins dimerize with the widely expressed bHLH products E12 and E47, and bind to cis-acting 'E boxes' (consensus sequence CANNTG, where N is any nucleotide) upstream of muscle-specific genes. Dimerization is mediated by the helix-loop-helix domains, and DNA binding by the adjacent basic domains of both proteins. In CH310T$\frac{1}{2}$ cells the myogenic differentiation process can be initiated by forced expression of some (but not all) muscle-specific HLH proteins. Myogenesis and cellular proliferation are mutually antagonistic, as indicated by a large number of observations (reviewed in [248a]). On the one hand, the overexpression of MyoD inhibits proliferation, even in cells in which it fails to initiate the myogenesis programme [249,250]. On the other, several agents including serum [251,252], protein kinases A and C [253,254], the phorbol ester 12-O-tetradecanoylphorbol 13-acetate (TPA) [255], transforming growth factor β (TGFβ) and fibroblast growth factor (FGF) [251,254,256,257], and the oncogenes ras [258,259], myc [260,261], fos and jun [76,262], interfere with the programme of muscle differentiation. Three mechanisms have been suggested for the inhibition of myogenesis: down-regulation of expression of myogenic regulators, interference with their DNA-binding activity, and interference with the transmission of their activating signal. Because myogenic factors auto-activate their own expression [263], the first of these effects may follow indirectly from either of the others.

TGFβ inhibits the trans-activation function of myogenin, MyoD, myf5 (myogenic factor 5) and MRF4 [256,257], but does not interfere with DNA-binding activity, at least of myogenin [251]. TGFβ exerts its negative effect through the bHLH region of myogenin, but shows no activity against the ubiquitous HLH protein E47 [256]. A muscle-specific cofactor has been suggested to contact the bHLH region of myogenic regulators [264], and this may be a target of TGFβ action.

FGF interferes with the regulatory activity of myogenic HLH proteins by a mechanism involving protein kinase C [254]. Protein kinase C phosphorylates myogenin at a conserved threonine residue within the basic domain, and in vitro this modification interferes with DNA binding by myogenin–E12 heterodimers. Inhibition of phosphatase activities also inhibits myogenesis [265], indicating that there may be a requirement for active dephosphorylation to initiate or maintain the regulatory activity of HLHs. Protein kinase A also interferes with myogenic HLH regulatory activity; however, this is not directly correlated with phosphorylation of the HLH proteins, and must involve an indirect mechanism [253]. Possible mediators of the effect of protein kinase A include c-Fos and c-Jun.

Activated ras oncoprotein interferes with the myogenesis programme [258], and this activity is mimicked by overexpression of Fos or Jun proteins [76,258,262]. Conversely, the overexpression of MyoD blocks transcriptional activation by AP1 at its cognate binding site [76]. Both phenomena are titratable, thus the ratio of AP1 components to myogenic factors is decisive, arguing against sequestration of an essential cofactor as the mechanism. The basic helix-loop-helix region of MyoD is a target of negative regulation by Jun, and again the ubiquitous HLH protein E47 is not affected [262]. An interaction between the leucine zipper of c-Jun and the helix-loop-helix region of MyoD has been demonstrated [76]. The simplest hypothesis is that this interaction mediates the mutual antagonism of MyoD and c-Jun by mutual inhibition of DNA binding, although this remains to be shown. A distinct negative regulatory function may reside within the N-terminal domain of c-Jun [262].

Treatment of CH310T$\frac{1}{2}$ cells with high concentrations of serum inhibits myogenesis. This blockage is accompanied by a decrease in the binding activity of myogenin, which may be explained by an increase in the expression of the factor Id [49,251,252]. Id belongs to a class of proteins which negatively regulate the activity of DNA-binding HLHs, and are more extensively discussed in the following section.

Clearly the mutual antagonism between myogenesis and proliferation involves a complex network of negative regulatory interactions, into which a degree of redundancy is built. The tension between proliferation and differentiation is a central theme in eukaryotic gene regulation, to which we will return in a later section.

HLH proteins in development

As well as being central to the process of myogenesis [246–248], HLH proteins participate in the tissue-specific and developmentally regulated expression of immunoglobulin [266,267], insulin [268,269] and exocrine pancreas-specific [270] genes, among others (see [271] for a review of HLH proteins in differentiation). In general, activation is mediated by a heterodimer of a tissue-specific bHLH (exemplified by MyoD) and a widely expressed bHLH (exemplified by E12 or E47 protein), binding to an E box (CANNTG) within the regulatory region of the gene.

A novel class of HLH proteins is exemplified by the mammalian Id and the Drosophila emc proteins [49–51], which contain the helix-loop-helix domain but lack an adjacent basic domain [referred to here as negative (n)HLHs]. They dimerize with widely expressed bHLHs, forming complexes which are unable to bind DNA [272], and will inhibit transcriptional activation by bHLH complexes, an effect entirely dependent upon abrogation of binding of the activatory species [273]. In vitro, such proteins are able to abrogate E box binding activity when present in low stoichiometry with bHLHs [49,274]. Thus in vivo E box binding activity may be highly sensitive to quite small changes in the ratio of bHLH to nHLH, pointing to a potential mechanism of regulation of E box-dependent promoters. In fact, a role for nHLH proteins in the developmental regulation of gene expression is suggested by several observations. The mammalian nHLHs Id1, Id2 and HLH426, as well as the Drosophila counterpart emc are expressed in complex temporally and spatially regulated patterns, often at higher levels in immature or undifferentiated tissues than in mature or differentiated tissues [76,275–278]. In Drosophila the emc and achaete–scute gene products are required for correct development of the peripheral nervous system. Dosage-altering experiments show the nHLH emc to be an important transducer of positional information in this process [277,279]. In several mammalian tissue culture systems for the study of differentiation, the induction of differentiation is accompanied by a transient or sustained decrease in mRNA encoding nHLH proteins, and the transient or sustained appearance of E box binding activity in nuclear extracts [76,272,275,280]. The overexpression of Id in transfected cells inhibits the activity of immunoglobulin, insulin or muscle cell-specific enhancers, and is able to block the execution of certain differentiation programmes [273,280–283].

It emerges that the nHLHs are negative regulators of certain differentiated states, namely those which depend upon transcriptional activation by bHLHs. They operate by sequestering their bHLH targets in a non-DNA-binding and therefore inactive form. A critical event in the developmental activation of gene expression appears to be a decrease in the effective concentration of nHLHs, probably coinciding with an

(a)

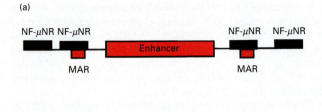

(b) Active

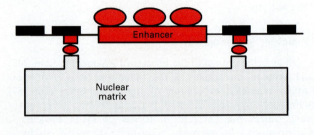

(c) Inactive

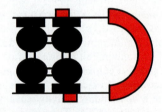

Figure 3   Model for negative regulation of the intronic immunoglobulin heavy chain (IgH) gene enhancer by NF-μNR

(a) Arrangement of binding sites within the IgH enhancer region. The enhancer contains binding sites for several factors. (b) Active transcription. Matrix attachment regions (MAR) may establish a domain of open chromatin structure, in which the enhancer is accessible to activators. (c) Transcriptional inactivity. The binding of NF-μNR inhibits enhancer activity through prevention of matrix attachment, and/or through the formation of NF-μNR tetramers, with structural alteration of intervening DNA.

increase in the concentration of the activatory bHLHs. These events themselves are as yet poorly understood.

The immunoglobulin genes

The intronic enhancer of the immunoglobulin heavy chain gene contains numerous binding sites for tissue-specific as well as widely distributed transcription factors [284]. Maximal enhancer activity is observed in mature lymphoid B-cells where the gene has undergone somatic rearrangement. However, non-tissue-specific enhancer activity can be detected in some non-lymphoid tissues, whilst in others (liver, foetal fibroblasts) the enhancer is completely silent [285]. Negative regulation of immunoglobulin enhancers has been suggested by extinction in B-cell–non-B-cell somatic hybrids [154,173,174,179,181,286]; by repression of *in vitro* transcription of immunoglobulin genes when non-B-cell nuclear extracts are added to B-cell nuclear extracts [287]; and by the cycloheximide induction of a transfected immunoglobulin gene in fibroblasts [288,289].

Repressor in excess

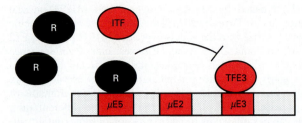

ITF in excess

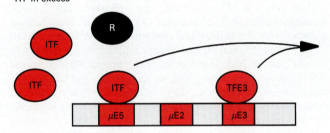

Figure 4   Negative regulation of the μE5–μE2–μE3 region of the immunoglobulin heavy chain enhancer

A repressor (R) and an activator (ITF) compete for access to the same binding site. The repressor prevents transcriptional activation by the ubiquitous factor TFE3, whilst the activator synergizes with TFE3 to activate transcription.

Mutagenic studies have revealed distinct mechanisms of negative regulation. Firstly, the enhancer is flanked by negative regulatory elements which down-regulate its activity in non-B-cells and in immature B-cells, but have no effect in mature B-cells [128,142]. Both of these elements are required for the negative effect, and their position flanking the enhancer is essential. Within the negative regulatory regions defined by mutagenesis, binding sites have been discovered for a factor NF-μNR which is most abundant in non-B and immature B-cells [128]. NF-μNR forms tetrameric complexes, suggesting that direct interactions between molecules at separated, enhancer-flanking sites may lead to a structural alteration of the intervening DNA, impeding the action of enhancer-binding factors ([128a]; Figure 3). Additionally, NF-μNR sites overlap potential nuclear matrix attachment and topoisomerase II sites [290], again hinting that its negative regulatory function may involve influences over local DNA structure. This mechanism may resemble the complete silencing of enhancer activity observed in some tissues, and may help to explain why factors present in numerous tissues footprint the enhancer only in the B lineage [266,267].

Other studies starting with smaller enhancer fragments have revealed further mechanisms of negative regulation. Non-tissue-specific activators are able to bind to the μE5 motif [291] and to the nearby μE3 motif [292]. It has been shown independently by Kadesch et al. [293] and by Weinberger et al. [294] that mutations or deletions within this region can increase enhancer activity in non-B-cells. In fact, a complex regulatory system exists to regulate the activity of μE3 and μE5 sites outside the B lineage ([295]; Figure 4). The μE3 site alone is able to stimulate transcription in non-B-cells, but this stimulation is blocked by an adjacent μE5

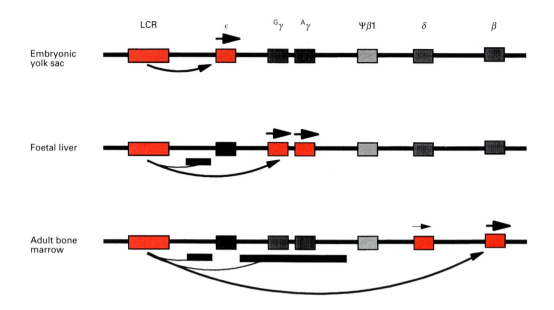

Figure 5   Developmental regulation of the human β globin locus

Active transcription is indicated by a horizontal arrow above the appropriate gene Ψβ1 is an unexpressed pseudogene, and δ is weakly expressed in the adult bone marrow. The LCR activates each of the genes in turn, and this requires the sequential inactivation of ε and γ genes at the correct developmental stages (indicated by horizontal bars below these genes).

motif. Negative regulation is presumably mediated by a *trans*-acting factor, since it can be overcome by the forced expression of the μE5 binding factor ITF-1. The overall activity of this region may be finely regulated by binding competition at the μE5 site between an activator which synergizes with TFE3 (transcription factor E3) and a repressor which blocks the activity of TFE3.

Finally, an element which mediates B-cell-specific activation of the human immunoglobulin heavy chain enhancer is a target of negative regulation in HeLa cells [296]. Also, as described above, octamer sites may not only activate transcription in B-cells, but also negatively regulate transcription in non-B-cells [173,179,181]. A strikingly common feature is the involvement of *cis*-acting sequences in both positive regulation within the B-cell lineage and negative regulation outside this lineage. Factors mediating such negative effects have so far proved difficult to identify, so that it remains unclear at what level of transcriptional regulation they operate.

The globin genes

The sequential expression of the genes in the α-like and β-like globin gene clusters during ontogeny has served as a paradigm for the study of developmental regulation in higher eukaryotes (reviewed in [297,298]). In humans the β-like genes are physically arranged and temporally expressed in the order ε (expressed in the embryonic yolk sac), Gγ and Aγ (in the foetal liver), then δ and β (in the adult bone marrow) (see Figure 5). This switching of expression is regulated at two levels. Firstly, the locus control region (LCR) upstream of the ε gene exerts a dominant influence over the entire cluster, and is required for the efficient expression of each of the genes (see [297] and references therein). Secondly, stage-specific control elements influence the activity of individual promoters.

Correct developmental regulation of the human β globin gene depends upon competition of *cis*-linked genes for the influence of the LCR. In the absence of other genes of the locus, the β promoter is deregulated (active at abnormally early stages of development) in transgenic mice [299,300]. In contrast, both human ε and γ promoters are correctly regulated in the absence of other genes [301–305], and therefore must be specifically switched off by stage-specific control elements. A silencer element has recently been identified upstream of the ε globin gene [100]. The deletion of this element partially deregulates ε gene expression in transgenic mice, leading to low levels of transcription in the adult bone marrow [301]. Within the silencer region both ubiquitous and erythroid-specific factors bind [100,306–309], the latter probably being GATA or a related protein [309]. This suggests a mechanism of developmental regulation of the silencer, namely that increasing concentrations of GATA lead to the displacement of silencer factor(s). It may also explain why the deletion of this region does not lead to high levels of gene expression in the adult bone marrow, GATA being a well-characterized powerful activator of transcription in erythroid cells.

Mechanisms of developmental switching of the human γ globin genes are less clear. However, some clues may come from analysis of HPFH (hereditary persistence of foetal haemoglobin) syndromes, in which γ globin genes fail to become inactivated (reviewed in [310]). Several forms of HPFH entail large deletions 3′ of the γ globin genes, suggesting the existence of negative regulatory elements in this region [311,312]. Other HPFH forms are associated with single base changes within the promoter sequences which have been shown to alter the binding of nuclear factors [313–316]. These alterations may increase the affinity for positively acting factors, making the promoter less susceptible to developmental silencing, or may directly interfere with the binding of factors involved in silencing expression. Because a single base change can influence binding of several factors, and because the high activity of mutant γ globin promoters has been difficult to reproduce in erythropoietic cell lines, it is presently difficult to distinguish between these hypotheses.

A number of other negative regulatory phenomena have been described for avian and mammalian globin genes [317–322];

however, their mechanisms and roles in developmental regulation remain largely unclear.

CELLULAR RESPONSES TO EXTERNAL SIGNALS

In the simplest terms two types of involvement of negative regulation in cellular signal responses can be imagined. Firstly, a cell may respond to an external cue by specifically down-regulating transcription of a gene or genes. Secondly, transcription may be up-regulated due to signal-mediated alleviation of repression. More complex negative regulatory mechanisms may be involved where it is essential to co-ordinate responses to different signalling pathways, or where the transcriptional response is necessarily transient. Negative regulation is required for transcriptional responses to a large number of diverse signalling agents including members of the steroid/thyroid hormone family (for references see later), peptide hormones such as insulin [323,324], growth factors [325], extracellular calcium [326] and double-stranded RNA [327]. Some selected examples are described here, beginning with the steroid/thyroid family of hormones.

The nuclear hormone receptors

The nuclear hormone receptors belong to a much-studied extended family of transcription factors with more than 70 members, and representatives in organisms as diverse as man and *Drosophila* (reviewed in [328–330]). An important distinction is made between the GR-related subfamily and the T_3R-related subfamily. The former are retained in the cytoplasm in the absence of the appropriate ligand, migrating to the nucleus when ligand-activated, and binding as homodimers at well-defined, palindromic hormone response elements which contain two hexameric half-sites separated by a 3 bp gap. In contrast, the members of the T_3R subfamily are found in the nucleus in the absence or presence of ligand, and may bind to DNA in the ligand-free state. Moreover, they show remarkable flexibility in terms of binding specificity and protein–protein interactions; binding sites may contain one, two, three or even four hexameric half-sites, varying in spacing and orientation, and may be recognized by homomeric or heteromeric receptor complexes [331,332]. In addition to the well-characterized hormone receptors, several 'orphan receptors' have also been described, for which no ligand is yet known. On the basis of their primary structure and other properties, these may also be defined as GR-related or T_3R-related. Very many negative regulatory processes involve the nuclear hormone receptors; several entail regulatory cross-talk with the *fos* and *jun* proto-oncogenes, and are dealt with as a separate subject (see later).

Receptor-dependent negative regulatory events may be divided into two classes, i.e. requiring and not requiring DNA binding by the receptor. The latter depend upon interactions of the receptor with DNA-bound transcription factors, or with transcription cofactors ('squelching'). Several promoters (including prolactin and glycoprotein hormone α subunit) are negatively regulated by activated GRs, in spite of the absence of high-affinity binding sites. In the case of the prolactin promoter the targets of this negative activity appear to be binding sites for the pituitary-specific activator Pit-1 [333], although it is not known whether this involves squelching or some other mechanism. Mutual interference between different nuclear hormone receptors does not require the DNA-binding domain of the negatively acting receptor, and is observed with endogenous as well as with trans-

fected proteins [79]. Thus squelching may provide a physiological mechanism of cross-talk between pathways of transcriptional regulation which involve different receptors co-expressed in a cell. DNA binding and *trans*-activation by T_3, retinoic acid and vitamin D receptors is enhanced by heterodimerization with the retinoid X receptors. Thus a further mechanism of crosstalk between nuclear hormone receptors involves competition for their common cofactor [333a]. This differs from squelching in that dimerization and not *trans*-activating domains of the negatively acting receptor species are required.

Several genes are transcriptionally repressed by ligand-activated GRs, apparently involving binding site competition between the GR and a positively acting factor. In some cases the positive site has not been characterized, but is inferred from the ability of a DNA fragment to stimulate transcription in the absence of ligand-activated GR [30–32]. A pro-opiomelanocortin promoter fragment containing a high-affinity GR-binding site activates a heterologous promoter and confers repressibility by the synthetic glucocorticoid dexamethasone [32,334]. Internal deletions removing the negative GRE (nGRE) reduce the basal activity of the pro-opiomelanocortin promoter and prevent repression by dexamethasone [32,335]. Deletions removing an upstream region which does not contain a GR-binding site also reduce basal expression and interfere with repression by dexamethasone [32,335]. It may be that efficient transcription depends upon co-operative interactions between proximal and distal positive sites, and that a function of the GR is to disrupt this interaction. As discussed earlier, a novel protein–DNA complex at the pro-opiomelanocortin nGRE has been described [33], and the formation of this unusual trimeric receptor complex may explain the failure of the receptor to activate transcription in this context.

A final word of caution is necessary here. As described above, negative regulation by nuclear hormone receptors may not require their binding to DNA. Because binding at atypical sites has often been determined using large excesses of purified receptor, and may not occur *in vivo*, it is necessary to consider both DNA-binding and non-DNA-binding mechanisms where such low-affinity sites are concerned [26,336–338].

Some investigators have reported negative regulation by the ligand-free T_3R at high-affinity binding sites [339–341]. T_3 converts the receptor into a transcriptional activator without any apparent change in DNA-binding affinity. Thus the transcriptional response to T_3 has two components: relief of repression and activation, potentially increasing the magnitude of the regulatory switch effected by this hormone. In general, sites which show the greatest activation by liganded T_3R show the greatest repression by the aporeceptor, perhaps reflecting higher affinity for both forms. Variations in the sequences of T_3 response elements may thus indicate selective pressures upon basal gene expression as well as on the extent of the T_3 response [342]. In the presence of the aporeceptor, T_3R binding sites show silencer-like properties of relative position/orientation-independence, an ability to operate upon heterologous promoters, and additive effects with increased copy number [106,341,343]. An active silencer domain is found in the T_3R, in the closely related oncogenic product v-ErbA and in the retinoic acid receptor [114,114a]. The silencing domain of the retinoic acid receptor acts only in some cell lines, apparently due to tissue-specific influences of an adjacent ligand-independent positive regulatory domain [114]. The v-ErbA product, a virally transduced thyroid hormone receptor protein, has lost its ability to bind ligand, and behaves as a constitutive negative regulator [341]. The cellular α_2 isoform of T_3R similarly fails to bind ligand or to activate transcription, and may interfere with transcriptional regulation by other receptor isoforms [344,345]. Tissue-specific variations in the relative levels of T_3R isoforms may permit modulation of cellular responses to T_3.

Nuclear hormone receptors have also been suggested to mediate ligand-dependent negative regulation of transcription by binding close to or overlapping the transcription start site, interfering with the formation or catalytic activity of the polymerase complex [12–16,346]. These sites are commonly very divergent from the binding consensus sequence. Effects of ligand upon the affinity of receptor for such sites might help to explain why they are not negatively regulated by the nuclear-localized ligand-free members of the T_3R family. Certainly their sequences, rather than simply their positions, are of critical importance [347,348].

The T_3, retinoic acid, retinoid and vitamin D receptors are related in structure and in DNA-binding specificity, with sites being constructed from (most often two) versions of the AGGTCA hexamer. The binding and regulatory activities of different receptors at an individual site are dependent upon the spacing and orientation of the hexamers, their precise sequence and possibly contextual sequences, complicating the task of constructing useful predictive rules for the specificity of hormonal action [331,349–352]. In any event a clear consequence of overlapping binding specificities and the nuclear localization of receptors is that one receptor may interfere with transcriptional activation by another, due to binding site competition [26,37,353]. The ability of ligand-free receptors to occupy binding sites may block the access of other members of the receptor superfamily, and contribute to the specificity of action of these sites in cells which co-express receptors [339,340,354].

The ability of a cell to respond to different ligands may also be modulated by the presence of the COUP (**C**hicken **O**valbumin **U**pstream **P**romoter) factors. These orphan receptors are structurally related to the T_3 family of receptors, recognize the same half-site sequence, and show flexibility in terms of the arrangement and spacing of pairs of half-sites which can be bound by dimers [355]. In co-transfection experiments COUP factors were shown to block the activation of transcription by ligand-bound vitamin D, T_3 and retinoic acid receptors [355–358], presumably by competition for the respective response elements, although the formation of inactive heterodimers is also a possible mechanism. This negative activity of the COUP factors suggests an important role in the regulation of development and differentiation, through regulation of transcriptional responses to various hormones.

Finally, nuclear hormone receptors may activate transcription indirectly by alleviating repression [359,360]. For instance, upstream oestrogen response elements of the *Xenopus laevis* vitellogenin B1 promoter co-operate with promoter-proximal elements to overcome the effect of an intervening negative regulatory element [361]. This co-operativity is potentiated by the precise positioning of a nucleosome between the two positively acting regions [362].

Cross-talk between regulatory pathways

The metazoan cell must co-ordinate its response to a large number of afferent signals by activating or switching off distinct programmes of gene expression. A case in point is that proliferation and differentiation frequently require the expression of discrete sets of genes. Thus a signal which activates the proliferative pathway must as rapidly and effectively shut down expression of genes active in the quiescent, differentiated state. Conversely, signals which induce differentiation must shut down the proliferative pathway. The ability to switch reversibly between

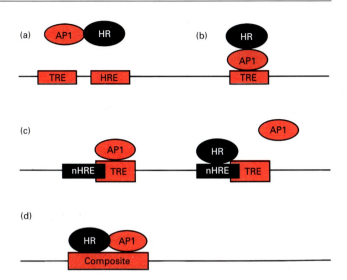

Figure 6   Mechanisms of cross-talk between AP1 complexes and nuclear hormone receptors (HRs)

(**a**) Mutual inhibition of DNA binding; (**b**) formation of a tertiary complex of AP1 and HR at an AP1-binding site (TRE); (**c**) binding competition between AP1 and HR at overlapping binding sites. The HR-binding site is referred to as a negative hormone response element (nHRE). (**d**) Action of HR and AP1 at a composite response element. The regulatory consequences depend upon the composition of the AP1 complex, as described in the text.

expression programmes is dependent upon extensive cross-talk between signal transduction machineries, constituting a 'regulatory lattice' of interactions. Some of the elements of this lattice are becoming increasingly well understood, in particular those which involve the AP1 family of transcriptional regulators (reviewed in [362–366]).

The AP1 factors are well-characterized, dimeric transcription factors which are commonly (but not exclusively) involved in proliferative processes [46,47]. Activation of the protein kinase C pathway, for instance by the phorbol ester TPA, leads to the activation of transcription by AP1 complexes, which bind at TPA response elements (TREs). In contrast to AP1, the steroid/thyroid hormones are commonly found to induce differentiation, or to maintain cells in a quiescent state. Antagonism between the proliferative function of AP1 factors and the differentiative function of the nuclear hormone receptor factors has been frequently noted, and a number of distinct mechanisms have been described (see Figure 6).

Several laboratories have independently demonstrated that nuclear hormone receptors interfere with transcriptional activation by AP1 complexes [74,75,367–369]. TREs linked to the thymidine kinase promoter render this heterologous promoter repressible by the nuclear receptors [75,368–370]. DNA binding of the receptor is not required for its negative regulatory activity; however, the zinc-finger DNA-binding domain is necessary [74,75]. Binding activity and repressor activity of the GR can be uncoupled, since a mutation in one of the DNA-binding zinc fingers retains binding but loses repressor activity [74]. Thus the DNA-binding domain must also be involved in protein–protein interactions which mediate its negative effect upon the AP1 complex. C-terminal regions may also mediate repression, as deletions in this region show diminished (GR [75]) or abolished (retinoic acid receptor and T_3R [367,370]) repressor activity.

AP1 components have also been shown to down-regulate the activity of nuclear hormone receptors [74,75,371–373]. The leucine zipper of c-Jun is implicated in negative regulation of GR function [75], whilst an N-terminal region of c-Fos is required for negative regulation of GR function. This N-terminal domain is absent from FosB, which fails to down-regulate GR function [374]. Interactions between receptors and AP1 components have been demonstrated *in vivo* and *in vitro* [74,371], the simplest hypothesis being that such interactions mutually inhibit DNA binding of the two classes of proteins (Figure 6a). Mutual inhibition of DNA binding is observed *in vitro* [74,75,367,370,372]; however, it is not clear that this occurs *in vivo* [369]. By genomic footprinting it has been shown that a negative effect of GR upon transcription activation by the AP1 complex is not accompanied by changes in the DNA-binding activity of the AP1 complex [375]. Because *fos* is reportedly a preferential target of the negative activity of GR [376], down-regulation of AP1 activity may be explained by a shift in AP1 composition from c-Jun/c-Fos heterodimers to c-Jun homodimers, which are weaker activators of transcription. Alternatively, the activity of the AP1 complex may be modulated through the formation of a tertiary complex of AP1, TRE and GR [369] (Figure 6b), although firmer evidence for the existence of such a complex *in vivo* is required.

A third mechanism of cross-talk involves competition between an activator and a negative regulatory factor for access to overlapping DNA-binding sites, and differs from the mechanism described above in that DNA binding is absolutely required for negative regulatory function (Figure 6c). Binding interference of this kind may be involved in the cross-talk of AP1 with steroid hormone receptors at the osteocalcin promoter, where a retinoic acid response element overlaps an AP1 binding site [377], and at the α-fetoprotein promoter, where a GRE overlaps an AP1-binding site [29,30].

A fourth cross-talk mechanism has been described [363,364], in which factors belonging to the AP1 and nuclear hormone receptor families co-occupy a so-called 'composite response element' (Figure 6d). A composite element of the proliferin gene confers GR-dependent stimulation where c-Jun homodimers occupy the AP1 site, GR-dependent repression where c-Jun/c-Fos heterodimers occupy the AP1 site, and confers no GR response in cells lacking AP1 activity [378,379]. Thus the balance of AP1 components acts as a cell-specific selector of the magnitude and direction of hormone response, and this is determined by the basic domains of the AP1 components [380]. Binding of GR to the proliferin composite response element has so far been demonstrated only with high concentrations of purified receptor [381]. Certain features of the action of GR at composite sites might be explained by selective interference with the DNA binding of different AP1 components [376]. It will thus be important to demonstrate the co-occupancy of composite sites by members of the two families of transcription factors, a central feature of this putative mechanism of cross-talk (cited in [363]).

Finally, a novel mechanism has been shown for AP1-receptor cross-talk in the negative regulation of the interleukin 2 promoter by the GR [382,383]. Nuclear factor of activated T-cells (NFAT) and an AP1-related complex co-operate to activate transcription. Neither site alone is a target of negative regulation by the receptor; however, the juxtaposition of the two sites creates a GR-repressible *cis*-acting element. The receptor, which does not bind to this sequence, down-regulates transcription by disturbing the synergy of the two positive sites [383].

An astonishing variety of both positive and negative cell-specific, factor-specific and promoter-specific interactions between nuclear hormone receptors and AP1 factors have been demonstrated [372,373,384–389]. Work is at a preliminary stage to sort out the mechanisms operating, and the consequences of different interactions, on and off DNA. Distinct receptor interactions are demonstrated by different members of the Fos and Jun families [364,374], perhaps providing a framework for understanding the apparent functional redundancy encoded within these families.

The rel family of transcription factors

The rel family of transcription factors includes the oncogene v-Rel, its cellular counterpart c-Rel, the *Drosophila* morphogen *dorsal*, and a nuclear factor binding to the immunoglobulin κ enhancer B motif (NFκB) (reviewed in [390–393]). The latter, à heterodimer of 50 and 65 kDa subunits [394], is involved in tissue-specific gene expression in mature B-lymphocytes, and also participates in transcriptional induction of some genes by external signals such as cytokines in non-lymphoid cells [395]. Various other homo- and hetero-dimeric complexes of rel-type proteins also bind to DNA, but differ in sequence specificity and in transcriptional regulatory activity [34,35,396–398].

The regulatory activity of rel family members is governed by a unique machinery which involves the IκB factors. The interactions between rel and IκB proteins are comprehensively reviewed elsewhere [392], and the reader is directed to this review for a more complete discussion and list of references.

Several IκB-type proteins have so far been identified, and shown to differ subtly in their properties ([399–425]; other references in [392]). All have in common a region of similarity to the structural protein ankyrin, which appears to mediate interactions with specific rel factors [408,412]. With possible exceptions [401,404], these interactions result in cytoplasmic retention [5,6,409], probably through masking of nuclear localization signals [400,405,409,425]. A second aspect of negative regulation by IκB is inhibition of the DNA-binding and thus the *trans*-activating function of rel factors [394,407,413, 422–424]. The rel family proteins p50 and p52 are synthesized as larger precursors, the C-terminal regions of which contain ankyrin-like repeats, and demonstrate IκB activity (referred to as IκBγ) [408,409,411,415–417,420]. The activation of p50 and p52 requires (as a first step) a proteolytic cleavage between the rel-related and IκB-related domains [403]. Free IκBγ may also be generated through alternative processing of primary transcripts encoding the precursors of p50 and p52 [411,415]. Cell surface signals lead to the dissociation of rel–IκB interactions, and the migration of the rel factors to the nucleus. This event is correlated with changes in the phosphorylation state of IκB [399,413,414,419,426], and although it is not yet known how such changes are effected, it is clear that different IκB species are inactivated on different signal transduction pathways [413,414,426].

The generation of active nuclear rel factors from an inactive cytoplasmic pool is a means of achieving rapid transcriptional induction in response to external signals. Furthermore, the rel-dependent activation of genes encoding IκB species constitutes a regulatory loop which may ensure rapid shut-down of transcription following a transient signal response [402,421]. Other features of the rel–IκB interaction contribute to signal, cell and promoter specificity of gene activation. These include the differential inactivation of IκB species by different signalling systems [413,414,426], the well-documented differences in specificity of IκB species for targets within the rel family [410,418,422], and the differential binding and activating properties of various homo- and hetero-meric rel complexes at alternative binding sites [34,35,396–398]. Additionally, activation by rel factors at certain promoters may be regulated by unrelated factors which bind in *cis*. In HeLa cells, activation of the Igκ

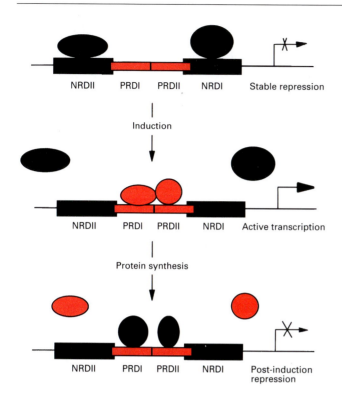

Figure 7　Phases of transcriptional regulation through the ISRE of the human βIFN gene

The identities of regulatory factors involved in this process are discussed in the text.

enhancer by NFκB is specifically blocked by a negative regulatory factor which binds in cis [427]. The Drosophila rel-related protein dorsal behaves as a repressor of zen gene expression, and this negative activity requires a co-repressor which binds adjacent to dorsal binding sites [220,221].

Transient responses: interferon genes and interferon responses

The type I (α and β) interferon genes are rapidly and transiently induced in response to external signals such as interferons, double-stranded RNA or virus infection (reviewed in [428]). In the human βIFN promoter this response is largely mediated by a complex region which contains two adjacent positively acting sites (PRDI and PRDII), the second (PRDII) overlapping with a negative regulatory domain NRDI ([429,430]; see Figure 7). De-repression of transcription by deletion of NRDI [429,431] or by competition with NRDI sequences [432] suggests that a role of this region is to block constitutive transcriptional activation, and to maintain a state of poised repression. Although its overlap with PRDII suggests a binding interference mechanism, NRDI retains its negative activity when widely separated from the PRDII motif upon which it acts [431].

PRDII is a target for binding of rel-related factors, and is thus subject to the negative regulatory mechanisms discussed above. In stimulated cells PRDII synergizes with the adjacent PRDI motif to activate transcription [433]. Following the transient response, both PRDI and PRDII sites participate in a post-induction switch-off event which, unlike the rapid induction of transcription [433], requires de novo protein synthesis [434–436].

Thus three phases of βIFN gene regulation, poised or stable repression, active transcription and post-induction repression, are controlled by the sequential binding of different classes of factors within a small but complex region (Figure 7).

The PRDI motif is highly related to the interferon-stimulated response elements (ISREs) which mediate transcriptional responses of several other genes to interferons. It is now clear that these sequences are targets for binding of several complexes which differ in precise binding specificity, inducibility and distribution [437–447]. One complex, interferon-responsive factor 1 (IRF1), behaves as an activator of transcription [438,446], whilst others may oppose transcriptional activation by competing for the binding site of IRF1 and/or other activators [439,440,442,444,447]. Negatively acting factors which are induced by stimulating treatments may contribute to the post-induction transcriptional repression of βIFN and other genes.

Like the NFκB–rel system, the extremely elaborate interferon response system appears to rely upon a large family of related factors which act through similar sequences but differ in their regulatory properties and patterns of induction. Cell-to-cell differences in the expression of members of this family are thought to contribute to differences in the magnitude, duration and signal specificity of transcriptional responses [440,444].

Transient responses: fos and immediate-early genes

The c-fos promoter is rapidly and transiently increased in response to several proliferative signals (references in [46]). Negative regulation is required for the repression of basal transcription, and for the post-induction shut-off which follows transient activation. Many regulatory processes involve the dyad symmetry element, which is 300 bp upstream of the transcription start site. The multifunctional nature of this cis-acting element is reflected by its complexity and by the large number of ubiquitous and tissue-specific factors by which it is recognized (reviewed in [448]). At the centre of the region of symmetry is the serum response element (SRE), which belongs to the CArG [CC(A/T)$_6$GG] class of regulatory sites. The SRE is constitutively bound by the serum response factor p67^{SRF}, or SRF [449,450], with other factors binding at the 5′ and 3′ boundaries of the SRF–SRE complex [451,452]. Adjacent to the SRE is the so-called FAP (fos AP1-binding sequence) site, which is able to bind AP1 complexes, ATF and CREB [453].

The FAP site is involved in the transcriptional response to protein kinase C [454], and also negatively regulates transcription in unstimulated cells [455,456]. This is revealed by the increase in basal expression caused by in vivo competition with FAP oligos [455], by mutation of the FAP site, or by altering its spacing from the adjacent SRE [456]. FAP thus appears to co-operate with the SRE in both induction of transcription by some external signals and the maintenance of low unstimulated expression. An involvement of AP1 complexes in the latter is implied by down-regulation of basal promoter activity in the presence of transfected Fos or Jun, and up-regulation when antisense Fos or Jun mRNAs are expressed [455].

The post-induction shut-off of serum-stimulated c-fos transcription is due to negative autoregulation, and constitutive Fos protein expression diminishes or inhibits the serum response of endogenous or transfected c-fos promoters [457,458]. The negative autoregulatory capacity of Fos protein involves its C-terminal domain [459,460], and may be controlled through its phosphorylation state [460]. Importantly, negative regulation of the serum response does not require the ability of Fos protein to interact with Jun, or to activate transcription [461], discounting the possibility of indirect repression by activation of expression of a negative regulatory factor.

The SRE alone is sufficient to act as a target of *trans*-repression [458,462–465], although in the intact c-*fos* promoter other sequences may also be involved in this process [457]. Other promoters including those of the Krox-20 and Krox-24 genes are transiently activated in response to serum refeeding. Like the c-*fos* promoter these contain CArG motifs which mediate *trans*-repression by Fos, hinting at a common mechanism for the post-induction shut-off of several immediate-early genes [458]. In mobility shift assays Fos protein is not present in complexes formed by SRF at the SRE [465], thus its mechanism of action is not known. It is not excluded that SRF and Fos may interact directly, this interaction being too weak or unstable for detection in mobility shift assays. Alternatively Fos may interact with a cofactor necessary for stimulation of transcription by SRF.

Because some factors are able to recognize sequences overlapping the SRE, the possibility arises that negative regulation may occur by competition for DNA binding. A zinc-finger protein YY1 competes for binding to the SRE *in vitro*, and in co-transfection experiments attenuates the response of the c-*fos* promoter to serum ([466]; also see [467,468]). However, because binding to the SRE does not appear to alter in the course of serum stimulation and post-induction repression, the significance of this binding site competition is uncertain.

The autoregulation of the c-*fos* gene reveals a general principle, i.e. the critical importance of transient responses to proliferative signals. As in the case of the c-*fos* gene, the activation of the c-*myc* gene appears to be limited by negative autoregulation ([469], references in [470]), and the disruption of such a feedback loop may play an important role in cellular transformation [471]. Transience of transcriptional responses may also be achieved through cross-regulatory interactions between immediate-early gene products; for instance, Fos and Jun have been described as negative regulators of the c-*myc* gene [472,473]. The following section describes a further class of regulatory events involved in controlling the cellular response to proliferative signals.

CONTROL OF GENE EXPRESSION IN THE CELL CYCLE

The strict regulation of proliferation is of obvious central importance to eukaryotic cells, because of the potentially oncogenic consequences of unregulated growth. The mechanisms controlling proliferation are increasingly well understood, and revolve around tumour suppressor proteins such as p53 and the retinoblastoma susceptibility gene product (Rb) (reviewed in [474–479]). The p53 and Rb proteins act as 'gatekeepers' of the cell cycle, blocking progression through the cycle unless proliferative signals converge appropriately. Their negative regulatory role is suggested by the discovery of mutations of the corresponding genes in a wide variety of tumours (references in [474,478]). Cell lines lacking functional tumour suppressor proteins are cell-cycle deregulated, and normal regulation can be restored by the introduction of wild-type p53 or Rb (e.g. [480,481]). Finally, oncoproteins of tumour viruses specifically interact with the tumour suppressor proteins, these interactions being essential for the oncogenic process (references in [474,477]). It appears that growth suppression by both p53 and Rb is at least partly accounted for by negative regulation of transcription, as we will now describe.

The Rb protein interacts with several cellular proteins [482], including Sp1 [483–485], ATF2 [486], c-myc [487] and the ets-like factor Elf-1 [488]. Its best understood interaction is with the transcription factor E2F [489–492], which was first characterized

as a regulator of the adenovirus E2 gene [493]. Binding sites for E2F are found in the promoters of several cell-cycle-regulated genes, including those for cdc2, thymidine kinase, DNA polymerase α, c-myb, c-myc and dihydrofolate reductase ([494–496]; other references in [497]). The developmentally regulated transcription factor DRTF1 [498,499] is extremely similar to E2F, and is not discussed separately here. E2F forms multicomponent DNA-binding complexes which alter in composition through the cell cycle [500], and contain Rb protein during the G1 phase [501]. Rb is subject to cyclical phosphorylation [502–504], and is hypophosphorylated in the G1 phase when complexed with E2F [489]. Two distinct lines of evidence indicate that this G1-specific E2F–Rb interaction may mediate negative effects of the tumour suppressor upon cell growth. Firstly, Rb arrests the cell cycle in G1 phase [480]. Secondly, tumour virus oncoproteins (SV40 T antigen, adenovirus E1a protein and human papillomavirus E7 protein) interact exclusively with hypophosphorylated Rb, suggesting that this is an important target of the oncogenic process [505,506]. The oncoproteins share an ability to disrupt the interaction between E2F and Rb [507,508], and mutant oncoproteins which fail to disrupt this interaction fail to deregulate growth. In co-transfection experiments E2F behaves as a potent, binding-site-dependent transcriptional activator. Wild-type Rb inhibits transcriptional activation by E2F [509,510], and a phosphorylation-defective mutant is a particularly potent repressor [509], in agreement with the proposed negative regulatory role of the hypophosphorylated form. Rb not only interferes with transcriptional activation by E2F, but may convert the E2F complex into a transcriptional repressor [78]. This suggests that Rb either possesses a silencer-like domain which can be anchored to DNA through the sequence-specific binding activity of E2F, or allosterically alters the regulatory activity of E2F. An E2F-binding site of the c-myb promoter does not activate transcription, but mediates negative regulation in the G_0 phase [496]. Such negative activity may be mediated by p107, an Rb-related tumour suppressor gene product, rather than by Rb itself. Nevertheless, this finding also indicates that E2F sites may be employed to present negative regulatory peptides to the basal transcription machinery, as well as to activate transcription. Finally, the negative effect of Rb upon transcriptional activation by E2F can be reversed in the presence of wild-type viral oncoproteins [511,512]. An undoubtedly oversimplified model for the role of Rb in cell-cycle control is illustrated in Figure 8. In the quiescent cell Rb is present in E2F complexes, down-regulating expression of target genes which are required for progression through the cell cycle. Proliferative signals lead to the phosphorylation of Rb, the dissociation of the E2F–Rb interaction, and the release of E2F as a transcriptional activator. This normal cell-cycle regulation is mimicked by two potentially oncogenic events: infection with tumour viruses, the oncoproteins of which sequester Rb, or mutations of Rb which prevent the interaction with E2F.

Unlike Rb, p53 possesses sequence-specific DNA-binding activity in its own right [513,514] and in several systems displays properties of a binding site-dependent transcription activator [515–518]. It also exerts negative regulatory control over many cellular and viral promoters [519–524]. The identification of *cis*-acting targets of negative regulation by p53 has proved difficult. In fact, minimal promoters containing little more than a TATA box can be down-regulated [522,525], repression possibly being mediated by direct interactions of p53 with the TATA-binding factors [525–527]. Several questions remain unanswered. What degree of specificity does p53 display for different promoters, and how is this achieved? How is its activity regulated in the cell

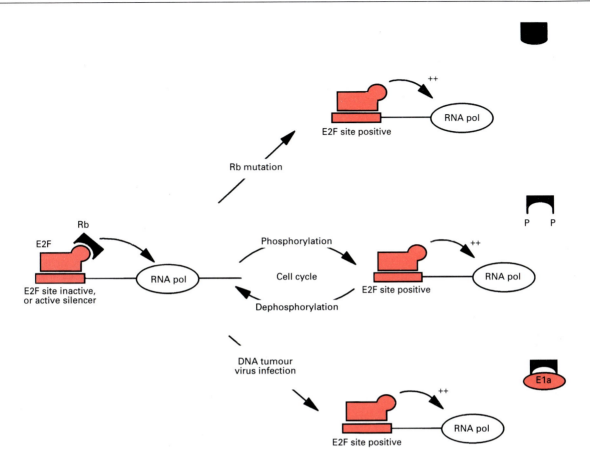

Figure 8   Cell cycle regulation of the activity of E2F complexes by Rb

The Figure also shows the disruption of this regulation through mutation of the Rb gene, or through the action of DNA tumour virus oncoproteins (here exemplified by the E1a gene product). RNA pol, RNA polymerase.

cycle? And what function is served by its positive and negative regulation of different target promoters?

The Wilms tumour suppressor gene product WT1 plays an important role in regulating proliferation in the developing kidney [528]. It is a zinc-finger DNA-binding protein which recognizes the same sites as the mitogen-inducible transcription activator Krox24 [529–531]. In transient transfection assays it represses transcription from the insulin-like growth factor II and platelet-derived growth factor A chain genes [532–534], both of which encode potent mitogens. The DNA-binding domain of WT1 is not sufficient for full repression, which therefore does not simply depend upon binding site competition between WT1 and Krox24 [118,533]. The N-terminus of WT1 contains a glutamine- and proline-rich active repressor domain, which can convert Krox24 into a repressor in domain swapping experiments [118].

The three tumour suppressor proteins discussed above share an ability to negatively regulate transcription, but otherwise differ in their properties. WT1 behaves as a tissue-specific silencer-like factor, regulating expression of growth factor genes in the kidney. p53 may be a broad-specificity negative regulator of transcription, and is also thought to control proliferation through regulation of DNA replication (references in [474]). Unlike the other two proteins, Rb possesses no specific DNA-binding activity in its own right, but behaves as a cell-cycle-regulated 'adaptor' for other transcription factors. It seems remarkably flexible in terms of its interactions, and the direction in which it

regulates transcription. It is likely that other tumour suppressor proteins, and other functions for the known tumour suppressor proteins, remain to be identified.

CONCLUDING REMARKS: REGULATORY FLEXIBILITY

It is clear that transcriptional repression is involved in the entire spectrum of eukaryotic regulatory events, and at this stage it would be surprising to find that any given gene was not subject to some form of negative regulation. It is also now commonplace to discover the participation of so-called positive transcription factors in negative regulatory events. Thus the description of any protein as a positive or negative transcription factor may have to be qualified by a description of the cellular and sequence context. It appears that the evolution of regulatory complexity in eukaryotic gene control has involved two separate processes: the growth of families of related but functionally distinct factors, and adaptive increases in the regulatory repertoire of single transcription factors. The first of these processes has occurred not only through duplication and divergence of genes encoding transcription factors, but also through alternative splicing and alternative translation initiation generating positive and negative factors from a single gene. Increases in the regulatory repertoires of transcription factors may involve several different effects: sequence-dependent variations in the higher-order structure and functional properties of the protein–DNA complex; signal-dependent post-translational modifications such as phosphorylation or ligand binding, which may alter the regulatory properties of the DNA-bound complex; the formation of various homo- and hetero-meric DNA-binding complexes which differ in *trans*-activating function; and context-dependent interactions with other complexes bound in *cis*. The finding of multiple positive and negative interactions between unrelated transcription factors, both on and off DNA, must signal a change in emphasis from the study of isolated regulatory events to a study of networks of overlapping and interconnecting pathways of transcriptional control.

A.R.C. was supported by a fellowship from the British Diabetic Association.

REFERENCES

1 Levine, M. and Manley, J. L. (1989) Cell **59**, 405–408
2 Goodbourn, S. (1990) Biochim. Biophys. Acta **1032**, 53–77
3 Renkawitz, R. (1990) Trends Genet. **6**, 192–197
4 Jackson, M. E. (1991) J. Cell Sci. **100**, 1–7
5 Baeuerle, P. A. and Baltimore, D. (1988) Cell **53**, 211–217
6 Baeuerle, P. A. and Baltimore, D. (1988) Science **242**, 540–546
7 Ghosh, S., Gifford, A. M., Riviere, L. R., Tempst, P., Nolan, G. P. and Baltimore, D. (1990) Cell **62**, 1019–1029
8 Pratt, W. B., Hutchison, K. A. and Scherrer, L. C. (1992) Trends Endocrinol. Metab. **3**, 326–333
9 Picard, D., Khursheed, B., Garabedian, M. J., Fortin, M. G., Lindquist, S. and Yamamoto, K. R. (1990) Nature (London) **348**, 166–168
10 Ptashne, M. (1986) A Genetic Switch, Cell Press and Blackwell Scientific Publications, San Francisco and Oxford
11 Rio, D., Robbins, A., Myers, R. and Tjian, R. (1980) Proc. Natl. Acad. Sci. U.S.A. **77**, 5706–5710
12 Carr, F. E., Burnside, J. and Chin, W. W. (1989) Mol. Endocrinol. **3**, 709–716
13 Crone, D. E., Kim, H.-S. and Spindler, S. R. (1990) J. Biol. Chem. **265**, 10851–10856
14 Krishna, V., Chatterjee, K., Lee, J.-K., Rentoumis, A. and Jameson, J. L. (1989) Proc. Natl. Acad. Sci. U.S.A. **86**, 9114–9118
15 Ray, A., LaForge, K. S. and Sehgal, P. B. (1990) Mol. Cell. Biol. **10**, 5736–5746
16 Strömstedt, P.-E., Poellinger, L., Gustaffson, J.-A. and Carlstedt-Duke, J. (1991) Mol. Cell. Biol. **11**, 33779–33783
17 Ackerman, S. L., Minden, A. G., Williams, G. T., Bobonis, C. and Yeung, C.-Y. (1991) Proc. Natl. Acad. Sci. U.S.A. **88**, 7523–7527
18 Barberis, A., Superti-Furga, G. and Busslinger, M. (1987) Cell **50**, 347–359
19 Karsenty, G., Ravazzolo, R. and deCrombrugghe, B. (1991) J. Biol. Chem. **266**, 24842–24848
20 Papazafiri, P., Ogami, K., Ramji, D. P., Nicosia, A., Monaci, P., Cladaras, C. and Zannis, V. I. (1991) J. Biol. Chem. **266**, 5790–5797
21 Roman, D. G., Toledano, M. B. and Leonard, W. J. (1990) New Biol. **2**, 642–647
22 Weissman, J. D. and Singer, D. S. (1991) Mol. Cell. Biol. **11**, 4217–4227
23 Yang, V. W., Christy, R. J., Cook, J. S., Kelly, T. J. and Lane, M. D. (1989) Proc. Natl. Acad. Sci. U.S.A. **86**, 3629–3633
24 Jaynes, J. B. and O'Farrell, P. H. (1988) Nature (London) **336**, 744–749
25 Han, K., Levine, M. S. and Manley, J. L. (1989) Cell **56**, 573–583
26 Oro, A. E., Hollenberg, S. M. and Evans, R. M. (1988) Cell **55**, 1109–1114
27 Masquilier, D. and Sassone-Corsi, P. (1992) J. Biol. Chem. **267**, 22460–22466
28 Lamph, W. W., Dwarki, V. J., Ofir, R., Montminy, M. and Verma, I. (1990) Proc. Natl. Acad. Sci. U.S.A. **87**, 4320–4324
29 Zhang, X.-K., Dong, J.-M. and Chiu, J.-F. (1991) J. Biol. Chem. **266**, 8248–8254
30 Guertin, M., LaRue, H., Bernier, D., Wrange, O., Chevrette, M., Gingras, M.-C. and Bélanger, L. (1988) Mol. Cell. Biol. **8**, 1398–1407
31 Sakai, D. D., Helms, S., Carlstedt-Duke, J., Gustaffsson, J.-A., Rottman, F. M. and Yamamoto, K. R. (1988) Genes Dev. **2**, 1144–1154
32 Drouin, J., Sun, Y. L., Chamberland, M., Gauthier, Y., DeLéan, A., Nemer, M. and Schmidt, T. J. (1993) EMBO J. **12**, 145–156
33 Drouin, J., Trifiro, M. A., Plante, R. K., Nemer, M., Eriksson, P. and Wrange, O. (1989) Mol. Cell. Biol. **9**, 5305–5314
34 Kieran, M., Blank, V., Logeat, F., Vandekerckhove, J., Lottspeich, F., Le Bail, O., Urban, M. B., Kourilsky, P., Baeuerle, P. A. and Israël, A. (1990) Cell **62**, 1007–1018
35 Fujita, T., Nolan, G. P., Ghosh, S. and Baltimore, S. (1992) Genes Dev. **6**, 775–787

36 Glass, C. K., Holloway, J. M., Devary, O. V. and Rosenfeld, M. G. (1988) Cell **54**, 313–323
37 Lipkin, S. M., Nelson, C. A., Glass, C. K. and Rosenfeld, M. G. (1992) Proc. Natl. Acad. Sci. U.S.A. **89**, 1209–1213
38 Leidig, F., Shepard, A. R., Zhang, W., Stelter, A., Cattini, P. A., Baxter, J. D. and Eberhardt, N. L. (1992) J. Biol. Chem. **267**, 913–921
39 Giese, K., Cox, J. and Grosschedl, R. (1992) Cell **69**, 185–195
40 Verrijzer, C. P., van Oosterhout, J. A. W. M., van Weperen, W. W. and van der Vliet, P. C. (1991) EMBO J. **10**, 3007–3014
41 Schreck, R., Zorbas, H., Winnacker, E.-L. and Baeuerle, P. A. (1990) Nucleic Acids Res. **18**, 6497–6502
42 Landschulz, W. H., Johnson, P. F. and McKnight, S. L. (1989) Science **243**, 1681–1688
43 Vinson, C. R., Sigler, P. B. and McKnight, S. L. (1989) Science **246**, 911–916.
44 Herr, W. and Sturm, R. A. (1988) Genes Dev. **2**, 1513–1516
45 Murre, C., McCaw, P. S. and Baltimore, D. (1989) Cell **56**, 777–783
46 Angel, P. and Karin, M. (1991) Biochim. Biophys. Acta **1072**, 129–157
47 Ransone, L. J. and Verma, I. M. (1990) Annu. Rev. Cell Biol. **6**, 539–557
48 Jones, N. (1990) Cell **61**, 9–11
48a Lamb, P. and McKnight, S. L. (1991) Trends Biochem. Sci. **16**, 417–422
49 Benezra, R., Davis, R. L., Lockshon, D., Turner, D. L. and Weintraub, H. (1990) Cell **61**, 49–59
50 Ellis, H. M., Spann, D. R. and Posakony, J. W. (1990) Cell **61**, 27–38
51 Garrell, J. and Modolell, J. (1990) Cell **61**, 39–48
52 Treacy, M. N., He, X. and Rosenfeld, M. G. (1991) Nature (London) **350**, 577–584
53 Ron, D. and Habener, J. F. (1992) Genes Dev. **6**, 439–453
54 Littlewood, T. D., Amati, B., Land, H. and Evan, G. I. (1992) Oncogene **7**, 1783–1792
55 Amati, B., Brooks, M. W., Levy, N., Littlewood, T. D., Evans, G. I. and Land, H. (1993) Cell **72**, 233–245
56 Amin, C., Wagner, A. J. and Hay, N. (1993) Mol. Cell. Biol. **13**, 383–390
57 Gu, W., Cechova, K., Tassi, V. and Dalla-Favera, R. (1993) Proc. Natl. Acad. Sci. U.S.A. **90**, 2935–2939
58 Ayer, D. E., Kretzner, L. and Eisenmann, R. N. (1993) Cell **72**, 211–222
59 Zervos, A. S., Gyuris, J. and Brent, R. (1993) Cell **72**, 223–232
60 Foulkes, N. S., Borrelli, E. and Sassone-Corsi, P. (1991) Cell **64**, 739–749
61 Foulkes, N. S., Laoide, B. M., Schlotter, F. and Sassone-Corsi, P. (1991) Proc. Natl. Acad. Sci. U.S.A. **88**, 5448–5452
62 Karpinski, B. A., Morle, G. D., Huggenvik, J., Uhler, M. D. and Leiden, J. M. (1992) Proc. Natl. Acad. Sci. U.S.A. **89**, 4820–4824
63 Descombes, P. and Schibler, U. (1991) Cell **67**, 569–579
64 Roman, C., Cohn, L. and Calame, K. (1991) Science **254**, 94–97
65 Reference deleted
66 Chiu, R., Angel, P. and Karin, M. (1989) Cell **59**, 979–986
67 Schütte, J., Viallet, J., Nau, M., Segal, S., Fedorko, J. and Minna, J. (1989) Cell **59**, 987–997
68 Mumberg, D., Lucibello, F. C., Schuermann, M. and Müller, R. (1991) Genes Dev. **5**, 1212–1223
69 Nakabeppu, Y. and Nathans, D. (1991) Cell **64**, 751–759
70 Yen, J., Wisdom, R. M., Tratner, I. and Verma, I. M. (1991) Proc. Natl. Acad. Sci. U.S.A. **88**, 5077–5081.
71 Treacy, M. N., Neilson, L. I., Turner, E. E., He, X. and Rosenfeld, M. G. (1992) Cell **68**, 491–505
72 Delmas, V., Laoide, B. M., Masquilier, D., DeGroot, R. P., Foulkes, N. S. and Sassone-Corsi, P. (1992) Proc. Natl. Acad. Sci. U.S.A. **89**, 4226–4230
73 Foulkes, N. S. and Sassone-Corsi, P. (1992) Cell **68**, 411–414
74 Yang-Yen, H.-F., Chambard, J.-C., Sun, Y.-L., Smeal, T., Schmidt, T. J., Drouin, J. and Karin, M. (1990) Cell **62**, 1205–1215
75 Schüle, R., Rangarajan, P., Kliewer, S., Ransone, L. J., Bolado, J., Yang, N., Verma, I. M. and Evans, R. M. (1990) Cell **62**, 1217–1226
76 Bengal, E., Ransone, L., Scharfmann, R., Dwarki, V. J., Tapscott, S. J., Weintraub, H. and Verma, I. M. (1992) Cell **68**, 507–519
77 Ma, J. and Ptashne, M. (1988) Cell **55**, 443–446
78 Weintraub, S. J., Prater, C. A. and Dean D. C. (1992) Nature (London) **358**, 259–261
79 Riggs, K. J., Merrell, K. T., Wilson, G. and Calame, K. (1991) Mol. Cell. Biol. **11**, 1765–1769
80 Kakkis, E., Riggs, K. J., Gillespie, W. and Calame, K. (1989) Nature (London) **339**, 718–721
81 Keleher, C. A., Goutte, C. and Johnson, A. D. (1988) Cell **53**, 927–936
82 Chen, S., West, R. W., Johnson, S. L., Gans, H., Kruger, B. and Ma, J. (1993) Mol. Cell. Biol. **13**, 831–840
83 Ptashne, M. and Gann, A. A. F. (1990) Nature (London) **346**, 329–331
84 Pugh, B. F. and Tjian, R. (1990) Cell **61**, 1187–1197
85 Martin, K. J., Lillie, J. W. and Green, M. R. (1990) Nature (London) **346**, 147–152
86 Gill, G. and Ptashne, M. (1988) Nature (London) **334**, 721–724

87 Meyer, M.-E., Gronemeyer, H., Turcotte, B., Bocquel, M.-T., Tasset, D. and Chambon, P. (1989) Cell **57**, 433–442

88 Braun, T., Bober, E. and Arnold, H. H. (1992) Genes Dev. **6**, 888–902

89 Hen, R., Borrelli, E. and Chambon, P. (1985) Science **230**, 1391–1394

90 Kalvakolanu, D. V. R., Liu, J., Hanson, R. W., Harter, M. L. and Sen, G. C. (1992) J. Biol. Chem. **267**, 2530–2536

91 Katoh, S., Ozawa, K., Kondoh, S., Soeda, E., Israel, A., Shiroki, K., Fujinaga, K., Itakura, K., Gachelin, G. and Yokoyama, K. (1990) EMBO J. **9**, 127–135

92 Nakajima, T., Nakamura, T., Tsunoda, S., Nakada, S. and Oda, K. (1992) Mol. Cell. Biol. **12**, 2837–2846

93 Offringa, R., Gebel, S., van Dam, H., Timmers, M., Smits, A., Zwart, R., Stein, R., Bos, J. L., van der Eb, A. and Herrlich, P. (1990) Cell **62**, 527–538

94 Rochette-Egly, C., Fromental, C. and Chambon, P. (1990) Genes Dev. **4**, 137–150

95 Stein, R. W. and Whelan, J. (1989) Mol. Cell. Biol. **9**, 4531–4534

96 Velcich, A. and Ziff, E. (1985) Cell **40**, 705–716

97 Webster, K. A., Muscat, G. E. O. and Kedes, L. (1988) Nature (London) **332**, 553–557

98 Horikoshi, N., Maguire, K., Kralli, A., Maldonardo, E., Reinberg, D. and Weinmann, R. (1991) Proc. Natl. Acad. Sci. U.S.A. **88**, 5124–5128

99 Brand, A. H., Breeden, L., Abraham, J., Sternglanz, R. and Nasmyth, K. (1986) Cell **41**, 41–48

100 Cao, S. X., Gutman, P. D., Dave, H. P. G. and Schechter, A. N. (1989) Proc. Natl. Acad. Sci. U.S.A. **86**, 5306–5309

101 Ip, Y. T., Kraut, R., Levine, M. and Rushlow, C. A. (1991) Cell **64**, 439–446

102 Kiss, I., Bösze, Z., Szabo, P., Altanchimeg, R., Barta, E. and Deák, F. (1990) Mol. Cell. Biol. **10**, 2432–2436

103 Mizuno, K., Goto, M., Masamune, Y. and Nakanishi, Y. (1992) Gene **119**, 293–297

104 Savagner, P., Miyashita, T. and Yamada, Y. (1990) J. Biol. Chem. **265**, 6669–6674

105 Hata, A., Ohno, S., Akita, Y. and Suzuki, K. (1989) J. Biol. Chem. **264**, 6404–6411

106 Baniahmad, A., Steiner, C., Köhne, A. J. and Renkawitz, R. (1990) Cell **61**, 505–514

107 Brand, A. H., Micklem, G. and Nasmyth, K. (1987) Cell **51**, 709–719

108 Imagawa, M., Osada, S., Okuda, A. and Muramatsu, M. (1991) Nucleic Acids Res. **19**, 5–10

109 Lee, J. W., Moffitt, P. G., Morley, K. L. and Peterson, D. O. (1991) J. Biol. Chem. **266**, 24101–24108

110 Winoto, A. and Baltimore, D. (1989) Cell **59**, 649–655

111 Farrell, F. X., Sax, C. M. and Zehner, Z. E. (1990) Mol. Cell Biol. **10**, 2349–2358

112 Nakabayashi, H., Hashimoto, T., Miyao, Y., Tjong, K.-K., Chan, J. and Tamaoki, T. (1991) Mol. Cell. Biol. **11**, 5885–5893

113 Roy, R. J., Gosselin, P., Anzivino, M. J., Moore, D. D. and Guérin, S. L. (1992) Nucleic Acids Res. **20**, 401–408

114 Baniahmad, A., Köhne, A. C. and Renkawitz, R. (1992) EMBO J. **11**, 1015–1023

114a Fondell, J. D., Roy, A. L. and Roeder, R. G. (1993) Genes Dev. **7**, 1400–1410

115 Licht, J. D., Grossel, M. J., Figge, J. and Hansen, U. M. (1990) Nature (London) **346**, 76–79

116 Jaynes, J. B. and O'Farrell, P. H. (1991) EMBO J. **10**, 1427–1433

117 Sauer, F. and Jäckle, H. (1991) Nature (London) **353**, 563–566

117a Sauer, F. and Jäckle, H. (1993) Nature (London) **364**, 454–457

118 Madden, S. L., Cook, D. M., Morris, J. F., Gashler, A., Sukhatme, V. P. and Rauscher, F. J. (1991) Science **253**, 1550–1553

119 Elgin, S. C. R. (1990) Curr. Opin. Cell Biol. **2**, 437–445

120 Rivier, D. H. and Rine, J. (1992) Curr. Opin. Genet. Dev. **2**, 286–292

121 Rivier, D. H. and Rine, J. (1992) Science **256**, 659–663

122 Miller, A. M. and Nasmyth, K. A. (1984) Nature (London) **312**, 247–251

123 Axelrod, A. and Rine, J. (1991) Mol. Cell. Biol. **11**, 1080–1091

124 Mahoney, D. J., Marquardt, R., Shei, G.-J., Rose, A. B. and Broach, J. R. (1991) Genes Dev. **5**, 605–615

125 Phi-Van, L. and Stratling, W. H. (1988) EMBO J. **7**, 655–664

126 Stief, A., Winter, D. M., Stratting, W. E. H. and Sippel, A. E. (1989) Nature (London) **341**, 343–345

127 Hofmann, J. R.-X., Laroch, T., Brand, A. H. and Gasser, S. M. (1989) Cell **57**, 725–737

128 Scheuermann, R. H. and Chen, U. (1989) Genes Dev. **3**, 1255–1266

128a Scheuermann, R. H. (1991) J. Biol. Chem. **267**, 624–634

129 Cedar, H. (1988) Cell **53**, 3–4

130 Enver, T., Zhang, J.-W., Papayannopoulou, T. and Stamatoyannopoulos, G. (1988) Genes Dev. **2**, 698–706

131 Becker, P. B., Ruppert, S. and Schütz, G. (1987) Cell **51**, 435–443

132 Antequara, F., Macleod, D. and Bird, A. P. (1989) Cell **58**, 5009–5017

133 Boyes, J. and Bird, A. (1991) Cell **64**, 1123–1134

134 Meehan, R. R., Lewis, J. D. and Bird, A. P. (1992) Nucleic Acids Res. **20**, 5085–5092

135 Meehan, R. R., Lewis, J. D., McKay, S., Kleiner, E. L. and Bird, A. P. (1989) Cell **58**, 499–507

136 Felsenfeld, G. (1992) Nature (London) **355**, 219–224

137 Han, M. and Grunstein, M. (1988) Cell **55**, 1137–1145

138 Kim, U.-J., Han, M., Kayne, P. and Grunstein, M. (1988) EMBO J. **7**, 2211–2219

139 Roth, S. Y., Shimizu, M., Johnson, L., Grunstein, M. and Simpson, R. T. (1992) Genes Dev. **6**, 411–425

140 Colantuoni, V., Pirozzi, A., Blance, C. and Cortese, R. (1987) EMBO J. **6**, 631–636

141 Hirsch, M.-R., Gaugler, L., Deagostini-Bazin, H., Bally-Cuif, L. and Goridis, C. (1990) Mol. Cell. Biol. **10**, 1959–1968

142 Imler, J.-L., Lemaire, C., Wasylyk, C. and Wasylyk, B. (1987) Mol. Cell. Biol. **7**, 2558–2567

143 Jackson, S. M., Keech, C. A., Williamson, D. J. and Gutierrez-Hartman, A. (1992) Mol. Cell. Biol. **12**, 2708–2719

144 Larsen, P. R., Harney, J. W. and Moore, D. D. (1986) Proc. Natl. Acad. Sci. U.S.A. **83**, 8283–8287

145 Shen, R., Goswami, S. K., Mascareno, E., Kumar, A. and Siddiqui, M. A. Q. (1991) Mol. Cell. Biol. **11**, 1676–1685

146 Seal, S. N., Davis, D. L. and Burch, J. B. E. (1991) Mol. Cell. Biol. **11**, 2704–2717

147 Simkevich, C. P., Thompson, J. P., Poppleton, H. and Raghow, R. (1992) Biochem. J. **286**, 179–185

148 Zhang, Z.-X., Kumar, V., Rivera, R. T., Chisholm, J. and Biswas, D. K. (1990) J. Biol. Chem. **265**, 4785–4788

149 Dynan, W. S. (1989) Cell **58**, 1–4

150 Weissman, J. D. and Singer, D. S. (1991) Mol. Cell. Biol. **11**, 4217–4227

151 Gourdeau, H. and Fournier, R. E. K. (1990) Annu. Rev. Cell Biol. **6**, 69–94

152 Chin, A. C. and Fournier, R. E. K. (1987) Proc. Natl. Acad. Sci. U.S.A. **84**, 1614–1618

153 Petit, C., Levilliers, J., Ott, M.-O. and Weiss, M. C. (1986) Proc. Natl. Acad. Sci. U.S.A. **83**, 2561–2565

154 Junker, S., Nielsen, V., Mattias, P. and Picard, D. (1988) EMBO J. **7**, 3093–3098

155 Baumhueter, S., Courtois, G. and Crabtree, G. R. (1988) EMBO J. **7**, 2485–2493

156 Tripputi, P., Guérin, S. L. and Moore, D. D. (1988) Science **241**, 1205–1207

157 Thayer, M. J., Lugo, T. G., Leach, R. J. and Fournier, R. E. K. (1990) Mol. Cell. Biol. **10**, 2660–2668

158 Bonapace, I. M., Sanchez, M., Obici, S., Gallo, A., Garofalo, S., Gentile, R., Cocozza, S. and Avvedimento, E. V. (1990) Mol. Cell. Biol. **10**, 1033–1040

159 Faraonio, R., Musy, M. and Colantuoni, V. (1990) Nucleic Acids Res. **18**, 7235–7242

160 Besnard, C., Monthioux, E., Loras., Jami, J. and Daegelen, D. (1991) J. Cell. Physiol. **146**, 349–355

161 Killary, A. M. and Fournier, R. E. K. (1984) Cell **38**, 523–534

162 Gourdeau, H., Peterson, T. C. and Fournier, R. E. K. (1989) Mol. Cell. Biol. **9**, 1813–1822

163 Boshart, M., Weih, F., Nichols, M. and Schütz, G. (1991) Cell **66**, 849–859

164 Jones, K. W., Shapero, M. H., Chevrette, M. and Fournier, R. E. K. (1991) Cell **66**, 861–872

165 Weiss, M. C. (1992) Nature (London) **355**, 22–23

166 Boshart, M., Weih, F., Schmidt, A., Fournier, R. E. K. and Schütz, G. (1990) Cell **61**, 905–916

167 Gonzalez, G. A. and Montminy, M. R. (1989) Cell **59**, 675–680

168 Chin, A. C. and Fournier, R. E. K. (1987) Proc. Natl. Acad. Sci. U.S.A. **84**, 1614–1618

169 Thayer, M. J. and Fournier, R. E. K. (1989) Mol. Cell. Biol. **9**, 2837–2846

170 Leshkowitz, D. and Walker, M. D. (1991) Mol. Cell. Biol. **11**, 1547–1552

171 Supowit, S. C., Ramsey, T. and Thompson, E. B. (1992) Mol. Endocrinol. **6**, 786–792

172 McCormick, A., Wu, D., Castrillo, J. L., Dana, S., Strobl, J., Thompson, E. B. and Karin, M. (1988) Cell **55**, 379–389

173 Junker, S., Pedersen, S., Schreiber, E. and Matthias, P. (1990) Cell **61**, 467–474

174 Bergman, Y., Strich, B., Sharir, H., Ber, R. and Laskov, R. (1990) EMBO J. **9**, 849–855

175 Bulla, G. A., DeSimone, V., Cortese, R. and Fournier, R. E. K. (1992) Genes Dev. **6**, 316–327

176 Cereghini, S., Yaniv, M. and Cortese, R. (1990) EMBO J. **9**, 2257–2263

177 Cereghini, S., Blumenfeld, M. and Yaniv, M. (1988) Genes Dev. **2**, 957–974

178 Yamada, T., Hitomi, Y., Shimizu, K., Ohki, M. and Oikawa, T. (1993) Mol. Cell. Biol. **13**, 1943–1950

179 Yu, H., Porton, B., Shen, L. and Eckhardt, L. A. (1989) Cell **58**, 441–448

180 Nitsch, D., Boshart, M. and Schütz, G. (1993) Genes Dev. **7**, 308–319

181 Shen, L., Lieberman, S. and Eckhardt, L. A. (1993) Mol. Cell. Biol. **13**, 3530–3540

182 Lawrence, P. A. (1992) The Making of a Fly, Blackwell Scientific Publications, Oxford

183 Akam, M. (1987) Development **101**, 1–22

184 Ingham, P. (1988) Nature (London) **335**, 25–34

185 Levine, M. and Hoey, T. (1988) Cell **55**, 537–540

186 Hoey, T. and Levine, M. (1988) Nature (London) **332**, 858–861

187 Tautz, D., Lehmann, R., Schnurch, H., Schuh, R., Seifert, E., Kienlin, A., Jones, K. and Jäckle, H. (1987) Nature (London) **327**, 383–389

188 Stanojevic, D., Hoey, T. and Levine, M. (1989) Nature (London) **341**, 331–335

189 Steward, R. (1987) Science **238**, 692–694

190 Capovilla, M., Eldon, E. D. and Pirrotta, V. (1992) Development **114**, 99–112

191 Krasnow, M. A., Saffman, E. E., Kornfeld, K. and Hogness D. S. (1989) Cell **57**, 1031–1043

192 Quian, S., Capovilla, M. and Pirotta, V. (1991) EMBO J. **10**, 1415–1425

193 Riley, G. R., Jorgensen, E. M., Baker, R. K. and Garber, R. L. (1991) Development Suppl. **1**, 177–185

194 Castelli-Gair, J. E., Capdevila, M.-P., Micol, J.-L. and Garcia-Bellido, A. (1992) Mol. Gen. Genet. **234**, 177–184

195 Appel, B. and Sakonju, S. (1993) EMBO J. **12**, 1099–1109

196 Pan, D. and Courey, A. J. (1992) EMBO J. **11**, 1837–1842

197 Jiang, J., Rushlow, C. A., Zhou, Q., Small, S. and Levine M. (1992) EMBO J. **11**, 3147–3154

198 Doyle, H. J., Kraut, R. and Levine, M. (1989) Genes Dev. **3**, 1518–1533

199 Leptin, M. (1991) Genes Dev. **5**, 1568–1576

200 Hoch, M., Seifert, E. and Jäckle, H. (1991) EMBO J. **10**, 2267–2278

201 Jäckle, H., Tautz, D., Schuh, R., Seifert, E. and Lehmann, R. (1986) Nature (London) **324**, 668–670

202 Kraut, R. and Levine, M. (1991) Development **111**, 611–621

203 Hoch, M., Gerwin, N., Taubert, H. and Jäckle, H. (1992) Science **256**, 94–97

204 Goto, T., Macdonald, P. and Maniatis, T. (1989) Cell **57**, 413–422

205 Howard, K. R. and Struhl, G. (1990) Development **110**, 1223–1231

206 Pankratz, M. J., Seifert, E., Gerwin, N., Billi, B., Nauber, U. and Jäckle, H. (1990) Cell **61**, 309–317

207 Riddihough, G. and Ish-Horowicz, D. (1991) Genes Dev. **5**, 840–854

208 Small, S., Kraut, R., Hoey, T., Warrior, R. and Levine, M. (1991) Genes Dev. **5**, 827–839

209 Warrior, R. and Levine, M. (1990) Development **110**, 759–767

210 Stanojevic, D., Small, S. and Levine, M. (1991) Science **254**, 1385–1387

211 Dearolf, C. R., Topol, J. and Parker, C. S. (1989) Genes Dev. **3**, 384–398

212 Brown, J. L., Sonoda, S., Ueda, H., Scott, M. P. and Wu, C. (1991) EMBO J. **10**, 665–674

213 Carroll, S. B. and Vavra, S. H. (1989) Development **107**, 673–683

214 Ish-Horowicz, D. and Pinchin, S. M. (1987) Cell **51**, 405–415

215 Topol, J., Dearolf, C. R., Prakash, K. and Parker, C. S. (1991) Genes Dev. **5**, 855–867

216 Hayashi, S. and Scott, M. P. (1990) Cell **63**, 883–894

217 Ohkuma, Y., Horikoshi, M., Roeder, R. G. and Despian, C. (1990) Cell **61**, 475–484

218 Biggin, M. D. and Tjian, R. (1989) Cell **58**, 433–440

219 Ten Harmsel, A., Austin, R. J., Savenelli, N. and Biggin, M. D. (1993) Mol. Cell. Biol. **13**, 2742–2752

220 Kirov, N., Zhelnin, L., Shah, J. and Rushlow, C. (1993) EMBO J. **12**, 3193–3199

221 Jiang, J., Cai, H, Zhou, Q. and Levine, M. (1993) EMBO J. **12**, 3201–3209

222 Hogan, B. L. M., Barlow, D. P. and Tilly, R. (1983) Cancer Surv. **2**, 115–140

223 Flanagan, J. R., Murata, M., Burke, P. A., Shirayoshi, Y., Appella, E., Sharp, P. A. amd Ozato, K. (1991) Proc. Natl. Acad. Sci. U.S.A. **88**, 3145–3149

224 Miyazaki, J.-I., Appella, E. and Ozato, K. (1986) Proc. Natl. Acad. Sci. U.S.A. **83**, 9537–9541

225 Sleigh, M. J. (1992) BioEssays **14**, 769–775

226 Vasios, G., Mader, S., Gold, J. D., Leid, M., Lutz, Y., Gaub, M.-P., Chambon, P. and Gudas, L. J. (1991) EMBO J. **10**, 1149–1158

227 Okazawa, H., Okamoto, K., Ishino, F., Ishino-Kaneko, T., Takeda, S., Toyoda, Y., Muramatsu, M. and Hamada, H. (1991) EMBO J. **10**, 2997–3005

228 Lenardo, M. J., Staudt, L., Robbins, P., Kuang, A., Mulligan, R. C. and Baltimore, D. (1989) Science **243**, 544–546

229 Kitabayashi, I., Kawakami, Z., Chiu, R., Ozawa, K., Matsuika, T., Toyashima, S., Umesono, K., Evans, R. M., Gachelin, G. and Yokayama, K. (1991) EMBO J. **11**, 167–175

230 DeGroot, R. P., Kruyt, F. A. E., van der Saag, P. T. and Kruijer, W. (1990) EMBO J. **9**, 1831–1837

231 Wasylyk, B., Imler, J. L., Chatton, B., Schatz, C. and Wasylyk, C. (1988) Proc. Natl. Acad. Sci. U.S.A. **85**, 7952–7956

232 Furukawa, K., Yamaguchi, Y., Ogawa, E., Shigesada, K., Satake, M. and Ito, Y. (1990) Cell Growth Differ. **1**, 135–147

233 Sleigh, M. J. (1987) Nucleic Acids Res. **15**, 9379–9395

234 Sleigh, M. J., Lockett, T. J., Kelly, J. and Lewy, D. (1987) Nucleic Acids Res. **15**, 4307–4324

235 Gorman, C. M., Rigby, P. W. J. and Lane, D. P. (1985) Cell **42**, 519–526

236 Sul, H. S. (1989) Curr. Opin. Cell Biol. **1**, 1116–1121

237 Swick, A. G. and Lane, M. D. (1992) Proc. Natl. Acad. Sci. U.S.A. **89**, 7895–7899

238 Distel, R. J., Ro, H.-S., Rosen, B. S., Groves, D. L. and Spiegelman, B. M. (1987) Cell **49**, 835–844

239 Rauscher, F. J., Sambucetti, L. C., Curran, T., Distel, R. J. and Spiegelman, B. M. (1988) Cell **52**, 471–480

240 Herrera, R., Ro, H. S., Robinson, G. S., Xanthopoulos, K. G. and Spiegelman, B. M. (1989) Mol. Cell. Biol. **9**, 5331–5339

241 Christy, R. J., Yang, V. W., Ntambi, J. M., Geiman, D. E., Landschulz, W. H., Friedman, A. D., Nakabeppu, Y., Kelly, T. J. and Lane, M. D. (1989) Genes Dev. **3**, 1323–1335

242 Samuelsson, L., Strömberg, K., Vikman, K., Bjursell, G. and Enerbäck, S. (1991) EMBO J. **10**, 3787–3793

243 Birkenmeier, E. H., Gwynn, B., Howard, S., Jerry, J., Gordon, J. I., Landschulz, W. H. and McKnight, S. L. (1989) Genes Dev. **3**, 1146–1156

244 Mischoulon, D., Rana, B., Bucher, N. L. and Farmer, S. R. (1992) Mol. Cell. Biol. **12**, 2553–2560

245 Umek, R. M., Friedman, A. D. and McKnight, S. L. (1991) Science **251**, 288–292.

246 Olson, E. N. (1990) Genes Dev. **4**, 1454–1461

247 Weintraub, H., Davis, R., Tapscott, S., Thayer, M., Krause, M., Benezra, R., Blackwell, K., Turner, D., Rupp, R., Hollenberg, S., Zhuang, Y. and Lassar, A. (1991) Science **251**, 761–766

248 Edmonson, D. G. and Olson, E. N. (1992) J. Biol. Chem. **268**, 755–758

248a Olson, E. N. (1992) Dev. Biol. **154**, 261–272

249 Crescenzi, M., Fleming, T. P., Lassar, A. B., Weintraub, H. and Aaronson, S. A. (1990) Proc. Natl. Acad. Sci. U.S.A. **87**, 8442–8446

250 Sorrentino, V., Pepperkok, R., Davis, R. L., Ansorge, W. and Philipson, L. (1990) Nature (London) **345**, 813–815

251 Brennan, T. J., Edmondson, D. G., Li, L. and Olson, E. N. (1991) Proc. Natl. Acad. Sci. U.S.A. **88**, 3822–3826

252 Edmondson, D. G., Brennan, T. J. and Olson, E. N. (1991) J. Biol. Chem. **266**, 21343–21346

253 Li, L., Heller-Harrison, R., Czech, M. and Olson, E. N. (1992) Mol. Cell. Biol. **12**, 4478–4485

254 Li, L., Zhou, J., James, G., Heller-Harrison, R., Czech, M. P. and Olson, E. N. (1992) Cell **71**, 1181–1194

255 Cohen, R., Pacifici, N., Rubinstein, N., Biehl, J. and Holtzer, H. (1977) Nature (London) **266**, 538–540

256 Martin, J. F., Li, L. and Olson, E. N. (1992) J. Biol. Chem. **267**, 10956–10960

257 Vaidya, T. B., Rhodes, S. J., Taparowsky, E. J. and Konieczny, S. F. (1989) Mol. Cell. Biol. **9**, 3576–3579

258 Lassar, A. B., Thayer, M. J., Overell, R. W. and Weintraub, H. (1989) Cell **58**, 659–667

259 Konieczny, S. F., Drobes, B. L., Menke, S. L. and Taparowsky, E. J. (1989) Oncogene **4**, 473–481

260 Denis, N., Blanc, S., Leibovitch, P., Nicolaiew, N., Dautry, F., Raymondjean, M., Kruh, J. and Kitzis, A. (1987) Exp. Cell Res. **172**, 212–217

261 Miner, J. H. and Wold, B. J. (1991) Mol. Cell. Biol. **11**, 2842–2851

262 Li, L., Chambard, J.-C., Karin, M. and Olson, E. N. (1992) Genes Dev. **6**, 676–689

263 Thayer, M. J., Tapscott, S. J., Davis, R. L., Wright, W. E., Lassar, A. B. and Weintraub, H. (1989) Cell **58**, 241–248

264 Schwarz, J. J., Chakraborty, T., Martin, J., Zhou, J. and Olson, E. N. (1992) Mol. Cell. Biol. **12**, 266–275

265 Park, K., Chung, M. and Kim, S.-J. (1992) Mol. Cell. Biol. **12**, 10810–10815

266 Church, G. M., Ephrussi, A., Gilbert, W. and Tonegawa, S. (1985) Nature (London) **313**, 798–801

267 Ephrussi, A., Church, G. M., Tonegawa, S. and Gilbert, W. (1985) Science **227**, 134–140

268 Karlsson, O., Edlund, T., Moss, J. B., Rutter, W. J. and Walker, M. D. (1987) Proc. Natl. Acad. Sci. U.S.A. **84**, 8819–8823

269 Moss, L. G., Moss, J. B. and Rutter, W. J. (1988) Mol. Cell. Biol. **8**, 2620–2627

270 Cockell, M., Stevenson, B. J., Staubin, M., Hagenbuchle, O. and Wellauer, P. K. (1989) Mol. Cell. Biol. **9**, 2464–2476

271 Kingston, R. E. (1989) Curr. Opin. Cell Biol. **1**, 1081–1087

272 Sun, X.-H., Copeland, N. G., Jenkins, N. A. and Baltimore, D. (1991) Mol. Cell. Biol. **11**, 5603–5611

273 Wilson, R. B., Kiledjian, M., Shen, C.-P., Benezra, R., Zwollo, P., Dymecki, S. M., Desiderio, S. V. and Kadesch, T. (1991) Mol. Cell. Biol. **11**, 6185–6191

274 Van Doren, M., Ellis, H. M. and Posakony, J. W. (1991) Development **113**, 245–255

275 Biggs, J., Murphy, E. V. and Israel, M. A. (1992) Proc. Natl. Acad. Sci. U.S.A. **89**, 1512–1516

276 Christy, B. A., Sanders, L. K., Lau, L. F., Copeland, N. G., Jenkins, N. A. and Nathans, D. (1991) Proc. Natl. Acad. Sci. U.S.A. **88**, 1815–1819

277 Cubas, P. and Modolell, J. (1992) EMBO J. **11**, 3385–3393

278 Duncan, M., DiCicco-Bloom, E. M., Xiang, X., Benezra, R. and Chada, K. (1992) Dev. Biol. **154**, 1–10

279 Vaessin, H., Caudy, M., Bier, E., Jan, L.-Y. and Jan, Y.-N. (1990) Cold Spring Harbor Symp. Quant. Biol. **55**, 239–245

280 Kreider, B. L., Benezra, R., Rovera, G. and Kadesch, T. (1992) Science **255**, 1700–1702

281 Jen, Y., Weintraub, H. and Benezra, R. (1992) Genes Dev. **6**, 1466–1479

282 Pongubala, J. M. R. and Atchison, M. L. (1991) Mol. Cell. Biol. **11**, 1040–1047

283 Cordle, S. R., Henderson, E., Masuoka, H., Weil, P. A. and Stein, R. (1991) Mol. Cell. Biol. **11**, 1734–1738

284 Nelsen, B. and Sen, R. (1992) Int. Rev. Cytol. **133**, 121–149

285 Jenuwein, T. and Grosschedl, R. (1991) Genes Dev. **5**, 932–943

286 Zaller, D. M. H., Yu, H. and Eckhardt, A. (1988) Mol. Cell. Biol. **8**, 1932–1939

287 Schöler, H. R. and Gruss, P. (1985) EMBO J. **4**, 3005–3013

288 Kitamura, D., Maeda, H., Araki, K., Kudo, A. and Watanabe, T. (1987) Eur. J. Immunol. **17**, 1249–1256

289 Ishihara, T., Kudo, A. and Watanabe, T. (1984) J. Exp. Med. **160**, 1937–1942

290 Cockerill, P. N., Yuen, M.-H. and Garrard, W. T. (1987) J. Biol. Chem. **262**, 5394–5397

291 Henthorn, P., Kiledjian, M. and Kadesch, T. (1990) Science **247**, 467–470

292 Beckmann, H., Su, L.-K. and Kadesch, T. (1990) Genes Dev. **4**, 167–179

293 Kadesch, T., Zervos, P. and Ruezinsky, D. (1986) Nucleic Acids Res. **14**, 8209–8221

294 Weinberger, J., Jat, P. S. and Sharp, P. A. (1988) Mol. Cell. Biol. **8**, 988–992

295 Ruezinsky, D., Beckmann, H. and Kadesch, T. (1991) Genes Dev. **5**, 29–37

296 Wang, J., Oketani, M. and Watanabe, T. (1991) Mol. Cell. Biol. **11**, 75–83

297 Orkin, S. H. (1990) Cell **63**, 665–672

298 Evans, R. M. (1988) Science **240**, 889–895

299 Enver, T., Raich, N., Ebens, A. J., Papayannopoulou, T., Costantini, F. and Stamatoyannopoulos, G. (1990) Nature (London) **344**, 309–313

300 Behringer, R. R., Ryan, T. M., Palmiter, R. D., Brinster, R. L. and Townes, T. M. (1990) Genes Dev. **4**, 380–389

301 Raich, N., Papayannopoulou, T., Stamatayannopoulos, G. and Enver, T. (1992) Blood **79**, 861–864

302 Lloyd, J. A., Krakowsky, J. M., Crable, S. C. and Lingrel, J. B. (1992) Mol. Cell. Biol. **12**, 1561–1567

303 Lindenbaum, M. H. and Grosveld, F. (1990) Genes Dev. **4**, 2075–2085

304 Shih, D. M., Wall, R. J. and Shapiro, S. G. (1990) Nucleic Acids Res. **18**, 5465–5472

305 Dillon, N. and Grosveld, F. (1991) Nature (London) **350**, 252–254

306 Peters, B., Merezhinskaya, N. and Noguchi, C. T. (1991) Blood **78** (Suppl. 1), 254a

307 Gutman, P. D., Cao, S. X., Dave, H. P. G., Mittelman, M. and Schechter, A. N. (1992) Gene **110**, 197–203

308 Wada-Kiyama, Y., Peters, B. and Noguchi, C. T. (1992) J. Biol. Chem. **267**, 11532–11538

309 Peters, B., Merezhinskaya, N., Diffley, J. F. X. and Noguchi, C. T. (1993) J. Biol. Chem. **268**, 3430–3437

310 Wood, W. G. (1989) in Hemoglobin Switching Part B, Cellular and Molecular Mechanisms (Stamatoyannopoulos, G. and Nienhuis, A. W., eds.), pp. 351–367, A. R. Liss, New York

311 Poncz, M., Henthorn, P., Stöckert, C. and Surrey, S. (1989) in Oxford Surveys on Eukaryotic Genes, vol. 5 (McLean, N., ed.), pp. 163–203, Oxford University Press, Oxford

312 Lumelsky, N. L. and Forget, B. G. (1991) Mol. Cell. Biol. **11**, 3528–3536

313 Superti-Furga, G., Barberis, A., Schaffner, G. and Busslinger, M. (1988) EMBO J. **7**, 3099–3107

314 Gumucio, D. L., Rood, K. L., Blanchard-McQuate, K. L., Gray, T. A., Saulino, A. and Collins, F. S. (1991) Blood **78**, 1853–1863

315 Gumucio, D. L., Rood, K. L., Gray, T. A., Riordan, M. F., Sartor, C. I. and Collins, F. S. (1988) Mol. Cell. Biol. **8**, 5310–5322

316 Berry, M., Grosveld, F. and Dillon, N. (1992) Nature (London) **358**, 499–502

317 Atweh, G. F., Liu, J. M., Brickner, H. E. and Zhu, X. X. (1988) Mol. Cell. Biol. **8**, 5047–5051

318 Berg, P. E., Williams, D. M., Qian, R.-L., Cohen, R. B., Cao, S-.X., Mittelman, M. and Schechter, A. N. (1989) Nucleic Acids Res. **17**, 8833–8852

319 Emerson, B. M., Nickol, J. M. and Fong, T. C. (1989) Cell **57**, 1189–1200

320 Jackson, P. D., Evans, T., Nickol, J. M. and Felsenfeld, G. (1989) Genes Dev. **3**, 1860–1873

321 Macleod, K. and Plumb, M. (1991) Mol. Cell. Biol. **11**, 4324–4332

322 Targa, F. R., Huesca, M. and Scherrer, K. (1992) Biochem. Biophys. Res. Commun. **188**, 416–423

323 O'Brien, R. M. and Granner, D. K. (1991) Biochem. J. **278**, 609–619

324 O'Brien, R. M., Lucas, P. C., Forest, C. D., Manguson, M. A. and Granner, D. K. (1990) Science **249**, 533–537

325 Kerr, L. D., Miller, D. B. and Matrisian, L. M. (1990) Cell **61**, 267–278

326 Okazaki, T., Ando, K., Igarashi, T., Ogata, E. and Fujita, T. (1992) J. Clin. Invest. **89**, 1268–1273

327 Visvanathan, K. V. and Goodbourn, S. (1989) EMBO J. **8**, 1129–1138

328 Evans, R. M. (1988) Science **240**, 889–895

329 Wahli, W. and Martinez, E. (1991) FASEB J. **5**, 2243–2249

330 Ham, J. and Parker, M. G. (1989) Curr. Opin. Cell Biol. **1**, 503–511

331 Green, S. (1993) Nature (London) **361**, 590–591

332 Brent, G. A., Williams, G. R., Harney, J. W., Forman, B. M., Samuels, H. H., Moore, D. D. and Larsen, P. R. (1991) Mol. Endocrinol. **5**, 542–548

333 Adler, S., Waterman, M. L., He, X. and Rosenfeld, M. G. (1988) Cell **52**, 685–695

333a Barettino, D., Bugge, T. H., Bartunek, P., Ruiz, M. D., Sonntag-Buck, V., Beug, H., Zenke, M. and Stunnenberg, H. G. (1993) EMBO J. **12**, 1343–1354

334 Charron, J. and Drouin, J. (1986) Proc. Natl. Acad. Sci. U.S.A. **83**, 8903–8907

335 Riegel, A. T., Lu, Y., Remenick, J., Wolford, R. G., Berard, D. S. and Hager, G. L. (1991) Mol. Endocrinol. **5**, 1973–1982

336 Akerblom, I. E., Slater, E. P., Beato, M., Baxter, J. D. and Mellon, P. L. (1988) Science **241**, 350–353

337 Keri, R. A., Andersen, B., Kennedy, G. C., Hamernik, D., Clay, C. M., Brace, A. D., Nett, T. M., Notides, A. C. and Nilson, J. H. (1991) Mol. Endocrinol. **5**, 725–733

338 Krishna, V., Chatterjee, K., Lee, J.-K., Rentoumis, A. and Jameson, J. L. (1989) Proc. Natl. Acad. Sci. U.S.A. **86**, 9114–9118

339 Brent, G. A., Dunn, M. K., Harney, J. W., Gulick, T., Larsen, P. R. and Moore, D. D. (1989) New Biol. **1**, 329–336

340 Graupner, G., Wills, K., Tzukerman, M., Zhang, X.-K. and Pfahl, M. (1989) Nature (London) **340**, 653–656

341 Damm, K., Thompson, C. C. and Evans, R. M. (1989) Nature (London) **339**, 593–597

342 Williams, G. R., Harney, J. W., Moore, D. D., Larsen, P. R. and Brent, G. A. (1992) Mol. Endocrinol. **6**, 1527–1537

343 Baniahmad, A., Muller, M., Steiner, C. and Renkawitz, R. (1987) EMBO J. **6**, 2297–2303

344 Rentoumis, A., Krishna, V., Chatterjee, K., Madison, L. D., Datta, S., Gallagher, G. D., Degroot, L. J. and Jameson, J. L. (1990) Mol. Endocrinol. **4**, 1522–1531

345 Izumo, S. and Mahdavi, V. (1988) Nature (London) **334**, 539–542

346 Wight, P. A., Crew, M. D. and Spindler, S. R. (1987) J. Biol. Chem. **262**, 5659–5663

347 Brent, G. A., Williams, G. R., Harney, J. W., Forman, B. M., Samuels, H. H., Moore, D. D. and Larsen, P. R. (1991) Mol. Endocrinol. **5**, 542–548

348 Carr, F. E., Kaseem, L. L. and Wong, N. C. W. (1992) J. Biol. Chem. **267**, 18689–18694

349 Näar, A. M., Boutin, J.-M., Lipkin, S. M., Yu, V. C., Holloway, J. M., Glass, C. K. and Rosenfeld, M. G. (1991) Cell **65**, 1267–1279

350 Umesono, K., Murakami, K. K., Thompson, C. C. and Evans, R. M. (1991) Cell **65**, 1255–1266

351 Williams, G. R. and Brent, G. A. (1992) J. Endocrinol. **135**, 191–194

352 Kim, H.-S., Crone, D. E., Sprung, C. N., Tillman, J. B., Force, W. R., Crew, M. D., Mote, P.L and Spindler, S. R. (1992) Mol. Endocrinol. **6**, 1489–1501

353 Harding, P. P. and Duester, G. (1992) J. Biol. Chem. **267**, 14145–14150

354 Graupner, G., Zhang, X.-K., Tzukerman, M., Wills, K., Hermann, T. and Pfahl, M. (1991) Mol. Endocrinol. **5**, 365–372

355 Cooney, A. J., Tsai, S. Y., O'Malley, B. W. and Tsai, M. J. (1992) Mol. Cell. Biol. **12**, 4153–4163

356 Tran, P., Zhang, X.-K., Salbert, G., Hermann, T., Lehmann, J. M. and Pfahl, M. (1992) Mol. Cell. Biol. **12**, 4666–4676

357 Kliewer, S. A., Umesono, K., Heyman, R. A., Mangelsdorf, D. J., Dyck, J. A. and Evans, R. M. (1992) Proc. Natl. Acad. Sci. U.S.A. **89**, 1448–1452

358 Liu, Y., Yang, N. and Teng, C. T. (1993) Mol. Cell. Biol. **13**, 1836–1846.

359 Gaub, M.-P., Dierich, A., Astinotti, D., Touitou, I. and Chambon, P. (1987) EMBO J. **6**, 2313–2320

360 Schmitt-Ney, M., Doppler, W., Ball, R. K. and Groner, B. (1991) Mol. Cell. Biol. **11**, 3745–3755

361 Corthesy, B., Cardinaux, J.-R., Claret, F.-X. and Wahli, W. (1989) Mol. Cell. Biol. **9**, 5548–5562

362 Schild, C., Claret, F.-X., Wahli, W. and Wolffe, A. P. (1993) EMBO J. **12**, 423–433

362 Cato, A. C. B., König, H., Ponta, H. and Herrlich, P. (1992) J. Steroid Biochem. Mol. Biol. **43**, 63–68

363 Miner, J. H., Diamond, M. I. and Yamamoto, K. R. (1991) Cell Growth Differ. **2**, 525–530

364 Miner, J. N. and Yamamoto, K. R. (1991) Trends Biochem. Sci. **16**, 423–426

365 Ponta, H., Cato, A. C. B. and Herrlich, P. (1992) Biochim. Biophys. Acta **1129**, 255–261

366 Schüle, R. and Evans, R. M. (1990) Trends Genet. **7**, 377–381

367 Zhang, X.-K., Wills, K. N., Husmann, M., Hermann, T. and Pfahl, M. (1991) Mol. Cell. Biol. **11**, 6016–6025

368 Nicholson, R. C., Mader, S., Nagpal, S., Leid, M., Rochette-Egly, C. and Chambon, P. (1990) EMBO J. **9**, 4443–4454

369 Jonat, C., Rahmsdorf, H. J., Park, K.-K., Cato, A. C. B., Gebel, S., Ponta, H. and Herrlich, P. (1990) Cell **62**, 1189–1204

370 Schüle, R., Rangarajan, P., Yang, N., Kliewer, S., Ransone, L. J., Bolado, J., Verma, I. M. and Evans, R. M. (1991) Proc. Natl. Acad. Sci. U.S.A. **88**, 6092–6096

371 Touray, M., Ryan, F., Jaggi, R. and Martin, F. (1991) Oncogene **6**, 1227–1234

372 Tzukerman, M., Zhang, X.-K. and Pfahl, M. (1991) Mol. Endocrinol. **5**, 1983–1992

373 Doucas, V., Spyrou, G. and Yaniv, M. (1991) EMBO J. **10**, 2237–2245

374 Lucibello, F. C., Slater, E. P., Jooss, K. U., Beato, M. and Müller, R. (1990) EMBO J. **9**, 2827–2834

375 König, H., Ponta, H., Rahmsdorf, H. J. and Herrlich, P. (1992) EMBO J. **11**, 2241–2246

376 Kerppola, T. K., Luk, D. and Curran, T. (1993) Mol. Cell. Biol. **13**, 3782–3791

377 Schüle, R., Umesono, K., Mangelsdorf, D. J., Bolado, J., Pike, J. W. and Evans, R. M. (1990) Cell **61**, 497–504

378 Diamond, M. I., Miner, J. N., Yoshinaga, S. K. and Yamamoto, K. R. (1990) Science **249**, 1266–1272

379 Yoshinaga, S. K. and Yamamoto, K. R. (1991) Mol. Endocrinol. **5**, 844–853

380 Miner, J. N. and Yamamoto, K. R. (1992) Genes Dev. **6**, 2491–2501

381 Mordacq, J. C. and Linzer, D. I. H. (1989) Genes Dev. **3**, 760–769

382 Northrop, J. P., Crabtree, G. R. and Mattila, P. S. (1992) J. Exp. Med. **175**, 1235–1245

383 Vacca, A., Felli, M. P., Farina, A. R., Martinotti, S., Maroder, M., Screpanti, I., Meco, D., Petrangeli, E., Frati, L. and Gulino, A. (1992) J. Exp. Med. **175**, 637–646

384 Shemshedini, L., Knauthe, R., Sassone-Corsi, P., Pornon, A. and Gronemeyer, H. (1991) EMBO J. **10**, 3839–3849

385 Weisz, A. and Rosales, R. (1990) Nucleic Acids Res. **18**, 5097–5106

386 Owen, T. A., Bortell, R., Yocum, S. A., Smock, S. L., Zhang, M., Abate, C., Ahalhoub, V., Aronin, N., Wright, K. L., van Wijnen, A. J., Stein, J. L., Curran, T., Lian, J. B. and Stein, G. S. (1990) Proc. Natl. Acad. Sci. U.S.A. **87**, 9990–9994

387 Tverberg, L. A. and Russo, A. F. (1992) J. Biol. Chem. **267**, 17567–17573

388 Gaub, M.-P., Bellard, M., Scheuer, I., Chambon, P. and Sassone-Corsi, P. (1990) Cell **63**, 1267–1276

389 Imai, E., Strömstedt, P.-E., Quinn, P. G., Carlstedt-Duke, J., Gustafsson, J.-A. and Granner, D. K. (1990) Mol. Cell. Biol. **10**, 4712–4719

390 Rushlow, C. and Warrior, R. (1992) Bioessays **14**, 89–95

391 Gilmore, T. D. (1990) Cell **62**, 841–843

392 Grimm, S. and Baeuerle, P. A. (1993) Biochem. J. **290**, 297–308

393 Blank, V., Kourilsky, P. and Israel, A. (1992) Trends Biochem. Sci. **17**, 135–140

394 Baeuerle, P. A. and Baltimore, D. (1989) Genes Dev. **3**, 1689–1698

395 Lenardo, M. J. and Baltimore, D. (1989) Cell **58**, 227–229

396 Nolan, G. P., Ghosh, S., Liou, H.-C., Tempst, P. and Baltimore, D. (1991) Cell **64**, 961–969

397 Schmitz, M. L. and Baeuerle, P. A. (1991) EMBO J. **10**, 3805–3817

398 Urban, M. B., Schreck, R. and Baeuerle, P. A. (1991) EMBO J. **10**, 1817–1825

399 Beg, A. A., Finco, T. S., Nantermet, P. V. and Baldwin, A. S. (1993) Mol. Cell. Biol. **13**, 3301–3310

400 Beg, A. A., Ruben, S. M., Scheinman, R. I., Haskill, S., Rosen, C. A. and Baldwin, A. S. (1992) Genes Dev. **6**, 1899–1913

401 Bours, V., Granzoso, G., Azarenko, V., Park, S., Kanno, T., Brown, K. and Siebenlist, U. (1993) Cell **72**, 729–739

402 Brown, K., Park, S., Kanno, T., Franzoso, G. and Siebenlist, U. (1993) Proc. Natl. Acad. Sci. U.S.A. **90**, 2532–2536

403 Fan, C.-M. and Maniatis, T. (1991) Nature (London) **354**, 395–398

404 Fujita, T., Nolan, G. P., Liou, H.-C., Scott, M. L. and Baltimore, D. (1993) Genes Dev. **7**, 1354–1363

405 Ganchi, P. A., Sun, S. C., Greene, W. C. and Ballard, D. W. (1992) Mol. Biol. Cell **3**, 1339–1352

406 Geisler, R., Bergmann, A., Hiromi, Y. and Nusslein-Volhard, C. (1993) Cell **71**, 613–621

407 Haskill, S., Beg, A. A., Tompkins, S. M., Morris, J. S., Yurochko, A. D., Sampson-Johannes, A., Mondal, K., Ralph, P. and Baldwin, A. S. (1991) Cell **65**, 1281–1289

408 Hatada, E. N., Nieters, A., Wulczyn, F. G., Naumann, M., Meyer, R., Nucifora, G., McKeithan, T. W. and Scheidereit, C. (1992) Proc. Natl. Acad. Sci. U.S.A. **89**, 2489–2493

409 Henkel, T., Zabel, U., van Zee, K., Müller, J. M., Fanning, E. and Baeuerle, P. A. (1992) Cell **68**, 1121–1133

410 Inoue, J.-I., Takahara, T., Akizawa, T. and Hino, O. (1993) Oncogene **8**, 2067–2073

411 Inoue, J.-I., Kerr, L. D., Kakizuka, A. and Verma, I. M. (1992) Cell **68**, 1109–1120

412 Inoue, J.-I., Kerr, L. D., Rashid, D., Davis, N., Bose, H. R. and Verma, I. M. (1992) Proc. Natl. Acad. Sci. U.S.A. **89**, 4333–4337

413 Kerr, L. D., Inoue, J.-I., Davis, N., Link, E., Baeuerle, P. A., Bose, H. R. and Verma, I. M. (1991) Genes Dev. **5**, 1464–1476

414 Link, E., Kerr, L. D., Schreck, R., Zabel, U., Verma, I. M. and Baeuerle, P. A. (1992) J. Biol. Chem. **267**, 239–246

415 Liou, H.-C., Nolan, G. P., Ghosh, S., Fujita, T. and Baltimore, D. (1992) EMBO J. **11**, 3003–3009

416 Mercurio, F., DiDonata, J., Rosette, C. and Karin, M. (1992) DNA Cell Biol. **11**, 523–537

417 Naumann, M., Nieters, A., Hatada, E. N. and Scheidereit, C. (1993) Oncogene **8**, 2275–2281

418 Naumann, M., Wulczyn, F. G. and Scheidereit, C. (1993) EMBO J. **12**, 213–222

419 Nolan, G. P., Fujita, T., Bhatia, K., Huppi, C., Liou, H.-C., Scott, M. L. and Baltimore, D. (1993) Mol. Cell. Biol. **13**, 3557–3566

420 Rice, N. R., MacKichan, M. L. and Israel, A. (1992) Cell **71**, 243–253

421 Sun, S. C., Ganchi, P. A., Ballard, D. W. and Greene, W. C. (1993) Science **259**, 1912–1915

422 Urban, M. B. and Baeuerle, P. A. (1990) Genes Dev. **4**, 1975–1984

423 Wulczyn, F. G., Naumann, M. and Scheidereit, C. (1992) Nature (London) **358**, 597–599

424 Zabel, U. and Baeuerle, P. A. (1990) Cell **61**, 255–265

425 Zabel, U., Henkel, T., dos Santos Silva, M. and Baeuerle, P. A. (1993) EMBO J. **12**, 201–211

426 Ghosh, S. and Baltimore, D. (1990) Nature (London) **344**, 678–682

427 Pierce, J. W., Gifford, A. M. and Baltimore, D. (1991) Mol. Cell. Biol. **11**, 1431–1437

428 DeMaeyer, E. and DeMaeyer-Guinard, J. (1988) Interferons and Other Regulatory Cytokines, John Wiley and Sons Inc., New York

429 Goodbourn, S., Burstein, H. and Maniatis, T. (1986) Cell **45**, 601–610

430 Goodbourn, S. and Maniatis, T. (1988) Proc. Natl. Acad. Sci. U.S.A. **85**, 1447–1451

431 Nourbakhsh, M., Hoffman, K. and Hauser, H. (1993) EMBO J. **12**, 451–459

432 Dirks, W., Mittnacht, S., Rentrop, M. and Hauser, H. (1989) J. Interferon Res. **9**, 125–133

433 Fan, C.-M. and Maniatis, T. (1989) EMBO J. **8**, 101–110

434 Dinter, H. and Hauser, H. (1987) EMBO J. **6**, 599–604

435 Whittemore, L.-A. and Maniatis, T. (1990) Mol. Cell. Biol. **10**, 1329–1337

436 Whittemore, L.-A. and Maniatis, T. (1990) Proc. Natl. Acad. Sci. U.S.A. **87**, 7799–7803

437 Dale, T. C., Rosen, J. M., Guille, M. J., Lewin, A. R., Porter, A. G. C., Kerr, I. M. and Stark, G. R. (1989) EMBO J. **8**, 831–839

438 Fujita, T., Kimura, Y., Miyamoto, M., Barsoumian, E. L. and Taniguchi, T. (1989) Nature (London) **337**, 270–272

439 Harada, H., Fujita, T., Miyamoto, M., Kimura, Y., Maruyama, M., Furia, A., Miyata, T. and Taniguchi, T. (1989) Cell **58**, 729–739

440 Harada, H., Willison, K., Sakakibara, J., Miyamoto, M., Fujita, T. and Taniguchi, T. (1990) Cell **63**, 303–312

441 Keller, A. D. and Maniatis, T. (1988) Proc. Natl. Acad. Sci. U.S.A. **85**, 3309–3313

442 Keller, A. D. and Maniatis, T. (1991) Genes Dev. **5**, 868–879

443 Levy, D. E., Kessler, D. S., Pine, R. and Darnell, J. E. (1988) Genes Dev. **2**, 383–393

444 Nelson, N., Marks, M. S., Driggers, P. H. and Ozato, K. (1993) Mol. Cell. Biol. **13**, 588–599

445 Pine, R., Decker, T., Kessler, D. S., Levy, D. E. and Darnell, J. E. (1990) Mol. Cell. Biol. **10**, 2448–2457

446 Reis, L., Harada, H., Wolchok, J. D., Taniguchi, T. and VilKek, J. (1992) EMBO J. **11**, 185–193

447 Whiteside, S. T., Visvanathan, K. V. and Goodbourn, S. (1992) Nucleic Acids Res. **20**, 1531–1538

448 Treisman, R. (1992) Trends Biochem. Sci. **17**, 423–426

449 Dey, A., Nebert, D. W. and Ozato, K. (1991) DNA Cell Biol. **10**, 537–544

450 Herrera, R. E., Shaw, P. E. and Nordheim, A. (1989) Nature (London) **340**, 68–70

451 Boulden, A. M. and Sealy, L. J. (1992) Mol. Cell. Biol. **12**, 4769–4783

452 Shaw, P. E., Schroter, H. and Nordheim, A. (1989) Cell **56**, 563–572

453 Treisman, R. (1990) Semin. Cancer Biol. **1**, 47–58

454 Gauthier-Rouvière, C., Basset, M., Lamb, N. J. C. and Fernandez, A. (1992) Oncogene **7**, 363–369

455 Schönthal, A., Büscher, M., Angel, P., Rahmsdorf, H. J., Ponta, H., Hattori, K., Chiu, R., Karin, M. and Herrlich, P. (1989) Oncogene **4**, 629–636

456 Morgan, I. M. and Birnie, G. D. (1992) Cell Prolif. **25**, 205–215

457 Sassone-Corsi, P., Sisson, J. C. and Verma, I. M. (1988) Nature (London) **334**, 314–319

458 Gius, D., Cao, X., Rauscher, F. J., Cohen, D. R., Curran, T. and Sukhatme, V. P. (1990) Mol. Cell. Biol. **10**, 4243–4255

459 Wilson, T. and Treisman, R. (1988) EMBO J. **7**, 4193–4202

460 Ofir, R., Dwarki, V. J., Rashid, D. and Verma, I. M. (1990) Nature (London) **348**, 80–82

461 König, H., Ponta, H., Rahmsdorf, U., Buscher, M., Schonthal, A., Rahmsdorf, H. J. and Herrlich, P. (1989) EMBO J. **8**, 2559–2566

462 Lucibello, F. C., Lowag, C., Neuberg, M. and Müller, R. (1989) Cell **59**, 999–1007

463 Subramaniam, M., Schmidt, L. J., Crutchfield, C. E. and Getz, M. J. (1989) Nature (London) **340**, 64–66

464 Shaw, P. E., Frasch, S. and Nordheim, A. (1989) EMBO J. **8**, 2567–2574

465 Rivera, V. M., Sheng, M. and Greenberg, M. E. (1990) Genes Dev. **4**, 255–268

466 Gualberto, A., lePage, D., Pons, G., Mader, S. L., Park, K., Atchison, M. L. and Walsh, K. (1992) Mol. Cell. Biol. **12**, 4209–4214

467 Lee, T.-C., Chow, K.-L., Fang, P. and Schwartz, R. J. (1991) Mol. Cell. Biol. **11**, 5090–5100

468 Park, K. and Atchison, M. L. (1991) Proc. Natl. Acad. Sci. U.S.A. **88**, 9804–9808

469 Penn, L. J. Z., Brooks, M. W., Laufer, E. M. and Land, H. (1990) EMBO J. **9**, 1113–1121

470 Lüscher, B. and Eisenman, R. N. (1990) Genes Dev. **4**, 2025–2035

471 Grignani, F., Lombardi, L., Inghirami, G., Sternas, L., Cechova, K. and Dalla-Favera, R. (1990) EMBO J. **9**, 3913–3922

472 Hay, N., Takimoto, M. and Bishop, J. M. (1989) Genes Dev. 3, 293–303

473 Takimoto, M., Quinn, J. P., Farina, A. R., Staudt, L. M. and Levens, D. (1989) J. Biol. Chem. **264**, 8992–8999

474 Levine, A. J., Momand, J. and Finlay, C. A. (1991) Nature (London) **351**, 453–456

475 Marshall, C. J. (1991) Cell **64**, 313–326

476 Sager, R. (1992) Curr. Opin. Cell Biol. **4**, 155–160

477 Cobrinik, D., Dowdy, S. F., Hinds, P. W., Mittnacht, S. and Weinberg, R. A. (1992) Trends Biochem. Sci. **17**, 312–315

478 Weinberg, R. A. (1991) Science **254**, 1138–1146

479 Horowitz, J. M. (1993) Genes Chromosom. Cancer **6**, 124–131

480 Goodrich, D. W., Wang, N. P., Quian, Y.-W., Lee, E. Y.-H. P. and Lee, W.-H (1991) Cell **67**, 293–302

481 Mercer, W. E., Shields, M. T., Amin, M., Sauve, G. J., Appella, E., Romano, J. W. and Ullrich, S. J. (1990) Proc. Natl. Acad. Sci. U.S.A. **87**, 6166–6170

482 Kaelin, W. G., Pallas, D. C., DeCaprio, J. A., Kaye, F. J. and Livingston, D. M. (1991) Cell **64**, 521–532

483 Robbins, P. D., Horowitz, J. M. and Mulligan, R. C. (1990) Nature (London) **346**, 668–671

484 Kim, S.-J., Lee, H.-D., Robbins, P. D., Busam, K., Sporn, M. B. and Roberts A. B. (1991) Proc. Natl. Acad. Sci. U.S.A. **88**, 3052–3056

485 Kim, S.-J., Onwuta, U. S., Lee, Y. I., Li, R., Botchan, M. R. and Robbins, P. D. (1992) Mol. Cell. Biol. **12**, 2455–2463

486 Kim, S.-J., Wagner, S., Liu, F., O'Reilly, M. A., Robbins, P. D. and Green, M. R. (1992) Nature (London) **358**, 331–334

487 Rustgi, A. K., Dyson, N. and Bernards, R. (1991) Nature (London) **352**, 541–543

488 Wang, C. Y., Petryniak, B., Thompson, C. B., Kaelin, W. G. and Leiden, J. M. (1993) Science **260**, 1330–1335

489 Chellappan, S. P., Hiebert, S., Mudryj, M., Horowitz, J. M. and Nevins, J. R. (1991) Cell **65**, 1053–1061

490 Chittenden, T., Livingston, D. M. and Kaelin, W. G. (1991) Cell **65**, 1073–1082

491 Kaelin, W. G., Krek, W., Sellers, W. R., DeCaprio, J. A., Ajchenbaum, F., Fuchs, C. S., Chittenden, T., Li, Y., Farnham, P. J., Blanar, M. A., Livingston, D. M. and Flemington, E. K. (1992) Cell **70**, 351–364

492 Helin, K., Lees, J. A., Vidal, M., Dyson, N., Harlow, E. and Fattaey, A. (1992) Cell **70**, 337–350

493 Yee, A. S., Raychaudhuri, P., Jakoi, L. and Nevins, J. R. (1989) Mol. Cell. Biol. **9**, 578–585

494 Mudryj, M., Hiebert, S. W. and Nevins, J. R. (1990) EMBO J. **9**, 2179–2184

495 Dalton, S. (1992) EMBO J. **11**, 1797–1804

496 Lam, E. W.-F. and Watson, R. (1993) EMBO J. **12**, 2705–2713

497 Nevins, J. R. (1992) Science **258**, 424–429

498 LaThangue, N. B., Thimmappaya, B. and Rigby, P. W. J. (1990) Nucleic Acids Res. **18**, 2929–2938

499 Shivji, M. K. K. and LaThangue, N. B. (1991) Mol. Cell. Biol. **11**, 1686–1695

500 Mudryj, M., Devoto, S. H., Hiebert, S. W., Hunter, T., Pines, J. and Nevins, J. R. (1991) Cell **65**, 1243–1253

501 Shirodkar, S., Ewen, M., DeCaprio, J. A., Morgan, J., Livingston, D. M. and Chittenden, T. (1992) Cell **68**, 157–166

502 Buchkovich, K., Duffy, L. A. and Harlow, E. (1989) Cell **58**, 1097–1105

503 Chen, P.-L., Scully, P., Shew, J.-Y., Wang, J. Y. J. and Lee, W.-H. (1989) Cell **58**, 1193–1198

504 Lin, B. T.-Y., Gruenwald, S., Morla, A. O., Lee, W.-H. and Wang, J. Y. J. (1991) EMBO J. **10**, 857–864

505 Ludlow, J. W., DeCaprio, J. A., Huang, C.-M., Lee, W.-H., Paucha, E. and Livingston, D. M. (1989) Cell **56**, 57–65

506 Imai, E., Strömstedt, P.-E., Quinn, P. G., Carlstedt-Duke, J., Gustafsson, J.-A. and Granner, D. K. (1990) Mol. Cell. Biol. **10**, 4712–4719

507 Chellappan, S., Kraus, V. B., Kroger, B., Munger, K., Howley, P. M., Phelps, W. C. and Nevins, J. R. (1992) Proc. Natl. Acad. Sci. U.S.A. **89**, 4549–4553

508 Bandara, L. R. and LaThangue, N. B. (1991) Nature (London) **351**, 494–497

509 Hamel, P. A., Gill, R. M., Phillips, R. A. and Gallie B. L. (1992) Mol. Cell. Biol. **12**, 3431–3438

510 Hiebert, S. W., Chellappan, S. P., Horowitz, J. M. and Nevins J. R. (1992) Genes Dev. **6**, 177–185

511 Arroyo, M. and Raychaudhuri, P. (1992) Nucleic Acids Res. **20**, 5947–5954

512 Zamanian, M. and LaThangue, N. B. (1992) EMBO J. **11**, 2603–2610

513 Kern, S. E., Kinzler, K. W., Bruskin, A., Jarosz, D., Friedman, P., Prives, C. and Vogelstein, B. (1991) Science **252**, 1708–1711

514 Weissker, S. N., Muller, B. F., Homfeld, A. and Deppert, W. (1992) Oncogene **7**, 1921–1932

515 Fields, S. and Jang, S. K. (1990) Science **249**, 1046–1049

516 Farmer, G., Bargonetti, J., Zhu, H., Friedman, P., Prywes, R. and Prives, C. (1992) Nature (London) **358**, 83–86

517 Zambetti, G. P., Bargonetti, J., Walker, K., Prives, C. and Levine, A. J. (1992) Genes Dev. **6**, 1143–1152

518 Schärer, E. and Iggo, R. (1992) Nucleic Acids Res. **20**, 1539–1545

519 Ginsberg, D., Mechta, F., Yaniv, M. and Oren, M. (1991) Proc. Natl. Acad. Sci. U.S.A. **88**, 9979–9983

520 Subler, M. A., Martin, D. W. and Deb, S. (1992) J. Virol. **66**, 4757–4762

521 Santhanam, U., Ray, A. and Sehgal P. B. (1991) Proc. Natl. Acad. Sci. U.S.A. **88**, 7605–7609

522 Kley, N., Chung, R. Y., Fay, S., Loeffler, J. P. and Seizinger B. R. (1992) Nucleic Acids Res. **20**, 4083–4087

523 Chin, K.-V., Ueda, K., Pastan, I. and Gottesman, M.M (1992) Science **255**, 459–462

524 Agoff, S. N., Hou, J., Linzer, D. I. H. and Wu, B. (1993) Science **259**, 84–87

525 Seto, E., Usheva, A., Zambetti, G. P., Momand, J., Horikoshi, N., Weinmann, R., Levine, A. J. and Shenk, T. (1992) Proc. Natl. Acad. Sci. U.S.A. **89**, 12028–12032

526 Liu, X., Miller, C. W., Koeffler, P. H. and Berk, A. J. (1993) Mol. Cell. Biol. **13**, 3291–3300

527 Ragimov, N., Krauskipf, A., Navot, N, Rotter, V., Oren, M. and Aloni, Y. (1993) Oncogene **8**, 1183–1193

528 Haber, D. A. and Housman, D. E. (1992) Cancer Surv. **12**, 105–117

529 Call, K. M., Glaser, T., Ito, C. Y., Buckler, A. J., Pelletier, J., Haber, D. A., Rose, E. A., Kral, A., Yeger, H., Lewis, W. H., Jones, C. and Housman, D. E. (1990) Cell **60**, 509–520

530 Rauscher, F. J., Morris, J. F., Tournay, O. E., Cook, D. M. and Curran, T. (1990) Science **250**, 1259–1262

531 Lemaire, P., Vesque, C., Schmitt, J., Stunneberg, H., Frank, R. and Charnay, P. (1990) Mol. Cell. Biol. **10**, 3456–3467

532 Drummond, I. A., Madden, S. L., Rohwer–Nutter, P., Bell, G. I., Sukhatme, V. P. and Rauscher, F. J. (1992) Science **257**, 674–678

533 Wang, Z. Y., Madden, S. L., Deuel, T. F. and Rauscher, F. J. (1992) J. Biol. Chem. **267**, 21999–22002

534 Gashler, A. L., Bonthron, D. T., Madden, S. L., Rauscher, F. J., Collins, T. and Sukhatme, V. P. (1992) Proc. Natl. Acad. Sci. U.S.A. **89**, 10984–10988

Biochem. J. (1993) **290**, 1–13 (Printed in Great Britain)

REVIEW ARTICLE
Stress response of yeast

Willem H. MAGER* and Pedro MORADAS FERREIRA†

* Department of Biochemistry and Molecular Biology, Vrije Universiteit, de Boelelaan 1083, 1081 HV Amsterdam, The Netherlands,
and † Departamento de Biologia Molecular, Instituto de Ciencias Biomedicas Abel Salazar, Universidade do Porto, 4000 Porto, Portugal

INTRODUCTION

All living cells display a rapid molecular response to adverse environmental conditions, a phenomenon commonly designated as the heat shock response. Since other kinds of stress have similar effects, this process can be considered as a general cellular response to metabolic disturbances. The most striking feature of the heat shock response is the induced synthesis of a set of proteins conserved during evolution, the heat shock proteins (hsps).

The heat shock response was first reported as a dramatic change in gene activity induced by a brief heat treatment of *Drosophila hydei* larvae (seen as changes in 'puffing patterns' of salivery gland polytene chromosomes; Ritossa, 1964). Until 1980 most attention was focused on the heat shock response in *Drosophila*, where, indeed, heat shock proteins were first discovered (Tissières et al., 1974). Similar findings with other eukaryotes and prokaryotes soon suggested that the response represents an evolutionarily conserved genetic system which might be beneficial for the living cell.

In yeast, in particular in the main species of investigations, *Saccharomyces cerevisiae*, a sudden temperature change generates considerable, temporary, alterations in the pattern of protein biosynthesis (Miller et al., 1979). Moreover, when yeast cells were shifted to a higher temperature, growth was transiently arrested at the G_1 phase of the cell division cycle (Johnston and Singer, 1980). Changes in the rate of protein synthesis were monitored by amino acid pulse-labelling, followed by two-dimensional gel analysis. These initial studies revealed a 10-fold or greater induction or repression of most proteins (Miller et al., 1979). Subsequent analysis demonstrated that, upon a shift of *S. cerevisiae* from 23 °C to 37 °C, out of 500 proteins examined more than 80 were transiently induced (20 of which could be classified as major heat shock proteins) and with more than 300 proteins the synthesis was reduced (Miller et al., 1982). The response to heat shock was temporary, with recovery occurring within a few hours after stress exposure.

The finding that the heat shock response is inhibited in mutants defective in RNA synthesis or processing/transport indicated that transcription plays a predominant role (Miller et al., 1979; McAlister and Finkelstein, 1980). Indeed, changes in the patterns of protein synthesis *in vivo* were found to parallel changes in the levels of translatable mRNAs (McAlister and Finkelstein, 1980). A major class of protein whose synthesis is strongly reduced are ribosomal proteins, which displayed a co-ordinated transient decrease upon shifting yeast cells from 23 °C to 36 °C (Gorenstein and Warner, 1979; Kim and Warner, 1983). The decrease in mRNA levels corresponding to repressed protein synthesis appeared to be much faster than could be explained merely by transcriptional arrest (McAlister and Finkelstein,

1980). In agreement with this finding, for ribosomal protein mRNAs a (heat-induced) temporarily enhanced decay rate was demonstrated (Herruer et al., 1985).

Although the early data suggested the universal nature of the heat shock response, differences were shown to exist in the way *Drosophila* and yeast achieve the fast changes in protein biosynthesis (Lindquist, 1981). In *Drosophila* a sudden heat shock induces a translational control mechanism, both specifically repressing the translation of pre-existing mRNAs and inducing the synthesis of mRNAs encoding hsps. In yeast such translational control of mRNAs does not occur.

The spectrum of hsps synthesized in yeast upon a stress challenge is similar to that produced in other cells. Several families can be distinguished which are designated, according to their average apparent molecular mass, hsp100 (in yeast hsp104), hsp90 (in yeast hsp83), hsp70, hsp60 (the chaperonin or groEL-family) and small-size hsps (in yeast hsp26 and hsp12); see Table 1. Several proteins homologous to hsps are synthesized constitutively, reflecting the important cellular functions performed by these proteins under normal circumstances. In addition, the rate of synthesis of several other proteins, e.g. ubiquitin, some glycolytic enzymes and a plasma membrane protein, is strongly

Table 1 Stress proteins of yeast

Designation	Cellular localization	Function
Hsp150	(Secretory)	Unknown
Hsp104	Nucle(ol)us	Stress tolerance
Hsp83	Cytosol/nucleus	Chaperone
Hsp70		
ssa1	Cytosol	Chaperone
ssa2	Cytosol	Chaperone
ssa3	Cytosol	Chaperone
ssa4	Cytosol	Chaperone
ssb1	Unknown	Unknown
ssb2	Unknown	Unknown
ssc1	Mitochondria	Chaperone
ssd1 (kar2)	Endoplasmic reticulum	Chaperone
Hsp60	Mitochondria	Chaperone
Hsp30	Plasma membrane	Unknown
Hsp26	Cytosol/nucleus	Unknown
Hsp12	Cytosol?	Unknown
Ubiquitin	Cytosol	Protein degradation
Enzymes		
Enolase	Cytosol	Glycolysis
Glyceraldehyde 3-phosphate dehydrogenase	Cytosol	Glycolysis
Phosphoglycerate kinase	Cytosol	Glycolysis
Catalase	Cytosol	Antioxidative defense

Abbreviations used: hsp, heat shock protein; GRE, glucocorticoid response element; HSG, heat shock granule; HSF, heat shock transcription factor; HSE, heat shock responsive element; URS, upstream repression site.
* To whom correspondence should be addressed.

induced upon exposing yeast cells to stress. These proteins, therefore, should also be considered as heat shock proteins.

The functional significance of the heat shock response is evident from the magnitude and the speed of the process. The ability of a cell to shift rapidly to heat shock protein synthesis suggests that it is pivotal for survival ('emergency response'; Lindquist and Craig, 1988). A shift from 23°C to 30 °C did not result in any major alterations in the pattern of protein synthesis (McAlister et al., 1979), but a shift to 36 °C (mild heat shock) does. Yet, the elevated temperature of 36 °C is within the normal growth range of yeast. Therefore, it has been assumed from the initial studies that the response might play a major role in the protection of cells from thermal injury. Indeed, a rapid shift in cultivation temperature of *S. cerevisiae* from 23 °C to 36 °C leads to protection from death due to a subsequent extreme (52 °C) heat treatment. It is a common feature of the response in all living organisms that it is evoked below lethal temperatures and, therefore, may provide the cell with the ability to withstand even higher, otherwise lethal, temperatures. The acquisition of stress tolerance, thus, is an important aspect of the stress response. A correlation between the cellular levels of hsps, transiently induced in yeast by a mild heat shock, and the level of acquired thermoresistance has been described as early as 1980 (McAlister and Finkelstein, 1980). However, a convincing body of data provide evidence that increased hsp levels are not sufficient for stress tolerance attainment (see below).

This Review deals with the cellular functions of hsps and other stress-induced proteins in yeast. In addition, stress-induced changes in yeast gene expression are discussed with particular emphasis on the specific *trans*-acting transcription factor HSF, its interaction with the corresponding *cis*-acting nucleotide element, HSE, and the regulation of transcription of some *HSP* genes. Also some stress-affected cellular processes are briefly discussed. Finally, data concerning the acquisition by yeast of tolerance against various kinds of environmental stress are reviewed. The concluding remarks address putative signals that trigger the stress response in yeast.

CELLULAR FUNCTIONS OF HEAT SHOCK (-INDUCED) PROTEINS

Heat shock proteins have been implicated in all major growth-related processes such as cell division, DNA synthesis, transcription, translation, protein folding and transport, membrane function; see Table 1 for a summary. The various members of the hsp subfamilies identified in yeast are discussed below.

Hsp70, mediator of protein folding

HSP70 proteins belong to the most highly conserved proteins in the cell. A 40–60% identity exists between hsp70 proteins found in higher eukaryotes and the *E. coli* hsp70 dnaK; among eukaryotes this percentage is even higher (reviewed by Lindquist, 1986; Lindquist and Craig, 1988). The genes encoding hsp70 in *S. cerevisiae* constitute a multigene family consisting of eight members which show amino acid sequence identities ranging from 50 to 97% (Ingolia et al., 1982; Craig et al., 1990). They are subdivided into four subfamilies, *SSA*, *SSB*, *SSC* and *SSD* ('Stress Seventy'). The *SSA* subfamily is indispensable for growth and encodes the ssa1, ssa2, ssa3, and ssa4 proteins, which are localized in the cytoplasm (Slater and Craig, 1989; Boorstein and Craig, 1990). Mutations in *SSB1* and *SSB2* result in a cold-sensitive phenotype and the cellular localization of the corresponding proteins is unknown (Craig et al., 1990). The protein

encoded by *SSC1* is a mitochondrial protein (Craig et al., 1987, 1989), whereas the *SSD1* (better known as *KAR2*) gene product occurs in the endoplasmic reticulum (Normington et al., 1989; Rose et al., 1989). Both genes are essential for growth. Expression of the family members is regulated differently upon changes in growth conditions. For instance, ssa4 is a characteristic heat shock protein displaying very low basal levels of expression and a strong induction upon heat treatment. In contrast, ssa1 and ssa2 show a rather high constitutive expression. This aspect will be discussed below, in the section on stress-induced changes in yeast gene expression. Also in other eukaryotic cells multiple genes coding for similar sets of related hsp70 proteins are present.

Sequence similarity between hsp70 proteins extends over the entire protein, but particularly conserved regions are present in the N-terminal part. In this region of the proteins an ATP-binding site is present displaying a weak ATPase activity (Chappell et al., 1987). A crystallographic study of a bovine hsp70 revealed a surprising similarity in tertiary structure and topology between this ATPase domain and the ATP-binding core of hexokinase (Flaherty et al., 1990). This shared property of hsp70 proteins is consistent with the current view that hsps70, in general, play a part in protein–protein interactions. More precisely, hsps70 may act by binding to certain polypeptide chains, hence modifying or maintaining their conformation or interaction with other proteins. According to this general model, release from such substrates depends on ATP (Rothman, 1989; Craig and Gross, 1991). The model of an 'ATP-driven detergent' action of hsp70 was proposed by Lewis and Pelham (1985). Their basic idea was that hsp70 binds to denatured, aggregated proteins and aids their solubilization, using the energy of ATP hydrolysis for release, with simultaneous (re)folding of the proteins. [See Gething and Sambrook (1992) for a recent review on all aspects of protein folding in the cell.]

All available data fit with such a chaperone function of hsp70. The first hsp70 whose cellular function was established is the mammalian uncoating enzyme, involved in ATP-dependent release of clathrin from coated vesicles (Rothman, 1989). Furthermore, the *E. coli* hsp70, encoded by the *dnaK* gene, was originally identified as a factor implicated in bacteriophage lambda DNA replication, being involved in the ATP-dependent disassembly of the primary nucleoprotein complex formed at the origin of replication (reviewed by Georgopoulos et al., 1990; Ang et al., 1991). However, dnaK plays also an important part in the normal growth of *E. coli*. Very recently, by *in vitro* reconstitution experiments, dnaK, dnaJ and groEL (the latter two also being stress proteins) were demonstrated to serve as protein-folding chaperones in a strictly sequential fashion (Langer et al., 1992). In HeLa cells, hsps70 have been implicated in the normal nuclear transport of proteins (Shi and Thomas, 1992).

Also in yeast, hsps70 were found to fulfil a major function under normal conditions. Evidence was obtained that hsps70 facilitate translocation of polypeptides across the endoplasmic reticulum and mitochondrial membranes (Deshaies et al., 1988; Chirico et al., 1988). Yeast cells depleted of the hsp70 proteins encoded by *SSA1* and *SSA2* appeared to accumulate in the cytosol precursor forms of proteins that are normally destined for import into the endoplasmic reticulum and mitochondria, indicating that hsps70 are involved in post-translational import pathways.

The protein encoded by *SSD1* (identical to *KAR2*) is an homologue of mammalian BiP (immunoglobulin heavy chain binding protein) which is most likely identical to grp78 (glucose-regulated protein), initially identified as a protein induced upon starvation for glucose (Shiu et al., 1977). *KAR2* was identified in a yeast mutant blocked in nuclear fusion after mating of haploid cells to form diploids (Normington et al., 1989; Rose et al., 1989). The function of BiP is uncertain but it appears to restrict transport of malfolded or aberrantly glycosylated secretory proteins from the endoplasmic reticulum to the Golgi body (reviewed by Deshaies et al., 1988; Pelham, 1989).

It is very likely that hsp70 proteins induced upon stress exposure perform functions similar to those under normal growth conditions. During stress, the cellular concentration of potential substrates, e.g. denatured proteins, is likely to increase, thus depleting the free pool of hsp70 and generating the need for an increase in the level of these proteins. Hsp70, therefore, has been considered as the cellular thermometer (DiDomenico et al., 1982; Craig and Gross, 1991). The intriguing possibility that hsp70 may also directly interact with the heat shock factor, thereby modulating its transcription activating potency, will be discussed below, in the section on the heat shock transcription factor HSF.

Hsp60, a chaperonin

Hsp60 proteins fulfil cellular functions that, presumably, are similar to those of hsp70. Hsp60 in yeast has been identified as a mitochondrial protein showing homology to the *E. coli* groEL protein (Cheng et al., 1989). *groEL* encodes a protein involved in bacteriophage head assembly (Sternberg, 1973; Georgopoulos and Hohn, 1978; reviewed by Zeilstra-Ryall et al., 1991). The yeast homologue was isolated as a nuclear mutation (mif4) preventing assembly of F_1-ATPase, cytochrome b_2 and the Rienke FeS protein of complex III (Cheng et al., 1989). The *HSP60* gene was found to be able to rescue the defect (reviewed by Pfanner et al., 1990). In mutants of *S. cerevisiae* defective in the constitutive expression of hsp60, incompletely processed proteins imported into the mitochondrial matrix appeared to accumulate (Cheng et al., 1989). Hsp60, therefore, most likely belongs to the proteins that facilitate post-translational assembly of polypeptides, commonly called molecular chaperones or 'chaperonins' (Ellis, 1987; Ellis and Vies, 1991). Proteins imported into the mitochondria do not fold spontaneously but need hsp60 function for proper folding (Ostermann et al., 1989; Neupert et al., 1990; Koll et al., 1992). Hsp60, consistently, acts in conjunction with ATP. Recently, another yeast protein, scj1, has been identified which is homologous to bacterial dnaJ (Blumenberg, 1991; Antencio and Yaffe, 1992), and may aid together with hsp70 in translocating proteins into the mitochondria.

Hsp83, cytoplasmic anchoring protein

Yeast hsp83 belongs to the family of hsp90 proteins. Also this family of heat shock genes encodes chaperone-like proteins. In eukaryotes, hsp90 is abundantly present in the cytoplasm; a small fraction translocates to the nucleus upon heat shock (reviewed by Lindquist and Craig, 1988; Schlessinger, 1990). Hsps90 have been found to interact with a variety of cellular proteins, including glucocorticoid receptors, several kinases and the cytoskeleton proteins actin and tubulin.

In particular, complex formation between hsp90 and the steroid receptors has been studied in detail (reviewed by Hunt, 1989). The receptor proteins are kept in the cytoplasm in an inactive conformation through their interaction with hsp90. Hsps90, therefore, serve as cytoplasmic anchoring proteins. Upon hormone binding to the receptor, hsp90 is released and the hormone–receptor complex moves into the nucleus and acts as a

transcription factor through specific responsive nucleotide elements, GREs.

The structure of hsps90 is highly conserved from bacteria to man and shows among eukaryotes at least 50 % sequence identity. Two notable regions with a very high content of negative charges are present in all eukaryotic hsp90 proteins. In addition, the C-terminal sequence of the protein, which in itself is rather divergent, uniformly ends with an EEVD motif of unknown function.

S. cerevisiae contains two genes encoding hsp90: *HSP83* and *HSC83* (Borkovich, 1989). *HSC83* ('heat shock cognate') is a constitutively expressed gene and is only weakly induced upon stress exposure. *HSP83* is expressed at a much lower basal level and is strongly activated upon heat treatment. Notably, expression of this gene is also induced when cells enter the stationary phase (Kurtz and Lindquist, 1984) or sporulate (Kurtz et al., 1986).

Though yeast probably does not naturally respond to steroid hormones, the glucocorticoid receptor from mammalian cells can act in yeast (Metzger et al., 1988; Schena and Yamamoto, 1988). Recently yeast strains were designed in which hsp90 expression can be regulated (Picard et al., 1991). At low levels of hsp90, aporeceptors appeared to be mostly hsp90-free. Yet, they failed to enhance transcription from a GRE-containing promoter. On hormone addition the receptors were activated, but with a markedly reduced efficiency as compared with cells possessing normal hsp90 levels. Therefore, hsp90 seems to facilitate the response of the receptor to the hormone signal. Apparently hsp90 acts in the signal transduction pathway for steroid receptors.

Hsp104, protector of nucleoli?

Much less is known with regard to the cellular function of heat shock proteins with a molecular mass greater than 100 kDa. Hsp110 in mouse has a nuclear localization, and is predominantly present in the nucleoli (Subjek, 1983). It was found to be associated with the fibrillar component of nucleoli, which, most likely, is the site of rDNA. This finding suggests that it may bind to rRNA. It has been proposed, therefore, that upon heat shock hsp110 is induced to protect ribosome formation, a process very sensitive to heat stress (Nover et al., 1986; see below, in the section on processes affected by stress).

A yeast gene, *HSP104*, belonging to this family has been isolated (Sanchez and Lindquist, 1990) and sequenced (Parsell et al., 1991). Hsp104 protein is not detectable at normal growth on fermentable carbon sources, is constitutively synthesized in respiring cells (Sanchez et al., 1992), and is strongly induced following a heat shock. Expression of this protein is also activated when cells enter the stationary phase or are induced to sporulate. Like in other eukaryotes, it may be a nucleolar protein to which an important function in the acquisition of stress tolerance has been assigned (see below). Two putative sites showing homology with a nucleotide consensus binding site have been identified in hsp104. By site-directed mutagenesis these sites were shown to be essential for the stress-protective function of this protein (Parsell et al., 1991). Hsp104 displays a striking similarity to the highly conserved ClpA/ClpB protein family, first identified in *E. coli* and supposed to possess ATP-dependent protease activity (Parsell et al., 1991).

Small hsps: hsp26 and hsp12

Yeast cells contain two small hsps: hsp26 and hsp12. The small hsps form a very diverse group which, nevertheless, display conserved structural elements (reviewed by Lindquist and Craig, 1988) and share the ability to form highly polymeric structures referred to as heat shock granules, HSGs (reviewed by Tuite et al., 1990). They show a notable and significant sequence similarity to the eye lens protein α-crystallin, in particular with respect to a highly conserved hydrophobic domain located at the C-terminus of these proteins (Tuite et al., 1990).

Hsp26 from yeast (Petko and Lindquist, 1986; Kurtz et al., 1986; Bossier et al., 1989) also is able to self-aggregate. The HSGs have a native molecular mass of 550 kDa, and contain about 20 copies of the protein (Tuite et al., 1990). Hsp26 aggregates probably accumulate in the perinuclear region of the cell (Nover et al., 1983; Leicht et al., 1986).

A universal property of the small hsps may also be their developmental regulation. Indeed, both yeast genes belonging to this group, *HSP26* and *HSP12* (Praekelt and Meacock, 1990) show, apart from a very strong stress-induction, also a dramatically increased expression following transition of cells to the stationary phase and upon induction of sporulation. The cellular role of these proteins, however, has yet to be elucidated.

Ubiquitin, mediator of proteolysis

Polyubiquitin, a protein encoded by the *UBI4* gene in yeast, is generally considered as a heat shock protein since it displays a strongly enhanced rate of synthesis under stress conditions (Finley et al., 1987). Selective, nonlysosomal proteolysis is mediated by the post-translational ubiquitination pathway (reviewed by Finley and Varshavsky, 1985; Hershko, 1988; Jentsch et al., 1990). Ubiquitin is a protein consisting of 76 amino acids which is found in eukaryotes either free or covalently bound to various other proteins. Examples of ubiquitinated proteins in mammalian cells are histones H2A and H2B, denatured globin and the platelet-derived growth factor receptor.

Attachment of ubiquitin to proteins occurs through its C-terminal glycine residue, in a series of steps: binding of ubiquitin to enzyme E1, transfer of activated ubiquitin to several conjugating enzymes E2 and, finally, joining of ubiquitin to specific proteins, sometimes through E3. Conjugation of ubiquitin to proteins can trigger their degradation. In yeast, *RAD6* has been identified as a gene encoding an E2-like enzyme, amongst others, involved in ubiquitination of histone H2B (Jentsch et al., 1987; Sung et al., 1988). A second ubiquitin carrier protein, showing sequence homology with rad6, has been identified as cdc34, a protein required for G_1 to S phase transition (Goebl et al., 1988).

Yeast ubiquitin differs in only three amino acids from mammalian ubiquitin (Finley and Varshavsky, 1985) and is encoded by a multigene family: *UBI1*, *UBI2*, *UBI3* and *UBI4*. *UBI1*, *2* and *3* encode hybrid proteins (Finley et al., 1989; Redman and Rechsteiner, 1989; Müller-Taubenberg et al., 1989; Özkaynak et al., 1987). Strikingly, in yeast, *UBI3* is fused to the gene for the small ribosomal subunit protein S37 (Finley et al., 1989), while the tail in the *UBI1*-hybrid gene encodes a so-far-unidentified protein of the large ribosomal subunit. It has been postulated that the ubiquitin moiety of the fusion protein stabilizes these ribosomal proteins during their synthesis and transfers them to the site of preribosome assembly in the nucleolus.

UBI4 (consisting of five head-to-tail arranged ubiquitin-encoding repeats) codes for a polyubiquitin precursor protein. Deletion of *UBI4* gives rise to mutants that are viable at vegetative growth conditions (Finley et al., 1987). These cells contain normal concentrations of free ubiquitin. However, these mutants were found to be very sensitive to high temperatures and other kinds of stress, such as starvation or the addition of amino acid

analogues. A single ubiquitin unit put under control of the *UBI4* promoter can complement for the defect. These results indicate that *UBI4* provides ubiquitin monomers after processing of the precursor. In addition, these data suggest that ubiquitin is an essential component of the stress response system. Indeed *UBI4* (but not *UBI1, 2* or *3*) was found to be a heat shock gene showing a rapid induction upon a temperature shift (Finley et al., 1987). Most likely, stress leads to a sudden increase in the level of damaged or denatured proteins which are very toxic for the cell. As a consequence of this effect there may be a need for excess ubiquitin against the depletion by the formation of ubiquitin–protein conjugates. Indeed, increased *UBI4* transcription was found to be triggered by the synthesis of abnormally high levels of aberrant polypeptides (Grant et al., 1989). At present it is unknown whether such ubiquitinated proteins are all degraded or are restored to their native conformation.

Other proteins involved in yeast stress response

Apart from the classical hsps discussed above, in yeast, as in other cells, several proteins have been shown or suggested to play a part in the stress response. Some of them exhibit significantly increased levels of expression following a stress treatment.

In the previous section the importance of proteolytic breakdown for cell survival has been indicated. It is not surprising, therefore, that a major protease in yeast, proteinase YscE, also appears to be involved in the stress response (Heinemeyer et al., 1991). YscE is an RNA-associated protein complex, composed of various subunits which mediate the non-lysosomal pathway of degradation of ubiquitinated proteins ('proteasome'). The gene *PRE1*, encoding a 22 kDa subunit of YscE has been isolated (Heinemeyer et al., 1991) and strains carrying a mutation in *PRE1* display enhanced sensitivity to stress. In these mutant cells protein degradation is decreased and ubiquitin–protein conjugates accumulated. It has not been reported whether expression of this gene may be induced upon stress.

Related to the occurrence of oxidative stress in yeast is the finding that catalase T is a protein whose rate of synthesis is controlled by heat shock (Belazzi et al., 1991). Catalase T is a cytosolic enzyme, encoded by the *CTT1* gene, which may contribute to protect cells against the adverse effects of heat shock. *CTT1* belongs to the genes that, in addition, are under negative cyclic AMP control (see below in this section).

In addition, several enzymes of the glycolytic pathway are induced upon heat treatment of yeast cells. One of the (three) genes encoding glyceraldehyde-3-phosphate dehydrogenase (hsp35) is induced following a temperature shock (Lindquist and Craig, 1988). Under normal conditions this enzyme is already abundantly present in yeast cells. The energy stress imposed onto cells by a heat shock may cause the observed drop in ATP levels (Findley et al., 1983). A further increased enzyme synthesis at high temperature, therefore, is beneficial, because it enables the cell to increase the rate of glycolysis, thereby restoring the intracellular level of ATP. Consistent with this assumption is the finding that two other glycolytic enzymes are also induced: enolase (eno, hsp48; Iida and Yahara, 1985) and phosphoglycerate kinase (pgk; Piper et al., 1986). An alternative explanation for the heat-induced increase of these glycolytic enzymes is that stress may damage membrane structures, resulting in a disruption of the normal coupling between electron transport and oxidative phosphorylation (Patriarca and Maresca, 1990). Notably, heat shock does not induce the gene for pyruvate kinase (pyk) which catalyses the second ATP-generating step in glycolysis (Piper et al., 1988).

An additional consequence of a stress challenge on yeast cells is the transient dissipation of the electrochemical pH gradient across the plasma membrane, leading to a decrease in the internal pH of the cell (Weitzel et al., 1987; Coote et al., 1991). The stress-induced intracellular acidification may play a (direct or indirect) role in triggering the stress response (Coote et al., 1991; see the final section). Under normal growth conditions the pH gradient is sustained by the action of an ATP-driven proton pump, the plasma membrane ATPase. Plasma membrane ATPase is an abundant trans-membrane protein showing a highly conserved structure among eukaryotes (Aaronson et al., 1988). Maintenance of the proton gradient over the plasma membrane is essential for control of intracellular pH and nutrient uptake. It is relevant, therefore, that the *PMA* gene encoding this protein displays a sustained expression upon heat shock (Panaretou and Piper, 1990). Analysis of membranes from stressed yeast cells revealed a 30 kDa heat shock protein that may be related to plasma membrane ATPase function (Piper et al., 1990; Panaretou and Piper, 1992). Recently another stress-induced gene, *TIP1*, was identified encoding a protein presumably located at the outside of the plasma membrane (Kondo and Inouye, 1991). This gene was isolated as a cold-shock inducible gene but it displays also a heat-stimulated transcription activation.

Several lines of evidence suggest that cyclic AMP metabolism is intimately involved in the stress response. Upon heat treatment, intracellular cyclic AMP levels increase (Boutelet et al., 1985; Canonis et al., 1986; reviewed by Piper et al., 1990). Increase in cyclic AMP is correlated with a stimulation of the plasma membrane ATPase discussed above (Goffeau and Slayman, 1981; Serrano, 1983). On the other hand, a decline in cyclic AMP (and probably a consequent decrease in cyclic AMP-dependent protein phosphorylation) was found to trigger the synthesis of several heat shock proteins (Iida and Yahara, 1984; Shin et al., 1987). As a consequence of the decrease in cyclic AMP when cells enter the stationary phase, for instance, the *SSA1* (Brazzell and Ingolia, 1984), *HSP26* (Tuite et al., 1990), *HSP12* (Praekelt and Meacock, 1990) and *UBI4* (Finley et al., 1987; Tanaka et al., 1987) genes show enhanced expression. These results indicate that the pertinent *HSP* genes are under transcriptional control of two opposed signals (see below). An additional noteworthy fact is the isolation of cyclic AMP-defective mutants that show remarkable increases in stress tolerance as compared with normal cells (Iida, 1988).

Another process closely related to the stress response in yeast cells is the synthesis of trehalose. Yeast cells exponentially growing on glucose contain little trehalose (Thevelein, 1984). In spores, however, this disaccharide is accumulated, which strongly suggested its role as an energy source. Strikingly however, an enormous accumulation of trehalose (up to 100-fold) also occurs in response to a heat shock from 27 °C to 40 °C (Hottiger et al., 1987), whereas a decrease occurred when cells were shifted back to the lower growth temperature. The biosynthesis of trehalose involves two steps (reviewed by Thevelein, 1984); UDP-glucose and glucose 6-phosphate form trehalose 6-phosphate, a reaction catalysed by trehalose-6-phosphate synthase. Phosphate is then cleaved off by trehalose-phosphatase. Degradation of trehalose is mediated by trehalase, which occurs in two forms: a regulatory and a non-regulatory enzyme. The activity of neutral trehalase is thought to be regulated by cyclic AMP-dependent protein phosphorylation. In agreement with the observed dramatic accumulation of trehalose upon stress exposure, the activity of trehalose-6-phosphate synthase was found to increase 6-fold under these conditions (Hottiger et al., 1989). Surprisingly, however, also the activity of neutral trehalase is increased by a factor of 3. Obviously, turnover of trehalose is a fast process at shift-conditions. For this reason a protective role of trehalose for

proteins and membranes and, thus, in maintaining the structural integrity of the cell, has been suggested (Hottiger et al., 1989; Wiemken, 1990).

It is likely that future investigations will uncover other stress-induced genes, in particular when other stress agents than heat are studied. For instance, recently metallothioneins, involved in heavy metal homeostasis and detoxification (Hamer, 1986) have been implicated in the heat shock response (Silar et al., 1991; Yang et al., 1991), and very recent evidence indicates the occurrence in yeasts of a heat-induced secretory protein (hsp150) with a so-far-unidentified function (Russo et al., 1992).

STRESS-INDUCED CHANGES IN YEAST GENE EXPRESSION

Heat shock transcription factor, HSF

In prokaryotes, the heat shock response is mediated by a specific heat-induced σ-factor, σ^{32}, that binds to core RNA polymerase and directs it to heat shock promoters (Grossman et al., 1984; Landick et al., 1984). Regulation of the level of σ^{32} is accomplished by several control mechanisms. The σ^{32} gene, *rpoH*, is constitutively transcribed by the normal (σ^{70}-containing) RNA polymerase, but the protein is extremely unstable. Upon heat shock, transcription of *rpoH* is induced (Tilly et al., 1989), the translation efficiency of the σ^{32} mRNA is increased and the protein is stabilized (Straus et al., 1987). The activity of σ^{32} may be negatively modulated by interaction with dnaK (Liberek et al., 1992), whereas, in addition, dnaJ and grpE were shown to be physically associated with this heat shock transcription factor (Gamer et al., 1992).

In eukaryotes, transcription of heat shock genes is regulated through the action of the heat shock transcription factor, HSF, which interacts with its cognate nucleotide element, the heat shock responsive element HSE (reviewed by Sorger, 1991). In higher eukaryotes, HSF binds to HSEs only after heat induction (Sorger et al., 1987), whereas, in contrast, yeast HSF was found to interact with HSEs irrespective of the transcriptional state of the *HSP* genes (Jakobsen and Pelham, 1988; Gross et al., 1990). HSF binds to the DNA as a trimer (Perisic et al., 1989; Sorger and Nelson, 1989) and, in addition, the binding of trimers to adjacent sites is highly co-operative, thus preferentially forming large complexes.

The gene encoding HSF has been isolated from the yeasts *S. cerevisiae* (Wiederrecht et al., 1988; Sorger and Pelham, 1988) and *Kluyveromyces lactis* (Jakobsen and Pelham, 1991) as well as from *Drosophila* (Clos et al., 1990), tomato (Scharf et al., 1990) and human (Schuetz et al., 1991; Rabindran et al., 1991). Tomato cells contain at least two HSFs. Human cells also contain two HSF genes but the relative levels of HSF1 and HSF2 are not yet known.

In contrast with the high conservation of the HSF-responsive element, HSFs from different organisms show little sequence similarity. The HSFs from the distantly related *S. cerevisiae* and *K. lactis*, for instance, share only 18 % amino acid identity, the similarity being mainly confined to the DNA-binding domain and the trimerization domain (see below).

HSF from *S. cerevisiae* is a protein consisting of 833 amino acids which is encoded by a single, essential gene (Sorger and Pelham, 1988; Wiederrecht et al., 1988; reviewed by Sorger, 1991). The DNA-binding region is located within residues 167–284 (Wiederrecht et al., 1988) and a further domain, encompassing residues 327–424, involved in trimerization, is required for high-affinity association with DNA (see Figure 1). The DNA-binding domain does not show a DNA-binding motif similar to other eukaryotic transcription factors; only a short match to the putative DNA-recognition helix of bacterial σ-

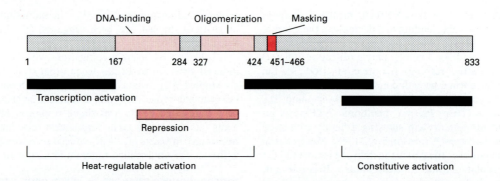

Figure 1   Domain structure of the heat shock factor HSF

The black bars indicate domains that can confer transcription activation to an heterologous, inactive DNA-binding domain (Sorger, 1990; Nieto-Sotelo et al., 1990).

factors can be distinguished (Clos et al., 1990). The trimerization domain, on the other hand, has been suggested to form a three-stranded α-helix coiled coil (Sorger and Nelson, 1989) on the basis of a leucine/isoleucine repeat structure.

Apparently, in higher eukaryotes activation of HSF to a potent transcription factor requires first induction of DNA-binding and subsequently activation, e.g. by phosphorylation (Larson et al., 1988). In yeast, however, unactivated HSF is already bound to DNA under non-shock conditions and, therefore, only the conversion of HSF into a form capable of efficiently stimulating transcription is required upon shock.

In order to unravel the mechanism by which transcription activation by HSF occurs, it should be recognized that HSF not only plays a part in heat shock-induced transcription but also in basal-level, constitutive ('sustained') transcription (Sorger, 1990). For instance, overexpression of HSF in yeast at normal growth conditions significantly increases the levels of a major hsp70 (Sorger and Pelham, 1988). In addition, as will be discussed below, 50–80 % of the basal expression of the hsp70 gene *SSA1* is mediated through HSEs.

Two studies have described the functional domains in yeast HSF (Sorger, 1990; Nieto-Sotelo et al., 1990). In one study, truncated HSF proteins were introduced into cells in the absence of wild-type factor and it was found that the heat shock-induced, characteristically transient, activity of HSF and its sustained activity are mediated by physically separable transcription-activating domains (Sorger, 1990; see Figure 1). The N-terminal region of the protein, residues 1–424, mediates the transient increase of HSF activity, required for growth at elevated temperatures. Analysis of lexA–HSF fusion proteins did not reveal a particular subregion of this domain capable of conferring the transcription activating function. Rather, integrity of the entire domain seems to be essential for proper function (Sorger, 1990). A C-terminal region consisting of amino acids 584–783 is essential for sustained increase. This activating region was unmasked by deletion of N-terminal residues, resulting in a 40-fold increase of HSF activity in the absence of a heat shock (Sorger, 1990). Both activities, transient as well as sustained, are associated with a rise in the extent of phosphorylation of HSF. Up to ten Ser/Thr sites are phosphorylated at 20 °C and an additional five at 39 °C (Sorger, 1990). It is doubtful, however, whether phosphorylation plays a major functional role in transcription activation. Recently evidence was obtained that phosphorylation of *trans*-activators may occur as a consequence of the formation of a transcription initiation complex (due to exposure of the target sites) rather than as a prerequisite for this formation (Sadowski et al., 1991).

In the second study (Nieto-Sotelo et al., 1990) residues 1–63 were demonstrated to be necessary for growth under heat shock conditions. Between amino acids 208 and 648 a transcription activation domain was found to be present. Consistently, fusion of this domain to the DNA-binding domain of yAP1 generated a hybrid temperature-regulatable transcription factor. A constitutive activation domain was shown to occur between residues 410 and 648. Notably, a third element could be distinguished, between residues 208 and 394, which is responsible for repression of the transcription-activating domain under non-shock conditions (Nieto-Sotelo et al., 1990). This region encompasses part of the DNA-binding domain and the Leu repeat.

HSF from *S. cerevisiae* can functionally substitute for HSF from *K. lactis* (Jakobsen and Pelham, 1991). The repressable activation domain of both proteins shows little sequence similarity. On the other hand, the region involved in masking the activity at low temperatures contains, apart from the evolutionarily conserved DNA-binding and oligomerization domains, an additional sequence similarity: a short conserved element, RXLLKNR, located near the activator region (Jakobsen and Pelham, 1991). Very recently, by domain-swapping, deletion, and mutagenesis experiments, the importance of the central evolutionarily conserved domain in keeping HSF unactivated under non-shock conditions was confirmed (Bonner et al., 1992): a hybrid HSF–VP16 factor appeared to be as temperature-regulatable as HSF itself. The putative masking element (Jakobson and Pelham, 1991) contributed to the regulation mechanism. Adjacent to this element a run of serine residues is present. Replacing these serines by alanine or aspartic acid, however, had no effect on regulation *in vivo*. The pertinent conserved element may either bind to the structural core of the protein or to another protein, thus keeping the activator in an unactivated conformation under non-shock conditions. An appealing model is that hsp70 may serve as a repressor. Upon heat shock the demand for hsps70 strongly increases as a consequence of the elevated level of 'thermally-damaged' proteins (see above). This may lead to the dissociation of the hsp70–HSF complex, thereby derepressing its transcription-activating activity (see Figure 2). Increased synthesis of hsp70 may result in re-association of the protein with HSF, which would explain the transient nature of the response (Morimoto et al., 1990; Sorger, 1991). As mentioned before, hsp70 has as such been proposed to serve as the cellular thermometer (Craig and Gross, 1991). However, direct evidence to support this model is still missing.

Within the context of transcription activation of heat-induced genes, three unexpected recent findings deserve some comment.

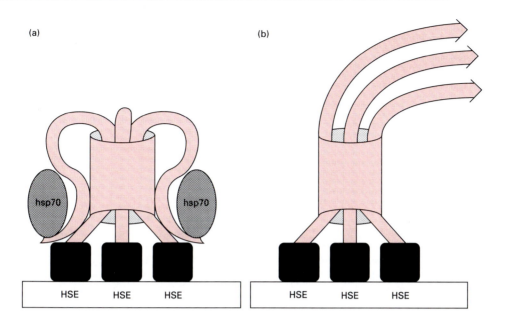

Figure 2   Model for the activation of the heat shock factor HSF

Under non-stress conditions, HSFs, bound to the HSEs as a trimer, are kept in an unactivated conformation through the interaction with hsps70 (**a**). Upon stress exposure hsp70 is released from the complex, thus enabling the factor to change into an activating-potent form (**b**).

First it was shown that HSF activates transcription of the yeast metallothionein gene (Silar et al., 1991; Yang et al., 1991). This gene, *CUP1*, is essential to prevent copper toxicity (Hamer, 1986). Metal (Cu or Ag)-induced transcription activation of *CUP1* is regulated by a specific *trans*-acting factor, ACE1, which binds to *cis*-acting metal-responsive elements in the promoter of *CUP1* (Furst et all., 1988; Furst and Hamer, 1989). Mutant strains have been isolated which are able to suppress the requirement for ACE1 in the activation of *CUP1*. Surprisingly, this suppressor gene appeared to encode a mutant form of HSF. A single amino acid substitution in the DNA-binding domain of HSF was found to strongly enhance transcription of *CUP1*, while it reduced *SSA3* gene expression. In the promoter of *CUP1* GAA repeats occur which, however, are spaced by three and four nucleotides. The results suggest HSF to be involved in transcriptional regulation of genes other than *HSP* genes. Secondly, a protein containing an amino acid sequence extensively similar to the DNA-binding domain of HSF has been identified as a suppressor of flocculation (*SFL1*; Fujita et al., 1989). This finding may reflect the presence in yeast cells of a family of (evolutionarily) related protein factors. Finally, evidence has been obtained for an HSF-independent mechanism of heat shock-induced transcription activation. In *S. cerevisiae* a set of genes has been identified which display high levels of expression after treatment with DNA-damaging compounds or heat. For one of those so-called DNA-damage response (*DDR*) genes, *DDR A2*, heat induction leads to a 20-fold increase in transcription (McClanahan and McEntee, 1986). The upstream region of the *DDR A2* gene harbours an element required for this stress regulation (Kobayashi and McEntee, 1990). Deletion of this element (−202 to −165) abolished heat-induced transcription and, in addition, this element can confer heat shock-induction onto a reporter gene. An element homologous to this region occurs in the promoter of *UBI4*, which also belongs to the *DDR* family. No sequence similarity with the HSE exists and it does not bind HSF. These data suggest the striking possibility

that, in addition to HSF, at least one other factor can mediate heat-induced transcription activation.

Heat shock responsive element, HSE

It has been well established that in yeast, as in other eukaryotes, a specific *cis*-acting promoter element mediates the heat shock response on transcription. This heat shock responsive element, HSE, represents the binding site for the *trans*-acting transcription factor HSF. The HSE was originally defined for *Drosophila HSP* genes as the sequence CNNGAANNTTCNNG, displaying a dyad symmetry characteristic for binding sites of multimeric DNA binding proteins (Parker and Topol, 1984). Later on, the definition has been revised on the basis of results obtained with site-directed mutagenesis (Xiao and Lis, 1988). Functionally-defined HSEs should be considered as modular elements encompassing at least three repeats of the sequence GAA in alternating orientations and separated from each other by two nucleotides. A gap of 5 bp between two modules is tolerated provided that the elements flanking the gap are direct repeats. *In vitro*, the minimal DNA sequence required for the formation of a stable complex with *Drosophila* HSF is an inverted 5 bp-repeat, GAA, irrespective of whether it has a tail-to-tail or a head-to-head arrangement. The aforementioned data have led to the model of trimerization of HSF binding to the DNA (Perisic et al., 1989; Sorger and Nelson, 1989).

Comparison of yeast HSEs indicates that they are composed of three to five appropriately positioned GAA-modules (Boorstein and Craig, 1990). On the other hand, a single HSE is sufficient to confer heat-inducible expression onto a reporter gene (Kirk and Piper, 1991), although in general multiple HSEs act co-operatively to mediate transcription activation of heat shock genes.

The conserved nature of the *cis/trans*-combination HSF–HSE emphasizes the general evolutionary conservation of transcriptional activation mechanisms among eukaryotes (reviewed by Guarente and Bermingham-McDonogh, 1992).

Transcription regulation of HSP-genes

Transcription of *HSP70*

The yeast hsp70-encoding genes *SSA1* and *SSA2* show a significant basal level of expression (reviewed by Craig et al., 1990). Only *SSA1* gene expression is strongly increased upon a mild heat shock; *SSA2* gene expression is almost unaffected by a stress treatment. On the other hand, *SSA3* and *SSA4* gene products are barely detectable at normal temperatures but display a very strong increase in the rate of transcription following a temperature upshift (Craig et al., 1990).

The upstream DNA region of each of these *SSA* genes contains multiple sites showing a close relationship to the HSE consensus sequence, which are sufficient to confer heat-inducibility onto a reporter gene (reviewed by Boorstein and Craig, 1990). In the *SSA1* promoter a number of HSEs are present but deletion analysis indicated that only one of these ('HSE2') is most important for the heat-inducible expression (Slater and Craig, 1987). Removal of 'HSE3', located further upstream and having a perfect match with the consensus, did not significantly affect heat-inducible expression of *SSA1*. 'HSE2' can confer heat shock-induced transcription onto a reporter gene. However, the basal level of expression conferred by this element was found to be considerably decreased when 'HSE2' was flanked by its original surrounding sequences (Park and Craig, 1989). This finding suggested the involvement of a negative regulatory element which, by deletions and point mutations, indeed could be shown to overlap this HSE (Park and Craig, 1989). This URS ('upstream repression site') has some sequence homology with a URS found in other yeast gene promoters (Luche et al., 1990). Consistently, mutations in the URS resulted in elevated levels of basal expression. For basal level expression of *SSA1* the heat shock factor HSF appeared to be essential (Park and Craig, 1989).

The HSE elements in the promoter of the *SSA3* gene are also involved in enhancement of the level of expression at heat shock conditions (Boorstein and Craig, 1990). In addition this gene is transcriptionally induced under conditions of lowered intracellular cyclic AMP, such as starvation.

SSA4 is the only canonical *HSP70* gene in yeast showing an extremely low level of expression at normal growth conditions and dramatically induced levels upon heat shock (Boorstein and Craig, 1990). Again an upstream region containing an extended match to the conserved HSE is necessary and sufficient for heat-inducible regulation.

Yeast cells bearing mutations in *SSA1* and *SSA2* display a high basal level of expression of other heat shock proteins even at a normal (23 °C) growth temperature (Craig and Jacobsen, 1984). In addition, it has been established that transcription activation of *SSA4* in these *ssa1/ssa2* mutants is mediated by the HSE in its promoter (Craig and Jacobsen, 1984).

These data lend support to the hypothesis that hsp70 may act as an autoregulator: it inhibits, directly or indirectly, its own expression. As discussed above, a tentative model presumes that hsp70 exerts its negative feed-back through the heat shock factor HSF.

Very recently, three independently acting *cis*-elements in the promoter of the *KAR2* gene were identified (Mori et al., 1992). In addition to a heat shock responsive element (HSE) and a GC-rich region functionally involved in the high constitutive rate of transcription of this gene, a 22 bp sequence was discovered which

mediates the response to unfolded proteins accumulating in the endoplasmic reticulum. This so called UPR-element is able to confer responsiveness to unfolded proteins onto an heterologous gene and its structure is similar to that of regulatory regions of the homologous mammalian genes (Mori et al., 1992). UPR turned out to represent a specific (*in vitro*) binding site of a protein factor. How the presence of unfolded proteins in the endoplasmic reticulum is sensed and may lead to the activation of a factor regulating *KAR2* gene transcription in the nucleus is an intriguing but as yet unsolved question.

Transcription regulation of *HSP83*

HSP83 is expressed at a low basal level at normal conditions and is highly induced following a mild heat shock (Borkovich et al., 1989). On the other hand, *HSC83* is expressed at a 10-fold higher constitutive level and is slightly induced upon a heat challenge. The promoter of the heat-inducible *HSP83* has been analysed in detail by genomic footprinting (Gross et al., 1990a,b). A combination of chemical and enzymic footprinting techniques was used to obtain a high resolution map. The *HSP83* promoter contains three HSEs that fully match the consensus sequence (Farelly and Finkelstein, 1984). Only one of these putative HSF-binding sites, the most proximal 'HSE1', appeared to be bound by a protein *in vivo*, within the major groove of the DNA (Gross et al., 1990a,b). In addition, the TATA box was found to be occupied, most likely by the basal transcription factor TFIID, and in this case binding occurs to the sugar-phosphate backbone of the DNA. Both protein-binding elements are on one side of the helix causing a marked local distortion of the DNA in the chromatin, which is in agreement with the general view that protein–protein interactions play a major part in transcription activation (Ptashne, 1988).

Binding of HSF to 'HSE1' was found to be irrespective of the transcriptional state – basal or induced – of the *HSP83* gene, which is in agreement with the finding that this HSE is also required for basal level expression of the *HSP83* gene (McDaniel et al., 1989). Moreover, mutational analysis of 'HSE1' confirmed that this element is absolutely required for both basal and induced expression (Gross et al., 1990a). On the other hand, some point mutations were preferentially found to reduce constitutive expression of *HSP83* and had no effect on the heat-inducible expression (Gross et al., 1990a). Therefore, HSEs may exist in functionally distinct subclasses. This is also emphasized by genomic footprinting showing that 'HSE2' and 'HSE3' are vacant, probably reflecting a lower affinity of HSF for these sequences.

Similar experiments performed for the *HSC83* gene revealed, in addition to a TATA box and a heat shock element ('C.HSE1') another factor-binding site 40 bp upstream of 'C.HSE1' (Gross et al., 1990a,b). Perhaps the 10-fold higher basal level of expression of *HSC83* as compared with *HSP83* is due to this (so far) unknown factor.

Transcription regulation of *HSP26*

HSP26 gene expression is strongly enhanced following a heat shock (Petko and Lindquist, 1986), during stationary phase growth (Kurtz et al., 1986) and upon induction to sporulation (Rossi and Lindquist, 1989). Several deletion and insertion mutations have been analysed (Susek and Lindquist, 1990) which demonstrated that regulation of *HSP26* gene expression primarily takes place at the transcriptional level. The promoter of *HSP26* has quite complex characteristics and, unlike the *HSP* promoters discussed above, appears to be composed of both repressing and activating elements. In the DNA region up to −500 matches to the consensus HSE sequence are present but none of them are essential for heat shock-induced transcription activation. They might act in a co-operative fashion (Susek and Lindquist, 1990).

Evidence for the presence of a transcriptional repressing element was provided by the finding that several upstream promoter deletions rendered *HSP26* expression strongly constitutive. Moreover, an upstream DNA fragment (−501 to −332) is able to confer transcriptional repression to an heterologous gene. This finding has led to the tempting proposal that the main features of the regulation of *HSP26* expression involve repression of basal transcription activity under normal conditions and derepression at heat shock conditions. This mechanism would significantly differ from the mode of transcription regulation of other heat shock genes in yeast (Susek and Lindquist, 1990). No distinction could be made so far between *cis*-acting elements involved in heat-shock control and those implicated in developmental regulation.

Transcription regulation of *HSP12*

HSP12 shows a pattern of gene expression that is quite similar to that of *HSP26* (Praekelt and Meacock, 1990). *HSP12* is also developmentally regulated and is strongly induced upon heat shock. During exponential growth, hsp12 mRNA levels are low, but following a temperature shift the concentration increases several-hundred-fold.

The gene may be regulated both by heat shock and by cyclic AMP, as evidenced by analyses of mutants defective in cyclic AMP-dependent protein phosphorylation (Praekelt and Meacock, 1990). In a *cyr1-2* mutant containing low intracellular levels of cyclic AMP, high levels of hsp12 mRNA are present during exponential growth. In contrast, in a *cyr1-2,bcy1* double mutant, whose protein kinase is constitutively active, these high levels do not occur. Cyclic AMP-dependent protein kinase may regulate both positive and negative transcription factors by phosphorylation (Tanaka et al., 1988) and, thus, control the level of *HSP12* expression. Notably, in minimal medium *HSP12* displays a low level of induction upon heat shock, which may be correlated with the low levels of cyclic AMP under those conditions. Evidence is available that, similar to the *HSP26* gene, also the *HSP12* promoter is activated upon a stress challenge by a derepression mechanism (P. A. Meacock and U. M. Praekelt, unpublished results; J. Varela and W. H. Mager, unpublished results).

Transcription regulation of the phosphoglycerate kinase gene

Levels of the phosphoglycerate kinase mRNA in *S. cerevisiae* are increased about 6-fold by transcriptional activation when fermentative cultures are subjected to a heat shock (Piper et al., 1986). The promoter of the *PGK* gene was searched for the presence of HSE elements and two imperfect matches with the consensus sequence were found at −165 and −365. The element at −365 seems to be functionally involved in heat-induced transcription activation (Piper et al., 1988). This HSE is located adjacent to the major upstream activation site (UAS) of the *PGK* gene, a binding site for the abundant protein factor RAP1 at about −460 (Chambers et al., 1990). The effect of the putative HSE on phosphoglycerate kinase mRNA synthesis is only evident when the UAS is deleted (Piper et al., 1988) since in the presence of the UAS, heat shock had a relatively small effect on phosphoglycerate kinase mRNA levels. The role of the HSE may be to sustain transcription of the *PGK* gene when, following a stress challenge, transcription of non-heat shock genes is suddenly arrested. The effect of heat shock on *PGK* gene expression appeared to be dependent on the carbon source as only fermentative cultures show elevated phosphoglycerate kinase mRNA levels (Piper et al., 1988). Using glycerol as a carbon source heat shock-induction did not occur, but addition of glucose rapidly induced the ability to enhance phosphoglycerate kinase mRNA levels. These data indicate a catabolite control of the induction phenomenon.

Processes affected by stress

Heat damages a wide variety of cellular processes and cellular structures. As stated in the introduction, stress exposure dramatically changes the pattern of gene expression. Apart from the increased levels of gene expression discussed above, transcription of many genes is transiently inhibited upon a temperature shock but it is unknown how this sudden arrest of transcription is brought about. Perhaps an essential component of the transcription machinery is extremely sensitive to stress. An alternative explanation is that conformational changes in chromatin occur which cause the arrest. It is striking that the HSF-mediated transcription activation can escape the inhibition.

Studies of two nuclear processes, ribosome formation and pre-mRNA splicing, point to drastic changes in nucle(ol)ar structure. It has been observed that ribosome assembly in mammalian cells is quite sensitive to a temperature stress and that rRNA processing slows down after heat shock leading to the accumulation of precursor rRNA (Sadis et al., 1988). In yeast, a strong inhibition of the rate of synthesis of rRNA has been found upon heat treatment (Veinot-Drebst et al., 1989). In addition, splicing of mRNA precursors was demonstrated to be disrupted by a severe heat shock (reviewed by Yost et al., 1990; Yost and Lindquist, 1991). A mild heat treatment prior to a shift to severe temperatures protects splicing. During such a pre-treatment no protein synthesis is required, but protein synthesis is needed for a rapid recovery of the splicing process after a sudden severe heat shock.

Notably, in *Drosophila*, *HSP83* is the only *HSP* gene that contains an intron. Indeed, following a stress challenge, pre-mRNA for hsp83 accumulates (Yost and Lindquist, 1985). The inhibition of splicing at high temperatures may explain why intron-less *HSP* genes have evolved. Perhaps *HSP83* has escaped from this selection because of the abundance of its gene product at normal conditions (Yost et al., 1990).

For mammalian cells a role of heat shock proteins in protection and repair of ribosome assembly and spliceosome assembly has been implicated. Upon heat shock, hsp70 migrates into the nucleus in order to associate with polypeptides that form insoluble complexes at increased temperatures (Pelham, 1990). Hsp70 also moves into the nucleolus to associate with partially assembled ribosomes (Munro and Pelham, 1985). Presumably nuclear proteins become partially denatured upon a temperature shock hence exposing hydrophobic regions that tend to interact and form insoluble aggregates (Pelham, 1990). The (auto-) regulatory role of hsp70 in recovery from stress may, thus, involve the promotion of disaggregation (Pelham, 1986). During the recovery process hsp70 moves back to the cytoplasm. Consistent with this proposed function of hsp70 is the finding that certain yeast hsp70 mutants that constitutively overproduce other heat shock proteins display protection of splicing at high temperatures without the need of a pre-treatment at mildly increased temperatures (Yost and Lindquist, 1991). Also hsp104 may play a part in the recovery of RNA splicing after a severe heat shock (Yost and Lindquist, 1991).

Finally, heat shock has a profound effect on RNA metabolism due to the selective degradation of mRNAs (reviewed by Yost et al., 1990). During shock conditions some normal cellular mRNAs display a transient rapid decay. In yeast, in particular ribosomal protein mRNAs have been shown to undergo a rapid degradation upon a mild temperature shift (Herruer et al., 1988; Mitsui and Tsurugi, 1991). On the other hand, during recovery from heat shock, while the levels of normal cellular mRNAs are being restored, hsp mRNAs are selectively degraded (Yost et al., 1990). This rapid decay is probably mediated by a *cis*-acting element present in the trailer regions. An appealing alternative explanation for the observed differences in turnover-rate of hsp mRNA is that these mRNAs are intrinsically unstable but are stabilized during stress exposure.

OTHER STRESS AGENTS AND THE ACQUISITION OF STRESS TOLERANCE

Studies on the stress response of living cells have so far mainly been focused on the effects of heat treatment. In particular for yeast, however, it has been demonstrated that, apart from heat, other stress agents can induce similar responses (reviewed by Watson, 1990). Heat shock protein synthesis has been reported to occur upon exposure of yeast cells to ethanol (Plesset et al., 1982; P. Moradas Ferreira, unpublished results), high salt (Varela et al., 1992), desiccation, heavy metals, arsenite (Chang et al., 1989), hydrogen peroxide (Collinson and Dawes, 1992) and amino acid analogues. It is clear, however, that the responses to the various stress agents are not identical (see below). The sensitivity of yeast for the induction of the stress response is manifest by the recent finding that conversion of yeast cells to spheroplasts (by incubating them in the presence of lyticase) evokes the expression of *HSP70* and *HSP83* genes (Adams and Gross, 1991). Treatment of cells with thiolutin (an inhibitor of all three RNA polymerases) induced a 5-, 25- and 50-fold increased transcription of *HSP83*, *SSA4* and *HSP26*, respectively (Adams and Gross, 1991). Another drug commonly used to inhibit transcription, phenanthroline, also caused elevated mRNA levels for some hsps, but in this case probably due to stabilization of the respective mRNAs (Adams and Gross, 1991).

Pre-treatment of yeast cells at mildly elevated temperatures leads to the attainment of tolerance against a severe heat shock. Thermotolerance develops rapidly in yeast after a shift from 23 °C to 37 °C reaching a maximum at 2 h and decreasing in the next 8–24 h (McAlister and Finkelstein, 1980). Consistent with the finding that other stress agents can induce similar responses as heat, it was demonstrated that heat treatment of *S. cerevisiae* results in a marked increase in ethanol tolerance as compared to control cells (Watson and Cavicchioli, 1983). The effect of ethanol on yeast has particularly been investigated since high ethanol concentrations inhibit fermentation and growth. This effect has been associated with plasma membrane ATPase activity (Rosa and Sá-Correia, 1992). Both ethanol and thermal stress lead to a slight but significant drop of internal pH which is related to the ATPase activity (Weitzel et al., 1987; Pampulha and Loureiro-Dias, 1989). Similarly, cells pre-exposed to osmotic stress were shown to become not only osmo-tolerant but also acquired thermotolerance (Trolmo et al., 1988; Varela et al., 1992). The reverse, however, is not true. Obviously, under both stress conditions no identical sets of proteins are synthesized.

From the initial studies (McAlister and Finkelstein, 1980), it has been suggested that heat shock proteins play an essential role in the acquisition of stress tolerance. In many aspects, indeed, the level of acquired thermal resistance shows a narrow correlation with induced cellular level of hsps. For instance, a yeast mutant, temperature-sensitive for growth at 39 °C by a defect in RNA transport from nucleus to cytoplasm and, therefore, incapable of synthesizing hsps at this temperature, failed to become resistant to a subsequent challenge at 55 °C (McAlister and Finkelstein, 1980). Since then, circumstantial evidence has become available that supports the idea that heat shock protein synthesis is a prerequisite for the development of stress-tolerance by yeast. A heat shock resistant mutant of *S. cerevisiae* (*hsr1*; Iida and Yahara, 1984) has been isolated with high constitutive levels of two hsps. Also *cyr1-2* mutants, having low intracellular cyclic AMP contents, constitutively synthesize hsps and are relatively resistant against lethal temperatures, while *bcy1*-mutants (deficient in cyclic AMP-dependent protein kinase) fail to attain thermotolerance (Shin et al., 1987). It is noteworthy, furthermore, that yeast cells in stationary phase are intrinsically more resistant to various stress agents than are exponential-phase cells (Schenberg-Frascino and Moustacchi, 1972; Parry et al., 1976; Walton et al., 1979). Under those growth conditions several hsps were shown to display enhanced rates of synthesis (see above). The heat shock response of *S. cerevisiae* cells during stationary growth differs from that occurring at exponential since the former shows long-term resistance in contrast with the transient nature of the latter (reviewed by Watson, 1990). Finally, studies performed with defective mutants of the *SSA1* and *SSA2* genes have been interpreted as evidence for a causal relationship between hsp levels and the acquisition of thermotolerance, since these double mutants are thermosensitive for growth at 37 °C (Craig and Jacobsen, 1984). On the other hand, a direct correlation between hsps and the attainment of stress-resistance so far has not been established. Rather, evidence is accumulating that hsps may not be needed for stress-tolerance acquisition but for a rapid recovery from the stress-affected situation, thus serving as components of the cellular defense mechanism.

The first argument against an obligatory role of hsps in the acquisition of stress tolerance was provided by the finding that *S. cerevisiae* cells pretreated with cycloheximide to block protein synthesis still develop thermotolerance (Hall, 1983). In addition, administration of inhibitors of both cytoplasmic and mitochondrial protein synthesis did not interfere with the heat-induced ethanol- or thermo-tolerance (Watson et al., 1984).

Analysis of *ssa1/ssa2* mutants demonstrated that their capability to attain thermotolerance is similar to that of control cells (Craig and Jacobson, 1984). These mutants appeared to be even more resistant against a short treatment at 52 °C than wild-type cells. Moreover, except for *HSP104*, analysis of disruption mutants did not reveal an important function of any hsp in the acquisition of stress-resistance. *HSP104* deletion mutants, however, do fail to acquire thermotolerance (Sanchez and Lindquist, 1990) and rescue of this capacity could be gained by transformation of the disruption strain with the wild-type gene. Very recently, a further analysis of the role of hsp104 in the acquisition of stress tolerance was reported (Sanchez et al., 1992). Respiring cells, displaying a constitutive expression of *HSP104*, were found to be basally more resistant against heat shock than cells in a fermentative culture. This selective advantage is absent in a *hsp104* deletion mutant. The attainment of tolerance against high ethanol concentrations (and to a minor extent also to arsenite) is similarly dependent upon a functional *HSP104* gene (Sanchez et al., 1992). These data strongly suggest the requirement of this heat shock protein for the acquisition of tolerance against stressful conditions.

So far, overproduction of hsps has not been found to render yeast cells thermoresistant. The *HSP83* gene, for instance, has been introduced into yeast on a multicopy plasmid. The resulting

3-fold increased induction of this protein upon heat shock did not lead to a corresponding increase in the level of thermotolerance normally attained (Finkelstein and Strausberg, 1983). Also a *HSP26* gene under control of the *GAL* promoter, allowing induction prior to the heat treatment, or the *PGK* promoter, did not alter the ability of cells to acquire tolerance (Tuite et al., 1990). Of course, this type of experiment does not exclude that hsps are involved in stress tolerance acquisition, but increased levels, obviously, are not sufficient to confer elevated levels of protection.

Yeast cells exposed to stress circumstances temporarily stop growing by an arrest at G_1 in the division cycle. Using cell cycle inhibitors it has been shown that arresting cells in G_1, S or G_2 phase of the mitotic cycle is not a stress condition that induces thermotolerance (Barnes et al., 1990). Arrested cells remained as sensitive to thermal death as growing cells, providing evidence for the conclusion that the full spectrum of hsps is not necessary for thermotolerance induction.

Convincing evidence that thermotolerance attainment does not require hsp synthesis was obtained by uncoupling both processes (Smith and Yaffe, 1991). A yeast strain bearing a mutation in the gene for the heat shock factor HSF (*hsfl-m3*) was isolated which causes a temperature-sensitive growth defect. The mutation prevents activation of HSF and therefore leads to a general block of heat shock-induced protein synthesis. However, it does not affect the acquisition of thermotolerance.

Finally, with regard to the development of stress tolerance by yeast cells, the synthesis of trehalose should be mentioned (see also above). Changes in intracellular trehalose levels have been correlated with the acquisition of tolerance, both against heat and desiccation (Hottiger et al., 1989; Wiemken, 1990). Again, however, it is unclear whether increased concentrations of trehalose are sufficient to render cells tolerant against stress conditions or just contribute to the protection of cellular structures at those adverse conditions.

WHAT IS THE TRIGGER?

The data summarized above 'stress' the pleiotropic nature of the heat shock response in yeast (and other living cells). Many cellular events reflect molecular consequences of the stress exposure and deal with protection, survival and repair. A remaining intriguing, so far unsolved question, however, is by which cellular component(s) the stress condition is sensed.

First, intracellular pH has been suggested to play an essential part in triggering the stress response, since the acquisition of thermotolerance is enhanced in cells having an acidic external environment (pH 4.0) as compared with a neutral one (Coote et al., 1991). Probably, plasma membranes become leaky for protons as a consequence of the stress challenge. At an external pH of 4.0 passive diffusion of protons may lead to internal acidification that, subsequently, may trigger the stress response. How changes in the pH_i might lead to, amongst others, stress protein synthesis is so far unknown.

A more classical view is that some proteins in the cell are very sensitive to stress-induced denaturation. (Although this may also be a pH-dependent process.) By exposure of cells to adverse circumstances, these denatured proteins might recruit hsps70 from their complex with the heat shock factor HSF. In this way hsps70 could positively regulate the expression of their own and other *HSP*-genes. Also in principle denatured nascent polypeptides may be signalled by the hsp70 'control-system'. For this reason in *E. coli* ribosomes have been implicated in sensing the stress response (van Bogelen and Neidhart, 1990). It is clear, on the other hand, that at least some *HSP* gene promoters, in addition to the HSF-mediated control, are regulated by other signals such as cyclic AMP. These data indicate the possible involvement in the stress response of protein phosphorylation. Another finding emphasizing that (de)phosphorylation may play a part, is the modification of HSF itself. Even if phosphorylation of HSF is a consequence of the formation of an active transcription initiation complex, kinases and phosphatases may play a regulatory part in the return to the normal growth conditions. Furthermore, it cannot be excluded that protein-(de)phosphorylation fulfils a key role in the expression of the response through a so far unidentified regulatory circuit.

By analysis of mutants defective in certain steps of well-known signal transduction pathways, the relevance of protein modifications will be elucidated. Moreover a comparative analysis of the responses evoked by different stress circumstances will contribute in revealing the nature of the actual trigger(s) of the process.

CONCLUDING REMARKS

In conclusion, the data reviewed in this article clearly demonstrate the pleiotropic nature of the stress response of yeast. Many questions with respect to the underlying molecular mechanisms remain to be answered. Evidently, the stress response is not limited to the action of the classical heat shock proteins and the heat-responsive transcription factor HSF, but involves many more factors and processes presently under study. These proteins and molecular events contribute to either protecting and repairing cells after exposure to stress (transient response) or their adaptation to prolonged stress (sustained response). It will be the challenge of future studies to unravel the functional links in this regulatory network, the more so since they are expected to play an important part under normal cellular growth conditions as well.

It is obvious that increased knowledge about the stress protection and adaptation mechanisms in yeast is of major both fundamental and biotechnological importance. Because of the universal nature of the stress response, further insight into the response of yeast is also relevant to improve the understanding of defense mechanisms in other cell types. For instance, a direct relationship exists between the expression of stress proteins and specific pathological conditions (Morimoto et al., 1990) and antigens from a wide variety of pathogens were identified as members of hsps (Young, 1992). Therefore, although the stress response certainly displays cell-specific features, the many shared characteristics render yeast an attractive model to investigate the response from bacteria to man.

REFERENCES

Aaronson, L. R., Hager, K. M., Davenport, J. W., Mandela, S. M., Chang, A., Speicher, D. W. and Slayman, C. W. (1988) J. Biol. Chem. **263**, 14552–14558

Adams, C. C. and Gross, D. S. (1991) J. Bacteriol. **173**, 7429–7435

Ang, D., Liberek, K., Skowyra, D., Zylicz, M. and Georgopoulos, C. (1991) J. Biol. Chem. **266**, 24233–24236

Atencio, D. P. and Yaffe, M. P. (1992) Mol. Cell. Biol. **12**, 283–291

Barnes, C. A., Johnson, G. C. and Singer, R. A. (1990) J. Bacteriol. **172**, 4352–4358

Belazzi, T., Wagner, A., Wieser, R., Schanz, M., Adam, G., Hartig, A. and Ruis, H. (1991) EMBO J. **10**, 585–592

Blumberg, H. and Silver, P. A. (1991) Nature (London) **349**, 627–630

Bonner, J. J., Heyward, S. and Fackenthal, D. L. (1992) Mol. Cell. Biol. **12**, 1021–1030

Boorstein, W. R. and Craig, E. A. (1990) J. Biol. Chem. **265**, 18912–18921

Borkovich, K. A., Farelly, F. W., Finkelstein, D. B., Taulien, J. and Lindquist, S. (1989) Mol. Cell. Biol. **9**, 3919–3930

Bossier, P., Fitch, I. A., Boucherie, H. and Tuite, M. F. (1989) Gene **78**, 323–330

Boutelet, F., Petitjean, A. and Hilger, F. (1985) EMBO J. **4**, 2635–2641

Brazzell, C. and Ingolia, T. D. (1984) Mol. Cell. Biol. **4**, 2573–2579

Canonis, J. H., Kalekine, M., Gondre, B., Garreau, H., Boy-Marcotte, E. and Jacquet, M. (1986) EMBO J. **5**, 375–380

Chambers, A., Stanway, C., Kingsman, A. J. and Kingsman, S. M. (1988) Nucleic Acids Res. **16**, 8245–8260

Chambers, A., Stanway, C., Tsang, J. S. H., Henry, Y., Kingsman, A. J. and Kingsman, S. M. (1990) Nucleic Acids Res. **18**, 5393–5399

Chang, E. C., Kosman, D. J. and Willsky, G. R. (1989) J. Bacteriol. **171**, 6349–6352

Chappell, T. G., Konforti, B. B., Schmid, S. L. and Rothman, J. E. (1987) J. Biol. Chem. **262**, 746–751

Cheng, M., Hartle, F., Martin, J., Pollock, R., Kalousek, F., Neupert, W., Hallberg, E., Hallberg, R. and Horwich, A. (1989) Nature (London) **337**, 620–625

Chirico, W., Waters, M. G. and Blobel, G. (1988) Nature (London) **332**, 805–810

Clos, J., Westwood, J. T., Becker, P. B., Wilson, S., Lambert, K. and Wu, C. (1990) Cell **63**, 1085–1097

Collinson, L. P. and Dawes, I. W. (1992) J. Gen. Microbiol. **138**, 329–335

Coote, P. J., Cole, M. B. and Jones, M. V. (1991) J. Gen. Microbiol. **137**, 1701–1708

Craig, E. A. and Gross, C. A. (1991) Trends Biochem. Sci. **16**, 135–139

Craig, E. A. and Jacobsen, K. (1984) Cell **38**, 841–849

Craig, E. A., Kramer, J. and Kosic-Smithers, J. (1987) Proc. Natl. Acad. Sci. U.S.A. **84**, 4156–4160

Craig, E. A., Kramer, J., Shilling, J., Werner-Washburne, M., Holmes, S., Kosic-Smither, J. and Nicolet, C. M. (1989) Mol. Cell. Biol. **9**, 3000–3008

Craig, E., Kang, P. J. and Boorstein, W. (1990) Anth. van Leeuwenhoek **58**, 137–146

Deshaies, R., Koch, B., Werner-Washburne, M., Craig, E. and Schekman, R. (1988) Nature (London) **332**, 800–805

Deshaies, R. J., Koch, B. D. and Schekman, R. (1988) Trends Biochem. Sci. **13**, 384–388

DiDomenico, B. J., Bugaisky, G. E. and Lindquist, S. (1982) Cell **31**, 593–603

Ellis, R. J. (1987) Nature (London) **328**, 378–379

Ellis, R. J. and van der Vies, S. (1991) Annu. Rev. Biochem. **60**, 321–347

Farrelly, F. W. and Finkelstein, D. B. (1984) J. Biol. Chem. **259**, 5745–5751

Findly, R. C., Gilies, R. J. and Schulman, R. G. (1983) Science **219**, 1223–1225

Finkelstein, D. B. and Strausberg, S. (1983) J. Biol. Chem. **258**, 1908–1913

Finley, D. and Varshavski, A. (1985) Trends Biochem. Sci. **10**, 343–346

Finley, D., Özkaynak, E. and Varshavsky, A. (1987) Cell **48**, 1035–1046

Flaherty, K. M., Deluca-Flaherty, C. and McKay, D. B. (1990) Nature (London) **346**, 623–628

Fujita, A., Kikuchi, Y., Kuhara, S., Misumi, Y., Matsumoto, S. and Kobayashi, H. (1989) Gene **85**, 321–328

Furst, P., Hu, S., Hackett, R. and Hamer, D. H. (1988) Cell **55**, 705–717

Furst, P. and Hamer, D. H. (1989) Proc. Natl. Acad. Sci. U.S.A. **86**, 5267–5271

Gamer, J., Bujard, H. and Bukau, B. (1992) Cell **69**, 833–842

Georgopoulos, C. P. and Hohn, B. (1978) Proc. Natl. Acad. Sci. U.S.A. **75**, 131–135

Georgopoulos, C., Ang, D., Liberek, K. and Zylicsz, M. (1990) Stress Proteins in Biology and Medicine, Cold Spring Harbor Laboratory Press, Cold Spring Harbor, NY

Gething, M.-J. and Sambrook, J. (1992) Nature (London) **355**, 33–45

Goebl, M. G., Yochem, J., Jentsch, S., McGrath, J. P., Varshavsky, A. and Byers, B. (1988) Science **241**, 1331–1335

Goffeau, A. and Slayman, C. W. (1981) Biochim. Biophys. Acta **639**, 197–223

Gorenstein, C. and Warner, J. R. (1976) Proc. Natl. Acad. Sci. U.S.A. **73**, 1547–1551

Grant, C. M., Firoozan, M. and Tuite, M. F. (1989) Mol. Microbiol. **3**, 215–220

Gross, D. S., Adams, C. C., English, K. E., Collins, K. W. and Lee, S. (1990a) Anth. van Leeuwenhoek **58**, 175–186

Gross, D. S., English, K. E., Collins, K. W. and Lee, S. (1990b) J. Mol. Biol. **216**, 611–631

Grossman, A. D., Erickson, J. W. and Gross, C. A. (1984) Cell **38**, 383–390

Guarente, L. and Bermingham-McDonogh, O. (1992) Trends Genet. **8**, 27–32

Hall, B. G. (1983) J. Bacteriol. **156**, 1363–1365

Hamer, D. H. (1986) Annu. Rev. Biochem. **55**, 913–951

Heinemeyer, W., Kleinschmidt, J. A., Saidowsky, J., Escher, C. H. and Wolf, D. (1991) EMBO J. **10**, 555–562

Herruer, M. H., Mager, W. H., Raué, H. A., Vreken, P., Wilms, E. and Planta, R. J. (1988) Nucleic Acids Res. **16**, 7917–7029

Hershko, A. (1988) J. Biol. Chem. **263**, 15237–15240

Hottiger, T., Boller, T. and Wiemken, A. (1987) FEBS Lett. **220**, 113–115

Hottiger, T., Schmutz, P. and Wiemken, A. (1987) J. Bacteriol. **169**, 5518–5522

Hottiger, T., Boller, T. and Wiemken, A. (1989) FEBS Lett. **255**, 431–434

Hunt, T. (1989) Cell **59**, 949–951

Iida, H. and Yahara, I. (1984) J. Cell Biol. **99**, 1441–1450

Iida, H. and Yahara, I. (1985) Nature (London) **315**, 688–690

Iida, H. (1988) Mol. Cell. Biol. **8**, 5555–5560

Ingolia, T. D., Slater, M. R. and Craig, E. A. (1982) Mol. Cell. Biol. **2**, 1388–1398

Jakobsen, B. K. and Pelham, H. R. B. (1988) Mol. Cell Biol. **8**, 5040–5042

Jakobsen, B. K. and Pelham, H. R. B. (1991) EMBO J. **10**, 369–375

Jentsch, S., McGrath, J. P. and Varshavsky, A. (1987) Nature (London) **329**, 131–134

Jentsch, S., Seufert, W., Sommer, T. and Reins, H. A. (1990) Trends Biochem. Sci. **15**, 195–198

Johnston, G. C. and Singer, R. A. (1980) Mol. Gen. Genet. **178**, 357–360

Kim, C. H. and Warner, J. R. (1983) Mol. Cell. Biol. **3**, 457–465

Kirk, N. and Piper, P. W. (1991) Yeast **7**, 539–546

Kobayashi, N. and McEntee, K. (1990) Proc. Natl. Acad. Sci. U.S.A. **87**, 6550–6554

Koll, H., Guiard, B., Rassow, J., Ostermann, J., Horwich, A. L., Neupert, W. and Hartl, F.-U. (1992) Cell **68**, 1163–1175

Kondo, K. and Inouye, M. (1991) J. Biol. Chem. **266**, 17537–17544

Kurtz, S. and Lindquist, S. (1984) Proc. Natl. Acad. Sci. U.S.A. **81**, 7323–7327

Kurtz, S., Rossi, J., Petko, L. and Lindquist, S. (1986) Science **231**, 1154–1157

Landick, R., Vaughn, V., Lau, E. T., Van Bogelen, R. A., Erickson, J. W. and Neidthardt, F. C. (1984) Cell **38**, 175–182

Langer, T., Lu, C., Echols, H., Flannagan, J., Hayer, M. K. and Hartl, F. U. (1992) Nature (London) **356**, 683–689

Larson, J. S., Schuetz, T. J. and Kingston, E. E. (1988) Nature (London) **335**, 372–375; erratum **336**, 184

Leicht, B. G., Biessman, H., Palter, K. B. and Bonner, J. J. (1986) Proc. Natl. Acad. Sci. U.S.A. **83**, 90–94

Lewis, M. J. and Pelham, H. R. B. (1985) EMBO J. **4**, 3137–3143

Liberek, K., Galitski, T. P., Zylicz, M. and Georgopoulos, C. (1992) Proc. Natl. Acad. Sci. U.S.A. **89**, 3516–3520

Lindquist, S. (1981) Nature (London) **293**, 311–314

Lindquist, S. (1986) Annu. Rev. Biochem. **55**, 1151–1191

Lindquist, S. and Craig, E. A. (1988) Annu. Rev. Genet. **22**, 631–677

Luche, R. M., Sumrada, R. and Cooper, T. G. (1990) Mol. Cell. Biol. **10**, 3884–3995

McAlister, L. and Finkelstein, D. B. (1980) J. Bacteriol. **143**, 603–612

McAlister, L. and Finkelstein, D. B. (1980) Biochem. Biophys. Res. Commun. **93**, 819–824

McAlister, L., Strausberg, S., Kulaga, A. and Finkelstein, D. B. (1979) Curr. Genet. **1**, 63–74

McClanahan, T. and McEntee, K. (1986) Mol. Cell. Biol. **6**, 90–96

McDaniel, D., Caplan, A. J., Lee, M.-S., Adams, C. C., Fishel, R. R., Gross, D. S. and Garrard, W. T. (1989) Mol. Cell. Biol. **9**, 4789–4798

Metzger, D., White, J. H. and Chambon, P. (1988) Nature (London) **334**, 31–36

Miller, M. J., Xuong, N.-H. and Geidusche, E. P. (1979) Proc. Natl. Acad. Sci. U.S.A. **76**, 5222–5225

Miller, M. J., Xuong, N.-H. and Geiduschek, E. P. (1982) J. Bacteriol. **151**, 311–327

Mitsui, K. and Tsurugi, K. (1991) Biochem. Archiv. **7**, 169–176

Mori, K., Sant, A., Kohno, K., Normington, K., Gething, M.-J. and Sambrook, J. F. (1992) EMBO J. **11**, 2583–2593

Morimoto, R. I., Tissières, A. and Georgopoulos, C. (1990) Stress Proteins in Biology and Medicine, pp. 1–36, Cold Spring Harbor Laboratory Press, Cold Spring Harbor, NY

Muller-Taubenberger, A., Graack, H. R., Grohmann, L., Schleicher, M. and Gerisch, G. (1989) J. Biol. Chem. **246**, 5319–5322

Munro, S. and Pelham, H. R. B. (1985) Nature (London) **317**, 477–478

Neupert, W., Hartl, F.-U., Craig, E. A. and Pfanner, N. (1990) Cell **63**, 447–450

Nieto-Sotelo, J., Wiederrecht, G., Okuda, A. and Parker, C. S. (1990) Cell **60**, 807–817

Normington, K., Kohno, K., Kozutsumi, Y., Gething, M. J. and Sambrook, J. (1989) Cell **57**, 1223–1236

Nover, L., Scharf, K. D. and Neumann, D. (1983) Mol. Cell. Biol. **3**, 1648–1655

Nover, L., Munsche, D., Neuman, D., Ohme, K. and Scharf, K. D. (1986) Eur. J. Biochem. **160**, 297–304

Ostermann, J. H., Horwich, A. L., Neupert, W. and Hartl, F.-U. (1989) Nature (London) **341**, 125–130

Özkaynak, E., Finley, D., Solomon, M. and Varshavsky, A. (1987) EMBO J. **6**, 1429–1439

Pampulha, M. E. and Loureiro-Dias, M. C. (1989) Appl. Microb. Biotechnol. **31**, 23–27

Panaretou, B. and Piper, P. W. (1990) J. Gen. Microbiol. **136**, 1763–1770

Panaretou, B. and Piper, P. W. (1992) Eur. J. Biochem. **206**, 635–640

Park, H.-O. and Craig, E. A. (1989) Mol. Cell. Biol. **9**, 2025–2033

Parker, C. S. and Topol, J. (1984) Cell **37**, 273–283

Parry, J. M., Davies, P. J. and Evans, W. E. (1976) Mol. Gen. Genet. **146**, 27–35

Parsell, D. A., Sanchez, Y., Stitzel, J. D. and Lindquist, S. (1991) Nature (London) **353**, 270–273

Patriarca, E. and Maresca, B. (1990) Exp. Cell Res. **190**, 57–64

Pelham, H. R. B. (1986) Cell **46**, 959–961

Pelham, H. R. B. (1989) EMBO J. **8**, 3171–3176

Pelham, H. R. B. (1990) Stress Proteins in Biology and Medicine, Cold Spring Harbor Laboratory Press, Cold Spring Harbor, NY

Perisic, O., Xiao, H. and Lis, J. T. (1989) Cell **59**, 707–806

Petko, L. and Lindquist, S. (1986) Cell **45**, 885–894

Pfanner, N., Ostermann, J., Rassow, J., Hartl, F.-U. and Neupert, W. (1990) Anth. van Leeuwenhoek **58**, 191–193

Picard, D., Khursheed, B., Garabedian, M. J., Fortin, M. G., Lindquist, S. and Yamamoto, K. R. (1990) Nature (London) **348**, 166–168

Piper, P. W., Curran, B., Davies, M. W., Lockheart, A. and Reid, G. (1986) Eur. J. Biochem. **161**, 525–531

Piper, P. W., Curran, B., Davies, M. W., Hirst, K., Lockheart, A., Ogden, J. E., Stanway, C., Kingsman, A. J. and Kingsman, S. M. (1988) Nucleic Acids. Res. **16**, 1333–1348

Piper, P. W., Curran, B., Davies, M. W., Hirst, K., Lockheart, A. and Seward, K. (1990) Mol. Microbiol. **2**, 353–361

Piper, P. (1990) Anth. van Leeuwenhoek **58**, 195–201

Plesset, J., Palon, C. and McLaughlin, C. S. (1982) Biochem. Biophys. Res. Commun. **108**, 1340–1345

Praekelt, U. M. and Meacock, P. A. (1990) Mol. Gen. Genet. **223**, 97–106

Ptashne, M. (1988) Nature (London) **335**, 683–689

Rabindran, S. K., Giorgi, G., Clos, J. and Wu, C. (1991) Proc. Natl. Acad. Sci. U.S.A. **88**, 6906–6910

Redman, K. L. and Rechsteiner, M. (1989) Nature (London) **338**, 438–440

Ritossa, F. M. (1964) Exp. Cell. Res. **35**, 601–607

Rosa, M. F. and Sá-Correia, L. (1992) Enzyme Microb. Technol. **14**, 23–27

Rose, M. D., Misra, L. M. and Vogel, J. P. (1989) Cell **57**, 1211–1221

Rossi, J. and Lindquist, S. (1989) J. Cell Biol. **108**, 425–439

Rothman, J. (1989) Cell **59**, 591–601

Russo, P., Kalkkinen, N., Sareneva, H., Paakkola, J. and Makarow, M. (1992) Proc. Natl. Acad. Sci. U.S.A. **89**, 3671–3675

Sadis, S., Hickey, E. and Weber, L. A. (1988) J. Cell Physiol. **135**, 377–386

Sadowski, I., Niedbala, D., Wood, K. and Ptashne, M. (1991) Proc. Natl. Acad. Sci. U.S.A. **88**, 10510–10514

Sanchez, Y. and Lindquist, S. L. (1990) Science **248**, 1112–1115

Sanchez, Y., Taulien, J., Borkovich, K. A. and Lindquist, S. (1992) EMBO J. **11**, 2357–2364

Scharf, K.-D., Rose, S., Zott, W., Schöff, F. and Nover, L. (1990) EMBO J. **9**, 4495–4501

Schena, M. and Yamamoto, K. R. (1988) Science **241**, 965–967

Schenberg-Frascino, A. and Moustacchi, E. (1972) Mol. Gen. Genet. **115**, 243–257

Schlessinger, M. J. (1990) J. Biol. Chem. **265**, 12111–12114

Schuetz, T. J., Gallo, G. J., Sheldon, L., Tempst, P. and Kingston, R. E. (1991) Proc. Natl. Acad. Sci. U.S.A. **88**, 6911–6915

Serrano, R. (1983) FEBS Lett. **156**, 11–14

Shi, Y. and Thomas, J. O. (1992) Mol. Cell. Biol. **12**, 2186–2192

Shin, D.-Y., Matsumoto, K., Iida, H., Uno, I. and Ishikawa, T. (1987) Mol. Cell. Biol. **7**, 244–250

Shiu, R. P. C., Pouyssegur, J. and Pastan, I. (1977) Proc. Natl. Acad. Sci. U.S.A. **74**, 3840–3844

Silar, P., Butler, G. and Thiele, D. J. (1991) Mol. Cell. Biol. **11**, 1232–1238

Slater, M. R. and Craig, E. A. (1987) Mol. Cell. Biol. **7**, 1906–1916

Slater, M. R. and Craig, E. A. (1989) Nucleic Acids Res. **17**, 805–806

Smith, B. J. and Yaffe, M. P. (1991) Proc. Natl. Acad. Sci. U.S.A. **88**, 11091–11094

Sorger, P. K. (1990) Cell **62**, 793–805

Sorger, P. K. (1991) Cell **65**, 363–366

Sorger, P. K. and Pelham, H. R. B. (1988) Cell **54**, 855–864

Sorger, P. K. and Nelson, H. C. M. (1989) Cell **59**, 807–813

Sorger, P. K., Lewis, M. J. and Pelham, H. R. B. (1987) Nature (London) **329**, 81–84

Sternberg, N. (1973) J. Mol. Biol. **76**, 25–44

Straus, D. B., Walter, W. A. and Gross, C. A. (1987) Nature (London) **329**, 348–351

Subjeck, J. R., Shyy, T., Shen, J. and Johnson, R. J. (1983) J. Cell Biol. **97**, 1389–1395

Sumrada, R. A. and Cooper, T. G. (1987) Proc. Natl. Acad. Sci. U.S.A. **84**, 3997–4001

Sung, P., Prakash, S. and Prakash, L. (1988) Genes Dev. **2**, 1476–1485

Susek, R. E. and Lindquist, S. (1990) Mol. Cell. Biol. **10**, 6362–6373

Tanaka, K., Matsumoto, K. and Toh-e, A. (1988) EMBO J. **7**, 495–502

Thevelein, J. M. (1984) Microbiol. Rev. **48**, 42–59

Tilly, K., Spence, J. and Georgopoulos, C. (1989) J. Bacteriol. **171**, 1585–1589

Tissières, A., Mitchell, H. K. and Tracy, U. (1974) J. Mol. Biol. **84**, 389–398

Trolmo, C., André, L., Blomberg, A. and Adler, L. (1988) FEMS Microbiol. Lett. **56**, 321–326

Tuite, M. F., Bossier, P. and Fitch, I. T. (1988) Nucleic Acids Res. **16**, 11845

Tuite, M. F., Bently, N. J., Bossier, P. and Fitch, I. T. (1990) Anth. van Leeuwenhoek **58**, 147–154

Van Bogelen, R. A. and Neidhardt, F. C. (1990) Proc. Natl. Acad. Sci. U.S.A. **87**, 5589–5593

Varela, J., van Beekvelt, C., Planta, R. J. and Mager, W. H. (1992) Mol. Microbiol. **6**, 2183–2190

Veinot-Drebst, L. M., Singer, R. A. and Johnston, G. C. (1989) J. Biol. Chem. **264**, 19473–19474

Vogel, J., Misra, L. and Rose, M. (1990) J. Cell Biol. **110**, 1885–1895

Walton, E. F., Carter, B. L. A. and Pringle, J. R. (1979) Mol. Gen. Genet. **171**, 111–114

Watson, K. (1990) Adv. Microbiol. Physiol. **31**, 183–223

Watson, K. and Cavicchioli, R. (1983) Biotechnol. Lett. **5**, 683–688

Watson, K., Dunlop, G. and Cavicchioli, R. (1984) FEBS Lett. **172**, 299–302

Weitzel, G., Pilatus, U. and Rensing, L. (1987) Exp. Cell Res. **170**, 64–79

Wiederrecht, G., Seto, D. and Parker, C. S. (1988) Cell **54**, 841–853

Wiemken, A. (1990) Anth. van Leeuwenhoek **58**, 209–217

Xiao, H. and Lis, J. T. (1988) Science **239**, 1139–1142

Yang, W., Gahl, W. and Hamer, D. (1991) Mol. Cell. Biol. **11**, 3676–3681

Yost, H. J., Petersen, R. B. and Lindquist, S. (1990) Trends Biochem. Sci. **6**, 223–227

Yost, H. J. and Lindquist, S. (1991) Mol. Cell. Biol. **11**, 1062–1068

Young, D. (1992) Curr. Opinion Immunol. **4**, 396–400

Zeilstra-Ryalls, J., Fayet, O. and Georgopoulos, C. (1991) Annu. Rev. Microbiol. **45**, 301–325

Biochem. J. (1993) **293**, 305–316 (Printed in Great Britain)

REVIEW ARTICLE
Regulated exocytosis

Robert D. BURGOYNE and Alan MORGAN

The Physiological Laboratory, University of Liverpool, P.O. Box 147, Liverpool L69 3BX, U.K.

INTRODUCTION

Exocytosis is the final vesicular transport step in the secretory pathway. In many cell types a class of secretory vesicles is formed that can only fuse with the plasma membrane following cell activation (Burgoyne, 1990, 1991; Lindau and Gomperts, 1991; Plattner, 1989). It is this regulated exocytosis, necessary for the secretion of neurotransmitters, hormones and many other molecules, that will be the subject of this Review. In recent years, considerable progress has been made in the identification of proteins involved in the various vesicular transport steps of the secretory pathway (Pryer et al., 1992; Rothman and Orci, 1992). This has been due, in part, to the availability of 'cell-free' assays using permeabilized (or semi-intact) cells or isolated organelles that allow transport to be reconstituted and manipulated. Regulated exocytosis was the first intracellular transport step to be studied by cell permeabilization (Baker and Knight, 1978; Bennett et al., 1981). Despite this it is notable that in a recent review of vesicular transport (Pryer et al., 1992) regulated exocytosis was not mentioned. The likely reason behind this is that much of the work on regulated exocytosis has followed a pharmacological approach and it has only been relatively recently that some of the proteins likely to be involved have been identified following the spectacular success for earlier steps in the secretory pathway.

A second, important, difference between the study of regulated exocytosis and the earlier transport steps is that a significant contribution towards our knowledge of vesicular transport has come from genetic studies on the yeast *Saccharomyces cerevisiae* (Novick et al., 1981; Pryer et al., 1992). Since this organism does not appear to possess a regulated exocytotic pathway, this particular transport step has not been amenable to the type of genetic analysis that has been so informative for the constitutive secretory pathway.

The extensively used pharmacological approach to regulated exocytosis using a variety of inhibitors and activators of potentially important intracellular proteins has, with one or two exceptions, proven to be strikingly unsuccessful in the generation of reliable insights into the proteins involved. We will concentrate in this Review, therefore, on the more recent findings on proteins involved in regulated exocytosis.

BASIC REQUIREMENTS FOR EXOCYTOSIS

Studies on a variety of cell types have shown that both constitutive (Edwardson and Daniels-Holgate, 1992; Helms et al., 1990; Miller and Moore, 1991) and regulated exocytosis require cytosolic proteins (Ali et al., 1989; Koffer and Gomperts, 1989; Martin and Walent, 1989; Sarafian et al., 1987) and ATP (Dunn and Holz, 1983; Knight and Baker, 1982; Wilson and Kirshner, 1983) to be optimal. Regulated exocytosis in some cells such as myeloid cells appears to be exceptional. In mast cells, for example, ATP is not required and activation of GTP-binding proteins can

be sufficient to trigger exocytosis (Lindau and Gomperts, 1991). Constitutive exocytosis differs from regulated exocytosis in that it is unimpaired even if the concentration of cytosolic free calcium ($[Ca^{2+}]_i$) is reduced to well below resting levels (Edwardson and Daniels-Holgate, 1992; Helms et al., 1990; Miller and Moore, 1991; Turner et al., 1992a,b) and it is blocked by activation of GTP-binding proteins by the non-hydrolysable GTP analogue GTPγS (Edwardson and Daniels-Holgate, 1992; Helms et al., 1990; Miller and Moore, 1991). In contrast, regulated exocytosis in both animal and plant (Zorec and Tester, 1992) cells is activated by a rise in $[Ca^{2+}]_i$ and GTP analogues have complex but often stimulatory effects (Gomperts, 1990; Lindau and Gomperts, 1991). The requirement for ATP and cytosolic proteins is common to all vesicular transport steps (Pryer et al., 1992). Many of the vesicular transport steps studied involve both formation and fusion of vesicles, whereas regulated exocytosis involves triggered fusion of already formed vesicles. This may be a partial explanation for the finding that certain vesicular transport steps are markedly blocked as temperature is reduced to 20 °C or below (Saraste and Kuismanen, 1984) whereas exocytotic fusion does not show such an acute temperature discontinuity and continues down to 4 °C (Oberhauser et al., 1992b).

One difficulty in the study of regulated exocytosis is that, in addition to essential components involved in the fusion machinery, there must be proteins involved in the regulation of exocytosis by second messengers. Regulated secretory vesicles must normally be prevented from fusing with the plasma membrane since in the same cells constitutive secretory vesicles do so readily. A mechanism must exist that allows a Ca^{2+} signal, for example, to be transduced into activation of the dormant fusion machinery. This could involve removal of inhibition as well as direct activation. Added complexity is due to variation between cell types in the regulation of exocytosis and in specialized features of the process. For example, in certain synapses neurotransmitter release has to be extremely rapid and it is likely that exocytosis is triggered and fusion complete within 100 μs or so of Ca^{2+} channels being opened to allow Ca^{2+} entry and may result from exocytosis of only synaptic vesicles docked to presynaptic Ca^{2+} channels (Almers, 1990; Augustine et al., 1991; Llinas et al., 1992). In contrast, exocytosis in neuroendocrine cells is triggered after a lag period of 3–50 ms (Chow et al., 1992; Neher and Zucker, 1993; Thomas et al., 1993) and in mast cells after a lag period of a minute or so (Fernandez et al., 1984). It is possible that in neuroendocrine cells, such as pituitary cells or adrenal chromaffin cells and in mast cells, there would be sufficient time for a protein fusion complex to assemble and trigger exocytosis. In contrast, in fast neurotransmitter release, a minimal conformational change in a protein at the site of exocytosis is likely to be the only event for which there is sufficient time. This would mean that the fusion complex is already assembled on docked synaptic vesicles.

Abbreviations used: GTPγS, guanosine 5′-[γ-thio]triphosphate; GDPβS, guanosine 5′-[β-thio]diphosphate; PKC, protein kinase C; RACK, receptor for activated C-kinase; PLA₂, phospholipase A₂.

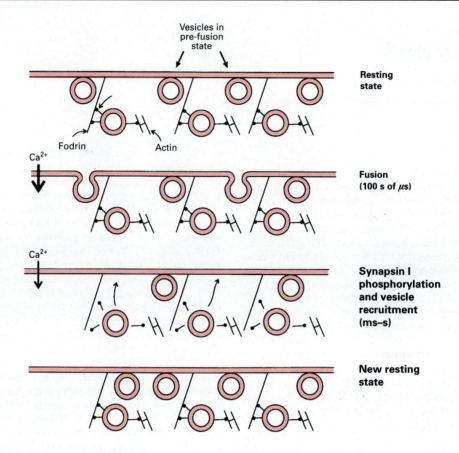

Figure 1 Exocytosis at the synapse

In the resting state some synaptic vesicles are closely associated with the presynaptic membrane and it is these that undergo exocytosis following depolarization and Ca^{2+} entry. Other vesicles are cross-linked by synapsin I to each other or to fodrin and are only released following phosphorylation of synapsin I. These vesicles can then move to the plasma membrane ready for the next depolarization.

To fully understand the mechanism of regulated exocytosis it is necessary to identify those proteins that act as regulatory as well as essential components of the exocytotic mechanism in addition to 'house-keeping' proteins that prepare the components for fusion. A comparison of the various cell types in which regulated exocytosis occurs should reveal whether there are universal protein components of the exocytotic machinery or whether the specialization required of regulated secretory cells has resulted in the evolution of a variety of distinct mechanisms for membrane fusion in regulated exocytosis.

STEPS IN REGULATED EXOCYTOSIS

As noted above, neurotransmitter release can be so fast that exocytosis must involve fusion of pre-docked synaptic vesicles tightly associated with the presynaptic release sites (Figure 1). Since the other synaptic vesicles are cross-linked within a cytoskeletal network (Hirokawa et al., 1989) by the extrinsic vesicle protein synapsin I (Sudhof et al., 1989), further events must occur subsequently in the synapse to allow recruitment of new vesicles to the plasma membrane in preparation for the next stimulus. These are believed to include phosphorylation of synapsin I leading to its dissociation from the vesicles and release of vesicles, bound by the cytoskeleton, that can then move to the presynaptic membrane. Synapsin I would then be rapidly dephosphorylated (DeCamilli and Greengard, 1986; Valtorta et al., 1992; Sudhof and Jahn, 1991). The initial activation of

exocytosis is believed to require a high (10–100 μM) $[Ca^{2+}]_i$ that is achieved locally at the presynaptic membrane following Ca^{2+} entry through plasma membrane channels (Augustine et al., 1991; Llinas et al., 1992). Synapsin I phosphorylation will be triggered by a lower rise in $[Ca^{2+}]_i$ due to activation of calmodulin-dependent kinase II (Valtorta et al., 1992). This kinase is also a synaptic vesicle protein that acts in the attachment of synapsin I to the vesicle (Benfenati et al., 1992). These ideas about the role of synapsin I in the regulation of synaptic vesicle availability have been supported by the findings that injected dephosphorylated synapsin I reduces neurotransmitter release from the squid giant synapse (Llinas et al., 1985) and rat brain synaptosomes (Nichols et al., 1992) and introduced calmodulin kinase II increases neurotransmitter release (Llinas et al., 1985; Nichols et al., 1990). Synapsin I is specific to nerve terminals and so this suggested mechanism does not occur in other cell types.

In other regulatory secretory cells, where exocytosis is triggered more slowly and can continue for prolonged periods, the situation is more complex and activation of the cells results in the sequential recruitment of multiple pools of secretory vesicles (Neher and Zucker, 1993; Thomas et al., 1993). In adrenal chromaffin cells (Aunis and Bader, 1988; Burgoyne and Cheek, 1987; Cheek and Burgoyne, 1986, 1987, 1992; Vitale et al., 1991), parotid salivary gland cells (Perrin et al., 1992), mast cells (Koffer et al., 1990) and various other cell types (reviewed in Cheek and Burgoyne, 1992; Trifaro and Vitale, 1993) there is an extensive actin network in the cell cortex that acts as a barrier to regulated secretory vesicles

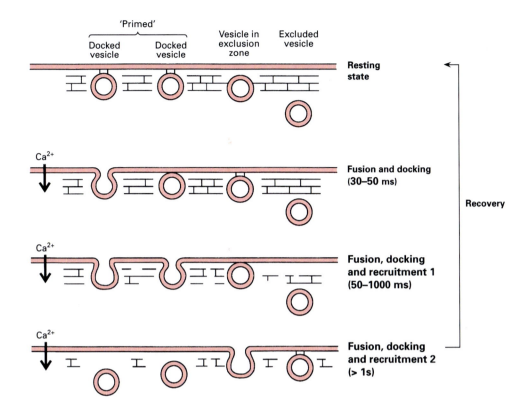

Figure 2   Discrete stages in exocytosis triggering in non-neuronal cells

Secretory vesicles or granules can be present at various stages in the resting state. In most non-neuronal cells, few vesicles would be in a pre-fusion state. Other docked vesicles could fuse with relatively short lag times. Disassembly of the cortical actin network would allow further vesicles to move to the plasma membrane for exocytosis at later times.

in the cell periphery (Figure 2). The difference in organization in these cells compared to synapses may be that such an actin network would not be sufficient to impede smaller synaptic vesicles, leading to the requirement for synapsin I for direct linkage of synaptic vesicles to the cytoskeleton. From electron microscopical observations on adrenal chromaffin cells, around 450 secretory granules are present in the cortical zone but the majority are excluded (Burgoyne et al., 1982; Burgoyne, 1991). In response to stimulation, many thousand granule fusions can be triggered in chromaffin cells and disassembly or reorganization of the actin network occurs reversibly after stimulation to allow secretory granules to move to the plasma membrane (Cheek and Burgoyne, 1986; Rodriguez Del Castillo et al., 1990; Vitale et al., 1991; Wu et al. 1992). Kinetic studies on digitonin-permeabilized chromaffin cells (Bittner and Holz, 1992a,b) and the use of patch-clamp analysis of exocytosis in chromaffin cells (Neher and Zucker, 1993) or pituitary cells (Thomas et al., 1993) have revealed that exocytosis can occur with several distinct kinetic components. The initial phases of release are ATP-independent but may already have been primed by ATP (Bittner and Holz, 1992a; Holz et al., 1989). ATP-dependent priming may involve vesicle binding through filamentous connections to the plasma membrane (Morimoto et al., 1990). The most likely interpretation of these kinetic data is that, in these cells, exocytotic fusion can be activated not only for secretory granules near the plasma membrane but also for other granules deeper within the cell following reorganization of the cytoskeleton. The first phase of release is fast and complete within 50 ms while later phases occur at slower rates (Neher and Zucker, 1993; Thomas et al., 1993). This fast burst of release involves an estimated 200 (Neher and

Zucker, 1993) or 450 (Thomas et al., 1993) vesicles from capacitance measurements or 1.3% of total catecholamine (around 390 secretory granules) from biochemical measurement (Bittner and Holz, 1992a). These values are similar to the estimated 450 granules near the plasma membrane in chromaffin cells (Burgoyne, 1991). Exocytosis can, therefore, be regulated at the level of secretory vesicle availability as well as by direct effects on membrane fusion. This again complicates the analysis of the components involved in the late steps leading to fusion.

IDENTIFICATION OF PROTEINS IN EXOCYTOSIS

Recent work has begun to identify cytosolic and membrane proteins involved in exocytosis. These proteins could be involved either in a regulatory function, or as essential components of the fusion machinery, but specific functions can not yet be assigned to the identified proteins. Amongst the proteins required for exocytosis must be at least one that acts as a Ca^{2+}-binding protein. It was originally believed that such a protein would be a high-affinity Ca^{2+}-binding protein with an affinity for Ca^{2+} in the range of $1-10\ \mu M$. The finding that fast exocytosis is triggered by high Ca^{2+} concentrations at the plasma membrane (Augustine et al., 1991; Augustine and Neher, 1992; Llinas et al., 1992; Neher and Zucker, 1993; O'Sullivan et al., 1989; Thomas et al., 1993) has led to the realization that a Ca^{2+}-binding protein with affinity in the range $10-100\ \mu M$ may be part of the mechanism. The various kinetic steps in exocytosis appear to have different Ca^{2+}-dependencies (Bittner and Holz, 1992a; Neher and Zucker, 1993), suggesting that exocytosis could be triggered or controlled by multiple Ca^{2+}-binding proteins with varying Ca^{2+} affinities. In the vesicular transport steps from endoplasmic reticulum to Golgi and between Golgi elements, multiple proteins are required including cytosolic, extrinsic and intrinsic membrane proteins (Pryer et al., 1992; Rothman and Orci, 1992). Even in constitutive exocytosis in yeast, 10 genes are known to be required and several of the gene products appear to assemble to form a cytosolic protein complex (Bowser et al., 1992). Therefore, we can expect to find a similar requirement for several different proteins in regulated exocytosis. Studies on mutants of *Paramecium* in which regulated exocytosis of trichocysts is impaired have already revealed a requirement for at least 13 distinct genes (Bonnemain et al., 1992).

Cytosolic and extrinsic membrane proteins

A requirement for soluble ('cytosolic') proteins in Ca^{2+}-dependent exocytosis has been shown for adrenal chromaffin cells (Ali et al., 1989; Morgan and Burgoyne, 1992a; Sarafian et al., 1987; Wu and Wagner, 1991), GH_3 cells (Martin and Walent, 1989) PC12 cells (Lomneth et al., 1991), mast cells (Howell and Gomperts, 1987; Koffer and Gomperts, 1989) and brain synaptosomes (Kish and Ueda, 1991). Two of the defective proteins in the *Paramecium* mutants mentioned above are cytosolic proteins (Bonnemain et al., 1992). Earlier pharmacological experiments using phorbol esters have suggested that one cytosolic protein, protein kinase C (PKC), is a regulator of Ca^{2+}-dependent exocytosis (Knight and Baker, 1983; Pocotte et al., 1985) and this has been confirmed by introduction of purified PKC into permeabilized chromaffin cells (Morgan and Burgoyne, 1992b), pituitary (Naor et al., 1989) and PC12 cells (Ben-Shlomo et al., 1991; Nishizaki et al., 1992). The role of soluble proteins has become apparent from the run-down of secretory responsiveness of permeabilized cells as they leak such proteins. This has formed the basis of an assay for the identification of certain essential or regulatory cytosolic and extrinsic membrane proteins.

Using reintroduction of proteins into digitonin-permeabilized chromaffin cells, several stimulatory proteins have been identified. The first to be identified in this way was annexin II (Ali et al., 1989) a member of the annexin family of Ca^{2+}- and phospholipid-binding proteins (Burgoyne and Geisow, 1989; Creutz, 1992). Annexin II is found on the inner surface of the plasma membrane of intact chromaffin cells (Nakata et al., 1990) and so is not normally a cytosolic protein but behaves as an extrinsic membrane protein at resting Ca^{2+} concentrations. It is extracted from the plasma membrane following permeabilization in the presence of EGTA (Burgoyne and Morgan, 1990). Annexin II has been suggested to be involved in exocytosis due to its location on the plasma membrane and the possibility that it forms filamentous cross-links between granules and plasma membranes (Nakata et al., 1990). Furthermore, it has the ability to aggregate isolated secretory granules in the presence of micromolar Ca^{2+} and to fuse them after addition of arachidonic acid (Drust and Creutz, 1988). Incubation with exogenous, purified annexin II resulted in an increase in Ca^{2+}- and ATP-dependent exocytosis in permeabilized chromaffin cells (Ali et al., 1989; Ali and Burgoyne, 1990). The stimulatory effect of annexin II was suggested to be dependent upon PKC-mediated phosphorylation of the protein (Sarafian et al., 1991) though other data do not agree with this conclusion (Ali and Burgoyne, 1990). The extreme N-terminus of annexin II, which contains sites for phosphorylation and for binding of the subunit p11, is critical for its ability to stimulate exocytosis (Ali and Burgoyne, 1990; Burgoyne and Morgan, 1990). Endogenous annexin II seems likely to be involved in exocytosis in chromaffin cells since exocytosis assayed without

added annexin II was partially inhibited by a synthetic peptide corresponding to the most conserved annexin domain (Ali et al., 1989). In contrast, a synthetic peptide corresponding to the N-terminal 15 residues of annexin II was without effect on exocytosis (Ali and Burgoyne, 1990).

As run-down is allowed to proceed for longer periods, the ability of annexin II to stimulate exocytosis is lost (Burgoyne and Morgan, 1990; Sarafian et al., 1991), suggesting that other required proteins leak from the permeabilized cells. In order to attempt to detect additional cytosolic proteins that regulate exocytosis, adrenal medullary and brain cytosols were fractionated and tested in a run-down/reconstitution assay. Three activities were detected, one was inhibitory and the other two (Exo1 and Exo2) were stimulatory (Morgan and Burgoyne, 1992a). Purified Exo1 consists of a family of proteins belonging to the 14-3-3 gene family (Morgan and Burgoyne, 1992a, Morgan et al., 1993b). These proteins were independently purified from adrenal medulla using a similar approach by another laboratory (Wu et al., 1992) and the Exo1/14-3-3 proteins were found to leak from permeabilized chromaffin cells (Morgan et al., 1993; Wu et al., 1992). Wu et al. (1992) used an immunodepletion approach to show that the 14-3-3 proteins were major stimulatory components of cytosol. Little is known about Exo2 except that it apparently behaves as a single 44 kDa protein. The ability of Exo1 to stimulate exocytosis is Ca^{2+}- and ATP-dependent, is blocked by tetanus toxin and is potentiated by activation of PKC using phorbol esters or co-introduction of purified rat brain PKC (Morgan and Burgoyne, 1992a,b; Morgan et al., 1993a). It did not, however, appear to be a good substrate itself for PKC. 14-3-3 proteins have been cloned from mammalian tissues and from *Xenopus*, *Drosophila*, plants and yeast. The proteins all show a high degree of sequence similarity with one another (Aitken et al., 1992; Isobe et al., 1992; Martens et al., 1992) but it is not clear whether all of the mammalian proteins are able to stimulate exocytosis. One of the domains conserved in the entire range of 14-3-3 proteins is homologous to the C-terminus of the annexins and is most similar to annexin II (Aitken et al., 1990). It seems possible that this domain could be necessary for interaction with a common target protein required for exocytosis stimulated by both classes of protein. The synaptic vesicle protein synaptotagmin (see below) and PKC have been shown to bind to target proteins known as RACK (receptor for activated C-kinase) proteins and this binding is inhibited by a synthetic peptide similar to this conserved sequence (Mochly-Rosen et al., 1992). Little is known about the RACK proteins but they may include certain annexins (Mochly-Rosen et al., 1991). Direct evidence that the common annexin/14-3-3 protein domain is important in exocytosis was shown by the finding that a 16-residue synthetic peptide, but not truncated peptides, based on the C-terminus of annexin II partially inhibited Ca^{2+}-dependent exocytosis in permeabilized chromaffin cells (Roth et al., 1993). It has been suggested that the 14-3-3 proteins have Ca^{2+}-dependent phospholipase A_2 (PLA_2) activity (Zupan et al., 1992). This could be significant since, as noted above, secretory granules cross-linked by annexin II fuse when arachidonic acid is added. The work of Zupan et al. (1992) did not actually demonstrate PLA_2 activity but only substrate binding and we and other laboratories were unable to detect PLA_2 activity in purified brain Exo1 using 1-palmitoyl-2-arachidonoylphosphocholine as substrate (Morgan et al., 1993b). Arachidonic acid generation and exocytosis in permeabilized chromaffin cells had previously been dissociated, suggesting that Ca^{2+}-dependent arachidonic acid production is not required for the activation of exocytosis (Morgan and Burgoyne, 1990).

Using an alternative cell permeabilization technique, known as cell-cracking, in which cells are sheared using a ball homogenizer and cytosol completely removed by washing, other cytosolic proteins have been identified which stimulate exocytosis in GH$_3$ and PC12 cells (Hay and Martin, 1992; Martin and Walent, 1989; Nishizaki et al., 1992; Walent et al., 1992). A protein in brain cytosol which stimulated exocytosis in GH$_3$ cells was partially characterized initially and then a 145 kDa protein which stimulated exocytosis in PC12 cells was purified from brain cytosol. The activity of this protein was Ca^{2+}- and ATP-dependent and it was found to be present in a range of tissues showing regulated secretion (Walent et al., 1992). Antiserum against this protein produced a marked inhibition of the ability of cytosol to stimulate exocytosis, indicating that it is a major contributor to the activity of crude cytosol. The protein did not appear to be related to any known proteins. The ability of p145 to stimulate exocytosis was increased by PKC-mediated phosphorylation and p145 is a substrate for the kinase (Nishizaki et al., 1992). The lack of detection of a requirement for this protein in the digitonin-permeabilized chromaffin cell system may be due to the fact that the native state of p145 is that of a dimer and it would be likely to leak slowly from digitonin-permeabilized cells, which have limited permeability, and thus p145 would not become rate-limiting.

Hay and Martin (1992) have recently developed a protocol that can resolve two stages in exocytosis in permeabilized PC12 cells similar to those previously detected in chromaffin cells (Bittner and Holz, 1992a; Holz et al., 1989). These are an ATP-dependent priming stage and a Ca^{2+}-dependent triggering stage. Each stage was stimulated by distinct protein fractions derived from brain or adrenal medullary cytosols and the 145 kDa protein isolated by this laboratory was functional in only the Ca^{2+}-triggering stage whereas a 20 kDa factor was active in priming.

These studies on permeabilized adrenal chromaffin and PC12 cells have revealed multiple cytosolic (or membrane-associated in the case of annexin II) proteins that stimulate Ca^{2+}-dependent exocytosis and that might function to accelerate distinct stages in the exocytotic process. The different permeabilization methods and protocols used to assay exocytosis would almost certainly result in different steps (or proteins) being rate-limiting in each cell type studied and there is no reason to believe that the fact that different cytosolic proteins were revealed is in any way contradictory. On the contrary, it merely reflects the complexity of regulated exocytosis. The interactions between the cytosolic proteins identified so far in chromaffin and PC12 cells and their exact function in exocytosis remains to be established.

Triggered exocytosis in *Paramecium* is accompanied by the rapid dephosphorylation of a protein, pp63, that behaves as a soluble cytosolic protein but is also localized within the cell cortex (Gilligan and Satir, 1982; Ziesness and Plattner, 1985; Momayezi et al., 1987; Hohne-Zell et al., 1992). This dephosphorylation event does not occur at the non-permissive temperature in various non-discharge mutants (Gilligan and Satir, 1982; Ziesness and Plattner, 1985) and microinjected anti-pp63 antibodies inhibit trichocyst discharge (Stecher et al., 1987), suggesting that pp63 is involved in exocytosis. A related protein has been detected in various species and mammalian tissues (Satir et al., 1989) but pp63 has not yet been characterized in any detail and no sequence data is available.

Secretory vesicle proteins

Vesicle docking at the plasma membrane must surely involve secretory vesicle proteins and these could act as key elements of the fusion mechanism. Over the past few years, extensive work

has been carried out in attempts to characterize secretory vesicle proteins and considerable information is now available on the membrane proteins of the synaptic vesicle (Sudhof and Jahn, 1991). Because the synaptic vesicle is so small (around 50 nm in diameter) and can contain limited amounts of protein (Bennett et al., 1992a), the protein composition of this vesicle is close to being fully characterized. Once transport or biosynthetic proteins are excluded, relatively few synaptic vesicle proteins remain as potential players in the exocytotic mechanism. One or more of these must be a key element in rapid exocytotic fusion at the synapse. Amongst the transmembrane vesicle proteins characterized so far, three have stood out as prime candidates—synaptophysin, synaptotagmin (p65) and synaptobrevin.

Synaptophysin is the major integral synaptic vesicle protein (Navone et al., 1986) and the first to be sequenced (Sudhof et al., 1987). It is believed to span the vesicle membrane four times and in some respects has the form of a channel protein. The reconstituted protein was shown to have channel activity in planar lipid bilayers (Thomas et al., 1988) and it was thought to be a Ca^{2+}-binding protein (Rehm et al., 1986). It was, therefore, suggested to be involved in initial pore formation (see the section on fusion pore below) prior to exocytosis (Thomas et al., 1988). It has been pointed out (Sudhof and Jahn, 1991), however, that the voltage-dependency of the synaptophysin channel would result in channel closure during exocytosis and the ability of synaptophysin to bind Ca^{2+} has been disputed (Brose et al., 1992). Studies on *Xenopus* oocytes support the idea that synaptophysin is required for exocytosis (Alder et al., 1992a). Total mRNA from brain was injected into oocytes and protein synthesis allowed to proceed. The oocytes developed the ability to release the neurotransmitter glutamate in response to a Ca^{2+} ionophore (Alder et al., 1992a) presumably due to the synthesis *de novo* of functional synaptic vesicles. Neurotransmitter release was inhibited by an antisense oligonucleotide designed to disrupt synaptophysin expression and also by anti-synaptophysin antibodies. These results suggest that synaptophysin may be necessary for synaptic vesicle exocytosis but do not distinguish between a purely structural role for this highly abundant vesicle protein and an essential mediatory role in exocytosis. The same comment applies to the demonstration that anti-synaptophysin antibodies inhibit neurotransmitter release at the neuromuscular junction (Alder et al., 1992b). In addition, all antibody experiments of this type can be criticized due to the possibility of a non-specific block of exocytosis by large antibody molecules decorating the synaptic vesicle.

Synaptotagmin was first discovered by chance as a widespread synaptic vesicle protein known as p65 (Matthew et al., 1981) and was found to be present in several secretory granules from endocrine tissues (Trifaro et al., 1989). Most of the protein projects into the cytoplasm (Perin et al., 1991; Tugal et al., 1991), it possesses two C$_2$-like domains (Figure 3) related to those of PKC (Perin et al., 1990) and it binds Ca^{2+} and phospholipid (Brose et al., 1992; Perin et al., 1990). Synaptotagmin has been found to bind to the receptor for α-latrotoxin, a spider venom component that acts extracellularly to activate exocytosis (Petrenko et al., 1991). The α-latrotoxin receptor belongs to a family of synaptic membrane proteins, the neurexins, and synaptotagmin binds in a Ca^{2+}-independent manner to the cytoplasmic C-termini of these proteins (Hata et al., 1993). Synaptotagmin has also been found associated with N-type Ca^{2+} channels (Bennett et al., 1992b; Leveque et al., 1992) which suggests an intriguing scenario in which synaptotagmin is responsible for docking of synaptic vesicles by binding to Ca^{2+} channels, thus holding the vesicle precisely at the site of Ca^{2+} entry for rapid exocytosis in synapses (Figure 4). Synaptotagmin

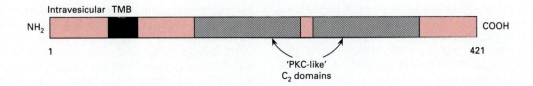

Figure 3   Schematic diagram of the domain structure of synaptotagmin (p65)

The figure is based on sequence data described by Perin et al. (1990).

is present on many synaptic vesicle and endocrine secretory granules (but not exocrine granules), supporting a widespread role in exocytosis in many but not all cell types. Recent work on PC12 cells has, however, argued against such a role. Shoji-Kasai et al. (1992) selected synaptotagmin-deficient PC12 cell lines. These cell lines not only secreted perfectly well but the extent of secretion was greater than that in the parent cell line. The exocytosis assayed would have involved dense core granules and so one possibility is that synaptotagmin is required for fast exocytosis of small synaptic vesicles but is not essential for slower, dense-core granule exocytosis. It is also possible that the role of synaptotagmin is not as a Ca²⁺ sensor in the essential fusion machinery but is simply in the docking of synaptic vesicles at exocytotic sites. Synthetic peptides based on the synaptotagmin sequence inhibit exocytosis in nerve terminals (Bonnert et al., 1993) and a partial reduction in the extent of exocytosis, mainly in a subpopulation of highly-secreting cells, was observed in a study in which antibodies against synaptotagmin or certain fragments of the cytoplasmic domain of synaptotagmin were microinjected into PC12 cells (Elferink et al., 1993). None of the studies have so far examined the kinetics of exocytosis in cells in which synaptotagmin function has been disrupted; if its role is in vesicle docking then such disruption might be manifest as a reduction in the initital rate of exocytosis or a loss of the first rapid phase of exocytosis.

Synaptobrevin is an evolutionarily conserved synaptic vesicle protein that exists in two isoforms in mammalian brain which are also known as VAMP 1 and 2 (Archer et al., 1990; Baumert et al., 1989; Elferink et al., 1989). Little had been known about synaptobrevin but now evidence has been obtained suggesting that the synaptobrevins are essential for neurotransmitter release. The neurotoxins tetanus toxin and the botulinum toxins potently produce long lasting blocks of neurotransmitter release. If the active fragments of these toxins are able to enter the cytosol of neurons or neuroendocrine cells, a specific inhibition of regulated exocytosis occurs. The toxins have been extensively studied since the view has been taken, almost certainly correctly, that the substrates for the toxin must be key components of the machinery for neurotransmitter exocytosis. It has now been shown that tetanus and botulinum B toxins have zinc-dependent protease activity (Schiavo et al., 1992a,b; Link et al., 1992) and in synaptic vesicles specifically cleave one of the synaptobrevin isoforms at a sequence motif apparently not found in any other known proteins (Figure 5). Botulinum toxins type A and E do not appear to act by precisely the same mechanism as type B and tetanus toxin and their mode of action is unknown. Unfortunately, no data is available on whether the tetanus and botulinum B toxins cleave any cytosolic, cytoskeletal or plasma membrane proteins nor on the exact substrate sequence motif that is sufficient to allow proteolysis. This limitation of the published data make it difficult to decide whether synaptobrevin 2 really is the sole substrate. If it is, then it is clearly essential for neurotransmitter release and it is important that more information on this point is now obtained. Tetanus and botulinum toxins inhibit exocytosis in neuroendocrine cells (Holz et al., 1992) but not exocrine secretory cells (Stecher et al., 1992); it is

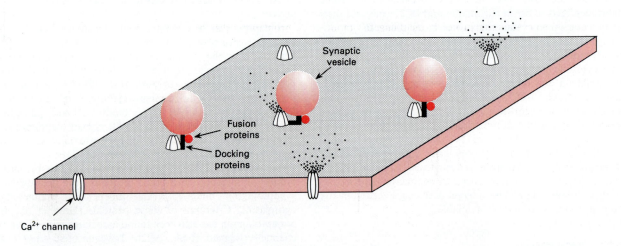

Figure 4   Organization of synaptic vesicles and calcium channels at the presynaptic membrane

In order to account for fast neurotransmitter release it is likely that synaptic vesicles are docked at the presynaptic membrane close to the Ca²⁺ channels, so that exocytosis is rapidly triggered by the cloud of Ca²⁺ at the mouth of the channel.

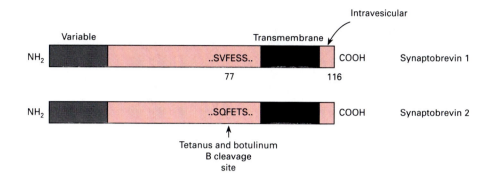

Figure 5   Schematic diagram of the domain structure of rat brain synaptobrevins

This is based on the data in Archer et al. (1990). Tetanus and botulinum B toxins (Schiavo et al., 1992b) specifically cleave only synaptobrevin 2 between Q and F in the amino acid sequence shown. The sequences are those of the rat synaptobrevin isoforms.

not known whether exocrine cells express synaptobrevin. The details of the tissue distribution of the synaptobrevins are sparse but clearly more information on this point is now essential in order to determine whether synaptobrevin is present on the secretory vesicles of all cell types sensitive to tetanus and botulinum B toxins and to enable assessment of whether synaptobrevin is a general or a synapse-specific component of the exocytotic machinery. In favour of a more general distribution is the discovery of synaptobrevin- (VAMP)-related proteins in adipocytes in secretory vesicles that are responsible for the insertion of the GLUT4 glucose transporter into the plasma membrane in a regulated fashion following exposure to insulin (Cain et al., 1992).

GTP-binding proteins

Many vesicular transport steps are regulated by GTP-binding proteins (Pryer et al., 1992; Rothman and Orci, 1992). The first GTP-binding protein to be identified that is involved in vesicular transport was the monomeric GTP-binding protein, Sec4, which is required for constitutive exocytosis in yeast (Salminen and Novick, 1987). A family of mammalian proteins related to Sec4, known as the rab proteins, have subsequently been found to be localized at distinct sites within the cell (Chavrier et al., 1990) and may each be involved in specific vesicular transport steps (Pfeffer, 1992). Additional GTP-binding proteins, including the monomeric ARF proteins (Taylor et al., 1992) and heterotrimeric G proteins (Barr et al., 1992) have been shown to control vesicle budding and transport. Indications of the importance of GTP-binding proteins has come from the finding that the non-hydrolysable GTP analogue GTPγS blocks many vesicular transport steps. This inhibition may be at the level of transport vesicle formation rather than fusion with the acceptor membrane (Barr et al., 1992).

In the case of regulated exocytosis, GTPγS and other GTP analogues were found to stimulate exocytosis, in some cases in the absence of Ca^{2+} (Barrowman et al., 1986; Gomperts, 1990; Lindau and Gomperts, 1991). In some cell types an additional effect can be demonstrated which is an inhibition of Ca^{2+}-dependent exocytosis if the permeabilized cells are first pre-incubated with GTPγS (Ahnert-Hilger et al., 1992; Knight and Baker, 1985; DeMatteis et al., 1991; Davidson et al., 1991; Smolen et al., 1991; Turner et al., 1992b). Synaptic vesicles and secretory granules possess both monomeric (Burgoyne and Morgan, 1989; Darchen et al., 1990; Fischer von Mollard et al.,

1991) and trimeric (Toutant et al., 1987) GTP-binding proteins. It is not clear, however, which type of GTP-binding protein is involved in the stimulatory or inhibitory effects of GTPγS. Antibody inhibition experiments on permeabilized chromaffin have suggested that G$_o$ normally exerts an inhibitory effect on Ca^{2+}-dependent exocytosis (Ohara-Imaizumi et al., 1992). In the nerve terminal of the squid giant axon, injected GTPγS and GDPβS produced a slow block of exocytosis of the docked synaptic vesicles (Hess et al., 1993). The data would be consistent with a role for a monomeric GTP-binding protein in vesicle docking and GTP hydrolysis having to occur before Ca^{2+} could trigger exocytosis.

One approach to investigating the function of the rab proteins has been to use synthetic peptides based on the postulated 'effector domain' of these proteins. One such 16-residue peptide known as rab3AL inhibited endoplasmic reticulum-to-Golgi and intra-Golgi transport (Plutner et al., 1990) and the interpretation of this finding was that the peptide bound to the downstream effector protein for the rab protein to antagonize the action of the GTP-bound form of the native rab protein. Surprisingly, the rab3AL peptide acts as an activator of exocytosis in pancreatic acinar cells (Padfield et al., 1992) and mast cells (Oberhauser et al., 1992a) in the absence of any other stimuli. It is not known whether rab3 is expressed in these cell types. Chromaffin cells do express rab3 (Darchen et al., 1990) but in digitonin-permeabilized chromaffin cells the rab3AL peptide produced only a small enhancement of Ca^{2+}-dependent exocytosis (Senyshyn et al., 1992). In the work on mast cells, patch clamp measurement demonstrated directly that the effect of the peptide was to stimulate exocytotic membrane fusion. Even more surprising was that a five-residue peptide corresponding to residues 33–37 of rab3 was sufficient to activate complete degranulation of the mast cells (Oberhauser et al., 1992a). The rab peptides required an extensive lag period of 10 min or so. The reason for this and the mechanism of action of the peptides and which rab proteins are expressed by mast cells are unknown. An earlier study has shown that injection of oncogenic ras protein resulted in mast cell degranulation after a prolonged lag period (Bar-Sagi and Gomperts, 1988). Much further work will be needed to determine the involvement of the rab proteins and exactly which GTP-binding proteins control regulated exocytosis.

Plasma membrane proteins

Little information is available on plasma membrane proteins in

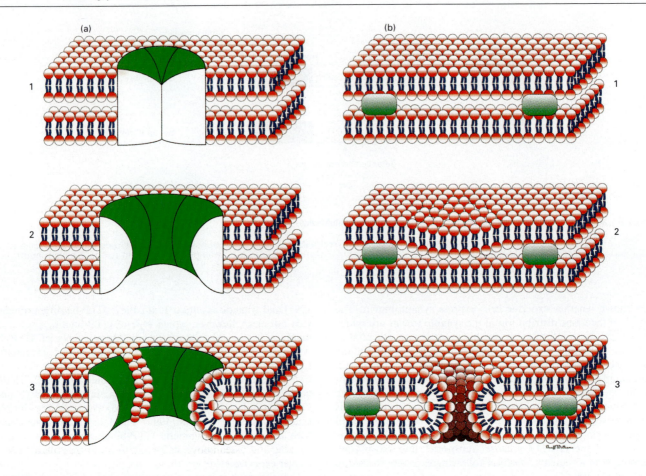

Figure 6   Models for the structure of the fusion pore in exocytosis

The two models shown are modified from Almers (1990) in (**a**) or based on the model of Monck and Fernandez (1992) as shown in (**b**). In the Almers model the fusion pore involved the formation of an oligomeric protein structure involving vesicle and plasma membrane proteins (1). This structure opens during stimulation to form the pore (2) and disassembly of the oligomeric subunits within the bilayer allows full exocytotic fusion to occur (3). In the Monck and Fernandez model proteins act to pull the bilayers into close apposition (1) and inward dimpling of the plasma membrane bilayer (2) is followed by formation of a lipid pore (3).

regulated exocytosis. Binding partners for the synaptic vesicle proteins synaptophysin (physophilin; Thomas and Betz, 1990) and synaptotagmin have been identified but their significance is unclear. Synaptotagmin was found to associate with the α-latrotoxin receptor (Petrenko et al., 1991) which forms part of a family of synaptic plasma membrane proteins known as the neurexins (Hata et al., 1993; Ushkaryov et al., 1992). The role of the neurexins in exocytosis is unclear since their structure is related to that of proteins involved in cell adhesion. One worrying possibility about the numerous interactions of synaptotagmin with other proteins that have been detected (Bennett et al., 1992a,b; Leveque et al., 1992; Petrenko et al., 1991) is that they simply represent non-specific interactions of a 'sticky' protein.

A rather specific interaction has been discovered involving synaptic plasma membrane proteins, the synaptic vesicle protein synaptobrevin and a set of proteins believed to be required for vesicle targeting in early stages of the secretory pathway (Sollner et al., 1993). From reconstitution studies on vesicular transport between Golgi cisternae, the protein NSF and the SNAP proteins (α, β and γ; Whiteheart et al., 1993) have been identified as essential components that interact in a 20 S particle (Rothman and Orci, 1992; Clary and Rothman, 1990; Wilson et al., 1992). In an attempt to identify integral membrane receptors for the 20 S particle an affinity chromatography approach was used with a detergent-solubilized extract of total brain membranes. Only four polypeptides were found to interact specifically, that is in an ATP-dependent manner, with the 20 S particle. Peptide sequencing revealed these to be synaptobrevin 2, syntaxin A and B and SNAP-25 (Sollner et al., 1993). Syntaxin A and B were originally identified as synaptic membrane proteins that co-immunoprecipitated with synaptotagmin (Bennett et al., 1992) but no synaptotagmin was detected associated with the 20 S particle. SNAP-25 has been little studied but is a previously identified synapse-specific protein (Oyler et al., 1989). No direct functional data is available on the role of the NSF/SNAP proteins in neurotransmitter release, but from what is known of their function in intra-Golgi transport it has been suggested that the interaction between the 20 S particle and the synaptic proteins could be a mechanism for vesicle targeting and docking on the presynaptic membrane. So far there is no evidence that NSF or the SNAPs actually act as membrane fusion proteins during vesicular transport or that these proteins interact with membrane lipids in any way and so it is not clear if they have any further function after docking on the target membrane. It is also important to note that NSF and SNAPs only function in Golgi transport in the presence of several additional cytosolic proteins (Clary and Rothman, 1990; Waters et al., 1992) and so its seems likely that additional cytosolic proteins would also be required for regulated exocytosis as discussed above. It is not known whether NSF and SNAPs are present in nerve terminals and this important point needs to be addressed to allow assessment of this recent intriguing data.

GAP-43 (B-50) is an extrinsic plasma membrane protein found predominantly in neurons. Antibodies against this protein inhibited Ca^{2+}-dependent exocytosis in permeabilized brain synaptosomes (Dekker et al., 1989) but the exact function of GAP-43, which is also highly expressed in developing neurons, is not known. Expression of an antisense construct that reduced GAP-43 levels in PC12 cells lead to an increase in basal secretion and a concomitant decrease in evoked secretion (Ivins, 1993). These results suggest that GAP-43 may exert an inhibitory influence to prevent exocytosis at resting Ca^{2+} concentration. A plasma membrane protein from adrenal chromaffin cells that binds to secretory granules has been identified. Antibodies against this 51 kDa protein inhibited exocytosis when introduced into chromaffin cells via a patch pipette (Schweizer et al., 1989) but relatively little is known about this protein.

THE FUSION PORE

One of the most powerful methods for the study of regulated exocytosis is the patch-clamp capacitance recording technique, since it allows direct measurement of membrane fusion of single secretory vesicles with high time resolution as well as single cell biochemistry (for more details see Monck and Fernandez, 1992). The method is based on the fact that addition of secretory vesicle membrane into the plasma membrane as fusion occurs will result in a step-wise increase in plasma membrane capacitance that can be detected electrophysiologically (Neher and Marty, 1982). The most important insights from capacitance measurement have come from studies of either normal or mutant mast cells from the beige mouse. The secretory vesicles of these cells are amongst the largest known, 0.8 μm or 1–5 μm in diameter for wild type and beige mice respectively, and exocytosis in mast cells is slow. For these reasons, not only can irreversible fusion events be measured but transient events, originally called capacitance flicker (Fernandez et al., 1984) can be detected. These appear to represent the initial events in exocytosis, the reversible formation of the 'fusion pore'. The fusion pore has the characteristics of a channel 1–2 nm in diameter with a fixed conductance (Breckenridge and Almers, 1987a) that can either close again or explosively open in a full exocytotic event (Breckenridge and Almers, 1987a,b; Monck et al., 1990; Zimmerberg et al., 1987). The lifetime of the fusion pore in mast cells is considerable, lasting up to several seconds. This is at least 10^4 times longer than the entire exocytotic process in a fast synapse. Therefore, there may be some very odd aspects of the mast cell that allow a fusion pore to open, remain stable in the fused lipid bilayers for such a prolonged period and then re-close. Two extreme models (Figure 6) have been suggested for the structure of the fusion pore as well as a third intermediate version (Zimmerberg et al., 1991). The first suggests that the pore is proteinaceous and must involve proteins in the secretory vesicle and plasma membrane that can initially act like a gap-junction channel (Almers, 1990). Synaptophysin was suggested to be able to act in this way in synapses by Thomas et al. (1988). This oligomeric channel would then have to spontaneously fall apart (Figure 6a), a scenario that seems thermodynamically improbable since considerable energy would need to be expended to split apart polypeptides that needed to be closely associated in the lipid bilayer in the first phase of the mechanism. The second extreme model, based on a lipid pore (Figure 6b), has been devised by Monck and Fernandez (1992). Previous work from this laboratory had suggested that lipid flow

could occur through the fusion pore while it was reversibly open (Monck et al., 1990). Secondly, a pore with all the characteristics of the exocytotic fusion pore could be induced in a pure lipid membrane and theoretical modelling demonstrated that closure or explosive opening of a fusion pore could be determined by the nature of the lipids flowing into the pore (Nanavati et al., 1992). In the lipidic fusion pore model of Monck and Fernandez the role of proteins would be to pull the plasma membrane and secretory vesicle membranes together. This idea is consistent with ultrastructural data on *Limulus* amoebocytes, *Paramecium* and mast cells, in which late stages in exocytosis could be visualized after rapid freezing (Chandler and Heuser, 1980; Knoll et al., 1991; Ornberg and Reese, 1981). In these studies widening fusion pores were seen but the most significant observation was that prior to pore formation the plasma membrane and vesicle membranes were linked by short filaments and the plasma membrane was pulled down on to the vesicle membrane at a focal point. This could then become the site of fusion pore formation (Monck and Fernandez, 1992). It is possible that this would require the presence, or the generation, of membrane-destabilizing lipids in this region. In *Paramecium* the site of fusion pore formation on the plasma membrane is within an ordered array of membrane particles (proteins) known as the rosette (Knoll et al., 1991) that could be important for fusion pore formation. The fusion pore that forms during haemagglutinin-mediated influenza virus fusion with the cell surface is preceded by the formation of an ordered array of haemagglutinin proteins within which the fusion pore is believed to form (White, 1992). Candidates for the proteins involved in cross-linking the bilayers and lipid reorganization during exocytosis are the annexins (discussed above, and see Creutz, 1992; Pollard et al., 1991). Growing evidence shows that the annexins can form multimeric structures on lipid bilayers (Zaks and Creutz, 1991) and interact with the bilayer sufficiently to acts as Ca^{2+} channels (Pollard et al., 1992).

The Monck and Fernandez model is fully consistent with all available data and seems intuitively reasonable. The correctness of this model will only become apparent when the essential exocytotic proteins are identified, but at the moment it provides an important pointer to the idea that these proteins do not necessarily need to be transmembrane vesicle or plasma membrane proteins as long as they can, alone or with other proteins, cross-link lipid bilayers in response to a rise in $[Ca^{2+}]_i$ (or the activation of GTP-binding proteins) and produce focal changes in lipid structure or organization.

The influence of lipid composition on membrane fusion in biological systems still remains to be fully explored, as does potential lipid changes during exocytosis. It had been suggested that lysolipids could promote membrane fusion (Poole et al., 1970) but more recent findings have clearly shown the opposite, that lysolipids inhibit biological membrane fusion (Chernomordik, 1993). Further work is needed in this area to define the role of lipids in the formation of the fusion pore.

SUMMARY AND FUTURE PROSPECTS

After many years of work that had given only limited insights into the proteins involved in regulated exocytosis, several soluble and membrane proteins that are essential or regulatory components of the exocytotic machinery have been identified recently. The picture that is emerging suggests that several proteins are likely to act in discrete steps in exocytosis or act together to form some kind of fusion machine. Further components are likely to be identified in the near future and a major area for investigation will be the nature of the interactions between these proteins.

Work on enveloped viruses has identified proteins with specific fusion peptide sequences that mediate viral fusion with the plasma membrane (White, 1992). These proteins act in binding the virus to the membrane as well as in membrane fusion. The fusion peptides from various viruses do not possess any marked sequence similarity, but all of them can potentially fold as a helix with a hydrophobic domain on one side. A protein that mediates sperm fusion with the egg has structural similarities to these viral proteins (Blobel et al., 1992). So far, no protein has been identified as being involved in exocytosis that has the properties of these fusion peptides. It will be interesting to see if intracellular bilayer fusion turns out to involve mechanisms related to or distinct from that in extracellular fusion, although one important point in these considerations will be the different lipid compositions of the inner and outer leaflets of the plasma membrane bilayer. It is interesting that viruses have evolved quite distinct fusion peptide sequences to solve the same biological problem and so it is conceivable that exocytosis could involve quite different proteins.

One important consideration is the relationship between constitutive and regulated exocytosis and between exocytosis and other intracellular transport vesicle steps. As described above, certain proteins known to be required for various intracellular vesicular transport steps (NSF and SNAPs) have been implicated in neurotransmitter release (Sollner et al., 1993) and in constitutive exocytosis (NSF; Sztul et al., 1993). In addition, annexins have been implicated in various membrane fusion events in the endocytic pathway (Emans et al., 1993; Lin et al., 1992). It is not yet clear to what extent the same or similar proteins are involved in membrane fusion of both constitutive and regulated secretory vesicles, but homology has been noted between nerve terminal proteins involved in Golgi-to-plasma-membrane transport in yeast (Bennett and Scheller, 1993). Since constitutive secretory vesicles can readily fuse with the plasma membrane immediately after their formation this suggests that fusion of regulated vesicles must be inhibited at resting $[Ca^{2+}]_i$. One possibility is that Ca^{2+} acts primarily to disinhibit the fusion mechanism. The inhibition could be due to the cytoskeleton acting as a cortical barrier, or holding regulated vesicles in a network (e.g. via synapsin I) to prevent exocytosis, or to additional mechanisms. Recent work on the biogenesis of small synaptic-like microvesicles in PC12 cells has suggested that synaptic vesicle proteins may be exported from the trans-Golgi network to the plasma membrane in constitutive secretory vesicles, recycled via the endosome pathway and only then become segregated to the regulated secretory vesicle (Cutler and Cramer, 1990; Linstedt and Kelly, 1991; Regnier-Vigouroux et al., 1991). The implications of this is that the synaptic vesicle proteins begin their life in vesicles (constitutive and recycling endocytic) that can spontaneously fuse with the plasma membrane but then become restricted to vesicles that can only fuse in response to a $[Ca^{2+}]_i$ rise. Does this mean that an inhibitory control becomes associated with the synaptic vesicle at a late stage in its biogenesis? This question may be significant for models in which synaptic vesicle membrane proteins are key elements in the fusion machinery.

Further progress towards solving the details of regulated exocytosis will require continued functional analysis, particularly of membrane proteins. An *in vitro* fusion system would be particularly useful for such studies and the *Xenopus* oocyte promises to provide an important experimental system. As we described above, synaptic vesicles can be assembled in oocytes from injected brain mRNA and the roles of particular vesicle proteins in neurotransmitter release assessed (Alder et al., 1992a). In addition, the oocytes possess a regulated secretory pathway

and Ca^{2+}-dependent exocytosis of chromaffin granules has been observed after their injection into *Xenopus* oocytes (Scheuner et al., 1992). This system could allow systematic manipulation of granule membrane proteins prior to the injection of the granules and thus provide functional data regarding these proteins.

Finally, future work on mutants of regulated exocytosis in *Tetrahymena* (Turkewitz et al., 1991) or *Paramecium* (Bonnemain et al., 1992) is likely to provide key insights into the nature of the proteins essential for exocytosis in the same way that the study of *Saccharomyces cerevisiae* has illuminated the constitutive secretory pathway. *Paramecium* may be a particularly useful organism for such studies, despite the difficulty of molecular genetics on this organism, because the biochemistry and morphology of exocytosis in *Paramecium* has been studied in detail (Hohne-Zell et al., 1992; Knoll et al., 1991; Momayezi et al., 1987; Stecher et al., 1987). The progress of biochemical studies on regulated exocytosis in the past year or so has been considerable, and the combination of this approach with a genetic approach should mean that within the next few years regulated exocytosis will be as well understood as other vesicular transport steps in the secretory pathway.

Work in the authors' laboratory was supported by grants from The Wellcome Trust. We thank Geoff Williams for his expert help in the preparation of the figures.

REFERENCES

Ahnert-Hilger, G., Wegenhorst, U., Stecher, B., Spicher, K., Rosenthal, W. and Gratzl, M. (1992) Biochem. J. **284**, 321–326

Aitken, A., Ellis, C. A., Harris, A., Sellers, L. A. and Toker, A. (1990) Nature (London) **344**, 594

Aitken, A., Amess, B., Howell, S., Jones, D., Martin, H., Patel, Y., Robinson, K. and Toker, A. (1992) Biochem. Soc. Trans. **20**, 607–611

Alder, J., Lu, B., Valtorta, F., Greengard, P. and Poo, M. (1992a) Science **257**, 657–661

Alder, J., Xie, Z.-P., Valtorta, F., Greengard, P. and Poo, M. (1992b) Neuron **9**, 759–768

Ali, S. M. and Burgoyne, R. D. (1990) Cell. Signalling **2**, 765–776

Ali, S. M., Geisow, M. J. and Burgoyne, R. D. (1989) Nature (London) **340**, 313–315

Almers, W. (1990) Annu. Rev. Physiol. **52**, 607–624

Archer, B. T., Ozcelik, T., Jahn, R., Francke, U. and Sudhof, T. C. (1990) J. Biol. Chem. **265**, 17267–17273

Augustine, G. J. and Neher, E. (1992) J. Physiol. (London) **450**, 247–271

Augustine, G. J., Adler, E. M. and Charlton, M. P. (1991) Ann. NY Acad. Sci. **635**, 365–381

Aunis, D. and Bader, M. F. (1988) J. Exp. Biol. **139**, 253–266

Baker, P. F. and Knight, D. E. (1978) Nature (London) **276**, 620–622

Bar-Sagi, D. and Gomperts, B. D. (1988) Oncogene **3**, 463–469

Barr, F. A., Leyte, A. and Huttner, W. B. (1992) Trends Cell Biol. **2**, 91–94

Barrowman, M. M., Cockcroft, S. and Gomperts, B. D. (1986) Nature (London) **319**, 504–507

Baumert, M., Maycox, P. R., Navone, F., De Camilli, P. and Jahn, R. (1989) EMBO J. **8**, 379–384

Ben-Shlomo, H., Sigmund, O., Stabel, S., Reiss, N. and Naor, Z. (1991) Biochem. J. **280**, 65–69

Benfenati, F., Valtorta, F., Rubenstein, J. L., Gorelick, F. S., Greengard, P. and Czernik, A. J. (1992) Nature (London) **359**, 417–420

Bennett, J. P., Cockcroft, S. and Gomperts, B. D. (1981) J. Physiol. (London) **317**, 335–345

Bennett, M. K. and Scheller, R. H. (1993) Proc. Natl. Acad. Sci. U.S.A. **90**, 2559–2563

Bennett, M. K., Calakos, N., Kreiner, T. and Scheller, R. H. (1992a) J. Cell Biol. **116**, 761–775

Bennett, M. K., Calakos, N. and Scheller, R. H. (1992b) Science **257**, 255–259

Bittner, M. A. and Holz, R. W. (1992a) J. Biol. Chem. **267**, 16219–16225

Bittner, M. A. and Holz, R. W. (1992b) J. Biol. Chem. **267**, 16226–16229

Blobel, C. P., Wolfsberg, T. G., Turck, C. W., Myles, D. G., Primakoff, P. and White, J. M. (1992) Nature (London) **356**, 248–252

Bommert, K., Charlton, M. P., De Bello, W. M., Chin, G. J., Betz, H. and Augustine, G. J. (1993) Nature (London) **363**, 163–165

Bonnemain, H., Gulik-Krywicki, T., Grandchamp, C. and Cohen, J. (1992) Genetics **130**, 461–470

Bowser, R., Muller, H., Govindan, B. and Novick, P. (1992) J. Cell Biol. **118**, 1041–1056

Breckenridge, L. J. and Almers, W. (1987a) Nature (London) **328**, 814–817

Breckenridge, L. J. and Almers, W. (1987b) Proc. Natl. Acad. Sci. U.S.A. **84**, 1945–1949

Brose, N., Petrenko, A. G., Sudhof, T. C. and Jahn, R. (1992) Science **256**, 1021–1025

Burgoyne, R. D. (1990) Annu. Rev. Physiol. **52**, 647–659

Burgoyne, R. D. (1991) Biochim. Biophys. Acta **1071**, 174–202

Burgoyne, R. D. and Cheek, T. R. (1987) Biosci. Rep. **7**, 281–288

Burgoyne, R. D. and Geisow, M. J. (1989) Cell Calcium **10**, 1–10

Burgoyne, R. D. and Morgan, A. (1989) FEBS Lett. **245**, 122–126

Burgoyne, R. D. and Morgan, A. (1990) Biochem. Soc. Trans. **18**, 1101–1104

Burgoyne, R. D., Geisow, M. J. and Barron, J. (1982) Proc. R. Soc. London Ser. B **216**, 111–115

Cain, C. C., Trimble, W. S. and Lienhard, G. E. (1992) J. Biol. Chem. **267**, 11681–11684

Chandler, D. E. and Heuser, J. E. (1980) J. Cell Biol. **86**, 666–674

Chavrier, P., Parton, R. G., Hauri, H. P., Simons, K. and Zerial, M. (1990) Cell **62**, 317–329

Cheek, T. R. and Burgoyne, R. D. (1986) FEBS Lett. **207**, 110–113

Cheek, T. R. and Burgoyne, R. D. (1987) J. Biol. Chem. **262**, 11663–11666

Cheek, T. R. and Burgoyne, R. D. (1992) in The Neuronal Cytoskeleton (Burgoyne, R. D., ed.), pp. 309–325, Wiley–Liss, New York

Chernomordik, L. V., Vogel, S. S., Sokoloff, A., Onaran, H. O., Leikina, E. A. and Zimmerberg, J. (1993) FEBS Lett. **318**, 71–76

Chow, R. H., von Ruden, L. and Neher, E. (1992) Nature (London) **356**, 60–63

Clary, D. O. and Rothman, J. E. (1990) J. Biol. Chem. **265**, 10109–10117

Creutz, C. E. (1992) Science **258**, 924–931

Cutler, D. F. and Cramer, L. P. (1990) J. Cell Biol. **110**, 721–730

Darchen, F., Zahraoui, A., Hammel, F., Monteils, M.-P., Tavitian, A. and Scherman, D. (1990) Proc. Natl. Acad. Sci. U.S.A. **87**, 5692–5696

Davidson, J., van der Merwe, P. A., Wakefield, I. and Millar, R. P. (1991) Mol. Cell. Endocrinol. **76**, C33–C38

De Matteis, M. A., Di Tullio, G., Buccione, R. and Luini, A. (1991) J. Biol. Chem. **266**, 10452–10460

DeCamilli, P. and Greengard, P. (1986) Biochem. Pharmacol. **35**, 4349–4357

Dekker, L. V., De Graan, P. N. F., Oestreicher, A. B., Versteeg, D. H. G. and Gispen, W. H. (1989) Nature (London) **342**, 74–76

Drust, D. S. and Creutz, C. E. (1988) Nature (London) **331**, 88–91

Dunn, L. A. and Holz, R. W. (1983) J. Biol. Chem. **258**, 4989–4993

Edwardson, J. M. and Daniels-Holgate, P. U. (1992) Biochem. J. **285**, 383–385

Elferink, L. A., Trimble, W. S. and Scheller, R. A. (1989) J. Biol. Chem. **264**, 11061–11064

Elferink, L. A., Peterson, M. R. and Scheller, R. A. (1993) Cell **72**, 153–159

Emans, N., Gorvel, J.-P., Walter, C., Gerke, V., Kellner, R., Griffiths, G. and Gruenberg, J. (1993) J. Cell Biol. **120**, 1357–1370

Fernandez, J. M., Neher, E. and Gomperts, B. D. (1984) Nature (London) **312**, 453–455

Fischer von Mollard, G., Mignery, G. A., Baumert, M., Perin, M. S., Hanson, T. J., Burger, P. M., Jahn, R. and Sudhof, T. C. (1990) Proc. Natl. Acad. Sci. U.S.A. **87**, 1988–1992

Fischer von Mollard, G., Sudhof, T. C. and Jahn, R. (1991) Nature (London) **349**, 79–81

Gilligan, D. M. and Satir, B. H. (1982) J. Biol. Chem. **257**, 13903–13906

Gomperts, B. D. (1990) Annu. Rev. Physiol. **52**, 591–606

Hata, Y., Davletov, B., Petrenko, A. G., Jahn, R. and Sudhof, T. C. (1993) Neuron **10**, 307–315

Hay, J. C. and Martin, T. F. J. (1992) J. Cell Biol. **119**, 139–152

Helms, J. B., Karrenbauer, A., Wirtz, K. W. A., Rothman, J. E. and Wieland, F. T. (1990) J. Biol. Chem. **265**, 20027–20032

Hess, S. D., Doroshenko, P. A. and Augustine, G. J. (1993) Science **259**, 1169–1172

Hirokawa, N., Sobue, K., Kanda, K., Harada, A. and Yorifuji, H. (1989) J. Cell Biol. **108**, 111–126

Hohne-Zell, B., Knoll, G., Riedel-Gras, U., Hofer, W. and Plattner, H. (1992) Biochem. J. **286**, 843–849

Holz, R. W., Bittner, M. A., Peppers, S. C., Senter, R. A. and Eberhard, D. A. (1989) J. Biol. Chem. **264**, 5412–5419

Holz, R. W., Senyshyn, J. and Bittner, M. A. (1992) Ann. NY Acad. Sci. **636**, 382–392

Howell, T. W. and Gomperts, B. D. (1987) Biochim. Biophys. Acta **927**, 177–183

Isobe, T., Hiyane, Y., Ichimura, T., Okuyama, T., Takahashi, N., Nakajo, S. and Nakaya, K. (1992) FEBS Lett. **308**, 121–124

Ivins, K. J., Neve, K. A., Feller, D. J., Fidel, S. A. and Neve, R. L. (1993) J. Neurochem. **60**, 626–633

Kish, P. E. and Ueda, T. (1991) Neurosci. Lett. **122**, 179–182

Knight, D. E. and Baker, P. F. (1982) J. Membr. Biol. **68**, 107–140

Knight, D. E. and Baker, P. F. (1983) FEBS Lett. **160**, 98–100

Knight, D. E. and Baker, P. F. (1985) FEBS Lett. **189**, 345–349

Knoll, G., Braun, C. and Plattner, H. (1991) J. Cell Biol. **113**, 1295–1304

Koffer, A. and Gomperts, B. D. (1989) J. Cell Sci. **94**, 585–591

Koffer, A., Tatham, P. E. R. and Gomperts, B. D. (1990) J. Cell Biol. **111**, 919–927

Leveque, C., Hoshino, T., David, P., Shoji-Kasai, Y., Leys, K., Omori, A., Lang, B., Far, E. L., Sato, K., Martin-Moutot, N., Newsom-Davis, J., Takahashi, M. and Seagar, M. J. (1992) Proc. Natl. Acad. Sci. U.S.A. **89**, 3625–3629

Lin, H. C., Sudhof, T. C. and Anderson, R. G. W. (1992) Cell **70**, 283–291

Lindau, M. and Gomperts, B. D. (1991) Biochim. Biophys. Acta **1071**, 429–471

Link, E., Edelman, L., Chou, J. H., Binz, T., Yamasaki, S., Eisel, E., Baumert, M., Sudhof, T. C., Niemann, H. and Jahn, R. (1992) Biochem. Biophys. Res. Commun. **189**, 1017–1023

Linstedt, A. D. and Kelly, R. B. (1991) Neuron **7**, 309–317

Llinas, R., McGuiness, T. L., Leonard, C. S., Sugimori, M. and Greengard, P. (1985) Proc. Natl. Acad. Sci. U.S.A. **82**, 3035–3039

Llinas, R., Sugimori, M. and Silver, R. B. (1992) Science **256**, 677–679

Lomneth, R., Martin, T. F. J. and DasGupta, B. R. (1991) J. Neurochem. **57**, 1413–1421

Martens, G. J. M., Piosik, P. A. and Danen, E. H. J. (1992) Biochem. Biophys. Res. Commun. **184**, 1456–1459

Martin, T. F. J. and Walent, J. H. (1989) J. Biol. Chem. **264**, 10299–10308

Matthew, W. D., Tsavaler, L. and Reichardt, L. F. (1981) J. Cell Biol. **91**, 257–269

Miller, S. G. and Moore, H.-P. H. (1991) J. Cell Biol. **112**, 39–54

Mochly-Rosen, D., Khaner, H. and Lopez, J. (1991) Proc. Natl. Acad. Sci. U.S.A. **88**, 3997–4000

Mochly-Rosen, D., Miller, K. G., Scheller, R. H., Khaner, H., Lopez, J. and Smith, B. L. (1992) Biochemistry **31**, 8120–8124

Momayezi, M., Lumpert, C. J., Kersen, H., Gras, U., Plattner, H., Krinks, M. H. and Klee, C. B. (1987) J. Cell Biol. **105**, 181–189

Monck, J. R. and Fernandez, J. M. (1992) J. Cell Biol. **119**, 1395–1404

Monck, J. R., Alvarez de Toledo, G. and Fernandez, J. M. (1990) Proc. Natl. Acad. Sci. U.S.A. **87**, 7804–7808

Morgan, A. and Burgoyne, R. D. (1990) Biochem. J. **271**, 571–574

Morgan, A. and Burgoyne, R. D. (1992a) Nature (London) **355**, 833–835

Morgan, A. and Burgoyne, R. D. (1992b) Biochem. J. **286**, 807–811

Morgan, A., Cenci de Bellow, I., Weller, U., Dolly, O. and Burgoyne, R. D. (1993a) in Botulinum and Tetanus Neurotoxins: Neurotransmission and Biomedical Aspects (Dasgupta, B., ed.), Plenum Press, New York, in the press

Morgan, A., Roth, D., Martin, H., Aitken, A. and Burgoyne, R. D. (1993b) Biochem. Soc. Trans. **21**, 401–405

Morimoto, T., Ogihara, S. and Takisawa, H. (1990) J. Cell Biol. **111**, 79–86

Nakata, T., Sobue, K. and Hirokawa, N. (1990) J. Cell Biol. **110**, 13–25

Nanavati, C., Markin, V. S., Oberhauser, A. F. and Fernandez, J. M. (1992) Biophys. J. **63**, 1118–1132

Naor, Z., Dan-Cohen, H., Herman, J. and Lima, R. (1989) Proc. Natl. Acad. Sci. U.S.A. **86**, 4500–4504

Navone, F., Jahn, R., Di Gioia, G., Stukenbrok, H., Greengard, P. and De Camilli, P. (1986) J. Cell Biol. **103**, 2511–2527

Neher, E. and Marty, A. (1982) Proc. Natl. Acad. Sci. U.S.A. **79**, 6712–6716

Neher, E. and Zucker, S. (1993) Neuron **10**, 21–30

Nichols, R. A., Sihra, T. S., Czernik, A. J., Nairn, A. C. and Greengard, P. (1990) Nature (London) **343**, 647–652

Nichols, R. A., Chilcote, T. J., Czernik, A. J. and Greengard, P. (1992) J. Neurochem. **58**, 783–785

Nishizaki, T., Walent, J. H., Kowalchyk, J. A. and Martin, T. F. J. (1992) J. Biol. Chem. **267**, 23972–23981

Novick, P., Fero, S. and Scheckman, R. (1981) Cell **25**, 461–469

O'Sullivan, A. J., Cheek, T. R., Moreton, R. B., Berridge, M. J. and Burgoyne, R. D. (1989) EMBO J. **8**, 401–411

Oberhauser, A. F., Monck, J. R., Balch, W. E. and Fernandez, J. M. (1992a) Nature (London) **360**, 270–273

Oberhauser, A. F., Monck, J. R. and Fernandez, J. M. (1992b) Biophys. J. **61**, 800–809

Ohara-Imaizumi, M., Kameyama, K., Kawae, N., Takeda, K., Muramatsu, S. and Kumakura, K. (1992) J. Neurochem. **58**, 2275–2284

Ornberg, R. L. and Reese, T. S. (1981) J. Cell Biol. **90**, 40–54

Oyler, G. A., Higgins, G. A., Hart, R. A., Battenberg, E., Billingsley, M., Bloom, F. E. and Wilson, M. C. (1989) J. Cell Biol. **109**, 3039–3052

Padfield, P. J., Balch, W. E. and Jamieson, J. D. (1992) Proc. Natl. Acad. Sci. U.S.A. **89**, 1656–1660

Perin, M. S., Fried, V. A., Mignery, G. A., Jahn, R. and Sudhof, T. C. (1990) Nature (London) **345**, 260–263

Perin, M. S., Brose, N., Jahn, R. and Sudhof, T. C. (1991) J. Biol. Chem. **266**, 623–629

Perrin, D., Möller, K., Hanke, K. and Söling, H.-D. (1992) J. Cell Biol. **116**, 127–134

Petrenko, A. G., Perin, M. S., Davletov, B. A., Ushkaryov, Y. A., Geppert, M. and Sudhof, T. C. (1991) Nature (London) **353**, 65–68

Pfeffer, S. R. (1992) Trends Cell Biol. **2**, 41–46

Plattner, H. (1989) Int. Rev. Cytol. **119**, 197–286

Plutner, H., Schwaninger, R., Pind, S. and Balch, W. E. (1990) EMBO J. **9**, 2375–2383

Pocotte, S. L., Frye, R. A., Senter, R. A., Terbush, D. R., Lee, S. A. and Holz, R. W. (1985) Proc. Natl. Acad. Sci. U.S.A. **82**, 930–934

Pollard, H. B., Rojas, E., Pastor, R. W., Rojas, E. M., Guy, H. R. and Burns, A. L. (1991) Ann. NY Acad. Sci. **635**, 328–351

Pollard, H. B., Gur, H. R., Ariope, N., de la Fuente, M., Lee, G., Rojas, E. M., Pollard, J. R., Srivastava, M., Zhang-Keck, Z.-Y., Merezhinskaya, N., Caohuy, H., Burns, A. L. and Rojas, E. (1992) Biophys. J. **62**, 15–18

Poole, A. R., Howell, J. I. and Lucy, J. A. (1970) Nature (London) **227**, 819–824

Pryer, N. K., Wuestehube, L. J. and Sheckman, R. (1992) Annu. Rev. Biochem. **61**, 471–516

Regnier-Vigouroux, A., Tooze, S. A. and Huttner, W. B. (1991) EMBO J. **10**, 3589–3601

Rehm, H., Wiedenmann, B. and Betz, H. (1986) EMBO J. **5**, 535–541

Rodriguez Del Castillo, A., Lemaire, S., Tchakarov, L. Jeyapragasan, M., Doucet, J. P., Vitale, M. L. and Trifaro, J. M. (1990) EMBO J. **9**, 43–52

Roth, D., Morgan, A. and Burgoyne, R. D. (1993) FEBS Lett. **320**, 207–210

Rothman, J. E. and Orci, L. (1992) Nature (London) **355**, 409–415

Salminen, A. and Novick, P. J. (1987) Cell **49**, 527–538

Sarafian, T., Aunis, D. and Bader, M.-F. (1987) J. Biol. Chem. **262**, 16671–16676

Sarafian, T., Pradel, L.-A., Henry, J.-P., Aunis, D. and Bader, M.-F. (1991) J. Cell Biol. **114**, 1135–1147

Saraste, J. and Kuismanen, E. (1984) Cell **38**, 535–549

Satir, B. H., Hamasaki, T., Reichman, M. and Murtaugh, T. J. (1989) Proc. Natl. Acad. Sci. U.S.A. **86**, 930–932

Scheuner, D., Logsdon, C. D. and Holz, R. W. (1992) J. Cell Biol. **116**, 359–365

Schiavo, G., Benfenati, F., Poulain, B., Rossetto, O., Polverino de Laureto, P., DasGupta, B. R. and Montecucco, C. (1992a) Nature (London) **359**, 832–835

Schiavo, G., Poulain, B., Rossetto, O., Benfenati, F., Tauc, L. and Montecucco, C. (1992b) EMBO J. **11**, 3577–3583

Schweizer, F. E., Schafer, T., Tapparelli, C., Grob, M., Karli, U. O., Heumann, R., Thoenen, H., Bookman, R. J. and Burger, M. M. (1989) Nature (London) **339**, 709–712

Senyshyn, J., Balch, W. E. and Holz, R. W. (1992) FEBS Lett. **309**, 41–46

Shoji-Kasai, Y., Yoshida, A., Sato, K., Hoshino, T., Ogura, A., Kondo, S., Fujimoto, Y., Kuwahara, R., Kato, R. and Takahashi, M. (1992) Science **256**, 1820–1823

Smolen, J., Stoehr, S. J., Kuczynski, B., Koh, E. K. and Omann, G. M. (1991) Biochem. J. **279**, 657–664

Sollner, T., Whiteheart, S. W., Brunner, M., Erdjument-Bromage, H., Geromanos, S., Tempst, P. and Rothman, J. E. (1993) Nature (London) **362**, 318–324

Stecher, B., Hohne, B., Gras, U., Momayezi, M., Glas-Albrecht, R. and Plattner, H. (1987) FEBS Lett. **223**, 25–32

Stecher, B., Ahnert-Hilger, G., Weller, U., Kemmer, T. P. and Gratzl, M. (1992) Biochem. J. **283**, 899–904

Sudhof, T. C. and Jahn, R. (1991) Neuron **6**, 665–677

Sudhof, T. C., Lottspeich, F., Greengard, P., Mehl, E. and Jahn, R. (1987) Science **238**, 1142–1144

Sudhof, T. C., Czernik, A. J., Kao, H.-T., Takei, K., Johnston, P. A., Horiuchi, A., Kanazir, S. D., Wagner, M. A., Perin, M. S., DeCamilli, P. and Greengard, P. (1989) Science **245**, 1474–1479

Sztul, E., Colombo, M., Stahl, P. and Samanta, R. (1993) J. Biol. Chem. **268**, 1876–1885

Taylor, T. C., Kahn, R. A. and Melancon, P. (1992) Cell **70**, 69–79

Thomas, L. and Betz, H. (1990) J. Cell Biol. **111**, 2041–2052

Thomas, L., Hartung, K., Langosch, D., Rehm, H., Bamberg, E., Franke, W. W. and Betz, H. (1988) Science **242**, 1050–1053

Thomas, P., Wong, J. G. and Almers, W. (1993) EMBO J. **12**, 303-306

Toutant, M., Aunis, D., Bockaert, J., Homburger, V. and Rouot, B. (1987) FEBS Lett. **215**, 339–344

Trifaro, J.-M. and Vitale, M. L. (1993) Trends Neurosci., in the press

Trifaro, J.-M., Fournier, S. and Novas, M. L. (1989) Neuroscience **29**, 1–8

Tugal, H. B., van Leeuwen, F., Apps, D. K., Haywood, J. and Phillips, J. H. (1991) Biochem. J. **279**, 699–703

Turkewitz, A. P., Madeddu, L. and Kelly, R. B. (1991) EMBO J. **10**, 1979–1987

Turner, M. D., Rennison, M. E., Handel, S. E., Wilde, C. J. and Burgoyne, R. D. (1992a) J. Cell Biol. **117**, 269–278

Turner, M. D., Wilde, C. J. and Burgoyne, R. D. (1992b) Biochem. J. **286**, 13–15

Ushkaryov, Y. A., Petrenko, A. G., Geppert, M. and Sudhof, T. C. (1992) Science **257**, 50–56

Valtorta, F., Benfenati, F. and Greengard, P. (1992) J. Biol. Chem. **267**, 7195–7198

Vitale, M. L., Rodriguez Del Castillo, A., Tchakarov, L. and Trifaro, J.-M. (1991) J. Cell Biol. **113**, 1057–1067

Walent, J. H., Porter, B. W. and Martin, T. F. J. (1992) Cell **70**, 765–775

Waters, M. G., Clary, D. O. and Rothman, J. E. (1992) J. Cell Biol. **118**, 1015–1026

White, J. M. (1992) Science **258**, 917–924

Whiteheart, S. W., Griff, I. C., Brunner, M., Clary, D. O., Mayer, T., Buhrow, S. A. and Rothman, T. J. (1993) Nature (London) **362**, 353–355

Wilson, S. P. and Kirshner, N. (1983) J. Biol. Chem. **258**, 4994–5000

Wilson, D. W., Whiteheart, S. W., Wiedmann, M., Brunner, M. and Rothman, J. E. (1992) J. Cell Biol. **117**, 531–538

Wu, Y. N. and Wagner, P. D. (1991) FEBS Lett. **282**, 197–199

Wu, Y. N., Vu, N.-D. and Wagner, P. D. (1992) Biochem. J. **285**, 697–700

Zieseniss, E. and Plattner, H. (1985) J. Cell Biol. **101**, 2028–2035

Zimmerberg, J., Curran, M., Cohen, F. S. and Brodwick, M. (1987) Proc. Natl. Acad. Sci. U.S.A. **84**, 1585–1589

Zimmerberg, J., Curran, M. and Cohen, F. S. (1991) Ann. NY Acad. Sci. **635**, 307–317

Zaks, W. J. and Creutz, C. E. (1991) Biochemistry **30**, 9607–9615

Zorec, R. and Tester, M. (1992) Biophys. J. **63**, 864–867

Zupan, L. A., Steffans, D. L., Berry, C. A., Landt, M. and Gross, R. W. (1992) J. Biol. Chem. **267**, 8707–8710

Biochem. J. (1993) **294**, 305–324 (Printed in Great Britain)

REVIEW ARTICLE

The structure, biosynthesis and function of glycosylated phosphatidylinositols in the parasitic protozoa and higher eukaryotes

Malcolm J. McCONVILLE and Michael A. J. FERGUSON
Department of Biochemistry, University of Dundee, Dundee DD1 4HN, U.K.

INTRODUCTION

The protozoa are the most diverse and amongst the most ancient group of organisms in the eukaryotic kingdom (Sogin et al., 1989). Many of their members are parasitic and some, like those belonging to the family Trypanosomatidae (African trypanosomes, *Trypanosoma cruzi*, *Leishmania*. spp.) and the genera *Plasmodium*, *Eimeria*, *Babesia*, *Theileria*, *Toxoplasma* and *Entamoeba*, are the cause of important diseases in humans and their domestic livestock. Glycoconjugates on the cell surface of these organisms frequently play a crucial role in determining parasite survival and infectivity. It has become clear over the last 5 years that many of these molecules are anchored to the plasma membrane via glycosyl-phosphatidylinositol (GPI) anchors. This type of anchor is not unique to the protozoa, but it does appear to be used with a much higher frequency in these organisms than in higher eukaryotes. In this article we review the structure, function and biosynthesis of GPI anchors in protozoan parasites and in higher eukaryotes. These data suggest that there may be significant differences in the function of GPI protein anchors in unicellular versus metazoan organisms. In addition, some of the parasitic protozoa, particularly those belonging to the Trypanosomatidae, synthesize a number of exotic GPI-related structures which are not attached to proteins. The structure, biosynthesis and role of these major cell surface glycoconjugates in parasite survival and infectivity are discussed, together with some speculations on the evolutionary aspects of the GPI-family. Previous general reviews on the subject of the GPI anchors and related structures can be found in Ferguson and Williams (1988), Low (1989), Cross (1990a), Thomas et al. (1990), Ferguson (1991, 1992b), McConville (1991) and Turco and Descoteaux (1992).

STRUCTURE OF PROTEIN GPI ANCHORS

Although some of the plasma membrane proteins of the parasitic protozoa use transmembrane polypeptide anchors, most of the major cell-surface proteins of these organisms are GPI-anchored (Table 1). These proteins are functionally diverse and include coat proteins, surface hydrolases and receptors. Some of these are known to be directly involved in parasite protection [e.g. *T. brucei* variant surface glycoprotein (VSG)] or specific host–parasite interactions (e.g. *T. cruzi* trans-sialidase and 35/50 kDa antigen, *P. falciparum* MSA-1 and *Leishmania* gp63).

Structures of parasite anchors

The complete or partial structures of the GPI anchors of four parasite proteins, all from the Trypanosomatidae, have been determined (Figure 1). In each case, the C-terminus of the protein is linked via ethanolamine phosphate to a glycan with the conserved backbone sequence Manα1–2Manα1–6Manα1–

4GlcNH$_2$, which in turn is linked to the 6-position of the *myo*-inositol ring of phosphatidylinositol (PI). The tetrasaccharide backbone may be substituted with other sugars in a species- and stage-specific manner, with the most elaborate side chains being found in the GPI anchors of the two *T. brucei* proteins; the VSG and the procyclic acidic repetitive protein (PARP or procyclin) (Table 1). The VSG anchor is substituted with branched side chains of α-galactose (Ferguson et al., 1988), while the PARP anchor has a large and complex side chain containing *N*-acetylglucosamine, galactose and sialic acid residues (Ferguson et al., 1993) (Figure 1). Preliminary studies on the latter side chains suggest that they contain sequences of poly(*N*-acetyllactosamine) that are substituted with sialic acid. In contrast, the GPI anchors of the *L. major* and *T. cruzi* antigens are either unsubstituted or only substituted with a single α-Man residue, respectively (Schneider et al., 1990; Güther et al., 1992) (Figure 1).

The lipid moieties of the parasite anchors can also vary in a species- and stage-specific manner. These anchors have been found to contain dimyristoylglycerol (in VSG), lyso-1-*O*-stearoylglycerol (in PARP) and alkylacylglycerol (in *Leishmania* gp63 and *T. cruzi* 1G7 antigen) (Figure 1). Some of these anchors may also contain an additional fatty acid (palmitate) on the inositol ring (Figure 1). In the PARP anchor, this fatty acid (acyl group) is attached to either the C-2 or C-3 position of the inositol (Field et al., 1991b; Ferguson, 1992a). This feature, first described by Rosenberry and colleagues, renders the anchor resistant to PI-specific phospholipase C (PI-PLC) hydrolysis, probably by blocking the formation of the inositol 1,2-cyclic phosphate during cleavage of the phosphodiester bond (Roberts et al., 1989a). Inositol acylation may be under developmental control in *T. cruzi*, as PI-PLC resistance of the 1G7 antigen increases as the cells develop from a non-infectious to an infectious 'metacyclic' stage (Schenkman et al., 1988).

Comparison with the anchors of other eukaryotes

All the GPI anchors which have been characterized to date (from protozoal, yeast, slime-mould, fish and mammalian sources) contain an identical ethanolamine-phosphate–Manα1–2Manα1–6Manα1–4GlcNα1–6*myo*-inositol backbone (Figure 1), suggesting that this sequence is likely to be conserved in all GPI anchors. These studies also indicate that the elaborate carbohydrate side chains and the distinctive mono- and diacylglycerol moieties of the *T. brucei* anchors are unique to this parasite. In contrast, the presence of additional α-mannose residues, as occurs in the *T. cruzi* anchor, also occurs on the anchors of yeast, slime-mould and mammalian proteins and appears to be a common substituent. It is of interest that all the anchors of higher eukaryotes (from slime moulds upward) are substituted

Abbreviations used: GPI, glycosyl-phosphatidylinositol; GIPL, glycoinositolphospholipid; LPG, lipophosphoglycan; LPPG, lipopeptidophosphoglycan; PARP, *Trypanosoma brucei* procyclic acid repeat protein; PI, phosphatidylinositol; PI-PLC, PI-specific phospholipase C; VSG, *T. brucei* variant surface glycoprotein.

Table 1 Occurrence of GPI-anchored proteins in the parasitic protozoa

Species	Protein	Properties	Key references
Trypanosoma brucei	Variant surface glycoprotein	Coat glycoprotein of bloodstream trypomastigotes	Ferguson et al., 1988
	Transferrin-binding protein	Surface of bloodstream trypomastigotes	Schell et al., 1991
	PARP/procyclin	Coat glycoprotein of procyclic forms	Clayton and Mowatt, 1989
	Procyclic trans-sialidase	Surface of procyclic forms	Engstler et al., 1992
T. congolense / T. equiperdum	Variant surface glycoprotein	Coat glycoprotein of bloodstream trypomastigotes	Lamont et al., 1987; Ross et al., 1987
Trypanosoma cruzi	Ssp-4	Major surface glycoprotein of amastigotes	Andrews et al., 1988
	90 kDa 1G7 antigen	Major surface glycoprotein of metacyclics	Schenkman et al., 1988; Güther et al., 1992
	gp50-55	Epimastigote/trypomastigote/amastigote antigen	Hernandez-Munain et al., 1991
	35/50 kDa/10D8 antigen	Epimastigote/metacyclic antigen; major acceptor of sialic acid	Schenkman et al., 1993
	TCNA/Shed Acute Phase Antigen (SAPA)	120–200 kDa, trans-sialidase/sialidase family (inactive	Pereira et al., 1991; Pollevick et al., 1991; Schenkman et al., 1992
	GP85 family	Trypomastigote trans-sialidase/sialidase family (inactive)	Fouts et al., 1991; Takle and Cross, 1991
	F1-160 (160 kDa)	160 kDa flagella antigen	van Voorhis et al., 1991
Leishmania spp.	Promastigote surface protease/gp63	Major surface glycoprotein on promastigote surface	Bouvier et al., 1985; Etges et al., 1986
	GP46/M2	Promastigote surface	Lohman et al., 1990
	PSA-2 (promastigote surface antigen-2)	Promastigote surface	Murray et al., 1989
Crithidia fasiculata	Protease	Homologue of *Leishmania* gp63	Zaretskia et al., 1989; Inverso et al., 1993
Herpetomonas samuelpessoai	Protease	Homologue of *Leishmania* gp63	Schneider and Glaser, 1993
Plasmodium falciparum	MSA-1 (merozoite surface antigen 1)	195 kDa, major merozoite surface glycoprotein	Haldar et al., 1985
	Transferrin binding protein	Merozoite surface	Haldar et al., 1986
	MSA-2 (merozoite surface antigen-2)	Merozoite surface	Smythe et al., 1988
	p76 proteinase	Merozoite/schizont surface	Braun-Breton et al., 1988
Toxoplasma gondii	P22, P23, P30, P35, P43	Major surface proteins on tachyzoite	Nagel and Boothroyd, 1989; Tomavo et al., 1992a
Giardia lamblia	GP49	Trophozoite surface	Das et al., 1991
Eimeria	Undefined antigens	Surface of sporozoites	Gurnett et al., 1990

with one or two additional ethanolamine phosphate residues. These residues have not been detected in the protozoan or yeast anchors, suggesting that they may be specific to metazoan eukaryotes. The nature of the lipid moieties in the non-protozoan anchors is also variable. Alkylacylglycerols and diacylglycerols have been found in the mammalian and fish anchors, whereas ceramide is the most common lipid in the anchors of yeast (Conzelmann et al., 1992; Fankhauser et al., 1993) and *Dictyostelium discoideum* (Stadler et al., 1989; Haynes et al., 1993). Many of the mammalian anchors also contain palmitate on the inositol ring (Roberts et al., 1989a; Luhrs and Slomiany, 1989; Walter et al., 1990; Lee et al., 1992).

GPI ANCHOR BIOSYNTHESIS

A protein destined to receive a GPI anchor must contain two pieces of information in its primary translation product. Firstly, it must contain an N-terminal signal sequence for entry into the lumen of the endoplasmic reticulum via the signal recognition particle and, secondly, it must contain a GPI-signal sequence at the C-terminus. This GPI-signal sequence is extremely degenerate but it is quite easily identified from cDNA sequences (reviewed in Ferguson and Williams, 1988; Cross, 1990a; Undenfriend et al., 1991; Caras et al., 1991; Kodukula et al., 1992). The most common feature is a run of 12–20 hydrophobic residues at the very C-terminus of the primary translation product, but there are exceptions. The GPI-signal sequence is cleaved and replaced by a preassembled GPI precursor in what appears to be an ATP- and GTP-independent (Mayor et al., 1991) transamidation reaction (Gerber et al., 1992, and references therein). However, both ATP and GTP appear to be required in some way immediately prior to the transamidation event (Amthauer et al., 1992). From kinetic and genetic data the addition of GPI to proteins is believed to occur in the endoplasmic reticulum (Bangs et al., 1985; Ferguson et al., 1986; Conzelmann et al., 1988). The biosynthetic pathways of GPI precursor formation are summarized in Figure 2. Previous reviews on this subject include Doering et al. (1990), Menon (1991), Tartakoff and Singh (1992), and Englund (1993).

The GPI biosynthetic pathway of bloodstream form African trypanosomes

The surface membrane of the *T. brucei* bloodstream form is covered by a dense coat of VSG molecules which protects the parasite from both non-specific and specific components of the host immune system (Cross, 1990b). The VSG coat is essential for the survival of the parasite and consequently a considerable amount of the parasite's metabolic effort is directed towards the synthesis and processing of VSG glycoprotein. Each trypanosome expresses about 10^7 copies of an individual VSG, all of which bear a GPI membrane anchor. For these reasons *T. brucei* has proved to be a convenient system for the analysis of GPI anchor biosynthesis. The pathway has been delineated by the use of a trypanosome lysates which, when fed with appropriate radio-labelled donor molecules such as UDP-[^{3}H]GlcNAc or GDP-[^{3}H]Man, produce the complete spectrum of intermediate species, all of which have been structurally characterized.

The GPI biosynthetic pathway in bloodstream form *T. brucei* parasites (Figure 2) may be summarized as follows: α-GlcNAc is transferred from UDP-GlcNAc to PI to form GlcNAc-PI, which is then de-N-acetylated to GlcN-PI (Doering et al., 1989). Subsequently, three α-Man residues are transferred in single steps from dolichyl phosphate mannose (Dol-P-Man) to form Man$_3$-GlcN-PI (Masterson et al., 1989; Menon et al., 1990a,b).

Ethanolamine phosphate is transferred from phosphatidylethanolamine to the terminal Man residue to form EtN-P-Man$_3$-GlcN-PI (known as glycolipid A') (Menon et al., 1993). This species then undergoes a complex series of fatty acid remodelling reactions (Masterson et al., 1990) as follows: the *sn*-2-fatty acid of glycolipid A' is removed to form a *lyso*-species called glycolipid θ. Glycolipid θ is myristoylated to form glycolipid A″ (which contains *sn*-1-stearoyl-2-myristoylglycerol). The *sn*-1-stearoyl group is removed from glycolipid A″ and replaced by myristic acid to form glycolipid A. The donor molecule for the two myristoyltransferase steps is myristoyl-CoA. Concomitant with the formation of glycolipid A is the formation of glycolipid C (the inositol-palmitoylated version of glycolipid A). The structures of glycolipids A and C (also known as glycolipids P2 and P3) have been rigorously determined (Krakow *et al.*, 1989; Mayor et al., 1990a,b). Both of these species have been shown to be competent for transfer to VSG polypeptide when added exogenously to a trypanosome cell-free system (Mayor et al., 1991), but there is no evidence of the transfer of glycolipid C (P3) *in vivo*. It has been suggested that glycolipid C is an obligate intermediate on the pathway to the glycolipid A GPI precursor (Doering et al., 1990; Masterson et al., 1990; Menon et al., 1990b; Menon, 1991), but recent data suggest that glycolipid C could be an end-product of the pathway (M. L. S. Güther, W. J. Masterson and M. A. J. Ferguson, unpublished work). Glycolipid C might therefore represent a reversible reservoir for glycolipid A and/or an intermediate in the catabolism of excess GPI precursors.

The transfer of GPI precursor to VSG polypeptide involves the removal of a hydrophobic C-terminal GPI signal sequence of 17–23 amino acids, depending on the VSG variant. The rapid kinetics of this reaction suggests that this occurs in the endoplasmic reticulum (Bangs et al., 1985; Ferguson et al., 1986). Finally, the GPI anchor becomes α-galactosylated. The first α-Gal residue is probably attached in the endoplasmic reticulum (Mayor et al., 1992) and the subsequent residues are added some 10–15 min later, most likely in the Golgi apparatus (Bangs et al., 1988). The α-galactosyltransferases involved in this processing appear to be unique to the African trypanosomes and may play a role in VSG coat function (see below). Recently a cell-free assay for VSG GPI α-galactosylation has been reported using UDP-[^{14}C]Gal as the donor (Pingel and Duszenko, 1992). Interestingly, small amounts of glycolipid A containing several α-Gal residues are observed in trypanosomes (Mayor et al., 1992), which suggests that some mono-galactosylated glycolipid A escapes the endoplasmic reticulum by bulk flow and becomes further galactosylated in the Golgi apparatus. The physiological significance of these structures is unknown.

GPI anchor biosynthesis in *T. brucei* procyclic forms

A structurally distinct GPI precursor (called PP1) has been isolated from the procyclic (insect-dwelling stage) of *T. brucei* and shown to have the structure EtN-P-Man$_3$-GlcN-*sn*-1-stearoyl-2-*lyso*-(palmitoyl)PI (Field et al., 1991b). These data suggest that the large side-chain found on the PARP anchor (Figure 1) is probably added after transfer of the anchor to the PARP polypeptide, possibly in the Golgi apparatus. The origin of the novel PI moiety is thought to arise from incomplete fatty acid remodelling such that, like the bloodstream form trypanosomes, the *sn*-2 fatty acid is removed by a specific phospholipase A$_2$ to yield a lyso species. However, unlike bloodstream forms, the *sn*-2 position is not subsequently reacylated with myristate nor is the *sn*-1-stearoyl group replaced by myristate. Experiments using a procyclic cell-free system suggest that the inositol

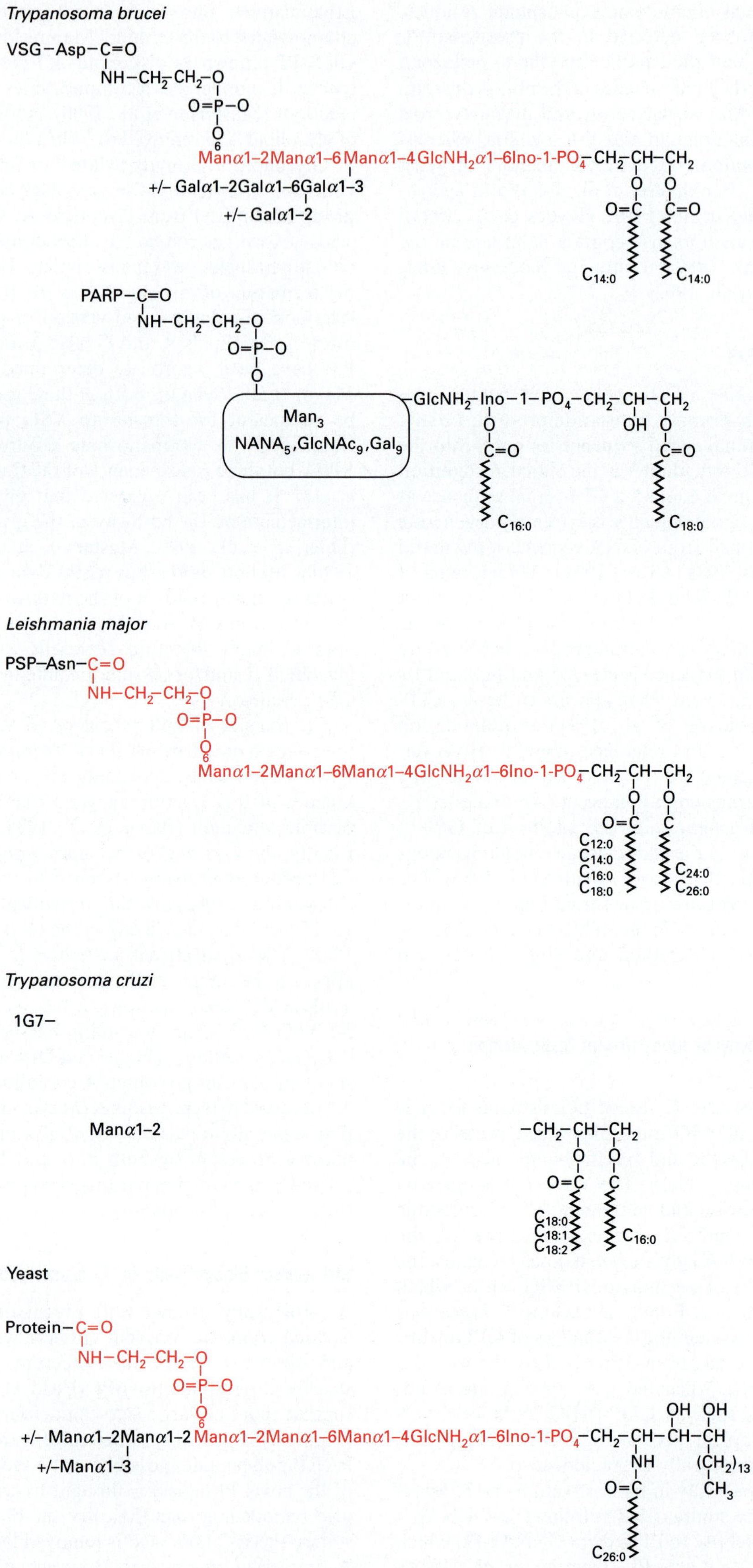

Trypanosoma brucei

VSG—Asp—C=O
        NH—CH$_2$—CH$_2$—O
                  O=P—O
                    O
                    6

Manα1–2Manα1–6Manα1–4GlcNH$_2$$\alpha$1–6Ino-1-PO$_4$—CH$_2$-CH-CH$_2$

+/– Galα1–2Galα1–6Galα1–3
          +/– Galα1–2

O=C   C=O

C$_{14:0}$ C$_{14:0}$

PARP—C=O
      NH—CH$_2$—CH$_2$—O
              O=P—O
                O

Man$_3$
NANA$_5$,GlcNAc$_9$,Gal$_9$ —GlcNH$_2$–Ino–1–PO$_4$–CH$_2$–CH–CH$_2$

O   OH   O

C=O   C=O

C$_{16:0}$ C$_{18:0}$

Leishmania major

PSP—Asn—C=O
        NH—CH$_2$—CH$_2$—O
               O=P—O
                 O
                 6

Manα1–2Manα1–6Manα1–4GlcNH$_2$$\alpha$1–6Ino-1-PO$_4$—CH$_2$-CH-CH$_2$

O=C   C

C$_{12:0}$
C$_{14:0}$
C$_{16:0}$ C$_{24:0}$
C$_{18:0}$ C$_{26:0}$

Trypanosoma cruzi

1G7–

Manα1–2

–CH$_2$–CH–CH$_2$

O=C   C

C$_{18:0}$
C$_{18:1}$
C$_{18:2}$ C$_{16:0}$

Yeast

Protein—C=O
       NH—CH$_2$—CH$_2$—O
              O=P—O
               O
              6

+/– Manα1–2Manα1–2 Manα1–2Manα1–6Manα1–4GlcNH$_2$$\alpha$1–6Ino-1-PO$_4$–CH$_2$–CH–CH–CH
    +/–Manα1–3

OH OH

NH   (CH$_2$)$_{13}$

O=C   CH$_3$

C$_{26:0}$

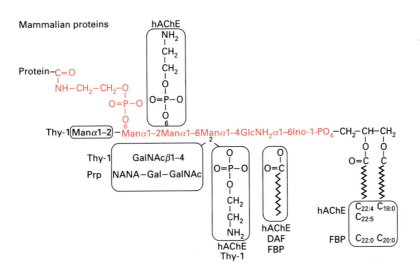

Figure 1   Structures of GPIs

The figure shows a comparison of the structures of the GPIs from four parasite proteins [*T. brucei* VSG (Ferguson et al., 1988) and PARP (Field et al., 1991a; Ferguson et al., 1992a, 1993), *Leishmania major* Gp63 (Schneider et al., 1990) and *T. cruzi* 1G7 antigen (Güther et al., 1992; N. Heise, M. L. Cardoso de Almeida and M. A. J. Ferguson, unpublished work)], a mixture of yeast glycoproteins (Conzelmann et al., 1992; Fankhauser et al., 1993), *Dictyostelium discoideum* PsA (Haynes et al., 1993), *Torpedo* acetylcholinesterase (AChE; Mehlert et al., 1993) and a number of mammalian proteins [rat brain Thy-1 (Homans et al., 1988), human erythrocyte acetylcholinesterase (hAChE; Roberts et al., 1989b; Deeg et al., 1992), scrapie prion protein (PrP; Stahl et al., 1992) and human folate binding protein (FBP; Luhrs and Slomiany, 1989; Lee et al., 1992)]. Components of the conserved GPI backbone, i.e. ethanolamine-phosphate–Manα1–2Manα1–6Manα1–4GlcNα1–6*myo*-inositol, are shown in red.

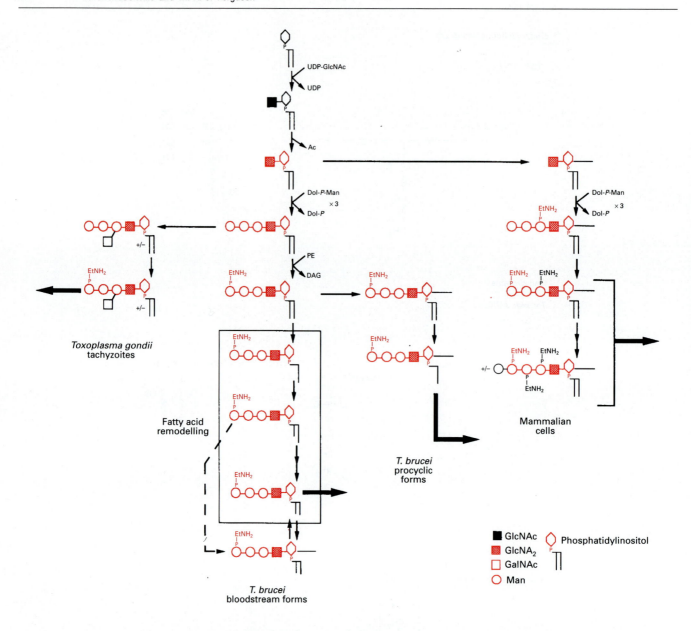

Figure 2   The GPI-anchor biosynthetic pathways in protozoan parasites and mammalian cells

The bold arrows indicate species which are transferred to protein. The donor species of the biosynthetic steps are shown where they are known. The boxed area represents the fatty acid remodelling reactions unique to bloodstream form *T. brucei*. The dashed arrows indicate alternative pathways. Components of the conserved GPI backbone, i.e. ethanolamine-phosphate–Manα1–2Manα1–6Manα1–4GlcNα1–6*myo*-inositol, are shown in red. Abbreviations: PE, phosphatidylethanolamine; EtNH₂, ethanolamine; DAG, diacylglycerol; P, phosphate.

palmitoylation event occurs earlier in the pathway than in bloodstream form trypanosomes (Field et al., 1992) (Figure 2). It is interesting that the GPI pathway in *T. brucei* appears to be under developmental control such that the timing of inositol palmitoylation, the extent of fatty acid remodelling and nature of protein-GPI carbohydrate processing are quite distinct. Presumably this represents developmental regulation of several GPI biosynthetic and processing enzymes during procyclic to bloodstream form differentiation, and vice versa.

GPI anchor biosynthesis in *Plasmodium* and *Toxoplasma*

The sporozoan parasites *Plasmodium* and *Toxoplasma* diverged from the kinetoplastid line at the earliest stage of eukaryotic evolution. Nevertheless, their GPI biosynthetic pathways are broadly similar to those of the African trypanosomes. In *Toxoplasma* the pathway produces exclusively diacyl- and *lyso*-acyl-PI GPI precursors (Tomavo et al., 1992b). A novel feature in this organism is the addition of a carbohydrate branch of β-GalNAc prior to the addition of the ethanolamine phosphate (Figure 2). Thus processing of the conserved trimannosyl backbone occurs before GPI transfer to protein. In the malaria parasite, *Plasmodium falciparum*, the majority of the GPI-intermediates are PI-PLC resistant and one of the major precursor species contains four Man residues (Gerold et al., 1992; P. Gerold et al., unpublished work). There is no evidence of fatty acid remodelling in either of these organisms. *Plasmodium* GPI anchors (or related structures) have been implicated in tumour necrosis factor and interleukin-1 production by macrophages during malarial infection (Taverne et al., 1990; Schofield and Hackett, 1993).

GPI biosynthesis in mammalian cells

Because of the emphasis of this article on parasite systems the data on the pathway in mammalian cells will be only briefly reviewed here. Essentially the pathway is the same as that found in parasites (i.e. sequential glycosylation and processing of PI). However, most but not all (Puoti and Conzelmann, 1993), of the intermediates (from GlcN-PI onwards) are inositol-palmitoylated (Urakaze et al., 1992; Hirose et al., 1992). Thus this reaction appears to be an early step in GPI biosynthesis in mammals. The occurrence of PI-PLC-sensitive mature protein-linked GPI anchors is, however, common in mammalian cells. The expression of either PI-PLC resistant or PI-PLC sensitive anchors appears to be cell-type specific (Toutant et al., 1990; Richier et al 1992) and the PI-PLC-resistant phenotype appears to be recessive in cell-fusion experiments (Singh et al., 1991). This suggests that mammalian cells transfer the mature inositol-palmitoylated GPI precursor to protein and then (in some cells) de-palmitoylate the inositol ring. Another notable difference in metazoan GPI anchors (from slime-moulds upwards) is the presence of at least one extra ethanolamine phosphate group (Figure 1). Recent studies show that these groups are added during GPI biosynthesis and not after GPI transfer to protein (Hirose et al., 1992; Kamitani et al., 1992; Puoti and Conzelmann, 1992) (Figure 2). The nature of the donor species for these extra groups is unknown.

The disease paroxysmal nocturnal haemoglobinuria (PNH) is caused by a defect in GPI anchor biosynthesis in blood cells as a result of a somatic mutation in one or more bone marrow progenitor cells. This leads to destruction of their progeny by autologous complement-mediated lysis since the GPI-anchored proteins decay accelerating factor (DAF) and CD59, which are responsible for preventing such autolysis, are not expressed on the cell surface. This phenotype appears to be due to a defect in the formation of GlcNAc-PI (Armstrong et al., 1992; Takahashi et al., 1993).

GPI biosynthesis in yeast

The mature protein-linked GPI anchors in yeast contain mostly ceramide-PI, with a phytosphingosine long chain base and an unusually long C_{26} fatty acid, although in one case (Gp125) the anchor is a *lyso*-acyl-PI (Frankhauser et al., 1993). The mechanism for this differential expression of lipid types is unknown. However, studies on newly synthesized GPI-anchored proteins show a time-dependent shift from base labile (acylglycerol-) GPIs to base stable (ceramide-) GPIs. This suggests that yeast may perform a novel form of lipid remodelling involving the exchange of glycerolipid for ceramide after transfer of the GPI precursors to protein (Conzelmann et al., 1992). Consistent with this, the putative early intermediates in the yeast GPI biosynthetic pathway, partially characterized as GlcNAc-PI, GlcN-PI and inositol-palmitoylated GlcN-PI, also appear to be base-labile (Costello and Orlean, 1992). These authors also present data to suggest that in yeast, palmitoyl-CoA is the donor for the inositol-palmitoylation event. However, this is in contrast to the situation in *T. brucei* procyclics where the palmitate donor appears to be a large pool of membrane-stable lipid, possibly phospholipid (Field et al., 1991b). The donor molecule for the addition of the ethanolamine phosphate bridge in yeast has been shown to be phosphatidylethanolamine (Menon and Stevens, 1992).

The topology and enzymology of the GPI pathway

Recent data, using sealed thymoma endoplasmic reticulum vesicles and permeabilized cells, strongly suggest that the first steps of the pathway (PI → GlcNAc-PI → GlcN-PI) occur on the cytoplasmic face of the endoplasmic reticulum (Vidugiriene and Menon, 1993). The topologies of the subsequent reactions are unknown, but it seems likely that one of the intermediates is able to 'flip' across the endoplasmic reticulum membrane such that the final GPI precursor can become attached to protein in the lumen of the endoplasmic reticulum (Amthauer et al., 1993). Such a translocation event is likely to be protein-mediated; a similar 'flip' has been postulated for the translocation of the Man_5-$GlcNAc_2$-PP-Dol intermediate involved in N-glycosylation (Abeijon and Hirschberg, 1992).

None of the GPI biosynthetic enzymes have been completely purified to date. However, the availability of thymoma mutants deficient in various stages of the pathway, and belonging to different complementation groups (Hyman, 1988), suggest that several of the genes may be cloned in the foreseeable future. So far only the cDNA which rescues the class A mutant has been cloned and sequenced (Miyata et al., 1993). The predicted protein of 484 amino acids has no homology with any known protein and its primary structure suggests that it is a transmembrane protein with a large cytoplasmic N-terminal domain (Miyata et al., 1993). The thymoma class A, C and H mutants are deficient in the synthesis of GlcNAc-PI (Stevens and Raetz, 1991), the class B mutant is deficient in the addition of the third Man residue (Puoti et al., 1991) and the class F mutant fails to add the ethanolamine phosphate bridge (Hirose et al., 1992; Kamitani et al., 1992; Puoti and Conzelmann, 1992). Some of these mutations may be in the genes encoding the respective transferases, or in genes encoding proteins involved in regulation and/or supply of donors. However, the possibility that several of the biosynthetic enzymes might form a complex requiring protein and/or RNA structural elements cannot be ruled out. Preliminary gel-filtration studies on partially purified GlcNAc-PI de-*N*-acetylase from *T. brucei* suggest that this enzyme is part of a high-molecular-mass complex or aggregate (Milne et al., 1993). A requirement for structural elements might explain the need for at least three gene products (A, C and H) for the formation of the first GPI intermediate GlcNAc-PI.

Although purified enzymes are not available, whole cell and cell-free system biosynthetic labelling experiments have been successful in identifying a few mechanistic aspects of the pathway. The inhibition of the addition of the ethanolamine phosphate bridge to the Man_3-GlcN-PI intermediate in *T. brucei*, by phenylmethanesulphonyl fluoride (PMSF) and di-isopropyl fluorophosphate, suggests that this enzyme has an active site serine residue which may form a serine–phosphoethanolamine covalent intermediate in the transfer reaction (Masterson and Ferguson, 1991). This is consistent with the data of Menon et al. (1993) who have shown that the ethanolamine and phosphate group of the phosphatidylethanolamine donor are transferred together. PMSF has also been shown to inhibit the formation of glycolipid C in *T. brucei* (Masterson and Ferguson, 1991; M. L. S. Güther, W. J. Masterson and M. A. J. Ferguson, unpublished work), presumably by inhibiting the palmitoyltransferase which adds the palmitate residue to the inositol ring.

The α-GlcNAc transferase, which forms GlcNAc-PI, is inhibited by thiol alkylating reagents such as *p*-chloromercuriphenylsulphonic acid, iodoacetamide and *N*-ethylmaleimide. The inhibition can be prevented by prior incubation with UDP-GlcNAc and UDP, indicating that there is a cysteine residue at or close to the UDP-GlcNAc donor binding site (Milne et al., 1992). In contrast, none of the downstream Dol-*P*-Man-dependent α-mannosyltransferases of the pathway are affected by thiol alkylating agents.

Mannosamine was first shown by Lisanti et al. (1990) to

Table 2   Functions of GPI in mammalian and protozoan cells

Function	Mammalian cells	Protozoan cells
1. Attachment of protein to plasma membrane	+	+
2. Association in membrane microdomains	+	−
3. Intracellular sorting	+	−
4. Transmembrane signalling via GPI clusters	+	−
5. Endocytosis via non- clathrin-coated pits (potocytosis)	+	+
6. High surface expression/low turnover rates	+	+
7. Selective release of protein by GPI-PLC	+	+
8. High surface packing	−	+
9. Contribution to surface glycocalyx	−	+

inhibit GPI biosynthesis in mammalian cells and procyclic trypanosomes . Experiments using mammalian cells have shown that this monosaccharide becomes incorporated into GPI intermediates and inhibits the formation of Man_3-GlcN-PI-containing species (Pan et al., 1992). Recent studies using bloodstream form trypanosomes have shown that the inhibition is less complete (about 80 %) in these cells. However, an accumulation of an intermediate with the structure $ManNH_2$-Man-GlcN-PI was observed. Trypanosome lysates made from cells preincubated with mannosamine (i.e. pre-loaded with $ManNH_2$-Man-GlcN-PI) were found to be significantly reduced in their capacity to make GPI precursors, suggesting that the $ManNH_2$-Man-GlcN-PI species may act as a competitive inhibitor of the $Man\alpha1$–$2Man$ α-mannosyltransferase (Ralton et al., 1993).

Certain myristic acid analogues (with oxygen replacing methylene groups in the acyl chain) become incorporated into GPI precursors and into the VSG GPI anchor in *T. brucei*, presumably via the fatty acid remodelling reactions. These compounds are toxic to bloodstream form trypanosomes but not to procyclic trypanosomes or to mammalian cells, which lack the fatty acid remodelling myristoyltransferases (Doering et al., 1991). These studies also suggest that the specificity for myristate shown by the remodelling enzymes is dependent on acyl-chain length rather than hydrophobicity of the acyl-CoA donor. *T. brucei* bloodstream forms are deficient in *de novo* fatty acid synthesis and they rely on exogenous (host serum) myristate. Recent studies have shown that they selectively incorporate myristate into GPI precursors and VSG, and not into other phospholipids, suggesting that this organism has developed an efficient system for directing this low abundance fatty acid specifically into the GPI pathway (Doering et al., 1993).

GPI ANCHOR FUNCTION

Various functions for GPI protein anchors have been described, particularly in mammalian and protozoan systems (reviewed in Cross, 1990b; Ferguson, 1991, 1992b); see Table 2. These data suggest that some basic functions are common to higher and lower eukaryotes, whereas others may represent specific adaptations that are advantageous to either unicellular or metazoan organisms, respectively.

Basic GPI functions and properties

The most fundamental function of GPI anchors is to afford the stable association of proteins with the lipid bilayer. The GPI anchor is an efficient and stable anchor, comparable with a hydrophobic polypeptide domain. Most GPI-anchored proteins are expressed at high levels on the outer leaflet of the plasma membrane and exhibit low turnover rates. This is true of both

mammalian (Lemansky et al., 1990) and protozoan (Bülow et al., 1989b; Seyfang et al., 1990) cells. However, there are exceptions; for example, GP-2 is a luminally disposed pancreatic secretory granule protein (Paul et al., 1991). One of the reasons for the low turnover rate of GPI-anchored proteins in mammalian cells is their exclusion from the clathrin-mediated endocytic pathway (Bretscher et al., 1980; Lemansky et al., 1990). In protozoan cells such as *T. brucei* and *T. cruzi*, which appear to lack clathrin-coated pits (Shapiro and Webster, 1989; Soares et al., 1992), GPI-anchored proteins are successfully endocytosed (Webster and Grab, 1988; Webster et al., 1990) and recycled (Seyfang et al., 1990). Clathrin-independent endocytosis also occurs in mammalian cells (see below).

Specialized GPI functions in higher eukaryotes

Membrane microdomains and transport

Over the past few years, GPI-anchored proteins in mammalian cells have been shown to perform some unexpected functions. These phenomena include intracellular targeting, transmission of transmembrane signals and clathrin-independent endocytosis. Recent data suggesting that GPI-anchored proteins become sequestered in specialized membrane microdomains in mammalian cells (reviewed by Brown, 1992), may be a unifying property that accounts for these functions. Following their transport from the endoplasmic reticulum to the Golgi, GPI-anchored proteins become insoluble in non-ionic detergents. This well-described phenomenon of neutral-detergent insolubility (Hooper and Turner, 1988) appears to correlate with the association of GPI-anchored proteins with sphingolipids and glycosphingolipids to form microdomains (Brown and Rose, 1992). The self-assembly of these microdomains may explain how GPI-anchored proteins are sequestered into specialized transport vesicles, for vectoral delivery exclusively to the apical membrane of polarized epithelial cells (Lisanti et al., 1990; Rodriguez-Boulan and Powell, 1992). Upon arrival at the plasma membrane the GPI-anchored proteins are immobile but become mobile with time (Hannan et al., 1993). It is possible that the delivered microdomains break up into smaller microdomain units with particular functions and properties, depending on their components.

Potocytosis

A good example of functional GPI-rich microdomains are the caveolae; these are membrane pits devoid of clathrin that are responsible for potocytosis (reviewed by Anderson et al., 1992 and Hooper, 1992). Potocytosis is a specialized type of pseudo-endocytosis, involving the occlusion of caveolae as 'tethered vesicles' to scavenge and concentrate small ligands such as folate. Caveolae are known to contain several GPI-anchored proteins (Ying et al., 1992), a cytoplasmic 'coat' of a 22 kDa protein called caveolin (Rothberg et al., 1992) and, probably, cholesterol (Rothberg et al., 1990). In some mammalian cells, as in the protozoa, GPI-anchored proteins are seen to be completely endocytosed and delivered to intracellular compartments in clathrin-free vesicles (Keller et al., 1992; Bamezai et al., 1992). Whether or not potocytosis and the latter endocytic events are variations on the same theme remains to be determined.

Transmembrane signalling

More evidence for GPI microdomains in mammalian cells has come from co-precipitation studies. Monoclonal antibodies against a variety of GPI-anchored proteins were shown to co-precipitate the intracellular protein tyrosine kinases p56^{lck} and p60^{fyn} from detergent lysates of T-cells (Stefanova et al., 1991; Thomas and Samuelson, 1992). Glycolipids and a 100 kDa membrane protein have also been associated with these microdomains (Cinek and Horejsi, 1992; Lehuen et al., 1992). These microdomains may have functional significance. It is known that the ligation of any GPI-anchored protein by a first antibody followed by a second antibody (i.e. the clustering of GPI-anchored proteins and their respective microdomains) produces a transmembrane signal, which in the presence of phorbol ester leads to mitogenesis in T-cells (Robinson, 1991). The nature of the signal is unclear, but it could involve the stimulation of the aforementioned intracellular tyrosine kinases and/or an increase in intracellular Ca^{2+} levels (Robinson, 1991). With regard to the latter effect, it is of interest that the cross-linking of surface glycosphingolipids with antibody (Dyer and Benjimins, 1990; Lund-Johansen et al., 1992) or pentavalent cholera toxin B-subunit (Masco et al., 1991) also leads to a rise in intracellular calcium in several different cell types. It is tempting to speculate that these transmembrane signalling events, brought about by artificially clustering GPI-anchored proteins and/or gangliosides, have some physiological counterpart. However, as yet, there are no physiological events which are known to operate this way.

GPI functions common to the lower and higher eukaryotes

Insulation of the cytoplasm

In the case of membrane proteins where their primary function resides exclusively in the ectoplasmic domain, such as hydrolases (e.g. alkaline phosphatase, 5'-nucleotidase, dipeptidase, etc) and cell adhesion molecules (e.g. contact site A, LFA-3, N-CAM$_{120}$), we find GPI anchors quite well represented in both lower and higher eukaryotes. This makes sense since the mode of anchorage is secondary to ectoplasmic domain function. However, the correct targeting of hydrolases to the apical membrane of epithelial cells, for example, could constitute a selective advantage to GPI usage in the higher eukaryotes. Another possible advantage, both in the protozoa and in epithelial cells, could be the physical isolation of the ectoplasmic domain from the cytoplasm and cytoskeleton. Thus proteins which are operating in a 'harsh environment' can be left to get on with their job in the correct location without compromising the intracellular environment. This may be a major functional advantage of GPI anchors to the unicellular protozoa, both free-living and parasitic, and a useful vestigial function for the apical surfaces of epithelia.

Fine tuning of protein function

It may be significant that some proteins are expressed in both transmembrane form and GPI-anchored form, either by differ-

ential mRNA splicing of a single gene transcript or encoded by distinct genes (reviewed by Ferguson, 1991). In this way a GPI-anchored version of a particular protein could be left to perform a purely ectoplasmic function (protozoa and higher eukaryotes), or specialized functions such as potocytosis or transmembrane signalling through microdomain effects (higher eukaryotes only). In contrast, the transmembrane version of the same protein could interact with a completely different set of cytoplasmic components, via the cytoplasmic domain. Thus ectoplasmic domain function could be fine-tuned to intracellular function via the alternative anchoring mechanisms. An example of this might include the FCγRIII receptor, which contains a transmembrane polypeptide domain in macrophages, where it can directly elicit cellular responses when ligated with immune complexes. This same receptor is found as a GPI-anchored form in neutrophils where it binds the same immune complexes, but only potentiates cellular responses via other receptors (Perussia and Ravetch, 1991; Anderson et al., 1990; Salmon et al., 1991).

In *Leishmania* parasites, the transcription of genes for GPI-anchored and non-GPI-anchored versions of the abundant metalloproteinase, gp63, is developmentally regulated (Ramamoorthy et al., 1992; Medina-Acosta et al., 1993). In the insect-dwelling promastigote stage, most or all of the protein contains a GPI anchor and is surface expressed, while in the amastigote stage, a low level of a non-GPI-anchored protein is expressed which is localized to the lysosomes (Medina-Acosta et al., 1989, 1993; Bahr et al., 1993). These results suggest that differences in the anchor type may regulate the subcellular localization of some proteins.

Surface release by phospholipases

The presence of a GPI anchor may allow the selective release of some surface proteins via the action of an endogenous PI-specific phospholipase. However, clear evidence for this type of release has been demonstrated in only a few instances, both in higher eukaryotes (Paul et al., 1991; Vogel et al., 1992) and in the protozoa. PI-PLC mediated release of surface proteins has been reported in *Trypanosoma cruzi*, for the major antigens SSp-4 and the polymorphic family of trans-sialidase/sialidases (Andrews et al., 1988; Rosenberg et al., 1991; Hall et al., 1992) and in *Plasmodium falciparum* and *P. chaboudi*, for the 76 kDa serine protease (Braun-Breton et al., 1988, 1992). In the latter case, cleavage by the phospholipase is apparently required for activation of the protease. In contrast to the situation in these parasites, there is no evidence for enzyme-mediated release of GPI-anchored molecules in either *Leishmania* or in the African trypanosomes, despite the identification and characterization of a GPI-PLC in the latter organism. The *T. brucei* enzyme has been cloned, sequenced and expressed in *E. coli* (Hereld et al., 1986; Carrington et al., 1989; Mensa-Wilmot and Englund, 1992) and localized by immuno-electron microscopy (Bülow et al., 1989b). It has no identifiable signal sequence and was localized to the cytoplasmic face of intracellular vesicles. The role of this enzyme therefore remains obscure since it is not on the same side of the membrane as its putative VSG substrate. One possibility is that it may be involved in the catabolism of excess GPI precursors in bloodstream form trypanosomes.

Specialized GPI anchor functions in the parasitic protozoa

There is no direct evidence that GPI-anchored proteins associate with membrane microdomains in the lower eukaryotes. The major GPI-anchored proteins of *T. brucei*, *Leishmania* spp. and yeast are completely soluble in neutral detergents such as Triton

X-100 and X-114 (Bouvier et al., 1985; Das et al., 1991; Schell et al., 1991; Frankhauser et al., 1993). This may reflect the radically different molecular architecture of the surfaces of these organisms, compared with higher eukaryotes. In the case of the kinetoplastid parasites (trypanosomes and *Leishmania*) the surfaces are dominated by GPI-anchored proteins and/or GPI-related glycolipids (see below). The need for membrane targeting in these organisms is obviated by the fact that there is only one small region of the plasma membrane, the flagellar pocket, that appears to be capable of supporting membrane fusion events and pinocytosis. In addition, the need for transmembrane signalling in these organisms is presumably relatively limited and probably served by small numbers of transmembrane proteins. Thus the GPI anchor has probably been widely adopted by these organisms because of its ability to insulate the cell interior from the harsh world of the insect vector gut and the mammalian host bloodstream (*T. brucei*) or the macrophage phagolysosome (*Leishmania*). Indeed, GPI anchors were probably common in their free-living ancestors for the same reason, i.e. protection rather than communication.

In considering the parasitic protozoa then, GPI anchors may be the predominant form of protein anchorage for the following reasons.

1. They isolate proteins with purely extracellular function, such as proteases and coat proteins, from the interior of the cell.

2. They allow very high levels of protein packing without using up membrane space for the inclusion of nutrient transporters. This is particularly relevant for the VSG coat of the African trypanosomes, where 10 million copies of VSG must occupy the cell surface. It may also be relevant for the circumsporozoite antigen, which forms a dense coat on the sporozoite form of *Plasmodium* parasites. Sporozoites are the form of the parasite which are first injected into the host by a mosquito bite, and which subsequently invade the liver. Although the circumsporozoite antigen has not been shown to contain a GPI anchor biochemically, the C-terminus predicted from cDNA sequencing looks like a GPI signal sequence (Ozaki et al., 1983).

3. They can be modified by complex oligosaccharide side chains to form 'glycocalyx' structures.

Space-filling 'glycocalyx' roles for the GPI side chains

The unique oligosaccharide side chains of the *T. brucei* anchors may themselves contribute to the surface glycocalyx of this parasite. Studies on the three-dimensional conformation of the galactosylated VSG anchor (Homans et al., 1989) suggests that the glycan portion lies plate-like along the plasma membrane and has a cross-sectional area of 6 nm². This is approximately the same cross-sectional area as the N-terminal domain of VSG (Metcalf et al., 1987), suggesting that the GPI glycan may itself contribute to the macromolecular diffusion barrier properties of the VSG coat (Figure 3). The extent of α-galactosylation may be influenced by the three-dimensional structure of the C-terminal domain of the VSG to which it is attached as there is a correlation between the size of the galactose side chains and VSG subtype (Ferguson and Homans, 1989). Most of the αGal side chain residues are added to the VSG anchor in the Golgi apparatus (Bangs et al., 1988), where the VSG molecules are packaged into coat arrays for transport to the cell surface. This raises the possibility that the α-galactosyltransferases act as spatial probes, filling space close to the membrane according to the steric constraints imposed by the three-dimensional structure of the VSG C-terminal domain. Similarly, the extensive side chains of the GPI anchor of PARP may influence the surface properties of the insect stage of *T. brucei*. These side chains are substituted

with charged sialic acid residues and probably form a protective glycocalyx over the surface of the procyclic (insect) stage, which is otherwise notable in lacking a significant glycolipid component (Ferguson et al., 1993; Figure 3). The sialic acid residues are probably added to the PARP anchor by a surface trans-sialidase that transfers sialic acid from serum sialoglycoconjugates (Pontes de Carvalho et al., 1993). These residues may have the more specific function of preventing complement activation on the procyclic surface (Tomlinson et al., 1992) when it is exposed to a blood meal in the insect gut.

PROTEIN-FREE GPIs

Several protozoa also synthesize free GPIs, which are not covalently linked to protein and which appear to be metabolic end-products. These structures are members of the GPI family by virtue of containing the core sequence Manα1–4GlcNα1–6-*myo*-inositol, but may diverge from the protein anchors beyond this sequence. In several trypanosomatid parasites these glycolipids are the major cellular glycoconjugates.

Protein-free GPIs in *Leishmania* spp.

All *Leishmania* synthesize two distinct classes of free GPI, the polydisperse lipophosphoglycans (LPGs) and the low-molecular-mass glycoinositolphospholipids (GIPLs) (reviewed in McConville 1991; Turco and Descoteaux, 1992). The expression of the LPGs is largely restricted to the promastigote (insect-dwelling) stage where it forms a major component in the densely organized surface glycocalyx (Handman et al., 1984; Tolsen et al., 1989; Pimenta et al., 1991) (Figure 3). It is present in very low or undetectable levels in the amastigote stage that infects mammalian macrophages (McConville and Blackwell, 1991; Moody et al., 1993; Bahr et al., 1993). In contrast, the GIPLs are abundant in both major developmental stages of the parasite and are expressed predominantly on the cell surface (McConville and Bacic, 1990; McConville and Blackwell, 1991; Rosen et al., 1989) (Figure 3).

Lipophosphoglycan structure

The primary structures of the LPGs from three species of *Leishmania* have now been elucidated (see Figure 4). They all contain a polydisperse phosphoglycan moiety of 4–40 kDa which is linked to the membrane via a complex GPI anchor. The phosphoglycan is made up of linear chains of repeat units, all of which contain the backbone sequence, P-6Galβ1–4Manα1–, where the 3-position of the galactose residue is either unsubstituted (as in *L. donovani* LPG) (Turco et al., 1987), partially substituted with βGlc residues (as in *L. mexicana* LPG) (Ilg et al., 1992) or highly substituted with monosaccharide or oligosaccharide side chains containing βGal, βGlc, or β-D-Ara$p$ (as in *L. major* LPG) (McConville et al., 1990a). The ends of the phosphoglycan chains, distal to the membrane, are capped by a number of neutral oligosaccharides, all of which contain the sequence Manα1–2Man (McConville et al., 1990a) (Figure 4). The GPI anchor of LPG is distinct from the protein GPI anchors in containing an unusual hexasaccharide core linked to *lyso*alkyl-PI with either $C_{24:0}$ or $C_{26:0}$ alkyl chains (Orlandi and Turco, 1987; Turco et al., 1989; McConville et al., 1987, 1990a; Ilg et al., 1992). The phosphoglycan chain is attached to the terminal Gal of this core, while one of the core Man residues is usually substituted with Glc-1-P (McConville and Homans, 1992; Thomas et al., 1992). The presence of both shared (i.e. the caps, the backbone disaccharide sequence and the core) and species-specific (i.e. the side chains of the repeat units) structures is

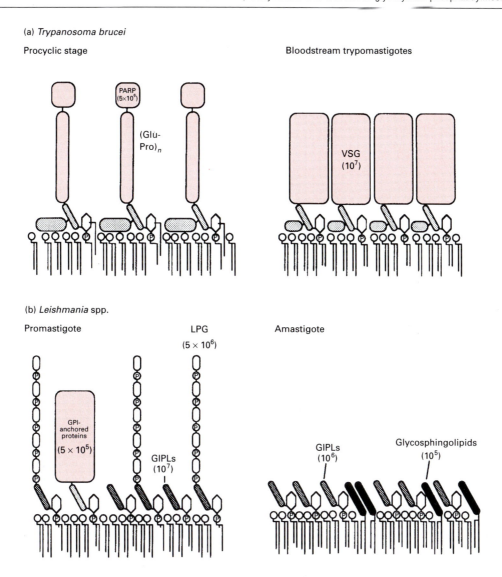

(a) *Trypanosoma brucei*

Procyclic stage

Bloodstream trypomastigotes

PARP
(5×10^6)

(Glu-Pro)$_n$

VSG
(10^7)

(b) *Leishmania* spp.

Promastigote

LPG
(5×10^6)

Amastigote

GPI-anchored proteins
(5×10^5)

GIPLs
(10^7)

GIPLs
(10^6)

Glycosphingolipids
(10^5)

Figure 3   Schematic representations of the cell surfaces of different developmental stages of *T. brucei* and *Leishmania* spp.

Only the major cell surface molecules are depicted (approximate copy numbers per cell are indicated in parentheses). The GPI anchors of the predominant cell surface proteins of *T. brucei* bloodstream and procyclic stages, VSG and PARP respectively, contain complex side chains (stippled area) which may form a glycocalyx over the plasma membrane. The cell surface of *Leishmania* parasites is coated by a more complex glycocalyx which also differs in the different developmental stages. The surface glycocalyx of the insect-dwelling promastigote stage contains several abundant GPI-anchored proteins, the polydisperse lipophosphoglycans (LPG) and the low-molecular-mass glycoinositolphospholipids (GIPLs). The surface expression of LPG and the major surface glycoprotein is greatly down-regulated in the intracellular amastigote stage. The plasma membrane of this stage also contains a number of glycosphingolipids which are apparently aquired from the mammalian host (McConville and Blackwell, 1991).

consistent with serological studies which predicted the presence of both conserved and species-specific epitopes (Greenblatt et al., 1983; Handman et al., 1984; Tolsen et al., 1989).

Marked variation in the size of the phosphoglycan and in the nature of the repeat unit side chains can occur in different developmental stages. During the development of *L. major* promastigotes in the insect midgut from a non-infective procyclic form to an infective 'metacyclic form', there is (1) an approximate doubling in the number of repeat units per molecule and (2) a decrease in the relative abundance of repeat unit side chains which terminate in βGal and a concomitant increase in un-substituted repeat units or repeat units with side chains that terminate in βAra (Sacks et al., 1990; McConville et al., 1992) (Figure 3). The expression of metacyclic LPG coincides with an increase in the thickness of the surface glycocalyx (Pimenta et al., 1989) and a loss in the binding of the Gal-specific lectin peanut

agglutinin to metacyclic promastigotes (Sacks, 1989). Similar developmental changes occur *in vivo* in the sandfly vector (Davies et al., 1990; Lang et al., 1991), and may be involved in regulating anterior migration of the parasite (see below). The intracellular amastigotes of *L. major* produce yet another form of LPG (Glaser et al., 1991; Turco and Sacks, 1991), albeit in much lower levels than in promastigotes. The majority of the repeat units in the amastigote LPG are unsubstituted, while the remainder are substituted with side chains of one to ten β-Gal residues (Moody et al., 1993).

A compelling image of the tertiary structure of LPG was provided by homo- and hetero-nuclear n.m.r. and molecular dynamics modelling of *L. donovani* LPG (Homans et al., 1992). These results suggest that the phosphoglycan chains have an extended helical conformation, although the length of the chains may vary considerably (from 180 Å to 320 Å, assuming 32 repeat

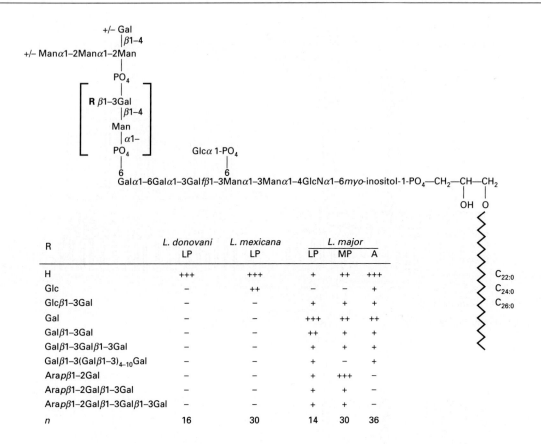

Figure 4   Generic structure of *Leishmania* LPG, showing common and species-specific structural features

The major species-specific differences (shown in inset) occur in the nature and frequency of side chain substitution (R) of the phosphorylated disaccharide repeat unit backbone. In the *L. major* LPG, these side chains also vary in different developmental stages (see text for references). LP, promastigotes in exponential (log) growth phase; MP, metacyclic promastigotes; A, amastigotes. The abundances of the different side chains [+ + + (> 50%); + + (10–50%); + (< 10%)] are indicated. It should be noted that the D-arabinopyranose residues in *L. major* LPG have been reassigned the β-configuration (S. W. Homans, unpublished work) and not the α-configuration as published (McConville et al., 1991a).

R	L. donovani LP	L. mexicana LP	L. major LP	L. major MP	L. major A
H	+++	+++	+	++	+++
Glc	–	++	–	–	+
Glcβ1–3Gal	–	–	+	+	+
Gal	–	–	+++	++	++
Galβ1–3Gal	–	–	++	+	+
Galβ1–3Galβ1–3Gal	–	–	+	+	+
Galβ1–3(Galβ1–3)$_{4–10}$Gal	–	–	+	–	+
Arapβ1–2Gal	–	–	+	+++	–
Arapβ1–2Galβ1–3Gal	–	–	+	+	–
Arapβ1–2Galβ1–3Galβ1–3Gal	–	–	+	+	–
n	16	30	14	30	36

units per molecule), reflecting the presence of several stable conformers that arise due to rotation around the flexible phosphodiester linkages. Transitions between these stable conformers may occur at physiological temperatures, allowing the LPG chains to contract or expand, somewhat like a slinky spring. The 3-position of the repeat unit galactose is always exposed in the different conformers, allowing the substitution of these residues without major conformation changes to the repeat backbone. These side chains would not only increase the cross-sectional area of the LPGs but would also be accessible for interaction with host receptors.

Lipophosphoglycan function

LPG-deficient strains of *Leishmania* are unable to survive in the sandfly vector (Schlein et al., 1990) or infect mammalian macrophages (Handman et al., 1986; Elhay et al., 1990; McNeely and Turco 1990), although both functions can be partly restored if exogenous LPG is inserted into the plasma membrane of these strains. These data suggest that LPG is an essential virulence factor for *Leishmania* parasites and that it is likely to perform a variety of functions (Table 3). The primary role of LPG is to form a coat over the promastigote surface which protects the plasma membrane from insect and mammalian hydrolases (El-On et al., 1980; McConville et al., 1990a; Schlein et al., 1990; Pimenta et al., 1991) and attack from the complement cascade. The resistance to complement-mediated lysis varies in different

developmental stages and appears to be critically dependent on the average chain length of the LPG. Only the infective metacyclic promastigotes, which contain the long (approximately 30 repeat units per molecule) LPG chains are resistant, while the non-infective procyclic forms which express shorter LPG chains (average 15 repeat units per molecule) are susceptible to complement lysis (Sacks, 1989; McConville et al., 1992). Although both forms activate complement to the same extent (Puentes et al., 1988), the longer LPG chains of metacyclic promastigotes appear to sterically hinder the stable insertion of the final C5–C9 complex into the plasma membrane and prevent cell lysis (Puentes et al., 1990). In addition, the LPG coat may mask underlying surface proteins from being opsonized by antibodies in either the insect bloodmeal or in the mammalian bloodstream (Karp et al., 1991). LPG also mediates a number of specific host–parasite interactions. In particular, it is required for binding of *L. major* promastigotes to the insect midgut during the early stages of insect colonization, where it is bound by receptors on insect epithelial cells that recognize the βGal terminating LPG side chains (Pimenta et al., 1992). Most of these side chains are capped with β-D-Arap residues on the LPG of metacyclic promastigotes (McConville et al., 1992) allowing this infective stage to detach from the gut wall and migrate to the insect foregut. These oligosaccharide side chains are unique to the LPG of *L. major* (see Figure 4) and may represent a specialized adaptation to allow this species to colonize the sandfly vector, *Phlebotiminae papatasi*, which is not a host for most other

Table 3   Structure/function relationships of *Leishmania* LPG

Function	Probable LPG domain involved	References
Surface coat		
(i) Protection against complement-mediated lysis		Puentes et al., 1990
(ii) Protection against insect/phagolysosomal hydrolases		El-On et al., 1980; Schlein et al., 1990
(iii) Masking of protein antigens		Karp et al., 1991
Cell–cell recognition and attachment		
(i) Binding to receptors in insect midgut	Repeat unit side chains	Pimenta et al., 1992
(ii) Binding to macrophage receptors	Repeat unit side chains	Handman and Goding, 1985; Talamas-Rohana et al., 1990; Kelleher et al., 1992
(iii) Activation of complement and binding to macrophages via the complement receptors CR1 or CR3	Repeat units/mannose cap structures	Puentes et al., 1988; da Silva et al., 1989; Mosser et al., 1992
Survival in mammalian macrophage		
(i) Scavenger of oxygen radicals	Repeat units	Chan et al., 1989
(ii) Chelation of Ca^{2+}	Repeat units	Eilam et al., 1985
(iii) Modulation of macrophage functions		
Inhibition of protein kinase C	Lipid and repeat units	McNeely et al., 1989; Descoteaux et al., 1992
Inhibition of oxidative burst	Lipid and repeat units	McNeely et al., 1990
Inhibition of chemotactic locomotion		Frankenburg et al., 1990
Down-regulation of tumour-necrosis-factor receptors		Descoteaux et al., 1991
Inhibition of IL-1 production		Frankenburg et al., 1990

Leishmania spp. LPG may also mediate the direct binding of promastigotes to cell surface receptors on the macrophage (Handman and Goding, 1985; Talamas-Rohana et al., 1990; Kelleher et al., 1992). Alternatively, opsonization of the LPG coat with complement components (particulary C3bi) may lead to the binding and uptake of promastigotes via the macrophage complement receptors CR1 and CR3 (da Silva et al., 1989; Mosser et al., 1992). This latter interaction seems to predominate with the physiologically relevant metacyclic forms. LPG also appears to be functionally important in protecting the parasite from hydrolyases and oxygen radicals (El-On et al., 1980; Chan et al., 1989) during the differentiation of the promastigote stage to the amastigote stage within the macrophage phagolysosome. Moreover, there is evidence that LPG, delivered exogenously or from an intracellular parasite, is able to prevent activation of the oxidative burst (McNeely and Turco, 1990), expression of the *c-fos* gene (Descoteaux et al., 1991) and chemotactic locomotion (Frankenburg et al., 1990) by macrophages. These functions are mediated by protein kinase C-dependent pathways, consistent with the finding that LPG is a potent inhibitor, both *in vitro* and *in vivo*, of mammalian protein kinase C (McNeely et al., 1989; Descoteaux et al., 1992). It is not known whether LPG released in the phagolysosome is able to gain access the host cell cytoplasm and directly inhibit the regulatory subunit of the kinase as a competitive inhibitor of the cofactor, diacylglycerol (McNeely and Turco, 1987), or whether it indirectly inhibits the kinase by chelating Ca^{2+}, another cofactor for the enzyme.

Glycoinositolphospholipids

The GIPLs are the major glycolipids synthesized by *Leishmania* parasites. Three distinct lineages of GIPLs have been identified, which are expressed in markedly different levels in different species or developmental stages (Figure 5). The type-1 and type-2 GIPLs have glycan headgroups which are structurally related to the GPI protein anchors and the LPG anchor, respectively (McConville et al., 1990b; McConville and Blackwell, 1991), while the hybrid-type GIPLs have branched glycan headgroups which share features in common with both types of GPI anchor (McConville and Blackwell, 1991; Sevlever et al., 1991) (Figure 5). Some of the hybrid-type GIPLs in *L. mexicana* are substituted

with an ethanolamine phosphate residue which is linked, unusually, to the core GlcN (McConville et al., 1993). The addition of this substituent appears to be restricted to these GIPLs as it is not present on either the type-2 GIPLs or protein anchors in the same strain. The lipid moieties of the *Leishmania* GIPLs are exclusively alkylacyl-PI or *lyso*alkyl-PI. Interestingly, the lipid compositions of the hybrid and type-1 GIPLs are distinct from those found in either the protein anchors or LPG in containing predominantly $C_{18:0}$ alkyl chains. In contrast, the type-2 GIPLs contain a more heterogeneous alkyl chain composition, which includes the longer alkyl chains ($C_{24:0}$ and $C_{26:0}$) found in the LPG anchor (McConville and Bacic, 1989; McConville et al., 1990b; McConville and Blackwell, 1991).

The GIPLs may coat a significant proportion of the plasma membrane and, in the case of the type-2 GIPLs, are highly immunogenic (McConville and Bacic, 1989; McConville et al., 1990b; Rosen et al., 1988; Avila et al., 1991). Although the function on these glycolipids is unknown, it is possible that, together with LPG, they are involved in protecting the parasite in the insect midgut and macrophage phagolysosome (McNeely et al., 1989; Descoteaux et al., 1992) and in mediating host–parasite interactions. Evidence that the GIPLs may play a role in parasite invasion of macrophages is suggested by the finding that *L. donovani* promastigotes and amastigotes, which both express mannose-terminating GIPLs, are able to utilize the mannose receptor on the macrophage surface (Blackwell et al., 1985). These functions may be crucial to the intracellular amastigote stage, which dramatically down-regulates the surface expression of the major macromolecules, LPG and gp63, leaving the GIPLs as the major components in the surface glycocalyx (Medina-Acosta et al., 1989; McConville and Blackwell, 1991; Schneider et al., 1992; Bahr et al., 1993).

Biosynthesis and metabolism of the GIPLs and LPG

Each of the GIPL lineages form a natural biosynthetic series suggesting that, like the protein GPI anchors, they are synthesized by the sequential addition of monosaccharides to PI (Figure 6). This is supported by metabolic labelling experiments (L. Proudfoot, P. Schneider, M. A. J. Ferguson and M. J. McConville, unpublished work), which also suggest that some of the type-2

Name	Structure	L. major (P)	L. mexicana (P)	L. donovani (P)	L. donovani (A)
M1	Manα1–4GlcN1–6PI	+	+	–	–
Type-1 GIPL					
M2	Manα1–6\Manα1–4GlcN1–6PI	–	–	–	++
M3	Manα1–2Manα1–6\Manα1–4GlcN1–6PI	–	–	–	++
Type-2 GIPLs					
iM2	Manα1–4GlcNα1–6PI / Manα1–3	+	++	++	–
GIPL-1	Manα1–4GlcNα1–6PI / Galβ1–3Manα1–3	++	–	–	–
GIPL-2	Manα1–4GlcNα1–6PI / Galα1–3Galβ1–3Manα1–3	++	+	–	–
GIPL-3	Manα1–4GlcNα1–6PI / Galα1–6Galα1–3Galβ1–3Manα1–3	++	+	–	–
GIPL-A	Manα1–4GlcNα1–6PI / Galβ1–3Galα1–3Galβ1–3Manα1–3	++	–	–	–
LPGp	Glcα1—PO4 / 6 / Galα1–6Galα1–3Galβ1–3Manα1–3 \ Manα1–4GlcN1–6lyso-PI	+	++	–	–
GIPL-4*	Manα1—PO4—6Galα1–6Galα1–3Galβ1–3Manα1–3 \ Manα1–4GlcNα1–6lyso-PI	++	–	–	–
GIPL-6*	Glcα1—PO4 / 6 / Manα1—PO4—6Galα1–6Galα1–3Galβ1–3Manα1–3 \ Manα1–4GlcNα1–6lyso-PI	++	–	–	–
Hybrid-type GIPLs					
iM3	Manα1–6\Manα1–4GlcN1–6PI / Manα1–3	–	++	++	–
iM4	Manα1–2Manα1–6\Manα1–4GlcN1–6PI / Manα1–3	–	++	++	–
EPiM3	$NH_2CH_2CH_2$—PO4 | Manα1–6\Manα1–4GlcN1–6PI / Manα1–3	–	++	–	–

Figure 5   Structures of the protein-free GPIs of *Leishmania* spp.

PI refers to alkylacylphosphatidylinositol. The predominant alkyl chain in type-1 and hybrid GIPLs is $C_{18:0}$, while type-2 GIPLs have a more heterogeneous lipid composition with $C_{18:0}$, $C_{22:0}$, $C_{24:0}$ and $C_{26:0}$ alkyl chains being the most abundant (from McConville et al., 1990a; McConville and Blackwell, 1991; McConville and Homans, 1992; McConville et al., 1993). Species marked with an asterisk have only been identified in an LPG-deficient *L. major* strain and probably represent truncated forms of LPG. –, not detected; +, minor species (< 5%); + +, major species. P, promastiogote; A, amastigote stage.

GIPLs may act as biosynthetic precursors to LPG (Figure 6). The cellular levels of the type-2 GIPLs varies greatly in different species, from less than 10^3 copies per cell in *L. donovani* to 10^5 and 10^7 copies per cell in *L. mexicana* and *L. major*, respectively (McConville and Blackwell, 1991; McConville et al., 1990a, 1993). As the levels of LPG synthesis are comparable in all these species, it is likely that only a minor population of these GIPLs (< 1 % in the high-expressing strains) will be utilized as LPG precursors, consistent with the finding that they are expressed in high copy number on the cell surface (McConville and Bacic, 1990). Recent studies have identified a polar GIPL species in *L. mexicana* which has the expected properties of an LPG precursor (LPGp in Figure 5; McConville et al., 1993). In particular it has a *lyso*alkyl-PI lipid moiety which is highly enriched for long alkyl chains ($C_{24:0}$ and $C_{26:0}$) and a glycan moiety which is substituted with Glc-1-*P*, suggesting that processing of the LPG anchor is

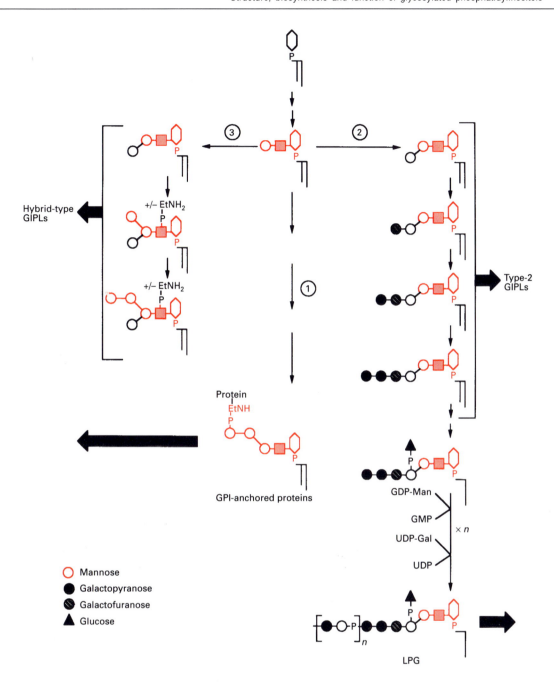

Figure 6   Pathways of GPI biosynthesis in *Leishmania* promastigotes

Pathways 1 and 2 lead to the formation of the protein and LPG GPI anchors, respectively, and probably occur in all species. Intermediates in pathway 1 are present in very low levels and have not been characterized. A putative protein anchor precursor species, identified in *L. mexicana* (Field et al., 1991c), probably corresponds to one of the ethanolamine phosphate-substituted hybrid-type GIPLs. Pathway 2 predominates in *L. major* promastigotes, while pathway 3 predominates in *L. donovani* and *L. mexicana* promastigotes. Large arrows indicate species that are expressed at the cell surface. Components of the conserved GPI backbone, i.e. ethanolamine-phosphate-Manα1–2Manα1–6Manα1–4GlcNα1–6*myo*-inositol, are shown in red. Symbols are defined in the keys to this Figure and to Figure 2.

finished prior to the addition of the repeat units (Figure 6). Studies by Turco and colleagues, using a cell-free system, suggest that the phosphoglycan chains of LPG are built up on this preformed anchor by the sequential addition of Man-1-*P* and Gal from GDP-Man and UDP-Gal, respectively (Carver and Turco, 1991,1992). An LPG-deficient strain of *L. major* has been identified which is unable to form the first repeat unit due to a defect in the transfer of the Gal residue (McConville and Homans, 1992). This strain expresses two highly truncated LPG structures

on its cell surface (GIPL-4 and -6 in Figure 5) and is unable to survive in mammalian macrophages (Handman et al., 1986).

The major GIPL species appear to have a relatively low turnover rate (L. Proudfoot, P. Schneider, M. A. J. Ferguson and M. J. McConville, unpublished work), in contrast to LPG, which is actively shed from the cell surface and has a high turnover (Handman et al., 1984; King et al., 1987). The rate of LPG shedding is increased when parasites are grown in the presence of serum albumin, which contains a hydrophobic pocket

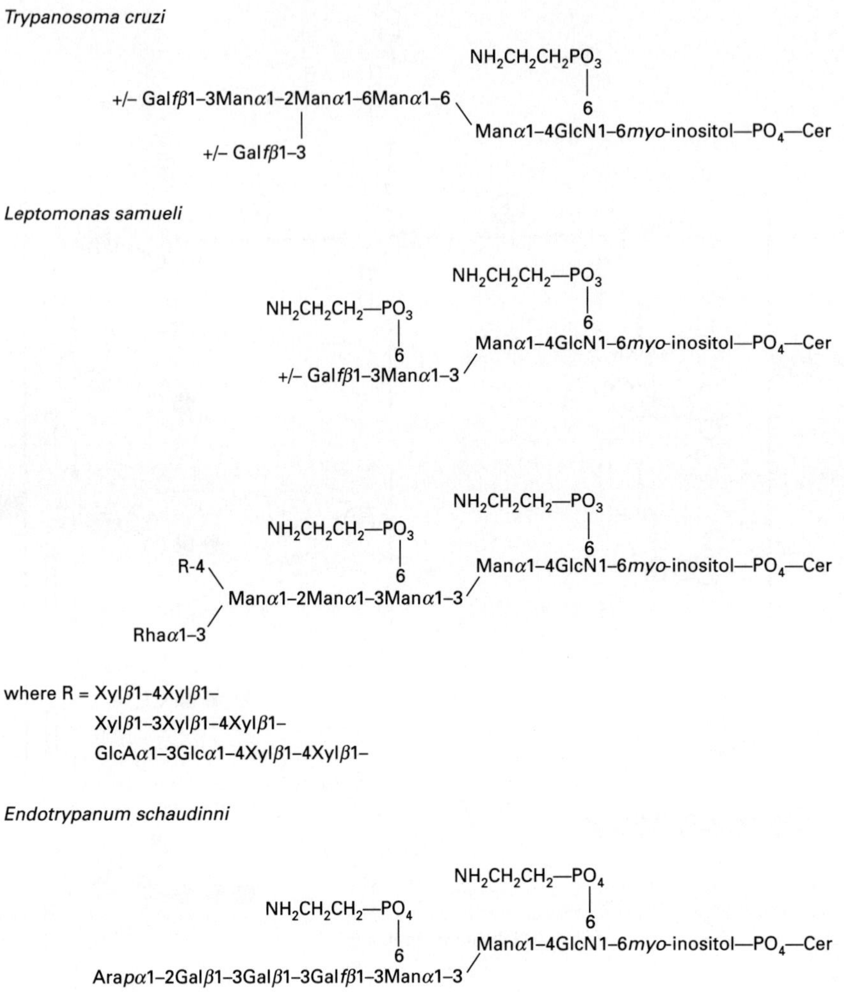

Trypanosoma cruzi

$$NH_2CH_2CH_2PO_3$$
$$|$$
$$6$$

+/– Gal$f\beta$1–3Manα1–2Manα1–6Manα1–6 ⟍
$$|$$
+/– Gal$f\beta$1–3

Manα1–4GlcN1–6myo-inositol—PO$_4$—Cer

Leptomonas samueli

$$NH_2CH_2CH_2—PO_3$$
$$|$$
$$6$$

$$NH_2CH_2CH_2—PO_3$$
$$|$$
$$6$$
+/– Gal$f\beta$1–3Manα1–3 ⟋

Manα1–4GlcN1–6myo-inositol—PO$_4$—Cer

$$NH_2CH_2CH_2—PO_3$$
$$|$$
$$6$$

$$NH_2CH_2CH_2—PO_3$$
$$|$$
R-4 ⟍ $$6$$
Manα1–2Manα1–3Manα1–3 ⟋
Rhaα1–3 ⟋

Manα1–4GlcN1–6myo-inositol—PO$_4$—Cer

where R = Xylβ1–4Xylβ1–
Xylβ1–3Xylβ1–4Xylβ1–
GlcAα1–3Glcα1–4Xylβ1–4Xylβ1–

Endotrypanum schaudinni

$$NH_2CH_2CH_2—PO_4$$
$$|$$
$$6$$

$$NH_2CH_2CH_2—PO_4$$
$$|$$
$$6$$
Ara$p\alpha$1–2Galβ1–3Galβ1–3Gal$f\beta$1–3Manα1–3 ⟋

Manα1–4GlcN1–6myo-inositol—PO$_4$—Cer

Figure 7 Structures of the protein-free GPIs of *Trypanosoma cruzi* (from Previato et al., 1990; Lederkremer et al., 1991), *Endotrypanum schaudinni* (Previato et al., 1993) and *Leptomonas samueli* (Previato et al., 1992; J. O. Previato, R. Wait, C. Jones and L. Mendonça-Previato, unpublished work)

(King et al., 1987), and the shed material retains the *lyso*alkyl-PI lipid moiety (Ilg et al., 1992). This is consistent with a non-enzymic mechanism of release whereby LPG partitions out of the plasma membrane as either monomers or micelles. This property is probably a reflection of the weak attachment of LPG to the outer leaflet of the plasma membrane via a single aliphatic chain and may be important in allowing the rapid expression of new LPG structures on the cell surface during parasite development. Phosphoglycan chains, which lack both a lipid and the core region of the LPG anchor, are also released into the culture medium (Greis et al., 1992). It is not known whether this material is derived from LPG due to the action of an endoglycosidase or phosphodiesterase or whether it is a distinct biosynthetic product. These shed/secreted antigens have been used to serotype different strains of *Leishmania* and are a valuable tool in diagnostic and demographic studies of the disease (Schnur, 1982; Greenblatt et al., 1983; Handman et al., 1987).

Protein-free GPIs in other kinetoplastid parasites

Protein-free GPI glycolipids are the major cellular glyco-conjugates in the insect stage of several other kinetoplastid parasites. In the digenetic parasite *T. cruzi*, the major molecule on the surface of the insect dwelling epimastigote stage is a heterogeneous lipopeptidophosphoglycan (LPPG) (Lederkremer et al., 1976). The expression of this molecule appears to be developmentally regulated as it is either absent from, or present in very low levels in, the stages that infect the mammalian host (Zingales et al., 1982). Recently, the complete primary structure of LPPG has been reported (Figure 7; Previato et al., 1990; Lederkremer et al., 1991). It contains the same tetrasaccharide backbone sequence as the protein GPI anchors of this species (Güther et al., 1992), but differs from these anchors in containing (1) up to two additional βGal$f$ residues, (2) a residue of 2-aminoethylphosphonic acid on the 6 position of glucosamine, and (3) a ceramide lipid moiety [containing sphinganine and N-linked lignoceric ($C_{24:0}$) acid], instead of alkylacylglycerol (Lederkremer et al., 1990; Previato et al., 1990). A small amount of peptide is frequently present in purified fractions, although it is not known if these residues are covalently linked to the glycan.

A structurally related family of GPI glycolipids have recently been characterized from the monogenetic parasite, *Leptomonas samueli* (Previato et al., 1992) and from the insect-dwelling stage of the digenetic parasite, *Endotrypanum schaudinni* (Previato et al., 1993). These glycolipids contain features in common with both the *Leishmania* type-2 GIPLs (in the glycan core) and the *T. cruzi* LPPG (in the inositolphosphoceramide lipid) (Figure 7). All these glycolipids appear to be substituted with either 2-aminoethylphosphonate (in *L. samueli*) or ethanolamine phosphate (in *T. schaudinni*) residues which are linked to both the core glucosamine and to the second mannose distal to the glucosamine. These structures may be extended with oligosaccharide sequences of variable length. In the *E. schaudinni* GPI, the terminal sequences are identical to those found in the side chains of *L. major* LPG (Figures 4 and 7), while in *L. samueli* these GPI glycolipids may be substituted with complex oligosaccharide chains containing rhamnose, xylose and glucuronic acid residues (Previato et al., 1992; J. Previato, R. Wait, C. Jones and L. Mendonça-Previato, unpublished work).

Protein-free GPIs in non-kinetoplastid parasites

There is evidence that some non-kinetoplastid protozoa may also synthesize non-protein linked GPIs. In particular, early studies identified a complex lipophosphonoglycan as the major cell surface glycoconjugate in *Acanthamoeba castellanii* (Dearborn et al., 1976). This glycoconjugate contained inositolphosphoceramide, and the monosaccharides mannose, glucosamine, galactose, xylose and galactosamine, although the structure of the glycan and the presence of the sequence Manα1–4GlcNα1–6*myo*-inositol has still to be determined. Similarly, an LPG-like structure has been identified in the human pathogen, *Entamoeba histolytica* (Stanley et al., 1992; Bhattacharya et al., 1992). Antibodies against this surface glycoconjugate inhibit adhesion of *E. histolytica* trophozoites to mammalian cells, suggesting that it may be involved in mediating host–parasite interactions. A GPI species with the structure Manα1–2Manα1–3Manα1–4GlcN-PI containing undefined phosphate constituents has also been described in the ciliate *Tetrahymena mimbres* (Weinhart et al., 1991).

EVOLUTION OF PROTEIN GPI ANCHORS AND FREE GPIs

Giardia lamblia and the kinetoplastid parasites are considered to be amongst the earliest diverging lineages in the eukaryotic line of descent, based on ribosomal RNA sequence analysis and morphological criteria (i.e. the absence of mitochondria, normal endoplasmic reticulum and Golgi in *G. lamlia*) (Sogin et al., 1989). The presence of GPI anchors in these organisms suggests that this type of anchor was present or evolved at the beginning of eukaryotic evolution. Indeed, the recent characterization of a novel phosphoglycerolipid with the structure, GlcNα1–6*myo*-inositol-*P*-dialkylglycerol, in the archaebacteria (Nishihara et al., 1992) raises the possibility of a relationship between the eukaryote GPIs and some prokaryote glycolipids. The presence of free GPIs as the major class of cellular glycolipid is, so far, unique to the protozoa. It is tempting to speculate that the high levels of GPI anchor expression in the ancestors of the kinetoplastid protozoa pre-adapted them for expanding their GPI metabolism for functions that have complemented their evolution, firstly to monogenetic parasitism (of annelids and insects) and subsequently to digenetic parasitism (of insects and mammals). The degree of structural relatedness in the core regions of the free GPIs of *Leishmania* spp., *T. cruzi*, *E. schaudinni* and *L. samueli* suggests that these glycolipids may have been inherited from a common monogenetic ancestor (Lake et al., 1988). Moreover, their continued expression in parasites such as *Leptomonas*, which only infects insects, suggests that their primary, or ancestral, function was related to parasite survival in the insect

gut. This is supported by structure–function studies on *Leishmania* LPG, which also indicate that some of the species-specific differences in LPG structure are adaptations to different sandfly vectors. It is surprising that free GPIs are apparently absent from the bloodstream or insect forms of *T. brucei* given that this species is thought to have evolved from the same ancestral organism as *Leishmania* and *T. cruzi* (Lake et al., 1988). In this case, it is possible that free GPI glycolipids have been down-regulated and replaced by the elaborate oligosaccharide side chains on the GPI anchors of the VSG and PARP surface proteins. Interestingly, the protein portion of PARP has similar physicochemical properties to *Leishmania* LPG, by virtue of containing an acidic (Asp-Pro)$_2$-(Glu-Pro)$_{22–29}$ repeat domain which is predicted to take up an extended conformation (Figure 3; Roditi et al., 1989). Thus LPG and PARP may represent examples of convergent evolution, giving rise to the expression of molecules of similar size, charge distribution and surface density on the insect stages of these two parasites.

CONCLUSIONS

The GPI anchors of plasma membrane proteins are ubiquitous throughout the eukaryotes. In higher eukaryotes, these anchors are only utilized by a minority of surface proteins, although there is evidence that GPI anchorage is frequently required for certain specific tasks related to multicellular existence. In the protozoa, GPI anchors appear to have been selected as the most useful form of surface protein anchorage. This is most pronounced in many parasitic protozoa, particularly the kinetoplastida, which have also expanded upon the conserved GPI biosynthetic pathway to produce a diverse array of unique structures. There is evidence that several of these parasite-specific structures are essential for survival of the parasite in the insect and mammalian hosts. We postulate that the evolution of these novel GPI metabolic pathways may have occurred in parallel with the evolution of these organisms to occupy new ecological niches as mono- and digenetic parasites. These studies predict that the expression of protein free GPIs may be quite common in other parasitic protozoa and that an improved understanding of their function will provide information on host–parasite interactions. Finally, some pathways of GPI biosynthesis are likely to be unique to these organisms and be potential targets for the development of antiparasite drugs.

We would like to thank Dr. Paul Englund, Dr. Chris Jones, Dr. Lucia Mendonça-Previato, Dr. José Osvaldo Previato, Dr. Anant Menon, Dr. Peter Overath, Dr. Ralph Schwarz and Dr. Robin Wait for providing unpublished work, Dr. Anant Menon for helpful discussions and Dr. Pascal Schneider for critical reading of the manuscript. This work was supported by the Wellcome Trust. M. J. M. is a Wellcome Trust Senior Research Fellow. M. A. J. F. is a Howard Hughes International Research Scholar.

REFERENCES

Abeijon, C. and Hirschberg, C. B. (1992) Trends Biochem. Sci. **17**, 32–36

Amthauer, R., Kodukula, K., Brink, L. and Udenfriend, S. (1992) Proc. Natl. Acad. Sci. U.S.A. **89**, 6124–6128

Amthauer, R., Kodukula, K., Gerber, L. and Udenfriend, S. (1993) Proc. Natl. Acad. Sci. U.S.A. **90**, 3973–3977

Anderson, C. L., Shen, L., Eicher, D. M., Wewers, M. D. and Gill, J. K. (1990) J. Exp. Med. **171**, 1333–1345

Anderson, R. G. W., Kamen, B. A., Rothberg, K. G. and Lacey, S. W. (1992) Science **255**, 410–411

Andrews, N. W., Robbins, E. S., Ley, V., Hong, K. S. and Nussenzweig, V. (1988) J. Exp. Med. **167**, 300–314

Armstrong, C., Schubert, E., Knez, J. J., Gelperin, D., Hirose, S., Silber, R., Hollan, S., Schmidt, R. E. and Medof, M. E. (1992) J. Biol. Chem. **267**, 25347–25351

Avila, J. L., Rojas, M. and Acosta, A. (1991) J. Clin. Microbiol. **29**, 2305–2312

Bahr, V., Stierhof, Y.-D., Ilg, T., Demar, M. and Overath, P. (1993) Mol. Biochem. Parasitol. **58**, 107–121

Bamezai, A., Goldmacher, V. S. and Rock, K. L. (1992) Eur. J. Immunol. **22**, 15–21

Bangs, J. D., Hereld, D., Krakow, J. L., Hart, G. W. and Englund, P. T. (1985) Proc. Natl. Acad. Sci. U.S.A. **82**, 3207–3211

Bangs, J. D., Doering, T. L., Englund, P. T. and Hart, G. W. (1988) J. Biol. Chem. **263**, 17697–17705

Bhattacharya, A., Prasad, R. and Sacks, D. L. (1992) Mol. Biochem. Parasitol. **56**, 161–168

Blackwell, J. M., Ezekowitz, A. B., Roberts, M. B., Channon, J. Y., Sim, R. B. and Gordon, S. (1985) J. Exp. Med. **162**, 324–331

Bouvier, J., Etges, R. J. and Bordier, C. (1985) J. Biol. Chem. **260**, 15504–15509

Braun-Breton, C., Rosenberry, T. L. and Da Silva, L. H. P. (1988) Nature (London) **332**, 457–459

Braun-Breton, C., Blisnick, T., Jouin, H., Barale, J. C., Rabilloud, T., Langsley, G. and Da Silva, L. H.P (1992) Proc. Natl. Acad. Sci. U.S.A. **89**, 9647–9651

Bretscher M. S., Thomson, J. N. and Pearse, B. M. F. (1980) Proc. Natl. Acad. Sci. U.S.A. **77**, 4156–4159

Brown, D. A. (1992) Trends Cell Biol. **2**, 338–343

Brown, D. A. and Rose, J. K. (1992) Cell **68**, 533–544

Bülow, R., Griffiths, G., Webster, P., Stierhof, Y.-D., Opperdoes, F. R. and Overath, P. (1989a) J. Cell Sci. **93**, 233–240

Bülow, R., Nonnengasser, C. and Overath, P. (1989b) Mol. Biochem. Parasitol. **32**, 85–92

Caras, I. W. (1991) Cell Biol. Int. Rep. **15**, 815–826

Carrington, M., Bülow, R., Reinke, H. and Overath, P. (1989) Mol. Biochem. Parasitol. **33**, 289–296

Carver, M. A. and Turco, S. J. (1991) J. Biol. Chem. **266**, 10974–10981

Carver, M. A. and Turco, S. J. (1992) Arch. Biochem. Biophys. **295**, 309–317

Chan, J., Fujiwara, T., Brennan, P., McNeil, M., Turco, S. J., Sibille, J. C., Snapper, M., Aisen, P. and Bloom, B. R. (1989) Proc. Natl. Acad. Sci. U.S.A. **86**, 2453–2457

Cinek, T. and Horejsi, V. (1992) J. Immunol. **149**, 2262–2270

Clayton, C. E. and Mowatt, M. R. (1989) J. Biol. Chem. **264**, 15088–15093

Conzelmann, A., Riezman, H., Desponds, C. and Bron, C. (1988) EMBO J. **7**, 2233–2240

Conzelmann, A., Puoti, A., Lester, R. L. and Desponds, C. (1992) EMBO J. **11**, 457–466

Costello, L. C. and Orlean, P. (1992) J. Biol. Chem. **267**, 8599–8603

Cross, G. A. M. (1990a) Annu. Rev. Cell Biol. **6**, 1–39

Cross, G. A. M. (1990b) Annu. Rev. Immunol. **8**, 83–100

Das, S., Traynor-Kaplan, A., Reiner, D. S., Meng, T. C. and Gillin, F. D. (1991) J. Biol. Chem. **266**, 21318–21325

da Silva, R. P., Hall, B. F., Joiner, K. A. and Sacks, D. L. (1989) J. Immunol. **143**, 617–622

Davies, C. R., Cooper, A. M., Peacock, C., Lane, R. P. and Blackwell, J. M. (1990) Parasitology **101**, 337–343

Dearborn, D. G., Smith, S. and Korn, E. D. (1976) J. Biol. Chem. **251**, 2976–2982

Deeg, M. A., Humphrey, D. R., Yang, S. H., Ferguson, T. R., Reinhold, V. N., and Rosenberry, T. L. (1992) J. Biol. Chem. **267**, 18573–18580

Descoteaux, A., Turco, S. J., Sacks, D. L. and Matlashewski, G. (1991) J. Immunol. **146**, 2747–2753

Descoteaux, A., Matlashewski, G. and Turco, S. J. (1992) J. Immunol. **149**, 3008–3015

Doering, T. L., Masterson, W. J., Englund, P. T. and Hart, G. W. (1989) J. Biol. Chem. **264**, 11168–11173

Doering, T. L., Masterson, W. J., Hart, G. W. and Englund, P. W. (1990) J. Biol. Chem. **265**, 611–614

Doering, T. L., Raper, J., Buxbaum, L. U., Adams, S. P., Gordon, J. I., Hart, G. W. and Englund, P. T. (1991) Science **252**, 1851–1854

Doering, T. L., Pessin, M. S., Hoff, E. F., Hart, G. W., Raben, D. M. and Englund, P. T. (1993) J. Biol. Chem. **268**, 9215–9222

Dyer, C. A. and Benjamins, J. A. (1990) J. Cell Biol. **111**, 625–633

Eilam, Y., El-On, J. and Spira, D. T. (1985) Exp. Parasitol. **59**, 161–168

El-On, J. Bradley, D. J. and Freeman, J. C. (1980) Exp. Parasitol. **49**, 167–174

Elhay, M. J., Kelleher, M., Bacic, A., McConville, M. J., Tolsen, D. L., Pearson, T. W. and Handman, E. (1990) Mol. Biochem. Parasitol. **40**, 255–268

Englund, P. T. (1993) Annu. Rev. Biochem. **62**, 121–138

Engstler, M., Reuter, G. and Schauer, R. (1992) Mol. Biochem. Parasitol. **54**, 21–30

Etges, R., Bouvier, J. and Bordier, C. (1986) EMBO J. **3**, 597–601

Fankhauser, C., Homans, S. W., Thomas-Oates, J. E., McConville, M. J., Desponds, C., Conzelmann, A. and Ferguson, M. A. J. (1993) J. Biol. Chem., in the press

Ferguson, M. A. J. (1991) Curr. Opin. Struct. Biol. **1**, 522–529

Ferguson, M. A. J. (1992a) Biochem. J. **284**, 297–300

Ferguson, M. A. J. (1992b) Biochem. Soc. Trans. **20**, 243–256

Ferguson, M. A. J. and Homans, S. W. (1989) in New Strategies in Parasitology (McAdam, K. P. W. J., ed.), pp. 121–143, Churchill-Livingstone, London

Ferguson, M. A. J. and Williams, A. F. (1988) Annu. Rev. Biochem. **57**, 285–320

Ferguson, M. A. J., Duszenko, M., Lamont, G., Overath, P. and Cross, G. A. M. (1986) J. Biol. Chem. **261**, 356–362

Ferguson, M. A. J., Homans, S. W., Dwek, R. A. and Rademacher, T. W. (1988) Science **239**, 753–759

Ferguson, M. A. J., Murray, P., Rutherford, H. and McConville, M. J. (1993) Biochem. J. **291**, 51–55

Field, M. C., Menon, A. K. and Cross, G. A. M. (1991a) EMBO J. **10**, 2731–2739

Field, M. C., Menon, A. K. and Cross, G. A. M. (1991b) J. Biol. Chem. **266**, 8392–8400

Field, M. C., Medina-Acosta, E. and Cross, G. A. M. (1991c) Mol. Biochem. Parasitol. **48**, 227–230

Field, M. C., Menon, A. K. and Cross, G. A. M. (1992) J. Biol. Chem. **267**, 5324–5329

Fouts, D. L., Ruef, B. J., Ridley, P. T., Wrightsman, R. A., Peterson, D. S. and Manning, J. E. (1991) Mol. Biochem. Parasitol. **46**, 189–200

Frankenburg, S., Lebovici, V., Mansbach, N., Turco, S. J. and Rosen, G. (1990) J. Immunol. **145**, 4284–4289

Gerber, L. D., Kodukula, K. and Udenfriend, S. (1992) J. Biol. Chem. **267**, 12168–12173

Gerold, P., Dieckmann-Schuppert, A. and Schwarz, R. T. (1992) Biochem. Soc. Trans. **20**, 297S

Glaser, T. A., Moody, S. F., Handman, E., Bacic, A. and Spithill, T. W. (1991) Mol. Biochem. Parasitol. **45**, 337–344

Greenblatt, C. L., Slutzky, G. M., de Ilbarra, A. A. L. and Snary, D. (1983) J. Clin. Microbiol. **18**, 191–193

Greis, K. D., Turco, S. J., Thomas, J. R., McConville, M. J., Homans, S. W. and Ferguson, M. A. J. (1992) J. Biol. Chem. **267**, 5876–5881

Gurnett, A., Dulski, P., Hsu, J. and Turner, M. J. (1990) Mol. Biochem. Parasitol. **41**, 177–186

Güther, M. L. S., Cardoso de Almeida, M. L., Yoshida, N. and Ferguson, M. A. J. (1992) J. Biol. Chem. **267**, 6820–6828

Haldar, K., Ferguson, M. A. J. and Cross, G. A. M. (1985) J. Biol. Chem. **260**, 4969–4974

Haldar, K., Henderson, C. L. and Cross, G. A. M. (1986) Proc. Natl. Acad. Sci. U.S.A. **83**, 8565–8569

Hall, B. F., Webster, P., Ma, A. K., Joiner, K. A. and Andrews, N. W. (1992) J. Exp. Med. **176**, 313–325

Handman, E. and Goding, J. W. (1985) EMBO J. **4**, 329–336

Handman, E., Greenblatt, C. L. and Goding, J. W. (1984) EMBO J. **3**, 2301–2306

Handman, E., Schnur, L. F., Spithill, T. W. and Mitchell, G. F. (1986) J. Immunol. **137**, 3608–3613

Handman, E., Mitchell, G. F. and Goding, J. W. (1987) Mol. Biochem. Med. **4**, 377–383

Hannan, L. A., Lisanti, M. P., Rodriguez–Boulan, E. and Edidin, M. (1993) J. Cell Biol. **120**, 353–358

Haynes, P. A., Ferguson, M. A. J., Gooley, A. A., Redmond, J. W. and Williams, K. L. (1993) Eur. J. Biochem., in the press

Hereld, D., Hart, G. W. and Englund, P. T. (1986) Proc. Natl. Acad. Sci. U.S.A. **85**, 8914–8918

Hernandez-Munain C., Fernandez, M. A., Alcina, A. and Fresno, M. (1991) Infect. Immun. **59**, 1409–1416

Hirose, S., Prince, G. M., Sevlever, D., Lakshmeswari, R., Rosenberry, T. L., Ueda, E. and Medof, M. E. (1992) J. Biol. Chem. **267**, 16968–16974

Homans, S. W., Ferguson, M. A. J., Dwek, R. A., Rademacher, T. W., Anand, R. and Williams, A. F. (1988) Nature (London) **333**, 269–272

Homans, S. W., Edge, C. J., Ferguson, M. A. J., Dwek, R. A. and Rademacher, T. W. (1989) Biochemistry **28**, 2881–2887

Homans, S. W., Melhert, A. and Turco, S. J. (1992) Biochemistry **31**, 654–661

Hooper, N. M. (1992) Curr. Biol. **2**, 617–619

Hooper, N. M. and Turner, A. J. (1988) Biochem. J. **250**, 865–869

Hyman, R. (1988) Trends Genet. **4**, 5–8

Ilg, T., Etges, R., Overath, P., McConville, M. J., Thomas-Oates, J. E., Homans, S. W. and Ferguson, M. A. J. (1992) J. Biol. Chem. **267**, 6834–6840

Inverso, J. A., Medina-Acosta, E., O'Connor, J., Russell, D. G. and Cross, G. A. M. (1993) Mol. Biochem. Parasitol. **57**, 47–54

Kamitani, T., Menon, A. K., Hallaq, Y., Warren, C. D. and Yeh, E. T. H. (1992) J. Biol. Chem. **267**, 24611–24619

Karp, C. L., Turco, S. J. and Sacks, D. L. (1991) J. Immunol. **147**, 680–684

Kelleher, M., Bacic, A. and Handman, E. (1992) Proc. Natl. Acad. Sci. U.S.A. **89**, 6–10

Keller, G.-A., Siegel, M. W. and Caras, I. W. (1992) EMBO J. **11**, 865–874

King, D. L., Chang, Y. D. and Turco, S. J. (1987) Mol. Biochem. Parasitol. **24**, 47–53

Kodukula, K., Cines, D., Amthauer, R., Gerber, L. and Udenfriend, S. (1992) Proc. Natl. Acad. Sci. U.S.A. **89**, 1350–1353

Krakow, J. L., Doering, T. L., Masterson, W. J., Hart, G. W. and Englund, P. T. (1989) Mol. Biochem. Parasitol. **36**, 263–270

Lake, J. A., de la Cruz, V. F., Ferreira, P. C. G., Morel, C. and Simpson, L. (1988) Proc. Natl. Acad. Sci. U.S.A. **85**, 4779–4783

Lamont, G. S., Fox, J. A. and Cross, G. A. M. (1987) Mol. Biochem. Parasitol. **24**, 131–136

Lang, T., Warburg, A., Sacks, D. L., Croft, S. L., Laine, R. P. and Blackwell, J. M. (1991) Eur. J. Cell Biol. **55**, 362–372

Lederkremer, R. M., Alves, M. J. M,. Fonseca, G. C. and Colli, W. (1976) Biochim. Biophys. Acta **444**, 85–96

Lederkremer, R. M., Lima, C., Ramirez, M. I. and Casal, O. L. (1990) Eur. J. Biochem. **192**, 337–345

Lederkremer, R. M., Lima, C., Ramirez, M. I., Ferguson, M. A. J., Homans, S. W. and Thomas-Oates, J. E. (1991) J. Biol. Chem. **265**, 19611–19623

Lee, H.-C., Shoda, R., Krall, J. A., Foster, J. D., Selhub, J. and Rosenberry, T. L. (1992) Biochemistry **31**, 3236–3243

Lehuen, A., Monteiro, R. C. and Kearney, J. F. (1992) Eur. J. Immunol. **22**, 2373–2380

Lemansky, P., Fatemi, S. H., Gorican, B., Meyale, S., Rossero, R. and Tartakoff, A. M. (1990) J. Cell Biol. **110**, 1525–1531

Lisanti, M. P., Caras, I. W., Gilbert, T., Hanzel, D. and Rodriguez-Boulan, E. (1990) Proc. Natl. Acad. Sci. U.S.A. **87**, 7419–7423

Lisanti, M. P., Field, M. C., Caras, I W., Menon, A. K. and Rodriguez-Boulan, E. (1991) EMBO J. **10**, 1969–1977

Lohman, K. L., Langer, P. J., and McMahon-Pratt, D. (1990) Proc. Natl. Acad. Sci. U.S.A. **87**, 8393–8397

Low, M. G. (1989) Biochim. Biophys. Acta **988**, 427–454

Luhrs, C. A. and Slomiany, B. L. (1989) J. Biol. Chem. **264**, 21446–21449

Lund-Johansen, F., Olweus, J., Horejsi, V., Skubitz, K. M., Thompson, J. S., Vilella, R. and Symington, F. W. (1992) J. Immunol. **148**, 3221–3229

Masco, D., Van de Walle, M. and Spiegel, S. (1991) J. Neurosci. **11**, 2443–2452

Masterson, W. J. and Ferguson, M. A. J. (1991) EMBO J. **10**, 2041–2045

Masterson, W. J., Doering, T. L., Hart, G. W. and Englund, P. T. (1989) Cell **56**, 793–800

Masterson, W. J., Raper, J., Doering, T. L., Hart, G. W. and Englund, P. T. (1990) Cell **62**, 73–80

Mayor, A., Menon, A. K. and Cross, G. A. M. (1990a) J. Biol. Chem. **265**, 6174–6181

Mayor, S., Menon, A. K., Cross, G. A. M., Ferguson, M. A. J., Dwek, R. A. and Rademacher, T. (1990b) J. Biol. Chem. **265**, 6164–6173

Mayor, S., Menon, A. K. and Cross, G. A. M. (1991) J. Cell Biol. **114**, 61–71

Mayor, S., Menon, A. K. and Cross, G. A. M. (1992) J. Biol. Chem. **267**, 754–761

McConville, M. J. (1991) Cell Biol. Int. Rep. **15**, 779–798

McConville, M. J. and Bacic, A. (1989) J. Biol. Chem. **264**, 757–766

McConville, M. J. and Bacic, A. (1990) Mol. Biochem. Parasitol. **38**, 57–68

McConville, M. J. and Blackwell, J. M. (1991) J. Biol. Chem. **266**, 15170–15179

McConville, M. J. and Homans, S. W. (1992) J. Biol. Chem. **267**, 5855–5861

McConville, M. J., Bacic, A., Mitchell, G. F. and Handman, E. (1987) Proc. Natl. Acad. Sci. U.S.A. **84**, 8941–8945

McConville, M. J., Thomas-Oates, J. E., Ferguson, M. A. J. and Homans, S. W. (1990a) J. Biol. Chem. **265**, 19611–19623

McConville, M. J., Homans, S. W., Thomas-Oates, J. E., Dell, A. and Bacic, A. (1990b) J. Biol. Chem. **265**, 7385–7394

McConville, M. J., Turco, S. J., Ferguson, M. A. J. and Sacks, D. L. (1992) EMBO J. **11**, 3593–3600

McConville, M. J., Collidge, T., Ferguson, M. A. J. and Schneider, P. (1993) J. Biol. Chem., in the press

McNeely, T. B. and Turco, S. J. (1987) Biochem. Biophys. Res. Commun. **148**, 653–657

McNeely, T. B. and Turco, S. J. (1990) J. Immunol. **144**, 2745–2750

McNeely, T. B., Rosen, G., Londner, M. V. and Turco, S. J. (1989) Biochem. J. **259**, 601–604

Medina-Acosta, E., Karess, R. E., Schwartz, H. and Russell, D. G. (1989) Mol. Biochem. Parasitol. **38**, 263–273

Medina-Acosta, E., Karess, R. E. and Russell, D. G. (1993) Mol. Biochem. Parasitol. **57**, 31–45

Mehlert, A., Silman, I., Homans, S. W. and Ferguson, M. A. J. (1992) Biochem. Soc. Trans. **21**, 43S

Menon, A. K. (1991) Cell Biol. Int. Rep. **15**, 1007–1021

Menon, A. K. and Stevens, V. L. (1992) J. Biol. Chem. **267**, 15277–15280

Menon, A. K., Mayor, S. and Schwarz, R. T. (1990a) EMBO J. **9**, 4249–4258

Menon, A. K., Schwarz, R. T., Mayor, S. and Cross, G. A. M. (1990b) J. Biol. Chem. **265**, 9033–9042

Menon, A. K., Eppinger, M., Mayor, S. and Schwarz, R. T. (1993) EMBO J. **12**, 1907–1914

Mensa-Wilmot, K. and Englund, P. T. (1992) Mol. Biochem. Parasitol. **56**, 311–322

Metcalf, P., Blum, M., Freymann, D., Turner, M. and Wiley, D. C. (1987) Nature (London) **325**, 84–86

Milne, K. G., Ferguson, M. A. J. and Masterson, W. J. (1992) Eur. J. Biochem. **208**, 309–314

Milne, K. G., Field, R. A. and Ferguson, M. A. J. (1993) in New Developments in lipid–protein interactions and receptor function (Gustafsson, J. A. and Wirtz, K. W. A., eds.), NATO/ASI Series, Plenum, London, in the press

Miyata, T., Takeda, J., Iida, Y., Yamamda, N., Inoue, N., Takahashi, M., Maeda, K., Katani, T. and Kinoshita, T. (1993) Science **259**, 1318–1320

Moody, S., Handman, E., McConville, M. J. and Bacic, A. (1993) J. Biol. Chem., in the press

Mosser, D. M., Springer, T. A. and Diamond, M. S. (1992) J. Cell Biol. **116**, 511–520

Murray, P. J., Spithill, T. W. and Handman, E. (1989) J. Immunol. **143**, 4221–4226

Nagel, S. D. and Boothroyd, J. C. (1989) J. Biol. Chem. **264**, 5569–5574

Nishihara, M., Utagawa, M., Akutsu, H. and Koga, Y. (1992) J. Biol. Chem. **267**, 12432–12435

Orlandi, P. A., Jr. and Turco, S. J. (1987) J. Biol. Chem. **262**, 10384–10391

Ozaki, L. S., Svec, P., Nussenzweig, V. and Godson G. N. (1983) Cell **34**, 815–822

Pan, Y.-T., Kamitani, T., Bhuvaneswaran, C., Hallaq, Y., Warren, C. D., Yeh, E. T. H. and Elbein, A. D. (1992) J. Biol. Chem. **267**, 21250–21255

Paul, E., Leblond, F. A. and LeBel, D. (1991) Biochem. J. **277**, 879–881

Pereira, M. E. A., Mejia, J. S., Ortega-Barria, E., Matzilevich, D. and Prioli, R. P. (1991) J. Exp. Med. **174**, 179–191

Perussia B. and Ravetch, J. V. (1991) Eur. J. Immunol. **21**, 425–429

Pimenta, P. F. P., da Silva, R. P., Sacks, D. L. and da Silva, P. P. (1989) Eur. J. Cell Biol. **48**, 180–190

Pimenta, P. F., Saraiva, E. M. B. and Sacks, D. L. (1991) Exp. Parasitol. **72**, 191–204

Pimenta, P. F. P., Turco, S. J., McConville, M. J., Lawyer, P. G., Perkins, P. V. and Sacks, D. L. (1992) Science **256**, 1812–1815

Pingel, S. and Duszenko, M. (1992) Biochem. J. **283**, 479–485

Pollevick, G. D., Affranchino, J. L., Frasch, A. C. C. and Sanchez, D. O. (1991) Mol. Biochem. Parasitol. **47**, 247–250

Pontes de Carvalho, L. C., Tomlinson, S., Vandekerckhove, F., Bienen, E. J., Clarkson, A. B., Jiang, M.-S., Hart, G. W. and Nussenzweig, V. (1993) J. Exp. Med. **177**, 465–474

Previato, J. O., Gorin, P. A., Mazurek, M., Xavier, M. T., Fournet, B., Wieruszesk, J. M. and Mendonca-Previato, L. (1990) J. Biol Chem. **265**, 2518–2526

Previato, J. O., Mendonca-Previato, L., Jones, C., Wait, R. and Fournet, B. (1992) J. Biol. Chem. **267**, 24279–24286

Previato, J. O., Mendonca-Previato, L., Jones, C. and Wait, R. (1993) in Structure, Function and Synthesis of Glycoconjugates, Royal Society of Chemistry/Biochemical Society Joint Symposium, Dundee (abstr.)

Puentes, S. M., Sacks, D. L., Da Silva, R. P. and Joiner, K. (1988) J. Exp. Med. **167**, 887–902

Puentes, S. M., da Silva, R. P., Sacks, D. L., Hammer, C. H. and Joiner, K. A. (1990) J. Immunol. **145**, 4311–4316

Puoti, A. and Conzelmann, A. (1992) J. Biol. Chem. **267**, 22673–22680

Puoti, A. and Conzelmann, A. (1993) J. Biol. Chem. **268**, 7215–7224

Puoti, A., Desponds, C., Fankhauser, C. and Conzelmann, A. (1991) J. Biol. Chem. **266**, 21051–21059

Ralton, J. E., Milne, K. G., Güther, M. L. S., Field, R. A. and Ferguson, M. A. J. (1993) J. Biol. Chem., in the press

Ramamoorthy, R., Donelson, J. E., Paetz, K. E., Maybodi, M., Roberts, S. C. and Wilson, M. E. (1992) J. Biol. Chem. **267**, 1888–1895

Richier, P., Arpagaus, M. and Toutant, J.-P. (1992) Biochim. Biophys. Acta **1112**, 83–88

Roberts, W. L., Myer, J. J., Kuksis, A., Low, M. G. and Rosenberry, T. L. (1989a) J. Biol. Chem. **263**, 18766–18775

Roberts, W. L., Santikarn, S., Reinhold, V. N. and Rosenberry, T. L. (1989b) J. Biol. Chem. **263**, 18776–18784

Robinson, P. J. (1991) Immunol. Today **12**, 35–41

Roditi, I., Schwarz, H., Pearson, T. W., Beecroft, R. P., Liu, M. K., Richardson, J. P., Bühring, H.-J., Pleiss, J., Bülow, R., Williams, R. O. and Overath, P. (1989) J. Cell Biol. **108**, 737–746

Rodriguez-Boulan, E. and Powell, S. K. (1992) Annu. Rev. Cell Biol. **8**, 395–427

Rosen, G., Londner, M. V., Sevlever, D. and Greenblatt, C. L. (1988) Mol. Biochem. Parasitol. **27**, 93–100

Rosen, G., Pahlsson, P., Londner, M. V., Westerman, M. E. and Nilsson, B. (1989) J. Biol. Chem. **264**, 10457–10463 (correction in **265**, 7708)

Rosenberg, J., Prioli, R. P., Ortega-Barria, E. and Pereira, M. E. A. (1991) Mol. Biochem. Parasitol. **46**, 303–305

Ross, C. A., Cardoso de Almeida, M. L. and Turner, M. J. (1987) Mol. Biochem. Parasitol. **22**, 153–158

Rothberg, K. G., Ying, Y.-S., Kamen, B. A. and Anderson, R. G. W. (1990) J. Cell Biol., **111**, 2931–2938

Rothberg, K. G., Heuser, J. E., Donzell, W. C., Ying, Y., Glenney, J. R. and Anderson, R. G. W. (1992) Cell **68**, 673–682

Sacks, D. L. (1989) Exp. Parasitol. **69**, 100–103

Sacks, D. L., Brodin, T. N. and Turco, S. J. (1990) Mol. Biochem. Parasitol. **42**, 225–234

Salmon, J. E., Brogle, N. L., Edberg, J. C. and Kimberly, R. P. (1991) J. Immunol. **146**, 997–1004

Schell, D., Evers, R., Preis, D., Ziegelbauer, K., Kiefer, H., Lottspeich, F., Cornelissen, A. W. C. A. and Overath, P. (1991) EMBO J. **10**, 1061–1066

Schenkman, S., Yoshida, N. and Cardoso de Almeida, M. L. (1988) Mol. Biochem. Parasitol. **29**, 141–152

Schenkman. S., Pontes de Carvalho, L. and Nussenzweig, V. (1992) J. Exp. Med. **175**, 567–575

Schenkman, S., Ferguson, M. A. J., Heise, N., Cardoso de Almeida, M. L. and Mortara, R. A. (1993) Mol. Biochem. Parasitol. **59**, 293–304

Schlein, Y., Schur, L. F. and Jacobson, R. L. (1990) Trans. R. Soc. Trop. Med. Hyg. **84**, 353–355

Schneider, P. and Glaser, T. A. (1993) Mol. Biochem. Parasitol. **58**, 277–282

Schneider, P., Ferguson, M. A. J., McConville, M. J., Mehlert, A., Homans, S. W. and Bordier, C. (1990) J. Biol. Chem. **265**, 16955–16964

Schneider, P., Rosat, J.-P., Bouvier, J., Louis, J. and Bordier, C. (1992) Exp. Parasitol. **75**, 196–206

Schnur, L. F. (1982) World Health Organization/TDR/WGS/IMMLEISH, pp. 25–45, WHO, Geneva

Schofield, L. and Hackett, F. (1993) J. Exp. Med. **177**, 145–153

Sevlever, D., Påhlsson, P., Rosen, G., Nilsson, B. and Londner, M. V. (1991) Glycoconjugate J. **8**, 321–329

Seyfang, A., Mecke, D. and Duszenko, M. (1990) J. Protozool. **37**, 546–552

Shapiro, S. Z. and Webster, P. (1989) J. Protozool. **36**, 344–349

Singh, N., Singleton, D. and Tartakoff, A. M. (1991) Mol. Cell. Biol. **11**, 2362–2374

Smythe, J. A., Coppel, R. L., Brown, G. V., Ramasamy, R., Kemp, D. J. and Anders, R. F. (1988) Proc. Natl. Acad. Sci. U.S.A. **85**, 5195–5199

Soares, M. J., Souto-Padron, T. and de Souza, W. (1992) J. Cell Sci. **102**, 157–167

Sogin, M. L., Gunderson, J. H., Elwood, H. J., Alonso, R. A. and Peattie, D. A. (1989) Science **243**, 75–77

Stadler, J., Keenan, T. W., Bauer, G. and Gerisch, G. (1989) EMBO J. **8**, 371–377

Stahl, N., Baldwin, M. A., Hecker, R., Pan, K.-M., Burlingame, A. L. and Prusiner, S. B. (1992) Biochemistry **31**, 5043–5053

Stanley, S. L., Jr., Huizenga, H. and Li, E. (1992) Mol. Biochem. Parasitol. **50**, 127–138

Stefanova, I., Horejsi, V., Ansotegui, I. J., Knapp, W. and Stockinger, H. (1991) Science **254**, 1016–1019

Stevens, V. L. and Raetz, C. R. H. (1991) J. Biol. Chem. **266**, 10039–10042

Takahashi, M., Takeda, J., Hirose, S., Hyman, R., Inoue, N., Miyata, T., Ueda, E., Kitani, T., Medof, M. E. and Kinoshita, T. (1993) J. Exp. Med. **177**, 517–521

Takle, G. D. and Cross, G. A. M. (1991) Mol. Biochem. Parasitol. **48**, 185–198

Talamas-Rohana, P., Wright, S. D., Lennartz, M. R. and Russell, D. G. (1990) J. Immunol. **144**, 4817–4824

Tartakoff, A. M. and Singh, N. (1992) Trends Biochem. Sci. **17**, 470–473

Taverne, J., Bate, C. A. W. and Playfair, J. H. L. (1990) Immunol. Lett. **25**, 207–212

Thomas, J. R., Dwek, R. A. and Rademacher, T. W. (1990) Biochemistry **29**, 5413–5422

Thomas, J. R., McConville, M. J., Thomas-Oates, J., Homans, S. W., Ferguson, M. A. J., Greis, K. and Turco, S. J. (1991) J. Biol. Chem. **267**, 6829–6833

Thomas, P. M. and Samuelson, L. E. (1992) J. Biol. Chem. **267**, 12317–12322

Tolsen, D. L., Turco, S. J., Beecroft, R. P. and Pearson, T. W. (1989) Mol. Biochem. Parasitol. **35**, 109–118

Tomavo, S., Dubremetz, J.-F. and Schwarz, R. T. (1992a) J. Biol. Chem. **267**, 11721–11728

Tomavo, S., Dubremetz, J.-F. and Schwarz, R. T. (1992b) J. Biol. Chem. **267**, 21446–21458

Tomlinson, S., Pontes de Carvaldo, L., Vandekerckhove, F. and Nussenzweig, V. (1992) Glycobiology **2**, 549–551

Toutant, J.-P., Richards, M. K., Drall, J. A. and Rosenberry, T. L. (1990) Eur. J. Biochem. **187**, 31–38

Turco, S. J. and Descoteaux, A. (1992) Annu. Rev. Microbiol. **46**, 65–94

Turco, S. J. and Sacks, D. L. (1991) Mol. Biochem. Parasitol. **45**, 91–95

Turco, S. J., Hull, S. R., Orlandi, P. A., Jr. and Shepherd, S. D. (1987) Biochemistry **26**, 6233–6238

Turco, S. J., Orlandi, P. A., Jr., Homans, S. W., Ferguson, M. A. J., Dwek, R. A. and Rademacher, T. W. (1989) J. Biol. Chem. **264**, 6711–6715

Udenfriend, S., Micanovic, R. and Kodukula, K. (1991) Cell Biol. Int. Rep. **15**, 739–759

Urakaze, M., Kamitani, T., DeGasperi, R., Sugiyama, E., Chang, H.-M., Warren, C. D. and Yeh, E. T. H. (1992) J. Biol. Chem. **267**, 6459–6462

van Voorhis, W. C., Schlekewy, L. and Le Trong, H. (1991) Proc. Natl. Acad. Sci. U.S.A. **88**, 5993–5997

Vidugiriene, J. and Menon, A. K. (1993) J. Cell Biol., **121**, 987–996

Vogel, M., Kowalewski, H., Zimmermann, H., Hooper, N. M. and Turner, A. J. (1992) Biochem. J. **284**, 621–624

Walter, E. I., Roberts, W. L., Rosenberry, T. L., Ratnoff, W. D. and Medof, M. E. (1990) J. Immunol. **144**, 1030–1036

Webster, P. and Grab, D. J. (1988) J. Cell Biol. **106**, 279–288

Webster, P. Russo, D. C. W. and Black, S. J. (1990) J. Cell Sci. **96**, 249–255

Weinhart, U., Thomas, J. R., Pak, Y., Thompson, G. A. and Ferguson, M. A. J. (1991) Biochem. J. **279**, 605–608

Ying, Y., Anderson, R. G. W. and Rothberg, K. G. (1992) Cold Spring Harbor Symp., in the press

Zingales, B., Martin, N. F., de Lederkremer, R. M. and Colli, W. (1982) FEBS Lett. **142**, 238–242

Biochem. J. (1993) **282**, 657–671 (Printed in Great Britain)

REVIEW ARTICLE
Molecular genetics of actin function

Emma S. HENNESSEY,* Douglas R. DRUMMOND† and John C. SPARROW‡
Department of Biology, University of York, York YO1 5DD, U. K.

INTRODUCTION

Actin is found in all eukaryotes and plays a fundamental role in many diverse and dynamic cellular processes. Despite being so ubiquitous, actin isoforms exhibit an unusually high degree of amino acid sequence similarity. This poses the question of how actin can fulfil such diverse functions in different organisms, yet retain such a high degree of sequence conservation? In this Review we examine the impact molecular genetics has made on our understanding of the relationship between actin amino acid sequence and function. We refer the reader to other reviews for more details of actin structure, biochemistry and cell biology than space allows us to include [1– 10].

Actin is a globular protein (G-actin) which polymerizes into filaments (F-actin) [6] for most of its biological functions [7]. F-actin is a major component of the cytoskeleton [3]. The control of F-actin assembly and turnover is important for many cellular processes such as motility, morphological changes, cell division and intracellular movements. Many proteins control actin polymerization and depolymerization within cells. These include proteins which sever or cap actin filaments, nucleate polymerization, cross-link and stabilise F-actin (reviewed in [3–5]. Others bind G-actin, maintaining up to 50 % of cytoplasmic actin as monomers [8] and can respond to extracellular signals by releasing the G-actin which then polymerizes [11]. Cytoskeletal actin may localize some metabolic pathways within the cell (reviewed in [12]). In addition to roles in non-muscle cell motility and contractile processes, such as cell division (reviewed in [9,10]), actin is a major component of myofibrils. In myofibrils, F-actin is assembled into the thin filaments which interact with the myosin heads of the thick filaments to produce contraction (for reviews see [13,14]).

Molecular genetic techniques can be used to study the relationships between a protein's functions and its amino acid sequence by examining the effects of sequence variation. Gene cloning and sequencing and, more recently, the polymerase chain reaction (PCR) have accelerated the accumulation of new actin sequences in the databases. These naturally occurring isoforms can be used to assess the significance of amino acid sequence variation for functional differences between actins. Information on the importance of specific amino acids in actin can also be obtained from induced 'man-made' mutations recovered from genetic studies. Actin mutations have been described in the yeast *Saccharomyces cerevisiae* [15,16], the fruitfly *Drosophila melanogaster* [17–22], the nematode *Caenorhabditis elegans* [23,24] and cultured human cells [25,26]. Mutations made in cloned actin genes have been studied after expression either by transcription and translation *in vitro* [27–36] or *in vivo* following transformation

of suitable organisms such as *S. cerevisiae* [37–43], *Dictyostelium discoideum* [44], *D. melanogaster* [20,33,45–49], or cultured cell lines [25,26,28,50].

A most significant development has been the determination of the atomic structure of actin from crystals of an actin–DNase I complex [51; reviewed in 1,2]. Atomic structures of complexes with the G-actin binding protein, profilin [52], the muscle protein, myosin [53] and gelsolin fragment-1 [54] will soon provide information on these physiologically important actin–ligand interactions. Most existing mutants were made without the benefit of the atomic structure of actin. It is therefore timely to consider the effects of actin mutants with reference to this atomic structure.

COMPARISON OF NATURALLY OCCURRING ACTINS

Conservation of actin sequences

Comparison of protein sequences can identify conserved regions and reveal evolutionary relationships. Sequence conservation implies evolutionary constraint, often identifying regions of functional importance. Actin sequences have been reviewed [55,56] and are much more conserved than other proteins, such as globins, found in the same species [56].

In an alignment of the 81 unique actin protein sequences taken from the recent OWL database version 15.1 [57], 17.4 % of positions in the consensus sequence are invariant in all actins, 35.6 % of positions are invariant or have a difference in only one of the 81 actins, and 40.4 % of positions are invariant or have what are considered conservative changes (D. R. Drummond, unpublished work). For most positions the amino acids are identical in the majority of actin sequences and at least 95 % of actins have the same amino acid at 62.9 % of positions. The most divergent sequences (44.9 %) are those from the protozoan *Euplotes crasus* and carrot actin 2. Skeletal muscle α-actins in human, mouse, rat, rabbit and chicken are identical, as are the cytoplasmic β-actins in human, mouse, rat, cow and chicken.

The high conservation of actin sequences means that, unlike other proteins, one cannot identify important sites as isolated regions with the most conserved sequences. Rather it suggests constraints throughout the actin structure. Thus regions of sequence variation may identify sequences where the constraints are reduced or which are important for isoform-specific functions.

Actin phylogenetic trees ([55,56,58,59]; Figure 1) produce relatively few groupings which can be considered significant [56,59], a consequence of the high sequence conservation of actins. The major divisions correspond to conventional taxonomic classifications. For example, plants and animals form distinct groups.

---

* Present address: Microbial Genetics Group, School of Biological Sciences, University of Sussex, Falmer, Brighton BN1 9QG, U. K.
† Present address: CRC Medical Oncology Unit, Southampton General Hospital, Southampton SO9 4XY, U. K.
‡ To whom correspondence and reprint requests should be addressed. The actin amino acid sequence alignment mentioned in the text as (D. R. Drummond, unpublished work), details of the sequences used, references etc. will be provided on request by J. C. S. Please make requests either by fax (to +44 904 432860) or by e-mail to JCS1@uk.ac.york.vaxa.

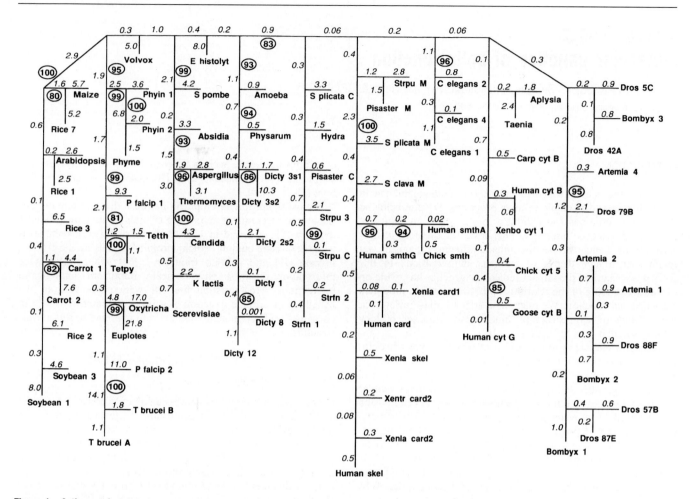

Figure 1   Actin protein tree

A phylogenetic tree of currently available actin protein sequences from the OWL database version 15.1 [57] was produced with the program Clustal V [191] which uses the neighbour-joining method of Saitou and Nei [192]. Actin names are described below. The percentage divergence is marked in italics on the branches. The reproducibility of the tree was determined using 1000 bootstrap replicas. The percentage of replicas containing a given branch is circled next to the branch. Only values greater than 80% are reported. The tree is unrooted but, as is common practice, the root can be placed on the midpoint of the longest branch, which is to *Euplotes*. Key: skel, skeletal; card, cardiac; smth, smooth; cyt, cytoskeletal. Xenla, *Xenopus laevis* (African clawed frog); Xentr, *X. tropicalis* (Western clawed frog); Xenbo, *X. borealis* (Kenyan clawed frog); S clava, *Styela clava*; S plicata, *Styela plicata* (sea squirts); *Artemia* sp. (brine shrimp); C elegans, *Caenorhabditis elegans* (nematode); Hydra, *Hydra attenuata*; Dros, *Drosophila melanogaster* (fruit fly); Bombyx, *Bombyx mori* (silk worm); Aplysia, *Aplysia californica* (California sea hare); Arabidopsis, *Arabidopsis thaliana* (mouse-ear cress); Pisaster, *Pisaster ochraceus* (sea star); Amoeba, *Acanthamoeba castellanii* (amoeba); E histolyt, *Entamoeba histolytica*; Strpu, *Strongylocentrotus purpuratus* (purple sea urchin); Strfn, *S. franciscanus* (sea urchin); Taenia, *Taenia solium*; Dicty, *Dictyostelium discoideum* (slime mould); Physarum, *Physarum polycephalum* (slime mould); S pombe, *Schizosaccharomyces pombe* (fission yeast); S cerevisiae, *Saccharomyces cerevisiae* (baker's yeast) and *S. carlsbergensis* (lager beer yeast); Absidia, *Absidia glauca* (zygomycete); Aspergillus, *Aspergillus nidulans*; Thermomyces, *Thermomyces lanuginosa* (*Humicola lanuginosa*); K lactis, *Kluyveromyces lactis* (yeast); Phyin, *Phytophthora infestans* (potato late blight fungus); Phyme, *P. megasperma* (potato pink rot fungus); Candida, *Candida albicans* (yeast); soybean, *Glycine max*; P falcip, *Plasmodium falciparum*; Tetth, *Tetrahymena thermophila*; Tetpy, *T. pyriformis*; T brucei, *Trypanosoma brucei*; Oxytricha, *Oxytricha nova*; Euplotes, *Euplotes crassus*. Details of sequences and references obtained from the OWL database [57] will be provided on request (see footnote). Actins listed as human are common to the following species; skel: mouse, rat, rabbit and chicken; card: mouse, chicken; smth A: mouse, rat, cow; smth G: mouse, rat; cyt G: mouse, rat, cow; cyt B: mouse, rat, cow, chicken.

Most species have several actin genes [60–71] with the largest numbers occurring in plants [68,69]. These multigene families encode different isoforms but in *Physarum polycephalum* [60], the protozoan *Taenia solium* [80] and *Caenorhabditis elegans* [71] multiple actin genes encode the same amino acid sequence. A number of lower eukaryotes, e.g. *Aspergillus nidulans* [72], *Saccharomyces cerevisiae* [73–75], and the protozoan *Tetrahymena thermophila* [76] have only a single actin gene, though some also contain genes for 'actin-like' proteins (see below). In metazoans actin multigene families have been maintained throughout evolution with a relatively low rate of change compared to many other proteins. Frequently members of actin gene families are more conserved between than within species. For example, rice actin 1 is more similar to *Arabidopsis* actin than any other rice actin [77]. The two carrot actins (dicot) are

closer to maize (monocot) than to soybean (dicot) [78], suggesting that the formation of different actin gene families predated the divergence of monocots and dicots [55,56]. Especially when actins from different species are more similar than within a multigene family the possibility of isoform-specific functional differences must be considered.

Muscle actins from several chordate species form a group distinct from their cytoplasmic actins, suggesting that the emergence of separate muscle and cytoplasmic actins predated chordate speciation [58]. Indeed, vertebrate cytoplasmic actins are more similar to arthropod actins than to vertebrate muscle actins [55], indicating a very ancient common origin for actin genes [59]. Within arthropods, *Drosophila melanogaster* (Diptera) and *Bombyx mori* (Lepidoptera) muscle actins form a group distinct from their cytoplasmic actins [59]. *Artemia* (crustacean) actins

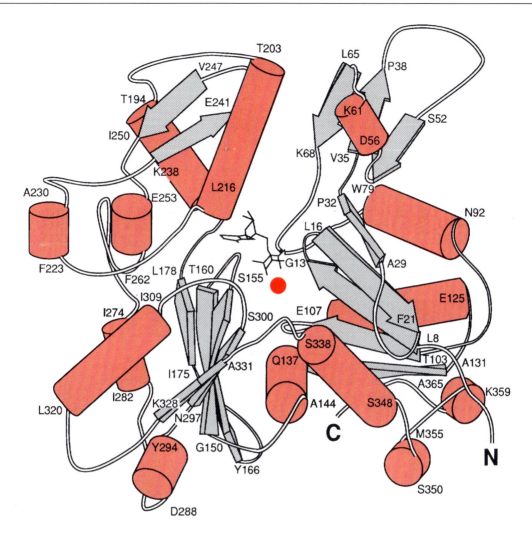

Figure 2   Ribbon and cylinder diagram of the actin structure, redrawn from [51]

The numbers indicate residues at either end of α-helices (cylinders) or β-sheet strands (arrows). A molecule of ATP is indicated by the wire diagram and the metal atom as a sphere. The orientation is the same as that shown in Figures 3(**a**) and 3(**b**). Note: the loop formed by amino acids 241–247 should be in front of the cylinder (α-helix) formed by amino acids 203–216.

divide between these two groups and it will be interesting to determine if these actins are muscle or cytoplasmic specific. Conserved amino acids that define different groups have been proposed for vertebrates [58], insect muscle [59] and plants [79]. The functional significance of these amino acids remains to be tested.

Intriguingly, the amino acids at which mammalian skeletal muscle actins differ from β- and γ-actins occur in three-dimensional clusters in the atomic structure [2]. The largest cluster, involving amino acids 1–6, 103, 129, 357 and 354, occurs in the lower right hand corner of subdomain 1 (Figures 2, 3a and 3b, coloured red). This region contains the proposed binding site of many actin binding proteins, including myosin [1,2]. Variations in this region may be adaptations to optimize the binding of different actin-binding proteins. Other clusters occur in the central β-sheet of subdomain 3 (residues 153, 162, 176 and 298) and in the region between subdomains 3 and 4 (residues 225, 259, 266 and 271). Subdomains 2 and 4 are conserved [2]. Of the nine amino acid substitutions which define the insect muscle group of actins, only four occur at positions included in those which define the mammalian skeletal muscle actins [59] and only Ile-76

is conserved in both groups. The insect muscle-specific substitutions do not cluster in the actin structure.

Although most actins are highly conserved in sequence and length (374–376 amino acids), proteins with more divergent sequences, known collectively as 'actin-like' proteins, have been identified recently in several species [80–83]. In addition to its single actin gene, *ACT1*, *Saccharomyces cerevisiae* contains the *ACT2* gene which encodes a protein of 391 amino acids that is 53.6 % divergent from *ACT1* actin [80]; *Schizosaccharomyces pombe* has an *ACT2* protein (427 amino acids) which is 52.6 % diverged from its *ACT1* actin [81]. The two *ACT2* proteins are 64.8 % diverged from each other. Genetic experiments show that *ACT1* cannot substitute for *ACT2*, or vice versa, arguing that both genes are essential for cell viability and have different functions [73,80,81]. Two identical actin-like sequences, with an identity of about 55 % compared to standard actins, but only 40 % identity to the *ACT2* genes of *S. cerevisiae* and *S. pombe*, have been isolated and studied in man [82] and dog [83]. These sequences, named respectively actin-RPV (related protein vertebrate) and centractin, are associated with microtubule based cellular motor systems. Actin-RPV is a component of the dynactin complex [82] and centractin is associated with the centrosomes [83]. Actin-like proteins have also been reported in *Aspergillus*, *Physarum*, *Caenorhabditis* and the prokaryotic cynanobacter *Spirulina platensis* [80]. Actin-like proteins are probably ubiquitous and the evidence so far indicates that they have different functions from standard actins.

In proteins with a conserved core of residues, the structure of one protein is usually a reasonable model of the main structural features of other members of the group [84,85]. Superimposition of the actin-RPV/centractin sequence onto the rabbit actin structure shows that regions corresponding to the actin core are more conserved [83]; similarly, the increased length of the two *ACT2* proteins derives from insertions in positions corresponding to surface loops in the rabbit actin structure [51,80,81]. This contrasts with substitutions in standard actins which occur with equal frequency in surface loops and other parts of the molecule. Like all actins, the 'actin-like' proteins conserve a region corresponding to the ATP binding site [51,80,81].

Comparison of actin isoform function

Functional differences between isoforms have been sought both in intact cells and by biochemical analysis [86]. Actin isoforms segregate into different structures within cells [87–94], suggesting they have specialized functions. Purified isoforms, and actins from different species, can differ in thermodynamic stability [31,95] and polymerization characteristics [32,96–99]. Actins from different species also differ in their affinity for profilin [100,101], DNase I [102–104], α-actinin and tropomyosin [104].

In binding assays of non-muscle actins [96,106–108] or rabbit muscle actin isoforms [91,109–111] with various rabbit muscle myosins, significant differences in affinities were found. The higher affinity of skeletal muscle actin compared to gizzard actin for both gizzard and enteric myosins suggests that the higher affinity is an intrinsic property of the actin, rather than the myosin [112]. In all cases the $V_{max.}$ for the actin-activated myosin ATPase was the same. Only in comparisons of actin and myosin from scallop and rabbit skeletal muscle was a difference in $V_{max.}$ found: homologous systems gave higher $V_{max.}$ values, but no difference in the affinity of each actin for either myosin [113]. Since such comparisons are often conducted in non-physiological conditions and any differences can often be eliminated by varying pH or ionic strength [111], it is unclear if any of these *in vitro* differences are relevant to the situation *in vivo* [111]. Furthermore, proteins which interact with actin will have coevolved [86,113]. Thus functional differences found with heterologous mixtures of actins and other proteins are probing not just the function of actin but also that of the other protein isoforms as well.

The physiological significance of isoform differences can be tested by substituting one actin for another in a whole cell, plant or animal. A mutation in the promoter sequence, which regulates expression, of the mouse cardiac actin gene reduces its expression to 16.5% of its normal level [114]. There is a corresponding increase in expression of the α-skeletal muscle actin gene. Despite containing predominantly the skeletal muscle actin isoform the heart functions normally, suggesting that the two isoforms are functionally equivalent. However, isoform substitution was incomplete and the low levels of cardiac actin may have been sufficient for any isoform-specific functions. Isoform substitution can be achieved by genetic transformation with cloned genes. Although no difference was found in the incorporation of cardiac and cytoplasmic actin isoforms into the cytoskeleton of cultured non-muscle cells [115], transformation of adult rat cardiomyocytes with different actin isoforms showed isoform-specific

effects on both their incorporation into the myofibrils and cell morphology [116]. In *Drosophila melanogaster*, of nine mutations encoding single amino acid substitutions in the *Act88F* actin gene corresponding to isoform-specific amino acid differences, only two of the mutations altered muscle structure [117]. However, two of three mutant *Act88F* gene constructs with multiple isoform-specific amino acid changes produced altered muscle structure, suggesting that as more changes are made, actins become functionally nonequivalent. Undoubtedly substitution of a complete actin isoform is desirable since it will contain multiple actin amino acid changes, which have co-evolved. Despite only 90% similarity the chicken β-actin is able to substitute for the yeast *ACT1* actin without loss of viability, although the cells exhibited an altered morphology, slower growth and temperature-sensitive lethality [118].

Functional isoform differences between some members of actin gene families offers a partial explanation for the evolution of actin multigene families. Although comparisons of isoform sequences and functions are a valuable source of information about actin, we need to look further than the study of naturally occurring actins to unravel the role of individual amino acids in actin function.

Actin processing

Actins are post-translationally modified. Most actins have a methylation of His-73 and modification of the N-terminus by removal of one or two residues followed by acetylation of the new N-terminus [3]. Actins are unusual in the type of amino acids which occur at the N-terminus [119] and the processing enzyme is specific for actin [34].

Methylation of His-73 occurs in all actins examined, except for that from *Naegleria gruberi*, whose actin binds DNase I less well than other actins and has a lower affinity compared to rabbit actin for scallop myosin S1 fragment but the same $V_{max.}$ for the actin-activated myosin ATPase activity [120]. His-73 is not close to the DNase I-binding loop [51], though both are in subdomain 2 (see Figures 2 and 3d). It is therefore unclear if these differences are due to a lack of methylation or to other amino acid differences in *Naegleria* actin. Substitution of His-73 with Arg or Tyr in human skeletal muscle actin expressed *in vitro* had no effect on the ability of the actins to co-polymerize or bind DNase I [28]. However, it is uncertain whether wild-type actin expressed *in vitro* and used as a control was methylated [32,121]. The same mutant actins, transformed into simian COS-1 cells, had no effect on cell growth [28]. No effects of His-73 methylation on actin functions are known.

Inhibition of N-terminal processing *in vitro* can alter actin polymerization [32]. However, yeast cells normally lack part of the N-terminal processing activity [38] and completely blocking all N-terminal processing does not affect the stability of actins expressed in yeast [38]. Non-yeast actins expressed in yeast are not fully processed yet polymerize *in vitro* [38,39]. Actin from *A. castellani* is also unprocessed at its N-terminus *in vivo* [122]. Indeed, although it is frequently inferred from sequence data that all actins can be N-terminally processed, this has been confirmed experimentally to occur *in vivo* in only a few cases.

The *Drosophila* flight muscle-specific *Act88F* actin undergoes two further modifications. First, the *mod*+ gene, which is not an actin gene, encodes a modifier which alters the charge of the actin [45] yet *mod*− flies, which lack this modification, are fully flighted and have a wild-type wingbeat frequency [32]. Second, a proportion of the *Act88F* actin forms a stable 1:1 conjugate with ubiquitin, called arthrin [123]. The roles of these actin modifications are unknown. In general, despite some small changes detected *in vitro*, no clear functions have emerged for post-translational processing.

ARTIFICIALLY INDUCED ACTIN MUTANTS

Mutants obtained by selection for flightlessness or uncoordinated movement have led to the recovery of actin mutants in *Drosophila melanogaster* [17–22] and *Caenorhabditis elegans* [23,24]. Serendipitous actin mutants have also been recovered in a melanoma [124] and transformed tissue culture cells [125]. However, these are inefficient ways to recover new actin mutants. First, mutants of other proteins often produce the same or similar phenotype. Second, severe actin mutations are often lethal, limiting selection of actin mutants to systems where the requirement for a specific isoform is not vital, or where conditional mutations, such as temperature-sensitive alleles, can be obtained. Finally, many cells and organisms express multiple actins so recovery of mutants may be difficult.

Protein engineering approaches remove many of these problems and, by *in vitro* mutagenesis of cloned actin genes, permit any amino acid substitution to be made at any point in actin. The only limits are our ability to formulate specific questions about actin at the level of individual amino acids. After mutagenesis the actin must be expressed. This has not been easy, in part because all eukaryotic cells contain actin leading to problems of separating mutant from endogenous actins. Only recently has the expression and purification of mutant actins in sufficient quantities for *in vitro* studies been achieved (see below).

An advantage of expression for *in vivo* analysis is that the mutant actin is in a normal environment where its ability to assemble into cellular structures and interact with a whole range of actin-binding proteins may be tested. However, it may be difficult to pinpoint which actin interactions are affected. For such studies it is useful if the mutant actin gene is the only one expressed since the expression of other actin isoforms may mask the effects of the mutation.

In the following section we describe the systems used for expression and study of actin mutants *in vivo*. Each has advantages and disadvantages but together they provide a wide choice for investigating different aspects of actin structure and function.

Drosophila melanogaster

The fruitfly genome contains six actin genes; two encode cytoplasmic actins and the remaining four are muscle-specific [62]. Mutations have been recovered only in the *Act88F* actin gene which is expressed specifically in the indirect flight muscles [62,126] and encodes all the muscle actin found in these tissues [see 123,127 for review]. *Act88F* mutations affect flight ability but not the viability of the flies, so the *Act88F* gene provides a unique system for studying the effects of actin mutations on muscle assembly and function.

Apart from *Act88F* mutants recovered as flightless flies following whole fly mutagenesis, others have been made by *in vitro* mutagenesis and inserted into the fly genome by P-element transformation [20,33,45–49]. In flies, the effects of the mutations can be analysed by flight testing, by microscopy of the flight muscles and by mechanical testing of muscle fibres [47,128]. Partially purified *Act88F* mutant actins have been used in biochemical studies [129].

Most *Act88F* mutations show a dominant flightless phenotype. This can be explained by a requirement for two wild-type gene copies for flight. However, there are many examples where this is clearly not the case. For instance, flies containing a mutant gene and two wild-type copies may still be flight defective and show aberrations of the myofibrillar filament lattice [20–22, 33,48]. This is a common feature of actin mutants in other organisms as well as *Drosophila*, and is known as 'antimorphism' or 'dominant negative complementation'. The usual explanation is that the mutant monomers interfere or 'poison' assembly of wild-type monomers into F-actin by retaining some capacity to bind wild-type monomers. Antimorphism thus indicates that the actin mutation has not destroyed all actin binding sites. The importance of actin polymerization for biological function undoubtedly explains the frequent antimorphism of actin mutants (Table 1) in flies and other systems. The incorporation of perhaps only one mutant monomer may be sufficient to 'cap' and prevent growth of an actin filament even in the presence of a large excess of wild-type monomers. The ease of genetically manipulating the gene copy number in organisms such as yeast and *Drosophila* means that antimorphic effects can be identified and studied, providing valuable information on mutant actin function *in vivo*.

Heat-shock proteins, a ubiquitous cellular response to stress and damage to proteins [130,131] are induced constitutively in *Drosophila* by certain *Act88F* mutants [19,20,33,126]. Actin is one of very few proteins for which heat-shock-protein-inducing mutants have been found [132]. Intriguingly, actin has some structural similarity with hsc70 (heat-shock cognate 70kDa protein) [133,134].

Caenorhabditis elegans

This nematode has been extensively used for molecular genetic studies of muscle development and function. Many muscle protein genes were identified by mutations with phenotypic effects of paralysis, twitching or uncoordinated movements [24,135]. The genome contains four actin genes but only regulatory mutants of these genes have been characterized to date [71]. The actin genes have been cloned and with a gene transformation system [136] *C. elegans* may provide another model system for the *in vivo* study of actin mutations.

Dictyostelium discoideum

The genetics of myosin and other cytoskeletal proteins is well developed in this slime mould [137–141]. No actin mutants have been recovered by selecting for abnormal phenotypes, probably because the genome contains at least 17 actin genes [67]. This system has been used to express and purify large quantities of actin following *in vitro* mutagenesis of the *Dictyostelium Act15* gene [44]. Charge changes were introduced into the *Act15* gene by site-directed mutagenesis to facilitate separation of the mutant from the endogenous actins.

Saccharomyces cerevisiae

The yeast *S. cerevisiae* provides a powerful system for the study of actin mutants and cell biology. Many mutations of the single actin gene, *ACT1*, are now known (see Table 1), although none were recovered by their phenotypic effects. *ACT1* mutants made by *in vitro* mutagenesis are easily transformed into the yeast genome and the phenomenon of homologous recombination can be used to disrupt the endogenous wild-type *ACT1* copy while inserting the mutant gene copy into the yeast genome [15,73,142]. This requires that the mutant alleles permit cell survival.

Mutant actins can be obtained from *S. cerevisiae* in sufficient quantities for biochemical studies either by replacing the wild-type *ACT1* copy [38,40] or using yeast expression vectors to express heterologous actins e.g chicken β-actin [37,39] in cells

Table 1 Actin mutants

Table 1 summarizes current actin mutants residue-by-residue from the N- to C- termini. The actin gene in which each mutation occurs, the system used to express it and any mutant effects are given. The mutants are named using the one letter code with the original amino acid first, its number, then the new amino acid; 'Term' denotes a translation stop codon; *in vitro*, expression in rabbit reticulocyte lysate; *ACT1*, yeast actin gene; *Act88F*, *Drosophila* flight-muscle specific gene; *Act15*, an actin gene from *Dictyostelium*, Hs α and Hs β, human α-skeletal and β-actin genes; and Ch β, chicken β-actin gene; pure, actin was purified after expression; NDG, mobility on non-denaturing gels; thermodynamic stability, as determined on urea gradient gels; $=$, as well as wild-type actin; $>$, better than wild-type actin; $<$, less than wild-type actin; $(+)$, normal; $(-)$, abnormal; structure, muscle appearance by light or electron microscopy; V_{max}, for myosin ATPase; K_m, for the binding of actin to myosin; ATPase, myosin ATPase; motility, velocity in the *in vitro* motility assay; mechanics, mechanical response of dissected myofibrils; DNase (DNase I), ATP, S1 (S1 fragment of myosin), profilin or tropomyosin, binding to each ligand relative to wild-type actin; F-actin, production of actin filaments; processing, N-terminal processing; heat, cold sensitive, yeast's inability to grow at high or low temperatures; antimorphic, mutant actin preventing wild-type actin from functioning normally *in vivo*; hypomorphic, produces normal actin but at a reduced level; hsp, induces heat-shock protein synthesis; ADP-ribosylation, actin is a substrate for *Clostridium* toxin ADP-ribosylating activity.

Position	Expression	Gene	Results	Reference
C0D,N,H	*in vitro*	Hs α	$(-)$ Processing	34
C0S,G,F,Y	*in vitro*	Hs α	Not processed	34
del0	*in vitro*	Hs α	$(+)$ Processed	30
D1H, D1H/D4H	*Dictyostelium*, pure	Act15	$<$ Motility; $< V_{max}$; $<$ ATPase	44
del1–12	*in vitro*	Act88F	$>$ Profilin; $=$ ATP; $=$ DNase; zero ADP-ribosylation	35,147
D2N/E4Q	Yeast, pure	ACT1	$<$ ATPase; $=$ polymerization	40
	Yeast, *in vivo*	ACT1	$(+)$ Growth; $=$ processing; $(-)$ F-actin	40
del2–4	Yeast, pure	ACT1	$<$ ATPase; $=$ polymerization	40
	Yeast, *in vivo*	ACT1	$(+)$ Growth; $(-)$ processing; $(-)$ F-actin	40
D2V	Yeast, *in vivo*	ACT1	$(+)$ Growth	142
D2A	Yeast, *in vivo*	ACT1	Recessive cold, heat sensitive; $<$ spore viability	42
D3A,H,N	*in vitro*	Hs α	$=$ Polymerization; $=$ S1; $=$ DNase; $=$ NDG	26
D3N	*in vitro*	Hs α	$(-)$ Processed	34
del3–4, D3K/D4K, D3A/D4A	Yeast, pure	Ch β	$=$ DNase; $=$ polymerization; $(-)$ S1	39
del3–4, D3A/D4A	Yeast, pure	Ch β	$<$ ATPase; $<$ velocity; $=$ tropomyosin	156
D3K/D4K	Yeast, pure	Ch β	$<$ ATPase; no velocity; $=$ tropomyosin	156
E4V,A	Yeast, *in vivo*	ACT1	$(+)$ Growth	42,142
D4E	*Drosophila, in vivo*	Act88F	$(+)$ Structure	117
G6A	*Drosophila, in vivo*	Act88F	$(-)$ Structure	117
G6A/A7T	*Drosophila, in vivo*	Act88F	Flightless; $(-)$ structure	46
I10V	*Drosophila, in vivo*	Act88F	$(+)$ Structure	117
D11E,N, D11N/N12D	*in vitro*	Hs α	$=$ Polymerization; $=$ S1; $<$ DNase; $<$ NDG; $<$ processing	29
D11H	*in vitro*	Hs α	$<$ Polymerization; $=$ S1; $<$ DNase; $<$ NDG; $<$ processing	29
D11Q,K	Yeast, *in vivo*	ACT1	Dominant lethal	142
D11A	Yeast, *in vivo*	ACT1	Partial dominant lethal	42
D11E	Yeast, *in vivo*	ACT1	$(+)$ Growth; viable spores	178
D11N	Yeast, *in vivo*	ACT1	$(+)$ Growth; no viable spores	178
D24H/D25H	*Dictyostelium*, pure	Act15	Zero motility	193
D24A/D25A	Yeast, *in vivo*	ACT1	Recessive cold, heat sensitive	42
R28C	*Drosophila, in vivo*	Act88F	Strong antimorph; weak hsp induction	17,19
R28L	Cells	Mouse A^x	Causes melanoma cell-line	50
P32L	Yeast, *in vivo*	ACT1	Temperature sensitive	15
G36E/E83D	*Drosophila, in vivo*	Act88F	$(-)$ Structure	49
G36E/E83D/G245D	*Drosophila, in vivo*	Act88F	$(+)$ Structure; flighted; hsp induction	49
	Cells	Hs β	$(+)$ Morphology; $<$ incorporation into structures; $<$ DNase	25
R37A/R39A	Yeast, *in vivo*	ACT1	Recessive cold, heat sensitive	42
K50A/D51A	Yeast, *in vivo*	ACT1	Recessive cold, heat sensitive	42
D56A/E57A	Yeast, *in vivo*	ACT1	Recessive heat sensitive; $=$ cold sensitive	42
A58T	Yeast, *in vivo*	ACT1	Temperature sensitive	15
K61A/R62A	Yeast, *in vivo*	ACT1	Partial dominant lethal	42
R68A/E72A	Yeast, *in vivo*	ACT1	$(+)$ Growth	42
H73Y	*in vitro*	Hs α	$<$ Polymerization; $=$ DNase; $=$ processing	28
H73Y,R	Cells	Hs α	$(+)$ Growth	28
H73R	*in vitro*	Hs α	$=$ Polymerization; $=$ DNase; $=$ processing	28
I76F	*Drosophila, in vivo*	Act88F	Strongly antimorphic; weak hsp induction; $(-)$ Z discs	17,19,117
T79Term	*Drosophila, in vivo*	Act88F	No actin accumulation	21
D80A/D81A	Yeast, *in vivo*	ACT1	Recessive cold, heat sensitive	42
E83A/K84A	Yeast, *in vivo*	ACT1	Recessive cold, heat sensitive	42
E93K	*Drosophila, in vivo*	Act88F	$(-)$ Structure; no Z-discs	22
E93A/R95A	Yeast, *in vivo*	ACT1	Recessive lethal	42
E99H/E100H	*Dictyostelium*, pure	Act15	$<$ Motility	193
E99A/E100A	Yeast, *in vivo*	ACT1	Recessive heat sensitive; $=$ cold	42
R116A/E117A/KII8A	Yeast, *in vivo*	ACT1	Recessive heat sensitive; $=$ cold	42
S129V	*Drosophila, in vivo*	Act88F	$(+)$ Structure	117
V139M	Cells, *in vivo*	β-actin	Partially responsible for cytochalasin resistance	185
D154A/D157A	Yeast, *in vivo*	ACT1	Partial dominant lethal	42
F169Y/A260S	*Drosophila, in vivo*	Act88F	$(-)$ Structure	117
F169Y/C257T	*Drosophila, in vivo*	Act88F	$(+)$ Structure	117
L176M	*in vitro*	Act88F	$>$ Profilin; $=$ ATP; $=$ DNase; $=$ NDG; $(+)$ ADP-ribosylation	35,147
R177Q	*in vitro*	Act88F	$>$ Profilin; $=$ ATP; $=$ DNase; $=$ NDG; zero ADP-ribosylation	35,147
R177A/D179A	Yeast, *in vivo*	ACT1	Recessive heat sensitive; $=$ cold	42

R183A/D184A	Yeast, *in vivo*	*ACT1*	(+) Growth	42
D187A/K191A	Yeast, *in vivo*	*ACT1*	(+) Growth	42
K191M, K191M/C374A	Yeast, *in vivo*	*ACT1*	(+) Growth	142
E195A/R196A	Yeast, *in vivo*	*ACT1*	(+) Growth	42
E205A/R206A/E207A	Yeast, *in vivo*	*ACT1*	Possibly dominant lethal	42
R210A/D211A	Yeast, *in vivo*	*ACT1*	Recessive weak heat sensitive; = cold	42
K213A/E214A/K215A	Yeast, *in vivo*	*ACT1*	Recessive cold; heat sensitive; low spore viability	42
D222A/E224A/E226A	Yeast, *in vivo*	*ACT1*	Recessive heat sensitive; = cold	42
T234S	*Drosophila, in vivo*	*Act88F*	(+) Structure	117
E237A/K238A	Yeast, *in vivo*	*ACT1*	Partial dominant lethal	42
E241A/D244A	Yeast, *in vivo*	*ACT1*	Partial dominant lethal	42
G245D	Cells	Hs β	< Polymerization	178
	Drosophila, in vivo	*Act88F*	Antimorphic	48
	Cells	Hs β	Stable; (−) morphology at high expression levels	25
	Yeast, pure	Ch β	< Polymerization; < velocity ; = S1; > K_m	151
G245K	Yeast, pure	Ch β	< Polymerization; = S1; = velocity	151
E253A/R254A	Yeast, *in vivo*	*ACT1*	Partial dominant lethal	42
R256A/E259A	Yeast, *in vivo*	*ACT1*	Recessive cold; heat sensitive	42
del269–272	Yeast, *in vivo*	*ACT1*	(+) Growth	179
E270A/D275A	Yeast, *in vivo*	*ACT1*	Recessive lethal	42
V278T	*Drosophila, in vivo*	*Act88F*	(+) Structure	117
D286A/D288A	Yeast, *in vivo*	*ACT1*	Recessive lethal	42
I289F	*Drosophila, in vivo*	*Act88F*	Hypomorphic; myofibrillar degeneration	49
R290A/K291A/E292A	Yeast, *in vivo*	*ACT1*	Recessive lethal	42
A295D	Cells, *in vivo*	β-actin	Partially responsible for cytochalasin resistance	185
T304G/M305T/Y307Term	*Drosophila, in vivo*	*Act88F*	Unstable; hsp inducing	21
E311A/R312A	Yeast, *in vivo*	*ACT1*	Recessive cold, heat sensitive; < spore viability	42
E316K	*Drosophila, in vivo*	*Act88F*	Flightless; almost = muscles; (−) mechanics; weak hsp induction	33,42
E316K	*in vitro*	*Act88F*	= Polymerization; = NDG; < thermodynamic stability	31,33
			> Profilin; = ATP; = DNase	35
K326A/K328A	Yeast, *in vivo*	*ACT1*	Possibly dominant lethal	42
E334K	*Drosophila, in vivo*	*Act88F*	(−) Structure; weak hsp induction	33
E334K	*in vitro*	*Act88F*	= Polymerization; = NDG; = thermodynamic stability	31,33
			> Profilin; = ATP; = DNase	35
E334A/R335A/K336A	Yeast, *in vivo*	*ACT1*	Recessive lethal	42
K336M,Q	Yeast, *in vivo*	*ACT1*	(+) Growth	142
V339I	*Drosophila, in vivo*	*Act88F*	No actin accumulation	33
V339I	*in vitro*	*Act88F*	= Polymerization; = NDG; < thermodynamic stablity	31,33
			> Profilin; = ATP; = DNase	35
Y356Term	*Drosophila, in vivo*	*Act88F*	Antimorph, some actin (?); strong hsp; nuclear swelling	18,21
W356Y,H	Yeast, *in vivo*	*ACT1*	(+) Growth	142
K359A/E361A	Yeast, *in vivo*	*ACT1*	(+) Growth	42
Q360E	*Drosophila, in vivo*	*Act88F*	(+) Structure	117
E360H/E361H	*Dictyostelium*, pure	*Act15*	= Motility	193
E361A/D363N	*in vitro*	*Act88F*	> Profilin; = ATP; = DNase; = NDG	35
Y362C,I	*in vitro*	*Act88F*	> Profilin; < ATP; < NDG; = DNase	35
Y362V,L	*in vitro*	*Act88F*	< Profilin; < ATP; < NDG; = DNase	35
Y362F	*in vitro*	*Act88F*	= Profilin; = ATP; = NDG; = DNase	35
D363H/E364H	*Dictyostelium*, pure	*Act15*	= Motility	193
D363Y	*in vitro*	*Act88F*	= Profilin; = ATP; = NDG; = DNase	35
D363H	*in vitro*	*Act88F*	> Profilin; = ATP; = NDG; < DNase	35
D363A/E364A	Yeast, *in vivo*	*ACT1*	Recessive heat sensitive; = cold	42
E364K	*Drosophila, in vivo*	*Act88F*	(−) Structure; strong hsp induction	33
E364K	*in vitro*	*Act88F*	= Polymerization; = thermodynamic stability; < NDG	31,33
			> Profilin; < ATP; < DNase	35
G366S	*Drosophila, in vivo*	*Act88F*	Antimorphic; strong hsp induction	20,21
G366D	*Drosophila, in vivo*	*Act88F*	(−) Structure; strong hsp induction	33
G366D	*in vitro*	*Act88F*	= Polymerization; = thermodynamic stability; < NDG	31,33
G366D,P,A	*in vitro*	*Act88F*	> Profilin; < ATP; = DNase	35
G368S	*Drosophila, in vivo*	*Act88F*	(+) Structure	117
G368E	*Drosophila, in vivo*	*Act88F*	Flightless; almost (+) muscles; (−) mechanics	47
G368E	*in vitro*	*Act88F*	= Polymerization; = NDG; = thermodynamic stability	31,33
			= Profilin; < ATP; = DNase	35
G368T,S,Q	*in vitro*	*Act88F*	= Profilin; = ATP; = DNase; = NDG	35
G368K	*in vitro*	*Act88F*	> Profilin; = ATP; = DNase; = NDG	35
G368Term	*in vitro*	*Act88F*	> Profilin; < ATP; = DNase; = NDG	35
R372H	*in vitro*	*Act88F*	= Polymerization; = NDG; = thermodynamic stability	31,33
			> Profilin; > ATP; = DNase	35
K373Term	Yeast, *in vivo*	*ACT1*	Lethal	142
C374A	Yeast, *in vivo*	*ACT1*	(+) Growth	142
C374Term	Yeast, *in vivo*	*ACT1*	Temperature sensitive	142
P375Term	Yeast, *in vivo*	*ACT1*	Temperature sensitive	142

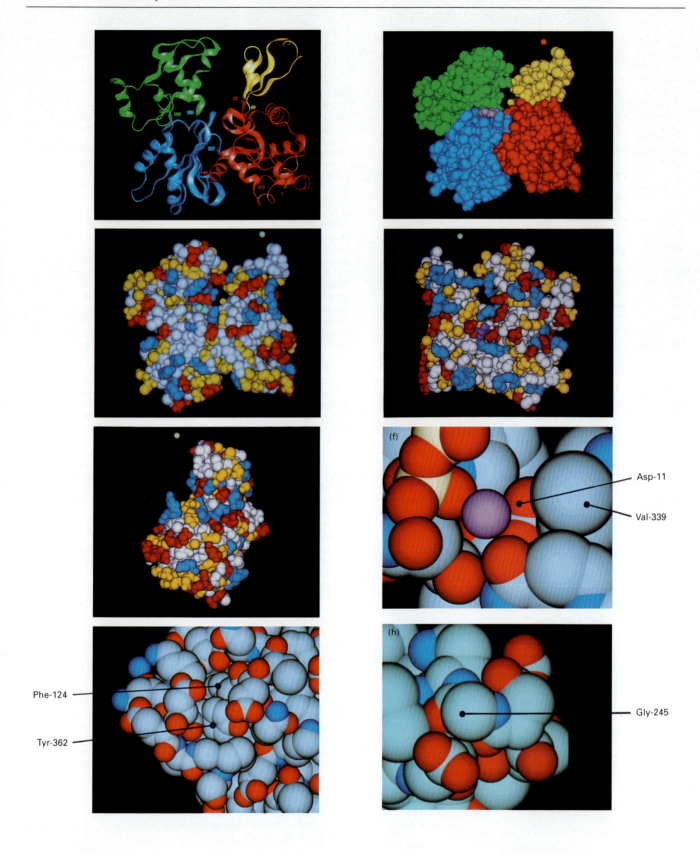

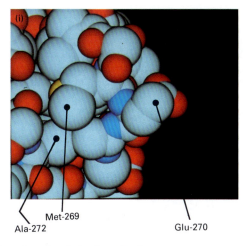
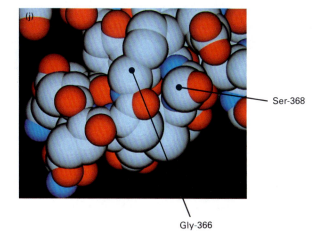

Figure 3   Three-dimensional structure of actin

Coloured ribbon diagram and spacefilling models of actin drawn using the coordinates of the actin–DNase I crystal structure [51] and displayed using the Quanta molecular graphics program (Molecular Simulations Inc.). Only the actin is shown. (**a**) and (**b**) show a ribbon diagram and spacefilling model respectively in which the four structural subdomains are shown in different colours. Subdomain 1 (residues 1–32, 70–144 and 338–375) is in red; subdomain 2 (residues 33–69) is yellow; subdomain 3 (residues 145–180 and 270–337) is blue; subdomain 4 (residues 181–269) is green. In (**b**) the isolated Ca^{2+} at the top right, shown in orange, is bound to DNase I and has been included to orientate the view. The ATP is shown in pink. Panels (**c**), (**d**) and (**e**) show the distribution of polarity on the 'front', 'back' and 'right-hand side' faces of actin respectively. The amino acid residues are coloured by polarity. Red, acidic; blue, basic; white, nonpolar; yellow, polar. Also shown are: pale green, ATP; orange, Ca^{2+}; purple, methylhistidine-73. The isolated Ca^{2+} from the DNase I is shown in pale green. Panel (**f**); the interior of actin around amino acids Asp-11 and Val-339. The atoms are coloured by type: red, oxygen; blue, nitrogen; green, carbon; pink, metal ion; yellow, phosphate. Arrows point to both Asp-11 and Val-339 and the approximate orientation is looking from the left-hand side of (**c**). Panel (**g**), a surface view of Tyr-362 and Phe-124. Atoms are coloured by type: red, oxygen; blue, nitrogen; green, carbon. Arrows point to both Tyr-362 and Phe-124 and the approximate orientation is looking towards (**e**). Panel (**h**), a surface view around amino acid Gly-245. The atoms are coloured by type: red, oxygen; blue, nitrogen; green, carbon. An arrow points to Gly-245. Approximate orientation is looking directly down on the top of the molecule. Panel (**i**), a surface view of amino acids 269–272. The atoms are coloured by type: red, oxygen; blue, nitrogen; green, carbon; yellow, sulphur. Arrows point to Met-269, Glu-270 and Ala-272. Orientation is as in (**d**). Panel (**j**), a surface view of Gly-366. The atoms are coloured by type: red, oxygen; blue, nitrogen; green, carbon. The approximate orientation is looking towards (**e**). Arrows point to Gly-366 and Ser-368.

with a wild-type genomic *ACT1* copy. Yields are limited in the latter system because actin overexpression is lethal [143] and must be regulated to maintain levels of the transformed actin equivalent to the endogenous actin [37,39]. The lack of N-terminal processing of actin in yeast [38] may be a disadvantage in the study of non-yeast actins.

Cultured cells

Cultured human fibroblasts [25,26,50] and simian COS-1 cells [28] have been transformed with mutant actin genes to look for effects on their growth and morphology. Partially purified preparations of a mutant actin from human fibroblasts have been used for *in vitro* polymerization studies [26]. Insect cells are commonly used to express foreign proteins following baculovirus transfection [144] but do not yet appear to have been used for expression of mutant actins.

Escherichia coli

The *Dictyostelium Act8* gene has been expressed in *E. coli* [145,146] but the actin forms inclusion bodies and only a small amount can be extracted in its native state. This is no doubt related to the ease with which native actin is irreversibly denatured. A cytoplasmic chaperone protein recently isolated from rabbit reticulocyte lysate catalyses the folding of denatured β-actin expressed in *E. coli* [147]. This may solve the problems of expressing native actin in *E. coli* for biochemical studies and allow suitable mutants to be used to study actin folding.

Expression *in vitro*

The expression of actin mutants from cloned and mutagenized DNA by *in vitro* transcription and translation in rabbit reticulo-

cyte lysate has been used to produce small quantities of radio-active actin. Although it may not be fully processed [32,121], such actin behaves the same as the product of the same gene expressed *in vivo* [32]. Mutant actins have been expressed from the genes for human α-skeletal muscle actin [28–30], rat brain cytoplasmic actin [27] and *Drosophila Act88F* [31–33,35,36,148] and used in co-polymerization and protein binding studies.

Although the costs are prohibitive for making large amounts of actin this way, expression is quickly and easily achieved and binding assays can be used to screen rapidly large numbers of mutations and pinpoint those of interest for more detailed study.

ACTIN MUTANTS

Table 1 summarizes the mutants residue-by-residue from the N- to C- termini, listing their effects *in vivo* and *in vitro*. We will consider actin functions in turn, describing the effects of mutants with reference to the actin structure where possible. Some mutants inevitably affect more than one function and will be found under more than one heading.

The atomic structure of actin [1,51] is an essential framework for interpreting mutant effects and is shown in Figures 2 and 3(a)–3(j). By comparing this structure with the extensive biochemical data from studies of actin by chemical modification, chemical crosslinking, spectroscopy etc. and electron microscope reconstructions of F-actin, F-actin–tropomyosin and filaments decorated with the myosin S1 fragment [149], potential binding sites can now be identified with some certainty [1,2,51].

Actin has four structural subdomains [51] (Figures 2, 3a and 3b). Subdomains 1 and 2 form what was previously known as the 'small domain'; subdomains 3 and 4 the 'large domain'. The chemical nature of amino acid side chains on the 'front', 'back' and 'right-hand side' surfaces of actin are shown in Figures 3(c), 3(d) and 3(e). These illustrate several interesting features. The

DNase I binding loop (top right in Figure 3c and top left in Figure 3d) contains several non-polar residues which form a hydrophobic contact with DNase I [51]. Long 'stripes' of non-polar residues in the middle of the front of the molecule suggest possible contact sites with other proteins (Figure 3c). There are also regions of stacked negative and positive charges visible on both the front and back surfaces. The N- and C-termini in subdomain 1 contain a cluster of negative charge (bottom right in Figure 3c and bottom left in Figure 3d). Methylhistidine-73 is located between two acidic groups and close to the conserved residue at Arg-177 (Figure 3d). The ATP is almost completely buried in the cleft between the so-called large and small domains, with only the purine ring visible (Figure 3c). Since there is ATP/ADP exchange some conformational changes must occur. Any opening of the cleft would involve breaking the salt bridge between Arg-62 and Glu-207 which links subdomains 2 and 4 (Figure 3c). Actin is a disc-like molecule and the side views present a very small aspect (Figure 3e).

STUDIES *IN VITRO*

Polymerization

In F-actin there are actin–actin contacts between the four structural subdomains [1,2,150] and it is not surprising to find extensive parts of the monomer surface involved in these contacts. From the model of F-actin [1,150] the contact sites have been identified as amino acids 41–50 (the DNase I loop), 110–112, 166–169, 195–197, 243–245, 264–273, 286–289, 322–325 and 375. The actin sequence alignment does not indicate that these sites are more highly conserved than other regions of the sequence. Only Gln-111, Gly-168, Asp-244, Gly-245, Asp-288, Arg-290 and Pro-322 are invariant. The conserved residues, with only occasional conservative substitutions, are Met-44, Pro-112, Asp-286, Ile-287, Ile-289, Lys-291, Thr-324, Met-325, Lys-326 and Phe-375. In the light of this it may seem surprising that actins from different species co- polymerize. However, this alignment is based on a large number of actins from a much broader range of species than have been used for co-polymerization studies. From their phenotypic effects *in vivo* it has been proposed that the alanine substitutions at residues 37, 39, 61, 62, 205, 244, 286 and 288 in the yeast *ACT1* gene affect the actin–actin contact surfaces [41,42] although these are not all within the contact sites proposed by modelling F-actin.

Two different actins containing the same G245D mutation both had reduced ability to polymerize [26,151] as did the G245K mutation [151]. Gly-245 is involved in actin–actin contacts [1,151] so the deleterious effects of these substitutions on polymerization are not unexpected.

Mutant actins, expressed *in vitro*, can be tested for co-polymerization with non-mutant carrier actin, added to achieve the critical concentration for polymerization. Although the interpretation of these experiments has been questioned [36], a number of mutants are able to co-polymerize as well as wild-type actin expressed *in vitro* [28,29,33,36] (Table 1). None of these mutants were in actin–actin contact sites.

Although there is no evidence of the N-terminus forming actin contacts in F-actin [53,150,152–154], complete N-terminal processing appears essential for normal polymerization [32]. Replacement of Asp-11 by non-acidic amino acids in a human α-skeletal muscle actin expressed *in vitro* prevented N-terminal processing, but in only one case did the actin have reduced ability to polymerize [29]. The lack of any polymerization effects of mutants at Asp-11 is surprising given their effect on protein conformation [155] and the involvement of this amino acid with metal binding (see below). The contradictory evidence for a role

of N-terminal processing in polymerization might arize from the use of heterologous or homologous actins for the co-polymerizations in the different studies [32]. Since there is no biochemical evidence for a direct role of the N-terminus in polymerization, these effects probably arise from a conformational change which affects a distant actin–actin contact site.

Motility *in vitro* and myosin binding

The ability of mutant actins to function as part of the actomyosin motor have been studied by mechanics experiments on de-membranated muscle fibres of *Drosophila* mutants [47], the use of *in vitro* motility assays [39–44,151,156] and assays of the actin-activated Mg-ATPase of myosin or myosin fragments [129,151,156,44]. The *in vitro* motility assays use purified actin and myosin which can be obtained from both muscle and non-muscle systems. By fixing one protein, usually myosin or myosin fragments, to a surface and adding F-actin stabilized with phalloidin and made fluorescent with a rhodamine label the velocity of the filaments can be measured, which is believed to correspond to the $V_{max.}$ of myosin ATPase [157].

Electron microscopic reconstruction of S1-decorated actin [149], the G- and F-actin structures [51,150] and biochemical data from many laboratories have been used to locate the myosin binding site, or at least that of the rigor state, on the outer face of subdomain 1 (Figures 2 and 3) [1,2]. The imminent publication of the atomic structure of the myosin S1 fragment means that modelling of actin and myosin S1 structures to the constraints of the electron microscopic reconstructions of S1–actin is now possible and the details of their interaction should soon be known.

Subdomain 1 is formed by residues 1–32, 70–144 and 338–372. While actin with the first 47 amino acids cleaved from the N-terminus is still able to bind myosin and activate its ATPase activity [154] albeit with reduced *in vitro* motility [153], the N-terminal peptide alone (amino acids 1–47) can also activate myosin ATPase [158,159]. The N-terminus forms contacts with myosin heavy chain [160,161] in the presence of nucleotide [161]. The proposed binding site includes the positions of known myosin- binding residues (in rabbit skeletal muscle) of Asp-1, Glu-2, Asp-3, Glu-4, Asp-24, Asp-25, Arg-28, Glu-93, Arg-95 and Lys-336. The hydrophobic surface of the α-helix 338–348 is probably a major contact site but due to the chemical nature of the residues these interactions would be refractory to most crosslinking chemicals. Crosslinking [162] and n.m.r. [163] studies implicate the 360–364 region in the binding of the A1 light chain from the vertebrate skeletal myosin. Actin mutants have been used to examine the importance of acidic groups at the N-terminus in myosin binding, a particularly variable region in different actin isoforms. Removal of acidic residues at positions within the first four amino acids reduced myosin S1 binding, myosin ATPase activation and *in vitro* motility [39,40,44,193]. Interestingly, substitutions with histidine at E360/E361 and D363/E364 resulted in wild-type motility suggesting the presence of acidic groups at the C-terminus is not necessary for motility [193]. This is surprising given the close proximity of the N-terminus.

Substitution of Gly-245 with aspartic acid in chicken β-actin gave an actin which bound S1 (decoration of F-actin filaments) but had a slower velocity in the motility assay and reduced affinity for myosin [151]. Replacement with lysine resulted in an actin with normal S1 binding and *in vitro* motility [151]. There is no evidence from biochemical studies of myosin-binding sites in this region, suggesting that the mutations cause long-range conformational effects on the whole actin molecule.

120 Sussman, D. J., Sellers, J. R., Flicker, P., Lai, E. Y., Cannon, L. E., Szent-Gyorgyi, A. G. and Fulton, C. (1984) J. Biol. Chem. **259**, 7349–7354

121 Redman, K. and Rubenstein, P. A. (1981) J. Biol. Chem. **256**, 13226–13229

122 Redman, K. L., Martin, D. J., Korn, E. D. and Rubenstein, P. A. (1985) J. Biol. Chem. **260**, 14857–14861

123 Ball, E., Karlik, C. C., Beall, C. J., Saville, D. L. Sparrow, J. C., Bullard, B. and Fyrberg, E. A. (1987) Cell **51**, 221–228

124 Taniguchi, S., Kawano, T., Kakunaga, T. and Baba, T. (1986) J. Biol. Chem. **261**, 6100–6106

125 Leavitt, J. and Kakunaga, T. (1980) J. Biol. Chem. **255**, 1650–1661

126 Hiromi, Y. and Hotta Y. (1985) EMBO J. **4**, 1681–1687

127 Sparrow, J., Drummond, D., Peckham, M., Hennessey, E. and White, D. (1991) J. Cell Sci. Suppl. **14**, 73–78

128 Peckham, M., Molloy, J. E., Sparrow, J. C. and White, D. C. S. (1990) J. Muscle Res. Cell Motil. **11**, 203–215

129 Geeves, M. A., Drummond, D. R., Hennessey, E. S. and Sparrow, J. C. (1991) Biophys. J. **59**, 419a

130 Burdon, R. H. (1986) Biochem J. **240**, 313–324

131 Lindquist, S. and Craig, E. A. (1988) Annu. Rev. Genet. **22**, 631–677

132 Parker-Thornburg, J. and Bonner, J. J. (1987) Cell **51**, 763–772

133 Flaherty, K. M., McKay, D. B., Kabsch, W. and Holmes, K. C. (1991) Proc. Natl. Acad. Sci. U.S.A. **88**, 5041–5045

134 Hightower, L. E. (1991) Cell **66**, 191–197

135 Anderson, P. (1989) Annu. Rev. Genet. **23**, 507–525

136 Fire, A. (1986) EMBO J. **5**, 2673–2680

137 DeLozanne, A. and Spudich, J. A. (1987) Science **236**, 1086–1091

138 Gerisch, G., Segall, J. E. and Wallraff, E. (1989) Cell Motil. Cytoskeleton **14**, 75–79

139 Noegel, A. A., Leiting, B., Witke, W., Gurniak, C., Harloff, C., Hartmann, E., Wiesmuller, E. and Schleicher, M. (1989) Cell Motil. Cytoskeleton **14**, 69–74

140 Manstein, D. J., Titus, M. A., DeLozanne, A. and Spudich, J. A. (1989) EMBO J. **8**, 923–932

141 Manstein, D. J., Ruppel, K. M., Kubalek, L. and Spudich, J. A. (1991) J. Cell Sci. Suppl. **14**, 63–65

142 Johannes, F.-J. and Gallwitz, D. (1991) EMBO J. **10**, 3951–3958

143 Huffaker, T. C., Hoyt, M. A. and Botstein, D. A. (1987) Annu. Rev. Genet. **21**, 259–284

144 Miller, L. K. (1988) Annu. Rev. Microbiol. **42**, 177–199

145 Frankel, S., Condeelis, J. and Leinwand, L. (1990) J. Biol. Chem. **265**, 17980–17987

146 Frankel, S., Sohn, R. and Leinwand, L. (1991) Proc. Natl. Acad. Sci. U.S.A. **88**, 1192–1196

147 Gao, Y., Thomas, J. O., Chow, R. L., Lee, G.-H. and Cowan, N. J. (1992) Cell **69**, 1043–1050

148 Just, I., Hennessey, E. S., Drummond, D. R., Aktories, K. and Sparrow, J. C. (1993) Biochem. J. **291**, 409–412

149 Milligan, R. A., Whittaker, M. and Safer, D. (1990) Nature (London) **348**, 217–221

150 Holmes, K. C., Popp, D., Gebhard, W. and Kabsch, W. (1990) Nature (London) **347**, 44–49

151 Aspenstrom, P., Engkvist, H., Lindberg, U. and Karlsson, R. (1993) Eur. J. Biochem. **207**, 315–320

152 DasGupta, G., White, J., Phillips, M., Bulinski, J. C. and Reisler, E. (1990) Biochemistry **29**, 3319–3324

153 Schwyter, D H., Kron, S. J., Toyoshima, Y. Y., Spudich, J. A. and Reisler, E. (1990) J. Cell Biol. **111**, 465–470

154 Kiessling, P., Schick, B., Polzar, B. and Mannherz, H. G. (1992) J. Muscle Res. Cell Motil. **13**, 263

155 Rubenstein, P. A., Solomon, L. R., Solomon, T. and Gay, L. (1989) Cell Motil. Cytoskeleton **14**, 35–39

156 Aspenstrom, P., Lindberg, U. and Karlsson, R. (1993) FEBS Lett. **303**, 59–63

157 Kron, S. J. and Spudich, J. A. (1986) Proc. Natl. Acad. Sci. U.S.A. **83**, 6272–6276

158 Kögler, H., Moir, A. J. G., Trayer, I. P. and Rüegg, J. C. (1991) FEBS Lett. **294**, 31–34

159 Van Eyk, J. E. and Hodges, R. S. (1991) Biochemistry **30**, 11676–11682

160 DasGupta, G., White, J., Cheung, P. and Reisler, E. (1990) Biochemistry **29**, 8503–8508

161 DasGupta, G. and Reisler, E. (1992) Biochemistry **31**, 1836–1841

162 Sutoh, K. (1982) Biochemistry **21**, 3654–3661

163 Trayer, I. P., Trayer, H. R. and Levine, B. A. (1987) Eur. J. Biochem. **164**, 259–267

164 Vandekerckhove, J. S., Kaiser, D. A. and Pollard, T. D. (1989) J. Cell Biol. **109**, 619–626

165 Asakura, S. (1961) Arch. Biochem. Biophys. **92**, 140–149

166 Kasai, M., Nakano, E. and Oosawa, F. (1965) Biochim. Biophys. Acta **94**, 494–503

167 Kasai, M. and Oosawa, F. (1968) Biochim. Biophys. Acta **154**, 520–528

168 Faustich, H., Merkler, I., Blackholm, H. and Stournaras, C. (1984) Biochemistry **23**, 1608–1612

169 Aktories, K., Bärmann, M., Ohishi, I., Tsuyama, S., Jakobs, K. H. and Habermann, E. (1986) Nature (London) **322**, 390–392

170 Aktories, K. and Just, I. (1990) in ADP-Ribosylating Toxins and G-Proteins (Moss, J. and Vaughan, M., eds), pp. 79–95, American Society for Microbiology, Washington DC

171 Aktories, K., Wille, M. and Just, I. (1992) Curr. Top. Microbiol. Immunol. **175**, 97–113

172 Schering, B., Bärmann, M., Chatwal, G. S., Geipel, U. and Aktories, K. (1988) Eur. J. Biochem. **171**, 225–229

173 Geipel, U., Just, I. and Aktories. K. (1990) Biochem. J. **266**, 335–339

174 Vandekerckhove, J., Schering, B., Bärmann, M. and Aktories, K. (1987) FEBS Lett. **225**, 48–52

175 Vandekerckhove, J., Schering, B., Bärmann, M. and Aktories, K. (1988) J. Biol. Chem. **263**, 696–700

176 Aktories, K. and Wegner, A. (1989) J. Cell Biol. **109**, 1385–1387

177 Mauss, S., Chapponnier, C., Just, I., Aktories, K. and Gabbiani, G. (1990) Eur. J. Biochem. **194**, 237–241

178 Cook, R. K. and Rubenstein, P. A. (1989) J. Cell Biol. **109**, 271a

179 Rubenstein, P. A., Cook, R. K. and Babcock, G. (1991) J. Cell Biol. **115**, 160a

180 Vandekerckhove, J., Leavitt, J., Kakunaga, T. and Weber, K. (1980) Cell **22**, 893–899

181 Leavitt, J., Bushar, G., Kakunaga, T., Hamada, H., Hirakawa, T., Goldman, D. and Merril, C. (1982) Cell **28**, 259–268

182 Friedman, E., Verderame, S., Winawer, S. and Pollack, R. (1984) Cancer Res. **44**, 3040–3050

183 Goldstein, D. and Leavitt, J. (1984) Cancer Res. **45**, 3256–3261

184 Leavitt, J., Gunning, P., Kedes, L. and Jariwalla, R. (1985) Nature (London) **316**, 840–842

185 Ohmori, H., Toyama, S. and Toyama, S. (1992) J. Cell Biol. **116**, 933–941

186 Sparrow, J. C., Drummond, D. R., Hennessey, E. S., Clayton, J. D. and Lindegaard, F. B. (1993) Soc. Exp. Biol. Symp. **46**, 111–129

187 Hitchcock-deGregori, S. E., Sampath, P. and Pollard T. D. (1988) Biochemistry **27**, 9182–9185

188 Weigt, C., Schoepper, B. and Wegner, A. (1990) FEBS Lett. **260**, 266–268

189 Mimura, N. and Asano, A. (1987) J. Biol. Chem. **262**, 4717–4723

190 Lebart, M.-C., Mejean, C., Roustan, C. and Benyamin, Y. (1993) J. Biol. Chem., in the press

191 Higgens, D. G. and Sharp, P. M. (1989) Comput. Appl. Biosci. **5**, 151–153

192 Saitou, N. and Nei, M. (1987) Mol. Biol. Evol. **4**, 406–425

193 Johara, M., Toyoshima, Y. Y., Ishijama, A., Kojima, H., Yanagida, T. and Sutoh, K. (1993) Proc. Natl. Acad. Sci. U.S.A. **90**, 2127–2131

Biochem. J. (1993) 295, 1–10 (Printed in Great Britain)

REVIEW ARTICLE
Plant metallothioneins

Nigel J. ROBINSON,*‡ Andrew M. TOMMEY,* Cheryl KUSKE† and Paul J. JACKSON†
*Department of Biological Sciences, University of Durham, Durham DH1 3LE, U.K., and †Genomics and Structural Biology Group, LS-2, M880, Los Alamos National Laboratory, Los Alamos, NM 87545, U.S.A.

INTRODUCTION

The name metallothionein was first used to describe a protein isolated from equine renal cortex in 1957 that contained large amounts of sulphur and cadmium (Margoshes and Vallee, 1957; Kägi and Vallee, 1960). During the ensuing three decades, structurally related proteins were identified in diverse organisms and shown to associate with several metal ions, most commonly Zn^{2+} and Cu^+ (or Cd^{2+} if Cd^{2+}-intoxicated) (Kägi and Kojima, 1987). Metallothioneins are thought to sequester excess amounts of certain metal ions although precise functions for most of these molecules remain the subject of debate. The specific metals sequestered by metallothioneins vary for the structurally distinct proteins/polypeptides occurring in different organisms (reviewed in Kägi, 1991). Roles in the regulation of gene expression have been proposed for some metallothioneins, in particular those in higher eukaryotes which co-ordinate Zn^{2+} and show programmed expression during development. Animal metallothionein genes respond to endogenous factors, which include a variety of hormones, second messengers, growth factors and cytokines, in addition to trace-metal levels. A proposed antioxidant role (Thornalley and Vasak, 1985) is supported by recent evidence that DNA strand-breakage, induced by oxidative stress, is reduced in the presence of elevated metallothionein levels and enhanced in Chinese hamster cells expressing a metallothionein antisense construct (Chubatsu and Meneghini, 1993). A wealth of information exists concerning the structure and regulation of expression of animal and fungal metallothioneins. These molecules are the subjects of a dedicated volume of Methods in Enzymology (Riordan and Vallee, 1991) and several articles thoroughly review the literature concerning animal and fungal metallothioneins and the genes encoding them (Karin, 1985; Hamer, 1986; Palmiter, 1987; Kägi and Schäffer, 1988; Thiele, 1992).

In 1985, it was reported that the major Cd^{2+} ligands in Cd^{2+}-intoxicated plant cells are composed of poly(γ-glutamyl-cysteinyl)glycine (Grill, 1985; Grill et al., 1985; Bernhard and Kägi, 1985; Robinson et al., 1985). These polypeptides, and other γ-glutamyl isopeptides in which Gly is either absent or substituted with β-alanine, are designated class III metallothioneins (Kojima, 1991). These compounds were first identified and characterized in the fission yeast Schizosaccharomyces pombe and termed cadystins (Murasugi et al., 1981; Kondo et al., 1984). Similar polypeptides were subsequently purified from plant cell cultures and termed phytochelatins (Grill et al., 1985). These class III metallothioneins have now been found in certain fungi and a broad spectrum of plant phyla (Grill, 1989). The structure, biosynthesis and proposed functions of these polypeptides have previously been reviewed (Tomsett and Thurman, 1988; Rauser, 1990; Steffens, 1990; Robinson, 1990; Jackson et al., 1990) and

only work which postdates these reviews is described in any detail here. The principal focus of this article is a growing body of literature describing gene-encoded plant metallothioneins (see Figure 1). A brief section on metallothioneins in other organisms is included to provide a context for discussion of plant metallo-thioneins.

METALLOTHIONEINS IN OTHER KINGDOMS

Mammalian metallothioneins are composed of approx. 61 amino acids with M_rs of 6000–7000. They contain no aromatic amino acids and 20 Cys residues that co-ordinate seven divalent metal ions (or 12 monovalent ions such as Cu^+; Nielson and Winge, 1984) in two distinct metal clusters. The locations of the Cys residues in mammalian metallothioneins are invariant, and proteins from any phyla (for example Neurospora crassa, quoted in Münger et al., 1987; and Agaricus bisporus, quoted in Münger and Lerch, 1985) that have similar primary structures are designated class I metallothioneins. Class II metallothioneins are low-M_r Cys-rich metal-binding proteins, but the distribution of Cys residues does not correspond to that in mammalian metallo-thioneins. These proteins have been identified in cyanobacteria, yeast, the nematode Caenorhabditis elegans and a higher plant (wheat germ E_c protein) (cited in Kägi, 1991).

Synthesis of metallothionein increases following exposure to elevated concentrations of Cu^+ and Ag^+ in fungal cells (Karin et al., 1984; Fürst et al., 1988), Cd^{2+} and Zn^{2+} in cyanobacteria (Olafson et al., 1988) [although other metals also stimulate increases in the abundance of metallothionein transcripts in cyanobacteria (Huckle et al., 1993)], and a range of trace metals including the ionic species of cadmium, zinc, copper, mercury, gold, silver, cobalt, nickel and bismuth in animals (cited in Kägi, 1991). Induction is primarily regulated at the level of gene transcription. Cis-acting metal-regulatory elements of animal and fungal metallothionein genes are known (for reviews see Hamer, 1986; Palmiter, 1987; Theile, 1992) and the first reports describing the use of such elements to control the expression of foreign genes in transgenic animals (Palmiter et al., 1982, 1983) are widely cited. Trans-acting metal-responsive factors have been identified (Labbé et al., 1991, and citations therein), purified (Labbé et al., 1993) and cloned from animals (Radtke et al., 1993), cloned from cyanobacteria (Huckle et al., 1993; Morby et al., 1993) and cloned and structurally characterized in yeasts (Dameron et al., 1991, and citations therein). Accumulation of metallothionein in response to elevated metal ion concentrations, combined with its association with these ions, indicates a role in the sequestration of excess metal.

Hypersensitivity to elevated trace-metal concentrations has been observed in fungal (cited in Hamer, 1986) and prokaryotic (Turner et al., 1993) cells with deleted metallothionein genes,

---

Abbreviations used: EDDHA, NN'-ethylenebis-[2-(2-hydroxyphenyl)glycine]; ABA, abscisic acid; GST, glutathione S-transferase; BPDS, bathophen-anthrolinedisulphonic acid; BCDS, bathocuproinedisulphonic acid.
‡ To whom correspondence should be addressed.

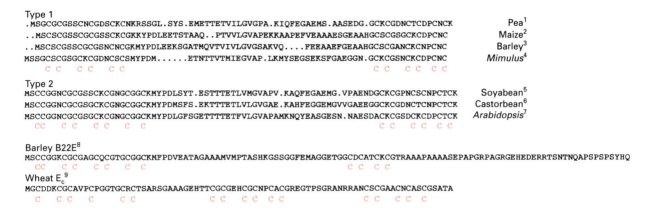

Figure 1   Sequences of the predicted products of plant genes with similarity to metallothionein genes, and of the E_c protein, emphasizing the distribution of Cys residues (appended in red).

Sources: [1]Evans et al. (1990); [2]de Framond (1991); [3]Okumura et al. (1991); [4]de Miranda et al. (1990); [5]Kawashima et al. (1991); [6]Weig and Komor (1992); [7]Takahashi (1991); [8]Klemsdal et al. (1991); [9]Kawashima et al. (1992). The predicted proteins are separated into two groups (type 1 and type 2) based on the arrangement of Cys residues. The E$_c$ protein and a second predicted protein from barley deviate from these two patterns.

while amplification of metallothionein genes has been observed in animal (Beach and Palmiter, 1981; Crawford et al., 1985) fungal (Fogel and Welch, 1982; Karin et al., 1984) and prokaryotic (Gupta et al., 1992, 1993) cells, selected for enhanced tolerance to certain trace metal ions. The phenotype of *Saccharomyces cerevisiae* cells lacking the metallothionein gene, *CUP1*, indicates that this gene performs no essential role(s) in cell growth, differentiation or 'normal' copper metabolism. These mutants grow with normal doubling times in standard low-copper media, and are capable of mating, diplophase growth, sporulation, germination, accumulation of copper and accumulation and activation of copper-dependent superoxide dismutase (cited in Hamer, 1986). *CUP1*-deficient cells are hypersensitive to elevated concentrations of copper. Sequestration of excess copper may be an exclusive role for metallothionein in *S. cerevisiae*. Similarly, cyanobacterial mutants deficient in the metallothionein locus, *smt*, are viable, but show reduced tolerance to Zn^{2+} and Cd^{2+} implying exclusive, but non-essential, roles in the homoeostasis/metabolism of Zn^{2+} and detoxification of Cd^{2+} (Turner et al., 1993).

The observations that animal metallothioneins show programmed expression during development and respond to certain endogenous signals suggest an undefined role in cellular regulation. This role may be mediated by the association of animal metallothionein with Zn^{2+}. More than 300 known enzymes, with representatives from all six recognized categories, and 200 known DNA-binding proteins require this divalent ion (cited in Vallee, 1991). While animal metallothionein has a high affinity for Zn^{2+}, the associated metal ions are also highly labile with rapid Zn^{2+} exchange between metallothionein molecules in solution (cited in Vallee, 1991). Animal metallothionein may thus serve as a Zn^{2+} store suited to the donation of metal ions when and where required.

Mammalian thionein (apo-metallothionein) can inactivate the Zn^{2+}-requiring transcription factor Sp1 (human) and can also acquire Zn^{2+} from *Xenopus laevis* transcription factor IIIA *in vitro* (Zeng et al., 1991a,b). In cultured adult rat hepatocytes, metallothionein accumulates in nuclei at early S-phase, but not at G_0 or G_1 when the protein is extranuclear (Tsujikawa et al., 1991). Modulation of animal thionein biosynthesis, or intracellular distribution (nuclear or extranuclear), could affect DNA binding by Zn^{2+}-requiring transcription factors and thereby regulate the activity of a large subset of genes (Zeng et al., 1991a,b).

THE E$_c$ PROTEIN FROM WHEAT GERM

At present, the wheat E$_c$ protein is the only plant protein that can be unequivocally designated a metallothionein (Hofmann et al., 1984; Lane et al., 1987; Kawashima et al., 1992). While 'metallothionein-like' plant genes have been isolated from other species (Figure 1), and data indicate that these genes have roles in metal metabolism (see below), their translational products remain to be purified from plant material and sequenced.

The E$_c$ protein is the dominant site of Cys incorporation during early germination, prior to degradation of mRNA species stored in the dry seed. A large proportion (20–25%) of radiolabelled [^{35}S]Cys is incorporated into a single protein following *in vitro* translation of bulk mRNA from dry wheat germ (Hanley-Bowdoin and Lane, 1983; Hofmann et al., 1984; Lane et al., 1987). The abundance of E$_c$ mRNA declines rapidly postimbibition and falls to imperceptible levels between 5 h and 10 h. In common with post-imbibition synthesis of other seed storage proteins, it is likely that synthesis of E$_c$ is 'residual', resulting from a role during embryogenesis and subsequent survival of E$_c$ transcripts in the desiccated seed (Hanley-Bowdoin and Lane, 1983).

The sequence of the E$_c$ protein, purified from wheat germ, is shown in Figure 1. The location of the Cys residues, notably the presence of Cys-Xaa-Cys motifs, suggested that the protein may be capable of binding metal ions. Purified E$_c$ protein was subsequently found to associate with Zn^{2+} at a stoichiometry (Zn^{2+}/protein) of approximately 5:1 (Lane et al., 1987). This led to its designation as a class II metallothionein (Kägi and Schäffer, 1988).

Recently, E$_c$ genes have been isolated (Kawashima et al., 1992). They are located in single copies on the long arms of chromosomes 1A, 1B and 1D in hexaploid wheat, unlike animal metallothionein genes which are contained in multigene clusters (cited in Kawashima et al., 1992). An element with similarity to known abscisic acid (ABA)-responsive elements is located in the 5′ flanking region of wheat E$_c$ genes but there are no clearly identifiable metal-responsive elements on the basis of sequence similarities (Figure 2). Northern blots confirm that E$_c$ transcripts

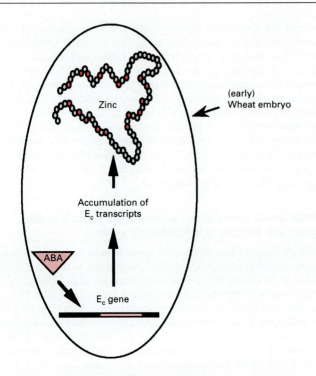

Figure 2   The E_c protein is expressed in early wheat embryos, binding Zn²⁺ and possibly modulating activity of Zn²⁺-requiring proteins (after Kawashima et al., 1991)

E_c transcripts persist in the desiccated seed but rapidly decline at germination unless supplemented with ABA. The E_c gene and a metallothionein-like gene from barley (Klemsdal et al., 1991) contain consensus ABA-responsive elements, but in the latter exogenous ABA is known to repress expression during seed development. Cysteine residues in the protein are shown in red.

accumulate in immature embryos and the highest levels of E_c mRNA were detected at the earliest stages (15 days post-anthesis) of embryogenesis, shortly after the onset of rapid cell division and differentiation (Kawashima et al., 1992). Furthermore, the abundance of E_c mRNA increases following addition of ABA, but not Zn²⁺, to germination media.

Kawashima et al. (1992) noted that wheat E_c genes are conspicuously expressed during embryogenesis under the control of endogenous factors, analogous to mammalian liver metallothionein genes. The hepatic concentration of metallothionein, primarily Zn²⁺-thionein, is 20-fold greater in neonatal than in adult rats. Kern et al. (1981) observed that the deposition of animal metallothionein is generally allied with a shift between proliferative and differentiating stages of embryo development. This also applies to wheat E_c. A homoeostatic role that engages Zn²⁺-E_c with Zn²⁺-dependent DNA and RNA polymerases, as well as with Zn²⁺-requiring *trans*-acting factors (Zn²⁺-fingers; Zn²⁺-twists; Zn²⁺-clusters; refer to Vallee et al., 1991) has been proposed for E_c (Kawashima et al., 1992). Only a small proportion (approx. 5%) of the total Zn²⁺ in mature wheat embryos is associated with E_c, suggesting that an alternative role in storage of Zn²⁺ for germination is less likely (Kawashima et al., 1992).

PLANT GENES WITH SIMILARITY TO METALLOTHIONEIN GENES

Genes that encode proteins with (some) sequence similarity to metallothioneins have been isolated from several plant species (Evans et al., 1990; de Miranda et al., 1990; Kawashima et al.,

1991; Takahashi, 1991; Okumura et al., 1991; de Framond, 1991; Robinson et al., 1992). All of these genes encode predicted proteins with two Cys-rich domains containing Cys-Xaa-Cys motifs (where Xaa is an amino acid other than Cys). Computer-based searches select metallothioneins as the most similar known proteins. Comparison (DNASTAR software) of the predicted protein from *Mimulus guttatus* with sequences in the NBRF protein database identified 19 of the top 23 matches as metallothioneins (de Miranda et al., 1990). Similar analysis (Fast P software) of the predicted translational product of the pea gene, *PsMT_A*, also selected metallothioneins as the top 10 best matches (Evans et al., 1990). Matrix comparisons of the amino acid sequences of the predicted PsMT_A protein and class I metallothionein from *Neurospora crassa* identify two regions of sequence similarity (amino acids 4–18 and 61–74 in the predicted pea protein), that correspond to the Cys-rich terminal domains. These two domains are linked by a central 'spacer' region of approx. 40 amino acids that is devoid of Cys residues. The Cys-rich terminal domains in the predicted sequences from different species are somewhat more conserved than the central 'spacer' regions (Figure 1).

Two categories of metallothionein-like proteins are proposed on the basis of the predicted locations of Cys residues and are designated types 1 and 2 (Figure 1). In type 1 there are exclusively Cys-Xaa-Cys motifs whereas in type 2 there is a Cys-Cys and a Cys-Xaa-Xaa-Cys pair within the N-terminal domain. Unlike the E_c protein, the translational products of these genes remain to be purified from plant tissue. Their discrimination from known Cys-rich plant proteins, such as leaf thioneins (Apel et al., 1990) and sulphur-rich prolamins (Shewry and Tatham, 1990), is based (primarily) upon the observed clustering of Cys residues to form these metallothionein-like domains although some other data support this distinction (see below). A second predicted protein sequence from barley (*Hordeum vulgare*) is also included in Figure 1.

Occurrence and isolation of plant genes with similarity to metallothionein genes

Several metallothionein-like plant genes have been isolated (by serendipity) via differential screening of cDNA libraries for root-abundant sequences (Evans et al., 1990; de Framond, 1991), sequences repressed by elevated copper (de Miranda et al., 1990) and sequences induced in response to depleted iron (Okumura et al., 1991). Sequences preferentially expressed in roots rather than other organs were isolated from cDNA libraries prepared from poly(A)⁺ RNA from roots of garden pea (*Pisum sativum*) and maize (*Zea mays*) (Evans et al., 1990; de Framond, 1991). The corresponding genes, *PsMT_A* and MT-L, were subsequently isolated from pea and maize genomic libraries respectively. These genes are members of small multi-gene families. Partial sequences of two further members of the pea gene family, *PsMT_B* and *PsMT_C*, have been obtained following PCR-mediated cloning (Robinson et al., 1992). A related cDNA from barley, *ids-1*, was identified in a library prepared from root poly(A)⁺ RNA isolated from plants grown under conditions of iron deficiency, following differential screening with probes prepared from poly(A)⁺ RNA from iron-deficient and iron-sufficient roots (Okumura et al., 1991). A soyabean (*Glycine max*) sequence was isolated from a cDNA library following hybridization to a 21-mer oligonucleotide (5′ ATGGACCCCAACTGCTCCTGC 3′) that corresponds to a conserved region found at the N-terminus of mammalian metallothionein genes (Kawashima et al., 1991). Several of the above sequences have been used as probes to isolate clones containing homologues from other

species including tobacco (*Nicotiana tabacum*) (Robinson et al., 1992), alfalfa (*Medicago sativum*) (Robinson et al., 1992) and *Arabidopsis thaliana* (Takahashi, 1991) and to reveal cognates in several other higher plant species by Southern analysis.

The identification of metallothionein-like genes in representative monocotyledonous (e.g. barley, maize) and dicotyledonous (e.g. pea, *Mimulus*, soyabean) species suggests a broad species distribution, although an extensive survey of genera, similar to that conducted for class III metallothioneins (cited in Grill, 1989), remains to be reported.

Attempts to detect the translational products of metallothionein-like genes within plants

Low-M_r, metal-induced, metal ligands whose structures remain to be elucidated have been isolated from plants (see citations in Robinson and Jackson, 1986). Many partly characterized Cd^{2+} complexes have amino acid compositions consistent with class III metallothioneins, while several partly characterized copper complexes appear to be unlike these and are more similar to the products of metallothionein genes (Robinson and Jackson, 1986). It has been proposed that either (i) the presence of copper may lead to increased oxidation of thiol groups during purification, resulting in isolates of class III metallothionein containing large amounts of impurities, or (ii) plants produce metallothionein-like proteins, as well as small metal-binding polypeptides, but the proteins are more important in the metabolism of an essential metal such as copper than in Cd^{2+} detoxification (Robinson and Jackson, 1986).

In addition to characteristic class III metallothioneins, a low-M_r copper complex was also isolated from *Mimulus guttatus* which was unlike these polypeptides (Tomsett and Thurman, 1988; Salt et al., 1989; de Miranda et al., 1990). The amino acid composition and size of the purified polypeptide was similar to an average of the two terminal domains of the predicted product of the *Mimulus guttatus* metallothionein-like gene. It was proposed that the internal 'spacer' region of the predicted product of this gene may be removed to generate smaller metal-binding polypeptides composed of the Cys-rich terminal domains (de Miranda et al., 1990). Following expression of the pea gene *PsMT*$_A$ in *Escherichia coli*, proteolysis was observed within the equivalent region of recombinant PsMT$_A$ protein (Kille et al., 1991; Tommey et al., 1991). Proteolytic cleavage within this region was further examined by exposure of purified PsMT$_A$ protein to proteinase K, followed by resolution of residual polypeptides by reversed phase f.p.l.c. (Kille et al., 1991). The amino acid composition of the residual polypeptides correlated with amino acids 2–21 and 56–75 from the Cys-rich N- and C-terminal domains of the full-length protein. Since cleavage within the PsMT$_A$ spacer region occurred in *E. coli*, it was proposed that this region may also act as a substrate for plant proteases (Kille et al., 1991). It is feasible that the spacer region could be important for folding the Cys-rich terminal domains into a conformation suitable for metal binding. Proteolysis of the spacer regions within plants may account for the failure of past attempts to isolate the native proteins.

The antigenic determinants for several vertebrate metallothioneins have been mapped close to the N-terminus within the region of residues 1–7 (cited in Katsuyuki et al., 1991). An epitope for many of these antibodies is a peptide region (amino acids 1–5) that includes the acetylated N-terminal Met. The epitope for antibody OAL-JI excludes the acetylated Met and is thought to include residues 3–7, i.e. Pro-Asn-Cys-Ser-Cys in vertebrate metallothionein. An equivalent heptapeptide occurs

within the C-terminal region of the predicted product of the soyabean metallothionein-like gene, and a related tetrapeptide Asn-Cys-Thr-Cys is present within the C-terminal domain of the predicted product of the pea gene *PsMT*$_A$. OAL-JI showed strong cross-reactivity with antigens present in soyabean tissues and somewhat less reactivity with material from pea. OAL-JI is reported not to cross-react with wheat germ E$_c$ protein or with protein extracted from *Mimulus guttatus* (Katsuyuki et al., 1991). However, full characterization of these antigens has not been reported and it thus remains to be established whether or not the antigens and the predicted products of the characterized metallothionein-like genes are synonymous.

Metal-binding characteristics of recombinant products of plant genes with similarity to metallothionein genes

The products of metallothionein genes from different organisms form conformations suitable for association with metal ions when expressed in *E. coli* and yeast (Romeyer et al., 1988, 1990; Jacobs et al., 1989; Kille et al., 1990; Silar and Wegnez, 1990). In the absence of purified native protein, the *PsMT*$_A$ gene was expressed in *E. coli* to facilitate examination of the metal-binding properties of its product (Tommey et al., 1991; Kille et al., 1991). Following growth in Cd^{2+}-supplemented media, cells expressing the *PsMT*$_A$ gene accumulated more Cd^{2+} than equivalent cells containing the vector (pPW1) alone (Kille et al., 1991). Increased accumulation of copper was also observed in *E. coli* cells expressing PsMT$_A$ as a C-terminal extension of glutathione S-transferase (GST) following growth in media supplemented with this metal (Evans et al., 1992). Such effects on metal accumulation have been used as a reliable indicator of intracellular production of metal–ligand complexes in *E. coli* (cited in Kille et al., 1991). The estimated Cd^{2+}/PsMT$_A$ stoichiometry ranged from 5.6 to 6.1 g-atoms of Cd^{2+} per mol of protein purified from *E. coli* (Kille et al., 1991). If all of these metal ions are associated with thiolate ligands (a total of 12 in the two Cys-rich domains) then the Cd^{2+}–Cys connectivities must differ from the pattern observed in other metallothioneins (Kille et al., 1991).

Estimations of the pH at which 50% of metal ions dissociate is a criterion used to distinguish metallothioneins from other metal-binding proteins (Vasák and Armitage, 1986). The pH of half-dissociation of Zn^{2+}, Cd^{2+} and copper ions from purified GST–PsMT$_A$ fusion protein was estimated to be 5.25, 3.95 and 1.45, respectively (Tommey et al., 1991). Comparison with equine renal metallothionein indicates that recombinant PsMT$_A$ protein has slightly lower affinities for Zn^{2+} and Cd^{2+}, but a slightly higher affinity for copper ions. Equivalent studies remain to be performed on the expressed products of the soyabean, castorbean and *Arabidopsis thaliana* sequences which contain Cys-Xaa-Xaa-Cys and Cys-Cys motifs that may modify metal binding. The metal-binding characteristics of the E$_c$ protein and the product of a second related gene from barley also require further examination.

It is well documented that association with metal ions protects proteins from proteolytic degradation and this is consistent with the observation that residues 2–21 and 56–75 in the PsMT$_A$ protein are refractory to proteinase K (Kille et al., 1991). Despite cleavage within the central spacer region, the two Cys-rich domains appear to remain associated with each other indicating that the metal–protein bonds are capable of holding this portion of the cleaved molecule together. Based upon these observations, Kille et al. (1991) (tentatively) proposed a model (illustrated in Figure 3) for association of metal to PsMT$_A$, followed by proteolysis of the spacer region. Computer analysis of the encoded protein from *Mimulus gutattus* predicts extensive folding

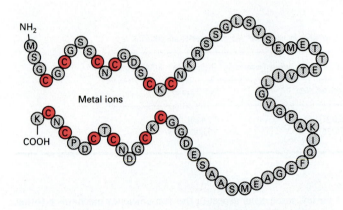

Figure 3 A putative structure for PsMT$_A$ associated with metal ions

Adapted from Kille et al. (1991). Cysteine residues are shown in red.

of the Cys-rich terminal domains but an extended configuration for the 39-amino-acid intervening spacer region (de Miranda et al., 1990).

Evidence that the products of these metallothionein-like genes bind metal ions within plants has been obtained from studies with *Arabidopsis thaliana* transformed with a construct containing *PsMT*$_A$ under the control of the cauliflower mosaic virus 35 S (CaMV 35 S) promoter (Evans et al., 1992). In a segregating progeny, derived from a single transgenic F$_1$ parent expressing *PsMT*$_A$, 75 % of individuals accumulated more copper than untransformed control plants.

Regulation of expression of plant genes with similarity to metallothionein genes

Information describing the expression of metallothionein-like plant genes is sparse (Table 1). Transcripts from type 1 genes are abundant in roots grown in nutrient solutions that have not been supplemented with supra-optimal concentrations of trace-metal ions (Evans et al., 1990; de Miranda et al., 1990; de Framond, 1991; Robinson et al., 1992). In pea plants *PsMT* transcripts were not detected in the embryonic radicle but increased in abundance during development of roots grown in hydroponic culture (Robinson et al., 1992). In soyabean and *Mimulus guttatus*, exposure to elevated concentrations of copper did not further increase transcript abundance in roots. Moreover, slight decreases in transcript abundance were observed in response to highly elevated copper concentrations (de Miranda et al., 1990; Kawashima et al., 1991). Transcripts encoded by the *ids-1* gene from barley were detected in roots grown under conditions of iron deficiency but not in roots supplemented with available iron (Okumura et al., 1991). The abundance of *PsMT* transcripts in pea roots also declines following supplementation with iron chelates (N. J. Robinson, unpublished work). The apparent constitutive expression of related genes observed in the roots of a number of species may correlate with iron deficiency since the stringent growth conditions required to maintain a supply of available iron, which depresses the activation of iron-efficiency mechanisms (Grusak et al., 1990), were not employed in these previous studies (Evans et al., 1990; de Miranda et al., 1990). Similarly, progressive increases in *PsMT* transcript abundance during root development will coincide with the activation of iron-efficiency mechanisms following depletion of iron stored (in plant ferritin; Laulhere and Briat, 1993) in the seed.

Table 1 Expression of genes with similarity to metallothionein genes in different plants

Refer to text for references.

Plant	Transcript abundance Localization			Elevated metal					Hormone ABA	CXC only	ABA element	Metal binding
	Root	Leaf	Seed	Zinc	Copper	Cadmium	Iron	Iron + copper				
Pea	High	Low	Low	–	No	–	Down	Up	–	Yes	No	Cu > Cd > Zn (recombinant) (Cu, transgenic plants)
Mimulus	High	Low	–	Down	Down	–	–	–	–	Yes	–	–
Maize	High	Low	Very low	–	–	Down	–	–	–	Yes	No	–
Barley	High	Low	–	–	–	–	Down	–	–	Yes	–	–
Arabidopsis	–	On	–	–	–	–	–	–	–	No	–	–
Castorbean	–	On	–	–	–	–	–	–	–	No	–	–
Soyabean	Low	High	High	–	Down	–	–	–	Down	No	Yes	–
Barley2	–	–	High	–	–	–	–	–	–	No	Yes	–
Wheat E$_c$	–	–	High	No	–	–	–	–	Up	No	Yes	Zn (native protein)

Table 2   Link between iron and copper metabolism in plants

There is an inverse correlation between iron availability, copper accumulation and copper and iron reduction by intact roots. Values in parentheses are S.D.s.

	Metal content of roots (nmol/g)		Metallo-reductase activity [μmol of Cu$^+$-BCDS (or Fe^{2+}-BPDS)/h per g]	
	Copper	Iron	Copper reduction	Iron reduction
Control	125(5)	89(5)	6.15(0.27)	3.29(0.3)
Fe-EDDHA	55(2)	160(11)	3.17(0.76)	1.36(0.11)

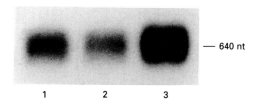

— 640 nt

1   2   3

Figure 4   Northern blot showing that the abundance of *PsMT* transcripts in pea roots correlates with the concentration of copper in media supplemented with available iron chelate

Equivalent amounts of total RNA were resolved in each track and the resultant Northern blot was probed with *PsMT$_A$*. Total RNA was isolated from pea seedlings grown hydroponically in the presence of added Fe^{3+}-EDDHA (2 μM), plus: no added copper (lane 2), micronutrient copper (40 nM) only (lane 1), highly elevated copper (1000 nM) (lane 3) (A. M. Tommey, L. V. Kochian, R. M. Welch, J. E. Schaff, S. C. Schaeffer and N. J. Robinson, unpublished work).

A root surface ferric-chelate reductase (Fe^{3+} to Fe^{2+}) is expressed in pea plants grown under conditions of low available iron (Buckhout et al., 1989; Grusak et al., 1990). Inducible *trans*-plasma-membrane ferric reductases have been extensively characterized in yeast, where electrons are transferred from cytosolic NAD(P)H to extracellular ferric ions (Lesuisse et al., 1991; Dancis et al., 1992). Ferrous ions are then imported. In yeast, transcription of the *FRE1* gene, which encodes ferric reductase, increases under iron deficiency. The FRE1 reductase also reduces Cu^{2+} to Cu$^+$ and appears to also be involved in the uptake of copper by *S. cerevisiae* (Lesuisse and Labbé, 1992). Correlative data indicating that pea root ferric-chelate reductase may similarly act as a cupric-chelate reductase and is implicated in copper acquisition includes the following: (1) increased cupric-chelate reductase activity coincides with increased ferric-chelate reductase activity under low-iron conditions (Table 2); (2) copper accumulation increases under low-iron conditions (Table 2; Evans et al., 1992); (3) cupric-chelate reductase activity increases in response to low copper (Welch et al., 1993); (4) ferric-chelate reductase activity increases coincident with the activation of cupric-chelate reductase activity under low-copper conditions (Welch et al., 1993); (5) iron accumulation increases under low-copper conditions (Welch et al., 1993). Expression of metallo-thionein-like genes under conditions of iron deficiency may be a direct response to concomitant increases in copper accumulation. Indeed, under conditions of high available iron, expression of *PsMT* transcripts correlates with the copper status of the growth media (Figure 4). Figure 5 summarizes these observations. According to this scheme, the products of type 1 metallothionein genes expressed under low-iron conditions chelate and thereby detoxify and store excess copper. However, if overaccumulation of copper in response to iron deficiency were also proposed to be the mediator of enhanced accumulation of *ids1* transcripts in

barley, some other aspect of the iron-efficiency mechanism (other than inducible ferric-chelate reductase) would have to account for increased copper uptake (at least in graminaceous species such as barley). Graminaceous species have alternative iron-efficiency systems which do not include inducible ferric-chelate reductases (cited in Grusack et al., 1990).

Overaccumulation of copper in transgenic *Arabidopsis thaliana* containing *PsMT$_A$* constructs (Evans et al., 1992) may be a reflection of constitutive expression of PsMT$_A$, whereas expression coupled to internal metal ion concentration may enhance tolerance without increasing metal accumulation. For example, net uptake of copper in wild type and *CUP1*-deleted *S. cerevisiae* is similar (Lin and Kosman, 1990). The metal tolerance of *CUP1*-deleted cells has additionally been examined following coupled (copper-induced) or uncoupled (constitutive) expression of the *Drosophila melanogaster* metallothionein gene, MTn (Silar and Wegnez, 1990). The correct coupling of metallothionein expression in this system is more effective in achieving a copper-resistant phenotype than uncoupled expression of the gene. In addition, a number of studies have shown that uncoupled expression of different metallothionein genes in *E. coli* mediates enhanced metal accumulation (Jacobs et al., 1989; Kille et al., 1990).

In soyabean, transcripts encoding the predicted metallo-thionein (type 2; Figure 1) are present in roots and leaves but are most abundant in leaves (Kawashima et al., 1991). Related cDNAs have been isolated from *Arabidopsis thaliana* and castor-bean libraries prepared from leaf poly(A)$^+$ RNA but the pattern of expression remains to be determined (Table 1). These observations raise the possibility that different members of a gene family, possibly encoding proteins with different metal-binding specificities, could be expressed in different organs or under different environmental conditions. Representatives of one group (type 1) appear to be expressed primarily in roots and are implicated in the metabolism or detoxification of copper, while others may be predominantly expressed in aerial tissues.

In maize, the abundance of MT-L transcripts in kernels is low (de Framond, 1991). In pea, *PsMT* transcripts are not detected in the embryonic radicle but transcripts of a slightly smaller size than those in roots are detectable in the embryonic cotyledon (Evans et al., 1990). Genes encoding analogues of the E$_c$ protein may be preferentially expressed in seeds and are implicated in Zn^{2+} metabolism, possibly regulating the activity of Zn^{2+}-requiring factors as proposed by Kawashima et al. (1992), or serving as a storage form of Zn^{2+}. Related cDNAs from barley were isolated from libraries constructed from poly(A)$^+$ RNA from the aleurone/pericarp and embryo of developing grains and also from germinating scutella (Klemsdal et al., 1991). The sequence of the predicted B22E product is shown in Figure 1. B22E transcripts are repressed by ABA in developing seeds (Olsen et al., 1990) and the corresponding gene contains consensus abscisic acid response elements (Klemsdal et al., 1991). It

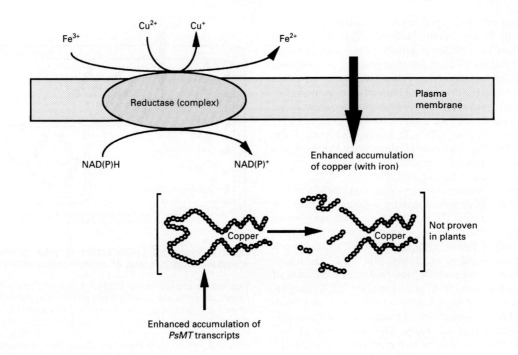

Figure 5 **Hypothetical scheme linking iron availability, expression of root surface ferric-chelate reductase, copper accumulation and the expression of (type 1) plant genes with similarity to metallothionein genes**

Under low-iron conditions there are coincident increases in root surface cupric-chelate reductase activity and copper accumulation. This scheme proposes that the ferric- and cupric-chelate reductases are synonymous. However, it is noted that the biochemical basis for enhanced copper accumulation under low-iron conditions remains to be established and that Cu^{2+} has previously been considered to be the transported form of copper in plants. Proposed proteolysis of the 'spacer' region of $PsMT_A$ within plants is also illustrated. Cysteine residues in metallothionein-like proteins are shown in red.

remains to be established whether or not the B22E product binds Zn^{2+} and is analogous to E_c in wheat.

In addition to an abscisic acid element, an element similar to that believed to be responsible for starchy-endosperm-specific expression of other cereal storage protein genes (Forde et al., 1985; Kreis et al., 1985) has also been detected in the B22EL8 gene, which encodes one class of B22E transcript (Klemsdal et al., 1991). Clearly, investigations are required to identify and characterize putative metallo-regulatory sequences in the flanking regions of type 1 genes, in addition to sequences responsive to endogenous signals (e.g. conferring spatial or temporal control of expression) such as the ABA elements in the E_c and B22EL8 genes (Klemsdal et al., 1991; Kawashima et al., 1992).

CLASS III METALLOTHIONEINS: γ-GLUTAMYL ISOPEPTIDES, CADYSTINS, PHYTOCHELATINS, POLY(γ-GLUTAMYLCYSTEINYL)GLYCINES

For more detailed discussions of these polypeptides, readers are referred to a number of reviews (Tomsett and Thurman, 1988; Rauser, 1990; Steffens, 1990; Robinson, 1990; Jackson et al., 1990). This section provides only a brief overview of these molecules and highlights some recent findings. Class III metallothioneins differ markedly from class I and II metallothioneins. They are enzymatically derived and are most commonly composed of poly(γ-glutamylcysteinyl)glycine, $(\gamma$-Glu-Cys$)_n$Gly, where $n = 2$–11 depending on the organism, although the most common forms have $n = 2$–4 (Grill et al., 1986a). However, class III metallothioneins isolated from the Fabaceae contain β-alanine in the C-terminal position and these species produce predominantly homoglutathione (γ-glutamylcysteinyl-β-alanine) rather than

glutathione (γ-glutamylcysteinylglycine) (Grill et al., 1986b). Des-glycine forms of class III metallothioneins have also been described (cited in Steffens, 1990).

Metal complexes containing these γ-glutamyl isopeptides have apparent native M_rs ranging from 2000 to 10000 depending upon the source and method of isolation, and include multiple polypeptides in a cluster (cited in Steffens, 1990; Rauser, 1990). Sulphide is sometimes present in Cd^{2+} complexes in varying amounts but has not been found in copper complexes (cited in Steffens, 1990; Rauser, 1990). High-M_r sulphide-containing complexes show enhanced affinity for metals (Reese and Winge, 1988). The structure of such complexes is of interest, being composed of a CdS quantum semiconductor crystallite core surrounded by polypeptides (Dameron et al., 1989). Sulphide-containing complexes have been described in *Candida glabrata* (Mehra et al., 1988), tomato (*Lycopersicon esculentum*) (Reese et al., 1992) and a selenium-tolerant wild mustard (*Brassica juncea*) (Speiser et al., 1992a). Metal-tolerant *Silene vulgaris* (Verkleij et al., 1990) and cell cultures of *Datura innoxia* (Robinson et al., 1990) incorporate greater amounts of S^{2-} into these complexes than their less tolerant counterparts. Most recently it has been observed that enzymes involved in purine biosynthesis are required for the introduction of S^{2-} into these complexes in *S. pombe* (Speiser et al., 1992b). The link between these two aspects of metabolism remains to be established.

Biosynthesis

Structural similarities between glutathione and class III metallothioneins suggest that the latter are synthesized from the former or its precursors. *In vivo* experiments demonstrate a significant

reduction of free glutathione upon exposure of plant cell cultures to Cd^{2+} (cited in Rauser, 1990; Steffens, 1990). Pulse–chase experiments, where the cellular glutathione pool is tagged with ^{35}S, show loss of radiolabel from glutathione with a concomitant increase in radiolabelled class III metallothioneins (Berger et al., 1989). Treatment of cell cultures with buthionine sulphoxamine, a potent inhibitor of γ-glutamylcysteine synthetase, results in the loss of metal tolerance and an inability to synthesize class III metallothioneins (Grill et al., 1987). In addition, mutants of the fission yeast deficient in enzymes of glutathione synthesis are unable to produce class III metallothioneins and are hypersensitive to Cd^{2+} (Mutoh and Hayashi, 1988).

There are several alternate pathways that might produce class III metallothioneins from glutathione or γ-glutamylcysteine. It has been reported that class III metallothioneins are synthesized from glutathione by the enzyme γ-glutamylcysteine dipeptidyl transpeptidase (phytochelatin synthase) in *Silene cucubalus* cell suspension cultures (Grill et al., 1989). The M_r of the native protein was reported to be 95000; the protein is composed of four subunits, each with an M_r of approximately 25000. Enzyme activity is dependent upon the presence of metal ions, with addition of EDTA or apopeptides to reaction mixtures terminating synthesis (Löffler et al., 1989). The mechanism of biosynthesis requires two glutathione molecules or one glutathione plus a previously synthesized class III metallothionein molecule. The transfer of the γ-glutamylcysteine moiety of glutathione to another glutathione molecule or to a previously synthesized class III metallothionein does not require additional ATP.

In the fission yeast *S. pombe*, two pathways for the biosynthesis of class III metallothioneins have been detected in a cell-free system (Hayashi et al., 1991). The first is similar to that described (Grill et al., 1989) for *Silene cucubalus* cell cultures except that either glutathione or class III metallothioneins can act as donors for γ-glutamylcysteine. The second involves the polymerization of γ-glutamylcysteine by the transfer of γ-glutamylcysteine from glutathione to $(\gamma$-Glu-Cys$)_n$ to produce $(\gamma$-Glu-Cys$)_{n+1}$ plus Gly. This is followed by the addition of Gly to poly(γ-Glu-Cys). In maize, Cd^{2+} has also been shown to inhibit glutathione biosynthesis, and γ-glutamylcysteine accumulates (Meuwly and Rauser, 1992; Rüegsegger and Brunold, 1992; Ric De Vos et al., 1992). Accumulation of this dipeptide could drive the biosynthesis of class III metallothioneins in the presence of Cd^{2+}. Metal activation of an enzyme for γ-glutamylcysteine transfer from glutathione was not observed in crude enzyme preparations from *S. pombe*, and enzyme extracts from Cd^{2+}-induced and uninduced cells showed no difference in γ-glutamylcysteine transfer.

There are several reports of rapid synthesis of class III metallothioneins being insensitive to inhibitors of *de novo* protein synthesis in plant cell cultures exposed to Cd^{2+} (Scheller et al., 1987; Robinson et al., 1988) indicating that the enzymes responsible for the synthesis of class III metallothioneins and their precursors are constitutive in cells in the absence of excess metal ions. Furthermore, enzyme activity was detected in cell-free extracts from cultures not exposed to elevated metal ion concentrations (Grill et al., 1989).

Localization

In Cd^{2+}-exposed hydroponically grown *Nicotiana rustica* var. Pavonii, the main Cd^{2+}-binding components were (γ-Glu-Cys$)_3$Gly and (γ-Glu-Cys$)_4$Gly. The location of these polypeptides was determined following isolation of protoplasts and vacuoles from leaves of Cd^{2+}-exposed seedlings. Both class III metallothioneins and Cd^{2+} were found in the vacuoles (Vögeli-Lange and Wagner, 1990).

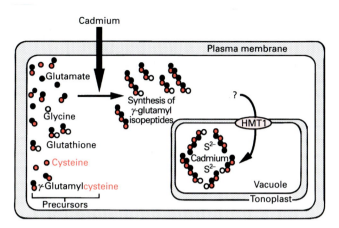

Figure 6   Plants and certain species of yeast accumulate γ-glutamyl isopeptides, designated class III metallothioneins, when exposed to Cd^{2+}

These thiol-rich compounds are also known as cadystins and phytochelatins. Cd^{2+} complexes contain multiple polypeptide molecules and some complexes also contain inorganic S^{2-} within the metal core. Vacuolar accumulation of such complexes has been observed in plants and in *S. pombe*. The product of the *hmt1* gene, localized to the vacuolar membrane, is required for the accumulation of these complexes in vacuoles of *S. pombe* (Ortiz et al., 1992). The predicted sequence of HMT1 is similar to those of ABC-type membrane transporters but it remains to be established which component(s) of the Cd^{2+} complexes (or their precursors) it transports. Red circles, cysteine residues; black circles, glutamate residues; white circles, glycine residues.

The recent observations of Ortiz et al. (1992) indicate that there is a specific transporter, designated HMT1, required for the accumulation of high-M_r CdS–(class III metallothionein) complexes in the vacuole of *S. pombe* cells. A Cd^{2+}-sensitive mutant of *S. pombe*, designated LK100, was isolated that accumulated less of these complexes than the wild type. LK100 cells transformed with *hmtF1* showed restored (increased) Cd^{2+} accumulation. The amino acid sequence deduced from *hmt1* cDNA suggests that its product is similar to ABC (ATP-binding cassette)-type membrane transport proteins (Ortiz et al., 1992). Subcellular fractionation of extracts from *S. pombe* containing an *hmt1–lacZ* fusion indicated that the encoded fusion protein is localized in the vacuolar membrane. At present it is not clear which of the components of the complex (Cd^{2+}, S^{2-}, polypeptides or their precursors) is transported into the vacuole via HMT1 (Figure 6). Delhaize et al. (1989) observed that Cd^{2+} tolerance in *Datura innoxia* cells correlated with the rapid assembly of class III metallothionein metal complexes, but not with modified rates of class III metallothionein synthesis. Modified activity of a plant analogue of HMT1 is one possible explanation for these observations.

CONCLUDING REMARKS: DIVERSITY OF PLANT METALLOTHIONEINS

Plants appear to contain a diversity of metal-binding metallothioneins with the potential to perform distinct roles in the metabolism of different metal ions. The E_c protein from wheat is implicated in the endogenous control of Zn^{2+} metabolism during embryogenesis (Figure 2) while the non-gene-encoded class III metallothioneins appear to play roles in the detoxification of Cd^{2+} (Figure 6) (and possibly of excesses of some essential metal ions). The putative type 1 products (Figure 1) of plant genes with similarity to metallothionein genes are implicated in the sequestration of copper in roots (Figure 5) while roles for type 2 products are even less certain. Clearly, there is an overwhelming necessity to isolate, from plants, the putative products shown in

Figure 1 and to establish whether or not species possess a complement of metallothioneins under the control of different exogenous and endogenous factors, fulfilling different biochemical requirements.

Support for work on plant metallothioneins from AFRC research grant PG12/519(PMB) is gratefully acknowledged. N. J. R. is a Royal Society University Research Fellow.

REFERENCES

Apel, K., Bohlmann, H. and Reimann-Philipp, U. (1990) Physiol. Plant. **80**, 315–321

Beach, L. R. and Palmiter, R. D. (1981) Proc. Natl. Acad. Sci. U.S.A. **78**, 2110–2114

Berger, J. M., Jackson, P. J., Robinson, N. J., Lujan, L. D. and Delhaize, E. (1989) Plant Cell Rep. **7**, 632–635

Bernhard, W. R. and Kägi, J. H. R. (1985) in Abstracts of the Second International Meeting on Metallothionein and Other Low Molecular Weight Metal Binding Proteins, Zurich, p. 25

Buckhout, T. J., Bell, P. F., Luster, D. G. and Chaney, R.L (1989) Plant Physiol. **90**, 151–156

Chubatsu, L. S. and Meneghini, R. (1993) Biochem. J. **291**, 193–198

Crawford, B. D., Enger, D. M., Griffith, B. B., Griffith, J. K., Hanners, J. L., Longmire, J. L., Munk, A. C., Stallings, R. L., Tesmer, J. G., Walters, R. A. and Hildebrand, C. E. (1985) Mol. Cell. Biol. **5**, 320–329

Dameron, C. T., Reese, R. N., Mehra, R. K., Korfan, A. R., Carroll, P. J., Steigerwald, M. L., Brus, L. E. and Winge, D. R. (1989) Nature (London) **338**, 596–597

Dameron, C. T., Winge, D. R., George, G. N., Sansone, M., Hu, S. and Hamer, D. H. (1991) Proc. Natl. Acad. Sci. U.S.A. **88**, 6127–6131

Dancis, A., Roman, D. G., Anderson, G. J., Hinnebusch, A. G. and Klausner, R. D. (1992) Proc. Natl. Acad. Sci. U.S.A. **89**, 3869–3873

de Framond, A. J. (1991) FEBS Lett. **290**, 103–106

Delhaize, E., Jackson, P. J., Lujan, L. D. and Robinson, N. J. (1989) Plant Physiol. **89**, 700–706

de Miranda, J. R., Thomas, M. A., Thurman, D. A. and Tomsett, A. B. (1990) FEBS Lett. **260**, 277–280

Evans, I. M., Gatehouse, L. N., Gatehouse, J. A., Robinson, N. J. and Croy, R. R. D. (1990) FEBS Lett. **262**, 29–32

Evans, K. M., Gatehouse, J. A., Lindsay, W. P., Shi, J., Tommey, A. M. and Robinson, N. J. (1992) Plant Mol. Biol. **20**, 1019–1028

Fogel, S. and Welch, J. W. (1982) Proc. Natl. Acad. Sci. U.S.A. **79**, 5342–5346

Forde, B. G., Heyworth, A., Pywell, J. and Kreis, M. (1985) Nucleic Acids Res. **13**, 7327–7339

Fürst, P., Hu, S., Hackett, R. and Hamer D. H. (1988) Cell **55**, 705–717

Grill, E. (1985) in Abstracts of the Second International Meeting on Metallothionein and Other Low Molecular Weight Metal Binding Proteins, Zurich, p. 24

Grill, E. (1989) in Metal Ion Homeostasis: Molecular Biology and Chemistry (Hamer, D. H. and Winge, D. R., eds.), pp. 282–300, A. R. Liss Inc., New York

Grill, E., Winnacker, E.-L. and Zenk, M. H. (1985) Science **230**, 674–676

Grill, E., Winnacker, E.-L. and Zenk, M. H. (1986a) FEBS Lett. **197**, 115–120

Grill, E., Gekeler, W., Winnacker, E.-L. and Zenk, M. H. (1986b) FEBS Lett. **205**, 47–50

Grill, E., Winnacker, E.-L. and Zenk, M. H. (1987) Proc. Natl. Acad. Sci. U.S.A. **84**, 439–443

Grill, E., Löffler, S., Winnacker, E.-L. and Zenk, M. H. (1989) Proc. Natl. Acad. Sci. U.S.A. **86**, 6838–6842

Grusak, M. A., Welch, R. M. and Kochian, L. V. (1990) Plant Physiol. **93**, 976–981

Gupta, A., Whitton, B. A., Morby, A. P., Huckle, J. W. and Robinson, N. J. (1992) Proc. R. Soc. London B **248**, 273–281

Gupta, A., Morby, A. P., Turner, J. S., Whitton, B. A. and Robinson, N. J. (1993) Mol. Microbiol. **7**, 189–195

Hamer, D. H. (1986) Annu. Rev. Biochem. **55**, 913–951

Hanley-Bowdoin, L. and Lane. B. G. (1983) Eur. J. Biochem. **135**, 9–15

Hayashi, Y., Nakagawa, C. W., Mutoh, N., Isobe, M. and Goto, T. (1991) Biochem. Cell. Biol. **69**, 115–121

Hofmann, T., Kélls, D. I. C. and Lane. B. G. (1984) Can. J. Biochem. Cell Biol. **62**, 908–913

Huckle, J. W., Morby, A. P., Turner, J. S. and Robinson, N. J. (1993) Mol. Microbiol. **7**, 177–187

Jackson, P. J., Unkefer, P. J., Delhaize, E. and Robinson, N. J. (1990) in Environmental Injury to Plants (F. Katterman, ed.), pp. 231–255, Academic Press, San Diego,

Jacobs, F. A., Romeyer, F. M., Beauchemin, M. and Brousseau, R. (1989) Gene **83**, 95–103

Kägi, J. H. R. (1991) Methods Enzymol. **205**, 613–626

Kägi, J. H. R. and Kojima, Y. (1987) Experientia Suppl. **52**, 25–61

Kägi, J. H. R. and Schäffer, A. (1988) Biochemistry **27**, 8509–8515

Kägi, J. H. R. and Vallee, B. L. (1960) J. Biol. Chem. **235**, 3460–3465

Karin, M. (1985) Cell **41**, 9–10

Karin, M., Najarian, R., Haslinger, A., Valenzuela, P., Welsh, J. and Fogel, S. (1984) Proc. Natl. Acad. Sci. U.S.A. **81**, 337–341

Katsuyuki, N., Suzuki, K., Otaki, N. and Kimura, M. (1991) Methods Enzymol. **205**, 174–189

Kawashima, I., Inokuchi, Y., Chino, M., Kimura, M. and Shimizu, N. (1991) Plant Cell Physiol. **32**, 913–916

Kawashima, I., Kennedy, T. D., Chino, M. and Lane, B. G. (1992) Eur. J. Biochem. **209**, 971–976

Kern, S. R., Smith, H. A., Fontaine, D. and Bryan, S. E. (1981) Toxicol. Appl. Pharmacol. **59**, 346–354

Kille, P., Stephens, P. E., Cryer, A. and Kay, J. (1990) Biochim. Biophys. Acta **1048**, 178–186

Kille, P., Winge, D. R., Harwood, J. L. and Kay, J. (1991) FEBS Lett. **295**, 171–175

Klemsdal, S. S., Hughes, W., Lönneborg, A., Aalen, R. B. and Olsen, O. A. (1991) Mol. Gen. Genet. **228**, 9–16

Kojima, Y. (1991) Methods Enzymol. **205**, 8–10

Kondo, N., Imai, K., Isobe, M., Goto, T., Murasugi, A., Wada-Nakagawa, C. and Hayashi, Y. (1984) Tetrahedron Lett. **25**, 3869–3872

Kreis, M., Shewry, P. R., Forde, B., Miflin, B. J. (1985) Oxford Surv. Plant Mol. Cell. Biol. **2**, 253–317

Labbé, S., Prevost, J., Remondelli, P., Leone, A. and Séguin, C. (1991) Nucleic Acids Res. **19**, 4225–4231

Labbé, S., Larouche, L., Mailhot, D. and Séguin, C. (1993) Nucleic Acids Res. **21**, 1549–1554

Lane, B., Kajioka, R. and Kennedy, T. (1987) Biochem. Cell Biol. **65**, 1001–1005

Laulhere, J.-P. and Briat, J.-F. (1993) Biochem. J. **290**, 693–699

Lesuisse, E. and Labbé, P. (1992) Plant Physiol. **100**, 769–777

Lesuisse, E., Horion, B., Labbe, P. and Hilger, F. (1991) Biochem. J. **280**, 545–548

Lin, C. M. and Kosman, D. J. (1990) J. Biol. Chem. **265**, 9194–9200

Löffler, S., Hochberger, A., Grill, E., Winnacker, E.-L. and Zenk, M. H. (1989) FEBS Lett. **258**, 42–46

Margoshes, M. and Vallee, B. L. (1957) J. Am. Chem. Soc. **79**, 4813–4814

Mehra, R. K., Tarbet, E. B., Gray, W. R. and Winge, D. R. (1988) Proc. Natl. Acad. Sci. U.S.A. **85**, 8815–8819

Meuwly, P. and Rauser, W. E. (1992) Plant Physiol. **99**, 8–15

Morby, A. P., Turner, J. S., Huckle, J. W. and Robinson, N. J. (1993) Nucleic Acids Res. **21**, 921–925

Münger, K. and Lerch, K. (1985) Biochemistry **24**, 6751–6756

Münger, K., German, U. A. and Lerch, K. (1987) J. Biol. Chem. **262**, 7363–7367

Murasugi, A., Wada, C. and Hayashi, Y. (1981) Biochem. Biophys. Res. Commun. **103**, 1021–1028

Mutoh, N. and Hayashi, Y. (1988) Biochem. Biophys. Res. Commun. **151**, 32–39

Nielson, K. B. and Winge, D. R. (1984) J. Biol. Chem. **259**, 4941–4946

Okumura, N., Nishizawa, N.-K., Umehara, Y. and Mori, S. (1991) Plant Mol. Biol. **17**, 531–533

Olafson, R. W., McCubbin, W. D. and Kay, C. M. (1988) Biochem. J. **251**, 691–699

Olsen, O.-A., Jakobsen, K. S. and Schmelzer, E. (1990) Planta **181**, 462–466

Ortiz, D. F., Kreppel, L., Speiser, D. M., Scheel, G., McDonald, G. and Ow, D. W. (1992) EMBO J. **11**, 3491–3499

Palmiter, R. D. (1987) Experientia Suppl. **52**, 63–80

Palmiter, R. D., Brinster, R. L., Hammer, R. E., Trumbauer, M. E., Rosenfeld, M. G., Birnberg, N. C. and Evans, R. M. (1982) Nature (London) **300**, 611–615

Palmiter, R. D., Norstedt, G., Gelinas, R. E., Hammer R. E. and Brinster, R. L. (1983) Science **222**, 809–814

Radtke, F., Heuchel, R., Georgiev, O., Hergersberg, M., Gariglio, M., Dembic, Z. and Schaffner, W. (1993) EMBO J. **12**, 1355–1362

Rauser, W. E. (1990) Annu. Rev. Biochem. **59**, 61–86

Reese, R. N. and Winge, D. R. (1988) J. Biol. Chem. **263**, 12832–12835

Reese, R. N., White, C. A. and Winge, D. R. (1992) Plant Physiol. **98**, 225–229

Ric De Vos, C. H., Vonk, M. J., Vooijs, R. and Schat, H. (1992) Plant Physiol. **98**, 853–858

Riordan, J. F. and Vallee, B. L. (1991) Methods Enzymol. **205**

Robinson, N. J. (1990) in Heavy Metal Tolerance in Plants (Shaw, A. J., ed.), pp. 195–214, CRC Press, Boca Raton

Robinson, N. J. and Jackson, P. J. (1986) Physiol. Plant **67**, 499–506

Robinson, N. J., Barton, K., Naranjo, C. M., Sillerud, J. O., Trewhella, J., Watt, K. and Jackson, P. J. (1985) in Abstracts of the Second International Meeting on Metallothionein and Other Low Molecular Weight Metal Binding Proteins, Zurich, p. 25

Robinson, N. J., Ratliff, R. L., Anderson, P. J., Delhaize, E., Berger, J. M. and Jackson, P. J. (1988) Plant Sci. **56**, 197–204

Robinson, N. J., Delhaize, E., Lindsay, W. P., Berger, J. M. and Jackson, P. J. (1990) in Sulfur Nutrition and Sulfur Assimilation in Higher Plants (Rennenberg, H., Brunold, C. H., DeKok, L. J. and Stulen, I., eds.), pp. 235–240, SPB Academic Publishing, The Hague

Robinson, N. J., Evans, I. M., Mulcrone, J., Bryden, J. and Tommey, A. M. (1992) Plant Soil **146**, 291–298

Romeyer, F. M., Jacobs, F. A., Masson, L., Hanna, Z. and Brousseau, R. (1988) J. Biotechnol. **8**, 207–220

Romeyer, F. M., Jacobs, F. A. and Brousseau, R. (1990) Appl. Environ. Microbiol. **56**, 2748–2754

Rüegsegger, A. and Brunold, C. H. (1992) Plant Physiol. **99**, 428–433

Salt, D. E., Thurman, D. A., Tomsett, A. B. and Sewell, A. K. (1989) Proc. R. Soc. London B **236**, 79–89

Scheller, H. V., Huang, B., Hatch, E. and Goldsbrough, P. B. (1987) Plant Physiol. **85**, 1031–1035

Shewry, P. R. and Tatham, A. S. (1990) Biochem. J. **267**, 1–12

Silar, P. and Wegnez, M. (1990) FEBS Lett. **269**, 273–276

Speiser, D. M., Abrahamson, S. L., Banuelos, G. and Ow, D. W. (1992a) Plant Physiol. **99**, 817–821

Speiser, D. M., Ortiz, D. F., Kreppel, L., Scheel. G., McDonald, G. and Ow, D. (1992b) Mol. Cell. Biol. **12**, 5301–5310

Steffens, J. C. (1990) Annu. Rev. Plant Physiol. Plant Mol. Biol. **41**, 533–575

Takahashi, K. (1991) OWL; accession number X62818, EMBL, Heidelberg, Germany

Thiele, D. J. (1992) Nucleic Acids Res. **20**, 1183–1191

Thornalley, P. J. and Vasak, M. (1985) Biochim. Biophys. Acta **827**, 36–44

Tommey, A. M., Shi, J., Lindsay, W. P., Urwin, P. E. and Robinson, N. J. (1991) FEBS Lett. **292**, 48–52

Tomsett, A. B. and Thurman, D. A. (1988) Plant Cell Environ. **11**, 383–394

Tsujikawa, K., Imai, T., Katutani, M., Kayamori, Y., Mimura, T., Otaki, N., Kimura, M., Fukuyama, R. and Shimizu, N. (1991) FEBS Lett. **283**, 239–242

Turner, J. S., Morby, A. P., Whitton, B. A., Gupta, A. and Robinson, N. J. (1993) J. Biol Chem. **268**, 4494–4498

Vallee, B. L. (1991) Methods Enzymol. **205**, 3–7

Vallee, B. L., Coleman, J. E. and Auld, D. S. (1991) Proc. Natl. Acad. Sci. U.S.A. **88**, 999–1003

Vasák, M. and Armitage, I. (1986) Environ. Health Perspect. **65**, 215–216

Verkleij, J. A. C., Koevoets, P., Riet, J. V., Bank, R., Nijdam, Y. and Ernst, W. H. O. (1990) Plant Cell Environ. **13**, 913–921

Vögeli-Lange, R. and Wagner, G. J. (1990) Plant Physiol. **92**, 1086–1093

Weig, A. and Komor, E. (1992) OWL; accession number L02306, EMBL, Heidelberg, Germany

Welch, R. M., Norvell, W. A., Schaeffer, S. C., Shaff, J. E. and Kochian, L. V. (1993) Planta **190**, 555–561

Zeng, J., Heuchel, R., Schaffner, W. and Kägi, J. H. R. (1991a) FEBS Lett. **279**, 310–312

Zeng, J., Vallee, B. L. and Kägi, J. H. R. (1991b) Proc. Natl. Acad. Sci. U.S.A. **88**, 9984–9988

Biochem. J. (1993) **289**, 17–24 (Printed in Great Britain)

REVIEW ARTICLE

Biology of the Rap proteins, members of the *ras* superfamily of GTP-binding proteins

Gary M. BOKOCH

Departments of Immunology and Cell Biology, IMM-14, Room 221, The Scripps Research Institute, 10666 N. Torrey Pines Road, La Jolla, CA 92037, U.S.A.

INTRODUCTION

A large superfamily (> 50 members) of monomeric GTP-binding proteins structurally related to the *ras* oncogene proteins has been described in the past few years (Bourne et al., 1991; Grand and Owen, 1991; Hall, 1990; Valencia et al., 1991). These proteins have been implicated in the regulation of a diverse array of cellular processes, utilizing their ability to bind and hydrolyse GTP as a means to regulate the reversible interactions and/or activities of macromolecules involved in these processes. Based upon sequence homology, the *ras* superfamily can be divided into four subfamilies, *ras*, *rho*, *rab* and *arf* (Valencia et al., 1991; Kahn et al., 1992). The *ras* proteins are essential components of receptor-mediated signaling pathways controling cell proliferation and differentiation; the *rho* proteins are involved in cytoskeletal assembly and NADPH oxidase regulation in phagocytes; and the *rab* and *arf* proteins regulate the transport of vesicles between intracellular compartments and/or the plasma membrane (see Hall, 1990).

The mammalian *rap* proteins, which are the focus of this Review, are members of the *ras* family of GTP-binding proteins, and share highly conserved structural motifs with the *ras* transforming proteins. Overall, the *rap* proteins share ~ 50% sequence similarity with *ras* and, like *ras*, are found in nearly all tissues. Two *rap* families, designated *rap1* and *rap2*, have been identified to date. Within each family are two members denoted as A and B, i.e. *rap1A* and *rap1B* (see Figure 1). As is the case with the majority of the low-molecular-mass GTP-binding proteins, the actual role(s) of the Rap proteins in cellular function have not yet been elucidated in detail. There is evidence however that Rap proteins exert biological activities in at least two cellular arenas: first in that of cellular growth and differentiation control, and secondly in a phagocyte-specific enzyme system responsible for the generation of microbicidal oxygen radicals. The latter suggests either that Rap can play very specific biological roles in certain cells or, alternatively, that Rap serves a very general function which can involve different proteins and different effector systems depending on the cell involved. In this Review, I examine current knowledge of the biology of the Rap proteins.

CLONING AND ISOLATION OF RAP PROTEINS

The Rap proteins were cloned and/or purified by a number of laboratories within a very short time period by using a variety of strategies. Pizon et al. (1988a) identified the *rap1A* and *rap2A* genes by screening a Raji human Burkitt lymphoma cell library with probes based upon the *Dras3* gene previously identified in *Drosophila* (Schejter and Shilo, 1985). This *Drosophila ras*-related protein differed from the other members of the *ras* superfamily known at that time in that it possessed a threonine-for-glutamine substitution at residue 61 of the highly conserved DTAGQE sequence found in positions 57–62 of *ras*. *rap1B* was

identified very soon afterward by Pizon et al. (1988b) using the same strategy. *rap1A* (termed Krev-1) was isolated by Kitayama et al. (1989) from a human fibroblast cDNA expression library based on its ability to cause reversion of the transformed phenotype of v-Ki-*ras*-transformed DT fibroblasts. Kawata et al. (1988) purified a 22 kDa GTP-binding protein (termed smg p21) from bovine brain, sequenced, and cloned the protein, and found it to be identical to *rap1B*. Takai's laboratory has subsequently purified Rap1B from human platelets (Ohmori et al., 1989) and bovine aortic smooth muscle (Kawata et al., 1989a). Rap1A has also been purified, sequenced, and cloned from human neutrophils (Bokoch et al., 1988; Quilliam et al., 1990). The identification of a fourth member of the *rap* family, *rap2B*, emerged through molecular cloning from a platelet cDNA library (Ohmstede et al., 1990).

rap1 and *rap2* proteins are 70% identical at the amino acid level (Figure 1). *rap1A* and 1B differ by only nine out of 184 amino acids (95% identity), with the sole region of substantial nonidentity being between positions 171–189 of the C-terminus. Similarly, *rap2A* and 2B differ by 18 out of 183 amino acids (90% identity), with the major area of divergence at residues 170–183 of the C-terminus.

POST-TRANSLATIONAL MODIFICATION OF RAP PROTEINS

Isoprenylation

It has been established that a diverse group of yeast and mammalian proteins are modified post-translationally by the covalent addition of an isoprenoid group (reviewed by Maltese, 1990; Cox and Der, 1992). This event is usually signalled by a C-terminal CAAX motif (C, cysteine; A, aliphatic; X, any residue) present in the proteins. The sequence of events involved in post-translational processing of the low-molecular-mass GTP-binding proteins includes the addition of a C_{15} (farnesyl) or a C_{20}

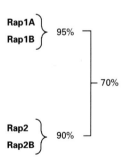

Figure 1 The Rap proteins, members of the *ras* superfamily

Four *rap* subtypes have been identified, as described in the text. The percentage of sequence similarity at the amino acid level between the indicated proteins is shown.

Abbreviations used: GTPγS, guanosine 5′-[γ-thio]triphosphate; GAP, GTPase-activating protein; GDS, GDP/GTP dissociation stimulator. Note that Rap2 is denoted as Rap2A in most of the literature cited.

(geranylgeranyl) isoprenyl group to the cysteine residue of the CAAX motif via a thioether linkage, proteolytic cleavage of the AAX residues, and carboxylmethylation of the now terminal and isoprenylated cysteine. While the specific contribution of isoprenoid modification to protein function is not known, it is apparent that such modifications are critical for the biological activities of isoprenylated proteins.

Like many of the other members of the *ras* superfamily, the Rap proteins contain a C-terminal CAAX consensus motif which directs post-translational isoprenylation (Maltese, 1990). Rap1A, ending in CLLL, has been shown to be modified by a geranylgeranyl group at Cys-181, with proteolytic truncation and subsequent carboxylmethylation also shown to take place (Buss et al., 1991). Rap1B and Rap2B, ending in CLQL and CVIL, appear to undergo similar modifications (Kawata et al., 1990; Winegar et al., 1991). Carboxylmethylation of Rap1B (Huzoor-Akbar et al., 1991) and Rap1A (Quilliam and Bokoch, 1992) has been found to be stimulated by guanosine 5'-[γ-thio]triphosphate (GTPγS); the significance of this interesting observation in terms of Rap modification and function *in vivo* is unknown, but suggests the possibility of regulatory significance for this modification.

It appears to be a general rule that isoprenylated proteins terminating in leucine are geranylgeranylated, while those ending in serine or methionine are farnesylated (Moores et al., 1991; Cox and Der, 1992). It is of note that Rap2B, which terminates in CVIL, is geranylgeranylated while the very closely related Rap2A, which terminates in CNIQ, is farnesylated (Farrell et al., 1992). These two structurally similar forms of Rap may thus partition into different membranes within cells, perhaps reflecting distinct biochemical activities. The concept that farnesylation versus geranylgeranylation of a low-molecular-mass GTP-binding protein may impart specific functions and/or subcellular localizations to that protein has been explored to a limited extent. A chimaeric protein containing the N-terminal half of oncogenic H-Ras linked to the C-terminal half of Rap1A was geranylgeranylated and retained the ability to transform cells (Buss et al., 1991). Cox et al. (1992) found that switching the C-terminal CAAX sequences of H-Ras and Rap1A did not prevent oncogenic Ras activity, nor the ability of Rap1A to antagonize Ras transforming activity. Farnesyl and geranylgeranyl moieties are thus functionally interchangeable for these biological activities. However, in this same study the expression of moderate levels of geranylgeranyl-modified normal Ras inhibited the growth of untransformed NIH 3T3 cells. These findings suggest that normal Ras function may specifically require protein modification by a farnesyl isoprenoid. Whether this reflects any changes in the subcellular localization of the chimaeric proteins however was not examined in this study.

The reported subcellular distribution of the Rap proteins seems to be distinct from that of Ras (see Grand and Owen, 1991). Both Beranger et al. (1991a) and Kim et al. (1990) observed Rap1 to be associated with subcellular fractions devoid of Ras. Beranger et al. (1991a) showed by immunofluoresence labelling and co-fractionation with Golgi-specific markers that Rap1 was associated with a Golgi-like structure in Rat-1 fibroblasts and HEP2 epidermoid carcinoma cells, while Ras was clearly detected in plasma membrane. Similarly, Rap2 was observed to associate with a low density structure that morphologically overlapped with the endoplasmic reticulum in Rat-1 cells (Beranger et al., 1991b). The structural and/or functional basis for this differential subcellular localization of Rap and Ras has not yet been determined. Rap1 and Rap2 proteins have been localized to both the plasma membrane and specific granules of human neutrophils (Maridonneau-Parini and de Gunzburg,

1992; Quinn et al., 1992a). This may relate to the role of Rap1 in the NADPH oxidase system of phagocytic cells (discussed below).

Phosphorylation

After the purification of Rap1A and Rap1B, these proteins were soon found to serve as substrates for phosphorylation by cyclic AMP-dependent protein kinase *in vitro* (Bokoch and Quilliam, 1990; Hoshijima et al., 1988; Kawata et al., 1989b; Lerosey et al., 1991; Quilliam et al., 1991). Both Rap proteins incorporate phosphate to a level of 1 mol/mol, suggesting a single site of phosphorylation. The site at which phosphorylation occurs has been shown to be Ser-180 in Rap1A (Quilliam et al., 1991). This is contained within the sequence KKKPKKKSC, which is similar to consensus cyclic AMP-dependent protein kinase phosphorylation sites (Kemp and Pearson, 1990). Rap1B is phosphorylated solely on Ser-179 within the sequence GKARKKSSC, again similar but not identical to "classical" cyclic AMP-dependent kinase motifs (Hata et al., 1991). In contrast, the Rap2 proteins lack such (potential) consensus motifs for phosphorylation near the C-terminus and do not serve as substrates for phosphorylation by cyclic AMP-dependent, or any known, protein kinases.

Phosphorylation is not influenced by whether Rap is in a GDP-bound versus GTP-bound state, and stoichiometric phosphorylation of Rap1A or 1B *in vitro* has no effect on the guanine nucleotide-binding or hydrolysis properties of the two proteins, nor the responsiveness to Rap-GAP (Bokoch and Quilliam, 1990; Hoshijima et al., 1988). The phosphorylation sites of both Rap1A and Rap1B are adjacent to the cysteine residue which is geranylgeranylated. Possible influences of the phosphorylation on post-translational processing of either Rap protein remain to be investigated. A role for Rap1 phosphorylation in regulating the interaction of Rap1 with other macromolecules will be discussed below.

Phosphorylation of Rap1A has been shown to occur in intact HL-60 cells which had been differentiated into neutrophil-like cells in response to dibutyryl cyclic AMP, forskolin, prostaglandin E$_1$ or isoprenaline (Quilliam et al., 1991). Phosphorylation of Rap1B occurs in human platelets in response to prostaglandin E$_1$ or the prostacyclin analogue iloprost (Kawata et al., 1989b; Lapetina et al., 1989; Siess et al., 1990). A form of Rap1 has also been found to be phosphorylated in intact fibroblasts by exposure to 8-bromo cyclic AMP (Lerosey et al., 1991). The possibility that Rap may mediate some of the cellular effects of cyclic AMP will be considered later in this Review.

The Rap proteins clearly do not serve as substrates for protein kinase C, myosin light chain kinase or insulin/EGF receptor tyrosine kinases *in vitro* (Bokoch and Quilliam, 1990; Hoshijima et al., 1988; Kawata et al., 1989b; Quilliam et al., 1991). Recently however, Rap1B has been reported to be phosphorylated by cyclic GMP-dependent protein kinase (Miura et al., 1992) and a neuronal calcium/calmodulin-dependent protein kinase (Sahyoun et al., 1991) *in vitro*. Both kinases modified the same serine residue in Rap1B that is phosphorylated by cyclic AMP-dependent kinase. However, phosphorylation of Rap by such kinases has not been demonstrated to occur in intact cells.

RAP1 REGULATORY PROTEINS

It is thought that the low-molecular-mass GTP-binding proteins, like the heterotrimeric signalling GTP-binding proteins, are regulated by a cycle of GTP binding and hydrolysis. The Rap proteins have clearly been shown to bind and hydrolyse GTP

(Bokoch and Quilliam, 1990; Kawata et al., 1988; Lerosey et al., 1991). The rate of GTP binding to Rap is limited at physiological concentrations of Mg^{2+} by the dissociation of GDP from the protein, suggesting that such exchange must be catalysed *in vivo* by other protein(s). Hydrolysis of GTP to GDP also occurs at a very low rate on the Rap proteins (0.0005–0.010 mol/min), and would be too low to terminate the action of Rap–GTP effectively without some means to enhance GTP hydrolysis. GTPase-activating proteins (GAPs) which stimulate the rate of GTP hydrolysis by Rap have been identified and are described below. A second regulatory protein, capable of stimulating the dissociation of GDP/GTP from Rap (Rap GDS), has also been purified and cloned (see below).

GTPase-activating proteins (GAPs)

Several GAP activities that appear to be specific for Rap1 have been detected in the plasma membrane and cytosol of a number of cell types. Polakis et al. (1991) purified a membrane-associated 88 kDa GAP from HL-60 cells. This GAP was subsequently cloned and shown to be a unique protein which did not exhibit similarity to any of the GAPs specific for Ras (Rubinfeld et al., 1991). This form of Rap-GAP was not ubiquitously expressed, and was most abundant in fetal tissues and certain tumour cell lines. Interestingly, the expression of the 88 kDa Rap-GAP was decreased in HL-60 cells which had been induced to differentiate by dimethyl sulphoxide.

Two chromatographically resolvable peaks of Rap-GAP activity have been observed in human platelets (Ueda et al., 1989) and bovine brain (Kikuchi et al., 1989) cytosol. One such cytosolic GAP was purified (Nice et al., 1992) as a 55 kDa protein. Limited amino acid sequence information indicates that this 55 kDa GAP is very closely related, if not identical, to the 88 kDa GAP purified by Polakis et al. (1991). It is thus not clear how many distinct forms of Rap-GAP exist and what the significance of the cytosolic versus membrane localization of this protein might be. The changes in Rap-GAP expression upon HL-60 cell differentiation and the identification of multiple, phosphorylated forms of this GAP (Polakis et al., 1992) suggest that Rap-GAP may be regulated in a very specific manner in order to control Rap activity.

GDP/GTP dissociation stimulator (GDS)

A guanine nucleotide exchange protein which is active on Rap1 has been identified and purified from bovine brain cytosol (Yamamoto et al., 1990). This 53 kDa protein has been cloned (Kaibuchi et al., 1991) and shown to be a unique protein with limited amino acid sequence similarity to the CDC25 and SCD25 proteins which may regulate the GDP/GTP exchange reaction of the yeast Ras 2 protein. Rap-GDS appears to interact with Rap1A and 1B as a 1:1 stoichiometric complex (Kawamura et al., 1991a). This interaction involves, at least in part, the C-terminal portion of Rap1, as indicated by: (a) the requirement for the post-translationally processed form of Rap1 for GDS binding and activity (Hiroyoshi et al., 1991); (b) the demonstration that proteolytic removal of an ∼ 1000 Da fragment of the C-terminus prevents binding and activation of Rap1 by GDS (Hiroyoshi et al., 1991); (c) geranylgeranylated synthetic peptides representing the C-terminus of Rap1 inhibit GDS action (Shirataki et al., 1991); and (d) phosphorylation of Rap1B at the C-terminus enhances its interaction with GDS (Hata et al., 1991; Itoh et al., 1991). It is of interest that this "Rap1-GDS" is also able to catalyse GDP/GTP exchange for other post-translationally processed low-molecular-mass GTP-binding proteins, including K-Ras, Rho A and Rac1 (Mizuno et al., 1991). These

proteins have in common a lysine/arginine-rich cationic region at their C-terminus which may enable them all to interact with the GDS.

A number of anionic phospholipids have been reported to antagonize the GDP/GTP exchange activity of Rap-GDS, and to reduce markedly the ability of GDS to stimulate Rap1B GTP binding (Kawamura et al., 1991b). The effect of these lipids was reduced when Rap1B was phosphorylated by cyclic AMP-dependent protein kinase (Itoh et al., 1991). Takai and associates (Itoh et al., 1991; Kawamura et al., 1991b) have suggested that Rap1B may bind to anionic lipids in the plasma membrane through the polycationic C-terminal domain, suppressing GDS action. Phosphorylation of Rap1B by cyclic AMP-dependent protein kinase would decrease the ionic interaction of the lipids with Rap1B in this region, sensitizing Rap1B to the action of the GDS. GDS binding to Rap1B is associated with the release of Rap1B from the membrane as a GDS complex (Kawamura et al., 1991a) and this may occur *in vivo* as a consequence of Rap1B phosphorylation (Lapetina et al., 1989). The operation of such a mechanism would implicate cyclic AMP as a critical regulator of Rap activity and could provide a link between this important hormonally-regulated second messenger system and biological activities regulated by Rap.

BIOLOGICAL ACTIVITIES OF RAP1 PROTEIN

Antagonism of Ras by Rap1

The isolation of *rap*1A by Noda and colleagues (Kitayama et al., 1989) as a cDNA which was able to suppress transformation of NIH3T3 cells by v-Ki *ras* led to the hypothesis that the Rap1A protein might directly antagonize Ras by competing for a common downstream target. Such a mechanism was postulated based upon the conservation in Rap1A and 1B of the putative "effector" domain region (amino acids 32–44) found to be crucial for the transforming activity of Ras and for GAP binding (Adari et al., 1988). It was possible however that Rap might antagonize Ras action by other mechanisms, such as directly activating a pathway which regulates cell growth and/or differentiation in a negative manner, or by indirect means, such as activation of other enzymes [such as kinases (Labadia et al., 1992) or phosphatases] able to inhibit Ras growth signals (see Figure 2).

In support of the "competition" hypothesis, Kitayama et al. (1990*a*) showed that an Asp-38 → Ala or Asn point mutation in the effector domain of *rap*1A markedly inhibited the ability of *rap*1A to cause phenotypic reversion of *ras*-transformed cells. Mutations at position 12 (Gly → Val) and at position 59 (Ala → Thr), which are thought to activate *rap*1 by maintaining it in the GTP-bound form (by analogy with *ras*), substantially enhanced its reversion-inducing activity. The latter observation suggested that *rap*1A was more "active" in the GTP-bound form.

More direct evidence that such a competitive model of Rap action was possible came from studies by Hata et al. (1990) and Frech et al. (1990). Both groups used purified components *in vitro* to demonstrate that Rap1A (Frech et al., 1990) or 1B (Hata et al., 1990) was able to compete with Ras for binding to Ras-GAP. Binding to Ras-GAP was more effective when Rap1 was bound with GTP than with GDP, with Frech et al. (1990) reporting that the GDP form of Rap1A had an affinity of at least 100-fold less than the GTP form, while Hata et al. (1990) only found a 2–3-fold difference between the two forms of Rap1B. Ras-GAP did not stimulate GTP hydrolysis by Rap1, as described in Quilliam et al. (1990), suggesting that the association of Rap1A with GAP produced a catalytically-inactive complex.

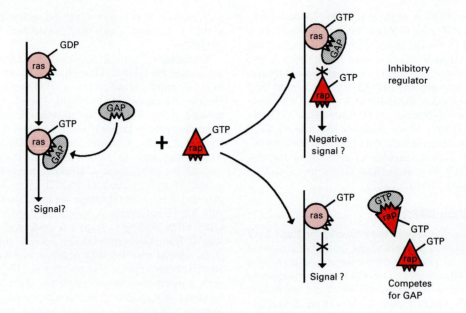

Figure 2   Hypothetical mechanisms through which Rap might antagonize Ras downstream signalling

· ʌʌʌ denotes the "effector" domain of Ras/Rap. GAP refers to Ras-GAP.

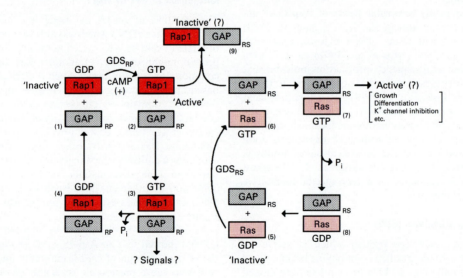

Figure 3   Model of Rap/Ras regulatory cycles

This Figure describes biochemical states of Ras and Rap1 and how the two GTP-binding proteins might interact based upon the concept that Rap acts as to bind Ras-GAP in a competitive manner, as described in the text. (1) In the basal or "inactive" state, Rap1 exists as the GDP form which is not associated with Rap GAP (GAP$_{RP}$). (2) Under the influence of a Rap-active GDS (GDS$_{RP}$), an interaction which may be regulated by cyclic AMP, GTP is exchanged for GDP on Rap1, producing the "active" form of Rap1. This active Rap1 may exert biological activities of its own, or may act in concert with GAP$_{RP}$ to pass on downstream signals. The Rap1-GTP form has a high affinity for Ras-GAP (GAP$_{RS}$), effectively binding to GAP$_{RS}$ to form a presumably inactive complex. This interaction would account for the anti-oncogenic action of Rap1 toward Ras. (3) Rap1-GTP binds to GAP$_{RP}$, which could represent an active signalling complex or (4) which catalyses the hydrolysis of Rap1-GTP to GDP, causing Rap1 to dissociate from GAP$_{RP}$ and return to the "inactive" state. (5) Ras with GDP bound also represents an "inactive" state which (6) is acted on by a Ras-active GDS (GDS$_{RS}$) to exchange bound GDP for GTP, producing a biologically "active" Ras. (7) Ras-GTP binds to GAP$_{RS}$, leading to the generation of signals which result in the biological manifestations of Ras activity. (8) The action of GAP$_{RS}$ converts Ras to the GDP form, which releases GAP$_{RS}$ and returns to the initial "inactive" state.

Interestingly, the affinity of Rap1A–GTP for Ras-GAP was 50–100-fold greater than that of Ras–GTP. Since the concentration of Rap1 in cells such as platelets (Ohmori et al., 1989) and neutrophils (Quilliam et al., 1991) seems to be at least 10 times greater than that of Ras, this suggests that Rap1A might be able to limit the amount of GAP available for interaction with Ras *in vitro*.

The ability of Rap1A to antagonize Ras action by a competitive mechanism in an intact membrane system was demonstrated by Yatani et al. (1991). This study utilized the M$_2$ muscarinic receptor-regulated K$^+$ channel of atrial membranes, which is inhibited by the action of Ras and Ras-GAP acting in concert (Yatani et al., 1990). Using a patch-clamp technique to manipulate the levels of Rap1A, Ras and GAP in the system, it was shown that: (1) Rap1A antagonized the effect of Ras-GAP on channel opening in a manner that was inversely proportional to the level of GAP added; (2) antagonism was dependent on an intact effector domain in Rap1A; (3) the inhibitory effect of Rap1A could be overcome by the addition of exogenous Ras; and (4) Rap1A did not antagonize a form of GAP (GAP32) whose ability to inhibit M_2 muscarinic receptor-regulated K^+ channels was independent of Ras. These results indicated that Rap1A was acting by a competitive mechanism, and that such inhibition could occur at picomolar concentrations of Rap1A, levels which are likely to occur in normal cells. The ability of Rap1A to suppress Ras-GAP action in this system, and also the antagonistic activity of Rap1B in *Xenopus* oocytes (Campa et al., 1991), indicates that this biological activity of Rap1 is not limited to the pathway leading to cell transformation by Ras. These studies also indicate that the antagonism of Ras action by Rap1 occurs via a competitive mechanism involving the "effector domain" of Rap1. A number of studies have localized the suppressive activity of Rap1 to the N-terminal portion of the molecule (Buss et al., 1991; Kitayama et al., 1990b; Zhang et al., 1990), and have identified residues 26, 27, 30, 31 and 45 as crucial in this regard (Marshall et al., 1991; Nur-e-Kamal et al., 1992). It is of interest then to note that both of the Rap2 proteins, as well as R-Ras, also contain an "effector domain" identical to Ras, yet do not suppress the ability of Ras to transform cells (Schweighoffer et al., 1990; Jimenez et al., 1991). Although R-Ras differs from Rap1 in the crucial amino acids adjacent to the "effector domain", Rap2 is identical to Rap1 in these positions. Consideration of these facts indicates that additional structural components need to be accounted for in order to understand antagonism of Ras by Rap1.

The physiological significance of the inhibition of Ras action by Rap1 is still uncertain. There is as yet little data to indicate that Rap1 plays such a regulatory role in normal cells. In fact, in some cell models of Ras action, Rap1 has been reported not to inhibit the actions of Ras (Schweighoffer et al., 1990). In yeast, mammalian *rap* genes can *stimulate* some of the same effector pathways as does H-*ras* (Xu et al., 1990; Ruggieri et al., 1992). It should be pointed out that most models we have discussed here to explain Rap suppression of Ras activity are based on the idea that Ras-GAP serves a downstream signalling function that can be disrupted by Rap (see Figure 3). If, in fact, Ras-GAP acts solely as a negative regulator of Ras activity in certain cells by keeping it in a GDP-bound state, then disruption of the binding of GAP to Ras could actually result in increased Ras activity!

It is tempting to speculate that, if Rap1 is indeed a physiological suppressor of Ras function, modulation of the guanine nucleotide state of Rap1 would play an important role in regulating the transduction of growth and differentiation signals via Ras (see Figure 3). One could thus envisage that changes in the activity of Rap1 GAPs and GDSs, etc., would also produce marked effects on Ras activity indirectly through their ability to regulate Rap1–GTP formation and thus binding to Ras-GAP(s). Mutations in Rap1 or in Rap1-associated regulatory components might then play a significant role in the pathogenesis of mammalian tumours. A marked decrease in the levels of expression of *rap*1 mRNA was reported in several types of tumours not normally associated with *ras* mutations (e.g. salivary gland fibrosarcomas and adenocarcinomas) by Culine et al. (1989). *rap*1B has been mapped to a chromosomal location near breakpoints associated with a number of malignant and benign neoplasms (Rousseau-Merck et al., 1990). Several investigations of Rap1 levels (Hong et al., 1990; Hsu and Gould, 1991) and loss of heterozygosity (Young et al., 1992) in certain types of tumours have proven negative. However, Kyprianou and Taylor-Papadi-

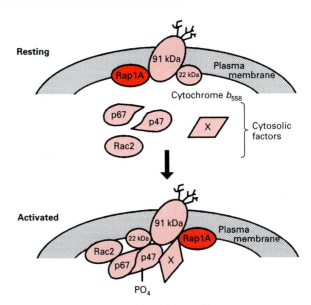

Figure 4   The phagocyte respiratory burst oxidase

Two GTP-binding proteins, Rac2 and Rap1A, appear to be involved in this enzyme system. 91 kDa and 22 kDa represent the subunits of the cytochrome b_{558}. Changes in the relative positions of these two subunits in the Figure are intended to represent schematically the potential conformational changes occuring in the cytochrome upon oxidase activation. p47, p67 and X are the cytosolic components of the NADPH oxidase. PO_4 indicates the known incorporation of phosphate groups into p47 during activation of the oxidase.

mitriou (1992) reported that Rap1 levels in azatyrosine-induced revertants of *ras*-transformed human mammary epithelial cells were significantly increased. Clearly, further studies along such lines are warranted.

Interaction of Rap1A with the phagocyte NADPH oxidase

The question of whether Rap1 serves as a physiological antioncogene remains an unsettled issue. The abundance of Rap1 in a number of untransformed cells, such as platelets and neutrophils, the existence of multiple Rap-specific GAPs which can potentially serve unique effector pathways, and the possible regulation of Rap1 and Rap-GAP by phosphorylation, all suggest that Rap may regulate other functions normally in cells. In support of this hypothesis, there is evidence that Rap1A may play some role in regulation of the NADPH oxidase system of human neutrophils.

Human neutrophils and other phagocytic cells respond rapidly to contact with opsonized micro-organisms by undergoing a "respiratory burst" in which molecular oxygen is reduced to form superoxide anion (O_2^-). O_2^- is subsequently converted to other toxic oxygen metabolites (Clark, 1990). This respiratory burst is catalysed by the NADPH oxidase, a multicomponent enzyme composed of at least four known proteins (see Figure 4). A cytochrome b_{558}, consisting of 91 kDa (gp91$_{phox}$) and 22 kDa (gp22$_{phox}$) subunits, resides in the plasma membrane and membrane of neutrophil-specific granules where it serves as the terminal electron carrier of the oxidase. Two additional cofactors, p47$_{phox}$ and p67$_{phox}$, are cytosolic and may exist as preformed complexes which translocate to the plasma membrane upon phagocyte activation, becoming integral parts of the active oxidase (Clark, 1990; Curnutte, 1992).

Guanine nucleotides, specifically GTP, have been shown to be absolutely required for NADPH oxidase activity in cell-free assay systems (Peveri et al., 1992; Uhlinger et al., 1991), supporting the idea of a GTP-binding protein being involved in oxidase regulation (Bokoch, 1990). Bokoch and Prossnitz (1992) have also shown that treatment of HL-60 cells with drugs able to block isoprenoid metabolism prevents a respiratory burst response in these cells after they have been differentiated into a neutrophil-like form by dimethyl sulphoxide. This isoprenoid requirement is presumably due to the need for isoprenoids in the post-translational processing of a low-molecular-mass GTP-binding protein. Indeed, Rac2 (Knaus et al., 1991) and possibly Rac1 (Abo et al., 1991) have been identified as important stimulatory regulators of the NADPH oxidase in human neutrophils. There are indications though that Rap1A must also play some modulatory role in this system. The first suggestion of this came when Quinn et al. (1989) observed the co-isolation of a 22 kDa protein distinct from the cytochrome b 22 kDa subunit during purification of the cytochrome b oxidase component. This protein had an N-terminal amino acid sequence identical to that of Rap1. Evidence that this co-purification was not merely fortuitous (since Rap1 could be separated from the cytochrome at a final sucrose gradient step) was provided by their observation that Rap1 present in neutrophil extracts would also bind to anti-cytochrome 91 kDa or 22 kDa antibody columns.

These observations were confirmed and extended by *in vitro* studies using a pure cytochrome preparation and purified baculovirus-Sf9 cell recombinant Rap1A or Rap1A purified from human neutrophils (Bokoch et al., 1991). Complexes between the two purified proteins were detected by gel filtration analysis. Formation of complex appeared to occur through the 1:1 association of Rap1A with the cytochrome. These results demonstrated that Rap1A binds directly to the cytochrome itself. The ability of Rap1A to interact with the cytochrome when it was in a GDP versus GTPγS-bound state was also examined. Rap1A bound to cytochrome in both the GTP and GDP forms; it appeared that the interaction was more stable when Rap1A was complexed with GTPγS, however, since the ability to observe the Rap1A–GDP–cytochrome complex was variable.

Agonists which elevate cyclic AMP in neutrophils are able to attenuate the respiratory burst in these cells (Rivkin et al., 1975; Sha'afi and Molski, 1988; Mueller et al., 1988). Since Rap1A can be phosphorylated by cyclic AMP-elevating agents in intact HL-60 cells (Quilliam et al., 1991), and stoichiometric phosphorylation of Rap1A *in vitro* does not affect its ability to bind or hydrolyse guanine nucleotides, nor to respond to a cytosolic Rap-GAP (Quilliam et al., 1990; Bokoch and Quilliam, 1990), the possibility of effects on Rap1A–cytochrome binding was examined. Bokoch et al. (1991) observed that the ability of phosphorylated Rap1A–GTPγS to form complexes with cytochrome b was markedly reduced. A potential mechanism by which elevations in neutrophil cyclic AMP could inhibit the respiratory burst response, therefore, is by modulation of Rap1A–cytochrome interactions.

It is of note that Rap1B phosphorylation has been shown to enhance its ability to bind and be stimulated by a GDS (Hata et al., 1991; Itoh et al., 1991). It is possible that Rap1 phosphorylation serves a general regulatory role for this low-molecular-mass GTP-binding protein. Thus, the interaction of Rap1A and 1B with other macromolecules at the C-terminus is postulated to occur, and to be modulated, either positively or negatively, by Rap1 phosphorylation. Such regulation could also extend to K-Ras, which has a similar cationic region at its C-terminus and which has been reported to undergo C-terminal phosphorylation as well (Ballester et al., 1987).

While it has been established that Rap1A can interact with the NADPH oxidase-associated cytochrome b, the significance of this interaction is not yet clear in terms of NADPH oxidase function. Eklund et al. (1991) have reported that an antiserum against a synthetic peptide corresponding to the effector region of Rap1A (amino acids 31–43) was able to totally inhibit oxidase activity in a cell-free system and that activity could be restored by addition of recombinant Rap1A. Similar experiments have proven unsuccessful in our laboratory (G. M. Bokoch, unpublished work). Recently, Mizuno et al. (1992) have also reported that Rap1 was inactive in supporting oxidase activity in a cell-free system. The reasons for these discrepancies are not yet clear. Several laboratories (Maridonneau-Parini and de Gunzburg, 1992; Quilliam et al., 1991; Quinn et al., 1992a) only find Rap1 in neutrophil membrane fractions when localized by specific immunblotting, in contrast to Eklund et al. (1991), who report Rap1 to be cytosolic and to translate from cytosol to membrane upon neutrophil activation by phorbol myristate acetate. The functional role of Rap1A in the NADPH oxidase system is thus not clear. It is possible that Rap1A (unlike Rac2) may play a more subtle role in the system which may not be evident in studies using cell-free assays, which clearly rely on non-physiological activators in a structurally disrupted system. For example, based upon studies of the BUD1 (or RSR1) Rap1A homologue in *Saccharomyces cerevesiae* which indicate it is crucial for positional information relevant to bud site selection (Ruggieri et al., 1992) and upon data which indicate an association of Rap1B with the platelet cytoskeleton upon thrombin activation (Fischer et al., 1990), it is possible that Rap1A in the neutrophil might direct translocation of cytochrome b to specific membrane sites upon cell activation or might mediate interactions between oxidase components and the cell cytoskeleton. Such interactions are thought to be important for oxidase activation and de-activation *in vivo* (Clark, 1990; Heyworth et al., 1991; Nauseef et al., 1991) and would be undetectable in normal cell-free oxidase assays. Recently, Quinn et al. (1992b) reported that the ability of partially purified Rap1-associated cytochrome b to reconstitute NADPH oxidase activity in CGD patient membranes lacking the cytochrome was decreased when the cytochrome was purified to essential homogeneity. This may reflect a requirement for Rap in order to allow the reconstitutive capacity of the cytochrome to occur. Other scenarios are also possible, as discussed in Quilliam and Bokoch (1992).

CONCLUSIONS AND FUTURE PROSPECTS

There are a number of inferences that can be drawn from the information just presented with regard to the biology of Rap. While it has not yet been shown that the ability of Rap1 to antagonize Ras action is of significance *in vivo*, such an action of Rap does occur *in vitro* and seems likely to be relevant to physiological growth regulation and/or tumourigenesis. Interactions of the Rap1 and Ras proteins at the biochemical level are depicted in Figure 3. The fact that Rap1A and 1B serve as excellent substrates for cyclic AMP-dependent protein kinase, as well as a number of other kinases, and that this phosphorylation can regulate the macromolecular interactions that Rap can undergo *in vitro*, also implicates Rap in cellular regulation. Activation of these kinases via cyclic AMP generation, etc. in intact cells is likely to exert significant effects then on the GTP state of Rap (via GDS) and, therefore, its suppression of Ras function. Hopefully, studies directed at such regulatory pathways involving Rap in intact cells will be forthcoming in the near future.

It is known that in some cell types elevations in cyclic AMP levels can lead to tumourigenesis (Dumont et al., 1989; Vallar et al., 1987), but the mechanism of this effect has not been established. A possible role for Rap activity in this process deserves investigation. Stimulation of Rap1 GTP binding via the activation of Rap GDS caused by increases in cyclic AMP would be expected to influence biological actions of Rap in other cellular systems as well. Rap may mediate some of the effects of cyclic AMP in human neutrophils (Quilliam et al., 1991), where cyclic AMP inhibits activity of the NADPH oxidase, and in human platelets (Lazarowski et al., 1990), where activation of phospholipase C is blocked by cyclic AMP. Finally, there are indications that Rap activity may be regulated by hormones *in vivo* through actions on its GAP (Marti and Lapetina, 1992). The concept that the activity of Rap and other low-molecular-mass GTP-binding proteins may be regulated by the seven-trans-membrane-spanning-domain receptors classically linked to the heterotrimeric G-proteins and/or that Rap might regulate the activity of these receptors in turn (Yatani et al., 1991) is an exciting one, with widespread implications for the integration of cell functions.

The association of Rap1A with the NADPH oxidase cyto-chrome *b* component in human neutrophils is more difficult to assess. Unlike oxidase regulation by Rac2, a GTP-binding protein which is expressed solely in cells of haematopoietic lineage (Didsbury et al., 1989), Rap1A is widely expressed and would seem unlikely to be solely a regulatory component for a phago-cyte-specific enzyme. One hypothesis is that Rap1 plays a more general function in the cell, and that this function involves different protein components in different cells. Regulation of cytoskeletal interactions or protein–protein interactions in gen-eral are possibilities that are consistent with the known functions of other low-molecular-mass GTP-binding proteins. This hy-pothesis predicts similar roles for Rap1 in other cells, but the biological consequences of its action would differ completely. Support for this hypothesis will require further work on the biochemical actions of Rap in a variety of cell types.

While at this time we do not fully understand the inter-relationships between biological events regulated by various members of the *ras* superfamily, there are indications that there may be a substantial amount of cross-talk between individual low-molecular-mass GTP-binding proteins. This is indicated by the recent identification of a variety of proteins which appear to contain multiple functional domains capable of GTP binding, GAP, or guanine nucleotide exchange activities directed against more than one GTP-binding protein (described in Hall, 1992; Marx, 1992). It is apparent from yeast genetic studies (Chant and Herskowitz, 1991) that the yeast homologue of Rap1A (BUD1) may interact with CDC42 to co-ordinate bud site selection and assembly. CDC42 is a low-molecular-mass GTP-binding protein which is closely related to Rac (Shinjo et al., 1990; Didsbury et al., 1989). It is very suggestive then that both Rap1A and Rac are associated with regulation of the NADPH oxidase system, which may involve very similar processes of site-directed protein assembly at the membrane. Hopefully, continuing progress in yeast genetic studies and in biochemical investigations of Rap1A, Rac, Ras, and CDC42Hs function in mammalian cells will provide us with a much clearer view of the complicated biological interactions involving these proteins and their associated regu-latory proteins.

REFERENCES

Abo, A., Pick, E., Hall, A., Totty, N., Teahan, C. G. and Segal, A. W. (1991) Nature (London) **353**, 668–670

Adari, H., Lowy, D. R., Williamsen, B. M., Der, C. J. and McCormick, F. (1988) Science **240**, 518–521

Ballester, R., Furth, M. E. and Rosen, O. M. (1987) J. Biol. Chem. **262**, 2688–2695

Beranger, F., Goud, B., Tavitian, A. and de Gunzburg, J. (1991a) Proc. Natl. Acad. Sci. U.S.A. **88**, 1606–1610

Beranger, F., Tavitian, A. & de Gunzburg, J. (1991b) Oncogene **6**, 1835–1842

Bokoch, G. M. (1990) Curr. Top. Membr. Transport **35**, 65–101

Bokoch, G. M. and Prossnitz, V. (1992) J. Clin. Invest. **89**, 402–408

Bokoch, G. M. and Quilliam, L. A. (1990) Biochem. J. **267**, 407–411

Bokoch, G. M., Parkos, C. A. and Mumby, S. M. (1988) J. Biol. Chem. **263**, 16744–16749

Bokoch, G. M., Quilliam, L. A., Bohl, B. P., Jesaitis, A. J. and Quinn, M. T. (1991) Science **254**, 1794–1796

Bourne, H. R., Sanders, D. A. and McCormick, F. (1991) Nature (London) **349**, 117–126

Buss, J. E., Quilliam, L. A., Kato, K., Casey, P. J., Solski, P. A., Wong, G., Clark, R., McCormick, F., Bokoch, G. M. and Der, C. J. (1991) Mol. Cell. Biol. **11**, 1523–1530

Campa, M. J., Chang, K.-J., Molina y Vedia, L., Reep, B. R. and Lapetina, E. G. (1991) Biochem. Biophys. Res. Commun. **174**, 1–5

Chant, J. and Herskowitz, I. (1991) Cell **65**, 1203–1212

Clark, R. A. (1990) J. Infect. Dis. **161**, 1140–1147

Cox, A. D. and Der, C. J. (1992) Crit. Rev. Oncogenesis, in the press

Cox, A. D., Hisaka, M. M., Buss, J. E. and Der, C. J. (1992) Mol. Cell. Biol. **12**, 2606–2615

Culine, S., Olofsson, B., Gosselin, S., Honore, N. and Tavitian, A. (1989) Int. J. Cancer **44**, 990–994

Curnutte, J. T. (1992) Immunodeficiency Rev. **3**, 149–172

Didsbury, J., Weber, R. F., Bokoch, G. M., Evans, T. and Snyderman, R. (1989) J. Biol. Chem. **264**, 16378–16382

Dumont, J. E., Jeuniaux, J. C. and Roger, P. P. (1989) Trends Biochem. Sci. **14**, 67–71

Eklund, E. A., Marshall, M., Gibbs, J. B., Crean, C. D. and Gabig, T. G. (1991) J. Biol. Chem. **266**, 13964–13970

Farrell, F., Torti, M. and Lapetina, E. G. (1992) J. Lab. Clin. Med. **120**, 533–537

Fischer, T. H., Gatling, M. N., Lacal, J.-C. and White, G. C., II (1990) J. Biol. Chem. **265**, 19405–19408

Frech, M., John, J., Pizon, V., Chardin, P., Tavitian, A., Clark, R., McCormick, F. and Wittinghofer, A. (1990) Science **249**, 169–171

Grand, R. J. A. and Owen, D. (1991) Biochem. J. **279**, 609–631

Hall, A. (1990) Science **249**, 635–640

Hall, A. (1992) Cell **69**, 389–391

Hata, Y., Kikuchi, A., Sasaki, T., Schaber, M. D., Gibbs, J. B. and Takai, Y. (1990) J. Biol. Chem. **265**, 7104–7110

Hata, Y., Kaibuchi, K., Kawamura, S., Hiroyoshi, M., Shirataka, H. and Takai, Y. (1991) J. Biol. Chem. **266**, 6571–6577

Heyworth, P. G., Curnutte, J. T., Nauseef, W. M., Volpp, B. D., Pearson, D. W., Rosen, H. and Clark, R. A. (1991) J. Clin. Invest. **87**, 352–356

Hiroyoshi, M., Kaibuchi, K., Kawamura, S., Hata, Y. and Takai, Y. (1991) J. Biol. Chem. **266**, 2962–2969

Hong, H. J., Hsu, L.-C. and Gould, M. N. (1990) Carcinogenesis **11**, 1245–1247

Hoshijima, M., Kikuchi, A., Kawata, M., Ohmori, T., Hashimoto, E., Yamamura, H. and Takai, Y. (1988) Biochem. Biophys. Res. Commun. **157**, 851–860

Hsu, L.-C. and Gould, M. N. (1991) Carcinogenesis **12**, 533–536

Huzoor-Akbar, X. X., Winegar, D. A. and Lapetina, E. G. (1991) J. Biol. Chem. **266**, 4387–4391

Itoh, T., Kaibuchi, K., Sasaki, T. and Takai, Y. (1991) Biochem. Biophys. Res. Commun. **177**, 1319–1324

Jimenez, B., Pizon, V., Lerosey, I., Beranger, F., Tavitian, A. and de Gunzburg, J. (1991) Int. J. Cancer **49**, 471–479

Kahn, R. A., Der, C. J. and Bokoch, G. M. (1992) FASEB J. **6**, 2512–2513

Kaibuchi, K., Mizuno, T., Fujioka, H., Yamamoto, T., Kishi, K., Fukumoto, Y., Hori, Y. and Takai, Y. (1991) Mol. Cell. Biol. **11**, 2873–2880

Kawamura, S., Kaibuchi, K., Hiroyoshi, M., Hata, Y. and Takai, Y. (1991a) Biochem. Biophys. Res. Commun. **173**, 1095–1102

Kawamura, S., Kaibuchi, K., Hiroyoshi, M., Fujioka, H., Mizuno, T. and Takai, Y. (1991b) Jpn. J. Cancer Res. **82**, 758–761

Kawata, M., Matsui, Y., Kondo, J., Hishida, T., Teranishi, Y. and Takai, Y. (1988) J. Biol. Chem. **263**, 18965–18971

Kawata, M., Kawahara, Y., Araki, S., Sumako, M., Tsuda, T., Fukuzaki, H., Mizoguchi, A. and Takai, Y. (1989a) Biochem. Biophys. Res Commun. **163**, 1418–1427

Kawata, M., Kikuchi, A., Hoshijima, M., Yamamoto, K., Hashimoto, E., Yamamura, H. and Takai, Y. (1989b) J. Biol. Chem. **264**, 15688–15695

Kawata, M., Farnsworth, C. C., Yoshida, Y., Gelb, M. H., Glomset, J. A. and Takai, Y. (1990) Proc. Natl. Acad. Sci. U.S.A. **87**, 8960–8964

Kemp, B. E. and Pearson, R. B. (1990) Trends Biochem. Sci. **15**, 342–346

Kikuchi, A., Sasaki, T., Araki, S., Hata, Y. and Takai, Y. (1989) J. Biol. Chem. **264**, 9133–9136

Kim, S., Mizoguchi, A., Kikuchi, A. and Takai, Y. (1990) Mol. Cell. Biol. **10**, 2645–2652

Kitayama, H., Sugimoto, Y., Matsuzaki, T., Ikawa, Y. and Noda, M. (1989) Cell **56**, 77–84

Kitayama, H., Matsuzaki, T., Ikawa, Y. and Noda, M. (1990a) Proc. Natl. Acad. Sci. U.S.A. **87**, 4284–4288

Kitayama, H., Matsuzaki, T., Ikawa, Y. and Noda, M. (1990b) Jpn. J. Cancer Res. **81**, 445–448

Knaus, U. G., Heyworth, P. G., Evans, T., Curnutte, J. T. and Bokoch, G. M. (1991) Science **254**, 1512–1515

Kyprianou, N. and Taylor-Papadimitriou, J. (1992) Oncogene **7**, 57–63

Labadia, M., Bokoch, G. M. and Huang, C.-K. (1992) FASEB J. **6**, abstr. no. 6490

Lapetina, E. G., Lacal, J.-C., Reep, B. R. and Molina y Vedia, L. (1989) Proc. Natl. Acad. Sci. U. S. A. **86**, 3131–3134

Lazarowski, E. R., Winegar, D. A., Nola, R. D., Oberdisse, E. and Lapetina, E. G. (1990) J. Biol. Chem. **265**, 13118–13123

Lerosey, I., Pizon, V., Tavitian, A. and de Gunzburg, J. (1991) Biochem. Biophys. Res. Commun. **175**, 430–436

Maltese, W. A. (1990) FASEB J. **4**, 3319–3328

Maridonneau-Parini, I. and de Gunzburg, J. (1992) J. Biol. Chem. **267**, 6396–6402

Marshall, M. S., Davis, L. J., Keys, R. D., Mosser, S. D., Hill, W. S., Scolnick, E. A. and Gibbs, J. B. (1991) Mol. Cell. Biol. **11**, 3997–4004

Marti, K. B. and Lapetina, E. G. (1992) Proc. Natl. Acad. Sci. U.S.A. **89**, 2784–2788

Marx, J. (1992) Science **257**, 484–485

Miura, Y., Kaibuchi, K., Itoh, T., Corbin, J. D., Francis, S. H. and Takai, Y. (1992) FEBS Lett. **297**, 171–174

Mizuno, T., Kaibuchi, K., Yamamoto, T., Kawamura, M., Sakoda, T., Fujioka, H., Matsuura, Y. and Takai, Y. (1991) Proc. Natl. Acad. Sci. U.S.A. **88**, 6442–6446

Mizuno, T., Kaibuchi, K., Ando, S., Musha, T., Hiroaka, K., Takaishi, K., Asada, M., Nunori, H., Matsuda, I. and Takai, Y. (1992) J. Biol. Chem. **267**, 10215–10218

Moores, S. L., Schober, M. D., Mosser, S. D., Rands, E., O'Hara, M. B., Garsky, V. M., Marshall, M. S., Pompliano, D. L. and Gibbs, J. M. (1991) J. Biol. Chem. **266**, 14603–14610

Mueller, H., Motulsky, H. J. and Sklar, L. A. (1988) Mol. Pharmacol. **34**, 347–353

Nauseef, W. M., Volpp, B. D., McCormick, S., Leidal, K. G. and Clark, R. A. (1991) J. Biol. Chem. **266**, 5911–5917

Nice, E. C., Fabri, L., Hammacher, A., Holden, J. and Simpson, R. J. (1992) J. Biol. Chem. **267**, 1546–1553

Nur-e-Kamal, M. S. A., Sizeland, A., D'Abaco, G. and Maruta, H. (1992) J. Biol. Chem. **267**, 1415–1418

Ohmori, T., Kikuchi, A., Yamamoto, K., Kim, S. and Takai, Y. (1989) J. Biol. Chem. **264**, 1877–1881

Ohmstede, C.-A., Farrell, F. X., Reep, B. R., Clemetson, K. J. and Lapetina, E. G. (1990) Proc. Natl. Acad. Sci. U.S.A. **87**, 6527–6531

Peveri, P., Heyworth, P. G. and Curnutte, J. T. (1992) Proc. Natl. Acad. Sci. U.S.A. **89**, 2494–2498

Pizon, V., Chardin, P., Lerosey, I., Olofsson, B. and Tavitian, A. (1988a) Oncogene **3**, 201–204

Pizon, V., Lerosey, I., Chardin, P. and Tavitian, A. (1988b) Nucleic Acids Res. **16**, 7719

Polakis, P. G., Rubinfeld, B., Evans, T. and McCormick, F. (1991) Proc. Natl. Acad. Sci. U.S.A. **88**, 239–243

Polakis, P. G., Rubinfeld, B. and McCormick, F. (1992) J. Biol. Chem. **267**, 10780–10785

Quilliam, L. A. and Bokoch, G. M. (1992) in Signal Transduction in Inflammatory Cells, Volume I (Cochrane, C. G. and Gimbrone, M. A., eds.), Academic Press, San Diego, in the press

Quilliam, L. A., Der, C. J., Clark, R., O'Rourke, E. C., Zhang, K., McCormick, F. P. and Bokoch, G. M. (1990) Mol. Cell. Biol. **10**, 2901–2908

Quilliam, L. A., Mueller, H., Bohl, B. P., Prossnitz, V., Sklar, L. A., Der, C. J. and Bokoch, G. M. (1991) J. Immunol. **147**, 1628–1635

Quinn, M. T., Parkos, C. A., Walker, L., Orkin, S. H., Dinauer, M. C. and Jesaitis, A. J. (1989) Nature (London) **342**, 198–200

Quinn, M. T., Mullen, M. L., Jesaitis, A. J. and Linner, D. G. (1992a) Blood **79**, 1563–1573

Quinn, M. T., Curnutte, J. T., Parkos, C. A., Mullen, M. L., Scott, P. J., Erickson, R. W. and Jesaitis, A. J. (1992b) Blood **79**, 2438–2445

Rivkin, I., Rosenblatt, J. and Becker, E. L. (1975) J. Immunol. **115**, 1126–1134

Rousseau-Merck, M. F., Pizon, V., Tavitian, A. and Berger, R. (1990) Cytogenet. Cell. Genet. **53**, 2–4

Rubinfeld, B., Munemitsu, S., Clark, R., Conroy, L., Watt, K., Crossier, W. J., McCormick, F. and Polakis, P. (1991) Cell **65**, 1033–1042

Ruggieri, R., Bender, A., Matsui, Y., Powers, S., Takai, Y., Pringle, J. R. and Matsumoto, K. (1992) Mol. Cell. Biol. **12**, 758–766

Sahyoun, N., McDonald, O. B., Farrell, F. and Lapetina, E. G. (1991) Proc. Natl. Acad. Sci. U.S.A. **88**, 2643–2647

Schejter, E. D. and Shilo, B. Z. (1985) EMBO J. **4**, 407–412

Schweighoffer, F., Rey, I., Barlot, I., Soubigou, P., Mayaux, J. F. and Tocque, B. (1990) in The Biology and Medicine of Signal Transduction (Nishizuka, Y. et al., eds.), pp. 329–334, Raven Press, New York

Sha'afi, R. I. and Molski, T. F. P. (1988) Prog. Allergy **42**, 1–64

Shinjo, K., Koland, J. G., Hart, M. J., Narasimhan, V., Johnson, D. I., Evans, T. and Cerione, R. A. (1990) Proc. Natl. Acad. Sci. U.S.A. **87**, 9853–9857

Shirataki, H., Kaibuchi, K., Hiroyoshi, M., Isomura, M., Osaki, S., Sasaki, T. and Takai, Y. (1991) J. Biol. Chem. **266**, 20672–20677

Siess, W., Winegar, D. A. and Lapetina, E. G. (1990) Biochem. Biophys. Res. Commun. **170**, 944–950

Ueda, T., Kikuchi, A., Ohga, N., Yamamoto, J. and Takai, Y. (1989) Biochem. Biophys. Res. Commun. **159**, 1411–1419

Uhlinger, D. J., Burnham, D. N. and Lambeth, J. D. (1991) J. Biol. Chem. **266**, 20990–20997

Valencia, A., Chardin, P., Wittinghofer, A. and Sander, C. (1991) Biochemistry **30**, 4637–4648

Vallar, L., Spoda, A. and Giannattasio, G. (1987) Nature (London) **330**, 566–568

Winegar, D. A., Molina y Vedia, L. and Lapetina, E. G. (1991) J. Biol. Chem. **266**, 4381–4386

Xu, H.-P., Wang, Y., Riggs, M., Rodgers, L. and Wigler, M. (1990) Cell. Regul. **1**, 763–769

Yamamoto, K., Kaibuchi, K., Mizuno, T., Hiroyoshi, M., Shirataki, H. and Takai, Y. (1990) J. Biol. Chem. **265**, 16626–16634

Yatani, A., Okabe, K., Polakis, P., Halenbeck, R., McCormick, F. and Brown, A. M. (1990) Cell **61**, 769–776

Yatani, A., Quilliam, L. A., Brown, A. M. and Bokoch, G. M. (1991) J. Biol. Chem. **266**, 2222–2226

Young, J., Searle, J., Stitz, R., Cowen, A., Ward, M. and Chenevix-Trench, G. (1992) Cancer Res. **52**, 285–289

Zhang, K., Noda, M., Vass, W. C., Papageorge, A. G. and Lowy, D. R. (1990) Science **249**, 162–165

Biochem. J. (1993) **291**, 329–343 (Printed in Great Britain)

REVIEW ARTICLE
Protein kinase C isoenzymes: divergence in signal transduction?

Hubert HUG* and Thomas F. SARRE†

Institute of Molecular Cell Biology, University of Freiburg, c/o Gödecke AG, Mooswaldallee 1–9, 7800 Freiburg, Federal Republic of Germany

INTRODUCTION

The development and life-time of multicellular eukaryotic organisms represents a complex interplay of numerous proliferation and differentiation events that proceed in a highly ordered manner. As a prerequisite for those events, cells must respond to extracellular *signals* with a specific set of mechanisms that regulate or modulate *gene expression*. Between the signal and the gene, a system of rather different cellular components is assembled to guarantee a specific and successful process of *signal transduction*. Pathways of signal transduction, though differing remarkably in their complexity and in the use of cellular components, seem to obey certain principles which are evolutionarily conserved and ubiquitously distributed amongst living organisms.

Extracellular signals, so-called *ligands*, either penetrate the cellular membrane or bind to the extracellular domain of *receptors*. Activated receptors as such, or in association with so-called *transducers*, are capable of activating *effectors*–either directly or by means of changing the amount or intracellular distribution of so-called *second messengers*. These second messengers activate target proteins which, as such, or by acting on further 'downstream' targets, finally modulate gene expression at both the transcriptional and translational levels. Target proteins at the same time act back on transducers and/or receptors to switch off the signal transduction in a kind of feedback inhibition. Figure 1 schematically summarizes the pathways of signal transduction (for review see Parker, 1991; Karin, 1992). In view of the complex interplay of signal transduction components that regulate proliferation and differentiation events, it is not surprising that most if not all known proto-oncogenes have turned out to represent proteins involved in signal transduction pathways at all levels, i.e. ligands, receptors, transducers and effectors (for a review see Hunter, 1991). Most of the components of signal transduction pathways are proteins whose activity is altered by either ligand or second messenger binding, by covalent modifications and by subsequent changes in conformation or subunit number. The majority of covalent modifications observed are phosphorylations on tyrosine or serine/threonine residues, and both Tyr- or Ser/Thr-specific kinases and their protein substrates are present amongst the numerous components of signal transduction. Moreover, some of these protein kinases are themselves substrates of other protein kinases or modulate their own activity by (multiple) autophosphorylation reactions.

Two prominent Ser/Thr-specific kinases, both activated by second messenger action, play a central role in signal

transduction: the cyclic AMP-dependent protein kinase A (PKA) (for a review see Taylor et al., 1990) and the Ca^{2+}/phospholipid-activated protein kinase C (PKC) (this Review). The latter emphasizes its role as a key enzyme in signal transduction by the fact that it represents the direct "receptor" protein of phorbol esters, substances known to interfere dramatically with proliferative and differentiation events by promoting oncogenic transformation of cells *in vivo* and *in situ*. (To avoid misunderstanding we define data from intact animals or tissue as *in*

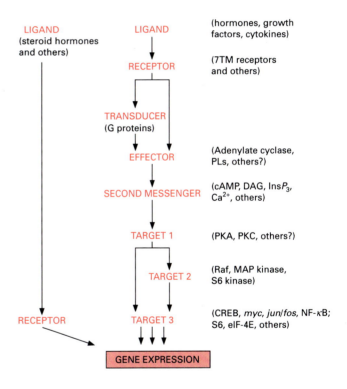

Figure 1 Schematic overview of the signal transduction pathway

Extracellular ligands bind to their cellular receptors. Hydrophobic ligands such as steroid hormones diffuse through membranes and bind to intracellular receptors that directly activate transcription. Activated receptors consisting of a structure with seven transmembrane domains (7TM receptors) interact with a transducer which in turn stimulates an effector. Activated receptors with Tyr kinase activity can directly activate several effector systems. The activated effectors generate second messengers which activate a target cascade leading to gene expression. In many cases PKC is the first target. For a complete description of signal transduction see Parker (1991), Hunter (1991) and Karin (1992).

Abbreviations used: AP, activator protein; CREB, cyclic AMP response element binding protein; DAG, *sn*-1,2-diacylglycerol; DMSO, dimethyl sulphoxide; EGF, epidermal growth factor; eIF, eukaryotic initiation factor of protein synthesis; GAP, GTPase activating protein; IFN, interferon; IL, interleukin; MAP, mitogen-activated protein (kinase); MBP, myelin basic protein; PKA, protein kinase A; PKC, protein kinase C; cPKC, "conventional" PKC; nPKC, "novel" PKC; PL, phospholipase; PMA, phorbol 12-myristate 13-acetate (= TPA); Sp1, stimulatory protein 1; SRE, serum-response element; S6, ribosomal protein 6 of the 40 S ribosomal subunit; TRE, TPA-response element.
*Present address: Osaka Bioscience Institute, Suita, Osaka 565, Japan.
†To whom correspondence should be sent, at present address: Institut für Biologie III, Universität Freiburg, Schänzlestr. 1, 7800 Freiburg, Federal Republic of Germany.

Table 1   PKC isoform cDNAs

All described PKC cDNAs are listed. " + " stands for the complete coding region. The isoform ϵ' is parenthesized because no functional protein could be identified.

PKC	Organism	Source	Region	Reference
α	Bovine	Brain	+	Parker et al., 1986a
	Human	Brain	Partial	Coussens et al., 1986
	Rat	Brain	Partial	Knopf et al., 1986
	Rabbit	Brain	+	Ohno et al., 1987
	Rat	Brain	+	Ono et al., 1988a
	Mouse	Fibroblasts	+	Rose-John et al., 1988
	Mouse	Brain	+	Megidish and Mazurek, 1989
	Human	T cells	+	Finkenzeller et al., 1990
βI	Rat	Brain	+	Ono et al., 1986
	Human	Spleen	+	Kubo et al., 1987a
	Rat	Brain	3'-end	Housey et al., 1987
	Rat	Brain	+	Housey et al., 1988
	Rabbit	Brain	+	Ohno et al., 1987
βII	Bovine	Brain	+	Coussens et al., 1986
	Human	Brain	+	Coussens et al., 1986
	Rat	Brain	+	Knopf et al., 1986
	Rat	Brain	+	Ono et al., 1986
	Human	Spleen	+	Kubo et al., 1987a
	Rabbit	Brain	+	Ohno et al., 1987
	Mouse	Brain	+	Tang and Ashendel, 1990
γ	Bovine	Brain	+	Coussens et al., 1986
	Human	Brain	5'-end	Coussens et al., 1986
	Rat	Brain	+	Knopf et al., 1986; Ono et al., 1988a
	Rabbit	Brain	+	Ohno et al., 1988b
δ	Rat	Brain	+	Ono et al., 1988b
	Mouse	Brain	+	Mizuno et al., 1991
	Mouse	Myeloid cells	+	Mischak et al., 1991a
ϵ	Rabbit	Brain	+	Ohno et al., 1988a
	Rat	Brain	+	Ono et al., 1988b
	Mouse	Brain	+	Schaap et al., 1989
(ϵ')	Rat	Brain	+	Ono et al., 1988b
ζ	Rat	Brain	+	Ono et al., 1989a
η	Mouse	Skin	+	Osada et al., 1990
	Human	Keratinocytes	+	Bacher et al., 1991, 1992
	Rat	Lung	+	Dekker et al., 1992
θ	Mouse	Skin	+	Osada et al., 1992

vivo, data from cell culture as *in situ*, and data obtained from experiments with cell-free components as *in vitro*.) The discovery that PKC represents a large gene family of isoenzymes differing remarkably in their structure and expression in different tissues, in their mode of activation and in substrate specificity may enable us to elucidate the key role of PKC isoenzymes in signal transduction and to link PKC isoenzyme action to the modulation of gene expression necessary for changes in the proliferative and differentiation status of eukaryotic cells.

In this Review, we have compiled current knowledge on the action of PKC isoenzymes with the goal of evaluating the possibility that signal transduction pathways lead to quite divergent responses by usage of the different PKC isoenzymes. Within this scope, we have refrained from citing numerous references that are compiled in detailed reviews on PKC which have appeared at regular intervals in the past (Kikkawa and Nishizuka, 1986; Nishizuka, 1988; Kikkawa et al., 1989; Bell and Burns, 1991; Stabel and Parker, 1991; Azzi et al., 1992; Clemens et al., 1992).

THE PROTEIN STRUCTURE OF PKC ISOENZYMES

Originally, PKC was discovered by Nishizuka and coworkers as a histone protein kinase from rat brain that could be activated by limited proteolysis (Inoue et al., 1977), Ca^{2+} and (phospho)lipids

(Takai et al., 1979) or phorbol esters and phospholipids (Castagna et al., 1982). From biochemical studies and purifications (Huang et al., 1986a), it soon became clear that PKC represented a group of at least three isoenzymes or isoforms (α, β, γ), but the major breakthrough emerged from cloning the cDNAs of an ever increasing number of PKC isoforms, mostly from brain cDNA libraries (see references in Table 1). So far, the cDNAs coding for nine different PKC isoenzymes have been cloned from different species and tissues or cell lines (Table 1; following the nomenclature of Nishizuka, 1988). They can be divided into two main groups: the Ca^{2+}-dependent or conventional PKCs (cPKCs) and the Ca^{2+}-independent or novel PKCs (nPKCs) (Ohno et al., 1991). The PKC isoforms α, βI, βII and γ belong to the Ca^{2+}-dependent group, and the isoforms δ, ϵ, ζ, η and θ to the Ca^{2+}-independent group. The recently identified murine (Osada et al., 1990) and rat (Dekker et al., 1992) PKC η is identical with the human PKC L (Bacher et al., 1991, 1992) and we refer to these PKCs as PKC η (Table 1). The PKC ϵ'-cDNA (Ono et al., 1988b) is very likely to represent a partial ϵ-cDNA clone since no ϵ'-protein could be identified by overexpressing the human ϵ'-cDNA in either mammalian or insect cells (H. Hug et al., unpublished work). For a comparison of the mammalian PKC isoenzymes with those from lower eukaryotes, which would exceed the scope of this Review, the reader should refer to Stabel and Parker (1991).

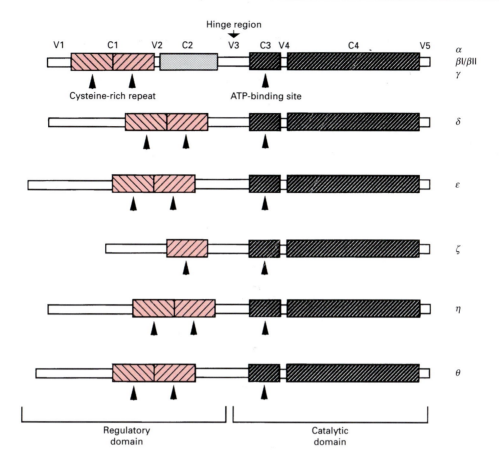

Figure 2   Domain structure of PKC isoenzymes

All PKC isoenzymes consist of constant (C) and variable (V) regions. The cysteine-rich repeats in the C1 region and the ATP binding site in the C3 region are indicated by arrowheads. The arrow points to the hinge region in the V3 domain which separates the regulatory from the catalytic domain.

The primary amino acid structure, deduced from the cDNA sequences available, can be divided into conserved and thus presumably functional domains (C1–C4) which are separated by variable regions (V1–V5), the function of which is not yet evident (Coussens et al., 1986). All PKC isoforms contain these constant and variable regions in a single subunit protein, although the nPKC isoenzymes differ in part from the structure of the cPKCs (Figure 2). The C-terminal regions C3–V5 have been defined in all PKC isoenzymes as the catalytic domain, which is separated by the V3 region from the N-terminal regulatory domain.

The N-terminal V1 region of the cPKC isoenzymes is a short stretch of approximately 20 amino acids and no function has been attributed to this region. In contrast, the V1 region of the nPKC isoenzymes is rather extended and may well influence or modulate the function of the conserved domains common to both c- and nPKC isoforms.

At the beginning of the C1 region, a sequence motif is located that is similar to the consensus sequence xRxxS/TxRx found in the phosphorylation sites of prominent PKC substrates (House. et al., 1987; Graff et al., 1989; Kemp and Pearson, 1990; House and Kemp, 1990). However, the serine or threonine residue found in the substrate motif is changed to alanine in all PKC isoenzymes (Figure 3). Thus, this motif cannot be phosphorylated and is very likely to represent a pseudosubstrate site that exhibits autoregulatory features (Soderling, 1990) by blocking the catalytic (substrate binding) site. In fact, pseudosubstrate peptides

are rather efficient inhibitors of PKC both *in vitro* and *in situ* (House and Kemp, 1987; Eicholtz et al., 1990; Shen and Buck, 1990); likewise, synthetic peptides containing this sequence with alanine replaced by a serine residue can be used as *in vitro* substrates (Marais and Parker, 1989; Schaap et al., 1989; Olivier and Parker, 1991; Dekker et al., 1992). A PKC α mutant containing a glutamic acid residue at position 25 (Figure 3) showed an increase in effector-independent kinase activity and in its sensitivity towards proteolytic activation (Pears et al., 1990), and anti-pseudosubstrate antibodies could be used to activate PKC *in vitro* (Makowske and Rosen, 1989). On the basis of the pseudosubstrate sequence, one can speculate on the substrate specificity of PKC isoenzymes; however, a comparison of these sites (Figure 3) does not reveal any significant differences and nor do the cPKC isoenzymes α, βI and γ show any distinct differences when probed with the respective substrate peptides (Marais and Parker, 1989). It should be noted, however, that within natural (holoprotein rather than peptide) substrates of PKC, structures of higher order may determine substrate specificity (Kemp and Pearson, 1990).

The Cys-rich region within the C1 domain (Figures 2 and 3) consists of two zinc finger motifs each with six cysteine residues, a DNA-binding motif found in transcription factors like GAL4 (Pan and Coleman, 1990). For both PKC and GAL4, it has been shown that two Zn^{2+} ions are co-ordinated between six cysteine residues (Quest et al., 1992) though no obvious role of Zn^{2+} ions

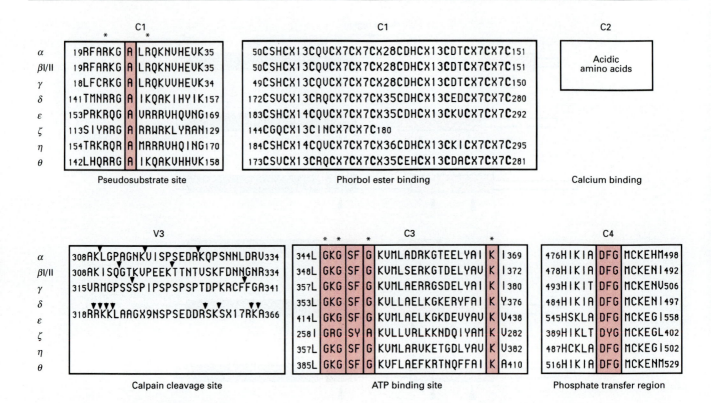

Figure 3   Sequence motifs of PKC isoenzymes

Sequences are taken from human PKC α (Finkenzeller et al., 1990), human PKC βI/II (Kubo et al., 1987); human PKC γ (Coussens et al., 1986; H. Hug, unpublished work), rat PKC δ and ϵ (Ono et al., 1988), rat PKC ζ (Ono et al., 1989), murine PKC η (Bacher et al., 1991, 1992) and murine PKC θ (Osada et al. 1992). Asterisks indicate conserved amino acids, arrowheads the calpain I and II cleavage sites in PKC α, β and γ (Kishimoto et al., 1989) or tryptic cleavage sites in murine PKC ϵ (Schaap et al., 1990). The murine PKC ϵ sequences shown in the V3 region are identical with the respective rat sequence. Longer stretches of non-homologous amino acids within the conserved sequences are marked with X followed by the number of amino acids not depicted.

is detectable. For PKC isoenzymes no DNA-binding activity has been demonstrated, but the regulatory subunit alone, generated by proteolytic cleavage in the V3 hinge region (see below), may bind. Three further proteins, diacylglycerol (DAG) kinase (Sakane et al., 1990; Schaap et al., 1990), c-Raf kinase isoenzymes (Bruder et al., 1992) and n-chimaerin (Ahmed et al., 1990), show no DNA-binding activity although they contain a zinc finger motif. The use of deletion mutants of different PKC isoenzymes revealed that the Cys-rich region is necessary for DAG and phorbol ester binding (Muramatsu et al., 1989; Kaibuchi et al., 1989; Burns and Bell, 1991). Moreover, the C1 domain, expressed in *Escherichia coli*, exhibited phospholipid-dependent phorbol ester binding (Ono et al., 1989b). PKC ζ contains only one zinc finger (Figure 3) and does not bind DAG or phorbol ester (Ono et al., 1989a; McGlynn et al., 1992); this is in agreement with the finding that, at least *in vitro*, PKC ζ exhibits a constitutive protein kinase activity (Liyanage et al., 1992; McGlynn et al., 1992; Nakanishi and Exton, 1992) which might be influenced *in vivo* by other unknown factors. It seems noteworthy that all PKC isoenzymes, even PKC ζ exhibit a distinct distance of 15 amino acids from the end of the pseudosubstrate box to the beginning of the zinc finger motif (see Figure 3).

The Ca^{2+}-independent nPKCs lack the C2 region (Figure 3), which is thought to represent the Ca^{2+}-binding domain of the cPKCs (Ono et al., 1988b; Ohno et al., 1988a). No sequence motif that represents a known Ca^{2+}-binding site, e.g. the classical E–F hand binding motif, could be identified, but the C2 region contains many acidic amino acids which are thought to par-

ticipate in Ca^{2+} binding (Ohno et al., 1987). Nevertheless, homologies of the C2 domain of the cPKCs with sequences of other Ca^{2+}-binding proteins have been reported, e.g. with phospholipase A$_2$ (Clark et al., 1991), phospholipase C-γ (Stahl et al., 1988) and two synaptic vesicle Ca^{2+}-dependent proteins (Perin et al., 1990; Geppert et al., 1991).

The V3 or hinge region separates the regulatory from the catalytic domain (Figure 2). This region is sensitive to proteolytic cleavage by trypsin or the Ca^{2+}-dependent neutral proteases calpain I and II which leads to a constitutively active kinase (Kishimoto et al., 1983; Huang and Huang, 1986; Schaap et al., 1990). PKC α is more resistant to proteolytic digestion than PKC β and γ (Huang et al., 1989; Kishimoto et al., 1989; Kochs et al., 1993). The cleavage sites for PKC α, β and γ by calpain I and II (Kishimoto et al., 1989) and for PKC ϵ by trypsin (Schaap et al., 1990) have been determined and are depicted in Figure 3, though no obvious consensus sequence can be found between those and other nPKC isoenzymes. Further experimental evidence will be necessary to determine whether or not the V3 region of the different PKC isoenzymes exhibits certain features that modulate the susceptibility to proteolytic cleavage. Recent findings (James and Olson, 1992) indicate that the hinge region (and the catalytic domain) may be involved in the nuclear targeting of PKC α (see below).

The C3 region contains the ATP-binding motif $xGxGx_2Gx_{16}Kx$ conserved in most protein kinases (Taylor et al., 1990; Kemp and Pearson, 1990). Only PKC ζ differs slightly from the consensus ATP-binding motif; it contains an alanine instead of a glycine at position 264 (Figure 3). Nevertheless, human PKC ζ purified from recombinant baculovirus-infected insect cells does show kinase activity (Liyanage et al., 1992; McGlynn et al., 1992; G. Kochs and H. Hug, unpublished work).

The C4 region contains the substrate binding site and the phosphate transfer region (Figure 3). The central element in the phosphate transfer region, again highly conserved amongst protein kinases (Taylor et al., 1990; Kemp and Pearson 1990), is the sequence DFG. The Asp residue is thought to be responsible for the transfer of the phosphate group to substrates. In all PKC isoenzymes, there is a conserved distance of 105–108 (113 in PKC γ) amino acids between the end of the ATP-binding site and the beginning of the phosphate transfer region. Again, PKC ζ exhibits a slight deviation from the consensus sequence DFG with a substitution of phenylalanine by tyrosine.

THE BIOCHEMICAL PROPERTIES OF PKC ISOENZYMES

Though certain features of the PKC isoenzymes can be deduced from the protein structure derived from the sequence of the cloned PKC cDNAs (see above), the greater body of information has been accumulated by detailed biochemical analyses. Those studies were carried out with isoenzymes purified to homogeneity mostly from rat, rabbit or bovine brain (Huang et al., 1986a; Jaken and Kiley, 1987; Sekiguchi et al., 1988; Marais and Parker, 1989; Leibersperger et al., 1990; Koide et al., 1992; Ogita et al., 1992; Saido et al., 1992). To date, expression of isoenzyme cDNAs in either COS cells or recombinant baculovirus-infected insect cells allows for the purification of distinct PKC isoenzymes for biochemical characterization (Knopf et al., 1986; Ohno et al., 1988a; Ono et al., 1989a; Patel and Stabel, 1989; Burns et al., 1990; Fiebich et al., 1990; Schaap and Parker, 1990; Akita et al., 1990; Osada et al., 1990; Olivier and Parker, 1991; Burns and Bell, 1991; Stabel et al., 1991; Liyanage et al., 1992; McGlynn et al., 1992; Osada et al., 1992; Dekker et al., 1992). The molecular mass values of these PKC isoenzymes, either calculated from the open reading frame of the respective cDNA sequence or reported from SDS/PAGE analyses, are listed in Table 2. The obvious deviations of these values are thought to be due to co- or post-translational modifications (see below); in several cases, a doublet of protein bands is observed.

Efficient *in vitro* assays have been developed to study partially or highly purified fractions of PKC isoenzymes derived from the sources mentioned above. As a major advance, the use of DAG/PtdSer vesicles (generated by sonication) as the classical activator of PKC (Takai et al., 1979; Castagna et al., 1982) has been substituted by so-called *mixed micelles* as the activating principle of PKC (Hannun et al., 1985). The latter, composed of (phospho)lipid activator(s) embedded in Triton X-100 micelles, have proven to virtually mimic the situation of the cellular membrane environment of PKC and has minimized artifacts such as damage to vesicle structure by the presence of Ca²⁺ (Bell and Burns, 1991). Originally, histones (H1 or III-S), myelin basic protein (MBP) or protamine were used as substrates for PKC isoenzymes (Jakes et al., 1988; Burns et al., 1990), but synthetic peptides derived from the Ala → Ser mutated pseudosubstrate sequence (Marais and Parker, 1989; Schaap et al., 1989; Olivier and Parker, 1991; Dekker et al., 1992) or from known PKC substrates, e.g. the EGF receptor or MBP (House et al., 1987; Yasuda et al., 1990) have proven to be adequate and even more specific substrates. This has been supported by the recent finding that some of the nPKCs, PKC δ, ε and η, display little or no

kinase activity on histone, MBP or protamine (Liyanage et al., 1992; Dekker et al., 1992).

By means of those *in vitro* systems, the biochemical properties of the different PKC isoenzymes have been investigated with respect to the activation, autophosphorylation, proteolytic activation/degradation and, last but not least, substrate specificity.

Activation

The finding that PKC represents a (phospho)lipid dependent protein kinase has drawn attention to the cellular glycerolipids, sphingolipids and their metabolic breakdown products (Berridge 1987, 1989) and the enzymes involved in lipid metabolism, i.e phospholipases and PtdIns kinase(s) (Meldrum et al., 1991). At least for the cPKCs, a model of activation was sufficient and convincing that included (i) the generation of DAG and Ins(1,4,5)P_3 from plasma membrane-associated PtdIns(4,5)P_2 by the action of phospholipase C, (ii) the release of Ca²⁺ from intracellular storage sites stimulated by Ins(1,4,5)P_3, (iii) the binding of Ca²⁺ to the C2 region of PKC and subsequent translocation of the enzyme to the plasma membrane, (iv) where it is activated, via its C1 region, by DAG and PtdSer, the latter being constitutively present in the membrane (reviewed by Bell and Burns, 1991; Gschwendt et al., 1991; Zidovetzki and Lester, 1992). In this model, phorbol ester would mimic the action of DAG and, by its persistence in the cellular membrane, lead to a long-term activation of PKC (Gschwendt et al., 1991; and references cited therein).

Activation of the cPKCs is thought to require DAG as activator and PtdSer as cofactor of activation, the presence of both reducing the Ca²⁺ requirement of PKC to the micromolar range (Lee and Bell, 1991, and references cited therein). For activation with phorbol esters in the case of cPKCs, the presence of Ca²⁺ is not required but lowers the concentration of phorbol ester necessary to obtain full PKC activity (Ryves et al., 1991). As outlined above, members of the nPKC group do not require Ca²⁺ for activation, but require either DAG/PtdSer or PMA/PtdSer, except for PKC ζ which exhibits a low but constitutive activator-independent kinase activity (Liyanage et al., 1992; McGlynn et al., 1992). However, Nakanishi and Exton (1992) have reported a marked stimulation of PKC ζ activity (purified from bovine kidney) by PtdSer or unsaturated fatty acids, e.g. arachidonic acid.

Recently it has been shown that other components of glycerolipid metabolism can be activators of PKC, at least *in vitro* (Lee and Bell, 1991; Chauhan et al., 1991; Orr and Newton, 1992; Kochs et al., 1993). Cardiolipin is able to activate PKC α, βI (Kochs et al., 1993) and ε (Saido et al., 1992). Arachidonic acid (Kikkawa et al., 1989; Burns et al., 1990; Ogita et al., 1992) and even better lipoxin A (Shearman et al., 1989), another lipoxygenase metabolite, are capable of activating PKC βII, γ (Shearman et al., 1989) and ε (Ogita et al., 1992), even in a Ca²⁺-independent fashion. PKC α can be activated by arachidonic acid only in the presence of Ca²⁺, whilst PKC βI and δ do not respond at all (Shearman et al., 1989; Burns et al., 1990). Surprisingly, PtdIns(4,5)P_2 can substitute for DAG (Chauhan et al., 1991) as an activator of PKC α, βI and γ (Lee and Bell 1991; Kochs et al., 1993), but not of βII (Lee and Bell, 1992), and PtdIns can replace PtdSer as cofactor of activation, at least for PKC α and βI (Kochs et al., 1993). These findings, together with observations that DAG can be generated in an inositol lipid-independent way, e.g. from phosphatidylcholine (Huang and Cabot, 1990; Cataldi et al., 1990) may indicate that some if not all PKC isoenzymes may be activated by different second

Table 2   Molecular mass values of PKC isoenzymes

Calculated molecular masses are from the cDNA sequence; apparent molecular masses are estimated from SDS/PAGE.

PKC	Molecular mass (kDa)		Reference
	Calculated	Apparent	
α	76.8	80–81	Parker et al., 1986; Marais and Parker, 1989; Burns et al., 1990; Borner et al., 1992
βI	76.8	79–80	Ono et al., 1987, Marais and Parker 1989
βII	76.9	80	Coussens et al., 1986, Ono et al., 1987, Burns et al., 1990
γ	78.4	82 + 84, 79–80, 77 + 80	Coussens et al., 1986; Patel and Stabel, 1989, Marais and Parker, 1989, Burns et al., 1990
δ	77.5	78, 76 + 78, 74 + 76, 77–79	Ono et al., 1988b; McGlynn et al., 1992; Olivier and Parker, 1991; Ogita et al., 1992; Borner et al., 1992; Liyanage et al., 1992
ϵ	83.5	89, 90, 90–91, 93 + 96, 96	Ohno et al., 1988a; Borner et al., 1992; Schaap and Parker, 1991; Liyanage et al., 1992; Koide et al., 1992; Saido et al., 1992
ζ	67.7	78, 76, 78–80	Ono et al., 1989a; Nakanishi and Exton, 1992; McGlynn et al., 1992; Liyanage et al., 1992
δ	77.9	82, 84, 86	Osada et al., 1992; Dekker et al., 1992; Greif et al., 1992
θ	81.6	79	Osada et al., 1992

messengers in a distinct way. Cytokines such as IFN-α, IL-1 and IL-3 have been reported to induce phosphatidylcholine hydrolysis, but not inositol phospholipid turnover (Duronio et al., 1989; Cataldi et al., 1990; Pfeffer et al., 1990), and recently, selective activation of PKC β and ϵ by IFN-α treatment of HeLa and Daudi cells, respectively, has been demonstrated (Pfeffer et al., 1990, 1991). Thus, the signal-induced production of a distinct second messenger or activator may actually decide which PKC isoenzyme becomes activated or not. Moreover, a detailed study on the activation potency of different phorbol ester derivatives on PKC isoenzymes reveals quite distinct differences, e.g. for PKC δ and PKC βI (Ryves et al., 1991).

Autophosphorylation

Most protein kinases described so far exhibit a rather pronounced autophosphorylation which is often but not necessarily linked to a modulation of kinase activity (Miller and Kennedy, 1986; Galabru and Hovanessian, 1987). Autophosphorylation has also been reported for all PKC isoenzymes known, e.g. PKC α (Pears et al., 1992), βI (Kochs et al., 1993), βII (Flint et al., 1990), γ (Patel and Stabel, 1989; Fiebich et al., 1990), δ (Ogita et al., 1992), ϵ (Koide et al., 1992; Saido et al., 1992), ζ (R. Hummel and G. Kochs, personal communication), η (Osada et al., 1990; Dekker et al., 1992) and θ (Osada et al., 1992). Detailed studies on the role of autophosphorylation have mostly been carried out with cPKC isoenzymes and have revealed that it is an intramolecular reaction at serine and threonine residues on both the regulatory and catalytic domains (Huang et al., 1986b; Mochly-Rosen and Koshland, 1987; Newton and Koshland, 1987). As expected, autophosphorylation has a K_m value for ATP about 10-fold lower than that for substrate phosphorylation and is strictly dependent on the presence of activators (Huang et al., 1986b; Newton and Koshland, 1989). Though it has been reported to affect PMA and Ca^{2+} binding (Huang et al., 1986b) and the K_m value for histone H1 as substrate (Mochly-Rosen and Koshland, 1987), autophosphorylation clearly seems not to be a prerequisite for PKC activity, but rather a concomitant event. A detailed study on PKC βII (Flint et al., 1990) nevertheless revealed certain interesting features of autophosphorylation: (i)

all phosphorylated residues were found in variable regions of PKC (see Figures 2 and 3), i.e. V1 (Ser-16, Thr-17) right before the C1 box, V3 (Thr-314, Thr-324) and V5 (Thr-634, Thr-641); (ii) the surrounding amino acid context did not exhibit any consensus sequence nor any similarity to the pseudosubstrate/substrate consensus motif (see above), though some selectivity must exist since threonine and serine residues next to autophosphorylated ones (e.g. Thr-314 and Thr-324) are not phosphorylated (Figure 3); (iii) the distribution of autophosphorylation sites over the N-terminus, the hinge region and the C-terminus indicates a close proximity of these regions to the catalytic centre of PKC, in spite of the fact that the current model of activator binding (a prerequisite of autophosphorylation) postulates a complete or partial unfolding of the PKC molecule (Bell and Burns, 1991; Zidovetzki and Lester, 1992). Whilst PKC βII phosphorylates on both serine and threonine residues (Flint et al., 1990), a comparative phospho-amino acid analysis of highly purified PKC α, βI and ζ revealed almost exclusive autophosphorylation of PKC α and ζ on serine; in contrast, PKC βI autophosphorylates primarily on threonine residues (S. Gaubatz and T. Sarre, unpublished work).

Reports on a potential role of autophosphorylation for the proteolytic activation and degradation of PKC (see below) have been controversial: whilst Ohno et al. (1990) found a severe impairment of PMA-induced downregulation for a kinase-negative mutant of PKC α expressed in COS cells or rat fibroblasts, kinase-negative mutants of PKC α and γ expressed in COS cells could be downregulated by PMA (Pears and Parker, 1991; Freisewinkel et al., 1991). Moreover, inhibition of endogenous PKC activity by the potent inhibitor K252a did not prevent downregulation in Swiss 3T3 cells (Lindner et al., 1991). Nevertheless, autophosphorylation, as one of the immediate early events of PKC activity, may be used as a potential measure of activation (Mitchell et al., 1989; Molina and Ashendel, 1991; Pfeffer et al., 1991; S. Gaubatz and T. Sarre, unpublished work) and may reveal isoenzyme-specific activation in distinct signalling pathways.

Several reports suggest a post- or even co-translational phosphorylation of PKC prior to activation (Borner et al., 1989; Pears et al., 1992). Without this initial phosphorylation, PKC is inactive and cannot be activated. This observation is supported by the finding that PKC expressed in *E. coli* is inactive (Dietrich et al., 1989) due to the lack of an as yet unidentified PKC kinase present only in eukaryotic cells. In this respect, protein phosphatases become relevant (Parker et al., 1986b), and it is notable that, at least with PKC α, potato acid phosphatase is capable of dephosphorylating PKC without loss of PKC activity, whilst protein phosphatases 1 and 2A completely abolish PKC activity (Pears et al., 1992). If autophosphorylation plays a role at all (if not for activity or proteolytic degradation then perhaps for translocation or substrate specificity), the localization of its target residues within the variable regions might indicate a role that is quite diverse between PKC isoenzymes.

Proteolytic activation and degradation

As mentioned above, PKC was first discovered as a protease-activated protein kinase (Inoue et al., 1977), and only later did it become evident that proteolytic cleavage actually followed activation (Kishimoto et al., 1983). The respective proteases *in vivo* are thought to be Ca^{2+}-dependent neutral proteases I and II (calpains) which are active in the micromolar and millimolar concentration range of Ca^{2+} respectively (Inoue et al., 1977; Kishimoto et al., 1983). Calpains cleave PKC in the V3 hinge region (Figure 3) and thus produce two distinct fragments, a protein comprising the regulatory domain and a protein containing the kinase domain which is catalytically active in the absence of any activators (Kishimoto et al., 1989; Saido et al., 1992). *In vitro*, this proteolytic activation can be achieved by a limited trypsin treatment (Huang et al., 1989; Newton and Koshland, 1989; Schaap et al., 1990; Kochs et al., 1993), though *in vivo*, activation of PKC and translocation to the cell membrane is thought to be a prerequisite for proteolytic cleavage. However, it is not yet clear whether further proteolytic degradation serves as a means of preventing continuing kinase activity or if the catalytic fragment, released from the membrane to the cytosol (and possibly to other cellular compartments), is capable of acting as a constitutive, activator-independent kinase. The latter possibility has found support in the recent observation that a PKC α mutant, devoid of the regulatory domain and expressed in COS cells, was selectively translocated to the nuclear envelope (James and Olson 1992). Moreover, one could speculate that the regulatory fragment may now bind – by its zinc finger domains – to DNA (Murray et al., 1987, and references cited therein).

In many cell types, prolonged treatment with phorbol esters results in (almost) complete depletion of cellular PKC (so-called downregulation) which is in favour of the first alternative, i.e. prevention of permanent kinase activity. In other cases, cellular responses involving PKC could be blocked by the protease inhibitor leupeptin, thus indicating a distinct role of proteolytic cleavage (Pontremoli et al., 1990, and references cited therein). With respect to downregulation, PKC isoenzymes exhibit quite extreme differences *in vivo* (see below) which is in agreement with observations *in vitro*. Compared to PKC β and γ, PKC α is relatively resistant to both calpain- and trypsin-mediated proteolysis (Kishimoto et al., 1989; Huang et al., 1989; Kochs et al., 1993). Moreover, tryptic activation of PKC ε not only rendered the enzyme lipid- and PMA-independent but increased its kinase activity towards histone about 10-fold (Schaap et al., 1990). Further work will be necessary to determine the role of proteolytic cleavage with respect to activation, degradation and relocalization to cellular compartments, and substrate specificity of PKC isoenzymes. It should be noted that recent reports indicate a specific role of the degradation of distinct PKC

isoenzymes during the process of cellular differentiation. A decrease of PKC ε seems to accompany (or promote?) the differentiation of mouse erythroleukaemia cells (Melloni et al., 1989; Powell et al., 1992), and the differentiation of a neuroblastoma cell line is promoted by an inactivation of PKC α and ε (Leli et al., 1992).

Substrate specificity

For quite a while, our insights into the substrate specificity of PKC have been constrained by the observation that at least the cPKC isoenzymes appeared to be non-specific Ser/Thr kinases *in vitro*. Thus, histones H1 or IIIS, MBP, protamine or any other basic protein or peptide could be used as efficient substrate as long as it contained the phosphorylation site motif xRxxS/TxRx (see above). This contrasts remarkably to several, but not all, Ser/Thr kinases and to the Tyr kinases known so far (for a review see Kemp and Pearson, 1990). Nevertheless, a comparison of PKC isoenzymes and their activities towards distinct substrates *in vitro* has revealed differences that may indicate an even more pronounced specificity towards natural substrates under physiological conditions. Kinase activities of PKC α, βI, βII and γ towards histone IIIS, protamine or MBP are very similar (Marais and Parker, 1989; Burns et al., 1990), though PKC γ has a 2–3-fold lower activity towards protamine or the pseudosubstrate site-derived synthetic peptide of PKC α or β (Marais and Parker, 1989). Members of the nPKC group differ significantly from cPKC isoenzymes in that they exhibit a rather poor kinase activity ($\delta > \eta > \zeta$) towards histone IIIS, MBP, protamine or protamine sulphate (Ono et al., 1989a; Schaap and Parker, 1990; Olivier and Parker, 1991; Saido et al., 1992; Liyanage et al., 1992; McGlynn et al., 1992; Dekker et al., 1992). Thus, it was sometimes necessary to use synthetic peptides derived from the respective pseudosubstrate sequence, the EGF receptor or MBP, to detect kinase activity of nPKC isoenzymes at all (Schaap et al., 1989; Saido et al., 1992; Ogita et al., 1992). This may well explain the difficulties in detecting new PKC isoenzymes by biochemical analysis, if the appropriate substrates are not available.

Several observations indicate that substrate specificity may be more complex than expected. For PKC ε, it could be demonstrated that proteolytic activation (see above) significantly increased its kinase activity towards histone (Schaap et al., 1990) indicating a distinct influence of the regulatory domain. Indeed, genetically engineered fusion of the PKC ε regulatory domain to the PKC γ catalytic domain imposed PKC ε substrate specificity onto the chimaeric enzyme (Pears et al., 1991). A similar observation has been made with PKC ζ: chymotrypsin-mediated proteolysis results in a catalytic fragment which – in contrast to the intact isoenzyme – now accepts the EGF receptor peptide as substrate (R. Hummel and T. Sarre, unpublished work). In some cases, the activator requirement seems to depend on the substrate used; for phosphorylation of protamine or protamine sulphate, neither cPKC nor nPKC isoenzymes require any activator (Bazzi and Nelsestuen, 1987; Liyanage et al., 1992). In order to phosphorylate an MBP-derived substrate peptide, but not the ε-specific pseudosubstrate site-derived peptide, PKC ε seems to require the presence of Ca^{2+} (Saido et al., 1992).

In situ studies by [32P]orthophosphate labelling of cells treated with PMA or other stimuli of PKC-involving signalling pathways have revealed such a large number of putative physiological substrates (for a review see Kikkawa and Nishizuka, 1986) that it is rather difficult to assess their significance in PKC-mediated cellular events. They can be arbitrarily divided into three major classes: (i) proteins involved in signal transduction and PKC activation (e.g. the EGF, T cell and insulin receptors, Ras and

GAP), (ii) proteins involved in metabolic pathways (channels, pumps), and last but not least (iii) proteins involved in regulatory functions concerning gene expression (transcription factors, translation factors). The phosphorylation of various transcription factors by PKC and other signal-transducing kinases has been reviewed recently (Meek and Street, 1992). It should be noted that amongst PKC substrates there might be several other protein kinases like S6 kinase or Raf kinase (see below).

Three prominent substrates, most likely involved in the control of cell proliferation mediated by PKC, should be mentioned, although little is known of the specificity of PKC isoenzymes towards them. The myristoylated, alanine-rich C kinase substrate (MARCKS) is phosphorylated by PKC under several conditions such as macrophage activation or growth-factor-dependent mitogenesis; its phosphorylation leads to a redistribution from the actin filaments of the membrane to the cytoplasm (Hartwig et al., 1992, and references cited therein). Two further PKC substrates, DNA topoisomerase I (Pommier et al., 1990) and lamin B (Hornbeck et al., 1988; Fields et al., 1988), are nuclear proteins and thought to be involved in the control of DNA synthesis; they gain interest in discussing a translocation of activated PKC isoenzymes to the cell nucleus (see below).

EXPRESSION OF PKC ISOENZYMES

The tissue distribution of the PKC isoenzymes has been determined mostly by Northern blot analyses and, more recently, by Western blotting using isoenzyme-specific antibodies. PKC α βI/II, δ, ϵ and ζ seem to be ubiquitously distributed, e.g. in brain, lung, spleen, thymus and skin (Nishizuka, 1988; Wada et al., 1989; Yoshida et al., 1989; Schaap et al., 1989; Ohno et al., 1991; Wetsel et al., 1992), whilst PKC γ is exclusively found in the central nervous system, e.g. brain (Nishizuka, 1988; Ohno et al., 1991; Wetsel et al., 1992), and PKC η is strongly expressed in skin and lung and only slightly in brain and spleen (Osada et al., 1990; Bacher et al., 1991). PKC θ is predominantly expressed in skeletal muscle and, to a clearly lower extent, in lung, spleen, skin and brain (Osada et al., 1992). Two remarkable deviations from the distribution of the ubiquitous isoenzymes should be mentioned: PKC α and ϵ seem not to be present in liver (Schaap et al., 1989; Rogue et al., 1990), where PKC β is the major isoenzyme (Rogue et al., 1990); in contrast, no PKC β could be detected in kidney, i.e. renal mesangial cells, though PKC α, δ, ϵ and ζ are present (Huwiler et al., 1992; see Table 3). Though it is still an open question why all PKC isoenzymes known so far are expressed in brain tissue, the remarkable differences in number and amount of PKC isoenzymes in other tissues make a concept of mere redundancy with respect to signal transduction rather unlikely. On the other hand, one or the other member of the ubiquitously distributed isoenzymes, e.g. PKC α or βI/II, may substitute for each other in different tissues with respect to identical or similar function. There may be even a distinction between isoenzymes for 'house-keeping' functions and those for distinct function in a differentiated, specialized cell, e.g. for PKC θ in skeletal muscle (Osada et al., 1992).

In Table 3, PKC isoenzyme expression patterns of (selected) cell lines from various origins have been accumulated, mostly determined by Northern or Western blot techniques. In accordance with the tissue distribution described above, PKC α, βI/II, δ, ϵ and ζ seem to be the most ubiquitous isoenzymes, whilst PKC γ is restricted to a neuronal cell line (PC12). Information on PKC η and θ is as yet too limited. Though both PKC α and β are present in most cell lines investigated, in none of them are both missing. Amongst cells of the haematopoietic

system, however, PKC α is absent from myeloid cells of different origin (Mischak et al., 1991b), megakaryocytes (HEL) and macrophages (Grabarek et al., 1992; Duyster et al., 1992), whilst PKC βI and βII are absent from mouse and rat fibroblasts, renal mesangial cells, murine erythroleukaemia and neuroblastoma cells. Moreover, the absence of PKC ϵ appears to be a feature of cells from the myeloid but not erythroid lineage of the haematopoietic system, i.e. megakaryocytes, macrophages, platelets and the promyelocytic leukaemia cell line HL 60. To date, these observations may represent just state of the art or may indicate an exciting script for distinct PKC isoenzymes within the complex network of proliferation and differentiation events of the haematopoietic system (Mischak et al., 1991b). Since a true picture of PKC isoenzyme expression in different tissues and cell lines is rapidly emerging, it can be anticipated that the role of a given PKC isoenzyme in a distinct signalling pathway may become evident once it is possible to define a given cell type unequivocally by its PKC isoenzyme expression pattern and by its signal response potential.

Little is known about how the expression of PKC isoenzymes is regulated, due to the fact that information on the gene and promoter structure is (yet) limited. So far, the promoters of the rat PKC γ (Chen et al., 1990) and the human PKC β (Niino et al., 1992; Obeid et al., 1992) gene have been cloned. The PKC γ promoter contains one binding site for the general transcription factor Sp1 (stimulatory protein 1) but lacks a TATA and CAAT box and therefore is similar to a housekeeping gene promoter. In addition, however, it contains several regulatory elements (reviewed by Locker and Buzard, 1990): two AP2 sites, one AP1 site, one cyclic AMP response element, one enhancer core element and one c-*myc* PRF site. Since AP1 and AP2 sites confer inducibility by phorbol esters, PKC γ transcription may be under positive control by itself or other PKC isoenzymes. The biological function could be that PKC mRNAs are immediately replenished if PKC is activated and subsequently degraded, e.g. during downregulation by PMA.

The PKC β promoter contains no TATA box, but a CAAT box, two Sp1-binding sites, one octamer binding motif site, one AP1 and one AP2 site are found (Niino et al., 1992, Obeid et al., 1992). By deletion analysis, three positive and two negative regulatory regions could be identified in the 1.9 kbp region upstream from the transcription start site of the PKC β promoter. Transcription under the control of these regions seems to be cell type specific (Niino et al., 1992). Comparison with further, not yet isolated, PKC promoters and a detailed analysis by reverse genetics should identify specific elements which could be responsible for the different expression patterns observed, especially for the nerve tissue-specific expression of PKC γ.

The PKC βI and βII isoforms differ in their 3-ends and originate by differential splicing. Genomic clones of the 3-end of the PKC β gene have been isolated (Ono et al., 1987; Kubo et al., 1987b; Coussens et al., 1987). According to the nomenclature of Kubo et al. (1987b) the βI-specific exon lies behind the βII-specific exon and is separated by an intron of 4–5 kbp (Kubo et al., 1987b; Coussens et al., 1987). If PKC β is expressed at all in a given tissue or cell line (Table 3), one splice form seems to be preferred, which indicates that the main regulation seems to occur at the level of alternative splicing. Differential splicing is also proposed for the PKC ϵ locus (Schaap et al., 1990); however, only one functional protein derived from the ϵ locus has been identified so far.

Soon after the detection of the various PKC isoenzymes it became clear not only that there are tissue-specific patterns of expression but that the amount and number of PKC isoenzymes varied within a given tissue depending on its developmental stage

Table 3   Expression of PKC isoenzymes in different cell lines

Cell lines: NIH 3T3, mouse fibroblast cell line; renal mesangial cells, rat renal mesangial cell line; HeLa, human carcinoma cell line; preB cells, B cells, plasmacytoma, myeloid cells, see Mischak et al. (1991b); T cells, human peripheral blood-derived T lymphocytes; HEL, human megakaryocyte-like cell line; macrophages, rat Kupffer cells; platelets, human peripheral blood-derived platelets; HL 60, U937, human promyelocytic leukaemia cell line; MEL, mouse erythroleukaemia cell line; GH_4C_1 pituitary, rat pituitary gland cells; neuroblastoma, mouse neuroblastoma cell line neuro 2a; PC12, pheochromocytoma cells. Abbreviation: n.d., not determined.

Cell line	α	$\beta I/\beta II$	γ	δ	ϵ	ζ	η	Reference
NIH 3T3	Yes	No	No	Yes?	Yes	Yes	No	H.Mischak (personal communication); T. Sarre (unpublished work)
Rat6 fibroblasts	Yes	No	No	Yes	Yes	Yes	n.d.	Borner et al., 1992
Renal mesangial cells	Yes	No	No	Yes	Yes	Yes	n.d.	Huwiler et al., 1991, 1992
HeLa	Yes	Yes‡	No	No?	Yes	n.d.	n.d.	Pfeffer et al., 1990
preB cells	Yes	Yes‡	No	Yes	Yes?	Yes	Yes?	Mischak et al., 1991a,b
B cells	Yes	Yes‡	No	Yes	Yes?	Yes	Yes	Mischak et al., 1991a,b
T cells	Yes	Yes*	No	Yes	Yes	Yes	Yes	Lucas et al., 1990; Mischak et al., 1991a,b
Plasmacytoma	Yes	Yes?	No	Yes	Yes?	Yes	Yes	Mischak et al., 1991a, b
Myeloid cells	No	Yes?	No	Yes	No	Yes	Yes	Mischak et al., 1991b
HEL	No	Yes‡	No	Yes	No	n.d.	n.d.	Grabarek et al., 1992
Macrophages	No	Yes‡	No	Yes	No	Yes?	n.d.	Duyster et al., 1992; J. Duyster (personal communication)
Platelets	Yes	Yes‡	No	Yes	No	Yes	n.d.	Grabarek et al., 1992; Cook et al., 1992; Crabos et al., 1992
HL 60	Yes	Yes†	No	Yes?	No	Yes	Yes?	Wada et al., 1989; Hashimoto et al., 1990; Hocevar and Fields, 1991; Ways et al., 1992
U937	Yes	Yes‡	No	n.d.	Yes	Yes	n.d.	Wada et al., 1989; Ways et al., 1992
MEL	Yes	No	No	No?	Yes	Yes	n.d.	Powell et al., 1992; T. Sarre (unpublished work)
GH_4C_1 pituitary	Yes	Yes†	n.d.	n.d.	Yes	n.d.	n.d.	Kiley et al., 1990; Akita et al., 1990
Neuroblastoma	Yes	No	No	Yes	Yes	Yes	n.d.	Wada et al., 1989; Bernards, 1991
PC12	Yes	Yes†	Yes	Yes	Yes	Yes	n.d.	Wooten et al., 1992

*βI; †βII; ‡βI or βII not determined.

(reviewed by Nishizuka, 1988; see also Yoshida et al., 1988; Wada et al., 1989). This again indicates that a certain set of PKC isoenzymes is necessary to guarantee the ordered sequence of proliferation and differentiation events which leads to and maintains the characteristics of a given tissue.

INTRACELLULAR DISTRIBUTION AND TRANSLOCATION OF PKC ISOENZYMES

More recently, the distribution of PKC isoenzymes within resting (unstimulated) cells of various origins has been determined. Upon treatment of these cells with certain stimuli, either specific ones (e.g. growth factors, hormones or cytokines) or unspecific ones (e.g. serum, phorbol esters), a redistribution of PKC isoenzymes can be observed. With prolonged treatment, the proteolytic degradation and downregulation (see above) could be investigated, too. The distribution, translocation and down-regulation of PKC could be visualized by analysis of the so-called particulate (i.e. membrane) and soluble (i.e. cytosolic) fractions of the cells, by means of PKC activity measurement or immunoblotting (using isoenzyme-specific anti-PKC antibodies).

Following the activation model outlined above, translocation to (the) cellular membrane(s) has been regarded as an equivalent of the activation of the respective PKC isoenzyme. However, in various cell types, rather significant portions of certain PKC isoenzymes are constitutively present in the particulate fraction, and it is hard to believe that this indicates a permanent and persistent activation or activity as has been proposed (Bazzi and Nelsestuen, 1988; Burgoyne, 1989). With cPKC isoenzymes, an increase in intracellular Ca^{2+} concentrations is thought to promote translocation to the cellular membrane where subsequent activation by DAG/PtdSer occurs. Following this line of thought, phorbol esters should not only be persistent activators of PKC within the cellular membrane, but should also exert a pleiotropic effect that leads to translocation of the respective isoenzymes. Moreover, phorbol ester treatment leads to the

translocation of the Ca^{2+}-independent nPKC isoenzymes, too, a mechanism which is not understood at all. Recent observations that phosphoinositides as such are capable of activating PKC *in vitro* (Chauhan et al., 1991; Lee and Bell, 1991; Kochs et al., 1993) allow for the possibility that translocation alone, without the additional production of second messengers and activators, may lead to the activation of the respective PKC isoenzyme.

In most cell lines investigated, e.g. those depicted in Table 3, PKC α seems to be located in the cytosol and is translocated to the cellular membrane and downregulated upon PMA treatment (Hocevar and Fields, 1991; Strulovici et al., 1991; Huwiler et al., 1991; Crabos et al., 1991; Borner et al., 1992). With more specific stimuli, however, remarkable differences are observed. In GH_4C_1 pituitary gland cells, PKC α is translocated to the particulate fraction, but not downregulated upon treatment with thyrotropin-releasing hormone (TRH) or PMA (Akita et al., 1990; Kiley et al., 1990). Likewise, the PKC α content of neuronal PC12 cells is unchanged over a 10 day period in the presence of nerve growth factor (NGF) (Wooten et al., 1992). In NIH 3T3 cells, PKC α is reported to translocate to the nucleus upon PMA treatment (Leach et al., 1989), and in Swiss 3T3 cells, PKC (presumably PKC α as the major isoenzyme in fibroblasts) seems to associate with the nuclear fraction upon treatment with insulin-like growth factor I, but with the cellular membrane in the presence of bombesin (Divecha et al., 1991). As reported recently (James and Olson, 1992), a PKC α mutant lacking the regulatory domain was associated primarily with the nuclear envelope.

PKC β also appears to be a cytosolic isoenzyme in unstimulated cells which is sensitive to downregulation by PMA (Huang et al., 1989; Akita et al., 1990; Kiley et al., 1990; Strulovici et al., 1991; Crabos et al., 1991; Cook et al., 1992; Duyster et al., 1992). It should be noted, however, that statements like that should not be generalized since, recently, Van den Berghe et al. (1992) reported the PMA-resistant localization of PKC βI in a human colonic cell line. In GH_4C_1 pituitary gland cells, PKC βII is down-regulated by PMA, but not by TRH treatment (Akita et al., 1990; Kiley et al., 1990). Stimulation of human platelets with IL-3 leads to a rapid translocation of PKC β that is more pronounced than that observed with PKC α (Cook et al., 1992), and IFN-α induces a selective translocation of PKC β to the particulate fraction in HeLa cells (Pfeffer et al., 1990). In HL 60 cells treated with retinoic acid (Makowske et al., 1988; Hashimoto et al., 1990) and in PC12 cells under long term treatment with NGF (Wooten et al., 1992), PKC βII is significantly accumulated whilst the other isoenzymes are downregulated or unchanged. Recently, Tanaka et al. (1992) reported an increase in the activities of PKC α and β during the differentiation of HL 60 cells and the appearance of a 'new' PKC isoenzyme when cells were induced to differentiation by retinoic acid. In rat liver and in HL 60 cells, a distinct portion of PKC β is reported to be found in the nuclear fraction (Kiss et al., 1988; Rogue et al., 1990; Hocevar and Fields, 1991).

Depending on the cell type, PKC δ seems to be differentially distributed within the cell. Whilst located in the cytosol in human platelets (Grabarek et al., 1992), the majority of PKC δ is found associated with the particulate fraction in rat6 fibroblasts (Borner et al., 1992) and in renal mesangial cells (Huwiler et al., 1992). In the latter cell lines, the isoenzyme can be completely down-regulated by PMA treatment (Borner et al., 1992; Huwiler et al., 1992).

Rather controversial observations have been made with PKC ϵ, which is mostly cytosolic in GH_4C_1 pituitary gland cells (Akita et al., 1990; Kiley et al., 1990) and human neuroblastoma cells (A. Javala and K. Akerman, personal communication), but membrane-associated to a certain extent in U937 cells (Ways et al., 1992b), rat6 fibroblasts (Borner et al., 1992) and renal mesangial cells (Huwiler et al., 1991). In the latter cell lines, downregulation by PMA is significantly diminished (Huwiler et al., 1991; Borner et al., 1992), whilst in the promonocytic cell line U937 as well as in thymocytes, PKC ϵ is resistant to down-regulation by prolonged PMA treatment (Strulovici et al., 1991). In contrast to PKC α and βII, PKC ϵ is downregulated in GH_4C_1 pituitary gland cells by both PMA and TRH treatment (Akita et al., 1990; Kiley et al., 1990). IFN-α treatment of Daudi cells has been reported to result in a selective and rapid activation of PKC ϵ (Pfeffer et al., 1991).

In several cell lines investigated so far, PKC ζ seems to be present as a cytosolic isoenzyme (Crabos et al., 1991; Borner et al., 1992; Huwiler et al., 1992); only in HL 60 cells, has it been reported to be mostly in the particulate fraction (Ways et al., 1992a). Again, translocation and downregulation by PMA appears to be dependent on the cell type. Whilst it is sensitive to PMA treatment in rat6 fibroblasts (Borner et al., 1992) and human platelets (Crabos et al., 1991), it is resistant in HL 60 (Ways et al., 1992a) and renal mesangial cells (Huwiler et al., 1992) and in two neuronal cell lines (Wada et al., 1989; Wooten et al., 1992). Sensitivity towards PMA treatment seems quite contradictory to the fact that PKC ζ neither binds to nor can be activated by phorbol esters (Ono et al., 1989a; Liyanage et al., 1992; McGlynn et al., 1992). However, especially for the nPKC isoenzymes, mechanisms of activation and degradation must be postulated which might involve other cellular (maybe cell type-specific) components; in fact, membrane-associated PKC-binding proteins have been reported which may serve to anchor or compartmentalize PKC isoenzymes to different intracellular membranes (Wolf and Baggiolini, 1990; Mochly-Rosen et al., 1991).

In several human tumour cell lines, PKC η has been reported to be present specifically in the cell nucleus; PMA treatment does not lead to downregulation (Greif et al., 1992). When PKC θ was expressed in COS cells, the majority of the protein was found in the particulate fraction (Osada et al., 1992).

Concerning the data accumulated above, we have to admit to being far away from a clear-cut picture of the distribution, translocation and degradation of PKC isoenzymes within distinct signalling pathways in different cells. Likewise, the concept of a "cell depleted of PKC by prolonged PMA treatment" has to be re-evaluated in the light of significant differences in PMA sensitivity between the PKC isoenzymes in various cells. However, a connection of the above findings with data on PKC-mediated proliferation and differentiation events may bring us closer to detailed insights.

In order to investigate the role of distinct PKC isoenzymes as such and, especially, in the control of cell proliferation, PKC isoenzyme cDNAs have been stably overexpressed in several mammalian cell systems under the control of various promoters. In general, the effects observed depend very much on the cell lines and the PKC isoenzyme cDNAs used.

Bovine PKC α has been overexpressed in Swiss 3T3 cells (Eldar et al., 1990) and murine PKC α in BALB/c-, rat6- and ψ2-fibroblasts (Borner et al., 1991). In all cases, no transformation could be observed by growing the cells in soft agar. Over-expression of human PKC α in NIH 3T3 fibroblasts resulted in a slightly transformed phenotype (Finkenzeller et al., 1992) which might be due to the parental cell line used.

Overexpression of rat PKC βI in rat6 fibroblasts gave rise to transformation, especially in the presence of PMA (Housey et al., 1988), and phospholipase D activity and DAG formation were increased (Pai et al., 1991). In contrast, overexpression of PKC βI in rat liver epithelial cells was not sufficient by itself for cell transformation (Hsieh et al., 1989). No alterations were observed by overexpressing rat PKC βI in C3H 10T1/2 cells (Krauss et al., 1989), whereas PMA exerted even an inhibitory effect on cell proliferation in HT29 colon cancer cells over-expressing the same PKC isoenzyme (Choi et al., 1990).

Rat PKC γ overexpression in NIH 3T3 fibroblasts may (Persons et al., 1988) or may not (Cuadrado et al., 1990) result in transformation. This may be due to different expression levels of rat PKC γ in these recombinant cell lines or to variations in the parental NIH 3T3 clone used and shows the general difficulty in interpreting effects of overexpressed proteins. When rat PKC γ was transiently cotransfected with a test gene under the control of the VL30 enhancer element it was shown that PKC γ could transactivate the murine VL30 enhancer element (Persons et al., 1991).

Overexpression of the murine PKC δ in NIH 3T3 cells gave rise to a slower cell growth rate, whereas overexpression of murine PKC ϵ in the same cell line led to an increased growth rate (H. Mischak, personal communication). Data on the overexpression of other nPKCs are not available at present.

Usually, overexpression of PKC isoenzymes in mammalian cells is correlated with an enhanced expression of early proto-oncogenes such as c-*jun*, c-*fos*, c-*myc* etc. Very often cells display an altered morphology such as an increase in refractility. In rat6 fibroblasts, it was possible to overexpress PKC βI (Housey et al., 1988) and α (Borner et al., 1991) 40–50-fold at the protein level, whereas the maximal expression level in other cell lines was up to 10-fold, so far. The PKC βI-overexpressing rat6 cell line (Housey et al., 1988) is the only PKC-overexpressing cell line that is known to give rise to tumours in nude mice. In general, it can be concluded that overexpression of PKC can lead to altered cell growth, but the type of alterations are dependent on other cellular factors.

Rather elegant experiments have been carried out by Ohno et al. (1991): rat fibroblasts were cotransfected with different PKC cDNAs and the cDNA for the bacterial chloramphenicol acetyl-transferase (CAT), the latter under the control either of a TPA-response element (TRE) or a serum-response element (SRE). PKC α, βII and ϵ were able to enhance significantly the expression of both the TRE/CAT and the SRE/CAT genes as well as of the endogenous c-*jun* gene. In contrast, PKC γ only enhanced the expression of the SRE/CAT gene. These findings represent a convincing demonstration of distinct roles of PKC isoenzymes at the level of the control of gene expression in a defined cellular environment.

MODULATION OF GENE EXPRESSION BY PKC ISOENZYMES

Besides analyses on the role of distinct PKC isoenzymes over-expressed in cells stably or transiently transfected with the respective cDNA (see above), several cellular systems have served to elucidate the role of endogenous PKC isoenzymes in the modulation of gene expression during differentiation and pro-liferation events *in situ*.

T cells

Activation of T cells by antigens, anti-receptor antibodies, IL-1, lectins, PMA, or a combination of these, leads to the expression of IL-2 and the 55 kDa α-subunit of the IL-2 receptor (IL-2Rα) which results in an autocrine stimulation of proliferation (reviewed by Berry and Nishizuka, 1990). In this classical model system, PKC is undoubtedly involved in the activation process, and although a rapid translocation of PKC has been observed, it is clear that PKC activation is required for a prolonged length of time (Berry et al., 1990). In this context, the observation that PKC β, but not PKC ϵ, is downregulated during thymocyte activation by PMA (Strulovici et al., 1991) might gain some significance.

During T cell activation, the expression of the c-*fos*, but not the c-*jun*, gene product is enhanced; these two proteins, as homo- or heterodimeric complexes, represent the transcription factor AP1 which confers inducibility by PMA to genes containing a TRE (TPA-response element) within their promoter DNA se-quence (for details see Berry and Nishizuka, 1990; Karin, 1992). Activation of the DNA-binding activity of AP1 seems to involve both PKC-mediated phosphorylation of c-Fos and de-phosphorylation of c-Jun (reviewed by Karin 1992; Meek and Street 1992); thus, activated PKC, via the modulation of AP1 activity, may enhance transcription of the IL-2 and IL-2Rα genes. Both genes also exhibit binding sites for the transcription factor NF-κB which, in its inactive cytosolic form, is complexed with an inhibitor of its DNA-binding activity named I-κB; the latter has been reported to be removed from the inactive complex by a phosphorylation event, most probably mediated by PKA and/or PKC by an indirect mechanism (Shirakawa and Mizel, 1989; Gosh and Baltimore, 1990; Link et al., 1992). Thus, PKC could be involved in transcriptional activation of the IL-2 and IL-2Rα genes in a dual way, via AP1 and NF-κB, maybe mediated by two different PKC isoenzymes (see below).

Recent reports have pointed out the possibility that two further components of the signal transduction pathway (see Figure 1) are involved in PKC-mediated T cell activation. A significant increase of the G protein p21^{ras} in its GTP-binding (active) form has been observed, which in turn indicates a diminished activity of the so-called GTPase activating protein GAP, due to (direct or indirect) phosphorylation of GAP by PKC (Downward et al., 1990). Another cytoplasmic Ser/Thr

kinase, c-Raf kinase, resembles PKC in that it represents a group of closely related isoenzymes also structured into a regulatory and catalytic domain, and containing a zinc finger motif (Morrison et al., 1991; Bruder et al., 1992). Its significant role within the signal transduction pathway has been unequivocally established, although details remain obscure (reviewed by Li et al., 1991).

As to T cell activation, Siegel et al. (1990) could demonstrate the concomitant phosphorylation and activation of the c-Raf kinase via a PKC-mediated pathway; this is supported by recent evidence that c-Raf kinase is a direct substrate of PKC *in situ* and *in vitro* (W.Kolch and G.Kochs, personal communication; Sözeri et al., 1992). Since expression of AP1-driven promotors and genes (see above) is thought to require c-Raf kinase (Bruder et al., 1992), this might indicate, at least for the case of T cell activation, the existence of a protein kinase cascade consisting of one (or more) PKC isoenzyme(s) and the Raf-1 kinase.

Erythroleukaemia cells

Mouse erythroleukaemia (MEL) cells represent a transformed cell line from the erythroid lineage of the haematopoietic system, most probably from the proerythroblast stage. MEL cells can be induced *in situ* to terminal erythroid differentiation marked by the cessation of proliferation, the onset of globin mRNA and haemoglobin synthesis and the expression of several other erythroblast marker genes (reviewed by Reuben et al., 1980). Recently, Melloni et al. (1989) convincingly demonstrated that PKC is involved in, if not responsible for, the onset of differentiation (so-called commitment). Moreover, in this system, the (transient?) presence of a proteolytically activated, cytosolic form of PKC (see above) seems to be important, as has been demonstrated by the inhibition of differentiation in the presence of the protease inhibitor leupeptin (Melloni et al., 1987). Since levels of PKC α remained unaltered, but PKC βI/βII decreased concomitantly with the onset of differentiation, the latter isoenzyme(s) seemed to exert the key role (Melloni et al., 1989). Re-evaluation of the PKC isoenzyme pattern of MEL cells, however, has revealed the absence of any PKC β isoenzyme, but a significant abundance of PKC ϵ (Powell et al., 1992; T. Sarre, unpublished work), which may now be the candidate isoenzyme during commitment. A potential role of PKC ϵ in T cell activation (see above) may be mere coincidence as well as the fact that cells from the non-erythroid myeloid lineage (see below), which cannot be induced to erythroid differentiation, lack this isoenzyme, but express PKC β which is not present in MEL cells (Table 3). Recently, a rapid elevation of DAG levels, followed by a significant decrease, has been reported as one of the earliest events during MEL cell differentiation (Michaeli et al., 1992).

Rather little is known of the modulation of gene expression linked to erythroid differentiation. In differentiating MEL cells, downregulation of both c-*myc* and c-*myb* gene expression has been reported (D. Eick, personal communication; Smith et al., 1990; Danish et al., 1992) which contrasts with the observation that erythroid differentiation achieved by erythropoietin in splenic erythroid cells is accompanied by an increase of c-*myc* RNA which seems to depend on the activation of PKC (Patel et al., 1992).

Involvement of PKC in the modulation of gene expression at the translational rather than the transcriptional level has been neglected for some time, though changes in both the proliferative and differentiation status of cells are accompanied by significant alterations of the overall rate and quality of protein biosynthesis (reviewed by Hershey, 1991). For example, differentiation of MEL cells leads to a decrease of the global protein synthesis rate to about 30%, though in parallel globin synthesis reaches a maximum (Bader and Sarre, 1986). The dominant mechanisms of translational control involve reversible phosphorylation reactions on components like ribosomal proteins, initiation or elongation factors (Hershey, 1991). The ribosomal protein S6 (of the 40 S subunit) as well as the initiation factors eIF-3, eIF-4E, eIF-4B and eIF-4F (the mRNA cap binding complex) have been reported to be phosphorylated in cells treated with insulin or phorbol esters (Morley and Traugh, 1990). Whilst S6 phosphorylation is exerted by a S6 kinase which seems to be downstream of PKC in a protein kinase cascade (Susa et al., 1989, and references cited therein), at least eIF-4F and -4E appear to be direct substrates of PKC in situ and in vitro (Morley et al., 1991; Smith et al., 1991). Interestingly, overexpression of the cDNA encoding eIF-4E in NIH 3T3 cells lead to cellular transformation (Lazaris-Karatzas et al., 1990) and expression of antisense cDNA diminished the proliferation rate of HeLa cells (De Benedetti et al., 1991).

Thus, it seems likely that translational control mechanisms involved in proliferation and differentiation events are exerted in part, directly or indirectly, by activated PKC isoenzymes.

Myeloid cells

A third in situ system used for the analysis of the role of PKC in the control of differentiation involves human myeloid or pro-myelocytic leukaemia cell lines (HL 60, U937, K562, HEL; see Table 3). These cells tend to differentiate into distinct cell types depending on the inducer used: upon exposure to phorbol esters, HL 60, K562 and U937 cells show a series of monocyte/macrophage properties, whilst HL 60 cells treated with retinoic acid or DMSO exhibit a neutrophil-like (granulocyte) phenotype (Hass et al., 1991; Okuda et al., 1991; Ways et al., 1992a,b; and references cited in Table 3). In contrast, K562 cells differentiate along the erythroid lineage upon treatment with haemin (Rutherford et al., 1979; Okuda et al., 1991). Following PMA treatment, the megakaryocyte-like HEL cells develop properties characteristic of platelets and megakaryocytes (Grabarek et al., 1992, and references cited therein). As is the case for T cells (Berry et al., 1990), HL 60 cells require a sustained activation of PKC for differentiation to the macrophage phenotype (Aihara et al., 1991). This is in agreement with the observation that high concentrations of PMA lead to differentiation of HL 60 cells whilst low doses of phorbol ester are just mitogenic (Trayner and Clemens, 1992), a finding which might reflect distinct roles in proliferation and differentiation events of PKC isoenzymes differing in their susceptibility towards PMA. The tremendous redistributions and changes in abundance of PKC isoenzymes observed during the differentiation processes in these cells (see above) indicate that PKC isoenzymes seem to play an important role, which will be worthwhile to elucidate.

PERSPECTIVES

An overwhelming amount of information on PKC has accumulated since its discovery in 1977. As outlined in this Review, successful work has been carried out with respect to the number of isoenzymes, their structure and biochemical proper-ties, their expression, localization and activation within resting, proliferating and differentiating cells. In order to understand the divergent role of PKC isoenzymes in distinct signal transduction pathways, however, several aspects will require special attention in the future and intensive investigations both in vitro and by the use of appropriate cellular systems.

Physiological inducers of a PKC response should get preference to artefactual inducers potentially exerting pleiotropic (or no) effects. Thus, in the light of the extremely different sensitivity of PKC isoenzymes to long-term PMA treatment in situ (see above), the experimental paradigm of 'PKC downregulation' should be handled with care (see also Varese et al., 1992). Instead, a detailed analysis of the activation of distinct PKC isoenzymes in cells which respond to a specific inducer with a distinct differentiation program will provide us with much more profound insights. Recently, Szamel and Resch (1992) demonstrated that early in T cell stimulation (see above), PKC α is transiently activated whilst a prolonged activation of PKC β is observed after a lag phase. PKC α activation appeared to be sufficient for the expression of IL-Rα, whereas the sustained activation of PKC β was a prerequisite for the expression of IL-2.

Although there are potent PKC inhibitors available now (reviewed by Tamaoki and Nakano, 1990), the development of isoenzyme-specific PKC inhibitors will be extremely useful. As described above, stable overexpression of PKC δ in NIH 3T3 cells leads to a diminished growth rate. Interestingly, in these cells normal proliferative behaviour could be restored by an inhibitor for both cPKC and nPKC, but not by a cPKC-specific inhibitor (G. Martiny-Baron and C. Schächtele, personal com-munication). Likewise, the inhibition of the synthesis of in-dividual PKC isoenzymes by the use of specific oligo-ribonucleotides as well as the intracellular use of isoenzyme-specific antibodies will allow for the depletion and inactivation respectively of distinct PKC isoenzymes. Recently, Leli et al. (1992) demonstrated that intracellular delivery of anti-PKC α and PKC ϵ antibodies into transiently permeabilized neuro-blastoma cells was sufficient to induce differentiation.

With regard to the group of nPKCs, the biochemical nature of second messengers generated by different incoming signals has to be determined. Correspondingly, the activator and cofactor requirements of the individual PKC isoenzymes both in vitro and in situ have to be elucidated. Northern and Western blot techniques with PKC isoenzyme-specific probes will permit the determination of the expression pattern, the localization and redistribution of PKC isoenzymes in different tissues and cells and thus complete our understanding of the complex PKC isoenzyme network.

Transient and stable (over)expression of PKC isoenzyme cDNAs under the control of constitutive and inducible promotors will surely enlarge our understanding, especially if more soph-isticated approaches are chosen: (i) expression of a PKC iso-enzyme not present in the respective cell type, (ii) co-expression of individual PKC isoenzymes with other (downstream or upstream) components of signal transduction pathways, (iii) expression of PKC isoenzyme mutants, e.g. chimerae of the regulatory and catalytic domains of different isoenzymes or mutants carrying deletions within the V1, the zinc finger and the hinge region, respectively. Promising work with respect to this goal has been initiated, e.g. the cotransfection experiments by Ohno et al. (1991), the PKC ϵ/γ chimera of Pears et al. (1991) and the PKC α devoid of the regulatory domain (James and Olson, 1992). Thus, the key role of PKC isoenzymes in signal transduction and its impact on the modulation of gene expression linked to proliferative and differentiation events will be further substantiated.

We acknowledge the work of numerous colleagues whose contributions to the field may not be mentioned in this Review, due to limits in space and to the scope of this article. We would like to thank Georg Kochs, Walter Kolch, Peter Parker and Marius Ueffing for helpful discussions, suggestions and criticism during the preparation of the manuscript. We are grateful to Justus Duyster, Dirk Eick, Georg Kochs, Walter Kolch, Georg Martiny-Baron and Harald Mischak for making data available to us prior to publication. The work in our laboratories is supported by the Bundesministerium für Forschung und Technologie of Germany.

REFERENCES

Ahmed, S., Kozma, R., Monfries, C., Hall, C., Lim, H. H., Smith, P. and Lim, L. (1990) Biochem. J. **272**, 767–773

Aihara, H., Asaoka, Y., Yoshida, K. and Nishizuka, Y. (1991) Proc. Natl. Acad. Sci. U.S.A. **88**, 11062–11066

Akita, Y., Ohno, S., Yajima, Y. and Suzuki, K. (1990) Biochem. Biophys. Res. Commun. **172**, 184–189

Azzi, A., Boscoboinik, D. and Hensey, C. (1992) Eur. J. Biochem. **208**, 547–557

Bacher, N., Zisman, Y., Berent, E. and Livneh, E. (1991) Mol. Cell. Biol. **11**, 126–133

Bacher, N., Zisman, Y., Berent, E. and Livneh, E. (1992) Mol. Cell. Biol. **12**, 1404 (erratum)

Bader, M. and Sarre, T. F. (1986) Eur. J. Biochem. **161**, 103–109

Bazzi, M. D. and Nelsestuen, G. L. (1987) Biochemistry **26**, 1974–1982

Bazzi, M. D. and Nelsestuen, G. L. (1988) Biochem. Biophys. Res. Commun. **152**, 336–343

Bell, R. M. and Burns, D. J. (1991) J. Biol. Chem. **266**, 4661–4664

Bellacosa, A., Testa, J. R., Staal, S. P. and Tsichlis, P. N. (1991) Science **254**, 274–277

Bernards, R. (1991) EMBO J. **10**, 1119–1125

Berridge, M. J. (1987) Annu. Rev. Biochem. **56**, 159–193

Berridge, M. (1989) Nature (London) **341**, 197–205

Berry, N. and Nishizuka, Y. (1990) Eur. J. Biochem. **189**, 205–214

Berry, N., Ase, K., Kishimoto, A. and Nishizuka, Y. (1990) Proc. Natl. Acad. Sci. U.S.A. **87**, 2294–2298

Borner, C., Filipuzzi, I., Wartmann, M., Eppenberger, U. and Fabbro, D. (1989) J. Biol. Chem. **264**, 13902–13909

Borner, C., Filipuzzi, I., Weinstein, I. B. and Imber, R. (1991) Nature (London) **353**, 78–80

Borner, C., Nichols-Guadagno, S., Fabbro, D. and Weinstein, I. B. (1992) J. Biol. Chem. **267**, 12892–12899

Bruder, J. T., Heidecker, G. and Rapp, U. R. (1992) Genes Dev. **6**, 545–556

Burgoyne, R. D. (1989) Trends Biol. Sci. **14**, 87–88

Burns, D. J. and Bell, R. M. (1991) J. Biol. Chem. **266**, 18330–18338

Burns, D. J., Bloomenthal, J., Lee, M.-H. and Bell, R. M. (1990) J. Biol. Chem. **265**, 12044–12051

Castagna, M., Takai, Y., Kaibuchi, K., Sano, K., Kikkawa, U. and Nishizuka, Y. (1982) J. Biol. Chem. **257**, 7847–7851

Cataldi, A., Miscia, S., Lisio, R., Rana, R. and Cocco, L. (1990) FEBS Lett. **269**, 465–468

Chauhan, A., Chauhan, V. P. S. and Brockerhoff, H. (1991) Biochem. Biophys. Res. Commun. **175**, 852–857

Chen, K.-H., Widen, S. G., Wilson, S. H. and Huang, K.-P. (1990) J. Biol. Chem. **265**, 19961–19965

Choi, P. M., Tchou-Wong, K.-M. and Weinstein, I. B. (1990) Mol. Cell. Biol. **10**, 4650–4657

Clark, J. D., Lin, L.-L., Kriz, R. W., Ramesha, C. S., Sultzman, L. A., Lin, A. Y., Milona, N. and Knopf, J. L. (1991) Cell **65**, 1043–1051

Clemens, M. J., Trayner, I. and Menaya, J. (1992) J. Cell Sci. **103**, 881–887

Cook, P. P., Chen, J. and Ways, D. K. (1992) Biochem. Biophys. Res. Commun. **185**, 670–675

Coussens L., Parker, P. J., Rhee, L., Yang-Feng, T. L., Chen, E., Waterfield, M. D., Francke, U. and Ullrich, A. (1986) Science **233**, 859–866

Coussens, L., Rhee, L., Parker, P. J. and Ullrich, A. (1987) DNA **6**, 389–394

Crabos, M., Imber, R., Woodtli, T., Fabbro, D. and Erne, P. (1991) Biochem. Biophys. Res. Commun. **178**, 878–883

Cuadrado, A., Molloy, C. J. and Pech, M. (1990) FEBS Lett. **260**, 281–284

Danish, R., El-Awar, O., Weber, B. L., Langmore, J., Turka, L. A., Ryan, J. J. and Clarke, M. F. (1992) Oncogene **7**, 901–907

De Benedetti, A., Joshi-Barve, S., Rinker-Schaeffer, C. and Rhoads, R. E. (1991) Mol. Cell. Biol. **11**, 5435–5445

Dekker, L. V., Parker, P. J. and McIntyre, P. (1992) FEBS Lett. **312**, 195–199

Dietrich, A., Rose-John, S. and Marks, F. (1989) Biochem. Int. **19**, 163–172

Divecha, N., Banfil, P. and Irvine, R. F. (1991) EMBO J. **10**, 3207–3214

Downward, J., Graves, J. D., Warne, P. H., Rayter, S. and Cantrell, D. A. (1990) Nature (London) **346**, 719–723

Duronio, V., Nip, L. and Pelech, S. L. (1989) Biochem. Biophys. Res. Commun. **164**, 804–808

Duyster, J., Hidaka, H., Decker, K. and Dieter, P. (1992) Biochem. Biophys. Res. Commun. **183**, 1247–1253

Eicholtz, T., Ablas, J., van Overveld, M., Moolenaar, W. and Poegh, H. (1990) FEBS Lett. **261**, 147–150

Eldar, H., Zisman, Y., Ullrich, A. and Livneh, E. (1990) J. Biol. Chem. **265**, 13290–13296

Fiebich, B., Marmé, D. and Hug, H. (1990) FEBS Lett. **277**, 15–18

Fields, A. P., Pettit, G. R. and May, S. W. (1988) J. Biol. Chem. **263**, 8253–8260

Finkenzeller, G., Marmé, D. and Hug, H. (1990) Nucleic Acids Res. **18**, 2183

Finkenzeller, G., Marmé, D. and Hug, H. (1992) Cell. Signalling **4**, 163–177

Flint, A. J., Paladini, R. D. and Koshland, D. E., Jr. (1990) Science **249**, 408–411

Freisewinkel, I., Riethmacher, D. and Stabel, S. (1991) FEBS Lett. **280**, 262–266

Galabru, J. and Hovanessian, A. (1987) J. Biol. Chem. **262**, 15538–15544

Geppert, M., Archer, B. T. and Sudhof, T. C. (1991) J. Biol. Chem. **266**, 13548–13552

Ghosh, S. and Baltimore, D. (1989) Nature (London) **344**, 678–682

Grabarek, J., Raychowdhury, M., Ravid, K., Kent, K. C., Newman, P. and Ware, J. A. (1992) J. Biol. Chem. **267**, 10011–10017

Graff, J. M., Stumpo, D. J. and Blackshear, P. J. (1989) J. Biol. Chem. **264**, 11912–11919

Greif, H., Ben-Chaim, J., Shimon, T., Bechor, E., Eldar, H. and Livneh, E. (1992) Mol. Cell. Biol. **12**, 1304–1311

Gschwendt, M., Kittstein, W. and Marks, F. (1991) Trends Biochem. Sci. **16**, 167–169

Hannun, Y. A., Loomis, C. R. and Bell, R. M. (1985) J. Biol. Chem. **260**, 10039–10043

Hartwig, J. H., Thelen, M., Rosen, A., Janmey, P. A., Nairn, A. C. and Aderem, A. (1992) Nature (London) **356**, 618–622

Hashimoto, K., Kishimoto, A., Aihara, H., Yasuda, I., Mikawa, K. and Nishizuka, Y. (1990) FEBS Lett. **263**, 31–34

Hass, R., Pfannkuche, H.-J., Kharbanda, S., Gunji, H., Meyer, G., Hartmann, A., Hidaka, H., Resch, K., Kufe, D. and Goppelt-Strübe, M. (1991) Cell Growth Differ. **2**, 541–548

Hershey, J. W. B. (1991) Annu. Rev. Biochem. **60**, 717–755

Hocevar, B. A. and Fields, A. P. (1991) J. Biol. Chem. **266**, 28–33

Hornbeck, P., Huang, K. P. and Paul, W. E. (1988) Proc. Natl. Acad. Sci. U.S.A. **85**, 2279–2283

House, C. and Kemp, B. E. (1987) Science **238**, 1726–1728

House, C. and Kemp, B. E. (1990) Cell. Signalling **2**, 187–190

House, C., Wettenhall, R. E. H. and Kemp, B. E. (1987) J. Biol. Chem. **262**, 772–777

Housey, G. M., OBrian, C. A., Johnson, M. D., Kirschmeier, P. and Weinstein, I. B. (1987) Proc. Natl. Acad. Sci. U.S.A. **84**, 1065–1069

Housey, G. M., Johnson, M. D., Hsiao, W. L. W., OBrian, C. A., Murphy, J. P., Kirschmeier, P. and Weinstein, I. B. (1988) Cell **52**, 343–354

Hsieh, L.-L., Hoshina, S. and Weinstein, I. B. (1989) J. Cell. Biochem. **41**, 179–188

Huang, C. and Cabot, M. C. (1990) J. Biol. Chem. **265**, 14858–14863

Huang, F. L., Yoshida, Y., Cunha-Melos, J. R., Beaven, M. A. and Huang, K.-P. (1989) J. Biol. Chem. **264**, 4238–4243

Huang, K.-P. and Huang, F. L. (1986) Biochem. Biophys. Res. Commun. **139**, 320–326

Huang, K.-P., Nakabayashi, H. and Huang, F. L. (1986a) Proc. Natl. Acad. Sci. U.S.A. **83**, 8535–8539

Huang, K.-P., Chan, K.-F. J., Singh, T. J., Nakabayashi, H. and Huang, F. L. (1986b) J. Biol. Chem. **261**, 12134–12140

Hunter, T. (1991) Cell **64**, 249–270

Huwiler, A., Fabbro, D. and Pfeilschifter, J. (1991) Biochem. Biophys. Res. Commun. **180**, 1422–1428

Huwiler, A., Fabbro, D., Stabel, S. and Pfeilschifter, J. (1992) FEBS Lett. **300**, 259–262

Inoue, M., Kishimoto, A., Takai, Y. and Nishizuka, Y. (1977) J. Biol. Chem. **252**, 7610–7616

Jaken, S. and Kiley, S. C. (1987) Proc. Natl. Acad. Sci. U.S.A. **84**, 4418–4422

Jakes, S., Hastings, T. G., Reimann, E. M. and Schlender, K. K. (1988) FEBS Lett. **234**, 31–34

James, G. and Olson, E. (1992) J. Cell Biol. **116**, 863–874

Jones, P. F., Jakubowicz, T., Pitossi, F. J., Maurer, F. and Hemmings, B. A. (1991) Proc. Natl. Acad. Sci. U.S.A. **88**, 4171–4175

Kaibuchi, K., Miyajima, A., Arai, K. and Matsumoto, K. (1989) Proc. Natl. Acad. Sci. U.S.A. **83**, 8172–8176

Karin, M. (1992) FASEB J. **6**, 2581–2590

Kemp, B. E. and Pearson, R. B. (1990) Trends Biochem. Sci. **15**, 342–346

Kikkawa, U. and Nishizuka, Y. (1986) Annu. Rev. Cell Biol. **2**, 149–178

Kikkawa, U., Kishimoto, A. and Nishizuka, Y. (1989) Annu. Rev. Biochem. **58**, 31–44

Kiley, S., Schaap, D., Parker, P. J., Hsieh, L.-L. and Jaken, S. (1990) J. Biol. Chem. **265**, 15704–15712

Kishimoto, A., Kajikawa, N., Shiota, M. and Nishizuka, Y. (1983) J. Biol. Chem. **258**, 1156–1164

Kishimoto, A., Mikawa, K., Hashimoto, K., Yasuda, I., Tanaka, S.-I., Tominaga, M., Kuroda, T. and Nishizuka, Y. (1989) J. Biol. Chem. **264**, 4088–4092

Kiss, Z., Deli, E. and Kuo, J. F. (1988) FEBS Lett. **231**, 41–46

Knopf, J. L., Lee, M.-L., Sultzman, L. A., Kriz, R. W., Loomis, C. R., Hewick, R. M. and Bell, R. M. (1986) Cell **46**, 491–502

Kochs, G., Hummel., R., Fiebich., B., Sarre, T. F., Marmé, D. and Hug, H. (1993) Biochem. J. **291**, 627–633

Kohlhuber, F., Strobl, L. and Eick, D. (1993) Oncogene, in the press

Koide, H., Ogita, K., Kikkawa, U. and Nishizuka, Y. (1992) Proc. Natl. Acad. Sci. U.S.A. **89**, 1149–1153

Kolch, W., Heidecker, G., Lloyd, P. and Rapp, U. R. (1991) Nature (London) **349**, 426–428

Krauss, R. S., Housey, G. M., Johnson, M. D. and Weinstein, I. B. (1989) Oncogene **4**, 991–998

Kubo, K., Ohno, S. and Suzuki, K. (1987a) FEBS Lett. **223**, 138–142

Kubo, K., Ohno, S. and Suzuki, K. (1987b) Nucleic Acids Res. **15**, 7179–7180

Lazaris-Karatzas, A., Montine, K. S. and Sonenberg, N. (1990) Nature (London) **345**, 544–547

Leach, K. L., Powers, E. A., Ruff, V. A., Jaken, S. and Kaufmann, S. (1989) J. Cell Biol. **109**, 685–695

Lee, M.-H. and Bell, R. M. (1991) Biochemistry **30**, 1041–1049

Leibersperger, H., Gschwendt, M. and Marks, F. (1990) J. Biol. Chem. **265**, 16108–16115

Leli, U., Parker, P. J. and Shea, T. B. (1992) FEBS Lett. **297**, 91–94

Levin, D. E., Fields, F. O., Kunisawa, R., Bishop, J. M. and Thorner, J. (1990) Cell **62**, 213–224

Li, P., Wood, K., Mamon, H., Haser, W. and Roberts, T. (1991) Cell **64**, 479–482

Lindner, D., Gschwendt, M. and Marks, F. (1991) Biochem. Biophys. Res. Commun. **176**, 1227–1231

Link, E., Kerr, L. D., Schreck, R., Zabel, U., Verma, I. and Baeuerle, P. A. (1992) J. Biol. Chem. **267**, 239–246

Liyanage, M., Frith, D., Livneh, E. and Stabel, S. (1992) Biochem. J. **283**, 781–787

Locker, J. and Buzard, G. (1990) DNA Sequence **1**, 3–11

Lucas, S., Marais, R., Graves, J. D., Alexander, D., Parker, P. J. and Cantrell, D. A. (1990) FEBS Lett. **260**, 53–56

Makowske, M. and Rosen, O. M. (1989) J. Biol. Chem. **264**, 16155–16159

Makowske, M., Ballester, R., Cayre, Y. and Rosen, O. M. (1988) J. Biol. Chem. **263**, 3402–3410

Marais, R. M. and Parker, P. J. (1989) Eur. J. Biochem. **182**, 129–137

McGlynn, E., Liebetanz, J., Reutener, S., Wood, J., Lydon, N. B., Hofstetter, H., Vanek, M., Meyer, T. and Fabbro, D. (1992) J. Cell. Biochem. **49**, 239–250

Meek, D. W. and Street, A. J. (1992) Biochem. J. **287**, 1–15

Megidish, T. and Mazurek, N. (1989) Nature (London) **342**, 807–811

Meldrum, E., Parker, P. J. and Carozzi, A. (1991) Biochim. Biophys. Acta **1092**, 49–71

Melloni, E., Pontremoli, S., Michetti, M., Sacco, O., Cakiroglu, A. G., Jackson, J. F., Rifkind, R. A. and Marks, P. A. (1987) Proc. Natl. Acad. Sci. U.S.A. **84**, 5282–5286

Melloni, E., Pontremoli, S., Viotti, P. L., Patrone, M., Marks, P. A. and Rifkind, R. A. (1989) J. Biol. Chem. **264**, 18414–18418

Michaeli, J., Busquets, X., Orlow, I., Younes, A., Colomer, D., Marks, P. A., Rifkind, R. A. and Kolesnick, R. N. (1992) J. Biol. Chem. **267**, 23463–23466

Miller, S. G. and Kennedy, M. B. (1986) Cell **44**, 861–870

Mischak, H., Bodenteich, A., Kolch, W., Goodnight, J., Hofer, F. and Mushinski, J. F. (1991a) Biochemistry **30**, 7925–7931

Mischak, H., Kolch, W., Goodnight, J., Davidson, W. F., Rapp, U., Rose-John, S. and Mushinski, J. F. (1991b) J. Immunol. **147**, 3981–3987

Mitchell, F. E., Marais, R. M. and Parker, P. J. (1989) Biochem. J. **261**, 131–136

Mizuno, K., Kubo, K., Saido, T. C., Akita, Y., Osada, S., Kuruoki, T., Ohno, S. and Suzuki, K. (1991) Eur. J. Biochem. **202**, 931–940

Mochly-Rosen, D. and Koshland, D. E., Jr. (1987) J. Biol. Chem. **262**, 2291–2297

Mochly-Rosen, D., Khaner, H. and Lopez, J. (1991) Proc. Natl. Acad. Sci. U.S.A. **88**, 3997–4000

Molina, C. A. and Ashendel, C. L. (1991) Cancer Res. **51**, 4624–4630

Morley, S. J. and Traugh, J. A. (1990) J. Biol. Chem. **265**, 10611–10616

Morley, S. J., Dever, T. E., Etchison, D. and Traugh, J. A. (1991) J. Biol. Chem. **266**, 4669–4672

Morrison, D. K., Kaplan, D. R., Escobedo, J. A., Rapp, U. R., Roberts, T. and Williams, L. T. (1989) Cell **58**, 649–657

Muramatsu, M.-A., Kaibuchi, K. and Arai, K.-I. (1989) Mol. Cell. Biol. **9**, 831–836

Murray, A. W., Fournier, A. and Hardy, S. J. (1987) Trends Biochem. Sci. **12**, 53–54

Nakanishi, H. and Exton, J. H. (1992) J. Biol. Chem. **267**, 16347–16354

Newton, A. C. and Koshland, D. E., Jr. (1987) J. Biol. Chem. **262**, 10185–10188

Newton, A. C. and Koshland, D. E., Jr. (1989) J. Biol. Chem. **264**, 14909–14915

Niino, Y. S., Ohno, S. and Suzuki, K. (1992) J. Biol. Chem. **267**, 6158–6163

Nishizuka, Y. (1988) Nature (London) **334**, 661–665

Obeid, L. M., Blobe, G. C., Karolak, L. A. and Hannun, Y. A. (1992) J. Biol. Chem. **267**, 20804–20810

Ogita, K., Miyamoto, S., Yamaguchi, K., Koide, H., Fujisawa, N., Kikkawa, U., Sahara, S., Fukami, Y. and Nishizuka, Y. (1992) Proc. Natl. Acad. Sci. U.S.A. **89**, 1592–1596

Ohno, S., Kawasaki, H., Imajoh, S., Suzuki, K., Inagaki, M., Yokokura, H., Sakoh, T. and Hidaka, H. (1987) Nature (London) **325**, 161–166

Ohno, S., Akita, Y., Konno, Y., Imajoh, S. and Suzuki, K. (1988a) Cell **53**, 731–741

Ohno, S., Kawasaki, H., Konno, Y., Inagaki, M., Hidaka, H. and Suzuki, K. (1988b) Biochemistry **27**, 2083–2087

Ohno, S., Konno, Y., Akita, Y., Yano, A. and Suzuki, K. (1990) J. Biol. Chem. **265**, 6296–6300

Ohno, S., Akita, Y., Hata, A., Osada, S., Kubo, K., Konno, Y., Akimoto, K., Mizuno, K., Saido, T., Kuroki, T. and Suzuki, K. (1991) Adv. Enzyme Regul. **31**, 287–303

Okuda, T., Sawada, H., Kato, Y., Yumoto, Y., Ogawa, K., Tashima, M. and Okuma, M. (1991) Cell Growth Differ. **2**, 415–420

Olivier, A. R. and Parker, P. J. (1991) Eur. J. Biochem. **200**, 805–810

Ono, Y., Kurokawa, T., Fujii, T., Kawahara, K., Igarashi, K., Kikkawa, U., Ogita, K. and Nishizuka, Y. (1986) FEBS Lett. **206**, 347–352

Ono, Y., Kikkawa, U., Ogita, K., Fujii, T., Kurokawa, T., Asaoka, Y., Sekiguchi, K., Ase, K., Igarashi, K. and Nishizuka, Y. (1987) Science **236**, 1116–1120

Ono, Y., Fujii, T., Igarashi, K., Kikkawa, U., Ogita, K. and Nishizuka, Y. (1988a) Nucleic Acids Res. **16**, 5199–5200

Ono, Y., Fujii, T., Ogita, K., Kikkawa, U., Igarashi, K. and Nishizuka, Y. (1988b) J. Biol. Chem. **263**, 6927–6932

Ono, Y., Fujii, T., Ogita, K., Kikkawa, U., Igarashi, K. and Nishizuka, Y. (1989a) Proc. Natl. Acad. Sci. U.S.A. **86**, 3099–3103

Ono, Y., Fujii, T., Igarashi, K., Kuno, T., Tanaka, C., Kikkawa, U. and Nishizuka, Y. (1989b) Proc. Natl. Acad. Sci. U.S.A. **86**, 4868–4871

Orr, J. W. and Newton, A. C. (1992) Biochemistry **31**, 4667–4673

Osada, S., Mizuno, K., Saido, T. C., Akita, Y., Suzuki, K., Kuroki, T. and Ohno, S. (1990) J. Biol. Chem. **265**, 22434–22440

Osada, S.-I., Mizuno, K., Saido, T. C., Suzuki, K., Kuroki, T. and Ohno, S. (1992) Mol. Cell. Biol. **12**, 3930–3938

Otte, A. P. and Moon, R. T. (1992) Cell **68**, 1021–1029

Pai, J.-K., Pachter, J. A., Weinstein, I. B. and Bishop, W. R. (1991) Proc. Natl. Acad. Sci. U.S.A. **88**, 598–602

Pan, T. and Coleman, J. E. (1990) Proc. Natl. Acad. Sci. U.S.A. **87**, 2077–2081

Parker, P. J. (1991) in Molecular Aspects of Cellular Regulation, Volume 6: The Hormonal Control of Gene Transcription (Cohen, P. and Foulkes, J. G., eds.), pp. 77–98, Elsevier/North Holland Biomedical Press, Amsterdam

Parker, P. J., Coussens, L., Totty, N., Rhee, L., Young, S., Chen, E., Stabel, S., Waterfield, M. D. and Ullrich, A. (1986a) Science **233**, 853–859

Parker, P. J., Goris, J. and Merlevede, W. (1986b) Biochem. J. **240**, 63–67

Patel, G. and Stabel, S. (1989) Cell. Signalling **1**, 227–240

Patel, H. R., Choi, H.-S. and Sytkowski, A. J. (1992) J. Biol. Chem. **267**, 21300–21302

Pears, C. and Parker, P. J. (1991) FEBS Lett. **284**, 120–122

Pears, C. J., Kour, G., House, B. E. and Parker, P. J. (1990) Eur. J. Biochem. **194**, 89–94

Pears, C., Schaap, D. and Parker, P. J. (1991) Biochem. J. **276**, 257–260

Pears, C., Stabel, S., Cazaubon, S. and Parker, P. J. (1992) Biochem. J. **283**, 515–518

Perin, M. S., Fried, V. A., Mignery, G. A., Jahn, R. and Sudhof, T. C. (1990) Nature (London) **345**, 260–263

Persons, D. A., Wilkinson, W. O., Bell, R. M. and Finn, O. J. (1988) Cell **52**, 447–458

Persons, D. A., Owen, R. D., Ostrowski, M. C. and Olivera, J. F. (1991) Cell Growth Differ. **2**, 7–14

Pfeffer, L. M., Strulovici, B. and Saltiel, A. R. (1990) Proc. Natl. Acad. Sci. U.S.A. **87**, 6537–6541

Pfeffer, L. M., Eisenkraft, B. L., Reich, N. C., Improta, T., Baxter, G., Daniel-Issakani, S. and Strulovici, B. (1991) Proc. Natl. Acad. Sci. U.S.A. **88**, 7988–7992

Pommier, Y., Kerrigan, D., Hartmann, K. D. and Glazer, R. I. (1990) J. Biol. Chem. **265**, 9418–9422

Pontremoli, S., Michetti, M., Melloni, E., Sparatore, B. Salamino, F. and Horecker, B. L. (1990) Proc. Natl. Acad. Sci. U.S.A. **87**, 3705–3707

Powell, C. T., Leng, L., Dong, L., Kiyokawa, H., Busquets, X., O'Driscoll, K., Marks, P. A. and Rifkind, R. A. (1992) Proc. Natl. Acad. Sci. U.S.A. **89**, 147–151

Quest, A. F. G., Bloomenthal, J., Bardes, E. S. G. and Bell, R. M. (1992) J. Biol. Chem. **267**, 10193–10197

Reuben, R. C., Rifkind, R. A. and Marks, P. A. (1980) Biochim. Biophys. Acta **605**, 325–346

Rogue, P., Labourdette, G., Masmoudi, A., Yoshida, Y., Huang, F. L., Huang, K.-P., Zwiller, J., Vincendon, G. and Malviya, A. N. (1990) J. Biol. Chem. **265**, 4161–4165

Rose-John, S., Dietrich, A. and Marks, F. (1988) Gene **74**, 465–471

Rutherford, T. R., Clegg, J. B. and Weatherall, D. J. (1979) Nature (London) **280**, 164–165

Ryves, W. J., Evans, A. T., Olivier, A. R., Parker, P. J. and Evans, F. J. (1991) FEBS Lett. **288**, 5–9

Saido, T. C., Mizuno, K., Konno, Y., Osada, S., Ohno, S. and Suzuki, K. (1992) Biochemistry **31**, 482–490

Sakane, F., Yamada, K., Kanoh, H., Yokoyama, C. and Tanabe, T. (1990) Nature (London) **344**, 345–348

Schaap, D. and Parker, P. J. (1990) J. Biol. Chem. **265**, 7301–7307

Schaap, D., Parker, P. J., Bristol, A., Kriz, R. and Knopf, J. (1989) FEBS Lett. **243**, 351–357

Schaap, D., Hsuan, J., Totty, N. and Parker, P. J. (1990) Eur. J. Biochem. **191**, 431–435

Sekiguchi, K., Tsukuda, M., Ase, K., Kikkawa, U. and Nishizuka, Y. (1988) J. Biochem. (Tokyo) **103**, 759–765

Shearman, M. S., Naor, Z., Sekiguchi, K., Kishimoto, A. and Nishizuka, Y. (1989) FEBS Lett. **243**, 177–182

Shen, S. S. and Buck, W. R. (1990) Dev. Biol. **140**, 272–280

Shirakawa, F. and Mizel, S. B. (1989) Mol. Cell. Biol. **9**, 2424–2430

Siegel, J. N., Klausner, R. D., Rapp, U. R. and Samelson, L. E. (1990) J. Biol. Chem. **265**, 18472–18480

Smith, M. J., Charron-Prochownik, D. C. and Prochownik, E. V. (1990) Mol. Cell. Biol. **10**, 5333–5339

Smith, M. R., Jaramillo, M., Tuazon, P. T., Traugh, J. A., Liu, Y., Sonenberg, N. and Kung, H. (1991) The New Biologist **3**, 601–607

Soderling, T. R. (1990) J. Biol. Chem. **265**, 1823–1826

Sözeri, O., Vollmer, K., Liyanage, M., Frith, D., Kour, G., Mark, G. E., III and Stabel, S. (1992) Oncogene **7**, 2259–2262

Stabel, S. and Parker, P. (1991) Pharmacol. Ther. **51**, 71–95

Stabel, S., Schaap, D. and Parker, P. J. (1991) Methods Enzymol. **200**, 670–673

Stahl, M., Ferez, C. R., Kelleher, K. L., Kriz, R. W. and Knopf, J. L. (1988) Nature (London) **332**, 269–272

Strulovici, B., Daniel-Issakani, S., Baxter, G., Knopf, J., Sultzman, L., Cherwinski, H., Nestor, J., Jr., Webb, D. R. and Ransom, J. (1991) J. Biol. Chem. **266**, 168–173

Sullivan, J. P., Connor, J. R., Tiffany, C., Shearer, B. G. and Burch, R. M. (1991) FEBS Lett. **285**, 120–123

Susa, M., Olivier, A. R., Fabbro, D. and Thomas, G. (1989) Cell **57**, 817–824

Szamel, M. and Resch, K. (1992) Biol. Chem. Hoppe–Seyler **372**, 828

Takai, Y., Kishimoto, A., Iwasa, Y., Kawahara, Y., Mori, T. and Nishizuka, Y. (1979) J. Biol. Chem. **254**, 3692–3695

Tamaoki, T. and Nakano, H. (1990) Bio/Technology **8**, 732–735

Tanaka, Y., Yoshihara, K., Tsuyuki, M., Itaya-Hironaka, A., Inada, Y. and Kamiya, T. (1992) J. Biochem. (Tokyo) **111**, 265–271

Tang, Y.-M. and Ashendel, C. L. (1990) Nucleic Acids Res. **18**, 5310

Taylor, S. S., Buechler, J. A. and Yonemoto, W. (1990) Annu. Rev. Biochem. **59**, 971–1005

Trayner, I. D. and Clemens, M. J. (1992) Exp. Cell Res. **199**, 154–161

Van den Berghe, N., Vaandrager, A. B., Bot, A. G., Parker, P. J. and de Jonge, H. R. (1992) Biochem. J. **285**, 673–679

Varese, R. V., Standaert, M. L. and Cooper, D. R. (1992) Nature (London) **360**, 305

Wada, H., Ohno, S., Kubo, K., Taya, C., Tsuji, S., Yonehara, S. and Suzuki, K. (1989) Biochem. Biophys. Res. Commun. **165**, 533–538

Ways, D. K., Cook, P. P., Webster, C. and Parker, P. J. (1992a) J. Biol. Chem. **267**, 4799–4805

Ways, D. K., Messer, B. R., Garris, T. O., Qin, W., Cook, P. P. and Parker, P. J. (1992b) Cancer Res. **52**, 5604–5609

Wetsel, W. C., Khan, W. A., Merchenthaler, I., Rivera, H., Halpern, A. E., Phung, H. M., Negro-Vilar, A. and Hannun, Y. A. (1992) J. Cell Biol. **117**, 121–133

Wolf, M. and Baggiolini, M. (1990) Biochem. J. **269**, 723–728

Wooten, M. W., Seibenhener, M. L., Soh, Y., Ewald, S. J., White, K. R., Lloyd, E. D., Olivier, A. and Parker, P. J. (1992) FEBS Lett. **298**, 74–78

Yasuda, I., Kishimoto, A., Tanaka, S., Tominaga, M., Sakurai, A. and Nishizuka, Y. (1990) Biochem. Biophys. Res. Commun. **166**, 1220–1227

Yoshida, Y., Huang, F. L., Nakabayashi, H. and Huang, K.-P. (1988) J. Biol. Chem. **263**, 9868–9873

Zidovetzki, R. and Lester, D. S. (1992) Biochim. Biophys. Acta **1134**, 261–272

Biochem. J. (1993) 292, 313-332 (Printed in Great Britain)

REVIEW ARTICLE
Cellular signalling mechanisms in B lymphocytes

William CUSHLEY and Margaret M. HARNETT
Department of Biochemistry, University of Glasgow, Glasgow G12 8QQ, Scotland, U.K.

INTRODUCTION

The B lymphocyte is the principal cellular mediator of the specific humoral immune response to infection. Activation of B cells occurs on selection of the appropriate clones by antigen leading to B cell proliferation and differentiation into antigen-specific, antibody-secreting plasma cells. After infection, the primed B lymphocyte serves as the cellular repository of specific immunological memory. The induction of antibody-secreting plasma cells and memory B cells is a complex process, integrating signals generated by a number of immunoregulatory receptors on several cell types, often in specialized environments: at each stage of development or differentiation, the fate of the B lymphocyte is determined not only by antigen but also by an array of soluble factors, including cytokines. In addition, signals generated by cell–cell contact and mediated by groups of complementary cell surface adhesion molecules also play a critical role in B cell development.

The B lymphocyte therefore provides an excellent model system in which to study not only cellular signalling processes but also cross-talk mechanisms by which distinct cell surface receptor-directed signal transduction pathways interact. This Review aims to detail the known structural and molecular pharmacological data relating to B cell antigen and cytokine receptors and, whilst identifying areas of current controversy, to set this molecular data in the context of the cell biology of B lymphocyte differentiation.

B LYMPHOCYTE DIFFERENTIATION

Antigen-independent differentiation

B lymphocytes arise from pluripotent stem cells in the bone marrow via a number of defined precursor B cell phenotypes (Figure 1). Once B cells emerge into the periphery, selection can lead to (i) cellular anergy and/or programmed cell death, (ii) activation, proliferation and differentiation into high rate anti-body-secreting plasma cells, or (iii) differentiation to memory B lymphocytes. The complex, multi-stage differentiation of normal B lymphocytes can be conveniently considered as divisible into two phases; antigen-independent and antigen-dependent (reviewed by Rolink and Melchers, 1991). The antigen-independent phase of differentiation occurs in the bone marrow, is dependent upon growth factors derived from stromal cells, and is concerned with providing the B cell with a functional cell surface receptor for antigen, [i.e., with a membrane immuno-globulin (mIg) molecule capable of binding an antigen]. The Clonal Selection Theory (Burnet, 1959) dictates that each B

lymphocyte expresses a receptor of a single antigen specificity; that is, antigen receptors are clonally distributed. The antigen-independent phase of B cell maturation proceeds via a series of stochastic rearrangements of the immunoglobulin heavy and light chain genes (Tonegawa, 1983), with the *Igh* locus being re-arranged first followed, in order, by *Igk* and *Igl* (Coleclough et al., 1981). Productive rearrangement of one of the two alleles at each locus inhibits recombination at the other allele, thus providing a molecular explanation for allelic exclusion of antigen receptors. Successful rearrangement of the heavy chain gene followed by that of one of the two light chain genes results in stable expression of IgM at the B cell surface, and the B lymphocyte can now be thought of as entering the antigen-dependent phase of B cell development.

Antigen-dependent differentiation

The mIgM$^+$ "immature" B cell is the first B cell which has the opportunity to respond to challenge with antigen. The great weight of available data, mostly from B cell lymphomas and transgenic mouse models, suggests that stimulation of the antigen receptor on immature B cells results in either clonal un-responsiveness (anergy) or in deletion of the clone (Scott et al., 1987; Goodnow et al., 1988; Nemazee and Bürki, 1989; Ales-Martinez et al., 1991). In the latter case, clonal deletion is accomplished by driving the antigen-stimulated B cells into apoptosis, or programmed cell death (Hasbold and Klaus, 1990). The biological significance of clonal deletion or anergy of immature mIgM$^+$ B cells is that it allows for removal of those B cells which possess antigen receptors which bind to self tissue components. If activated, such B cells would secrete self-reactive antibodies, potentially leading to autoimmune disease.

The immature IgM$^+$ B cell next develops to express mIgD; the mIgM$^+$/mIgD^{++} B cell can be regarded as a "mature" B lymphocyte. The mature B cell reacts positively to ligation of the antigen receptor, but in the case of a thymus-dependent antigen (which includes the majority of protein antigens), the precise nature of the response is shaped by T cell-derived cytokines. The first possibility is for the antigen-activated B cell to become a high rate IgM antibody-secreting plasma cell, which expresses essentially no mIg and whose energies are devoted to production of the secretory form of IgM. Alternatively, the antigen-activated cell can undergo isotype switch, V region somatic mutation and emerge as a memory B cell, a series of events manifest in the plasma of the host as a more rapid response to re-challenge with the same antigen together with appearance of higher affinity antibodies of, for example, the IgG class.

Abbreviations used: mAb, monoclonal antibody; AP-1, activated protein-1; CD, cluster of differentiation; EGF, epidermal growth factor; FcR, Fc receptor; FDC, follicular dendritic cell; GH, growth hormone; GM-CSF, granulocyte/monocyte colony-stimulating factor; GPI, glycophosphatidylinositol; GTPγS, guanosine 5-[γ-thio]triphosphate; HRS, haematopoietin receptor superfamily; ICAM-1, intercellular adhesion molecule-1; mIg, membrane immunoglobulin; IFN, interferon; IL, interleukin; InsP$_3$, inositol trisphosphate; LFA, leukocyte functional antigen; MAP, mitogen-activated protein (kinase); MHC, major histocompatibility complex; NF-BRE, nuclear factor–B cell response element; NF-κB, nuclear factor κB; PDGF, platelet-derived growth factor; PKA, protein kinase A; PKC, protein kinase C; PLC, phospholipase C; PTK, protein tyrosine kinase; PTPase, protein tyrosine phosphate phosphatase; TcR, T cell receptor.

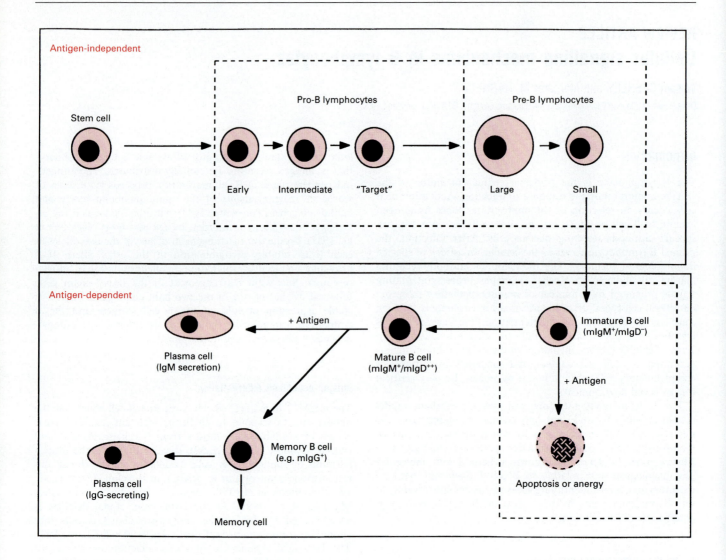

Figure 1   The B lymphocyte differentiation pathway

B lymphocyte differentiation is divisible into two phases, with antigen-independent steps which occur in the mammalian bone marrow and the antigen-dependent phase which takes place in the periphery (i.e. secondary lymphoid organs). The pathway illustrated is based on data from murine models. In the antigen-independent phase, the *Early* pro-B cell is CD45$^-$ but possesses a functionally rearranged heavy chain locus and is the first B cell precursor to express the *mb*-1 protein (despite being mIg$^-$). This cell also expresses the surrogate light chains V$_{pre-B}$ and λ_5 which are important in allelic exclusion and are present in all B cell precursor cells. The *Intermediate* pro-B cell is the first to express the CD45 antigen, and the *Target* pro-B cell is so called because it appears to be the target cell for the Abelson murine leukaemia virus. Pre-B cells are distinguished on the basis of size, with large cells being dividing cells. The small pre-B cells express cytoplasmic μ chains and are non-dividing cells. Each differentiation step from the Early pro-B cell to the small pre-B cell involves one round of division; this means that the productively rearranged VDJ gene in the early pro-B cell is found in 16 daughter small pre-B cells prior to light chain gene re-arrangement. Functional light chain rearrangement leads to expression of mIgM and the B cell is now reactive to antigen and enters the antigen-dependent phase of differentiation. Contact of an immature (mIgD$^-$) B cell with antigen leads to anergy or apoptosis of that cell, while a mature B cell (mIgM$^+$/mIgD^{++}) will respond to antigen by differentiating to a plasma cell or memory B cell depending upon the nature of T cell help available.

Cellular events in generation of memory B cells

The cellular events which give rise to memory B cells, the follicular reaction and germinal centre formation, have been the subject of study for many years and have recently been reviewed (Liu et al., 1992). Follicles are found in secondary lymphoid tissues (e.g. tonsil, spleen, lymph nodes) and consist of a network of follicular dendritic cells (FDCs) and, in the case of follicles where an antigen-driven response to a T-cell-dependent antigen is occurring, large numbers of lymphoblasts. The key property of FDCs is that they "fix" antigen on their cell surface in an unprocessed form in immune complexes which can later be recognized by newly-formed memory B lymphocytes. Germinal

centre formation is a characteristic feature of a secondary immune response to antigen and the initial step in this process is the exponential growth of mIg$^+$ B cell blasts within the follicle. Once the FDC network is filled, a polarization of cells within the follicle occurs which gives rise to the characteristic histological appearance of a germinal centre (Figure 2). The B cell blasts move to one edge of the germinal centre, called the dark zone, lose expression of mIg and continue to proliferate; these cells are now referred to as centroblasts. The proliferating mIg$^-$ centroblasts give rise to centrocytes which are non-dividing B cells which express mIg of isotypes other than IgM. The centrocytes move into the light zone of the germinal centre which possesses a dense network of FDCs, in contrast to the finer FDC network

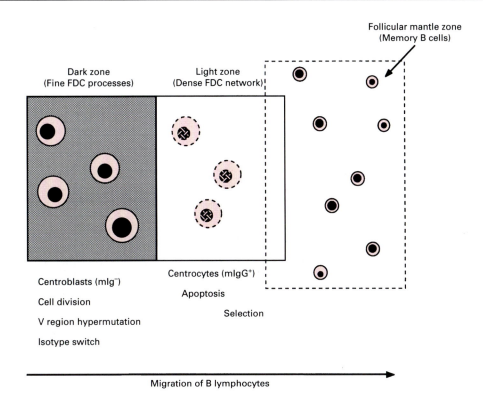

Figure 2 Cellular events in the generation of B cell memory

Memory B cells arise in germinal centres in secondary lymphoid organs. Antigen-activated B cells are driven to proliferate, move to the dark zone of the germinal centre and cease mIgM/mIgD expression; at the genetic level V region genes undergo somatic hypermutation, and the pattern of C region gene expression changes from μ and δ to, for example, Cγ. The centroblasts in the dark zone give rise to the centrocytes, smaller non-dividing B cells expressing mIgG (for example), which migrate to the light zone where they encounter antigen fixed as immune complexes on follicular dendritic cells (FDCs) and CD40 (the division of the light zone into basal and apical regions is not shown on the Figure). Centrocytes which have (for example) mIgG receptor for antigen fixed on the FDCs are selected and migrate to the follicular mantle zone as memory B cells.

processes found in the dark zone, and shows evidence of massive cell death by apoptosis. The light zone is divisible into two areas, the basal light zone adjacent to the dark zone and the apical light zone closer to the follicular mantle. Surviving memory cells emerging from the light zone reside in the follicular mantle zone.

In molecular terms, it seems likely that the mIg⁻ centroblasts undergo heavy chain isotype switch and V gene somatic hypermutation to give rise to the non-proliferating centrocytes which express mIg of isotypes other than mIgM and mIgD. The centrocytes are now programmed to undergo apoptosis and die rapidly unless actively rescued from this fate. *In vitro* studies of human tonsillar centrocytic cells indicate that two signals enable the centrocyte to escape apoptosis; one is generated by the cross-linking of the antigen receptors (*in vivo*, presumably by antigen fixed on the surface of FDCs), whilst the other is provided by ligation of the CD40 antigen on the centrocyte cell surface (Liu et al., 1989). This rescue of centrocytes from apoptosis appears to occur in the basal light zone. Thus, newly-generated centro-cytes have the opportunity to interrogate their somatically mutated and isotype-switched antigen receptors with the original antigen in these complexes and, if specific binding occurs, the cells are rescued from apoptosis. The main consequences of this process are that redundant specificities are not selected and only the highest affinity clones are rescued if antigen concentrations are limiting.

Further centrocyte maturation occurs in the apical light zone (Liu et al., 1992) where the cells can be induced to become either non-cycling memory cells via interaction with CD40 ligand (Liu

et al., 1991a) or can be driven to differentiate to plasmablast-type cells following stimulation with soluble CD23 and IL-1α (Liu et al., 1991a, 1992) or with IL-2 (Holder et al, 1992). This finding correlates well with the observation that apical light zone FDCs produce large amounts of CD23 and soluble CD23 (sCD23), the latter presumably acting upon the centrocytes via a paracrine mechanism. The signals delivered to the centrocyte as a consequence of encounter with each of these ligands and the importance of each signal in determining the survival and differentiation of the centrocyte remain to be elucidated.

The antigen receptor: death, life and suicide

The above brief account of B cell differentiation indicates that antigen plays a critical role in determining the fate of the B cell in the antigen-dependent phase of B cell development. Thus, immature B cells are clonally deleted, or at least functionally tolerised, if their antigen receptor is stimulated. This is in striking contrast to the situation with mature B cells or the immediate precursors of memory cells, the centrocytes, which respond to antigen stimulation by activation or rescue from cell suicide, respectively. However, it appears that, at least in models of immature and mature B cells, the signals transduced via the B cell antigen receptor are very similar and involve protein tyrosine kinase (PTK) activation, phospholipase C (PLC)-mediated hydrolysis of the inositol phospholipid phosphatidylinositol 4,5-bisphosphate (PtdInsP_2) and elevation of intracellular Ca²⁺ concentration (Monroe and Cambier, 1983 ; Bijsterbosch et al.,

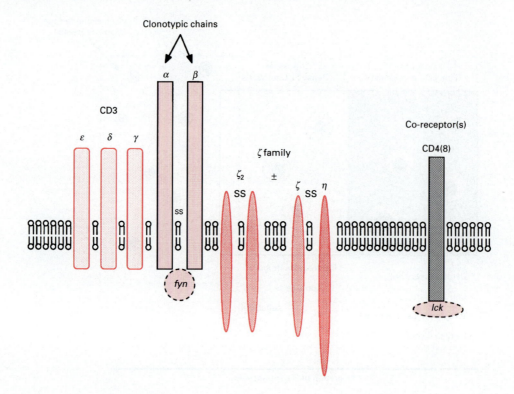

Figure 3   The T lymphocyte antigen receptor complex

Mature T cells possess antigen receptor complexes containing seven or nine components. All functional receptors possess the clonotypic α and β subunits (or γ and δ chains in the case of TcR1$^+$ cells) plus the γ, δ and ϵ chains of the CD3 complex. Mature T cells can possess ζ_2 homodimers either alone or in conjunction with the $\zeta\eta$ heterodimer. The CD4 (or CD8) co-receptor is shown in association with its associated non-receptor tyrosine kinase, the pp56^{lck} product (broken circle). The pp59^{fyn} product is shown in association with the TcR/CD3 complex; the precise point of interaction with these receptor components is undefined.

1986; Ales-Martinez et al., 1991). Thus, although there is some controversy as to the extent of protein kinase C (PKC) activation in immature B cell lines, the general consensus is that the negative signals generated by ligation of the antigen receptors are likely to be due to coupling of sIg to additional (as yet unknown) transmembrane signalling pathways. The mechanisms by which the same receptor can elicit three different biological responses in a maturation state-dependent manner is one of the central problems in B cell differentiation which remains to be resolved unequivocally.

THE B LYMPHOCYTE ANTIGEN RECEPTOR

General considerations

It is intuitively obvious that mIg occupies a central position in the B cell antigen receptor, and great efforts have been made to define the structural basis for the function of mIg in this context. Analysis of the genomic organisation of the constant region of the μ heavy chain gene revealed that two exons, the M exons, contributed unique codons to the mRNA molecule directing the synthesis of the membrane form of the μ heavy chain (μ_m) (Rogers et al., 1981; Alt et al., 1981). The membrane form of the μ chain contained 25–26 hydrophobic amino acids which traversed the bilayer in a helical arrangement, and a charged cytoplasmic tail of only three amino acids, Lys-Val-Lys. The paradox presented by this information was that although this charged tail could serve to facilitate stable membrane insertion of mIg, it was evidently too small to possess any intrinsic catalytic activity. However, stimulation of B cells with anti-Ig, or in specialized antigen-specific systems with antigen itself (Grupp et

al., 1987), leads to activation of the calcium-mobilizing second messenger system. Moreover, recent data indicate that tyrosine kinases, phosphatidylinositol 3-kinases (Yamanishi et al., 1992) and mitogen-activated protein kinases (MAP kinases) are also activated by perturbation of the B cell antigen receptor (Gold et al., 1992; Casillas et al., 1991). Bearing these observations in mind, several groups reasoned that the mIg molecule must be associated with other cell surface proteins to form a multi-component complex which serves to transmit signals to the B cell interior upon contact with antigen.

Lessons from the T lymphocyte antigen receptor complex

A paradigm for the existence of multi-component antigen receptor complexes involved in signal transduction in lymphocytes was provided by studies of the T cell receptor (TcR). Using mild detergent solublization techniques followed by immunoprecipitation, it was demonstrated that the TcR contained chains concerned with antigen recognition and other non-covalently associated structures (e.g. reviewed by Frank et al., 1990). The complete TcR structure at the surface of mature T lymphocytes contains either seven or nine components (Figure 3). Mature T lymphocytes possess one of two types of clonally distributed receptor; the most abundant of these, TcR2, contains α and β chains (transmembrane glycoproteins, each of approx 45 kDa) as its clonotypic components, while the less widely expressed receptor, TcR1, possesses γ and δ chains. Whilst the clonotypic heterodimer is structurally homologous to mIg (i.e., possesses Ig superfamily domains) and is the product of genes which undergo somatic recombination and allelic exclusion, the other chains of the TcR are invariant transmembrane proteins, composed of the CD3 ($\gamma\delta\epsilon$) complex and the ζ(ζ and η) family of proteins (Frank et al., 1990). The invariant chains of the TcR complex play roles in intracellular trafficking of the newly-synthesized receptor, and in transduction of the transmembrane signals generated by ligation of the clonotypic receptor (Frank et al., 1990; Finkel et al., 1991).

Optimal signalling via the TcR complex appears to require additional associations with appropriate co-receptors and their accessory transducing molecules. CD4 and CD8 are "co-receptors" in the antigen receptor complex. The TcR recognizes a peptide located in a peptide binding groove of a class I or II MHC antigen with low affinity and the CD4 or CD8 molecule confers stability upon the TcR–MHC complex. Non-receptor protein tyrosine kinases (PTKs) are found in association with this activated TcR–co-receptor complex. The two best character-ized kinases are members of the *src* family of tyrosine kinases and are encoded by the *lck* and *fyn* cellular proto-oncogenes (Veillette et al., 1988; Samelson et al., 1990). The pp56*lck* kinase is non-covalently associated with the CD4 structure on helper T cells and the CD8 molecule on cytotoxic T cells. The second tyrosine kinase is pp59*fyn*, which co-immunoprecipitates with the TcR–CD3-ζ complex (Dasgupta et al., 1992). These tyrosine kinases appear to be crucial in the coupling of the T cell antigen receptor complex to inositol phospholipid signalling events (June et al., 1990; Mustelin et al., 1990) and their activity appears to be further regulated by recruitment of the transmembrane protein tyrosine phosphatase (PTPase) CD45 to the TcR receptor complex (Koretzky et al., 1990; Alexander, 1990). That this structure plays a crucial rôle in regulation of signalling in lymphocytes is evidenced by the observation in T cells that signal transduction via the antigen receptor, but not via transfected G-protein-coupled muscarinic receptors, is abrogated in a CD45⁻ variant of the human HPB-ALL leukaemic T cell line (Koretzky et al., 1990); similar results were observed for a plasmacytoma B cell line expressing a transfected mIg receptor (Justement et al., 1991). The transfection of a CD45-expressing plasmid into the CD45⁻ HPB-ALL cell line corrects the defect, presumably via dephosphorylation and activation of p56*lck* (Ostergaard et al., 1989; Mustelin et al., 1989), and restores the capacity for functional signalling via the TcR. Thus, in a fashion analogous to that described for the intrinsic protein tyrosine kinase growth factor receptors, perturbation of the TcR complex leads to PTK-mediated activation of PLC (PLC-γ1) (Park et al., 1991; Secrist et al., 1991; Weiss et al., 1991), generation of InsP_3 and diacylglycerol, elevation of intracellular Ca²⁺ levels and activation of protein kinase C. There is, as yet, no evidence of involvement of classical heterotrimeric G proteins in linking the TcR to PLC (Graves and Cantrell, 1991; Phillips, 1991). However, recent evidence has indicated that a novel G-protein (p32) is associated with the CD4 (CD8)–pp56*lck* complex which may play a role in modulating these key early signalling events (Telfer and Rudd, 1991). Furthermore, a role for G-proteins has also been impli-cated by the recent finding that ligation of the TcR complex is coupled to p21*ras* activation resulting from a rapid decrease in the levels of GTPase activating protein (GAP) activity (Downward et al., 1990). The *ras* family of proteins are believed to be involved in the regulation of cellular proliferation and, thus, TcR coupling to *ras* activation would be consistent with the ability of antigen receptor-mediated signals to induce T cell proliferation.

The B lymphocyte antigen-receptor complex

Mild detergent lysis studies in the B cell system have revealed the consistent presence of two glycoproteins, found as a disulphide-

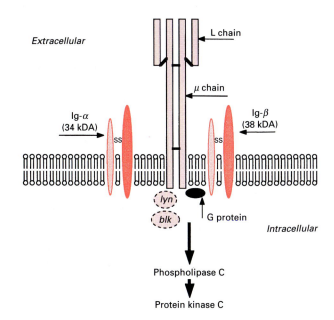

Figure 4   The B lymphocyte antigen receptor complex

The B cell antigen receptor complex is shown for a mIgM molecule; a similar layout is likely to apply to other Ig isotypes. The clonotypic chains of the receptor are provided by the μ heavy chain and light chain molecules, and the Ig-α and Ig-β associated proteins are shown as shaded ellipses. The associations with the *src*-like kinases, *blk* and *lyn*, and with a G protein are also shown. No definitive stoichiometric data are available for numbers of $\alpha\beta$ heterodimers per clonotypic unit.

bonded heterodimer, which are non-covalently associated with the mIg complex (Campbell and Cambier, 1990; Chen et al., 1990; Hombach et al., 1990; Parkhouse, 1990) (Figure 4). Both of these mIg-associated proteins have been demonstrated to be encoded by genes expressed only in B lymphocytes. The Ig-β chain, encoded by the B29 gene (Hermanson et al., 1988), has a molecular mass of 38 kDa and is a transmembrane glycoprotein. The electrophoretic properties of Ig-β are essentially identical for all isotypes studied, leading to the conclusion that mIg molecules of all classes can associate with Ig-β (Wienands et al., 1990). The Ig-α chain, also a transmembrane glycoprotein and encoded by the B cell-specific *mb*-1 gene (Sakaguchi et al., 1988), displays a range of molecular masses when analysed by two-dimensional electrophoresis. This observation initially attracted great interest as it suggested the possibility that multiple forms of the Ig-α chain existed which associated with heavy chains in an isotype-specific manner. This notion has been shown to be incorrect by transfection of "tagged" genes which demonstrate that the differences in electrophoretic mobility of Ig-α molecules are entirely attributable to variability in N-glycosylation patterns, and that there is only one Ig-α form which can associate with all mIg isotypes (Venkitaraman et al., 1991). The stoichiometry of the complex formed between the clonotypic chains of the receptor and the Ig-α–Ig-β heterodimer remains to be defined.

Considerable excitement surrounds the possible functions of Ig-α and Ig-β. The two most likely functions of these mIg-associated glycoproteins are in signal transduction and in as-sembly and transport of the B cell antigen-receptor complex. Early evidence for a role for Ig-α in transport was provided by transfection studies in the J$_{558}$ plasmacytoma (these B cells do not synthesize heavy chains, but do produce λ_1 light chains necessary for assembly of complete Ig molecules). Introduction of the secretory form (μ_s) of the μ heavy chain resulted in efficient secretion of IgM (Hombach et al., 1988). However, transfection of μ_m plasmids resulted in synthesis of the μ_m heavy chain, but no appearence of mIgM at the surface of the cells: surface expression could only be achieved if mb-1 (i.e. Ig-α) was also transfected into the J_{558} cells together with the μ_m constructs (Hombach et al., 1990). This result led to the proposal that Ig-α was mandatory for surface expression of all Ig isotypes. While this maxim is true for IgM, it is not an universally applicable rule, since detailed transfection experiments have demonstrated that IgD (Wienands et al., 1990; Venkitaraman et al., 1991) and certain IgG subclasses can be efficiently expressed at the cell surface in the absence of the Ig-α protein (Venkitaraman et al., 1991).

Both Ig-α and Ig-β are transmembrane glycoproteins with sizeable intracellular domains. However, inspection of the sequences of the cytoplasmic domains reveals an absence of any motifs which suggest that either Ig-α or Ig-β possesses intrinsic catalytic activity. Indeed, the available data suggest no definitive role for Ig-α and Ig-β as direct couplers of mIg to the B cell signal transduction machinery. The cytoplasmic sequences of Ig-α and Ig-β do, however, possess motifs which are also found in the cytoplasmic domains of certain of the CD3 complex components. Moreover, there is abundant evidence that both Ig-α and Ig-β are substrates for protein kinases, including protein tyrosine kinases (Campbell and Cambier 1991; Gold et al., 1991). The involvement of PTKs and PTPases in the activation of T lymphocytes is well established, and studies in B cells have also strongly implicated activation of PTKs, both in intact cell and isolated membrane systems (Campbell and Sefton, 1990; Gold et al., 1990). Thus, src-like kinases such as fyn , and others which are apparently unique to B cells, the blk and lyn kinases, have recently been shown to be associated with the mIg complex and coupled to mIg activation (Yamanishi et al., 1991; Burkhardt et al., 1991; Gold et al., 1991). One or more of these PTKs is likely to be responsible for tyrosine phosphorylation of the mIg-associated molecules following receptor crosslinking, and it is possible that CD45 (Justement et al., 1991) is involved in the dephosphorylation, and hence modulation, of cellular signalling via Ig-α and Ig-β.

SIGNAL TRANSDUCTION VIA THE B LYMPHOCYTE ANTIGEN RECEPTOR

In a normal individual, each B cell possesses an unique receptor for antigen and, even in an on-going response to deliberate immunization, the frequency of B cells specific for a given antigen is of the order of 10^{-7}. Except in specialized normal B cell systems (Snow et al., 1983; Pike and Nossal, 1985; Grupp et al., 1987), or in B cell lymphomas such as the CH12 series whose antigen receptor has a defined specificity (Mercolino et al., 1986), the use of antigen to stimulate the B cell antigen receptor is quite unrealistic. Consequently, the majority of studies have been performed using anti-Ig reagents to "mimic" the effect of antigen. As will be noted below, this approach is not without inherent complications of its own, but consistent patterns of data have emerged in both human and murine models and these are detailed below.

G proteins and tyrosine kinases

The antigen receptors (mIgM and mIgD) on B cells are coupled to the activation of at least two early transmembrane signalling events: (i) the PLC-mediated hydrolysis of PtdInsP_2 to generate the intracellular second messengers InsP_3 and diacylglycerol (Bijsterbosch et al., 1986) and (ii) activation of PTK activity

leading to the tyrosyl-phosphorylation of target proteins (Campbell and Sefton, 1990). The identity of the PTK(s) involved has not yet been defined, but is likely to be one or more of the src-related PTKs, blk, lyn and fyn recently shown to be associated with the mIg receptors (Yamanishi et al., 1991; Burkhardt et al., 1991). Another early event in B cell activation is the redistribution of PKC to the plasma membrane where up to nine proteins have been identifed as substrates, including class I, but not class II, MHC antigens (Burke et al., 1989).

There is, at present, considerable debate concerning the coupling of the antigen receptors to PLC activation in B cells. The observation that the kinetics of PLC activation following crosslinking of the antigen receptors are typical of classical G-protein-coupled calcium-mobilizing receptors and experiments demonstrating disruption and functional GTP-dependent reconstitution of the mIg-signalling pathway in permeabilized cells support the hypothesis that both classes of antigen receptors are regulated by a toxin-insensitive G protein (Gold et al., 1987, Harnett and Klaus, 1988), which may belong to the Gq subclass of toxin-insensitive G-proteins recently shown to be involved in regulation of PLC activity (Smrcka et al., 1991; Taylor et al., 1991). However, recent data detailing the structural properties of the G-protein-coupled receptor superfamily, notably the pre-eminence of a seven membrane-spanning loop structure, sparked off a controversy concerning the role of classical heterotrimeric G-proteins in the coupling of the lymphoid antigen receptors to PLC. Thus, despite evidence that other classes of receptors with single transmembrane-spanning regions, such as the TNFα receptor, can be G-protein regulated (Yanaga et al., 1992), together with the possibility that the antigen receptor-associated accessory molecules would provide candidates for coupling elements, many workers began to consider it unlikely that the antigen receptors, with their pairs of single transmembrane regions and short cytoplasmic tails, would be able to couple to a G-protein. The debate was further fuelled by the consistent failure to demonstrate classical G-protein coupling of the TcR (Graves and Cantrell, 1991; Phillips et al., 1991), which led to the acceptance of the proposal that the TcR is coupled to PLC activation by PTK activity in a manner similar to that described for receptors such as those for epidermal growth factor (EGF) and platelet-derived growth factor (PDGF) which possess intrinsic tyrosine kinase activities.

The discovery of sIg-mediated PTK activation (Gold et al., 1990; Campbell and Sefton, 1990) and studies using PTK inhibitors to investigate mIg/PLC coupling (Carter et al., 1991; Lane et al., 1991), suggested that similar events could also occur in some B cell lines. Studies with PTK inhibitors, such as genistein, herbimycin and tyrphostin, showed that these reagents could inhibit anti-Ig-mediated InsP_3 generation and calcium mobilization in human B cells and lymphoblastoid cell lines such as the Daudi cell line [generally following a long-term pre-incubation (16–20h) of cells with the inhibitor]. In addition, Carter et al. (1991) demonstrated that ligation of mIg could lead to phosphorylation of PLC-γ1 in a human B lymphoblastoid cell line. These findings, although interesting, should be assessed carefully, as Hempel and DeFranco (1991) have shown that whilst the predominant forms of PLC expressed in a range of murine B cell lines are the α and γ2 isoforms, PLC-γ1 levels of expression are generally very low or, indeed, absent in some lines. In addition, the kinetics of Ca^{2+} mobilization by PTK-coupled receptors are typically slower than those observed when signals are transduced via classical G-protein-coupled calcium-mobilizing receptors (Margolis et al., 1990), and the rapid kinetics of mIg-activated normal B cells would therefore be more consistent with coupling via a classical G protein system rather

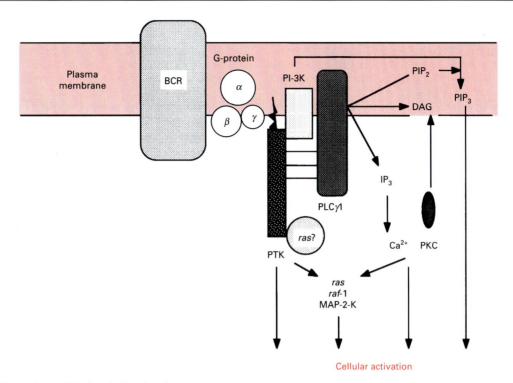

Figure 5   Signalling pathways linked to the B cell antigen receptor

"BCR" in this Figure represents the clonotypic and associated structures depicted in Figure 4. The spatial linkage between the associated heterotrimeric G protein and the effector systems [PTK, phosphatidylinositol 3-kinase (PI-3K) and PLCγ1] are detailed, together with their relationship to likely downstream effectors necessary for cellular activation. Abbreviations: PIP$_2$, phosphatidylinositol bisphosphate; PIP$_3$, phosphatidylinositol trisphosphate; DAG, diacylglycerol; IP$_3$, inositol trisphosphate.

than via a PTK system. Furthermore, under the conditions necessary to induce inhibition of mIg-mediated InsP_3 production by PTK inhibitors in normal murine B cells, these reagents also induce toxic and/or nonspecific effects on other cellular kinases such as PKC and the phosphatidylinositol kinases (M. M. Harnett, unpublished work). Indeed, recent data demonstrate that those effects of the PTK inhibitors which have been interpreted in the above studies in terms of the disruption of receptor–PLC coupling could simply reflect the ability of these reagents to block production of PtdInsP_2, the PLC substrate (M. M. Harnett, unpublished work). Thus, mIg-mediated PTK activation may play an important role(s) in B cell activation, although it does not appear to be the sole or even major regulatory element underlying coupling of the antigen receptors to PLC in normal resting B cells. However, in view of the increasing evidence supporting G-protein regulation of intrinsic PTK growth factors in primary cells, rather than the PTK coupling observed in transformed cells or in cell lines that overexpress receptors (Liang and Garrison, 1991; Yang et al, 1991), it is possible that the antigen receptors are differentially coupled to various PLC isoforms via G-proteins and/or PTKs depending on their state of activation or differentiation. Finally, PTKs and G-proteins need not play mutually exclusive roles in the regulation of mIg/PLC coupling (Figure 5). Indeed, recent studies by Roifman and co-workers may finally reconcile the conflicting evidence supporting roles for both G-proteins and PTKs in mIg/PLC coupling. These authors now have evidence that in human peripheral blood B lymphocytes, mIg is coupled to PTK activation, tyrosine phosphorylation of PLC-γ1 and inositol phosphate generation via a pertussis toxin-sensitive G-protein (Melamed et al., 1992) in a manner reminiscent of the

involvement of Gi in EGF-mediated activation of PLC-γ1 in primary rat hepatocytes (Yang et al., 1991).

Models of T cell-dependent B cell activation

Anti-Ig antibodies are most likely to provide a model for polyclonal B cell activation by type-2 T cell-independent (TI-2) antigens, which are typically large polymers, with repeating epitopes, capable of efficient crosslinking of sIg receptors. It is still a matter of debate how these effects of anti-Ig relate to those of soluble T cell-dependent antigens which are likely to be rather poor crosslinking agents and which do not cause substantial B cell activation in the absence of T cells. However, evidence from studies using hapten–protein conjugates suggests that that T cell-dependent antigens may produce a limited inositol phospholipid turnover leading to abortive activation of B cells. This abortive entry into the cell cycle may be important since the resultant highly-efficient internalization of antigen followed by proteolysis and re-expression of antigen-derived peptides in association with MHC Class II molecules (Germain and Hendrix, 1991) is likely to promote B cell–T cell co-operation; that is, abortive activation promotes the conditions necessary for "cognate" recognition of antigen (Noelle and Snow, 1991). Secondary intercellular interactions between adhesion and accessory molecules (see below) may then generate 'bi-directional' signalling in which the T helper cell is stimulated to proliferate and release cytokines which, in turn, act to amplify the B cell response. The finding that the T cell-derived cytokine IL-4 could act, *in vitro*, as a co-mitogen with submitogenic concentrations of anti-Ig (Howard et al., 1982) (which induce a low level of InsP_3 and diacylglycerol generation) led to the current hypothesis that IL-4 (and possibly other lymphokines generated by antigen-presenting cells?) interacts with the mIg signalling cascade to "prime" the B cell for additional signals generated during B cell–T cell co-operation. This is consistent with the recent demonstration that crosslinking of the MHC Class II molecules (to mimic interactions with the TcR–CD3 complex and CD4/CD8 receptors on T cells during B cell–T cell co-operation *in vitro*) on IL-4/anti-Ig-primed murine cells, drives all resting B cells into S phase (Cambier and Lehman, 1989). Provision of additional lymphokines (e.g. IL-4 and IL-5) at this stage drives the differentiation of B cells into antibody-secreting plasma cells (Cambier and Lehman, 1989). This murine experimental model of B cell activation by T cell-dependent antigens has been strongly supported by reconstitution experiments which show that incubation of B cells with fixed, pre-activated T cells (or even membranes from pre-activated T cells) in the presence of IL-4 and IL-5 is sufficient to induce B cell proliferation and differentiation into antibody-secreting cells (Brian, 1988; Hodgkin et al., 1990; Noelle et al., 1991).

Patterns of gene expression

Stimulation of B cells with antigen or anti-Ig causes several changes in the pattern of gene expression in the cell. The most pronounced of these is the massive up-regulation of class II MHC antigens which, as noted above, is important in facilitating T cell–B cell co-operation. Other rapid changes include increases in levels of c-*fos*, *Egr*-1 and *c-myc* mRNA (Cambier and Campbell, 1990), all of which appear to be linked to the calcium-mobilizing second messenger system. Two inducible genes are particularly noteworthy in terms of the nature of the B cell response to stimulation of the antigen receptor.

Immature B cells, which are mIgM$^+$/mIgD$^-$, respond negatively to stimulation of the antigen receptor (i.e. they become anergic or enter apoptosis), while the next differentiation stage, the mature B cell (mIgM$^+$ / mIgD^{++}) becomes activated by perturbation of the antigen receptor. A possible molecular explanation for this phenomenon, in the murine model, lies in the capacity of mature B lymphocytes to express the *Egr*-1 (*early growth response*) gene in response to anti-Ig stimulation. The *Egr*-1 gene product is a transcriptional regulatory factor, which is induced by anti-Ig (or phorbol ester) stimulation in mature but not immature B cells; it is also absent from B cell precursor cells in the bone marrow, and cannot be induced in IgM$^+$/IgD$^-$ B cell lymphomas such as the WEHI-231 line. The inability to express *Egr*-1 is associated with a lack of proliferation in response to antigen receptor stimulation (Seyfert et al., 1991). The mechanism by which immature B cells fail to respond to anti-Ig stimulation at the level of expression of *Egr*-1 does not involve the presence or absence of differentiation-stage-specific *trans*-acting factors (as is the case for control of MHC class II expression by IL-4 in murine B cells; see section below), but is dependent upon the state of methylation of the promoter region of the *Egr*-1 gene itself (Seyfert et al., 1991). Thus, in immature B lymphocytes, the *Egr*-1 promoter is heavily methylated, while B cell lines representative of the mature B cell phenotype (e.g. BAL-17) and normal splenic B lymphocytes possess a hypomethylated *Egr*-1 promoter region. Interestingly, treatment of the immature B lymphoma line WEHI-231 with 5-azacytidine, a non-specific inhibitor of methylation, leads to hypomethylation of the *Egr*-1 promoter region in these cells which results in the cells synthesizing *Egr*-1 mRNA upon anti-Ig stimulation. Thus, the methylation status of the *Egr*-1 gene is a reflection of the state of maturity of early antigen-dependent B cells, and this gene must be in a hypomethylated state before B cells can respond to anti-Ig stimulation by cellular activation.

Once mature mIgM$^+$ / mIgD^{++} mature B lymphocytes have been activated by antigen, they face a second selection step if they are driven to differentiate into memory B cells. This occurs in the germinal centre where centrocytes are programmed to die by apoptosis unless they encounter antigen (and other appropriate stimuli). In this instance, data from human tonsillar B cell models strongly suggest that the gene which is necessary for escape from apoptosis is the *bcl*-2 oncogene (Liu et al., 1991a,b, 1992). Thus, stimulation of centrocytes with anti-Ig causes induction of expression of *bcl*-2 within 4 h (Liu et al., 1991b): other signals which can promote *bcl*-2 expression in centrocytes include the combination of sCD23 and IL-1α (Liu et al., 1991a) and ligation of the CD40 antigen on the centrocyte surface (Liu et al., 1991b). Thus, the stimuli which induce *bcl*-2 expression in centrocytes therefore correlate exactly with those which rescue the same cells from apoptosis. Interestingly, certain lymphomas of centrocytic cells display a chromosomal translocation which places a section of chromosome 18 containing the *bcl*-2 oncogene under the influence of the *Igh* enhancer on chromosome 14 (Tsujimoto et al., 1985). Centrocytic cells bearing this translocation are neoplastic and survive without the need for signals from antigen receptors or FDCs by virtue of their constitutive expression of the *bcl*-2 protein. Finally, transfection of *bcl*-2 into *bcl*-2$^-$ B cell lymphomas renders such cells refractory to apoptosis as a consequence of serum starvation (Henderson et al., 1991).

CYTOKINES AND B LYMPHOCYTE SIGNALLING

Introduction

All stages of B lymphocyte activation, growth and differentiation are regulated by cytokines. Thus, in the bone marrow, the antigen-independent phase of B cell differentiation is heavily dependent upon cytokines derived from stromal cells (e.g. IL-7 and IL-11) and, to a lesser extent, by cytokines derived from T cells (reviewed by Callard, 1990). Once a functional antigen receptor is expressed upon the B cell surface and the cell is capable of responding to antigen, a range of cytokines tightly regulates the precise immunological characteristics of the behaviour of the B cell clones recruited to the response. The majority of the cytokines which are important in the antigen-dependent phase of B cell differentiation are derived from helper T lymphocytes: for example, whilst IL-2 is involved in regulating growth of cycling B cells (Jelinek et al., 1986) and accelerating antibody secretion [as is IL-6 (Hirano et al., 1986)], other cytokines such as IL-4 and IFNγ direct molecular processes in the B cell, for example, by positively influencing heavy chain isotype switching (Lutzker et al., 1988; Rothman et al., 1988). Many B cell responses to cytokines are not a reflection of stimulation with one cytokine alone, but are indicative of the combined effects of several cytokines. Thus, certain cytokines can interact synergistically upon B cells, while others, such as IL-4 and IFNγ (Rabin et al., 1987) or IL-4 and IL-2 (Jelinek and Lipsky, 1987; Lorente et al., 1990), appear to be mutually inhibitory. All of these phenomena pose interesting questions at the level of the nature of the cellular signals generated upon interaction of cytokines with their complementary cell surface receptors, and set further challenges for an understanding of how such cellular signals might interact, positively or negatively, within the B lymphocyte to elicit the appropriate biological response.

Recent advances in cloning of the receptors for a number of cytokines has improved our knowledge of the structural biochemistry of the receptors, but has not yet yielded definitive data to explain their interaction with cellular signalling systems. The data suggest that cytokine receptors may be conveniently grouped into four main categories. The first are members of the Ig-like superfamily, and are exemplified by the IL-1 receptors which are discussed below. A second group are classical G-protein coupled receptors. An example of such a cytokine receptor is that for the 9 kDa cytokine IL-8 (Oppenheim et al., 1991), which is present as a single class of high-affinity receptors (Samanta et al., 1989) linked to phospholipase C via a pertussis toxin-sensitive G-protein, with ligation of the receptor triggering Ca^{2+} mobilization and activation of PKC (Dewald et al., 1988). The cloning of the IL-8 receptor from a cDNA library of neutrophil-specific genes (Holmes et al., 1991; Murphy and Tiffany, 1991; Thomas et al., 1991) reveals it to be a seven-transmembrane-domain type, prototypic G-protein-coupled receptor. The third group of receptors expressed on B cells are those which possess intrinsic tyrosine kinase catalytic domains in their intracellular domains, as opposed to the antigen receptors and some cytokine receptors which appear to be coupled to non-receptor PTK activities. Thus, activated murine T and B cells have been shown to mount a mitogenic response to insulin stimulation (Snow et al., 1980), and functional receptors for stem cell factor are present on some B lymphocyte types. The final group of cytokine receptors, to which many interleukin receptors belong, is the haematopoietin receptor superfamily (HRS) (Bazan, 1989). While the signal transduction pathways for a few systems have been rigorously characterized (e.g. for the IL-8 receptor), a consistent pattern has not emerged for signal transduction within or amongst the cytokine receptor families. Heated debate regarding lineage-specific (IL-1) and species-specific (IL-4) cellular signalling mechanisms surrounds certain individual cytokine/receptor systems.

Receptors of the Ig-like superfamily: IL-1

Members of the Ig-like superfamily include the receptor(s) for IL-1. The key feature of these receptors is that they contain structures characteristic of the prototypic Ig domain. Thus, members of this family have one or more protein domains which are approximately 110 amino acids in size and display good primary structural similarity to immunoglobulin molecules, including a single disulphide bond enclosing 65–75 residues. At the higher structural levels, such domains possess seven anti-parallel β strands as their principal secondary structural motif, and have the β strands arranged in two faces of four and three strands linked by a single disulphide bond (Williams and Barclay, 1988). This structural motif is widely represented in the cell surface molecules of the immune and neural systems.

IL-1 is a highly pleiotropic cytokine, influencing the growth and differentiation of a wide range of cell types (reviewed by Dinarello et al., 1989). The cytokine is found in two forms, IL-1α (159 amino acids) and IL-1β (153 amino acids), both of which are initially found as cell-surface-associated molecules before being cleaved to generate soluble cytokine (Martin and Resch, 1988). Both IL-1α and IL-1β display activity towards a similar range of target cells. Both the T cell, type I (Sims et al., 1989), and the B cell, type II (McMahan et al., 1991), IL-1 receptors have now been cloned. The IL-1 receptor found on human T lymphocytes and fibroblasts comprises 552 amino acids (557 in the mouse), and the layouts of the domains in both species are similar (Sims et al., 1989). Thus, there is an extracellular binding region composed of three Ig-like domains, a single trans-membrane sequence, and a large (215 residues) cytoplasmic tail, with some features conserved in many nucleotide binding proteins; there are no motifs consistent with the IL-1 receptor cytoplasmic domain possessing intrinsic protein kinase activities (Sims et al., 1989). The B cell type receptor has a similar

extracellular region, with three Ig-like domains, also possesses a single transmembrane region, but has a small 29-amino-acid cytoplasmic tail region (McMahan et al., 1991). Both types of receptor can bind both IL-1α and IL-1β. Interestingly, type I receptors show a single class of binding sites for IL-1α, but two sites for IL-1β, while the type II receptor displays the reverse characteristics (McMahan et al., 1991). Both receptors are linked to pertussis toxin-sensitive G proteins (Mizel, 1990; O'Neill et al., 1990) and, whilst they activate the Na^+/H^+ antiporter, neither receptor mediates an increase in intracellular Ca^{2+} levels. Finally, both receptors undergo receptor-mediated endocytosis following ligand binding, and there is strong evidence to suggest that the receptor–ligand complex may ultimately translocate to the nucleus via an endosomal compartment (Grenfell et al., 1989; Curtis et al., 1990).

Lineage-specific IL-1-triggered signalling pathways

Controversy surrounds the precise downstream signal transduction mechanisms activated in IL-1-sensitive cells (Mizel, 1990; O'Neill et al., 1990). The IL-1-driven transcriptional activation in 70Z/3 pre-B cells is mediated by the NF-κB *trans*-acting factor, and can be mimicked by exposure of the cells to forskolin via protein kinase A activation. Together with studies indicating that IL-1 receptor activation leads to elevations in cyclic AMP levels in other lymphocyte models, the data from the 70Z/3 pre-B cell model suggest that IL-1 mediates its effect upon NF-κB activity via elevation of cyclic AMP and subsequent activation of protein kinase A (Mizel, 1990). This suggestion is not, however, supported by data from other studies of the mechanism of action of IL-1 upon fibroblasts (O'Neill et al., 1990). Thus, fibroblasts respond to IL-1 stimulation by phosphorylation, upon serine residues, of the 80 kDa type I IL-1 receptor itself and the EGF receptor. This effect cannot be mimicked by forskolin, suggesting a lack of involvement of protein kinase A, and also fails to be induced by exposing the fibroblasts to phorbol esters. The hypothesis that PKC is not involved in the signal transduction pathways activated by IL-1 in fibroblasts is further supported by the observation that chronic treatment with phorbol ester, which down-regulates protein kinase C activity (O'Neill et al., 1990), fails to inhibit the ability of IL-1 to stimulate phosphorylation of the IL-1 and EGF receptors. This suggests the involvement of a distinct kinase(s) linked to the IL-1 receptor, and precedents for this are found in the IL-2 receptor system discussed below.

The data from analysis of IL-1 receptor-mediated signal transduction in pre-B cells and fibroblasts raises the possibility that distinct cell lineages possess structurally-different IL-1 receptors which are, in turn, linked to non-identical signal transduction pathways. Further circumstantial support for such a notion is provided by the observation that in certain IL-1-sensitive cell types, diacylglycerols are generated from different lipid sources (O'Neill et al., 1990). Thus, phosphatidylcholine appears to be the principal source of diacylglycerol in T cells, phosphatidylinositol serves the equivalent function in macrophages, while in mesangial cells phosphatidylethanolamine is the predominant source of ligand-sensitive diacylglycerol.

Cytokine receptors of the haematopoietin receptor superfamily

The final group of cytokine receptors is the haematopoietin receptor superfamily (HRS). This superfamily was initially described by Bazan (1989), D'Andrea et al. (1989) and Cosman et al. (1989). The HRS is divisible into two subgroups which, while having many features in common, also possess characteristic structural differences. Type I receptors (which for con-

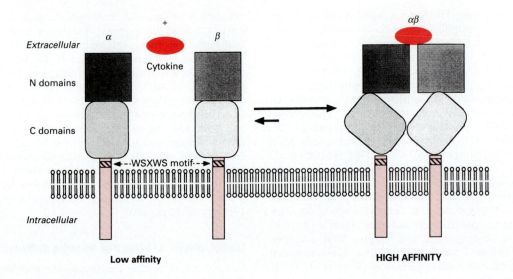

Figure 6   A prototypic HRS cytokine receptor

A prototypic HRS receptor is shown. The N domains are shown as squares, the C domains as oblongs, and the cytoplasmic tails as open rectangles; the WSXWS motif is shown as a hatched box. Based on the IL-3 or IL-6 receptor models (see the text for details), the cytokine (filled ellipse) binds to the α chain of the receptor driving association with the β subunit. The association is via the C domains. After cytokine binding the equilibrium of the reaction lies to the right, and signalling proceeds principally via the high-affinity receptor complex.

venience we shall refer to as "HRS" receptors) include receptors, or one or more components of receptors, for IL-2, IL-3, IL-4, IL-5, IL-6, IL-7, GM-CSF, erythropoietin, prolactin and growth hormone (GH). Type II receptors includes receptors for the interferons and for tissue factor.

In general terms, all HRS receptors are transmembrane glycoproteins, possessing two extracellular domains, a short transmembrane region and a large cytoplasmic tail. The recent crystallization and determination of the structure of a complex of GH and its receptor at a resolution of 2.8 Å (De Vos et al., 1992) has suggested many interesting properties for the HRS receptors and their ligands in terms of conserved primary and higher structural features. For example, the extracellular surface of the receptor comprises two domains, N and C (Bazan, 1989), which are composed of anti-parallel β strands arranged in faces of three and four strands. The N domain of the GH receptor possesses three disulphide bonds, and the placement of four of the cysteines participating in formation of these disulphide bonds is highly conserved in all HRS receptors; no disulphide bond is found in the C domain. The HRS receptors also possess structural characteristics akin to type III fibronectin domains, which could suggest a adhesion function of these molecules or may simply reflect the evolutionary origins of members of this receptor group (Bazan et al., 1990). The final characteristic feature of note in the extracellular surface of the HRS receptors is the finding of a very highly conserved pentapeptide motif (Trp-Ser-Xaa-Trp-Ser, the WSXWS motif) located close to the transmembrane region in the primary structure. The intracellular domain is of variable size and shows little conservation within the HRS. In terms of activation of cellular signalling pathways, there are few notable structural features in the intracellular domains of HRS receptors, and these display no motifs which indicate that they possess catalytic domains. There are, however, several good candidate residues which could be substrates for protein kinases, and there is evidence that ligation of several of the cytokine receptors of this family leads to p21^{ras} activation (Satoh et al., 1991).

Perhaps the most important finding from the GH model is that the stoichiometry of the complex is one ligand molecule to two receptor molecules. The data also indicate that ligand binding is accomplished by the N domains, and the binding causes a very close association between C domains to take place (Figure 6). Moreover, binding appears to show a degree of co-operativity, with binding to one receptor site being necessary for binding to a second site (Cunningham et al., 1991). In addition, occupation of the two receptor chains seems to confer stability on the receptor–ligand complex which may have important consequences for initiation of signal transduction via the dimerized receptors: this is consistent with observations on the EGF, insulin and PDGF receptors, where maximal activation of their intrinsic tyrosine kinase activities depends on receptor dimerization. The following examples of cytokine receptors functioning in B cells serve to illustrate this point. However, unlike the growth factor and GH systems, the cytokine receptor systems currently best understood show receptor heterodimerization.

IL-2: receptor heterodimerization

The IL-2 receptor is a true cytokine receptor complex, possessing two defined ligand binding subunits, and a range of associated proteins (Figure 7). The two ligand-binding chains are both transmembrane glycoproteins, and both make intimate contact with the ligand itself (Teshigawara et al., 1987; Robb et al., 1987; Smith, 1989). The best characterized of the two subunits is a 55 kDa glycoprotein originally defined on activated T cells (variously called Tac antigen, CD25 or IL-2Rα), which is not an HRS receptor and which is inducible by a variety of stimuli on T lymphocytes including perturbation of the antigen receptor and the cytokines IL-1, IL-2, IL-6 and TNFα (reviewed by Greene et al., 1989; Smith, 1989; Ullman et al., 1990). IL-2Rα is also inducible in B cells, but with an apparently distinct range of cytokines. Thus, IL-4 appears to be the crucial cytokine in human tonsillar B lymphocytes (Butcher et al., 1990; Butcher and Cushley, 1991; Tomizawa et al., 1991; Zola et al., 1991), while IL-5 occupies a central role in murine B cells (Loughnan and Nossal, 1989). The second component is the 75 kDa IL-2Rβ subunit (Tsudo et al., 1986; Takeshita et al., 1989) which is

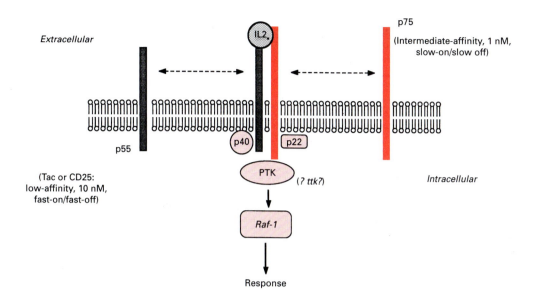

Figure 7   The IL-2 receptor complex

The properties of the two ligand-binding chains, p55 (α) and p75 (β), are indicated at either end of the Figure. In the presence of ligand, both chains bind IL-2 (see Figure 6 for the general case) and signals are transduced. The complex of α and β has a fast-on/slow- off character with high affinity (10 pM). The relationship of the high-affinity receptor to defined associated chains and enzymes involved in the signalling cascade [i.e. protein kinases (possibly the novel *ttk* activity) and c-*raf*-1] are also indicated.

constitutively expressed on lymphocytes, and is an HRS receptor (Hatakayama et al., 1989). A range of studies have illustrated a number of other proteins associated with the IL-2 receptor. One of these membrane proteins has been unequivocally identified as intracellular adhesion molecule-1 (ICAM-1, CD54) in both human and murine systems (Burton et al., 1990; Sharon et al., 1990). Radioligand cross-linking experiments have identifed the existence of two other proteins of 22 kDa and 40 kDa in murine B cells which are of unknown function (Saragovi and Malek, 1990). Scatchard analyses of IL-2 receptors on T and B cells reveal three distinct families of IL-2 binding subunits which can be explained in terms of the binding characteristics of IL-2 for each of the two receptor subunits (Teshigawara et al., 1987; Smith, 1989; Greene et al., 1989). Thus, IL-2Rα binds IL-2 with low affinity and IL-2Rβ with intermediate affinity, but the non-covalent complex of the two chains gives rise to the high-affinity receptor complex which is most likely to bind IL-2 at physiological concentrations. By analogy with the GH receptor system, IL-2Rα would bind ligand first, making the primary interaction necessary for binding of a second site on the ligand to IL-2Rβ. In this instance, distinct sites on the IL-2 molecule would be expected to make contact with unique motifs on the α and β subunits. Finally, there are recent data from human T cell lines which report the involvement of a 65 kDa component of the IL-2 receptor complex which is necessary for maximal ligand binding and activation of signal transduction (Arima et al., 1992).

At present, no consistent pattern has emerged for activation of cellular signalling pathways by IL-2 in lymphocytes (Mills et al., 1991). For example, although it is generally accepted that activation of the IL-2R does not cause the hydrolysis of PtdInsP_2 leading to the generation of the intracellular second messengers, InsP_3 and diacylglycerol, IL-2 has been reported to induce the membrane translocation, and hence activation, of PKC (Saltzmann et al., 1989: Merida and Gaulton, 1990; Farrar and Anderson, 1985). It has also been reported that IL-2 receptor activation may lead to induction of signal transduction pathways involving cyclic AMP (Wickremasinghe et al., 1987) and glyco-

phosphatidylinositol (GPI) (Eardley and Koshland, 1991). The loss of GPI molecules, presumably by action of a PLC, was paralleled by an increase in intracellular levels of inositol-phosphoglycan: diacylglycerol, in the form of myristoyl diacyl-glycerol, also accumulated, but was converted to myristoyl phosphatidic acid. GPI lipid release appears to be physiologically relevant to IL-2-driven T lymphocyte proliferation, since (i) IL-2 has been shown to promote loss of GPI molecules within 30 s of ligand binding at concentrations where only the high-affinity IL-2 receptor would be predicted to be occupied (Merida et al., 1990) and (ii) addition of one of the breakdown products, inositolphosphoglycan, to T cells synergized with IL-2 in promotion of cellular growth without altering the levels of IL-2 receptors expressed (Merida et al., 1990).

There is now good evidence to show that, in T cells, the IL-2 receptor is coupled to tyrosine kinases including pp56[lck] via the IL-2Rβ subunit (Hatekayama et al., 1991) and possibly pp59[fyn] (Mills et al., 1991). Recent data, in a *lck*-negative pre-B cell line, have demonstrated that the IL-2Rβ subunit can activate the B cell-specific pp53/56[lyn] kinase (Torigoe et al., 1992), suggesting lineage specificity in activitation of *src* family non-receptor associated PTKs by IL-2. In T cells, the IL-2 receptor also seems to be associated with a member of the recently described *ttk* family (*tyrosine-threonine kinase*), although it remains unclear if this is a direct coupling, or if another system links the receptor structure to the *ttk* activity (Mills et al., 1991). IL-2 binding is also linked to increases in the activity of phosphatidylinositol bisphosphate 3-kinase (PI-3-kinase) within 1 min of ligand binding in human peripheral blood T lymphoblasts (Remillard et al., 1991), and to activation of p21[ras], possibly either via direct tyrosine phosphorylation of p120[GAP] or by phosphorylation of intermediary regulatory proteins such as p62 and p190 (Downward et al., 1992). IL-2 stimulation of T cells also leads to tyrosine phosphorylation of the product of the c-*Raf*-1 oncogene, a 72–74 kDa protein which possesses serine-threonine protein kinase activity (Turner et al., 1991). The precise sequence and interaction of the above tyrosine kinase-mediated signalling events is, as yet, not clear. For example, it is conceivable that c-*raf* acts downstream of p21^{ras} since in fibroblasts, p21^{ras} has been shown to modulate the function of c-*raf* (Downward et al., 1992). The central role of the tyrosine kinases in IL-2-driven cellular signalling events is underscored by the finding that exposure of cells to tyrphostins or genistein prior to IL-2 stimulation abrogates phosphorylation of IL-2Rβ and expression of other cellular signalling activities (Mills et al., 1991). It remains to be determined if the B cell IL-2 receptor is linked to the same cellular signalling effectors.

B lymphocytes respond to IL-2 by proliferation and by increased secretion of immunoglobulin. While it is unclear whether these two biological responses are mutually exclusive in a single lymphocyte, available data indicate that B cells may use the same cellular signalling pathway to couple cytokine binding to both responses (Tigges et al., 1989). The effect of IL-2 upon Ig secretion, particularly IgM, is explained by its capacity to elevate the biosynthesis of the J chain protein necessary for successful assembly and secretion of polymeric IgM. This effect is at the transcriptional level (Tigges et al., 1989) and is controlled by a *cis* promoter element, J$_B$, which is located at position -75 to -45 in the J chain gene promoter (Lansford et al., 1992). The J$_B$ element is acted upon by a B cell-specific, IL-2-regulated *trans*-acting factor, NF-J$_B$. By analogy with the T cell system, the capacity of IL-2 to promote the growth of activated B cells may be explained by the induction of *ras*, and subsequent gene regulation by nuclear *trans*-activating factors such as serum response factor, c-*jun* and NF-κB (Downward et al., 1992). Indeed, in a transfected pre-B cell line, recent data have shown that IL-2 specifically induces the expression of the c-*fos* oncogene, an event which will be critical in facilitating expression of other genes which possess AP-1 sites in their promoter sequences (Hatakayama et al., 1992).

A further question which must be addressed is that of regulation of signals via the IL-2 receptor. Several mechanisms can exist to account for this, including rapid down-regulation of receptors following ligand binding by receptor mediated endocytosis. It is not unreasonable to expect that PTPases may be implicated in cytokine signalling regulation. Finally, it is noteworthy that in human B cells, IL-4 completely abrogates all of the effects of IL-2 upon GPI-signalling, cellular proliferation and Ig secretion (Eardley and Koshland, 1991; Jelinek and Lipsky, 1987; Galanaud et al., 1990; Lorente et al., 1990), indicating that the cellular signalling pathways activated by the two cytokines interact at some point. Molecular pharmacological explanations for the inhibition of IL-2-driven signalling by IL-4 remain to be fully elucidated.

IL-3, IL-5 and GM-CSF: shared receptor subunits

Early lymphoid cell progenitors are sustained in the bone marrow by haematopoietic factors such as GM-CSF and IL-3 (Kinashi et al., 1990), while eosinophil differentiation is regulated by IL-5 (Sanderson et al., 1989). Recent data have illustrated that IL-3, IL-5 and GM-CSF interact with receptors which have two components: an α chain which is specific for the individual cytokine, and a β chain which is shared by all three cytokine receptors (Tavernier et al., 1991). The α subunits of the IL-3, IL-5 and GM-CSF receptors have molecular masses of 70 kDa, 60 kDa and 80 kDa, respectively, while the common β subunit has a molecular mass of 120 kDa (reviewed in Miyajima et al., 1992). Transfection studies have illustrated that the β subunit is essential both for formation of high-affinity receptor complexes and for signal transduction (Metcalf et al., 1990).

IL-3, IL-5 and GM-CSF all induce PTK activities in cytokine-sensitive cells (Koyasu et al., 1988; Kanakura et al., 1990; Miyajima et al., 1992) although, in common with other HRS members, it is clear that neither the common β subunit nor the ligand-specific α subunits possess intrinsic tyrosine kinase domains. Since all three cytokines induce similar patterns of tyrosine phosphorylation of cellular proteins (Isfort and Ihle, 1990; Murata et al., 1990) it is possible that the common β subunit of the receptors associates with the same PTK and phosphorylates the same pool of substrates regardless of the ligand bound. The substrates for PTK activities associated with the IL-3 and GM-CSF receptors include c-*raf*-1 (Carroll et al., 1990; Kanakura et al., 1991), p21^{ras} (Satoh et al., 1991) and, in the case of IL-3, the β subunit of the IL-3 receptor itself (Sorensen et al., 1989).

There is also evidence that IL-3 (Linnekin and Farrar, 1990) and IL-5 (Murata et al., 1990) induce serine phosphorylation and, in the case of IL-3, that this is due to activation of PKC. Thus, IL-3 promotes a redistribution of PKC from the cytosol to the plasma membrane (Farrar et al., 1985), and drives cellular proliferation via a PKC-dependent mechanism (Whetton et al., 1988). Since IL-3 fails to promote inositol lipid hydrolysis, elevated PKC activity appears not to be a consequence of PLC action (Whetton et al., 1988); phosphatidylcholine may be the source of ligand-sensitive diacylglycerol in the IL-3 receptor system (Duronio et al., 1989).

IL-6: ligand-driven assembly of a functional receptor complex

IL-6 is a highly pleiotropic cytokine which elicits a range of responses in many different cell types, including B and T cells, hepatocytes, neural cells and fibroblasts (Wong and Clark, 1988; Kishimoto and Hirano, 1988; Van Snick, 1990). All sensitive cell types possess apparently identical IL-6 receptors. Little is known in detail about signal transduction mechanisms activated via the IL-6 receptor complex, but there are no reports of the cytokine stimulating lipid hydrolysis, PKC activation or Ca^{2+} mobilization. The IL-6 receptor complex comprises two distinct transmembrane glycoproteins, gp80 and gp130, both of which are members of the HRS. The gp80 component possesses a redundant Ig-like domain and binds ligand but is, by itself, unable to activate cellular signalling (Taga et al., 1989). However, the binding of IL-6 to gp80 promotes binding of gp80 to the second component of the receptor complex, gp130, and the transmission of cellular signalling information. Thus, gp130, which cannot bind IL-6, is the signal transduction element of the IL-6 receptor complex (Hibi et al., 1990), although it is unclear how these signals are transmitted given the lack of catalytic domains in the gp130 intracellular region. The sole signal for formation of the receptor complex capable of initiating second messenger generation is binding of IL-6, and this presumably is mediated allosterically via IL-6-induced exposure of an evanescent binding site for gp130 on the gp80 molecule. A final point of note in this system is that generation of a soluble, secreted form of gp80 can bind IL-6 and this complex of cytokine and soluble receptor retains its capacity to interact with gp130 and provide stimulatory signals to sensitive cells (Taga et al., 1989). Thus, the molecular motifs which participate in the interaction of gp80 and gp130 are located in the extracellular domains of the two glycoproteins. More interestingly, the soluble IL-6 receptor is an agonistic receptor, an apparently unique property amongst the soluble cytokine receptors since other well-characterized secreted cytokine receptors, for example IL-4 (Mosley et al., 1989) and IL-7 (Goodwin et al., 1990) behave as antagonistic soluble receptors.

Other cytokine receptors also possess gp130-like structures as part of their receptor elements. Thus, the receptor for Leukaemia Inhibitory Factor (LIF) displays sequence similarity to the gp130 element of the IL-6 receptor, particularly in its transmembrane and cytoplasmic domains (Gearing et al., 1991). The gp130 glycoprotein may be a member of a family of such molecules, all of which are involved in signal transduction via cytokine receptors. Sharing of a gp130-like element may account for cross-competition of IL-6 and LIF for ligand binding sites and their partially shared spectra of biological activities. This situation has a parallel with the shared β subunits in the IL-3/IL-5/GM-CSF model discussed above.

IL-7: PTK-coupled activation of PLC activity?

IL-7 is a stromal cell-derived cytokine (Henney, 1989) which is active upon B cell precursors (Namen et al., 1988) and certain immature T cell subsets (Chazen et al., 1989). However, the IL-7 receptor occupies an unusual position in this discussion inasmuch as there is little evidence to suggest a need for receptor dimerization. Thus, the sole component of the receptor characterized to date is a 75 kDa ligand-binding transmembrane glycoprotein which displays molecular motifs consistent with membership of the HRS. Studies on fetal thymocytes and leukaemic precursor B cell lines indicated that the IL-7 receptor was coupled to PtdInsP_2 hydrolysis via PTK-mediated activation of PLCγ1 (Uckun et al., 1991a,b) in a manner analogous to that reported for the T cell antigen-receptor complex. However, in striking contrast to these results, a recent report demonstrates that although the IL-7 receptor does indeed mediate tyrosine phosphorylation it is not coupled to PLC-γ1 phosphorylation or indeed any inositol phosphate generation or calcium mobilization in thymocytes, mature T cells and pre-pre-B cells (from a patient with acute lymphoblastic leukaemia) (Roifman et al., 1992). This major discrepancy may reflect differential coupling of the IL-7 receptor during differentiation and oncogenic transformation.

IL-4: species-specific signalling mechanisms?

The debate surrounding the biochemical mechanisms underlying IL-4 action on B lymphocytes has attracted perhaps the greatest controversy of all of the cytokines which regulate B cell development. IL-4 is a T cell-derived cytokine which elicits a range of effects upon B cell differentiation (Paul and Ohara, 1987). Thus, in the murine B cell system, it can sustain the growth of pre-B cells, promote early activation of quiescent B lymphocytes, which is manifest as increased expression of MHC class II antigens (Noelle et al., 1984; Roehm et al., 1984) and of low-affinity Fc$_\varepsilon$ receptors (Hudak et al., 1987). Finally, IL-4 mediates directed isotype switching in mitogen-activated B cell blasts (Coffman et al., 1987; Lutzker et al., 1988; Rothman et al., 1988; Bergstedt-Lindqvist et al., 1988). A similar array of activities is evident in the human model, although increased levels of expression of membrane IgM and IgD are more pronounced, relative to murine models, at low doses of IL-4 (Rigley et al., 1991). This latter datum has implications for the existence of multiple IL-4 receptors.

The most compelling data currently available for signalling via the IL-4 receptor come from studies of human tonsillar B lymphocytes. In such cells, IL-4 promotes generation of InsP_3, diacylglycerol, and a rise in intracellular Ca^{2+} immediately after ligand binding. This is followed after several minutes by a sustained rise in intracellular cyclic AMP levels (Finney et al.,

1990). In pharmacological mimicry experiments, the ability of IL-4 to stimulate CD23 expression, a marker of IL-4 action on human B cells, was achieved by a combination of a brief pulse with phorbol ester and ionomycin, followed by exposure to dibutyryl cyclic AMP. However, studies in murine B cells failed to find similar effects. Thus, stimulation of quiescent murine B cells with IL-4 fails to trigger inositol lipid hydrolysis and changes in intracellular Ca^{2+} levels, and also does not promote translocation of protein kinase C to the plasma membrane (Mizuguchi et al., 1986). Protein kinase activities do seem to be activated by IL-4 in B cells, however, as a 42 kDa protein of unknown identity is phosphorylated in isolated membranes in the presence of IL-4 (Justement et al., 1986; McGarvie and Cushley, 1989a). A similar protein kinase substrate has been reported in human tonsillar B cells, although the apparent molecular mass in this model is 38 kDa, slightly smaller than that reported in murine splenocytes (Finney et al., 1991).

The biochemistry of the IL-4 receptor also remains to be fully characterized. Initial experiments employing cross-linking of radioactive IL-4 to B cells suggested that, in murine and human models, IL-4 was principally associated with a 75 kDa cell surface component. However, similar studies of human B cells in other laboratories implicated a binding component of some 130–140 kDa (Galizzi et al., 1989). The IL-4 binding component was finally isolated by molecular cloning and shown to be of 140 kDa in the mature, fully glycosylated form in both murine (Mosley et al., 1989) and human (Idzerda et al., 1991) B cells. It remains as yet unclear if the 140 kDa component is the sole molecular species which makes up the IL-4 receptor. There are reports from cross-linking experiments which indicate the existence of species of 42 kDa and 110 kDa in murine B cells (Fernandez-Botran et al., 1990). Moreover, culture of murine B cells with IL-4 leads to the appearance of a 75 kDa tyrosine-phosphorylated protein in the membranes of such lymphocytes (McGarvie and Cushley, 1989b). No definitive cross-linking experiments have been reported for the human IL-4 receptor, but the crystallographic data for the growth hormone receptor (De Vos et al., 1992) and the dose-dependent nature of specific responses to IL-4 in human B cells (Rigley et al., 1991) suggests that other molecules are likely to be involved in modulation of the affinity and, possibly, component composition of the IL-4 receptor in B lymphocytes.

B lymphocytes respond to IL-4 in characteristic ways, one of which, in the murine model, is to greatly up-regulate the expression of class II MHC antigens (Noelle et al., 1984). This effect is not uniquely driven by IL-4 stimulation, and can also be elicited by gross stimulation with lipopolysaccharide and by anti-Ig treatment. However, in the case of class II antigen expression, the response is regulated at the molecular level by a specific DNA binding protein (NF-BRE) which interacts with a defined dodecanucleotide sequence element in the far upstream region of the promoter region of class II α chains (Boothby et al., 1989). The NF-BRE recognition element is distinct from the conserved W, X and Y boxes found in the 5 promoter region of all MHC class II genes. The precise biochemical requirements for activation of NF-BRE , and the cellular signalling pathways which deliver these, remain undefined.

ADHESION MOLECULES AND OTHER SURFACE STRUCTURES MEDIATING CELLULAR SIGNALLING IN B LYMPHOCYTES

The vast majority of data relating to cellular signalling in B cells has been derived from studies of the antigen and cytokine receptors. However, other cell surface structures, some with defined ligands and others with unknown ligands also initiate signal transduction in B lymphocytes with consequences for activation, growth and differentiation of the B cell. Such structures include MHC antigens, adhesion molecules, receptors for complement components and Ig molecules, and members of the CD differentiation antigen family (Table 1). Signals delivered through these structures can influence the response of the B cell to antigen.

Adhesion molecules

The antigen receptors of B and T cells differ in one crucial respect; the B cell antigen receptor can bind antigen which is free in solution, and can do so with high affinity, whereas the T cell receptor cannot bind soluble antigen, requiring peptides derived from the intact antigen to be presented to it in association with MHC antigen; this binding is of several orders of magnitude lower affinity. In the case of T cells, adhesion molecules serve to promote the overall *avidity* of the multi-component recognition complex in order to facilitate specific binding of antigen.

The interaction of class II MHC antigen with CD4 could be regarded as a reaction between two adhesion molecules, although the interactions between CD2 on T cells and Leucocyte Functional Antigen-3 (LFA-3) on presenting cells (including B cells) would be considered a more classical pair of adhesion molecules, as would the ICAM-1 (CD54)–LFA-1 interaction (Figdor et al., 1990). These interactions have consequences for cellular signalling. Thus, it has been suggested that LFA-1 may be a Ca^{2+}-mobilizing receptor (Pardi et al., 1989), and it is well documented that anti-CD2 antibody alone triggers a range of cellular signalling processes including PtdInsP_2 hydrolysis, mobilization of calcium, activation of PKC, PTK(s) and *ras*, all of which are generally believed to be activated via the CD3-$\zeta\eta$ complex (Downward et al., 1992). Moreover, simultaneous administration of anti-CD3 and anti-CD2 mAbs to T cells provides a powerful activatory signal, particularly if these are in the form of a bi-specific antibody complex (Tutt et al., 1991). Thus, CD2 provides a further conduit for signal transduction in the T lymphocyte, and there is the possibility that the generation of the adhesion molecule pair sends a signal in both directions; in this example, the LFA-3$^+$ presenting cell could also be stimulated via LFA-3 as a consequence of the CD2–LFA-3 interaction. While the binding of antigen to the antigen receptor in B cells, even in low affinity and potentially polyreactive mIgM$^+$/mIgD$^+$ mature B cells, is of sufficiently high affinity to obviate the need for adhesion molecules in formation of a functional antigen recognition unit, B cells nonetheless possess a range of classical adhesion molecules which could potentially deliver cellular signals to the B cell during cognate recognition of thymus-dependent antigen. Additionally, the B lymphocyte possesses potentially unique adhesion molecule pairs, namely CD5/CD72 and B7/CD28 (De Franco, 1991) which could transduce activation signals to the B cell. In this regard, CD72 ligation is known to increase intracellular Ca^{2+} concentrations and to enhance proliferation induced by anti-Ig stimulation of B lymphocytes.

Cellular signalling via B lymphocyte differentiation antigens

B lymphocytes possess a range of cell surface differentiation CD (*C*luster of *D*ifferentiation) antigens, in addition to antigen and cytokine receptors, which are capable of transmitting information to the cell interior (reviewed in detail by Clark and Lane, 1991). It has already been noted that adhesion molecules have a role to play in signal transduction, and other molecules currently defined

Table 1 Summary of signals transmitted via B lymphocyte differentiation antigens

Surface antigen	Size (kDa)	Function	Signals	Effect of ligation on anti-Ig-stimulated events	References
CD19	95	Pan B cell marker	Possible increase in [Ca^{2+}]$_i$. PKC translocation to plasma membrane	Decreases c-*myc* mRNA levels	Clark and Ledbetter, 1989; Barrett et al., 1990; Rigley et al., 1991
CD20	35	Ca^{2+} channel	Increased c-*myc* and MHC class II via PKC activation. No PtdInsP_2 or Ca^{2+} effects. Tyrosine kinase activation.	Stimulates quiescent B cells and inhibits activated B cells	Clark and Ledbetter, 1989; Clark and Lane, 1991; Brown et al., 1989
CD21	145	Receptor for C3d. Receptor for Epstein–Barr virus	Differentially phosphorylated in B cells. EBV causes PtdInsP_2 hydrolysis and increases [Ca^{2+}]$_i$	Monomeric C3d lowers Ca^{2+} response, but anti-CD21 mAbs enhance.	Changelian and Fearon, 1986; Dugas et al., 1988; Tsokos et al., 1990
CD22	150	Adhesion molecule	Unknown	Boosts Ca^{2+} response	Knapp et al., 1989
CD23	45	Low-affinity IgE receptor. sCD23 has autocrine activity. Adhesion molecule	Increased [Ca^{2+}]$_i$ and PtdInsP_2 hydrolysis		Kolb et al., 1990; Letellier et al., 1988; Kikutani et al., 1986
CD32	40	B cell Fc receptor for IgG	Increased [Ca^{2+}]$_i$ with anti-CD32	Inhibitory. Uncouples mIg from its G protein	Hunziker et al., 1990; Rigley et al., 1989
CD40	50	Adhesion molecule	Ser/Thr phosphorylation, but no Tyr phosphorylation	Enhances	Einfeld et al., 1988; Gruber et al., 1989
CD45	200	Protein tyrosine phosphatase		Blocks proliferation and increases [Ca^{2+}]$_i$	Tonks et al., 1988; Gruber et al., 1989; Ledbetter et al., 1988
CD72	85	Adhesion molecule	Increased [Ca^{2+}]$_i$	Enhances proliferation	Clark and Lane, 1991; Subbarao et al., 1988

by monoclonal antibodies also influence B cell activation, growth and differentiation. Many of these differentiation antigens influence, either positively or negatively, the cellular signalling pathways activated by anti-Ig stimulation of B cells. A brief mention of the signalling capabilities of some of the differentiation antigens is appropriate at this point; the list is not exhaustive (Table 1). The elucidation of cellular signalling via such molecules is in its infancy, and depends upon use of mAbs to these structures since the normal ligands for many CD antigens remain undefined.

However, many CD antigens are receptors for known ligands. Thus, CD21 is the receptor for the complement component C3d and for Epstein–Barr virus, while CD23 functions as a low affinity receptor for IgE. Other CD structures, however, do not have defined ligands, and much of the information regarding their signalling functions comes from studies using mAbs directed against the structures. Thus, stimulation of B cells with anti-CD19 mAbs causes increases in intracellular Ca²⁺ levels and a translocation of protein kinase C to the plasma membrane (Clark and Ledbetter, 1989), although other groups fail to note Ca²⁺ mobilization following anti-CD19 stimulation (Rigley et al., 1991). In addition, while anti-CD20 mAbs fail to influence intracellular Ca²⁺ levels (Lane et al., 1991; Clark and Lane, 1991), both anti-CD19 and anti-CD20 mAbs can influence the signals transmitted via the antigen receptor, although the precise nature of the modulatory effect can be determined by the activation status of the B cells (Clark and Ledbetter, 1989; Brown et al., 1989; Barrett et al., 1990; Rigley et al., 1991).

Analysis of the data from the CD21 system suggest that due caution should be applied when interpreting data from signalling experiments in complex cellular systems using mAbs as stimulatory agents. Thus, anti-CD21 mAbs fail to mobilize intracellular Ca²⁺(Dugas et al., 1988), whereas Epstein–Barr virus (one of the physiological ligands) does so (Carter et al., 1988). Moreover, monomeric C3d inhibits anti-Ig stimulated Ca²⁺ levels, but anti-CD21 mAbs enhance this response (Tsokos et al., 1990). A final point of note with respect to CD21 is the recent finding that CD23 binds to this differentiation antigen (Aubry et al., 1992). The reaction can be blocked by some, but not all, anti-CD21 mAbs, indicating that CD21 functions as a receptor for CD23 as well as for Epstein–Barr virus and C3d. This also raises the possibility of signalling in both directions as a result of CD21–CD23 interaction as both these differentiation antigens are known to drive Ca²⁺ mobilization following ligation. The interaction of the CD5 and CD72 molecules provides another example of potential bi-directional signalling.

One particular B cell differentiation antigen which has attracted particular attention is CD40. Ligation of CD40 by antibody leads to a number of effects on B cells and influences signals delivered via other receptors (notably the IL-4 and antigen receptors). Thus, use of mAbs to cross-link CD40 on the B cell surface leads to elevated release of sCD23 (Cairns et al., 1988) and also promotes synthesis and secretion of IL-6 (Clark and Shu, 1990). Use of anti-CD40 and IL-4 in conjunction with CD32⁺ adherent cells leads to the establishment of long-term B cell lines from normal primary human B cells (Banchereau et al., 1991). Studies of cellular signalling via CD40 suggest a role for PTK activities (Uckun et al., 1991c) and, indeed, CD40 shows some sequence similarity to members of the Nerve Growth Factor family. In germinal centre cells, ligation of CD40 is one of the signals required to allow centrocytes to escape from apoptosis, an effect mediated via induction of *bcl*-2 expression in the centrocytes (Liu et al., 1992). The CD40 ligand has recently been cloned, and is a 39 kDa structure expressed on activated T lymphocytes. Anti-39 kDa mAbs block the capacity of this

structure to activate B cells for growth and differentiation, showing that this molecule has a critical role in B cell activation (Noelle et al., 1992).

RECEPTOR CROSS-TALK AND REGULATION OF SIGNAL TRANSDUCTION IN B LYMPHOCYTE PROLIFERATION AND DIFFERENTIATION

The foregoing discussion illustrates that the B lymphocyte provides a range of ligand receptor systems which are interesting and worthy of investigation as individual models. However, there are numerous interactions between receptor systems in B cells which are particularly important for the insights which they may yield into mechanisms of receptor desensitization and in the positive and negative interaction of cellular signalling networks within the B cell. Some examples of interacting signal transduction systems are now discussed.

Homologous and heterologous desensitization of antigen receptors

Perturbation of the antigen receptor on the B cell is the critical event in the initiation of clonal expansion; all subsequent ligands deliver essentially regulatory signals. The cellular signalling systems activated by stimulation of the antigen receptor have been documented above. However, it has been well-documented that following ligand binding, the antigen receptor, like many other receptors, enters a period where it is refractory to further stimulation. In the case of a mature mIgM⁺/mIgD⁺ B lymphocyte, stimulation of the cell with anti-μ-specific reagents causes not only desensitization of mIgM-containing antigen receptors, but also results in the desensitization of the heterologous, mIgD antigen receptor complexes; the same is true when anti-δ reagents are employed as primary stimulus (Klaus et al., 1985; Cambier et al., 1988). Inositol phospholipid signalling via the mIgM and mIgD receptors appear to be regulated by a common form of G-protein, and desensitization appears to be at the level of this G-protein–PLC coupling (Harnett et al., 1989). Neither homologous nor heterologous desensitization processes are permanent, and the B cell regains its capacity to mobilize Ca²⁺ in response to anti-Ig after 8–20 h of initial stimulation (Cambier et al., 1988; Harnett et al., 1989).

Stimulation of membrane Ig leads not only to homologous and heterologous desensitization of mIg molecules, but also influences the generation of signals via other cell surface receptors. Thus, in human B cells, both CD21 (Carter et al., 1988) and, more controversially (Uckun et al., 1988; Rigley et al., 1989), CD19 mobilize Ca²⁺ when stimulated by mAbs. Exposure of B cells to anti-μ or anti-δ antibodies prevents subsequent mobilization of Ca²⁺ in response to anti-CD19 or anti-CD21 stimulation (Rijkers et al., 1990).

Antigen, Fc and IL-4 receptors: three-way receptor cross-talk

In an on-going immune response, activated B cells are subject to negative regulation by circulating immunoglobulin, which is important in feedback inhibition of the antibody response. In B lymphocytes, such negative feedback regulation of activation and antibody secretion is mediated by the 40 kDa Fc receptor, CD32 (Hunziker et al., 1990). Occupation of the B cell CD32 structure sends a profoundly negative signal to the B lymphocyte rendering it refractory to signals received via the antigen receptor and resulting in abortive activation upon antigenic stimulation (Klaus et al., 1984). However, the negative influence of occupied CD32 can be overcome in B cells by IL-4 (O'Garra et al., 1987). Thus, exposure of B cells to IL-4 prior to treatment with intact anti-Ig molecules prevents the abortive activation caused by simultaneous occupation of antigen and Fc receptors by the intact IgG antibodies. Thus, IL-4 over-rides and/or negates the signal delivered via CD32, or maintains a key element of the antigen receptor signal transduction pathway in an "active" condition refractory to the negative influence of occupied CD32.

The biochemical mechanism underlying the FcR-dependent inhibition of mIg signalling has been at least partially elucidated. Thus, whilst $F_{(ab)2}$ fragments of anti-Ig provoke a sustained hydrolysis of $PtdInsP_2$, intact anti-Ig causes only a transient release of $InsP_3$. Moreover, intact anti-Ig, which co-crosslinks mIg and Fc receptors, profoundly inhibits the release of $InsP_3$ provoked by the mitogenic $F_{(ab)2}$ fragments. Thus, the effect of occupation of CD32 is to reduce PLC activity, such that little Ca^{2+} can be mobilized and activation is abortive (Bijsterbosch and Klaus, 1985). Mechanistically, this effect is explained in terms of the functional coupling of the antigen receptor complex to its G protein (Rigley et al., 1989). In permeabilized B cells, $F_{(ab)2}$ anti-Ig reagents plus GTPγS can reconstitute the mIg-mediated inositol lipid hydrolysis observed in intact cells. Co-stimulation of these permeabilized B cells with intact anti-Ig antibodies abrogates the functional reconstitution, but does not disrupt the GTPγS-induced PLC activity (Rigley et al., 1989). These data indicate that the occupied FcR mediates its effect by disrupting the function of the B cell antigen receptor complex at a point proximal to the G protein. Thus, whilst the G protein itself is functionally linked to PLC, as demonstrated by the lack of effect of anti-Ig upon GTPγS-induced lipid hydrolysis, the G protein itself appears to be uncoupled from the antigen receptor (reduction of GTPγS plus anti-Ig response to that of GTPγS alone). Thus, the G protein can transmit signalling information, but cannot receive stimulatory input from the B cell antigen receptor complex (Rigley et al., 1989). Although the mechanism leading to this uncoupling is unknown, one possibility is that engaging CD32 could disrupt receptor–Gp contact by modifying the receptor contact site with Gp, perhaps by phosphorylation (Rall and Harris, 1987; Sullivan et al., 1987). Alternatively, it has been shown previously that capping of sIg also leads to co-capping of FcR, suggesting that the uncoupling might be brought about by physical dissociation, perhaps by disrupting the sIg/accessory transducing molecules complex. IL-4 reverses the capacity of FcR to inhibit stimulatory signals delivered via the antigen receptor (O'Garra et al., 1987), and although the molecular basis for this effect has not been elucidated, it is clear that IL-4 does not overcome the inhibition of $PtdInsP_2$ hydrolysis in murine B cells (O'Garra et al., 1987).

Receptor cross-talk mechanisms underlying T cell-dependent cell activation

At present, the molecular mechanisms underlying the complex interactions of signals generated by mIg, IL-4R (and other cytokine receptors such as IL-5R in the mouse), CD40, and class II molecules which lead to B cell proliferation and differentiation are poorly understood (Cambier and Lehmann, 1989; Cambier et al., 1991). However, although it has not yet been possible to assign a particular signalling pathway to the IL-4 receptors on murine B cells, some progress has been made towards understanding cross-talk mechanisms underlying the interactions between the antigen and IL-4 receptors during the priming phase (Harnett et al., 1991). Thus, although IL-4 does not appear to induce the release of $InsP_3$, Ca^{2+} mobilization, PKC translocation (Justement et al., 1986; Mizuguchi et al., 1986), or indeed to modify signalling via the phosphoinositide pathway induced by

ligation of the sIg receptors in murine B cells (Klaus and Harnett, 1990), recent evidence demonstrates that IL-4 synergizes with non-mitogenic concentrations of anti-Ig to provoke translocation of PKC from the cytosol to the plasma membrane (Harnett et al., 1991). Thus, the lymphokine upregulates PKC levels and activity and also acts to prevent PKC downregulation in B cells (Harnett et al., 1991). These data therefore suggest that signals generated via IL-4 receptors potentiate and/or prolong sIg-induced PKC activation and may provide a biochemical basis for explaining how IL-4 and anti-Ig synergize to induce B cell activation. The proposal that strong, prolonged PKC signals are sufficient to induce DNA synthesis in murine B cells is also likely to be consistent with the other models (Cambier and Lehmann, 1989; Cambier et al., 1991) of T cell-dependent B cell activation: Cambier's group were initially puzzled by the finding that mIg, IL-4R and Class II ligation synergised to induce B cell entry to S phase as earlier studies reported that (i) ligation of Class II molecules could induce cyclic AMP elevation and inhibit mitogen-driven proliferation, and (ii) it was widely accepted that elevation of cyclic AMP levels would be antagonistic for lymphocyte proliferation (Forsgren et al., 1984; Cambier et al., 1987; Cohen and Rothstein, 1989). This paradox now appears to have been resolved by the recent finding that cross-linking of Class II molecules on murine B cells pretreated with both IL-4 and anti-Ig generates an inositol phospholipid (and presumably PKC) response rather than the cyclic AMP response observed in resting cells. Furthermore, the inositol phospholipid response resulting from crosslinking Class II on IL-4/anti-Ig "primed" cells is of greater magnitude than that observed when anti-Ig alone is employed as primary stimulus (Cambier et al., 1991). Although this model of murine B cell activation now appears to be highly dependent on calcium and PKC signals, it may not be prudent to totally discard a role for Class II-mediated cyclic AMP signals at some stage in this process as the likely signals generated by mIg, IL-4R and Class II molecules on resting B cells are reminiscent of the PKC/cyclic AMP response associated with the IL-4R on human B lymphocytes (Finney et al., 1990; Rigley and Harnett, 1990).

CONCLUSION AND FUTURE PERSPECTIVES

Great advances have been made in the understanding of the structures and functions of receptors for antigen and cytokines on B lymphocytes, and the biological consequences of the signals delivered via these receptors. It is clear that the antigen receptors of both T and B lymphocytes are multi-component assemblies comprising both clonotypic components and conserved elements involved in complex assembly, transport and signal transduction. Many cytokine receptors, particularly those of the haematopoietin receptor superfamily, also appear to be multi-component complexes, but in this instance additional interest is provided by the fact that assembly of functional complexes may be driven by ligand (e.g., IL-6 and IL-2 receptors), and a novel receptor–ligand stoichiometry (2:1) seems to prevail in these systems.

Interesting questions do remain to be resolved, including the precise mechanisms via which the antigen receptors are coupled to the PLC signalling effector. Is the critical coupling event mediated via a PTK, G protein or both? Evidence to support either mechanism is available in abundance, but it is entirely possible that both systems may operate, to greater or lesser extents, in B cells at different stages of differentiation or depending upon other stimuli which the cells may have previously met. It could be argued that virgin and memory B cells have the capacity to encounter and respond to antigen in qualitatively different ways. For example, soluble anti-Ig reagents are generally adequate in induction of activation of mIgM$^+$/mIgD^{++} B cells, while the same reagents fail to rescue centrocytic cells from apoptosis, anti-Ig immobilized on Sepharose being mandatory to achieve this outcome. Thus, virgin cells respond to soluble antigen while recently formed memory B cells (of non-mIgM/mIgD isotypes) require to encounter insoluble antigen, at least on the first occasion, in order to mount a positive response. Given that there are little or no apparent differences in the complement of receptor complex components, might this reflect subtle differences in the coupling of original and somatically mutated B cell antigen receptors?

Good experimental models for the study of B lymphocyte activation processes are not abundant. Cell lines are transformed and are therefore, by definition, not quiescent. Moreover, the transformation process may involve expression or over-expression of cellular proto-oncogenes which ablate a normal cellular regulatory function, thereby compromising the normal physiological linkage of a receptor to its signalling effector system(s). Abnormal expression of any kinase (e.g. *lyn*, *raf*), GTP binding protein (e.g. *ras*) or transcription factor (e.g. *fos*) could easily bias the activation system used by a particular receptor in a transformed cell. Therefore, data from normal B cell systems and those from B cell lines should not be regarded as strictly comparable.

Moreover, while lessons can be learned from data in the T cell receptor system, it is worth considering that there is no *a priori* reason why the antigen receptors of the two lymphocyte types should be linked to their signalling effectors by the same coupling intermediates. Thus, clonotypic T cell receptors experience antigen at very low affinity (relative to B lymphocyte receptors) and the functional activation signal may result from the net avidity of multiple receptor–ligand systems rather than simply stimulation of the TcR itself. In this context, the relevance of use of anti-TcR (or CD3ϵ) mAbs as a stimulus to the physiological state should be regarded cautiously; for example, the concentration of anti-TcR reagents used to drive PtdInsP_2 hydrolysis frequently induces cell death rather than activation in some T cell models (Cherwinski et al., 1992).

While several of the downstream signalling events associated with many cytokine receptors are well defined, including PTK activation, c-*raf* and p21^{ras} activation, the initial coupling events are less well understood. Moreover, interpretation of data from cytokine systems faces the additional complication of receptors of different affinities. Taking human IL-4 as a particular example, low doses of cytokine promote elevation of membrane Ig levels, while higher doses appear to drive directed isotype switching; IL-4 causes inositol lipid hydrolysis and cyclic AMP production in human B cells but it is not known which of these two signalling pathways, *if either*, is linked to the biological responses observed. The same questions arise for other cytokines where multiple cellular responses can be elicited by a single cytokine. Since B cells are likely to encounter high local doses of cytokines within secondary lymphoid organs during an on-going response to antigen, it may be that responses to ligation of low affinity receptors will prove to be more physiologically relevant than at first thought.

It is clear than B lymphocytes can mount characteristically distinct responses to antigen and/or cytokine stimulation depending upon their differentiation status. In a few cases, such as the expression of *egr*-1 in mature B cells being correlated with a positive response to stimulation via mIg, the basis for differentiation state-dependent responses is beginning to be understood. However, data of this kind suggest that substrates for signalling effector systems linked to many B cell receptors may differ in a differentiation stage-dependent manner, and it is the range of substrates available at the time of contact with ligand (antigen, cytokine or adhesion structure) which determines the ultimate biological response of the B lymphocyte. The identification and characterization of such putative substrates and their cellular functions awaits elucidation, and will provide the next step forward in our understanding of regulation of cellular signalling systems in the B lymphocyte system.

M. M. H. is an M. R. C. Senior Fellow. The authors thank Dr. M. J. O. Wakelam for constructive criticism of the manuscript.

REFERENCES

Ales-Martinez, J. E., Cuende, E., Martinez, A. C., Parkhouse, R. M. E., Pezzi, L. and Scott, D. W. (1991) Immunol. Today **12**, 201–205

Alexander, D. R. (1990) The New Biologist **2**, 1049–1062

Alt, F. W., Bothwell, A. L.M, Knapp, M., Siden, E., Mather, E., Koshland, M. and Baltimore, D. (1981) Cell **20**, 293–301

Arima, N., Kamio, M., Imada, K., Hari, T., Hattori, T., Tsudo, M., Okuma, M. and Uchiyama, T. (1992) J. Exp. Med. **176**, 1265–1272

Aubry, J.-P., Pochon, S., Graber, P., Jansen, K. U. and Bonnefoy, J.-Y. (1992) Nature (London) **358**, 505–507

Banchereau, J., De Paoli, A., Vallé, A., Garcia, E. and Rousset, F. (1991) Science **251**, 70–72

Barrett, T. S., Shu, G. L., Draves, K. E., Pezzutto, A. and Clark, E. A. (1990) Eur. J. Immunol. **20**, 1053–1059

Bazan, H. F. (1990) Proc. Natl. Acad. Sci. U.S.A. **87**, 6934–6938

Bergstedt-Lindqvist, S., Moon, H.-B., Persson, U., Möller, G., Heusser, C. and Severinson, E. (1988) Eur. J. Immunol. **18**, 1073–1077

Bijsterbosch, M. K. and Klaus, G. G. B. (1985) J. Exp. Med. **162**, 1825–1836

Bijsterbosch, M., Meade, J. C., Turner, G. A. and Klaus, G. G. B. (1985) Cell **41**, 999–1005

Boothby, M., Gravallesse, E., Liou, H.-G. and Glimcher, L. H. (1989) Science **242**, 1559–1562

Brian, A. A. (1988) Proc. Natl. Acad. Sci. U.S.A. **85**, 564–570

Brown, A. N., Thurstan, S. M. and Callard, R. E. (1989) in Leukocyte Typing IV; White Cell Differentiation Antigens, pp. 203–205, Oxford University Press, Oxford

Burke, T., Pollok, K., Cushley, W. and Snow, E. C. (1989) Mol. Immunol. **26**, 1095–1104

Burkhardt, A. L., Brunswick, M., Bolen, J. B. and Mond, J. J. (1991) Proc. Natl. Acad. Sci. U.S.A. **88**, 7410–7414

Burnet, M. F. (1959) The Clonal Selection Theory of Acquired Immunity, Vanderbilt University Press

Burton, J., Goldman, C. K., Rao, P., Moos, M. and Waldmann, T. A. (1990) Proc. Natl. Acad. Sci. U.S.A. **87**, 7329–7333

Butcher, R. D. J. and Cushley, W. (1991) Immunology **74**, 511–518

Butcher, R. D. J., McGarvie, G. M. and Cushley, W. (1990) Immunology **69**, 57–64

Cairns, J. L., Flores-Romo, L., Millsum, M. J., Guy, G. R., Gillis, S., Ledbetter, J. A. and Gordon, J. (1988) Eur. J. Immunol. **18**, 349–356

Callard, R. E., Smith, S. H., Shields, J. G. and Levinsky, R. J. (1986) Eur. J. Immunol. **16**, 1037–1042

Callard, R. E. (1990) Cytokines and B Lymphocytes, Academic Press, London

Cambier, J. C. and Campbell, K. S. (1990) Semin. Immunol. **2**, 139–149

Cambier, J. C. and Lehman, K. R. (1989) J. Exp. Med. **170**, 877–886

Cambier, J. C., Newell, M. K., Justement, L. B., McGuire, J. C., Leach, K. L. and Chen, Z. Z. (1987) Nature (London) **327**, 629–632

Cambier, J. C., Chen, Z. Z., Pasternak, J., Ransom, J., Sandoval, V. and Pickles, H. (1988) Proc. Natl. Acad. Sci. U.S.A. **85**, 6493–6497

Cambier, J. C., Morrison, D. C., Chien, M. M. and Lehmann, K. R. (1991) J. Immunol. **146**, 2075–2082

Campbell, K. S. and Cambier, J. C. (1990) EMBO J. **9**, 441–448

Campbell, M. A. and Sefton, B. (1990) EMBO J. **9**, 2125–2131

Carroll, M. P., Clark-Lewis, I., Rapp, U. R. and May, W. S. (1990) J. Biol. Chem. **265**, 19812–19817

Carter, R. H., Park, D. J., Rhee, S. G. and Fearon, D. T. (1991) Proc. Natl. Acad. Sci. U.S.A. **88**, 2745–2749

Casillas, A., Hanekom, C., Williams, K., Katz, R. and Nel, A. (1991) J. Biol. Chem. **266**, 19088–19094

Changelian, P. S. and Fearon, D. T. (1986) J. Exp. Med. **163**, 101–115

Chazen, G. D., Pereira, G. M. B., LeGros, G., Gillis, S. and Shevach, E. M. (1989) Proc. Natl. Acad. Sci. U.S.A. **86**, 5923–5927

Chen, J., Stall, A. M., Herzenberg, L. A. and Herzenberg, L. A. (1990) EMBO J. **9**, 2117–2124

Cherwinski, H. M., Semenuk, G. T. and Ransom, J. T. (1992) J. Immunol. **148**, 2996–3003

Clark, E. A. and Lane, P. J. L. (1991) Annu. Rev. Immunol. **9**, 97–127

Clark, E. A. and Ledbetter, J. A. (1989) Adv. Cancer Res. **52**, 81–149

Clark, E. A. and Shu, G. L. (1990) J. Immunol. **145**, 1400–1411

Coffman, R. L., Ohara, J., Bond, M. W., Carty, J., Zlotnik, A. and Paul, W. E. (1986) J. Immunol. **136**, 4538–4541

Cohen, D. P. and Rothstein, T. L. (1989) Cell. Immunol. **121**, 113–124

Coleclough, C., Perry, R. P., Karjalainen, K. and Weigert, M. (1981) Nature (London) **290**, 372–378

Cosman, D., Lyman, S. D., Idzerda, R. L., Beckmann, M. P., Park, L. S., Goodwin, R. G. and March, C. J. (1990) Trends Biochem. Sci. **15**, 265–269

Cunningham, B. C., Ultsch, M., De Vos, A. M., Mulkerrin, M. G., Clauser, K. R. and Wells, J. A. (1991) Science **254**, 821–825

Curtis, B. M., Widmer, M. B., DeRoos, P. and Qwarnstrom, E. E. (1990) J. Immunol. **144**, 1295–1303

D'Andrea, A. D., Fasman, G. D. and Lodish, H. F. (1989) Cell **58**, 1023–1024

Dasgupta, J. D., Granja, C., Druker, B., Lin, L.-L., Yunis, E. J. and Relias, V. (1992) J. Exp. Med. **175**, 285–288

De Franco, A. L. (1991) Nature (London) **353**, 603–604

De Vos, A. M., Ultsch, A. and Kossiakoff, A. A. (1992) Science **255**, 306–312

Dewald, B., Thelen, B. and Baggiolini, M. (1988) J. Biol. Chem. **263**, 16179–16184

Dinarello, C. A., Clark, B. D., Puren, A. J., Savage, N. and Rosoff, P. M. (1989) Immunol. Today **10**, 49–51

Downward, J., Graves, J. D., Warne, P. H., Rayter, S. and Cantrell, D. A. (1990) Nature (London) **346**, 719–723

Downward, J., Graves, J. and Cantrell, D. A. (1992) Immunol. Today **13**, 89–92

Dugas, B., Delfraissy, J. F., Calenda, A., Peuchmaur, M., Wallon, C., Rannou, M. T. and Galanaud, P. (1988) J. Immunol. **141**, 4344–4351

Duronio, V., Nip, L. and Pelch, S. L. (1989) Biochem. Biophys. Res. Commun. **164**, 804–808

Eardley, D. D. and Koshland, M. E. (1991) Science **251**, 78–81

Einfeld, D. A., Brown, J. P., Valentine, M. A., Clark, E. A. and Ledbetter, J. A. (1988) EMBO J. **7**, 711–717

Farrar, W. L. and Anderson, W. B. (1985) Nature (London) **315**, 233–234

Farrar, W. L., Thomas, T. P. and Anderson, W.B (1985) Nature (London) **315**, 235–237

Fernandez-Botran, R., Uhr, J. W. and Vitetta, E. S. (1989) Proc. Natl. Acad. Sci. U.S.A. **86**, 4235–4239

Figdor, C. G., van Kooyk, Y. and Keizer, G. D. (1990) Immunol. Today **11**, 277–283

Finkel, T. H., Kubo, R. T. and Cambier, J. C. (1991) Immunol. Today **12**, 79–85

Finney, M., Guy, G., Michell, R. H., Gordon, J., Dugas, B., Rigley, K. P. and Callard, R. E. (1990) Eur. J. Immunol. **20**, 151–156

Finney, M. J., Michell, R. H., Gillis, S. and Gordon, J. (1991) Biochem. Soc. Trans. **19**, 287–291

Forsgren, S., Pobor, G., Coutinho, A. and Pierres, M. (1984) J. Immunol. **133**, 2104–2110

Frank, S. J., Samelson, L. E. and Klausner, R. D. (1990) Semin. Immunol. **2**, 89–98

Galanaud, P., Karray, D. S. and Lorente, L. L. (1990) Eur. Cytokine Network **1**, 57–64

Galizzi, J.-P., Castle, B., Djossou, O., Harada, N., Cabrillat, H., Yahia, S. A., Barrett, R., Howard, M. and Banchereau, J. (1990) J. Biol. Chem. **265**, 439–444

Gearing, D. P., Thut, C. J., Vanden Bos, T., Gimpel, S. D., Delaney, P. B., King, J., Price, V., Cosman, D. and Beckmann, M. P. (1991) EMBO J. **10**, 2839–2848

Germain, R. and Hendrix, L. A. (1991) Nature (London) **353**, 134–139

Gold, M. R., Jakway, J. P. and DeFranco, A. L. (1987) J. Immunol. **139**, 3604–3613

Gold, M. R., Law, D. and DeFranco, A. L. (1990) Nature (London) **345**, 810–813

Gold, M. R., Matsuuchi, L., Kelly, R.B and De Franco, A. L. (1991) Proc. Natl. Acad. Sci. U.S.A. **88**, 3436–3440

Gold, M. R., Sanghera, J. S., Stewart, J. and Pelech, S. L. (1992) Biochem. J. **287**, 269–276

Goodnow, C. C., Crosbie, J., Adelstein, S., Lavoie, T. B., Smith-Gill, S., Brink, R. A., Pritchard-Bristoe, H., Wotherspoon, J. S., Loblay, R. H., Raphael, K., Trent, R. J. and Basten, A. (1988) Nature (London) **334**, 676–682

Goodwin, R. G., Friend, D., Ziegler, S. F., Jerzy, R., Falk, B. A., Gimpel, S., Cosman, D., Dower, S. K., March, C. J., Namen, A. E. and Park, L. S. (1990) Cell **60**, 941–951

Graves, J. and Cantrell, D. (1991) J. Immunol. **146**, 2102–2107

Greene, W. C., Bohnlein, E. and Ballard, D. W. (1989) Immunol. Today **10**, 272–278

Grenfell, S., Smithers, N., Miller, K. and Solari, R. (1989) Biochem. J. **264**, 813–822

Gruber, M. F., Bjorndahl, J. M., Nakamura, S. and Fu, S. M. (1989) J. Immunol. **142**, 4142–4152

Grupp, S. A., Snow, E. C. and Harmony, J. A. K. (1987) Cell. Immunol. **109**, 181–191

Harnett, M. M. and Klaus, G. G. B. (1988) J. Immunol. **140**, 3135–3139

Harnett, M. M., Holman, M. and Klaus, G. G. B. (1989) Eur. J. Immunol. **19**, 1933–1939

Harnett, M. M., Holman, M. J. and Klaus, G. G. B. (1991) J. Immunol. **147**, 3831–3835

Hasbold, J. and Klaus, G. G. B. (1990) Eur. J. Immunol. **20**, 1685–1690

Hatakayama, M., Tsudo, M., Minamoto, S., Kono, T., Doi, T., Miyata, T., Miyusaka, M. and Taniguchi, T. (1989) Science **244**, 551–556

Hatakayama, M., Kono, T., Kobayashi, N., Kawahara, A., Levin, S. D., Perlmutter, R. M. and Taniguchi, T. (1991) Science **252**, 1523–1528

Hatakayama, M., Kawahara, A., Mori, H., Shibuya, H. and Taniguchi, T. (1992) Proc. Natl. Acad. Sci. U.S.A. **89**, 2022–2026

Hempel, W. M. and DeFranco, A. L. (1991) J. Immunol. **146**, 3713–3720

Henderson, S., Rowe, M., Gregory, C., Croom-Carter, C., Wang, F., Longnecker, R., Kieff, E. and Rickinson, A. (1991) Cell **65**, 1107–1115

Henney, C. S. (1989) Immunol. Today **10**, 170–173

Hermanson, G. G., Eisenberg, D., Kincade, P. W. and Wall, R. (1988) Proc. Natl. Acad. Sci. U.S.A. **85**, 6890–6894

Hibi, M., Murakami, M., Saito, M., Hirano, T., Taga, T. and Kishimoto, T. (1990) Cell **63**, 1149–1157

Hirano, T., Taga, T., Nakano, N., Yasukawa, K., Kashiwamura, S., Shimizu, K., Nakajima, K., Pyun, K. H. and Kishimoto, T. (1985) Proc. Natl. Acad. Sci. U.S.A. **82**, 5490–5494

Hodgkin, P. D., Yamashita, L. C., Coffman, R. L. and Kehry, M.R (1990) J. Immunol. **145**, 2025–2034

Holmes, W. E., Lee, J., Kuang, W.-J., Rice, J. C. and Wood, W. I. (1991) Science **253**, 1278–1280

Hombach, J., Sablitzky, F., Rajewsky, K. and Reth, M. (1988) J. Exp. Med. **167**, 652–657

Hombach, J., Tsubata, T., Leclercq, L., Stappert, H. and Reth, M. (1990) Nature (London) **343**, 760–762

Howard, M., Farrar, J., Hilfiker, M., Johnson, B., Takatsu, K., Hamaoka, T. and Paul, W. E. (1982) J. Exp. Med. **155**, 914–924

Hudak, S., Gollnick, S. O., Conrad, D. H. and Kehry, M. R. (1987) Proc. Natl. Acad. Sci. U.S.A. **84**, 4606–4610

Hunziker, W., Koch, T., Whitney, S. A. and Mellman, I. (1990) Nature (London) **345**, 628–632

Idzerda, R. L., March, C. J., Mosley, B., Lyman, S. D., Vanden Bos, T., Gimpel, S. D., Din, W. S., Grabstein, K. H., Widmer, M. B., Park, L. S., Cosman, D. and Beckmann, M. P. (1990) J. Exp. Med. **171**, 861–873

Isfort, R. J. and Ihle, J.N (1990) Growth Factors **2**, 213–220

Jelinek, D. F. and Lipsky, P. E. (1987) J. Immunol. **141**, 164–173

Jelinek, D. F., Slawski, J. and Lipsky, P. (1986) Eur. J. Immunol. **16**, 925–932

June, C. H., Fletcher, M. C., Ledbetter, J. A., Schieven, G. L., Siegel, J. N., Phillips, A. F. and Samelson, L. E. (1990) Proc. Natl. Acad. Sci. U.S.A. **87**, 7722–7726

Justement, L. B., Chan, Z., Harris, L., Ransom, J., Sandoval, V., Smith, C., Rennick, D., Roehm, N. and Cambier, J. C. (1986) J. Immunol. **137**, 456–464

Justement, L. B., Campbell, K. S., Chien, N. C. and Cambier, J. C. (1991) Science **252**, 1839–1842

Kanakura, Y., Druker, B., Cannistra, S. A., Furukawa, Y., Torimoto, Y. and Griffin, J. D. (1990) Blood **76**, 706–715

Kanakura, Y., Druker, B., Wood, K. W., Mamon, H. J., Okuda, K., Roberts, T. M. and Griffin, J. D. (1991) Blood **77**, 243–248

Kikutani, H., Suemura, M., Owaki, H., Nakamura, H., Sato, R., Yamasaki, K., Barsumian, E. L., Hardy, R. R. and Kishimoto, T. (1986) J. Exp. Med. **164**, 1455–1469

Kinashi, T., Tashiro, K., Lee, K. H., Inaba, K., Toyama, K., Palacios, R. and Honjo, T. (1990) Philos. Trans. R. Soc. London Ser. B **327**, 117–125

Kishimoto, T, and Hirano, T. (1988) Bioessays **9**, 11–15

Klaus, G. G. B. and Harnett, M. M. (1990) Eur. J. Immunol. **20**, 2301–2307

Klaus, G. G. B., Hawrylowicz, C. M., Holman, M. J. and Keeler, M. D. (1984) Immunology **53**, 693–698

Klaus, G. G. B., Bijsterbosch, M. and Parkhouse, R. M. E. (1985) Immunology **54**, 677–683

Knapp, W., Dorken, B., Gilks, W. R., Rieber, E. P., Schmidt, R. E., Stein, H. and van dem Borne, A. E. G. (1989) Leukocyte Typing IV: White Cell Differentiation Antigens, Oxford University Press, Oxford

Kolb, J.-P., Renard, D., Dugas, B., Genot, E., Petit-Koskas, E., Sarfati, M., Delespesse, G. and Poggioli, J. (1990) J. Immunol. **145**, 429–437

Koretzky, G. A., Picus, J., Thomas, M. L. and Weiss, A. (1990) Nature (London) **346**, 66–68

Koyasu, S., Tojo, A., Miyajima, A., Akiyama, T., Kasuga, M., Urabe, A., Schreurs, J., Arai, K., Takaku, F. and Yahara, I. (1987) EMBO J. **6**, 3979–3984

Lane, P. J., Ledbetter, J. A., McConnell, F. M., Draves, K., Deans, J., Schieven, G. and Clark, E. A. (1991) J. Immunol. **146**, 715–722

Lansford, R. D., McFadden, H. J., Siu, S. T., Cox, J. S., Cann, G. M. and Koshland, M. E. (1992) Proc. Natl. Acad. Sci. U.S.A. **89**, 5966–5970

Letellier, M., Sarfati, M. and Delespesse, G. (1989) Mol. Immunol. **26**, 1105–1112

Liang, M. and Garrison, J. C. (1991) J. Biol. Chem. **266**, 13342–13349

Linnekin, D. and Farrar, W. L. (1990) Biochem. J. **271**, 317–324

Liu, Y.-J., Joshua, D. E., Williams, G. T., Smith, C. A., Gordon, J. and MacLennan, I. C. M. (1989) Nature (London) **342**, 929–931

Liu, Y.-J., Cairns, J. A., Holder, M., Abbot, S. D., Jansen, K. H., Bonnefoy, J.-Y., Gordon, J. and MacLennan, I. C. M. (1991a) Eur. J. Immunol. **21**, 1107–1114

Liu, Y.-J., Mason, D. Y., Johnson, G. D., Abbot, S., Gregory, C. D., Hardie, D. L., Gordon, J. and MacLennan, I. C. M. (1991b) Eur. J. Immunol. **21**, 1905–1910

Liu, Y.-J., Johnson, G. D., Gordon, J. and MacLennan, I. C. M. (1992) Immunol. Today **13**, 17–21

Lorente, L. L., Mitjavila, F., Crevon, M.-C. and Galanaud, P. (1990) Eur. J. Immunol. **20**, 1887–1892

Loughnan, M. and Nossal, G. J. V. (1989) Nature (London) **340**, 76–79

Lutzker, S., Rothman, P., Pollock, R., Coffman, R. and Alt, F. W. (1988) Cell **53**, 177–184

Margolis, B., Zilberstein, A., Franks, C., Felder, S., Kremer, S., Ullrich, A., Rhee, S. G., Skorecki, K. and Schlessinger, J. (1990) Science **248**, 607–610

Martin, M. and Resch, K. (1988) Trends Pharmacol. Sci. **9**, 171–177

McGarvie, G. M. and Cushley, W. (1989a) Cell. Signalling **1**, 447–460

McGarvie, G. M. and Cushley, W. (1989b) Immunol. Lett. **22**, 221–226

McMahan, C. J., Slack, J. L., Mosley, B., Cosman, D., Lupton, S. D., Brunton, L. L., Grubin, C. E., Wignall, J. M., Jenkins, N. A., Brannan, C. I., Copeland, N. G., Huebner, K., Croce, C. M., Cannizzarro, L. A., Benjamin, D., Dower, S. K., Spriggs, S. K. and Sims, J. E. (1991) EMBO J. **10**, 2821–2832

Melamed, I., Wang, G. and Roifman, C. M. (1992) J. Immunol. **149**, 169–174

Mercolino, T. J., Arnold, L. W. and Haughton, G. (1986) J. Exp. Med. **163**, 155–165

Merida, I. and Gaulton, G. N. (1990) J. Biol. Chem. **265**, 5690–5694

Merida, I., Pratt, J. C. and Gaulton, G. N. (1990) Proc. Natl. Acad. Sci. U.S.A. **87**, 9421–9425

Metcalf, D., Nicola, N. A., Gearing, D. P. and Gough, N. M. (1990) Proc. Natl. Acad. Sci. U.S.A. **87**, 4670–4674

Mills, G. B., Zhang, N., Schmandt, R., Fung, M., Greene, W., Mellors, A. and Hogg, D. (1991) Biochem. Soc. Trans. **19**, 277–287

Miyajima, A., Kitamura, T., Harada, N., Yokota, T. and Arai, K. (1992) Annu. Rev. Immunol. **10**, 295–331

Mizel, S. B. (1990) Immunol. Today **11**, 390–391

Mizuguchi, J., Beaven, M. A., Ohara, J. and Paul, W. E. (1986) J. Immunol. **137**, 2215–2220

Monroe, J. G. and Cambier, J. C. (1983) J. Exp. Med. **157**, 2073–2086

Monroe, J. G. and Haldar, S. (1989) Biochim. Biophys. Acta **1013**, 372–378

Mosley, B., Beckmann, M. P., March, C. J., Idzerda, R. L., Gimpel, S. D., VandenBos, T., Friend, D., Alpert, A., Anderson, D., Jackson, J., Wignall, J. M., Smith, C., Gallis, B., Sims, J. E., Urdal, D., Widmer, M. B., Cosman, D. and Park, L. S. (1989) Cell **59**, 335–348

Murata, Y., Yamaguchi, N., Hitoshi, Y., Tominaga, A. and Takatsu, K. (1990) Biochem. Biophys. Res. Commun. **173**, 1102–1108

Murphy, P. M. and Tiffany H. L. (1991) Science **253**, 1280–1283

Mustelin, T., Coggeshall, K. M. and Altman, A. (1989) Proc. Natl. Acad. Sci. U.S.A. **86**, 6302–6306

Mustelin, T., Coggeshall, K. M., Isakov, N. and Altman, A. (1990) Science **247**, 1584–1587

Namen, A. E., Lupton, S., Hjerrild, K., Wignall, J., Mochizuki, D. Y., Schmierer, A., Mosley, B., March, C. J., Urdal, D., Gillis, S., Cosman, D. and Goodwin, R. G. (1988) Nature (London) **333**, 571–573

Nemazee, D. A. and Bürki, K. (1989) Nature (London) **337**, 562–566

Noelle, R. J. and Snow, E. C. (1990) Immunol. Today **11**, 361–368

Noelle, R. J., Krammer, P. H., Ohara, J., Uhr, J. W. and Vitetta, E. S. (1984) Proc. Natl. Acad. Sci. U.S.A. **81**, 6149–6153

Noelle, R. J., Daum, J., Bartlett, W. C., McCann, J. and Shepherd, D. M. (1991) J. Immunol. **146**, 1118–1124

O'Garra, A., Warren, D. J., Holman, M., Popham, A. M., Sanderson, C. J. and Klaus, G. G. B. (1986) Proc. Natl. Acad. Sci. U.S.A, **83**, 5228–5232

O'Garra, A., Rigley, K. P., Holman, M., McLaughlin, J. B. and Klaus, G. G.B (1987) Proc. Natl. Acad. Sci. U.S.A. **84**, 6254–6258

O'Neill, L. A. J., Bird, T. A. and Saklatvala, J. (1990) Immunol. Today **11**, 392–394

Oppenheim, J. J., Zachariae, C. O. C., Mukaida, N. and Matsushima, K. (1991) Annu. Rev. Immunol. **9**, 617–648

Ostergaard, H. L., Shackelford, D. A., Hurley, T. R., Johnson, P., Hyman, R., Sefton, B. M. and Trowbridge, I. M. (1989) Proc. Natl. Acad. Sci. U.S.A. **86**, 8959–8963

Pardi, R., Bender, J. R., Dettori, C., Giannazza, E. and Engleman, E. G. (1989) J. Immunol. **143**, 3157–3166

Park, D. J., Rho, H. W. and Rhee, S. G. (1991) Proc. Natl. Acad. Sci. U.S.A. **88**, 5453–5456

Parkhouse, R. M. E. (1990) Immunology **69**, 289–302

Paul, W. E. and Ohara, J. (1987) Annu. Rev. Immunol. **5**, 429–459

Phillips, R. J., Harnett, M. M. and Klaus, G. G. B. (1991) Int. Immunol. **3**, 617–621

Pike, B. L. and Nossal, G. J. V. (1985) Proc. Natl. Acad. Sci. U.S.A. **82**, 3395–3399

Rabin, E. M., Mond, J. J., Ohara, J. and Paul, W. E. (1986) J. Immunol. **137**, 1573–1576

Rall, T. and Harris, B. A. (1987) FEBS Lett. **224**, 365–371

Ravetch, J. V. and Kinet, J.-P. (1991) Annu. Rev. Immunol. **9**, 457–492

Remillard, B., Petrillo, R., Maslinski, W., Tsudo, M., Strom, T. B., Cantley, L. and Varticovski, L. (1991) J. Biol. Chem. **266**, 14167–14170

Rigley, K. P. and Harnett, M. M. (1990) in Cytokines and B Lymphocytes (Callard, R. E., ed.), pp. 39–63, Academic Press, London

Rigley, K. P., Harnett, M. M. and Klaus, G. G. B. (1989) Eur. J. Immunol. **19**, 481–485

Rigley, K. P., Thurston, S. M. and Callard, R. E. (1991) Int. Immunol. **3**, 197–203

Rijkers, G. T., Griffioen, A. W., Zegers, B. J. M. and Cambier, J. C. (1990) Proc. Natl. Acad. Sci. U.S.A. **87**, 9766–8770

Robb, R. J., Rusk, C. M., Yodoi, J. and Greene, W. C. (1987) Proc. Natl. Acad. Sci. U.S.A. **84**, 2002–5658

Roehm, N. W., Leibson, H. J., Zlotnik, A., Kappler, J., Marrack, P. and Cambier, J. C. (1984) J. Exp. Med. **160**, 679–687

Rogers, J., Early, P., Carter, C., Calame, K., Bond, M., Hood, L. and Wall, R. (1981) Cell **20**, 303–312

Roifman, C. M., Wang, G., Freedman, M. and Pan, Z. (1992) J. Immunol. **148**, 1136–1142

Rolink, A. and Melchers, F. (1991) Cell **66**, 1081–1094

Rothman, P., Lutzker, S., Cook, W., Coffman, R. and Alt, F. W. (1988) Proc. Natl. Acad. Sci. U.S.A. **85**, 7704–7708

Sakaguchi, N., Kashiwamura, S., Kimoto, M., Thalmann, P. and Melchers, F. (1988) EMBO J. **7**, 3457–3464

Saltzman, E. M., White, K. and Casnellie, J. E. (1990) J. Biol. Chem. **265**, 10138–10142

Samanta, A. K., Oppenheim, J. J. and Matsushima, K. (1989) J. Exp. Med. **169**, 1185–1189

Samelson, L., Phillips, A. F., Luong, E. T. and Klausner, R. D. (1990) Proc. Natl. Acad. Sci. U.S.A. **87**, 4358–4362

Sanderson, C. J., Campbell, D. H. and Young, I. G. (1988) Immunol. Rev. **102**, 29–50

Saragovi, H. and Malek, T. R. (1990) Proc. Natl. Acad. Sci. U.S.A. **87**, 11–15

Satoh, T., Nakafuku, M., Miyajima, A and Kaziro, Y. (1991) Proc. Natl. Acad. Sci. U.S.A. **88**, 3314–3318

Scott, D. W., Chace, J. H., Warner, G. L., O'Garra, A., Klaus, G. G. B. and Quill, H. (1987) Immunol. Rev. **99**, 153–172

Secrist, J. P., Karnitz, L. and Abraham, R. T. (1991) J. Biol. Chem. **266**, 12135–12139

Seyfert, V. L., McMahon, S. B., Glenn, W. D., Yellen, A. J., Sukhatme, V. P., Cao, X. and Monroe, J. G. (1990) Science **250**, 797–800

Sharon, M., Gnarra, J. R. and Leonard, W. J. (1990) Proc. Natl. Acad. Sci. U.S.A. **87**, 4869–4873

Sherman, P. A., Basta, P. V., Heguy, A., Wloch, M. K., Roeder, R. G. and Ting, J. P.-Y. (1989) Proc. Natl. Acad. Sci. U.S.A. **86**, 6739–6743

Sidman, C. L. and Unanue, E. R. (1976) J. Exp. Med. **144**, 882–888

Sims, J. E., Acres, R. B., Grubin, C. E., McMahan, C. J., Wignall, J. M., March, C. J. and Dower, S. K. (1989) Proc. Natl. Acad. Sci. U.S.A. **86**, 8946–8950

Smith, K. A. (1989) Annu. Rev. Cell Biol. **5**, 397–425

Smrcka, A. V., Hepler, J. R., Brown, K. O. and Sternweis, P. C. (1991) Science **251**, 804–807

Snow, E. C., Feldbush, T. L. and Oaks, J. A. (1980) J. Immunol. **124**, 739–744

Snow, E. C., Noelle, R. J., Uhr, J. W. and Vitetta, E. S. (1983) J. Immunol. **130**, 614–618

Sorensen, P., Mui, A. L.-F. and Krystal, G. (1989) J. Biol. Chem. **263**, 19203–19209

Subbarao, B., Morris, J. and Baluyut, A. R. (1988) Cell. Immunol. **112**, 329–342

Sullivan, K. A., Tyler Miller, R., Masters, S. B., Beiderman, B., Heidaman, W. and Bourne, H. R. (1987) Nature (London) **330**, 758–760

Taga, T., Hibi, M., Hirata, Y., Yamasaki, K., Yasukawa, K., Matsuda, T., Hirano, T. and Kishimoto, T. (1989) Cell **58**, 573–581

Takeshita, T., Goto, Y., Tada, K., Nagata, K., Asao, H. and Sugamura, K. (1989) J. Exp. Med. **169**, 1323–1332

Tavernier, J., Devos, R., Cornelis, S., Tuypens, T., Van der Heyden, T., Fiers, W. and Plaetinck, G. (1991) Cell **66**, 1175–1184

Taylor, S. J., Chae, H. Z., Rhee, S. G. and Exton, J. H. (1991) Nature (London) **350**, 516–518

Telfer, J. C. and Rudd, C. E. (1991) Science **237**, 439–441

Teshigawara, K., Wang, H. M., Kato, K. and Smith, K. A. (1987) J. Exp. Med. **165**, 223–238

Thomas, K. M., Taylor, L. and Navarro, J. M. (1991) J. Biol. Chem. **266**, 14839–14841

Tigges, M. A., Casey, L. and Koshland, M. E. (1989) Science **243**, 781–786

Tomizawa, K., Ishizaka, A., Kojima, K., Nakanishi, M., Sakiyama, Y. and Matsumoto, S. (1991) Clin. Exp. Immunol. **83**, 492–496

Tonegawa, S. (1983) Nature (London) **302**, 575–581

Tonks, N. K., Charbonneau, H., Diltz, C. D., Fischer, E. H. and Walsh, K. A. (1988) Biochemistry **27**, 8695–8701

Torigoe, T., Saragovi, H. U. and Reed, J. C. (1992) Proc. Natl. Acad. Sci. U.S.A. **89**, 2674–2678

Tsokos, G. C., Lambris, J. D., Finkelman, F. D., Anastassiou, E. D. and June, C. H. (1990) J. Immunol. **144**, 1640–1645

Tsudo, M., Kozak, R. W., Goldman, C. and Waldmann, T. A. (1986) Proc. Natl. Acad. Sci. U.S.A. **83**, 9694–9698

Tsujimoto, Y., Cossman, J., Jaffe, E. and Croce, C. (1985) Science **228**, 1440–1443

Turner, B., Rapp, U., App, H., Greene, M., Dobashi, K. and Reed, J. (1991) Proc. Natl. Acad. Sci. U.S.A. **88**, 1227–1231

Tutt, A., Stevenson, G. T. and Glennie, M. J. (1991) J. Immunol. **147**, 60–69

Uckun, F. M., Jaszcz, W., Ambrus, J. L., Fauci, A. S., Gajl-Peczalska, K., Song, C. W., Wick, M. R., Myers, D. E., Waddick, K. and Ledbetter, J. A. (1988) Blood **71**, 13–29

Uckun, F. M., Tuel-Ahlgren, L., Obuz, V., Smith, R., Dibirdik, I., Honson, M., Langlie, M.-C. and Ledbetter, J. A. (1991a) Proc. Natl. Acad. Sci. U.S.A. **88**, 6323–6327

Uckun, F. M., Dibirdik, I., Smith, R., Tuel-Ahlgren, L., Langlie, C.-M., Schieven, G. L., Waddick, K. G., Hanson, M. and Ledbetter, J. A. (1991b) Proc. Natl. Acad. Sci. U.S.A. **88**, 3589–3593

Uckun, F. M., Schieven, G. L., Dibirdik, I., Chandan-Langlie, M., Tuel-Ahlgren, L. and Ledbetter, J. A. (1991c) J. Biol. Chem. **266**, 17478–17485

Ullman, K., Northrop, J. P., Verweij, C. L. and Crabtree, G. R. (1990) Annu. Rev. Immunol. **8**, 421–452

Van Snick, J. (1990) Annu. Rev. Immunol. **8**, 253–278

Veillette, A., Bookman, M. A., Horak, E. M. and Bolen, J. B. (1988) Cell **55**, 301–308

Venkitaraman, A. R., Williams, G. T., Dariavach, P. and Neuberger, M. S. (1991) Nature (London) **352**, 777–781

Weiss, A., Koretzky, G., Schatzman, R. C. and Kadlecek, T. (1991) Proc. Natl. Acad. Sci. U.S.A. **88**, 5484–5488

Whetton, A. D., Monk, P. N., Consalvey, S. D., Huang, S. J., Dexter, T. M. and Downes, C. P. (1988) Proc. Natl. Acad. Sci. U.S.A. **85**, 3284–3288

Wickremasinghe, R. G., Mire-Sluis, A. R. and Hoffbrand, A. V. (1987) FEBS Lett. **220**, 52–56

Wienands, J., Hombach, J., Radbruch, A., Riesterer, C. and Reth, M. (1991) EMBO J. **9**, 449–455

Williams, A. F. and Barclay, A. N. (1988) Annu. Rev. Immunol. **6**, 381–405

Wong, G. G. and Clark, S. C. (1988) Immunol. Today **9**, 137–139

Yamanishi, Y., Kakiuchi, T., Mizuguchi, J., Yamamoto, T. and Toyoshima, K. (1991) Science **251**, 192–194

Yanaga, F., Abe, M., Koga, T. and Hirata, M. (1992) J. Biol. Chem. **267**, 5114–5121

Yang, L., Baffy, G., Rhee, S. G., Manning, D., Hansen, C. A. and Williamson, J. R. (1991) J. Biol. Chem. **266**, 22451–22458

Zola, H., Weedon, H., Thompson, G. R., Fung, M. C., Ingley, E. and Hapel, A. J. (1991) Immunology **72**, 167–173

Biochem. J. (1993) **292**, 617–629 (Printed in Great Britain)

REVIEW ARTICLE
Platelet-activating factor: receptors and signal transduction

Wei CHAO and Merle S. OLSON*
Department of Biochemistry, The University of Texas Health Science Center, San Antonio, TX 78284-7760, U.S.A.

INTRODUCTION

In 1970, during a study of immunological mechanisms involving histamine and serotonin release from platelets in immunized rabbits, Henson [1] proposed that 'a soluble factor' was released from leukocytes which stimulated platelets to release vasoactive amines. This observation was confirmed independently by Siraganian and Osler [2] in 1971. In 1972, Benveniste, Henson and Cochrane [3] demonstrated that the antibody involved in the immunological reaction described above was an IgE class antibody and coined the term 'platelet-activating factor (PAF)' for the soluble factor released from basophils following IgE stimulation. Several reports followed describing the lipid character of PAF [4–6]. However, it was not until 1979 that Demopoulos, Pinckard and Hanahan [7] demonstrated that a semisynthetic phosphoacylglycerol, 1-O-alkyl-2-acetyl-sn-glycero-3-phosphocholine (AGEPC), had physiochemical as well as biological properties (i.e. aggregation of platelets and secretion of serotonin) indistinguishable from those of naturally-generated PAF [6]. At the same time, Blank et al. [8] reported independently the preparation of 1-O-alkyl-2-acetyl-sn-glycero-3-phosphocholine from choline plasmalogens isolated from bovine heart by using the same semisynthetic approach used by Demopoulos et al. [7]. The compound synthesized by Blank et al. possessed profound antihypertensive properties in the rat. Shortly after these events, Benveniste et al. [9] reported the chemical preparation of 1-O-alkyl-2-acetyl-sn-glycero-3-phosphocholine, which had similar biological properties to those of naturally-occuring PAF.

After these initial attempts to characterize the chemical structure of PAF, Hanahan et al. [10] isolated PAF from sensitized rabbit basophils and characterized the chemical structure of the naturally-occuring substance by using gas–liquid chromatography and mass spectral analysis and demonstrated that naturally-occuring PAF was indeed AGEPC. The structure of this novel ether lipid is shown in Figure 1.

Several important features of this phosphoacylglycerol mediator were revealed through evaluation of the functional capacities of structural analogues of PAF (reviewed in [11,12]). PAF has an O-alkyl ether residue at the sn-1 position and a short acyl chain, i.e. an acetyl moiety, at the sn-2 position. The sn-3 position is occupied by the polar head group O-phosphocholine. Several modifications, such as (a) acyl analogues at the sn-1 position, (b) different chain lengths of the ester group beyond three carbon atoms or a hydroxyl group at the sn-2 position, and (c) ethanolamine substituted for choline at the sn-3 position, greatly diminish or even abolish biological activity.

During the past decade, the elucidation of the biological properties of PAF has indicated that this molecule is involved extensively in intercellular signalling in a variety of patho-physiological situations. Numerous cell types and tissues have been shown to synthesize and release PAF upon stimulation and at the same time to exhibit biological responses to this compound (Table 1). There are two metabolic pathways involved in the biosynthesis of PAF, the remodelling and the de $novo$ pathways. The details of the synthetic and degradative pathways for PAF

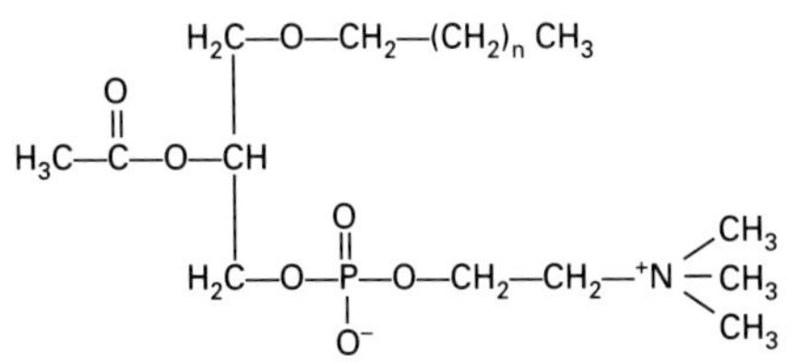

Figure 1   Chemical structure of PAF (1-O-alkyl-2-acetyl-sn-glycero-3-phosphocholine); $n = 14$–16

Table 1   Biosynthesis of PAF in various cells in response to different stimuli

Cell and tissue	Stimuli	Reference
Endothelial cells	Thrombin	152,153
	Bradykinin, A23187	154
	Tumour necrosis factor	155,156
	Interleukin 1α	156
	Leukotrienes C_4 and D_4	157
	Interleukin 1	158
	Histamine, bradykinin, ATP	159
	A23187	160
Neutrophils		
Human	Zymosan	161
	A23187	162–165
	fMet-Leu-Phe	166
	Zymosan and A23187	167
Rat	A23187	168
	Tumour necrosis factor	155
Rabbit	A23187	169,170
Platelets	Thrombin	171
Macrophages	Zymosan, A23187	172,173
	A23187	174
	Zymosan	175–178
	Tumour necrosis factor	155
HL-60 cells	A23187	179
Rat Kupffer cells	A23187	142
Exocrine cells	Carbachol	180
Eosinophils	A23187	165
Rat kidney cells	A23187	181

Abbreviations used: PAF, platelet-activating factor; AGEPC, 1-O-alkyl-2-acetyl-sn-glycero-3-phosphocholine; alkylacylGPC, 1-O-alkyl-2-acyl-sn-glycero-3-phosphocholine; lysoPAF, 1-O-alkyl-2-lyso-sn-glycero-3-phosphocholine; LPS, lipopolysaccharide; PG, prostaglandin; TX, thromboxane; LT, leukotriene; HPETE, hydroperoxyeicosatetraenoic acid; HETE, hydroxyeicosatetraenoic acid; PMA, phorbol 12-myristate 13-acetate; GTPγS, guanosine 5-[3-O-thio]triphosphate; DAG, diacylglycerol; ROI, reactive oxygen intermediates.
*To whom reprint requests should be sent.

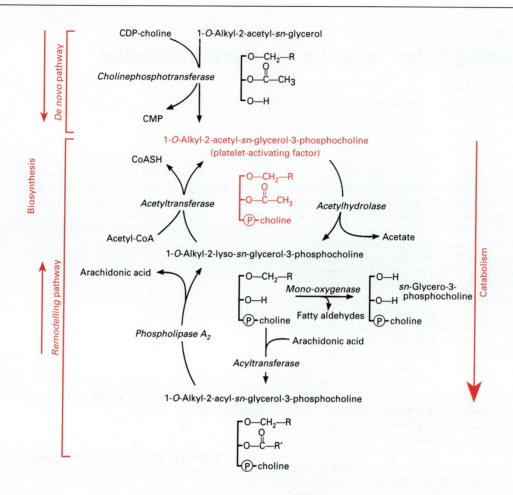

Figure 2   Metabolic pathways of PAF (1-O-alkyl-2-acetyl-sn-glycero-3-phosphocholine)

Table 2   Physiological and pathophysiological effects of PAF

Platelet aggregation and secretion, thrombosis
Stimulation of neutrophils and macrophages
Acute inflammation
Asthma and systemic anaphylaxis
Endotoxin and immune factor-induced shocks
Gastrointestinal ulceration
Glycogenolysis and increased portal pressure
Pancreatitis
Cardiac anaphylaxis (negative inotropic effect and increased heart beating rate)
Pregnancy and ovoimplantation
Ovulation
Acute lung injury

are summarized in Figure 2. Ligand binding studies indicate that specific PAF receptors can be identified in various cells and tissues (reviewed in [13]). Such a wide distribution of PAF receptors suggests that PAF must be an important mediator involved in cellular responses to trauma. Several studies have indicated that the PAF receptor is under stringent regulation as it functions in the cellular signalling mechanisms in which it plays a role. Successful cloning of the PAF receptor gene from guinea pig lung and several other types of cells represents a significant breakthrough which will allow further detailed examination of the PAF receptor and the attendant signalling mechanisms which depend upon its participation [14–16]. Following the

activation of specific PAF receptors, diverse biochemical effects are elicited, including activation of phospholipases C and A_2 leading to hydrolysis of phosphoinositide and release of arachidonic acid, respectively, an increased cytosolic calcium concentration, activation of protein kinase C, protein tyrosine phosphorylation, and proto-oncogene expression. These biochemical studies provide insight into the complex intracellular signalling mechanisms of PAF. There is substantial evidence that PAF plays an important role in various physiological and pathophysiological episodes. Table 2 summarizes some of these situations. Several comprehensive reviews have focused on this rapidly developing area [11,12,17–19]. This article will discuss the characterization and regulation of the PAF receptor and PAF receptor-mediated transmembrane signalling mechanisms.

PAF RECEPTORS

Identification and characterization

Specific receptors for PAF have been identified in numerous tissues and cells (Table 3). The first binding experiment utilizing [3H]PAF was conducted in human platelets in 1982 [20]. Using [3H]PAF (a mixture of 1-O-hexadecyl- and 1-O-octadecyl-2-acetyl-sn-glycero-3-phosphocholine) in the absence or presence of excess unlabelled PAF at 20 °C, two distinct types of binding sites were revealed. One binding site for PAF on platelets exhibited a high affinity with a K_d value of 37 ± 13 nM and had a low capacity of 1399 ± 498 sites/platelet. The other binding site possessed nearly infinite binding capacity with a low affinity for

Table 3 Specific PAF receptors in various tissues and cells

Cell and tissue	K_d (nM)	$B_{max.}$ (sites/cell, or as indicated)	Reference
Human platelet	37 ± 13	1399 ± 498	20
Human platelet	1.58 ± 0.36	1983 ± 391	22
Human platelet	0.05	242 ± 64	21
Rabbit platelet	0.5	400	66
Rabbit platelet membrane	1.36	150–300	25
Rat platelet	No specific PAF binding		22
Neutrophil	0.11 ± 0.02	5×10^6	26
Neutrophil	45	2.8×10^4	27
Neutrophil membrane	0.2	1100	28
Macrophage	0.08	7872	29
	0.25	117 fmol/mg of protein	69
Mononuclear leukocyte	5.7	1.11×10^4	31
Lung membrane	0.49	140 fmol/mg of protein	182
Liver membrane	0.5 ± 0.14	140 ± 18 fmol/mg of protein	183
Gerbil brain membrane	3.66 ± 0.92 (high)	0.83 pmol/mg of protein	40
	20.4 ± 0.5 (low)	1.1 pmol/mg of protein	40
Synaptic plasma membrane	0.023 (high)	8.75 fmol/mg of protein	41
	25 (low)	0.96 pmol/mg of protein	41
Rat retina	2.9	0.85 pmol/mg of protein	42
Eosinophil	1.6	3.5×10^4	32
Kupffer cells	0.12–0.45	1.06×10^4	33

PAF. The high-affinity binding sites were responsible for PAF-elicited platelet aggregation. By comparing the capacity of several analogues of PAF (i.e. lysoPAF and 1-hexadecyl-2-benzoyl-*sn*-glycero-3-phosphocholine) to inhibit the specific binding of [³H]PAF and to induce platelet aggregation, it was demonstrated that several features of the PAF structure described above (an ether linkage at the *sn*-1 position, an acetyl moiety at the *sn*-2 position, and a polar head group containing choline at the *sn*-3 position) were critical both for the specific PAF binding to human platelets and for the initiation of platelet aggregation [20]. Similar to human platelets [20,21], rabbit platelets possess high-affinity binding sites for PAF of K_d 0.9 ± 0.5 nM [22], while rat platelets show only non-specific binding, explaining perhaps the observation that the functional responses of the rat platelet are insensitive to PAF [23]. Platelets from patients with septicaemia exhibit fewer specific [³H]PAF binding sites compared with platelets from normal humans or from patients with respiratory or cardiovascular disturbances [24]. In addition, platelets from patients with sepsis contained significant amounts of PAF, whereas this mediator could not be found in platelets from patients with either respiratory or cardiovascular disturbances with negative blood cultures, or from normal individuals. The observation that a large amount of PAF is associated with one of its target cells under certain pathophysiological conditions such as sepsis may provide the rationale for using PAF antagonists in patients with severe shock or multiple organ dysfunction.

Specific binding sites for PAF are present on smooth muscle cells [25], neutrophils [26–28], macrophages [29,30], mononuclear leukocytes [31], eosinophils [32], and Kupffer cells [33]. In neutrophils, there exist two classes of binding sites [26–28,34], high- and low-affinity (Table 3). The high-affinity binding sites are believed to mediate PAF-induced cellular responses. The typical number of PAF receptors found on neutrophils ranges from several hundred to thousands. Both murine [29,30] and rat [33] macrophages possess high-affinity receptors for PAF. PAF receptors following ligand binding mediate several biochemical activities including phosphoinositide and arachidonic acid

metabolism, intracellular calcium changes, and protein phosphorylation.

A problem with some of the early studies of PAF binding to its receptors in intact cells was that data on [³H]PAF metabolism under the binding conditions were not presented. The lack of information on PAF metabolism renders the binding data difficult to interpret since it is known that many cells actively metabolize PAF [28,35–37]. O'Flaherty et al. [28] characterized the binding and metabolism of [³H]PAF by human neutrophils and showed that neutrophils rapidly metabolize [³H]PAF to its alkylacyl derivative at 37 °C. Subcellular fractionation of cells pretreated with radiolabelled PAF on Percoll gradients revealed that most [³H]PAF associated with alkaline phosphatase-rich membranes was converted rapidly to alkylacyl-GPC and was transferred slowly to specific intracellular granules. In contrast, human neutrophils did not metabolize [³H]PAF at 4 °C, but rather accumulated PAF in plasma membrane subfractions. Under non-metabolizing conditions, [³H]PAF binding experiments indicated that human neutrophil plasma membranes possess two classes of binding sites, high-affinity and low-affinity, with K_d values of 0.2 nM and 500 nM, respectively. The potency of several structural analogues in inhibiting binding of [³H]PAF to membranes correlated closely with their respective potency in stimulating degranulation responses.

A possible physiological and pathophysiological role for PAF in tissue or cells of the central nervous system was first suggested by Kornecki and Ehrlich [38] and by Kumar et al. [39]. It was found that incubation of cultured NG108-15 neuroblastoma cells for 3–4 days with low concentrations of PAF caused neuronal differentiation while higher concentrations of PAF were neurotoxic. In addition, PAF caused an almost immediate increase in intracellular free Ca^{2+} in both NG108-15 and PC12 cells. This effect was dependent upon extracellular calcium and was inhibited by the PAF receptor antagonist CV-3988. The functional consequence of the increase in intracellular Ca^{2+} was a secretion of ATP from PC12 cells [38]. PAF caused an accelerated turnover of ^{32}P-labelled phosphoinositides in a synaptosome preparation from rat brain [39]. Most interesting

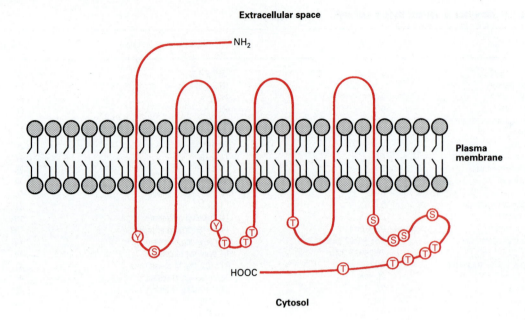

Figure 3   Schematic representation of PAF receptor

Guinea pig lung PAF receptor has 342 amino acids and a molecular mass of 38982. There are seven transmembrane segments. The possible intracellular phosphorylation sites of the PAF receptor [14] are illustrated. S, serine residues; T, threonine residues; Y, tyrosine residues.

was the observation that both electrical and chemical stimulation of either the isolated perfused rat brain or the brain of an intact animal resulted in a substantial increase in PAF levels in the brain [39]. Such observations suggest an important role(s) for PAF in the brain under various physiological and patho-physiological situations. Supporting this suggestion was an investigation conducted by Domingo et al. [40] in which specific [^{3}H]PAF binding sites on membrane preparations from gerbil brain were characterized. Scatchard analysis revealed two apparent populations of binding sites (Table 3). A study of the distribution of [^{3}H]PAF binding revealed that maximal binding was present in the midbrain and hippocampus. Marcheselli et al. [41] found three distinct classes of binding sites for PAF in synaptic plasma membranes and in intracellular membranes of rat cerebral cortex. The significance of intracellular PAF receptors has not been clarified but it has been postulated that intracellular PAF receptors may mediate PAF-induced proto-oncogene expression in this type of tissue [42].

Early attempts to purify and to characterize PAF receptor proteins from different sources, mainly from rabbit and human platelets [43–45] have been unsuccessful. Using photoaffinity labelling techniques, PAF binding proteins in rabbit platelet membranes have been characterized [46]. A photoreactive, radio-iodinated derivative of PAF, 1-O-(4-azido-2-hydroxy-3-iodo-benzamido)undecyl-2-O-acetyl-sn-glycero-3-phosphocholine ([^{125}I]AAGP), was synthesized and was used to label specific PAF binding proteins. This derivative of PAF maintained the biological effects of authentic PAF, inducing aggregation of rabbit platelets. Photoaffinity labelling of platelet membranes with [^{125}I]AAGP revealed several ^{125}I-labelled components in SDS/polyacrylamide-gel electrophoresis. One labelled component exhibiting an apparent molecular mass of 52000 was observed consistently and its labelling was inhibited significantly by unlabelled PAF at nanomolar concentrations and by the specific PAF antagonists SRI-63,675 and L652,731, but not by lysoPAF. A significant advance was made in this area when

Shimizu and his colleagues cloned a cDNA for a PAF receptor from guinea pig lung [14]. The cloning strategy involved the construction of a cDNA library from size-fractionated poly(A) RNA, the synthesis of a transcript of the cDNA *in vitro* using phage DNA as a template, the expression of the transcript in *Xenopus* oocytes, and the electrophysiological detection of a PAF-induced response in the oocytes. The PAF receptor cDNA analysis indicated that the PAF receptor has 342 amino acids and a molecular mass of 38982 Da. A hydropathy profile analysis suggested the existence of seven transmembrane segments and the cytoplasmic tail of the PAF receptor contained four serine and five threonine residues as possible phosphorylation sites (Figure 3). Subsequently, the PAF receptor was cloned from human leukocytes [15,47] and from HL-60 granulocytes [16]. The cloned PAF receptor from human leukocytes has similar characteristics to that of guinea pig lung with 83% identity in the amino acid sequence. Using the expressed PAF receptor in *Xenopus laevis* oocytes and COS-7 cells, it was demonstrated that the PAF receptor is linked functionally to phosphoinositide metabolism via a G-protein [15].

Linkage with guanine nucleotide regulatory proteins

Numerous studies have indicated that PAF receptor-induced transmembrane signalling mechanisms involve guanine nucleotide regulatory proteins (G-proteins). Although it is clear that the PAF receptor is coupled to various cellular effector systems such as phospholipase A$_2$ and phospholipase C through G-proteins, the identities of the G-proteins involved have not been characterized.

The initial observations concerning the involvement of G-proteins in PAF action were made in the early 1980s. It was found that synthetic PAF inhibited adenylate cyclase activity in the particulate fraction of platelets [48,49] and stimulated GTPase activity as well [49–53]. In an effort to elucidate the identity and properties of G-protein(s) involved in the activation of platelets,

Brass et al. [54] examined the relationship between various receptors (including the PAF receptor) and G-proteins in human platelets by comparing the ability of various agonists (e.g. PAF) to stimulate phospholipases (via G_p) and to inhibit adenylate cyclase (via G_i) with their ability to induce structural changes in G_p and G_i which would preclude subsequent [^{32}P]ADP-ribosylation. Using permeabilized platelets and platelet membranes, it was found that thrombin, which elicited responses that were mediated by both G_i and G_p, decreased ^{32}P-radiolabelling of G-proteins by > 90%. PAF and vasopressin, which were found to couple only to G_p to stimulate phosphoinositide breakdown, decreased the radiolabelling by 50%, as did adrenaline, which was coupled functionally only to G_i. On the other hand, an agonist that neither inhibited cyclic AMP formation nor caused pertussis toxin-sensitive phosphoinositide hydrolysis, such as the thromboxane analogue U46619, had no effect on [^{32}P]ADP-ribosylation. The [^{32}P]ADP-ribosylation catalysed by pertussis toxin in permeabilized platelets or platelet membranes labelled a protein(s) (α_{41}) which migrated in SDS/polyacrylamide-gel electrophoresis slightly below rabbit and bovine α_i (M_r 41000). Based on experiments involving proteolytic digestion of the G-protein and subsequent two-dimensional electrophoresis, it was concluded that in platelets a single pertussis toxin-sensitive, α_{41}-containing G-protein was invoved in the regulation of both adenylate cyclase and phospholipase C.

In rabbit platelets, PAF receptor stimulation increased GTPase activity [51,52] and this stimulatory effect of PAF on GTPase activity was attenuated in cells pretreated with phorbol ester, dibutyryl cyclic AMP, and by PAF-induced desensitization [51]. It is likely that both homologous desensitization [55] and dibutyryl cyclic AMP [56] modify PAF receptors causing functional dissociation of PAF receptors from their G-proteins. On the other hand, phorbol ester was found to suppress PAF-mediated signal transduction through modification of the GTP-binding proteins, as phorbol ester was found to abolish PAF-stimulated GTPase activity [51]. This observation suggests that protein kinase C selectively inhibits PAF effects by inactivating a GTP-binding protein coupled with PAF receptors, although activation of protein kinase C also down-regulates surface PAF receptors [57–59].

Although evidence has been presented that GTP-binding protein(s) are involved in signal transduction from PAF receptors to a receptor-stimulated phospholipase A_2, the identity and properties of the G-protein regulating the activity of phospholipase A_2 remain uncertain. Pertussis toxin inhibits PAF-elicited arachidonic acid release, PGE_2 formation, and inositol trisphosphate production, but it does not alter significantly the rise in intracellular calcium [60]. Further information concerning the G-protein regulation of phospholipase A_2 was presented by Nakashima et al. [61] in neutrophils and Kajiyama et al. [62] in rabbit platelets. In neutrophils permeabilized with saponin, the guanine nucleotide analogue guanosine 5-[3-O-thio]triphosphate (GTPγS) and NaF, which bypass receptors and directly activate G-proteins, induced the release of [^{3}H]arachidonic acid. The effect of GTPγS was inhibited by pretreatment with pertussis toxin. Similar information was obtained from a human neutrophil homogenate and membrane preparation [63] and in platelets [62]. GTPγS, guanosine 5-[βγ-imido]triphosphate and $NaAlF_4$ all caused a significant release of arachidonic acid in digitonin-permeabilized platelets. The stimulatory effect of the GTP analogues was inhibited by guanosine 5-[2-O-thio]diphosphate and by pertussis toxin.

Interestingly, GTP-binding proteins may be involved in PAF-stimulated release of PAF. PAF enhanced the release of newly synthesized PAF in human neutrophils, as assessed by [^{3}H]acetate incorporation into PAF. The non-metabolizable bioactive PAF analogue 1-O-hexadecyl-2-(N-methyl)-carbamoyloxy-sn-glycero-3-phosphocholine [64], but not lyso-PAF, enhanced the release of newly-synthesized PAF. The PAF-stimulated PAF release was inhibited in pertussis toxin-treated neutrophils. The biosynthesis of PAF involves activation of at least two critical enzymes, phospholipase A_2 and the PAF acetyltransferase. The observation that pertussis toxin inhibits the PAF-stimulated synthesis of PAF suggests that a G-protein is likely to be involved in the signal transduction from PAF receptors to the PAF-synthesizing enzyme system.

Regulation

Knowledge of the regulatory factors that affect specific PAF receptors and subsequently control PAF-elicited cellular responses remains in the initial stages of development. Generally, any factor affecting the process of PAF binding to its receptor or subsequent PAF receptor-mediated signal transduction is a likely candidate as a regulator of specific PAF receptors.

It has been found that PAF down-regulates its own receptors in platelets [20,21] and in cultured Kupffer cells [55]. Valone et al. [20] showed that pretreatment of human platelets with PAF led to a decrease in specific [^{3}H]PAF binding. The ligand-mediated loss of specific PAF receptors was very rapid, with only 50% of the specific binding sites remaining after 90 s of incubation at 37 °C. Kloprogge and Akkerman [21] confirmed this finding that desensitized human platelets failed to respond to PAF because of a loss of available binding sites for PAF. However, Chesney et al. [65] reported that pretreatment of human platelets with PAF caused a decrease in the binding affinity of specific PAF receptors rather than the loss of binding sites. In contrast, Homma et al. [66] found, in an investigation of the mechanism for the desentization of PAF responses, that functionally desensitized rabbit platelets did not lose their capacity for specific binding of [^{3}H]PAF. Rather, PAF-treated platelets internalized more [^{3}H]PAF than control platelets. A potential problem with the investigations regarding homologous down-regulation of the PAF receptor was that very little attention was given to the extent to which PAF molecules added during pretreatment associate with the platelet membrane after the pretreatment and any subsequent wash procedure. The presence of residual unlabelled PAF from the pretreatment could decrease subsequent specific [^{3}H]PAF binding. The lack of such information renders data concerning the loss of specific [^{3}H]PAF binding sites difficult to interpret since PAF is very hydrophobic and exhibits high non-specific binding to the lipid bilayer of the plasma membrane. A washing procedure employing a high concentration (1%) of bovine serum albumin can remove PAF efficiently from the outer leaflet of cells [55].

In neutrophils [57,58] and rat Kupffer cells [59] the PAF receptor is modulated by protein kinase C. In neutrophils, specific [^{3}H]PAF receptors were modulated by the protein kinase C activator phorbol ester [57,58], as well as by LTB4 [58]. The regulatory effect of protein kinase C on specific PAF receptors was bidirectional, increasing specific [^{3}H]PAF binding at low concentrations and decreasing binding at high concentrations. The protein kinase C regulation of PAF receptors seems to be cell type-dependent since the protein kinase C activator phorbol 12-myristate 13-acetate (PMA) had no effect on specific binding of PAF in rabbit platelets [51]. Although dibutyryl cyclic AMP was reported to have profound inhibitory effects on PAF-stimulated biological responses such as platelet aggregation and GTPase activity, cyclic AMP analogues at concentrations as high as 2 mM had little apparent effect on [³H]PAF binding in rabbit platelets [51]. However, in cultured Kupffer cells, prolonged incubation with cyclic AMP analogues or forskolin decreased the surface expression of PAF receptors [56]. In U937 cells, dibutyryl cyclic AMP was found to down-regulate PAF receptor mRNA levels after a 72 h treatment [47]. These observations indicate that cyclic AMP down-regulates surface expression of PAF receptors through a mechanism involving decreased PAF receptor gene expression and/or subsequent receptor protein synthesis.

It is evident that various monovalent and divalent cations exert regulatory effects on PAF receptors. Na^+ specifically inhibited [³H]PAF binding at 0 °C, primarily due to a decrease in the affinity of specific PAF receptors while Li^+ was 25-fold less effective and K^+, Cs^+ and Rb^+ enhanced PAF binding. Mg^{2+}, Ca^{2+} and Mn^{2+} enhanced the specific PAF receptor binding 8–10-fold. Scatchard analysis of the binding data suggested that the Mg^{2+}-induced enhancement of specific PAF binding was attributable to an increase in the affinity of the receptor for PAF and an increase in the available specific PAF binding sites. The specific mechanism involved in these cation effects on PAF receptors has not been elucidated [52]. Zinc ions have been shown to have profound inhibitory effects on PAF-induced activation of rabbit [67] and human [68] platelets, as well as [³H]PAF binding [68]. The decrease in specific [³H]PAF binding caused by zinc resulted primarily from a decreased affinity of the specific binding sites rather than a reduced number of binding sites. Compared with Zn^{2+}, Cd^{2+} and Cu^{2+} had much weaker inhibitory effects on specific PAF binding [68].

Recent studies have demonstrated that bacterial lipopolysaccharide (LPS) induces an increase in the surface expression of PAF receptors in macrophages [69]. Consequently, the PAF-induced increase in intracellular Ca^{2+} is enhanced [69,70]. The receptor regulatory effect of LPS is time-dependent with a maximal effect (150–200 %) observed within 5 h and 8 h. Cycloheximide and actinomycin D can abolish the effect of LPS, suggesting the involvement of enhanced receptor protein synthesis and mRNA production in this event [69]. The alteration of PAF receptor expression in response to LPS treatment may represent one of the mechanisms for LPS priming of PAF-induced responses such as prostaglandin E_2 production [71].

PAF-MEDIATED BIOCHEMICAL EFFECTS

Arachidonic acid metabolism

During the past decade, a vast literature has been elaborated to define the relationship between PAF action and arachidonic acid metabolism. Various types of cells release arachidonic acid when PAF is synthesized as well as in response to PAF stimulation (Figures 2 and 4). Moreover, it is known that both PAF and arachidonic acid can be released from a common precursor, 1-O-alkyl-2-arachidonoyl-sn-glycero-3-phosphocholine (Figure 2).

Numerous types of cells and tissues release arachidonic acid and its metabolites in response to PAF stimulation. Responsive cells include platelets [72,73], neutrophils [63,74–76], eosinophils [77,78], macrophages [33,56,59,79–81], smooth muscle cells [82], epithelial cells [83], and the perfused heart [84] and liver [85], to name a few. The released arachidonic acid and its metabolites in response to PAF are believed to play an important role in various physiological and pathophysiological processes.

In an *in vivo* study, McManus et al. [73] found that intravenous antigen challenge of IgE-sensitized rabbits caused a significant elevation of plasma TXB_2. Simultaneously, basopenia, thrombocytopenia and neutropenia occurred. The cyclo-oxygenase inhibitor aspirin, when given to rabbits 18 h before

antigen challenge, did not prevent the development of thrombocytopenia, neutropenia, basopenia, or the release of platelet factor 4, but reduced significantly the release of TXB_2 and resulted in mortality in IgE-sensitized rabbits. Based on these observations and data from *in vitro* experiments where PAF was found to stimulate platelets to synthesize TXB_2, serotonin and platelet factor 4, it was suggested that TXB_2 release into the circulation during IgE-induced anaphylaxis in the rabbit may result, in part, from PAF stimulation of thromboxane synthesis by circulating rabbit platelets. However, this arachidonic acid metabolite was not necessary to alter circulating blood cells or platelet factor 4 secretion but may have served to reduce the IgE-induced anaphylactic reaction.

In cultured rat mesangial cells, PAF stimulated arachidonic acid release as well as PGE_2 synthesis. At the same time, PAF caused contraction of mesangial cells with a dose–response and time-course parallel to that for PGE_2 production. The PAF-elicited contraction of mesangial cells was enhanced when PGE_2 synthesis was inhibited. It was suggested that the production of glomerular prostaglandins may be an important facet of glomerulonephritis [79].

Arachidonic acid metabolites also play an important role in PAF-induced coronary vasoconstriction and reduced cardiac contractility in the isolated perfused rat heart [84]. Both cyclo-oxygenase- and lipoxygenase-derived arachidonic acid metabolites were released into the cardiac effluent perfusate in response to PAF challenge. These metabolites of arachidonic acid included $PGF_{2\alpha}$, PGE_2, 6-ketoPGF$_{1\alpha}$, TXB_2, LTB_4 and LTC_4. Through the use of inhibitors of lipoxygenase and cyclo-oxygenase and a leukotriene receptor antagonist, these authors demonstrated that LTC_4 released during PAF challenge was largely responsible for the coronary vasoconstriction induced following infusion of PAF. There was no information presented, however, as to whether a PAF receptor antagonist inhibited the cardiac effects of PAF. The possibility that PAF may act directly on target cells causing cardiac haemodynamic changes in the perfused heart was not ruled out [84].

In human [77] and guinea pig eosinophils [78], isolated from either the peritoneal cavity or bronchoalveolar lavages, PAF has been found to stimulate synthesis of LTC_4, LTB_4 and TXB_2. Human eosinophils produce primarily LTC_4, a powerful airway smooth-muscle constrictor [86], and 15-HETE, a substance thought to be responsible for excessive airway mucus production [87]. The finding that PAF induces synthesis of leukotrienes from lung cells may explain some of the pathophysiological roles of PAF in allergic respiratory diseases. Supporting this contention was an early observation [88] that PAF-induced rapid pulmonary vasoconstriction and oedema in isolated lungs was mediated through the action of LTD_4 and LTC_4, which were identified in the lung effluent perfusate after stimulation with PAF. Lipoxygenase-derived products of arachidonic acid also may mediate PAF-induced neutrophil aggregation and release of granule-associated enzymes [74–76].

The mechanisms by which PAF stimulates arachidonic acid release have been investigated extensively in various types of cells [56,63,80,89–93]. Using the relatively non-specific phospholipase A_2 inhibitors mepacrine and 2-(p-amylcinnamoyl)amino-4-chlorobenzoic acid [63,71] and using [¹⁴C]arachidonic acid-labelled membranes as endogenous substrate and dioleoyl-phosphatidyl[¹⁴C]ethanolamine as an exogenous substrate [80], it was demonstrated that phospholipase A_2 was responsible for the release of arachidonic acid in response to PAF. Through the use of EGTA and TMB-8 (an inhibitor of intracellular calcium mobilization), it was demonstrated that PAF-induced activation of phospholipase A_2 in neutrophils was dependent upon intra-

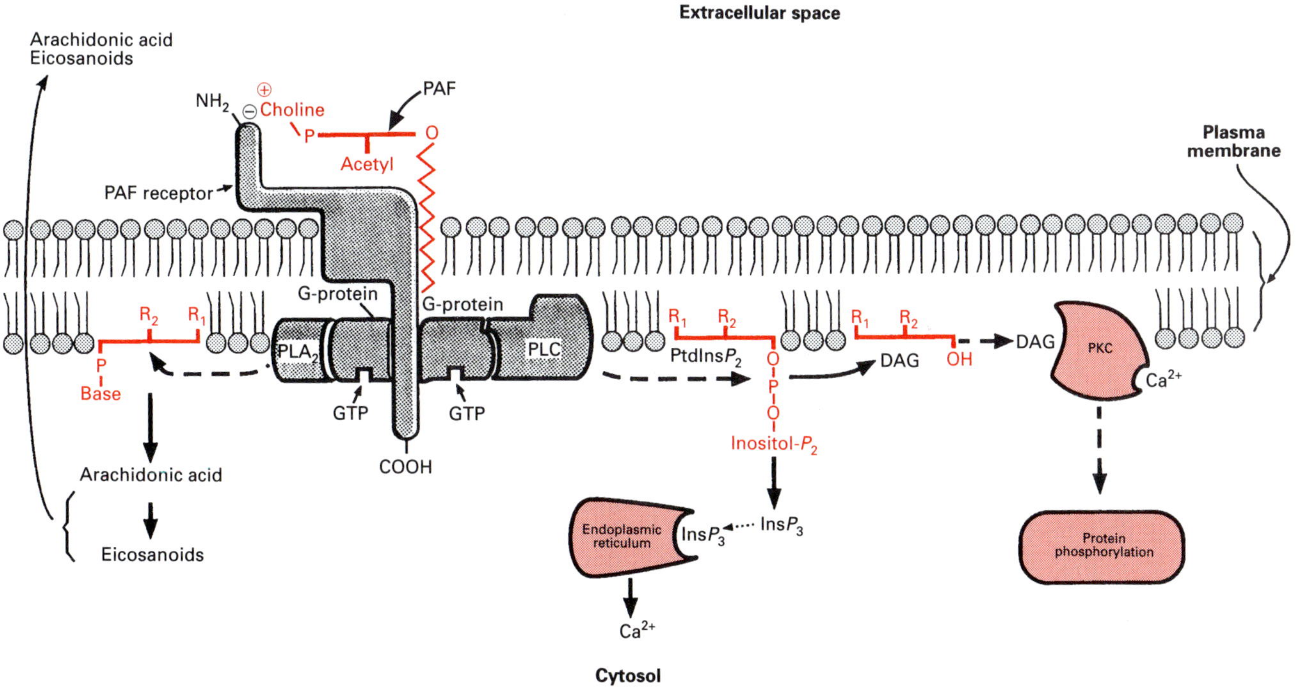

Figure 4　Schematic representation of PAF receptor-mediated intracellular signalling mechanisms

Recent studies have suggested that protein tyrosine phosphorylation (not shown) may also play a role in PAF signalling mechanisms [71,93,130–133].

cellular but not extracellular Ca^{2+} [63]. In cultured Kupffer cells, however, PAF-induced activation of phospholipase A_2 depended on extracellular Ca^{2+} [33,93]. In addition, PAF induces activation of phospholipase A_2 through a protein kinase C-dependent mechanism [59,90,91,93]. Pertussis toxin pretreatment abolished PAF-induced arachidonic acid release, suggesting a G-protein involvement in this event [60,63,90]. The stimulatory effect of PAF on phospholipase A_2 may also be regulated by intracellular cyclic AMP levels [56,92].

PAF not only stimulates arachidonic acid release from membrane phospholipids but also promotes incorporation of arachidonic acid into phospholipids. PAF stimulated the incorporation of [1-^{14}C]arachidonic acid in a Ca^{2+}-independent fashion most significantly into phosphatidylinositol and phosphatidylcholine in guinea pig [94] as well as in human neutrophils [95,96]. Moreover, PAF did not alter the distribution of [1-^{14}C]arachidonic acid in the various molecular species of phosphatidylcholine (diacyl, alkylacyl and alkenylacyl species) after brief incubation intervals, suggesting that the increased formation of [1-^{14}C]arachidonylphosphatidylcholine was not derived from the added PAF, e.g. alkylacyl-phosphatidylcholine formation with an ether linkage at the sn-1 position. The mechanism of the stimulatory effect of PAF on acylation of phospholipids has not been elucidated completely. Both increased fatty acid uptake and increased availability of lysophospholipids following phospholipase A_2 activation may contribute to the increased phospholipid acylation induced by PAF [96]. However, other mechanisms such as the increased activity of fatty acyl-CoA synthetase and acyltransferase, two enzymes essential for the acylation of phospholipids during PAF stimulation, deserve to be considered.

Arachidonic acid can be generated directly or indirectly by agonist activation of phospholipases and may act as a signalling element in various cellular reactions. In addition to serving as a precursor for eicosanoid synthesis, a function which has considerable pathophysiological significance, arachidonic acid has been implicated in a variety of biochemical responses such as activation of protein kinase C and adenylate cyclase, and the regulation of intracellular calcium concentrations [97]. Also, arachidonic acid was found to play an important role in the regulation of K^+ channel opening in cardiac and smooth muscle [98]. Moreover, it has been suggested that there exists a regulatory relationship between arachidonic acid and phosphoinositide hydrolysis. It has been demonstrated that arachidonic acid causes a dose-dependent increase in the accumulation of inositol phosphates, including $InsP_3$, $InsP_2$, and $InsP$, in cultured astrocytes [99]. Inositol phosphate formation following application of carbachol or noradrenaline was additive with arachidonic acid, whereas a similar response evoked by PAF or ATP was not additive with arachidonic acid.

Phosphoinositide turnover

It has become clear that one of the key events in PAF-mediated signalling mechanisms is the hydrolysis of phosphatidylinositol 4,5-bisphosphate, by a specific phospholipase C, yielding two second messengers, diacylglycerol (DAG) and inositol 1,4,5-trisphosphate. DAG was found to activate protein kinase C leading to phosphorylation of various substrates [100,101] whereas $InsP_3$ mobilizes intracellular calcium [102,103] (Figure 4). PAF has been found to stimulate the hydrolysis of phosphatidylinositol in a wide variety of cell types including rabbit [104,105], human [106], and horse [107,108] platelets, smooth muscle cells [109], hepatocytes [110,111], macrophages [30,112,113], neutrophils [114], endothelial cells [115], human keratinocytes [116], glomerular mesangial cells [60,117] and

Kupffer cells [118,119]. It was first reported that PAF stimulated metabolism of inositol phospholipids and phosphatidic acid in washed rabbit platelets [104]; PAF caused a 15–20 % decrease in the PtdIns level within 15 s with a dramatic four-fold increase in phosphatidic acid. The effect of PAF on the metabolism of phosphoinositide was quite specific since other major classes of phospholipids were not affected. If $[^{32}P]P_i$ was present in the medium, PAF enhanced significantly the incorporation of radio-activity into the PtdIns$P$, PtdInsP_2 and phosphatidic acid fractions within 1 min while the incorporation of $[^{32}P]P_i$ into phosphatidylinositol increased thereafter. In ^{32}P-labelled hepato-cytes [110], PAF, at a concentration of 5×10^{-10} M, caused a 30–40 % decrease in $[^{32}P]$PtdInsP_2 within 10 s. The ^{32}P content of the PtdIns$P$ and PtdIns fractions also decreased but at a slower rate. With horse platelets prelabelled with $[^{32}P]P_i$, it was shown that PAF initiated a rapid formation of labelled phos-phatidic acid followed by an increase in phosphatidylinositol.

PAF has potent stimulatory effects on macrophages [30,112]. Macrophages stimulated with PAF produced several inositol phosphates, including Ins(1,4,5)P_3 and Ins(1,3,4,5)P_4, and as a consequence, the intracellular level of calcium was elevated up to $290 \pm 27\%$ of the basal level (82.7 ± 12 nM) [30]. It should be pointed out that the metabolites [e.g. Ins(1,4,5)P_3] of phosphoinositide metabolism may not always be responsible for the PAF-induced increase in cytosolic calcium. In Kupffer cells, for example, more than 90 % of cytosolic free Ca^{2+} is due to extracellular Ca^{2+} influx rather than mobilization of Ca^{2+} from intracellular sites in response to PAF stimulation [93] although PAF also stimulates metabolism of phosphoinositide within the same time frame [118,119]. On the other hand, as a consequence of metabolism of phosphoinositide, PAF increases DAG levels [30], stimulates protein kinase C activity [93], and causes protein phosphorylation [30]. In an attempt to investigate the relationship between inositol lipid metabolism and the production of reactive oxygen intermediates (ROI), Huang et al. [112] found that PAF treatment led to a rapid increase in $[^3H]$InsP_3 levels and 1,2-diacyl$[^3H]$glycerol levels in bone marrow-derived macrophages labelled with $[^3H]$inositol or $[^3H]$glycerol, respectively. This response to PAF was followed by an increase in the production of ROI. Pretreatment of macrophages with phorbol ester or pertussis toxin attenuated both PAF-induced $[^3H]$inositol phos-phate production and ROI production. Phorbol ester, a protein kinase C activator, but not calcium ionophore A23187, stimulated ROI production. It was proposed, therefore, that the DAG formed and consequent protein kinase C activation fol-lowing PAF stimulation was responsible for the increased production of ROI. Protein kinase C activation also may mediate other PAF-induced cellular responses such as arachidonic acid release and eicosanoid production and protein tyrosine phosphorylation [93].

Using a double labelling technique, Okayasu et al. [111] demonstrated in primary cultured rat hepatocytes that PAF stimulated the breakdown of phosphoinositides via phospholipase A_2. Addition of PAF to cells labelled with $[^{14}C]$glycerol and $[^3H]$arachidonic acid caused a transient decrease in $[^{14}C]$glycerol-labelled PtdIns and an increase in $[^{14}C]$glycerol-labelled lysoPtdIns. $[^3H]$Arachidonic acid-labelled PtdIns decreased in a time-dependent fashion and the radioactivity in phosphatidylcholine, phosphatidylethanolamine and other major phospholipids was not affected by the addition of PAF. The $[^3H]$arachidonate/$[^{14}C]$glycerol ratio decreased significantly in PtdIns and $[^3H]$arachidonic acid appeared within 10 s upon stimulation with PAF. Also, PAF increased $[^3H]$inositol-labelled lysoPtdIns in myo-$[^3H]$inositol-labelled hepatocytes in the ab-sence of an accumulation of $[^3H]$inositol-labelled inositol

phosphates. In addition, a precursor–product relationship was detected between PtdIns and lysoPtdIns in $[^{32}P]P_i$-labelled hepatocytes stimulated with PAF. These observations suggested that PAF stimulated the metabolism of phosphoinositides via activation of phospholipase A_2 rather than via the PtdIns cycle or a polyphosphoinositide turnover mechanism in primary cultured hepatocytes.

Calcium flux

PAF causes an elevation of cytosolic free calcium in various cells such as platelets [120,121], neutrophils [90,114,122,123], macrophages [30,69,124], mesangial cells [111,117], vascular smooth muscle cells [109,125], endothelial cells [126], Kupffer cells [93,118] and neuronal cells [38]. There are at least two mechanisms involved in PAF-induced increases in cytosolic free calcium: (a) calcium influx occurs through a membrane-associated calcium channel regulated by PAF receptors or by signalling molecules generated intracellularly (e.g. metabolites of arachidonic acid) and (b) mobilization of calcium is instigated from intracellular stores in response to the intracellular second messenger InsP_3 produced during PAF receptor stimulation.

Lee et al. [120,121] first reported that PAF induced a calcium influx in rabbit platelets in a dose-dependent manner. The PAF-stimulated $^{45}Ca^{2+}$ influx in rabbit platelets could be blocked by verapamil, a calcium channel blocker, and was dependent upon extracellular $[Ca^{2+}]$. Calcium mobilization was independent of cyclo-oxygenase products of arachidonic acid metabolism but was inhibited significantly by mepacrine, p-bromophenacyl bro-mide, eicosatetraynoic acid and nordihydroguaiaretic acid. These observations suggest that lipoxygenase-derived metabolites of arachidonic acid produced in response to PAF stimulation may mediate the PAF-induced calcium uptake in rabbit platelets [121]. In mesangial cells [117], PAF increased cytosolic free calcium within 10 s, due to both a release of Ca^{2+} from intracellular storage sites as well as an influx of extracellular Ca^{2+}. The rise in cytosolic free $[Ca^{2+}]$ in response to PAF stimulation was attributed to an increased production of InsP_3.

In vascular smooth muscle cells, PAF stimulated the hydrolysis of phosphoinositide and a rapid efflux of $^{45}Ca^{2+}$ from preloaded cells [109]. In an attempt to elucidate the mechanism for desensitization of the PAF response, it was demonstrated that PAF elicited a transient, dose-dependent increase in cytosolic free Ca^{2+} in vascular smooth muscle cells preloaded with fura-2 and an increase in InsP_3 and InsP_4 levels [125]. Pretreatment of the cells with PAF or PMA attenuated subsequent PAF- and angiotensin II-induced Ca^{2+} mobilization but not vasopressin-stimulated Ca^{2+} mobilization. The authors proposed that both homologous and heterologous desensitization is mediated by PAF-stimulated phosphoinositide hydrolysis and DAG formation [125]. The PAF-induced increase in cytosolic free Ca^{2+} seemed to be independent of InsP_3 since the peak of Ca^{2+} concentration was reached before there was a significant increase in the amount of InsP_3. The effect of PAF was dependent upon extracellular calcium. In vascular endothelial cells preloaded with fura-2, however, it was demonstrated that PAF elicited an elevation of cytosolic free Ca^{2+} released primarily from intra-cellular stores [126].

A PAF-elicited increase in cytosolic free Ca^{2+} in rabbit neutrophils preloaded with fura-2 was inhibited by pretreatment of the cells with fMet-Leu-Phe [123]. This finding was explained as an increased production of endogenous PAF which bound to and inactivated the PAF receptor. Pretreatment of the cells with PMA, a potent protein kinase C activator, abolished completely the rise in intracellular free Ca^{2+} induced by PAF. This effect of

PMA appeared to be mediated by protein kinase C activation, since the protein kinase C inhibitor H-7 attenuated the inhibitory effect of PMA [123]. In an attempt to assess the relative contribution of Ca^{2+} released from intracellular stores and Ca^{2+} influx from the extracellular medium to the elevation of intracellular Ca^{2+} during neutrophil activation, it was found that Ca^{2+} release from intracellular stores was rate-limiting for the PAF- and fMet-Leu-Phe-induced increase in intracellular free Ca^{2+} [122].

With murine peritoneal macrophages preloaded with fura-2/AM, Prpic et al. [30] demonstrated that PAF elevated intracellular levels of Ca^{2+} to $290 \pm 27\%$ of basal levels (82.7 ± 12 nM). Using colour-enhanced computer images of the 340nm:380 nm fluoresence ratio of a single macrophage, Prpic showed that increases in intracellular Ca^{2+} were observed first in a submembranous area of the macrophage. In single mouse macrophages preloaded with fura-2, Randriamampita and Trautmann [124] observed that a pulse administration of PAF caused a biphasic increase in intracellular free $[Ca^{2+}]$, including an initial transient and then a more sustained increase in intracellular Ca^{2+}. The initial transient phase, which lasted for a few seconds, was independent of extracellular Ca^{2+} concentration, suggesting a release of Ca^{2+} from intracellular stores. The second phase of this response, which lasted for several minutes, was sensitive to extracellular Ca^{2+} concentrations and was probably due to an influx of Ca^{2+} through the plasma membrane. Also, similar biphasic responses to PAF were observed in cultured rat mesangial cells [127]. Human monocytic leukaemic U-937 cells [128] and HL-60 cells [129], when differentiated with dimethyl sulphoxide to a macrophage-like state, express specific PAF receptors and respond to PAF stimulation by increasing their intracellular free Ca^{2+} levels [128]. This response to PAF depends upon extracellular Ca^{2+} concentrations and can be blocked by the receptor antagonist CV-3988 but not by calcium channel blockers such as nifedipine or verapamil. Both PAF-induced calcium mobilization and phosphoinositide metabolism were insensitive to pertussis toxin, but sensitive to the phospholipase C inhibitor, manoalide.

Protein tyrosine phosphorylation

Recent studies have demonstrated that PAF stimulates tyrosine phosphorylation of numerous cellular proteins in platelets [130,131], neutrophils [132], and liver macrophages [93,133]. PAF-induced tyrosine phosphorylation is extracellular Ca^{2+}-dependent and may be mediated by G-protein(s) and PKC activation [93,132]. Although it has been suggested that a tyrosine kinase is involved in the PAF-stimulated phosphoinositide turnover in platelets [130] and arachidonic acid metabolism in macrophages [71], the molecular mechanism for the interaction between the tyrosine kinase and the phospholipases is not understood. Characterization of the tyrosine-phosphorylated proteins in platelets led to the identification of a 60 kDa phosphoprotein as the proto-oncogene product $pp60^{c\text{-}src}$ [131]. It will be of interest to determine the functional consequence of the tyrosine phosphorylation of $pp60^{c\text{-}src}$ protein in the PAF-stimulated platelets. Also, PAF may play a role in gene expression. It was found that PAF is capable of inducing transcription of the nuclear proto-oncogenes c-*fos* and c-*jun* in B cells [134,135]. Since the cloned PAF receptor contains several tyrosine residues in its intracellular loops and tail (Figure 3) [14,16], it will be of interest to examine whether homologous down-regulation of PAF receptors [55] and PAF-induced protein tyrosine phosphorylation are related. Vanadate, an inhibitor of protein

tyrosine phosphatase, stimulates tyrosine phosphorylation of numerous cellular proteins and induces a decrease in the number of the surface PAF receptors [133]. Both effects of vanadate can be inhibited by genistein, a putative tyrosine kinase inhibitor, suggesting that protein tyrosine phosphorylation plays a role, directly or indirectly, in the regulation of the surface expression of PAF receptors.

Hepatic actions of PAF

PAF, when infused into the perfused rat liver, resulted in a several-fold increase in glucose output in the effluent perfusate [110,136]. In contrast, when a 500-fold higher concentration of 1-*O*-alkyl-2-lyso-*sn*-glycero-3-phosphocholine or the stereoisomer 3-*O*-alkyl-2-acetyl-*sn*-glycero-1-phosphocholine was infused, no increased glycogenolysis was observed. The glycogenolytic effect of PAF depends upon both ligand and the Ca^{2+} concentration in the perfusate. Surprisingly, PAF failed to stimulate glycogenolysis in isolated hepatocytes although adrenaline and glucagon elicited glucose output in the same preparation [137]. In an attempt to elucidate the mechanism by which PAF causes glycogenolysis in the perfused liver, it was observed that infusion of PAF caused a transient increase in portal vein presure concomitant with the increase in glycogenolysis [138]. The vascular and metabolic responses were correlated closely, displaying similar dose dependence and similar attenuation in response to a reduction in perfusate Ca^{2+} concentration. The activity of glycogen phosphorylase *a* and the tissue ADP level were increased significantly in the perfused liver in response to PAF. Furthermore, nitric oxide, a compound which relaxes vascular smooth muscle, was found to inhibit or to abolish PAF-induced increases in portal vein pressure, oxygen consumption and, most importantly, hepatic glucose output in the perfused rat liver [139]. In contrast to its effect on PAF-induced hepatic responses, nitric oxide inhibited only the haemodynamic but not the glycogenolytic effects of phenylephrine, which acts directly on hepatocytes to induce glycogenolysis and glucose production [139]. Based on these observations, it was proposed that the glycogenolytic effect of PAF in the perfused liver was a result of the haemodynamic effects of PAF, rather than a direct effect of the agonist on the hepatocyte (Figure 5). Further evidence suggested that hepatic reticuloendothelial cells, e.g. endothelial and/or Kupffer cells, may play an important role in the PAF-induced hepatic responses observed in the perfused liver: heat-aggregated IgG [140], which stimulates reticuloendothelial cells in the liver, causes hepatic responses similar to those of PAF in the perfused liver; [³H]PAF infused into the perfused liver was localized specifically in small portal venules instead of in parenchymal cells [141]; appropriately stimulated isolated Kupffer cells synthesize PAF and actively metabolize this potent phospholipid [37,142]; and, finally, isolated rat hepatocytes, the glycogen-storing cells in the liver, lack detectable specific PAF-binding sites (W. Chao and M. S. Olson, unpublished work).

In an effort to elucidate the important role of Kupffer cells in the hepatic actions of PAF, the specific receptor for PAF and the regulatory characteristics of the receptor have been investigated [33,55,56,59]. It was found that isolated Kupffer cells possess a large number of high-affinity receptors for PAF [33]. The PAF receptor identified in Kupffer cells is functionally active since it mediates arachidonic acid release and eicosanoid production [33,56,59,143]. It has been proposed that non-parenchymal cells, in response to certain stimuli, release biologically active eicosanoids which then act on parenchymal cells to stimulate glycogenolysis [144]. The release of the biologically active eicosanoids from non-parenchymal cells was suggested to

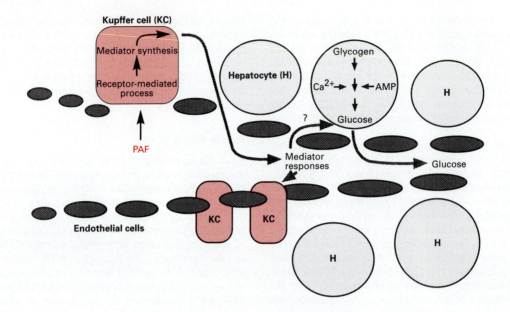

Figure 5   Schematic representation of the intercellular signalling mechanism for PAF in the liver

When infused into the rat liver through the portal vein, PAF induces glycogenolysis and vasoconstriction as illustrated by the narrowing sinusoid. The glycogenolytic effect of PAF is indirect, requiring interaction of PAF with Kupffer cells and/or endothelial cells resulting in severe hepatic vasoconstriction. It has been proposed that this mediator-induced haemodynamic response produces a transient ischaemia which initiates the glycogenolytic response, probably through an elevation of tissue AMP levels and possibly of intracellular free calcium levels in hepatocytes.

mediate both PAF- [143] and endotoxin- [145] stimulated glycogenolysis in the perfused liver since the cyclo-oxygenase inhibitors aspirin and indomethacin blocked the glycogenolytic action of these factors [143,145,184] (Figure 5). In addition, media obtained from aspirin-treated Kupffer cells or endothelial cells had no effect on glucose production by parenchymal cells [144]. However, this proposal has been questioned by the observation that ibuprofen (50 mM), a specific cyclo-oxygenase inhibitor, does not inhibit PAF-induced glycogenolysis and vasoconstriction significantly although it abolishes eicosanoid production in the perfused liver [85]. Nevertheless, it is probable that the hepatic glycogenolytic action of PAF observed in the perfused liver is indirect, requiring interaction between the parenchymal cells, i.e. hepatocytes, and the non-parenchymal cells, i.e. Kupffer cells and endothelial cells (Figure 5).

The haemodynamic and hyperglycaemic effects of PAF can be regulated by β-adrenergic receptor stimulation [146,147]. Infusion of isoproterenol, a β-adrenergic receptor agonist, into the perfused rat liver attenuated the glycogenolytic stimulation caused by PAF. The regulatory effect of the β-adrenergic agonist was believed not to be mediated by a cyclic AMP increase in 'parenchymal' cells since glucagon, which increases hepatic cyclic AMP levels to a far greater extent than does isoproterenol, had no effect on the glycogenolytic response of the liver to PAF [146]. It was suggested that the mechanism by which isoproterenol regulates the glycogenolytic effect of PAF in the perfused liver may involve interaction of the β-agonist with non-parenchymal cells [56,146,147]. Further studies demonstrated that isoproterenol attenuates the subsequent PAF-stimulated biological effects, including arachidonic acid release and cyclo-oxygenase-derived eicosanoid production in isolated Kupffer cells. The regulatory effect of isoproterenol is highly specific and involves a β_2-adrenergic receptor- and a cyclic AMP-mediated mechanism [56]. Long-term incubation of Kupffer cells with cyclic AMP analogues or forskolin down-regulates the surface expression of PAF receptors [56].

Ligand binding studies indicated that PAF down-regulates the surface expression of its own receptor in cultured Kupffer cells [55]. Both the rate of loss and the maximal extent of loss of the receptors were dependent upon PAF concentration. With receptor synthesis inhibited by cycloheximide in the absence of PAF, the half-life of the surface PAF receptor was about 4 h, suggesting that the turnover of the PAF receptors on the plasma membrane is continuous. Through the use of cycloheximide [55] or actinomycin D (our unpublished work), it was demonstrated that PAF receptors are not recycled and that the restoration of lost or inactivated PAF receptors requires newly synthesized protein. The fact that both cycloheximide and actinomycin D prevent the restoration of PAF receptors suggests that tissue responsiveness to PAF may be regulated by both transcription and translation. Also, surface expression of PAF receptors in Kupffer cells is down-regulated by protein kinase C activation [59,93]. The effect of protein kinase C is specific and transient and, as a consequence, PAF-mediated arachidonic acid release and eicosanoid production are attenuated. Also, protein kinase C may be involved in the stimulatory signal transduction between the PAF receptor and the phospholipase(s) responsible for the release of arachidonic acid and subsequent production of eicosanoids. This contention is based on the observations that both down-regulation of protein kinase C and a protein kinase C inhibitor attenuate PAF-stimulated arachidonic acid release as well as eicosanoid production [59,93].

SUMMARY

During the past two decades, studies describing the chemistry and biology of PAF have been extensive. This potent phosphoacylglycerol exhibits a wide variety of physiological and pathophysiological effects in various cells and tissues. PAF acts, through specific receptors and a variety of signal transduction systems, to elicit diverse biochemical responses. Several important future directions can be enumerated for the

characterization of PAF receptors and their attendant signalling mechanisms. The recent cloning and sequence analysis of the gene for the PAF receptor will allow a number of important experimental approaches for characterizing the structure and analysing the function of the various domains of the receptor. Using molecular genetic and immunological technologies, questions relating to whether there is receptor heterogeneity, the precise mechanism(s) for the regulation of the PAF receptor, and the molecular details of the signalling mechanisms in which the PAF receptor is involved can be explored. Another area of major significance is the examination of the relationship between the signalling response(s) evoked by PAF binding to its receptor and signalling mechanisms activated by a myriad of other mediators, cytokines and growth factors. A very exciting recent development in which PAF receptors undoubtedly play a role is in the regulation of the function of various cellular adhesion molecules [148,185,186]. Finally, there remain many incompletely characterized physiological and pathophysiological situations in which PAF and its receptor play a crucial signalling role. Our laboratory has been active in the elucidation of several tissue responses in which PAF exhibits major autocoid signalling responses, e.g. hepatic injury and inflammation [149], acute and chronic pancreatitis [150,151], and cerebral stimulation and/or trauma [39]. As new experimental strategies are developed for characterizing the fine structure of the molecular mechanisms involved in tissue injury and inflammation, the essential role of PAF as a primary signalling molecule will be affirmed. Doubtless the next 20 years of experimental activity will be even more interesting and productive than the past two decades.

Our work is supported by grants from the National Institutes of Health (DK-33538 and DK-19473) and the Robert A. Welch Foundation (AQ-728).

REFERENCES

1   Henson, P. M. (1970) J. Exp. Med. **131**, 287–304
2   Siraganian, R. P. and Osler, A. G. (1971) J. Immunol. **106**, 1244–1251
3   Benveniste, J., Henson, P. M. and Cochrane, C. G. (1972) J. Exp. Med. **136**, 1356–1377
4   Benveniste, J. (1974) Nature (London) **249**, 581–582
5   Benveniste, J., Le Couedic, J. P., Polonsky, J. and Taence, M. (1977) Nature (London) **269**, 170–171
6   Pinckard, R. N., Farr, R. S. and Hanahan, D. J. (1979) J. Immunol. **123**, 1847–1857
7   Demopoulos, C. A., Pinckard, R. N. and Hanahan, D. J. (1979) J. Biol. Chem. **254**, 9355–9358
8   Blank, M. L., Snyder, F., Byers, L. W., Brooks, B. and Muirhead, E. E. (1979) Biochem. Biophys. Res. Commun. **90**, 1194–1200
9   Benveniste, J., Tence, M., Varenne, P., Bidault, J., Boullet, C. and Polonsky, J. (1979) C. R. Acad. Sci. Paris Ser. D **289**, 1037–1040
10 Hanahan, D. J., Demopoulos, C. A., Liehr, J. and Pinckard, R. N. (1980) J. Biol. Chem. **255**, 5514–5516
11 Hanahan, D. J. (1986) Annu. Rev. Biochem. **55**, 483–509
12 Snyder, F. (1985) Med. Res. Rev. **5**, 107–140
13 Hwang, S.-B. (1990) J. Lipid Mediators **2**, 123–158
14 Honda, Z.-i., Nakamura, M., Miki, H., Minami, M., Watanabe, T., Seyama, Y., Okado, H., Toh, H., Ito, K., Miyamoto, T. and Shimizu, T. (1991) Nature (London) **349**, 342–346
15 Nakamura, M. Honda, Z.-i., Izumi, T., Sakanaka, C., Mutoh, H., Minami, M., Bito, H., Seyama, Y., Matsumoto, T., Noma, M. and Shimizu, T. (1991) J. Biol. Chem. **266**, 20400–20405
16 Ye, R. D., Prossnitz, E. R., Zou, A. and Cochrane, C. G. (1991) Biochem. Biophys. Res. Commun. **180**, 105–111
17 Prescott, S. M., Zimmerman, G. A. and McIntyre, T. M. (1990) J. Biol. Chem. **265**, 17381–17384
18 Snyder, F. (1990) Am. J. Physiol. **259**, c697–c708
19 Braquet, P., Touqui, L., Shen, T. Y. and Vargaftig, B. B. (1987) Pharmacol. Rev. **39**, 97–145
20 Valone, F. H., Coles, E., Reinhold, V. R. and Goetzl, E. J. (1982) J. Immunol. **129**, 1639–1641
21 Kloprogge, E. and Akkerman, W. N. (1984) Biochem. J. **223**, 901–909
22 Inarrea, P., Gomez-Cambronero, J., Nieto, M. and Sanchez-Crespo, M. (1984) Eur. J. Pharmacol. **105**, 309–315
23 Sanchez-Crespo, M., Alonso, F., Inarrea, P. and Egido, J. (1981) Agents Actions **11**, 565–566
24 Diez, F. L., Nieto, M. L., Fernandez-Gallardo, S., Gijon, M. A. and Sanchez-Crespo, M. (1989) J. Clin. Invest. **83**, 1733–1740
25 Hwang, S.-B., Lee, C.-S. C., Cheah, M. J. and Shen, T. Y. (1983) Biochemistry **22**, 4756–4763
26 Valone, F. H. and Goetzl, E. J. (1983) Immunology **48**, 141–149
27 Bussolino, F., Breviario, F., Tetta, C., Aglietta, M., Mantovani, A. and Dejana, E. (1986) J. Clin. Invest. **77**, 2027–2033
28 O'Flaherty, J. T., Surles, J. R., Redman, J., Jacobson, D., Piantadosi, C. and Wykle, R. L. (1986) J. Clin. Invest. **78**, 381–388
29 Valone, F. H. (1988) J. Immunol. **140**, 2389–2394
30 Prpic, V., Uhing, R. J., Weiel, J. E., Jakoi, L., Gawdi, G., Herman, B. and Adams, D. O. (1988) J. Cell Biol. **107**, 363–372
31 Ng, D. S. and Wong, K. (1988) Biochem. Biophys. Res. Commun. **155**, 311–316
32 Ukena, D., Krogel, C., Dent, G., Yukawa, T., Sybrecht, G. and Barnes, P. J. (1989) Biochem. Pharmacol. **38**, 1702–1705
33 Chao, W., Liu, H., DeBuysere, M. S., Hanahan, D. J. and Olson, M. S. (1989) J. Biol. Chem. **264**, 13591–13598
34 Hwang, S.-B. (1988) J. Biol. Chem. **263**, 3225–3233
35 Kramer, R. M., Patton, G. M., Pritzker, C. R. and Deykin, D. (1984) J. Biol. Chem. **259**, 13316–13320
36 Malone, B., Lee, T.-c. and Snyder, F. (1985) J. Biol. Chem. **260**, 1531–1534
37 Chao, W., Siafaka-Kapadai, A., Hanahan, D. J. and Olson, M. S. (1989) Biochem. J. **261**, 77–81
38 Kornecki, E. and Ehrlich, Y. H. (1988) Science **240**, 1792–1794
39 Kumar, R., Harvey, S. A. K., Kester, M., Hanahan, D. J. and Olson, M. S. (1988) Biochim. Biophys. Acta **963**, 375–383
40 Domingo, M. T., Spinnewyn, B., Chabrier, P. E. and Braquet, P. (1988) Biochem. Biophys. Res. Commun. **151**, 730–736
41 Marcheselli, V. L., Rossowska, M. J., Domingo, M. T., Braquet, P. and Bazan, N. G. (1990) J. Biol. Chem. **265**, 9140–9145
42 Thierry, A. T., Doly, M., Braquet, P., Cluzel, J. and Meyniel, G. (1989) Eur. J. Pharmacol. **163**, 97–101
43 Valone, F. H. (1984) Immunology **52**, 1169–1174
44 Nishihira, J., Ishibashi, T., Imai, Y. and Muramatsu, T. (1985) J. Exp. Med. **147**, 145–152
45 Chau, L.-Y. and Jii, Y.-J. (1988) Biochim. Biophys. Acta **970**, 103–112
46 Chau, L.-Y., Tsai, Y.-M. and Cheng, J.-R. (1989) Biochem. Biophys. Res. Commun. **161**, 1070–1076
47 Kunz, D., Gerard, N. P. and Gerard, C. (1992) J. Biol. Chem. **267**, 9101–9106
48 Haslam, R. J. and Vanderwel, M. (1982) J. Biol. Chem. **257**, 6879–6885
49 Avdonin, P. V., Svitina-Ulitina, I. V. and Kulikov, V. I. (1985) Biochem. Biophys. Res. Commun. **131**, 307–313
50 Houslay, M. D., Bojanic, D. and Wilson, A. (1986) Biochem. J. **234**, 737–740
51 Homma, H. and Hanahan, D. J. (1988) Arch. Biochem. Biophys. **262**, 32–39
52 Hwang, S.-B., Lam, M.-H. and Pong, S.-S. (1986) J. Biol. Chem. **261**, 532–537
53 Avdonin, P. V., Svitina-Ulitina, I. V. and Tkachuk, V. A. (1989) J. Mol. Cell. Cardiol. **21**, 139–143
54 Brass, L. F., Woolkalis, M. J. and Manning, D. R. (1988) J. Biol. Chem. **263**, 5348–5355
55 Chao, W., Liu, H., Hanahan, D. J. and Olson, M. S. (1989) J. Biol. Chem. **264**, 20448–20457
56 Chao, W., Liu, H., Zhou, W., Hanahan, D. J. and Olson, M. S. (1990) J. Biol. Chem. **265**, 17576–17583
57 O'Flaherty, J. T., Jacobson, D. P. and Redman, J. F. (1989) J. Biol. Chem. **264**, 6836–6843
58 Yamazaki, M., Gomez-Cambronero, J., Durstin, M., Molski, T. F., Becker, E. L. and Sha'afi, R. I. (1989) Proc. Natl. Acad. Sci. U.S.A. **86**, 5791–5794
59 Chao, W., Liu, H., Hanahan, D. J. and Olson, M. S. (1990) Arch. Biochem. Biophys. **282**, 188–197
60 Schlondorff, D., Singhal, P., Hassid, A., Satriano, J. A. and DeCandido, S. (1989) Am. J. Physiol. **256**, F171–F178
61 Nakashima, S., Nagata, K.-I., Ueeda, K. and Nozawa, Y. (1988) Arch. Biochem. Biophys. **261**, 375–383
62 Kajiyama, Y., Murayama, T. and Nomura, Y. (1989) Arch. Biochem. Biophys. **274**, 200–208
63 Nakashima, S., Suganuma, A., Sato, M., Tohmatsu, T. and Nozawa, Y. (1989) J. Immunol. **143**, 1295–1302
64 Gomez-Cambronero, J., Durstin, M., Molski, T. F. P., Naccache, P. H. and Sha'afi, R. I. (1989) J. Biol. Chem. **264**, 21699–21704

65 Chesney, C. M., Pifer, D. D. and Huch, K. M. (1985) Biochem. Biophys. Res. Commun. **127**, 24–30

66 Homma, H., Tokumura, A. and Hanahan, D. J. (1987) J. Biol. Chem. **262**, 10582–10587

67 Huo, Y., Ekholm, J. and Hanahan, D. J. (1988) Arch. Biochem. Biophys. **260**, 841–846

68 Nunez, D., Kumar, R. and Hanahan, D. J. (1989) Arch. Biochem. Biophys. **272**, 466–475

69 Liu, H., Chao, W. and Olson, M. S. (1992) J. Biol. Chem. **267**, 20811–20819

70 Aepfelbacher, M., Ziegler-Heitbrock, H. W., Lux, I. and Weber, P. C. (1992) J. Immunol. **148**, 2186–2193

71 Glaser, K. B., Asmis, R. and Dennis, E. A. (1990) J. Biol. Chem. **265**, 8658–8664

72 Shaw, J. O., Printz, M. P., Hirabayashi, K. and Henson, P. M. (1978) J. Immunol. **121**, 1939–1945

73 McManus, L. M., Shaw, J. O. and Pinckare, R. N. (1980) J. Immunol. **125**, 1950–1954

74 Smith, R. J. and Bowman, B. J. (1982) Biochem. Biophys. Res. Commun. **104**, 1495–1501

75 Lin, A. H., Morton, D. R. and Gorman, R. R. (1982) J. Clin. Invest. **70**, 1058–1065

76 Chilton, F. H., O'Flaherty, J. T., Walsh, C. E., Thomas, M. J., Wykle, R. L., DeChartelet, L. R. and Waite, B. M. (1982) J. Biol. Chem. **257**, 5402–5407

77 Bruijnzeel, P. L. B., Kok, P. T. M., Hamelink, M. L., Kijne, A. M. and Verhagen, J. (1987) Prostaglandins **34**, 205–214

78 Sun, F. F., Czuk, C. I. and Taylor, B. M. (1989) J. Leukocyte Biol. **46**, 152–160

79 Schlondorff, D. and Satriano, J. A. (1984) J. Clin. Invest. **73**, 1227–1231

80 Bachelet, M., Masliah, J., Vargaftig, B. B., Bereziat, G. and Colard, O. (1986) Biochim. Biophys. Acta **878**, 177–183

81 Kadiri, C., Masliah, J., Bachelet, M., Vargaftig, B. B. and Berezilat, G. (1989) J. Cell. Biochem. **40**, 157–164

82 Yousufzai, S. Y. K. and Abdel-Latif, A. A. (1985) Biochem. J. **228**, 697–706

83 Kawaguchi, H. and Yasuda, H. (1986) Biochim. Biophys. Acta **875**, 525–534

84 Piper, P. J. and Stewart, A. G. (1986) Br. J. Pharmacol. **88**, 595–605

85 LaPointe, D. S. and Olson, M. S. (1989) J. Biol. Chem. **264**, 12130–12133

86 Dahlen, S., Hedqvist, P., Hammarstrom, S. and Samuelsson, B. (1980) Nature (London) **288**, 484–486

87 Zvi, M., Shelhamer, J. H., Sun, F. and Kaliner, M. (1983) J. Clin. Invest. **72**, 122–127

88 Voelkel, N. F., Worthen, S., Reeves, J. T., Henson, P. M. and Murphy, R. C. (1982) Science **218**, 286–288

89 Haslam, R. J., Williams, K. A. and Davidson, M. M. L. (1985) Adv. Exp. Med. Biol. **192**, 265–280

90 Tao, W., Molski, F. P. and Sha'afi, R. I. (1989) Biochem. J. **257**, 633–637

91 O'Flaherty, J. T. and Nishihira, J. (1987) J. Immunol. **138**, 1889–1895

92 Bachelet, M., Adolfs, M. J. P., Masliah, J., Bereziat, G., Vargaftig, B. B. and Bonta, I. L. (1988) Eur. J. Pharmacol. **149**, 73–78

93 Chao, W., Liu, H., Hanahan, D. J. and Olson, M. S. (1992) J. Biol. Chem. **267**, 6725–6735

94 Tou, J.-s. (1985) Biochem. Biophys. Res. Commun. **127**, 1045–1051

95 Tou, J.-s. (1987) Lipids **22**, 333–337

96 Tou, J.-s. (1989) Lipids **24**, 812–817

97 Axelrod, J., Burch, R. M. and Jelsema, C. L. (1988) Trends Neurosci. **11**, 117–123

98 Bourne, H. R. (1989) Nature (London) **337**, 504–505

99 Murphy, S. and Welk, G. (1989) FEBS Lett. **257**, 68–70

100 Nishizuka, Y. (1984) Nature (London) **308**, 693–698

101 Nishizuka, Y. (1986) Science **233**, 305–312

102 Berridge, M. J. (1984) Biochem. J. **220**, 345–360

103 Berridge, M. J. and Irvine, R. F. (1984) Nature (London) **312**, 315–321

104 Shukla, S. D. and Hanahan, D. J. (1982) Biochem. Biophys. Res. Commun. **106**, 697–703

105 Mauco, G., Chap, H. and Douste-Blazy, L. (1983) FEBS Lett. **153**, 361–365

106 Tysnes, L.-B., Verhoeven, A. J. M. and Holmsen, H. (1987) Biochem. Biophys. Res. Commun. **144**, 454–462

107 Lapetina, E. G. (1982) J. Biol. Chem. **257**, 7314–7317

108 Billah, M. M. and Lapetina, E. G. (1983) Proc. Natl. Acad. Sci. U.S.A. **80**, 965–968

109 Doyle, V. M., Creba, J. A. and Ruegg, U. T. (1986) FEBS Lett. **197**, 13–16

110 Shukla, S. D., Buxton, D. B., Olson, M. S. and Hanahan, D. J. (1983) J. Biol. Chem. **258**, 10212–10214

111 Okayasu, T., Hasegawa, K. and Ishibashi, T. (1987) J. Lipid Res. **28**, 760–767

112 Huang, S. J., Monk, P. N., Downes, C. P. and Whetton, A. D. (1988) Biochem. J. **249**, 839–845

113 Stephens, L., Hawkins, P. T., Carter, N., Chahwala, S. B., Morris, A. J., Whetton, A. D. and Downes, P. C. (1988) Biochem. J. **249**, 271–282

114 Rossi, A. G., McMillan, R. M. and MacIntyre, D. E. (1988) Agents Actions **24**, 272

115 Kawaguchi, H., Sawa, H. and Yasuda, H. (1990) Biochim. Biophys. Acta **1052**, 503–508

116 Fisher, G. J., Talwar, H. S., Ryder, N. S. and Voorhees, J. J. (1989) Biochem. Biophys. Res. Commun. **163**, 1344–1350

117 Bonventre, J. V., Weber, P. C. and Gronich, J. H. (1988) Am. J. Physiol. **254**, F87–F94

118 Fisher, R. A., Sharma, R. V. and Bhalla, R. C. (1989) FEBS Lett. **251**, 22–26

119 Gandhi, C. R., Hanahan, D. J. and Olson, M. S. (1990) J. Biol. Chem. **265**, 18234–18241

120 Lee, T.-c., Malone, B., Blank, M. L. and Snyder, F. (1981) Biochem. Biophys. Res. Commun. **102**, 1262–1268

121 Lee, T.-c., Malone, B. and Snyder, F. (1983) Arch. Biochem. Biophys. **223**, 33–39

122 Von Tscharner, V., Prod'hom, B., Baggiolini, M. and Reuter, A. (1986) Nature (London) **324**, 369–372

123 Molski, T. F. P., Tao, W., Becker, E. L. and Sha'afi, R. I. (1988) Biochem. Biophys. Res. Commun. **151**, 836–843

124 Randriamampita, C. and Trautmann, A. (1989) FEBS Lett. **249**, 199–206

125 Schwertschlag, U. S. and Whorton, A. R. (1988) J. Biol. Chem. **263**, 13791–13796

126 Hirafuji, M. H., Maeyama, K., Watanabe, T. and Ogura, Y. (1988) Biochem. Biophys. Res. Commun. **154**, 910–917

127 Kester, M., Mené, P., Dubyak, G. R. and Dunn, M. J. (1987) FASEB J. **1**, 215–219

128 Barzaghi, G., Sarau, H. M. and Mong, S. (1989) J. Pharmacol. Exp. Ther. **248**, 559–566

129 Vallari, D. S., Austinhirst, R. and Snyder, F. (1990) J. Biol. Chem. **265**, 4261–4265

130 Dhar, A., Paul, A. K. and Shukla, S. D. (1990) Mol. Pharmacol. **37**, 519–525

131 Dhar, A. and Shukla, S. D. (1991) J. Biol. Chem. **266**, 18797–18801

132 Gomez-Cambronero, J., Wang, J., Johnson, G., Huang, C.-K. and Sha'afi, R. I. (1991) J. Biol. Chem. **266**, 6240–6245

133 Chao, W., Liu, H., Hanahan, D. J. and Olson, M. S. (1992) Biochem. J. **288**, 777–784

134 Schulam, P. G., Kuruvilla, A., Putcha, G., Mangus, L., Franklin-Johnson, J. and Shearer, W. T. (1991) J. Immunol. **146**, 1642–1648

135 Mazer, B., Domenico, J., Sawami, H. and Gelfand, E. W. (1991) J. Immunol. **146**, 1914–1920

136 Buxton, D. B., Shukla, S. D., Hanahan, D. J. and Olson, M. S. (1984) J. Biol. Chem. **259**, 1468–1471

137 Fisher, R. A., Shukla, S. D., DeBuysere, M. S., Hanahan, D. J. and Olson, M. S. (1984) J. Biol. Chem. **259**, 8685–8688

138 Buxton, D. B., Fisher, R. A., Hanahan, D. J. and Olson, M. S. (1986) J. Biol. Chem. **261**, 644–649

139 Moy, J. A., Bates, J. N. and Fisher, R. A. (1991) J. Biol. Chem. **266**, 8092–8096

140 Buxton, D. B., Hanahan, D. J. and Olson, M. S. (1984) J. Biol. Chem. **259**, 13758–13761

141 Hill, C. E., Miwa, M., Sheridan, P. J., Hanahan, D. J. and Olson, M. S. (1988) Biochem. J. **253**, 651–657

142 Chao, W., Siafaka-Kapadai, A., Olson, M. S. and Hanahan, D. J. (1989) Biochem. J. **257**, 823–829

143 Kuiper, J., De Rijke, Y. B., Zijlstra, F. J., Van Waas, M. P. and Van Berkel, T. J. C. (1988) Biochem. Biophys. Res. Commun. **157**, 1288–1295

144 Casteleijn, E., Kuiper, J., Van Rooij, H. C. J., Kamps, J. A. A. M., Koster, J. F. and Van Berkel, T. J. C. (1988) J. Biol. Chem. **263**, 2699–2703

145 Casteleijn, E., Kuiper, J., Van Rooij, H. C. J., Kamps, J. A. A. M., Koster, J. F. and Van Berkel, T. J. C. (1988) J. Biol. Chem. **263**, 6953–6955

146 Fisher, R. A., Kumar, R., Hanahan, D. J. and Olson, M. S. (1986) J. Biol. Chem. **261**, 8817–8823

147 Steinhelper, M. E., Fisher, R. A., Revtyak, G. E., Hanahan, D. J. and Olson, M. S. (1989) J. Biol. Chem. **264**, 10976–10981

148 Lorant, D. E., Patel, K. D., McIntyre, T. M., McEver, R. P., Prescott, S. M. and Zimmerman, G. A. (1991) J. Cell Biol. **115**, 223–234

149 Zhou, W., Chao, W., Levine, B. A. and Olson, M. S. (1992) Am. J. Physiol. **263**, G587–G592

150 Zhou, W., Chao, W., Levine, B. A. and Olson, M. S. (1990) Am. J. Pathol. **137**, 1501–1508

151 Zhou, W., Levine, B. A. and Olson, M. S. (1993) Am. J. Pathol. **142**, 1–9

152 Prescott, S. M., Zimmerman, G. A. and McIntyre, T. M. (1984) Proc. Natl. Acad. Sci. U.S.A. **81**, 3534–3538

153 Hirafuji, M., Mencia-Huerta, J. M. and Benveniste, J. (1987) Biochim. Biophys. Acta **930**, 359–369

154 Whatley, R. E., Nelson, P., Zimmerman, G. A., Stevens, D., Parker, C. J., McIntyre, T. M. and Prescott, S. M. (1989) J. Biol. Chem. **264**, 6325–6333

155 Camussi, G., Bussolino, F., Salvidio, G. and Baglioni, C. (1987) J. Exp. Med. **166**, 1390–1404

156 Bussolino, F., Camussi, G. and Baglioni, C. (1988) J. Biol. Chem. **263**, 11856–11861

157 McIntyre, T. M., Zimmerman, G. A. and Prescott, S. M. (1986) Proc. Natl. Acad. Sci. U.S.A. **83**, 2204–2208

158 Bussolino, F., Breviario, F., Tetta, C., Aglietta, M., Mantovani, A. and Dejana, E. (1986) J. Clin. Invest. **77**, 2027–2033

159 McIntyre, T. M., Zimmerman, G. A. and Prescott, S. M. (1985) J. Clin. Invest. **76**, 271–280

160 Camussi, G., Aglietta, M., Malavasi, F., Tetta, C., Piacibello, W., Sanavio, F. and Bussolino, F. (1983) J. Immunol. **131**, 2397–2403

161 Alonso, F., Gil, M. G., Sanchez-Crespo, M. and Mato, J. M. (1982) J. Biol. Chem. **257**, 3376–3378

162 McIntyre, T. M., Reinhold, S. L., Prescott, S. M. and Zimmerman, G. A. (1987) J. Biol. Chem. **262**, 15370–15376

163 Billah, M. M., Bryant, R. W. and Siegel, M. I. (1985) J. Biol. Chem. **260**, 6899–6906

164 Jouvin-Marche, E., Nino, E., Beaurain, G., Tence, M., Niaudet, P. and Benveniste, J. (1984) J. Immunol. **133**, 892–898

165 Lee, T.-c., Malone, B., Wasserman, S. I., Fitzgerald, V. and Snyder, F. (1982) Biochem. Biophys. Res. Commun. **105**, 1303–1308

166 Ludwig, J. C., Hoppens, C. L., McManus, L. M., Mott, G. E. and Pinckard, R. N. (1985) Arch. Biochem. Biophys. **234**, 337–347

167 Nieto, M. L., Velasco, S. and Sanchez-Crespo, M. (1988) J. Biol. Chem. **263**, 4607–4611

168 Ninio, E., Mencia-Huerta, J. M. and Benveniste, J. (1983) Biochim. Biophys. Acta **751**, 298–304

169 Swendsen, C. L., Ellis, J. M., Chilton, F. H., III, O'Flaherty, J. T. and Wykle, R. L. (1983) Biochem. Biophys. Res. Commun. **113**, 72–79

170 Mueller, H. W., O'Flaherty, J. T. and Wykle, R. L. (1983) J. Biol. Chem. **258**, 6213–6218

171 Coeffer, E., Nino, E., Le Couedic, J. P. and Chignard, M. (1986) Br. J. Haematol. **62**, 641–651

172 Albert, D. H. and Snyder, F. (1984) Biochim. Biophys. Acta **796**, 92–101

173 Albert, D. H. and Snyder, F. (1983) J. Biol. Chem. **258**, 97–102

174 Wey, H. E. (1989) J. Cell. Biol. **39**, 305–313

175 Dulioust, A., Vivier, E., Meslier, N., Roubin, R., Haye-Legrand, I. and Benveniste, J. (1989) Biochem. J. **263**, 165–171

176 Mencia-Huerta, J.-M., Roubin, R., Morgat, J.-L. and Benveniste, J. (1982) J. Immunol. **129**, 804–808

177 Roubin, R., Dulioust, A., Haye-Legrand, I., Nino, E. and Benveniste, J. (1986) J. Immunol. **136**, 1796–1802

178 Roubin, R., Mencia-Huerta, J.-M., Landes, A. and Benveniste, J. (1982) J. Immunol. **129**, 809–813

179 Suga, K., Kawasaki, T., Blank, M. L. and Snyder, F. (1990) J. Biol. Chem. **265**, 12363–12371

180 Domenech, C., Domenech, E. M.-D. and Soling, H.-D. (1987) J. Biol. Chem. **262**, 5671–5676

181 Pirotzky, E., Ninio, E., Bidault, J., Pfister, A. and Benveniste, J. (1984) Lab. Invest. **51**, 567–572

182 Hwang, S.-B., Lam, M.-H. and Shen, T. Y. (1985) Biochem. Biophys. Res. Commun. **128**, 972–979

183 Hwang, S.-B. (1987) Arch. Biochem. Biophys. **257**, 339–344

184 Mendlovic, F., Corvera, S. and Garcia-Sainz, J. A. (1984) Biochem. Biophys. Res. Commun. **123**, 507–514

185 Gasic, A. C., McGuire, G., Krater, S., Farhood, A. I., Goldstein, M. A., Smith, C. W., Entman, M. L. and Taylor, A. A. (1991) Circulation **84**, 2154–2166

186 Zimmerman, G. A., Prescott, S. M. and McIntyre, T. M. (1992) Immunol. Today **13**, 93–100

Biochem. J. (1993) **296**, 1–14 (Printed in Great Britain)

REVIEW ARTICLE
Calcium: its modulation in liver by cross-talk between the actions of glucagon and calcium-mobilizing agonists

Fyfe L. BYGRAVE*‡ and Angelo BENEDETTI†
*Division of Biochemistry and Molecular Biology, Faculty of Science, Australian National University, Canberra, ACT 2601, Australia, and †Istituto di Patologia Generale, Universita di Siena, 53100 Siena, Italy

INTRODUCTION

Few would dispute the necessity to study cellular Ca^{2+} movements – indeed, in higher organisms, changes in cytosolic free $[Ca^{2+}]$ are a vital physiological event from the very beginning of life at the moment of fertilization, to its very end at the final beat of the heart. These changes are brought about either as a result of mobilizing intracellular Ca^{2+} or as a result of inflow of the ion from the extracellular medium. The means by which cells control Ca^{2+} fluxes, and particularly Ca^{2+} inflow, is increasingly becoming a focus of attention. Changes in intracellular $[Ca^{2+}]$, especially that following Ca^{2+} inflow across the cell membrane induced by a range of different stimuli, are the trigger not only in the short-term but also in the long-term of many vital physiological responses in cells in a range of tissues and species. Moreover, aberrant control of Ca^{2+} fluxes appears to be associated with many pathological situations.

During the past decade in particular, a large number of articles have been published about many aspects of cellular Ca^{2+} fluxes: about the discovery and generation of the new second messenger $Ins(1,4,5)P_3$ and of the related enzymic reactions that generate and dispose of it; about the range of agonists able to induce changes in cellular Ca^{2+} fluxes; about the cellular compartments that appear to contain agonist-sensitive Ca^{2+} pools; about the continual development of new techniques to measure and locate the ion within cells; and about how 'cross-talk' between second messenger systems, such as those generating cyclic AMP and $Ins(1,4,5)P_3$ for instance, are able to induce/modulate Ca^{2+} fluxes. Numerous offerings and ideas have in turn evolved from these studies as to the nature of the mechanisms involved in the control of cell Ca^{2+}. The aim of this article is to contribute further to this knowledge by integrating a number of relevant observations and reports obtained over the years by numerous investigators and to synthesize a general scheme that we hope will be useful in the current debates about the control of cellular Ca^{2+} fluxes. We focus on liver tissue, drawing upon the large body of information available about such fluxes derived from studies from both single cells (hepatocytes) and the intact organ (perfused liver). We also focus particularly on the issue of how glucagon (and cyclic AMP) might regulate/modulate Ca^{2+} fluxes induced by calcium-mobilizing agonists. This is an aspect of the control of Ca^{2+} fluxes in liver that is not only important in many aspects of liver function, but also one that we believe holds clues as to the mechanism of hormone-induced Ca^{2+} inflow and outflow in this tissue.

BRIEF HISTORY OF RELEVANT EVENTS

Ringer (1883) and Heilbrunn (1952) were among the first to recognize the ubiquitous role of Ca^{2+} in physiological events. Features of their work were later extended by others who envisaged the ion as having a regulatory role in cell metabolism (Lehninger, 1950; Bygrave, 1966, 1967; Gevers and Krebs, 1966; Rasmussen, 1970). During the 1970s, the numerous studies on cellular Ca^{2+} fluxes were confined largely to movements across the membranes of isolated organelles, in particular those of the mitochondria and endoplasmic reticulum (e.g. Lehninger et al., 1967; Bygrave, 1978a; Williamson et al., 1981; Åkerman and Nicholls, 1983). Initially, little attention was paid to the possible role of hormones in the control of these intracellular events, but towards the end of that decade a realization emerged that certain hormones and also cyclic AMP might play a role in the control of intracellular Ca^{2+} (Bygrave, 1978a; Rasmussen, 1981; Borle, 1981; Exton, 1981). Others were advocating a possible controlling role of cyclic AMP in Ca^{2+} fluxes across the cell (plasma) membrane (Friedmann and Park, 1968; Borle, 1972) and that Ca^{2+} and cyclic AMP could have an interactive role in the control of cell metabolism (Rasmussen and Barrett, 1984).

A further significant landmark occurred towards the end of the 1970s. This was the realization that cyclic AMP generation and action (Robison et al., 1971) on their own could not account for the mechanism by which α-adrenergic agonists, vasopressin and angiotensin induce metabolic events such as glycogenolysis; the evidence was mounting that Ca^{2+} was involved in some (at that time unknown) way (Van de Werve et al., 1977; Assima-copoulos-Jeanett et al., 1977; Keppens et al., 1977; Keppens and de Wulfe, 1977; reviewed by Exton, 1981, 1985, 1988). In studies quite separate, but which were later to be very relevant to this subject, knowledge was being obtained about the role of Ca^{2+} in the hydrolysis of phosphatidylinositol (Michell, 1975). The role of $Ins(1,4,5)P_3$ as second messenger in the action of Ca^{2+}-mobilizing agonists was reviewed by Berridge (1987), and the fascinating story of the concepts leading up to it is outlined by Michell (1992).

During the 1980s a number of reports appeared in which the general question was raised as to whether glucagon (or cyclic AMP) was itself able to induce cellular Ca^{2+} fluxes in isolated organelles, in intact cells or in the perfused rat liver. In many instances the responses induced by glucagon were compared with those induced by Ca^{2+}-mobilizing hormones which, it was realized early on, are far more potent than glucagon in their ability to mobilize Ca^{2+}. However, what a number of those reports revealed, and others continue to reveal, is that glucagon is able to exert effects in liver such that the Ca^{2+} flux-inducing action of Ca^{2+}-mobilizing agonists is greatly modified. It is this aspect, the means by which glucagon is able to modify the action of Ca^{2+}-mobilizing agonists, that this review primarily addresses.

First, we briefly review the patterns of Ca^{2+} inflow and outflow induced by Ca^{2+}-mobilizing agonists that occur in the perfused rat liver and in hepatocytes, and where possible we integrate information gained from each of the two experimental systems.

Abbreviation used: EGF, epidermal growth factor.
‡ To whom correspondence should be addressed.

We show how these patterns are altered by prior administration or co-administration of glucagon with these agonists. Bearing in mind current views about how the inflow of Ca^{2+} might take place, we attempt to identify cellular sites at which glucagon might be acting so as to modify cellular Ca^{2+} fluxes induced in turn by the action of Ca^{2+}-mobilizing agonists. Finally, we attempt to draw this information together in a form that will permit further experimentation both in respect of the basic mechanisms involved and in respect of some of the physiological events affected by the combined action of glucagon and Ca^{2+}-mobilizing agonists.

FEATURES OF Ca^{2+} FLUXES INDUCED BY Ca^{2+}-MOBILIZING AGONISTS

Prior to 1980, the majority of studies on Ca^{2+} flux changes in hepatocytes and intact perfused liver employed ^{45}Ca. This radionuclide served as a tracer for Ca^{2+} movements and distribution and provided information about the rates of movement of the ion across cell and intracellular membranes, about the nature of the intracellular Ca^{2+} pools and, more particularly, about how hormones effect these events (see Berry et al., 1991). From the earliest studies it was soon apparent that the hormones mobilizing Ca^{2+} induced multiple effects. For example it was shown that α-adrenergic agonists could stimulate the rate of Ca^{2+} inflow (e.g. Assimacoupolous-Jeanett et al., 1977) as well as Ca^{2+} outflow (e.g. Blackmore et al., 1978; Poggioli et al., 1980) in hepatocytes. Around this time an extensive analysis of hormone-induced Ca^{2+} mobilization in the perfused rat liver, also employing ^{45}Ca, appeared (Claret-Berthon et al., 1977). Later, Barritt et al. (1981) showed that under steady-state conditions adrenaline is able to stimulate the rate of inflow, the exchange and the loss of ^{45}Ca from hepatocytes. An important technical advance was the synthesis and development of fluorescent indicators that could be loaded into hepatocytes and used to report on the changes in intracellular $[Ca^{2+}]$ following administration to cells of an appropriate agonist (see, e.g., Tsien et al., 1982). Although other intracellular probes are available (see, e.g., Borle and Snowdowne, 1986), the fluorescent indicators quin2 and more recently fura2 have been used by most researchers to measure intracellular free Ca^{2+} in hepatocytes, largely on account of their ease of handling and non-invasive properties. The application of these indicators has provided an important step forward in enabling a further understanding of the effects of α-agonists and glucagon on Ca^{2+} fluxes in cells.

As implied above, the two principal experimental systems that have contributed to our current understanding of cellular Ca^{2+} fluxes in liver are the perfused rat liver and single cells (hepatocytes) prepared therefrom. While each system inherently has particular advantages and disadvantages (see Exton, 1975; Berry et al., 1991), information gained from both, described below, is necessary in order to fully appreciate how such Ca^{2+} fluxes operate under the influence of the various agonists in question.

STUDIES ON Ca^{2+} FLUXES IN THE PERFUSED RAT LIVER INDUCED BY Ca^{2+}-MOBILIZING AGONISTS

Early studies on Ca^{2+} movements in the perfused rat liver relied on measurements of perfusate Ca^{2+} by using ^{45}Ca, atomic absorption spectroscopy or a Ca^{2+}-selective electrode. The latter two sets of measurements were often restricted to experimental situations where the perfusate Ca^{2+} needed to be maintained at a relatively low concentration so as to increase the sensitivity of measurement (see, e.g., Althaus-Salzmann et al., 1980; Sies et al.,

1981; Morgan et al., 1983). As is now known and shown below, such a situation can have a large bearing on the nature of the responses observed. Many of these problems were overcome by the application of a sensitive Ca^{2+}-selective electrode to the outflow medium (detailed in Reinhart et al., 1982; see also Sies et al., 1983) that, together with appropriate backoff instrumentation, enabled the detection of a $+1\ \mu M\ Ca^{2+}$ change in extracellular $[Ca^{2+}]$ when the extracellular $[Ca^{2+}]$ was varied from approx. zero to 1.3 mM. This has provided a sensitive system through which much information has been gained about a number of aspects of the hormonal control of cellular Ca^{2+} fluxes in liver from studies of the perfused rat liver system alone (reviewed in Reinhart et al., 1984d). Other relevant features of this system are that the extracellular $[Ca^{2+}]$ is able to be rapidly altered and (in the flowthrough, non-recirculating set-up) any agent or agonist can be rapidly administered and, most importantly, rapidly removed. It should be appreciated that in the description which follows, it is the information gained at higher extracellular $[Ca^{2+}]$ that is physiologically relevant. Work with lower extracellular $[Ca^{2+}]$, on the other hand, can be useful experimentally.

In the perfused liver, the α_1-adrenergic agonist phenylephrine induces a sizable efflux of Ca^{2+} within 3–4 s, the amount extruded depending on the concentration and the time for which the hormone is administered. Details about the properties of the hormone-sensitive Ca^{2+} pools in rat liver are reviewed in Reinhart et al. (1984d). The amount of Ca^{2+} efflux caused by phenylephrine, vasopressin and angiotensin II is less at higher extracellular $[Ca^{2+}]$. Conversely, the amount of Ca^{2+} taken up increases as the extracellular $[Ca^{2+}]$ is increased. Influx of Ca^{2+} is induced only after removal of the agonist. This latter feature, which clearly distinguishes the actions of phenylephrine from those of other Ca^{2+}-mobilizing agonists (Altin and Bygrave, 1985), cannot be seen in experiments with hepatocytes unless a perifusion system is used.

Ca^{2+} influx promoted by vasopressin exhibits a pattern similar to that with phenylephrine only when the extracellular $[Ca^{2+}]$ is relatively low, i.e. approx. 500 μM or less (see also Benedetti et al., 1989). At higher concentrations of Ca^{2+}, vasopressin induces a small efflux of the ion but, and while still being infused, spontaneously induces a large influx followed by a second small efflux. A second pulse of vasopressin administered later, even when the previously depleted intracellular pools have refilled (see below), induces only Ca^{2+} efflux. When the length of the time of administration of the (first) vasopressin pulse is varied between 15 and 45 s, the extent of efflux is the same but the extent of influx is more pronounced as the period of administration is increased. Responses to angiotensin are essentially similar to those induced by vasopressin. A number of these general features of agonist-induced Ca^{2+} fluxes in the perfused rat liver are illustrated in Figure 1. The oscillatory phenomena observed following the administration of a Ca^{2+}-mobilizing agonist to the perfused rat liver seen at low $[Ca^{2+}]$ in the Figure have been studied in some detail by Graf et al. (1987).

One serious limitation to the study of Ca^{2+} fluxes in the perfused rat liver has been the inability to date to directly monitor changes in intracellular $[Ca^{2+}]$. Some success in perfused heart has been achieved, however, using intracellular indicators (Levine et al., 1990). On the other hand, qualitative information about the extent to which these changes occur can be gauged from the appearance of glucose (from glycogen breakdown) in the medium, particularly following alterations to the extracellular $[Ca^{2+}]$ (see Reinhart et al., 1984b) and by assessing the degree to which oxygen uptake is stimulated. This point is amplified in the data illustrated in Figure 2. Here, changes in perfusate Ca^{2+},

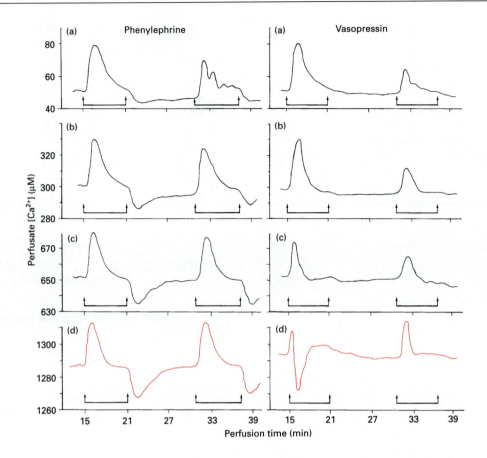

Figure 1   Extracellular [Ca²⁺] influences the Ca²⁺ flux responses in perfused rat liver induced by phenylephrine or vasopressin

The data, obtained with a Ca²⁺-selective electrode and perfused rat liver system, show that Ca²⁺ inflow (downwards deflection) is minimal for both agonists at low extracellular [Ca²⁺]. They also show that at extracellular [Ca²⁺] > 500 μM, repeated pulses of phenylephrine (2 μM) are able to induce repeated outflow/inflow cycles; by contrast, vasopressin (10 nM) will induce only one 'typical' cycle and thereafter only Ca²⁺ outflow. The curves shown in red illustrate the cycles that occur at a physiological Ca²⁺ concentration (1.3 mM). For further general details see the text and Reinhart et al. (1982), Altin and Bygrave (1986) and Benedetti et al. (1989). See Graf et al. (1987) for further information concerning the oscillatory behaviour observed at low perfusate [Ca²⁺].

oxygen uptake and glucose output following administration of phenylephrine were assessed in situations where the extracellular [Ca²⁺] was either maintained at 1.3 mM or reduced to nominally zero in the presence of EGTA. The data reveal two basic types of response; one transient and the other sustained.

Transient responses are characterized as being rapid in onset, occurring within seconds, and once the maximum response is attained they decay quite quickly; in addition, in the case of perfusate Ca²⁺ measurements (Figure 2a), no influx is observed when the agonist phenylephrine is withdrawn (see above). The responses occur independently of extracellular [Ca²⁺] and hence are dependent on the mobilization of intracellular Ca²⁺. While not shown here (but see Reinhart et al., 1984b), the response can be rapidly re-activated by providing an excess of Ca²⁺ to the perfusate medium.

Sustained responses are equally rapid in onset but once the maximum is attained, little or no decay is observed. Clearly these responses are dependent on Ca²⁺ inflow from the extracellular medium. As will be shown later in this review, much of this information is able to be obtained also with hepatocytes. Stimulation of oxygen uptake by Ca²⁺-mobilizing agonists (see, e.g., Figure 2c) most probably reflects movement of Ca²⁺ into the mitochondria which then stimulates enzymes of the citrate cycle (see, e.g., Denton and McCormack, 1985; Assimacoupolos-Jeanett et al., 1986). These aspects will not be discussed further here. As a final point at this juncture, it needs to be appreciated

that changes in glucose output and oxygen consumption are a good index of changes in intracellular [Ca²⁺] for studies when only one of the glycogenolytic inducing agonists is used. Their use in examining synergism (see below), like that of phosphorylase activity, is somewhat limited.

EFFECTS OF GLUCAGON ON Ca²⁺ FLUXES INDUCED BY Ca²⁺-MOBILIZING AGONISTS IN THE PERFUSED RAT LIVER

Administration of glucagon alone to the perfused rat liver induces a reproducibly significant efflux of Ca²⁺. This was shown in early studies by Friedmann and Park (1968), Friedmann (1972), Friedmann and Dambach (1973) and Kraus-Friedmann (1986), and has since been confirmed in other laboratories (e.g. Morgan et al., 1985; Altin and Bygrave, 1986; Rashed and Patel, 1987; Benedetti et al., 1989). Where examined, the effects of glucagon could be reproduced when dibutyryl cyclic AMP was the agonist. The amount of Ca²⁺ released, however, and the rate of release are much less than that mobilized by phenylephrine, vasopressin or angiotensin; these glucagon-induced responses appear to be increased substantially when the extracellular [Ca²⁺] is lowered (Benedetti et al., 1989). On the other hand, when pre-administered, glucagon greatly alters the characteristics of Ca²⁺ mobilization induced by all three of the above-mentioned Ca²⁺-mobilizing agonists (Morgan et al., 1985; Altin and Bygrave, 1986) as well as by ADP, ATP and epidermal growth factor

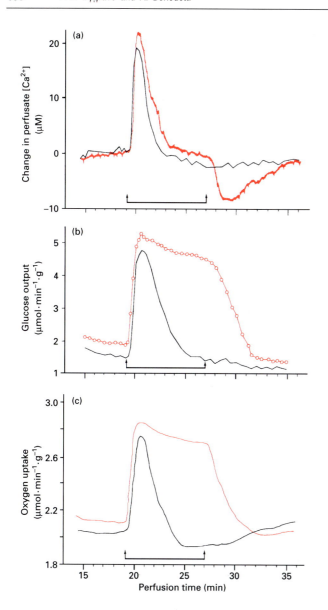

Figure 2   Extracellular [Ca²⁺] sustains agonist-stimulated, Ca²⁺-sensitive, intracellular responses in the perfused rat liver

Phenylephrine (2 μM) was administered for the time indicated by the bar in the presence of 1.3 mM Ca²⁺ (red line) or nominally zero Ca²⁺ by the addition of EGTA. It is seen (1) that Ca²⁺ outflow (upwards deflection), in terms of both magnitude and rapidity of response, is unaffected by variations in extracellular [Ca²⁺]; (2) that phenylephrine-induced inflow will not occur following removal of the agonist when extracellular [Ca²⁺] is approx. zero; and (3) that both glucose output and oxygen uptake induced by the action of phenylephrine require high extracellular [Ca²⁺] and therefore Ca²⁺ inflow in order to exhibit a sustained response to the hormone. For further details see the text and Reinhart et al. (1984b,d). Similar information can be gleaned with hepatocytes, for example from the studies of Charest et al. (1985a) and Combettes et al. (1986).

(EGF) (Altin and Bygrave, 1987a) and by phosphatidic acid and arachidonic acid (Altin and Bygrave, 1987b). Preadministration of glucagon to the perfused rat liver was shown to also suppress the oscillations induced by vasopressin alluded to above (Graf et al., 1987). Earlier Jenkinson and Koller (1977) had shown that the combined addition of α₁- and β-agonists increased synergistically the efflux of ⁴⁵Ca from liver slices of the guinea pig. Similar findings were made by Cocks et al. (1984).

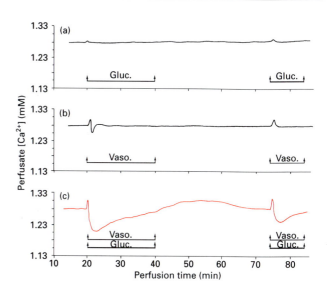

Figure 3   Glucagon modulates vasopressin-induced Ca²⁺ fluxes in the perfused rat liver

The data show that, in the presence of 1.3 mM extracellular Ca²⁺, whereas glucagon (Gluc.; 10 nM) alone (**a**) induces a slight outflow of the ion, its co-administration with vasopressin (Vaso.; 10 nM) (**c**) results in little effect on Ca²⁺ outflow induced by vasopressin (cf. **b**) but significant (synergistic) increases in Ca²⁺ inflow. A second co-administration of the two hormones also induces Ca²⁺ inflow. Note that the scale of perfusate [Ca²⁺] is different to that shown in Figure 1. For further details see the text and Altin and Bygrave (1986)

A major effect of pre-administered glucagon is to significantly attenuate the initial Ca²⁺ efflux induced by these agonists. Its other major effect is to greatly accelerate Ca²⁺ influx. The magnitude of this synergistic effect on Ca²⁺ influx is dependent on the period of preincubation; maximal effects in the perfused rat liver system are seen after approx. 4 min of preincubation. An important observation relating to the issue of the possible mechanisms involved, however, is that while co-administration of glucagon with the Ca²⁺-mobilizing agonists does not influence the extent of efflux induced by the agonists alone, glucagon still enhances Ca²⁺ inflow to a considerable extent (Altin and Bygrave, 1986). This indicates that the effects of glucagon in the perfused rat liver are virtually immediate and that its earliest effects are on Ca²⁺ inflow and later, either directly or indirectly as a consequence of some presently unknown action, on Ca²⁺ outflow. As mentioned above, the effects of glucagon on agonist-induced inflow are not observed when the extracellular [Ca²⁺] is relatively low (Benedetti et al., 1989), similar to the action of vaspressin alone on Ca²⁺ inflow also mentioned above. Glucagon will, however, promote Ca²⁺ inflow induced by a second pulse of vasopressin if the extracellular [Ca²⁺] is above approx. 500 μM. These experiments thus reveal that, despite the hormone regime under study, the Ca²⁺ gradient across the plasma membrane is a significant factor determining whether the synergistic effects of glucagon with Ca²⁺-mobilizing agonists on Ca²⁺ inflow will (or will not) be manifest. In the perfused liver the acceleration of Ca²⁺ influx seen with vasopressin and angiotensin in these conditions is much greater than that seen with phenylephrine. A number of the features described above of the effect of glucagon on vasopressin-induced Ca²⁺ fluxes in the perfused rat liver are illustrated in Figure 3.

The information in Figure 3 reveals another important aspect about the massive Ca²⁺ influx induced by the synergistic effects of the two hormones. That is, despite the very large amount of

the ion taken up by the liver, there is no sign of damage having occurred, at least over the time course of the experiment. This is reflected in the fact that a second administration of the hormones at 74 min into the perfusion will produce a second response only slightly less than the first.

In summarizing to this point, several features are evident. The first is that glucagon alone is able to induce a small but significant outflow of Ca^{2+} from the perfused rat liver. Secondly, the onset of this outflow follows a short lag period, and the magnitude of it is less than that resulting from the action of Ca^{2+}-mobilizing agonists. Thirdly, glucagon-induced Ca^{2+} outflow appears to increase as the extracellular $[Ca^{2+}]$ is decreased. Finally, the major effect of glucagon in the perfused rat liver is (a) when co-administered with Ca^{2+}-mobilizing agonists, to immediately promote Ca^{2+} inflow without altering Ca^{2+} outflow, (b) when administered before Ca^{2+}-mobilizing agonists, to attenuate the ability of α-agonists to promote Ca^{2+} outflow, and (c) to greatly augment the ability of these agonists to promote Ca^{2+} inflow.

STUDIES ON Ca^{2+} FLUXES IN HEPATOCYTES INDUCED BY Ca^{2+}-MOBILIZING AGONISTS

As pointed out by Crofts and Barritt (1989) and Berry et al. (1991), accurate measurement of rates of hormone-stimulated Ca^{2+} inflow in hepatocytes has proved difficult. The main methods employed to measure rates of Ca^{2+} inflow are the use of ^{45}Ca under steady-state conditions (see Barritt et al., 1981), (indirect) measurement of increases in intracellular Ca^{2+} using fluorescent indicators following the addition of Ca^{2+} to a medium containing negligible Ca^{2+}, measurement of the rates of quenching of fluorescence of intracellular quin2 following Mn^{2+} addition (see, e.g., Crofts and Barritt, 1990), and measurement of the initial rate of activation of glycogen phosphorylase following the addition of Ca^{2+} to a medium containing negligible Ca^{2+}. A further complication to studies of Ca^{2+} inflow is that there is mounting evidence that more than one system facilitates Ca^{2+} entry in the plasma membrane of liver cells (Barritt et al., 1981; Hughes et al., 1986; Altin and Bygrave, 1987a; Barritt and Hughes, 1991; Llopis et al., 1992). The potentially useful technique of patch-clamping appears to be technically difficult in liver plasma membranes (see, e.g., Sawanobori et al., 1989; Barritt and Hughes, 1991).

These potential complications may in large part account for some of the apparent discrepancies between groups on the characteristics of Ca^{2+} inflow in hepatocytes. For instance, Crofts and Barritt (1989) found that vasopressin-induced Ca^{2+} inflow measured in quin2-loaded cells exhibits a biphasic curve as the extracellular $[Ca^{2+}]$ is increased; the second phase showed no signs of saturation even at 5 mM Ca^{2+}. By contrast, the data of Mauger et al. (1984) and Joseph et al. (1985) showed Ca^{2+} inflow induced by noradrenaline, vasopressin and angiotensin to be saturated at 1–2 mM external Ca^{2+}. Mauger et al. (1985) also showed that glucagon alone can stimulate the initial rate of Ca^{2+} inflow.

Both the basal and hormone-stimulated Ca^{2+} inflow systems exhibit a broad specificity for bivalent metal ions (Crofts and Barritt, 1990), with Mn^{2+}, Co^{2+}, Ni^{2+} and Zn^{2+} all able to quench the fluorescence of Ca-quin2 loaded into hepatocytes.

As intimated above, a serious drawback to studies with the perfused rat liver has been the inability to gain quantitative information about changes in intracellular $[Ca^{2+}]$ following agonist presentation. Now, using hepatocytes loaded with appropriate fluorescent indicators that monitor changes in intracellular $[Ca^{2+}]$ (Tsien et al., 1982), it is possible to gauge with

a reasonable degree of accuracy the magnitude of and the rapidity with which intracellular $[Ca^{2+}]$ is altered under a variety of experimental conditions, for example. Such measurements in turn provide information about the changes occurring to Ca^{2+} inflow (Berry et al., 1991). Because of the lack of appropriate inhibitory agents, however, it is not always possible at the present time to strictly distinguish between contributions to raising the cytoplasmic $[Ca^{2+}]$ from Ca^{2+} inflow on the one hand, and intracellular Ca^{2+} mobilization on the other.

The following observations most pertinent to this review have been made in studies of this nature. (1) The Ca^{2+}-mobilizing agonists vasopressin, angiotensin, phenylephrine, adrenaline and noradrenaline all rapidly increase intracellular $[Ca^{2+}]$ (Charest et al., 1983, 1985a; Sistare et al., 1985; Combettes et al., 1986). The onset of the increase occurs at approx. 2–3 s, with maximal changes attained by approx. 10 s. The order of potency is adrenaline, phenylephrine, vasopressin. A similar order of potency was observed from measurements of ^{45}Ca inflow (Mauger et al., 1984) and this in turn is similar to what is observed with activation of phosphorylase a (Charest et al., 1985a). (2) Glucagon alone is also able to induce an increase in intracellular $[Ca^{2+}]$. However, the time of onset lags by some 10 s and the rapidity and magnitude of the change are not quite as great (Charest et al., 1983; see also Mauger and Claret, 1986). (3) Reducing the extracellular $[Ca^{2+}]$ does not affect the time of onset nor the magnitude of the change in intracellular $[Ca^{2+}]$, but does reduce the time for which the response is sustained (Charest et al., 1985a). This group showed that varying the extracellular $[Ca^{2+}]$ between 30 and 500 μM had little effect on either the initial rate or the maximal extent of the intracellular Ca^{2+} increase. However, whereas 500 μM extracellular $[Ca^{2+}]$ prolonged the rise in intracellular $[Ca^{2+}]$, 30 μM extracellular $[Ca^{2+}]$ allowed intracellular $[Ca^{2+}]$ to return to basal levels by 4 min following vasopressin addition to the cells. These experiments again illustrate the importance of extracellular Ca^{2+} in maintaining the sustained responses alluded to in an earlier section of this review (see text and Figure 2). (4) Parallel with rapid increases in intracellular $[Ca^{2+}]$ and other metabolic responses following vasopressin addition to hepatocytes, Charest et al. (1983, 1985b) demonstrated that $Ins(1,4,5)P_3$ plus $Ins(1,3,4)P_3$ production caused by adrenaline, vasopressin and angiotensin occurred very rapidly (within seconds), attaining a maximum by 5 s and remaining maximal for at least 10 min. Of interest is that by comparison with their abilities to induce Ca^{2+} inflow, vasopressin and angiotensin were better able to induce $Ins(1,4,5)P_3$ plus $Ins(1,3,4)P_3$ formation. Seemingly at odds with these findings, Hansen et al. (1991) have shown that phenylephrine-induced accumulation of $Ins(1,4,5)P_3$ in rat hepatocytes, while attaining a maximum also by 5 s, rapidly decreased thereafter; by 30 s a new steady-state level was reached that was 30% of the maximal level. The apparent discrepancy between these two groups, however, is possibly explained by the fact that the former group were examining both isoforms of myo-inositol tris-phosphate, i.e. $Ins(1,4,5)P_3$ and $Ins(1,3,4)P_3$, whereas the latter group examined the turnover of $Ins(1,4,5)P_3$. Studies by Tennes et al. (1987) have shown that guinea pig hepatocytes stimulated with angiotensin produce $Ins(1,4,5)P_3$ maximally by 10 s and that this declines rapidly during the next 10 min. By contrast, $Ins(1,3,4)P_3$ gradually rises over 10 min.

Mauger et al. (1984) examined the initial rate of Ca^{2+} inflow into hepatocytes and found that Ca^{2+} inflow was activated for as long as the hormones occupied their receptors. Increasing the extracellular $[Ca^{2+}]$ led, in contrast with the report of Barritt et al. (1981), to a saturation of Ca^{2+} inflow; but see Barritt and Hughes (1991) who elaborate on this point.

Studies on single hepatocytes (see, e.g., Woods et al., 1986, 1987) have shown that Ca^{2+}-mobilizing hormones induce oscillations in intracellular Ca^{2+}, the frequency of which depends on the agonist concentration. It is of interest that in this single-cell system the responses to vasopressin are different from those to phenylephrine (Sanchez-Bueno and Cobbold, 1993), as in the intact perfused rat liver (Altin and Bygrave, 1985; see Figure 1).

EFFECTS OF GLUCAGON OR OF DIBUTYRYL CYCLIC AMP ON Ca^{2+} FLUXES IN HEPATOCYTES INDUCED BY Ca^{2+}-MOBILIZING AGONISTS

As intimated above, glucagon alone appears able to induce an increase in intracellular $[Ca^{2+}]$ as well as having a number of other Ca^{2+}-flux-related effects in hepatocytes. It is able to increase ^{45}Ca movement into hepatocytes (Keppens et al., 1977), but Studer et al. (1984) report that glucagon at physiological concentrations has no effect on intracellular $[Ca^{2+}]$. The interesting observation of Assimacoupoulus-Jeanett et al. (1982) and Morgan et al. (1983) that increases in dibutyryl cyclic AMP are able to reverse vasopressin-induced Ca^{2+} mobilization (see also Crane et al., 1982) appears to be consistent with the finding (Keppens and De Wulf, 1984) that vasopressin inhibits the maintenance of cyclic AMP levels in hepatocytes by activating phosphodiesterase. Charest et al. (1985b) have shown that glucagon is able to potentiate Ca^{2+} inflow induced by extracellular ATP.

In kinetic studies of the influence of glucagon on ^{45}Ca inflow induced by Ca^{2+}-mobilizing agonists, Mauger et al. (1985) and Poggioli et al. (1986) reported that co-addition of glucagon with vasopressin greatly stimulated Ca^{2+} inflow. The dose–response curves for either glucagon or vasopressin were unaffected by the other agent, i.e. the EC_{50} did not change but the maximal effects of each were enhanced. The effects of glucagon were seen also with dibutyryl cyclic AMP. These workers concluded that both hormones were activating the same gating mechanism.

Burgess et al. (1986), using guinea pig hepatocytes, showed that the β-adrenoceptor agonist isoprenaline alone induced Ca^{2+} efflux when the extracellular $[Ca^{2+}]$ was low but not when it was high. They also observed a significant potentiation by this agent of angiotensin-induced Ca^{2+} mobilization. They suggested that the interacting effects did not involve steps subsequent to the generation of signals by angiotensin. Because there appeared to be no apparent increase in the total intracellular pool size induced by isoprenaline, it was concluded that the agonist may have increased the sensitivity of the endoplasmic reticular pool to $Ins(1,4,5)P_3$.

Studies by Kass et al. (1990) on Mn^{2+} inflow showed that glucagon alone had no effect on Mn^{2+} inflow but that its prior administration to hepatocytes enhanced vasopressin-induced Mn^{2+} inflow and abolished the latency period observed between vasopressin addition and the onset of Mn^{2+} uptake. The effects of glucagon could be mimicked by dibutyryl cyclic AMP. Crofts and Barritt (1990), however, provided evidence that the effects of co-addition of glucagon and vasopressin on Mn^{2+}-induced quenching were the same as the sum of their individual effects.

The oscillations in intracellular $[Ca^{2+}]$ induced by vasopressin or phenylephrine mentioned earlier are also able to be modulated by elevated levels of intracellular cyclic AMP induced by glucagon, dibutyryl cyclic AMP or forskolin (Somogyi et al., 1992; Sanchez-Bueno et al., 1993; see also Capiod et al., 1991). While the elevated cyclic AMP does not itself induce the oscillations, it enhances both the peak free Ca^{2+} and the frequency of 'spikes' induced by phenylephrine. Again, and consistent once

more with differences seen between the effects of elevated cyclic AMP levels on Ca^{2+} fluxes in the perfused rat liver (Altin and Bygrave, 1986; Figure 3), the effects on vasopressin-induced spikes are different to those induced by phenylephrine; they do not affect peak free Ca^{2+} but induce an eventual fall in spiking frequency (Sanchez-Bueno et al., 1993). Of additional interest is the finding of Somogyi et al. (1992) that protein kinase C activation by diacylglycerol, generated together with $Ins(1,4,5)P_3$ during phosphatidylinositol breakdown, inhibits the cytoplasmic Ca^{2+} oscillations.

There is a feature of the synergism between glucagon and Ca^{2+}-mobilizing hormones that appears to have been neglected from the experimental viewpoint, yet may hold clues to the mechanism(s) involved. Some years ago, Blackmore et al. (1984) reported that when the pH of the incubation is increased to values near 8, the ability of glucagon and Ca^{2+}-mobilizing agonists to increase hepatocyte cell Ca^{2+} is greatly enhanced. Indeed, at this pH, glucagon appeared to be a more potent Ca^{2+}-mobilizing agonist than vasopressin, adrenaline or angiotensin. These findings were observed also in the perfused rat liver (Altin and Bygrave, 1987a), where direct measurements of the synergism (as described in Figure 3) showed little glucagon-induced influx at pH 7.7 but a large influx at higher pH values. Ca^{2+} influx was virtually immediate upon hormonal stimulation. Of some interest also was the finding that, at pH 8, vasopressin alone induced no Ca^{2+} efflux, but rather a much greater (than at pH 7.4) influx of the ion. It would appear that further exploration of this system should provide useful information in the future.

EFFECTS OF GLUCAGON AND Ca^{2+}-MOBILIZING AGONISTS ON THE Ca^{2+} EXTRUDING PUMP, THE Ca^{2+}-ATPase

It will be evident that the plasma membrane, like the mitochondrial and endoplasmic reticular membranes, possesses separate Ca^{2+} transporter systems that permit routes of Ca^{2+} inflow which are different from those of Ca^{2+} outflow (see Barritt, 1992). This enables a 'cycling' of the ion between the two compartments into and out of which it is moving. It has been suggested, following the principles of substrate cycling laid down earlier by Newsholme and Crabtree (1976), that this form of control may be important in the regulation of Ca^{2+} fluxes induced by hormones as discussed here (Bygrave, 1978a,b). This has obvious implications for the regulation of metabolic processes in different cell compartments that are sensitive to Ca^{2+} (Bygrave, 1967; Rasmussen, 1981). In this context it is relevant to consider whether Ca^{2+} cycling across the plasma membrane might be altered by the action of hormones, not only on Ca^{2+} inflow mechanisms but also on the Ca^{2+}-ATPase.

It has been reported that the purified Ca^{2+}-ATPase is inhibited by the action *in vitro* of glucagon (Lotersztajn et al., 1981, 1985; Lin et al., 1983) and more recently it has been suggested that the Ca^{2+}-ATPase is coupled to a G_s-like protein (Lotersztajn et al., 1992; Jouneaux et al., 1993). In examining the influence of glucagon and Ca^{2+}-mobilizing agonists in an enriched plasma membrane fraction obtained from a low-speed centrifugation of the homogenate, Prpic et al. (1982) observed a degree of inhibition (some 30% after 3 min) of Ca^{2+}-ATPase by vasopressin, adrenaline and angiotensin; glucagon was without effect. In these experiments, the agonist was first administered to the perfused rat liver for a short time; the plasma membrane was subsequently prepared and Ca^{2+}-ATPase activity determined. The effects of vasopressin, the most potent of the agonists studied in these experiments, did not appear to be significant after 1 min of administration. Epping and Bygrave (1984), using a plasma membrane-enriched liver cell fraction preparation derived from an initial centrifugation at medium to high speed, could not detect any hormone sensitivity of Ca^{2+}-ATPase or Ca^{2+} uptake when the perfused liver was first administered with phenylephrine.

By combining the Ca^{2+}-electrode and ^{45}Ca techniques in the perfused rat liver, Reinhart et al. (1984c) were able to show that net Ca^{2+} fluxes across the plasma membrane could be described in terms of the activities of a Ca^{2+} cycle comprising separate Ca^{2+} inflow and outflow components. The α_1-adrenergic agonists phenylephrine, vasopressin and angiotensin were found to stimulate both inflow and outflow in a time-dependent manner; the latter occurring before the former. By comparison, glucagon had a relatively small but significant effect in this study. As mentioned above, Altin and Bygrave (1986) observed that following co-administration of glucagon and vasopressin to perfused rat livers, Ca^{2+} outflow was unaffected but Ca^{2+} inflow was greatly enhanced. Finally, and again using the Ca^{2+}-electrode and ^{45}Ca techniques, Altin and Bygrave (1987c) were able to demonstrate in the perfused rat liver that the main effect of these hormones was to activate a Ca^{2+} inflow pathway rather than to inhibit Ca^{2+} outflow via the Ca^{2+}-ATPase. On the other hand, a recent study has concluded that dibutyryl cyclic AMP stimulates Ca^{2+} efflux in rat hepatocytes (F. L. Bygrave, A. Gamberucci, R. Fulceri, R. Giunti and A. Benedetti, unpublished work).

Thus the extent to which hormones, particularly glucagon, activate the plasma membrane Ca^{2+}-ATPase appears to remain a debatable issue. Clarification of the possible role of G proteins in Ca^{2+}-ATPase activity is needed (see Jouneaux et al., 1993). We also should not lose sight of the fact that the high sensitivity of the Ca^{2+}-ATPase to intracellular $[Ca^{2+}]$ (see Lotersztajn et al., 1981), independent of any hormone action, is likely to be a prevailing factor determining Ca^{2+} outflow in liver.

DOES GLUCAGON EMPTY THE (Ca²⁺-MOBILIZING) AGONIST-SENSITIVE INTRACELLULAR Ca²⁺ STORES?

Mechanisms by which Ca^{2+} entry might be evoked in non-excitable cells in general have been considered recently by Irvine (1992) [see also Putney (1990) and Neher (1992)]; the general issue will not be elaborated upon here. What is clear from these considerations, however, is that Ca^{2+} entry appears to be able to be controlled to some (possibly quite large) degree by the extent to which intracellular Ca^{2+} stores themselves are empty or full. Presumably the greater the extent of emptying of these stores, the greater the potential ability to exert a positive effect on plasma membrane Ca^{2+} inflow. In the present context, the question arises as to whether evidence exists linking glucagon action to the emptying (or refilling) of these stores and whether this purported action has any bearing on altering intracellular $[Ca^{2+}]$.

Studies with both hepatocytes and the perfused rat liver have provided evidence that glucagon releases Ca^{2+} from the same intracellular stores as does vasopressin. Combettes et al. (1986), studying the role of agonists in altering intracellular $[Ca^{2+}]$, observed that when the extracellular $[Ca^{2+}]$ is high, low glucagon concentrations stimulate both vasopressin-induced Ca^{2+} inflow and the final intracellular $[Ca^{2+}]$ reached. When the extracellular $[Ca^{2+}]$ is low, on the other hand, glucagon and vasopressin both mobilize Ca^{2+} from a common pool. Of further interest, and similar to results in experiments with perfused rat liver (see above), was the finding that under conditions where the extracellular $[Ca^{2+}]$ is low, glucagon was unable to potentiate the effect of vasopressin in inducing an increase in intracellular $[Ca^{2+}]$. Combettes et al. (1986) concluded that glucagon accelerates the fast phase of vasopressin-induced inflow and elevates the final intracellular $[Ca^{2+}]$ attained, by (a) releasing Ca^{2+} from the same

internal store as that affected by vasopressin, and (b) potentiating the inflow of external Ca^{2+} induced by vasopressin. In a related study, Mauger and Claret (1986) showed that glucagon mobilizes Ca^{2+} from a pool that is also sensitive to the action of Ca^{2+}-mobilizing agonists.

Kass et al. (1990), examining Mn^{2+} inflow into hepatocytes, showed that prior emptying of the endoplasmic reticular store with the inhibitor of endoplasmic reticular Ca^{2+} inflow 2,5-di-(tert-butyl)-1,4,-benzohydroquinone enhanced the action of glucagon or of dibutyryl cyclic AMP on vasopressin-induced Mn^{2+} inflow. Of additional interest was their finding that the effects of glucagon on vasopressin-induced Mn^{2+} inflow were unaffected by membrane depolarization and by pertussis toxin treatment.

Several reports from studies with the perfused rat liver following the sequential administration of the agonists provide evidence that glucagon mobilizes Ca^{2+} from the same store as that mobilized by phenylephrine (Reinhart et al., 1982) or by vasopressin (Kimura et al., 1982; Kraus-Friedmann, 1986; Mine et al., 1988a; Benedetti et al., 1989).

Related to these issues is the question of which intracellular store the Ca^{2+} enters as a result of the synergistic action of glucagon and Ca^{2+}-mobilizing agonists on hepatocytes and the perfused rat liver as discussed above. The studies of Morgan et al. (1985) provided evidence that mitochondria were the intracellular site into which this Ca^{2+} enters. The evidence was based on both invasive analyses of the mitochondrial Ca^{2+} content following hormonal treatment of hepatocytes and non-invasive analyses of Ca^{2+} movements following treatment of intact hepatocytes with specific inhibitors of mitochondrial energy transduction.

In a study examining the influence of vasopressin and/or glucagon on mitochondrial Ca^{2+} and Ca^{2+}-sensitive oxidative enzymes in the perfused rat liver, Assimacopoulos-Jeannet et al. (1986) also showed that these hormones additively induce increases in intramitochondrial $[Ca^{2+}]$. This was reflected also in the stimulation of the activities of pyruvate dehydrogenase and 2-oxoglutarate dehydrogenase.

In agreement with these conclusions, Altin and Bygrave (1986) rapidly isolated mitochondria and endoplasmic reticulum from the perfused rat liver following co-administration of glucagon with Ca^{2+}-mobilizing agonists. Their data revealed an accumulation of Ca^{2+} into the subsequently isolated organelles, particularly mitochondria, although significant increases in the endoplasmic reticulum-enriched fraction also were evident. Although considerable care was taken in those studies to minimize post-isolation Ca^{2+} redistribution (see Reinhart et al., 1984a), it is possible that some redistribution did take place especially involving Ca^{2+} release from the microsomes and its re-uptake into the mitochondria. Benedetti et al. (1989), for example, have found that lowering the temperature to approx. 4 °C, as occurs in the course of the invasive organellar isolation technique, induces a rapid efflux of Ca^{2+} from the microsomes that could move into mitochondria which readily take up Ca^{2+} at these temperatures (Fulceri et al., 1991; Banhegyi et al., 1991). Banhegyi et al. (1991) have shown that a non-invasive technique using hepatocytes provided evidence that it is a non-mitochondrial pool into which the bulk of the Ca^{2+} enters following co-administration of glucagon and vasopressin. It is significant however that, at low temperatures, the majority of the Ca^{2+} entered the mitochondria (Banhegyi et al., 1991). In this regard Assimacopoulos-Jeannet et al. (1986) reported that, provided Na^+ is kept to a minimum and the mitochondria are maintained in ice-cold media, the (mitochondrial) preparation remains stable for the duration of the isolation procedure.

Kleineke and Soling (1985) concluded from a non-invasive study of perfused rat liver that the endoplasmic reticulum and not the mitochondria is the hormone-sensitive intracellular store. Further evidence for a reticular component for Ca^{2+} deposition is the observation that destruction of the endoplasmic reticulum with bromotrichloromethane prevented the synergistic accumulation of Ca^{2+} by perfused rat liver (Benedetti et al., 1989). Bond et al. (1987) had previously examined this issue by employing electron probe analyses of the intact liver and also had concluded that the Ca^{2+} had entered the endoplasmic reticulum (but see Altin and Bygrave, 1988).

Mine et al. (1988a) examined Ca^{2+} mobilization in rat hepatocytes loaded with aequorin and concluded that the intracellular Ca^{2+} pool mobilized by glucagon is different from that mobilized by the vasoactive agents phenylephrine and angiotensin II. This conclusion was based on the observation that brief prior treatment of hepatocytes with dinitrophenol prevented glucagon but not angiotensin II from inducing an increase in cytoplasmic Ca^{2+}. Kraus-Friedmann (1986) had previously shown that glucagon releases Ca^{2+} from a carbonyl cyanide p-(trifluoromethoxy)phenylhydrazone (FCCP)-releasable pool.

As implied above, it seems that the experimental conditions can influence the extent to which mitochondria or the endoplasmic reticulum (or both) provide the agonist-sensitive pool of Ca^{2+}. Most workers seem to be of the view, however, that the bulk of this pool is non-mitochondrial.

OTHER Ca^{2+} FLUX-RELATED SIGNALS GENERATED BY GLUCAGON

During the 1970s a number of reports appeared indicating that administration of glucagon to the intact rat, or to the perfused liver, or to hepatocytes, induced changes in the ability of the subsequently isolated organelle to transport Ca^{2+} (for reviews see Bygrave, 1978a; Soboll and Sies, 1989). A large body of work showed, for instance, that administration of glucagon to intact rats will enhance the ability of the subsequently isolated microsomes to accumulate Ca^{2+} in vitro (e.g. Bygrave and Tranter, 1978; Reinhart and Bygrave, 1981; Andia-Waltenbaugh et al., 1981). With mitochondria too, it was established that prior glucagon administration to the intact rat led to an increased ability of the subsequently isolated organelle to retain Ca^{2+} (e.g. Prpic et al., 1978). While it is somewhat unclear how and if these phenomena relate to the control by glucagon of hepatic Ca^{2+} fluxes, they provide a framework for a more recent set of observations indicating that Ca^{2+} uptake and release can be modulated by metabolites whose cytosolic concentrations can increase very rapidly following glucagon action.

EFFECTORS GENERATED BY GLUCAGON ACTION THAT MIGHT SERVE AS MODULATORS OF Ca^{2+} FLUXES

There is increasing evidence that intracellular Ca^{2+} fluxes can be affected not only by intracellular second messengers like $Ins(1,4,5)P_3$ and cyclic AMP but also by variations in the (cytosolic) concentration of metabolites resulting from the action of glucagon. For instance, it is well known that glucagon rapidly increases the flux into and the rate of hydrolysis of glucose 6-phosphate within the endoplasmic reticulum since this hormone stimulates glucose production. This event could promote Ca^{2+} inflow and storage in these organelles; physiological concentrations of glucose 6-phosphate have been observed to stimulate Ca^{2+} uptake by isolated liver microsomes and by digitonin-permeabilized hepatocytes (Benedetti et al., 1985, 1987). In this

respect, a recent study by Banhegyi et al. (1991) showed that the extent of Ca^{2+} inflow into a reticular pool(s) of isolated hepatocytes caused by co-administration of glucagon and vasopressin (see above) can be correlated with the extent of glucose production. It seems possible, therefore, that increased fluxes of glucose 6-phosphate and concomitant stimulation of reticular active Ca^{2+} translocation (see above) could be involved in the synergistic Ca^{2+} inflow induced by glucagon and Ca^{2+}-dependent agonists. It is notable that glucose 6-phosphate appeared to promote Ca^{2+} storage into both $Ins(1,4,5)P_3$-sensitive and -insensitive endoplasmic reticular pools (Benedetti et al., 1988), and therefore a sensitization by glucagon to $Ins(1,4,5)P_3$-dependent agonists can be envisaged. The effect on the $Ins(1,4,5)P_3$-insensitive Ca^{2+} pool also could contribute to the termination of the effect of $Ins(1,4,5)P_3$-dependent agonists by reaccumulating released Ca^{2+} (Benedetti et al., 1986). However, the relative roles of the activation of these two pools need further clarification.

Inorganic P_i is another metabolite able to induce changes in Ca^{2+} fluxes. The ability of P_i to promote mitochondrial Ca^{2+} inflow and storage in vitro under conditions of high Ca^{2+} loading is well known. At concentrations that exist in the cytosol, P_i also has been shown to promote Ca^{2+} inflow into (smooth) liver microsomal vesicles and the reticular pool(s) of permeabilized hepatocytes (Fulceri et al., 1990). Glucagon stimulates P_i uptake into perfused livers and increases the cytosolic $[P_i]$ (e.g. Bygrave et al., 1990). Hence, variations in cytosolic $[P_i]$ may also play a role in the modulation of intracellular reticular Ca^{2+} fluxes by glucagon.

Other metabolic effects induced by glucagon that are potentially relevant to the regulation of intracellular Ca^{2+} fluxes in liver are the variation in redox state (e.g. Rusinko and Lee, 1989) and the increase in acyl-CoA cellular content (Comerford and Dawson, 1993). Both events have been shown to affect cellular Ca^{2+} fluxes.

Few reports appear to have addressed the issue of whether Ca^{2+} fluxes induced by glucagon, with or without vasopressin, are influenced by the nutritional status of the animal. The work of Sistare et al. (1985) examining glucagon-induced Ca^{2+} fluxes in rat hepatocytes was carried out with 24-h-starved animals and the data appear qualitatively similar to those obtained by others using hepatocytes from fed animals. On the other hand, Rashed and Patel (1987) found that in 24-h-fasted animals, glucagon failed to induce ^{45}Ca efflux, but that induced by phenylephrine was unaffected. They suggested that the cellular redox potential regulates the ability of glucagon to promote Ca^{2+} efflux, probably through altering the cyclic AMP concentration. It was also reported (Banhegyi et al., 1991) that the synergistic inflow of Ca^{2+} induced by glucagon and Ca^{2+}-mobilizing agonists was reduced in hepatocytes from fasted rats (as compared to those from fed rats). As in the study of Rashed and Patel (1987), Ca^{2+} inflow was restored in fasted hepatocytes in the presence of gluconeogenic substrates. The extent of Ca^{2+} inflow appeared to correlate with the extent of glucose production; see also Taylor et al. (1983).

OTHER ASPECTS OF GLUCAGON-STIMULATED MODULATION OF (Ca^{2+}-MOBILIZING) AGONIST-INDUCED Ca^{2+} FLUXES: HORMONE SENSITIVITY AND RAPIDITY OF ONSET

An important observation by two groups in studies with hepatocytes and with the perfused rat liver is that glucagon, besides potentiating the changes in Ca^{2+} fluxes, appears also to increase the sensitivity of the cells to the agonists in question. Thus Morgan et al. (1985) noted that while very low doses of noradrenaline would not increase intracellular $[Ca^{2+}]$, preincubation of the cells with glucagon allowed noradrenaline to induce a large increase. This change in sensitivity was interpreted to be correlatable with large increases in binding of noradrenaline to sites on the hepatocyte surface. These workers suggested that a modification by cyclic AMP at the level of the α_1-adrenergic receptor mediated this effect. Altin and Bygrave (1986), making dose–response measurements with the perfused rat liver, observed that the sensitivity of vasopressin and angiotensin (but not so much that of phenylephrine) was apparently increased by the presence of a low glucagon concentration. It should be noted that Poggioli et al. (1986) did not observe an alteration in the sensitivity to the agonists. As found by Morgan et al. (1985), the increase in sensitivity is such that the concentrations of hormones that are inducing the effects on Ca²⁺ fluxes is well within the physiological range. Buhler et al. (1978), for example, have analysed the circulating glucagon and noradrenaline concentrations, finding these to be 0.1–0.01 nM. Conditions of severe stress, like some other conditions *in vivo*, cause a concomitant increase in plasma adrenaline and glucagon (Eigler et al., 1979). Elevated levels of glucagon in the blood appear to accompany hyperglycaemia in diabetes mellitus (Unson et al., 1989).

Studer et al. (1984) have called into question the physiological relevance of a number of reported effects of glucagon both *in vivo* and *in vitro*, on account of the potentially apparently high (unphysiological) concentrations of glucagon employed in the experiments. It is difficult to argue against this point in many instances; the message is clear.

Several reports have now appeared indicating that an early action of glucagon is on Ca²⁺ inflow. For instance, Mauger et al. (1985) showed that the initial rate of ⁴⁵Ca uptake by hepatocytes was stimulated by glucagon, and Staddon and Hansford (1986) using the Ca²⁺-free/Ca²⁺ re-admission technique with fura2-loaded hepatocytes also found that glucagon could stimulate Ca²⁺ inflow. More recently, Bygrave et al. (1993a), using the same technique, have confirmed and extended the finding of Staddon and Hansford (1986) to show that the effects of glucagon on Ca²⁺ inflow are already maximal by 5–10 s after its addition to hepatocytes and therefore are far more rapid in onset than those on mobilization of intracellular Ca²⁺ pools. In this latter study, dibutyryl cyclic AMP could reproduce the effects on Ca²⁺ inflow seen with glucagon.

POSSIBLE LOCI IN THE SIGNALLING TRANSDUCTION PATHWAYS WHERE CROSS-TALK MIGHT TAKE PLACE AND THE RELATED ISSUE OF WHETHER GLUCAGON IS ABLE TO GENERATE INS(1,4,5)P_3 IN THE ABSENCE OR PRESENCE OF Ca²⁺-MOBILIZING AGONISTS

As we have seen above, there is mounting evidence that practically from the moment of its administration to a suspension of hepatocytes, glucagon itself is able to rapidly induce the mobilization of Ca²⁺. Such rapidity of action raises the issue of the extent to which cross-talk might be taking place at or near the level of the G proteins that couple receptors to various effector molecules (see Figure 4), or at points distal to the site of the plasma membrane-located agonist–receptor interactions (Figure 5). It will also have become apparent that one potential mechanism whereby glucagon could influence the action of Ca²⁺-mobilizing agonists is through altering the ability of the cell to generate Ins(1,4,5)P_3.

Earlier studies from a number of laboratories indicated that glucagon alone has no significant effect on Ins(1,4,5)P_3 generation in hepatocytes (Prpic et al., 1982; Kaibuchi et al., 1982; Charest et al., 1985b) and fails to increase Ins(1,4,5)P_3 breakdown (Creba et al., 1983). Moreover, Poggioli et al. (1986) and Williamson

and Hansen (1987) showed that glucagon, while enhancing adrenaline-induced Ca²⁺ inflow in hepatocytes, had no effect on the generation of Ins(1,4,5)P_3. Burgess et al. (1986, 1991) found (like Cocks et al., 1984) that isoprenaline administration to guinea pig hepatocytes caused little, if any, increase in Ins(1,4,5)P_3 formation in the absence or presence of angiotensin II.

As will now be discussed, this issue has been re-opened by several sets of data. First, Wakelam et al. (1986) have provided evidence that glucagon is able to activate two signal-transduction systems in hepatocytes depending on the agonist concentrations employed. At low (physiological; K_a approx. 0.25 nM) concentrations, glucagon activates a system that leads to the generation of inositol phosphates and little, if any, cyclic AMP. Adenylate cyclase is activated only at higher concentrations ($> 10^{-9}$ M) with concomitant production of cyclic AMP. Wakelam et al. (1986) also showed that by 5 s after administration of glucagon to hepatocytes, Ins(1,4,5)P_3 is already increased by some 15%, by 10 s it is maximally generated (by some 35% over basal) and by 30 s it has returned to near-basal levels. The amount of Ins(1,4,5)P_3 generated was, however, considerably less than that generated by the action of α_1-adrenoceptors or by vasopressin.

Secondly, vom Dahl et al. (1988), working with the perfused rat liver, showed that while vasopressin and glucagon each were able to promote [³H]inositol release from the liver of a rat previously administered [³H]inositol, with vasopressin the response was Li-sensitive but with glucagon it was Li-insensitive.

Thirdly, and consistent with the conclusions of Wakelam et al. (1986), Mine et al. (1988b) have shown that the cytoplasmic [Ca²⁺] can be increased following glucagon administration to aequorin-loaded rat hepatocytes, through an action apparently largely independent of cyclic AMP generation. The action of glucagon was very rapid in onset, the response was transient, and the increase in cytoplasmic [Ca²⁺] was accompanied by only a very small elevation of cyclic AMP. Mine et al. (1988b) concluded that at least part of the action of glucagon on Ca²⁺ mobilization, especially stimulation of Ca²⁺ inflow, is a process that occurs independently of cyclic AMP generation.

Fourthly, Unson et al. (1989) have described the biological properties of the glucagon antagonist des-His¹-[Glu⁹]glucagon amide. This analogue bound to liver membranes but did not stimulate cyclic AMP production. It did, however, activate a pathway that led to the generation of inositol phosphates. The authors concluded that either there are two distinct glucagon receptors on the hepatocyte plasma membrane, or there is one binding site but two G proteins.

Finally, a group of recent reports have appeared that concern the cloning and expression of the glucagon receptor and other closely related receptors. Jelinek et al. (1993) recently described the cloning, expression and signalling properties of the rat glucagon receptor. The cloned receptor was reported to increase intracellular cyclic AMP following the binding of glucagon. Moreover, and most relevant to the present discussion, it transduced a signal that led to a very rapid elevation (i.e. within seconds) of the cytoplasmic [Ca²⁺]. Thus the cloned glucagon receptor is able to transduce signals that lead to the accumulation of two different second messengers, cyclic AMP and Ca²⁺. In a related study, Abou-Samra et al. (1992) reported the cloning and expression of the parathyroid hormone receptor. Upon activation, both adenylate cyclase and phospholipase C were stimulated. This led to intracellular accumulation of cyclic AMP and inositol trisphosphates with concomitant rapid increases in the cytoplasmic [Ca²⁺] (Abou-Samra et al., 1992). The calcitonin receptor also has been cloned and expressed (Chabre et al., 1992). Upon activation, it too stimulated the generation of cyclic

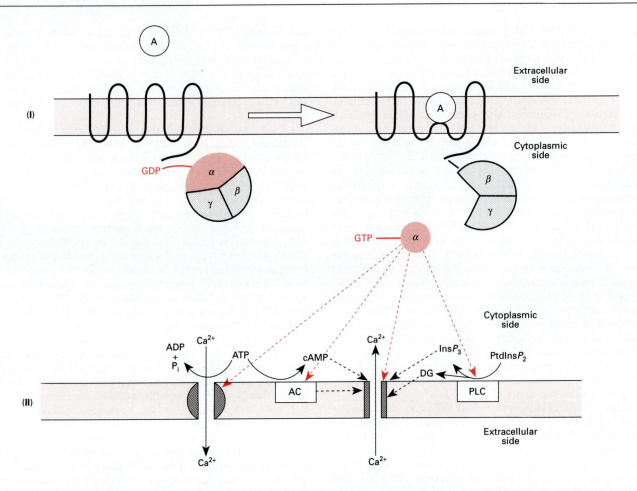

Figure 4   Possible interactions associated with the hepatocyte plasma membrane that might be involved in the cross-talk between the actions of glucagon and agonists that mobilize Ca²⁺

The interaction of glucagon or the agonist (both indicated by A) with its receptor (R) activates a G protein (shown in **I**). This activated G protein, located in the plasma membrane, in turn produces a range of possible effects on the effector molecules shown (see **II**) that lead directly or indirectly to modulation of Ca^{2+} fluxes (indicated by the broken lines). The membrane environment would provide rapid and sensitive interactions without the relatively slow diffusion of generated messengers through the cytoplasmic medium. α, β and γ are subunits of the G protein located in the plasma membrane. AC, adenylate cyclase; PLC, phospholipase C; cAMP, cyclic AMP. For further explanation, see the text.

AMP and inositol trisphosphates, again with rapid increases in intracellular [Ca²⁺].

Thus all three of these closely related receptors are able to be activated in such a way that the individual (single) receptor efficiently couples to multiple effector systems, i.e. those leading to generation of both cyclic AMP and inositol trisphosphates. In the light of this, it is of some interest that Yamaguchi (1991) showed that calcitonin was able to stimulate basal Ca²⁺ inflow in rat hepatocytes as well as potentiate adrenaline-induced Ca²⁺ inflow.

All of these observations thus raise the issue of the nature of the locus/loci at which the signals generated by glucagon and by Ca²⁺-mobilizing agonists interact so as to permit cross-talk. A number of articles have appeared reviewing the evidence for the role of G proteins in signal transduction (see, e.g., Freissmuth et al., 1989; Taylor, 1990; Simon et al., 1991; Cockroft and Thomas, 1992; Milligan, 1992) and in coupling to ion channel pathways (Brown, 1991). What is clear is that G proteins allow for great diversity in signal transduction, acting as switches by being able to rapidly activate or desensitize effector molecules. Moreover, a single receptor can in principle activate multiple G protein molecules, thus amplifying the ligand binding event (Simon et al.,

1991). Similar G proteins may also be able to generate signals with different time constants.

Cross-activation, it has been suggested (Taylor, 1990; Simon et al., 1991), can be an essential part of the G-protein-mediated information-transducing circuit, co-ordinating and integrating signal cross-talk. In situations where a G protein is closely linked spatially with an ion channel in a membrane, there is an increased likelihood of an increase in the time constant (Brown, 1991). Thus directly coupled G protein–ion channel pathways have faster time constants and new steady states are reached more quickly.

All of these events can be further modulated in turn by the actions of cyclic AMP-dependent protein kinases and by diacylglycerol and protein kinase C. Phosphorylation of both the G protein receptors and the G proteins themselves (see Brown, 1991) will add to the possibilities for amplification, memory, changes in sensitivity and sharing of pathways in signal transduction (Taylor, 1990). In this context, the report of Dasso and Taylor (1992) in which evidence was provided that α_1-adrenoceptors and V_1-vasopressin receptors share the same pool of G proteins in rat hepatocytes is of interest. Also relevant to this general issue is the extent to which any cyclic AMP generated

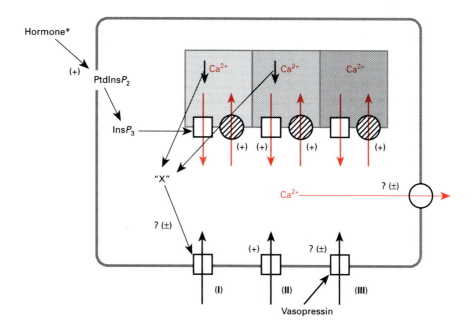

Figure 5   **Scheme illustrating current knowledge of the effects of glucagon or elevated cyclic AMP on events affecting Ca²⁺ mobilization in liver cells**

Three Ca²⁺ inflow pathways in the plasma membrane are shown: I, the pathway activated as a result of depletion of the intracellular Ca²⁺ stores; II, the pathway activated by direct action of glucagon or cyclic AMP; and III, the putative receptor (vasopressin)-operated pathway (this pathway is shared by Mn²⁺ and Ca²⁺). Three pools of endoplasmic reticular Ca²⁺ are shown; one discharged by Ins(1,4,5)P_3, one discharged by cyclic AMP action and a third insensitive to known Ca²⁺-discharging messengers. Although for clarity separate pumps are shown for each pool, a common active uptake transport system may operate. This system(s) is activated by cyclic AMP. X represents the putative messenger generated by the emptying of intracellular agonist-sensitive stores; + or − represent activation or inhibition respectively by glucagon or elevated intracellular cyclic AMP of the pathway in question; * represents the event in which glucagon sensitizes Ca²⁺-mobilizing hormones to their receptors.

through glucagon action is metabolized by cyclic AMP phosphodiesterases and the degree to which adenylate cyclase itself is activated/desensitized (Houslay et al., 1992). Another factor having a potentially critical bearing on this is the extent to which intracellular cyclic AMP is free or is bound to intracellular ligands (see, e.g., Sammak et al., 1992). Many of these points are collected together in the scheme shown in Figure 4.

Two further points can now be made concerning the locus/loci at which signalling cross-talk occurs. The first is that it is clear that cyclic AMP can potentiate Ca²⁺ mobilization induced by Ca²⁺-mobilizing agonists at loci distal to the point at which the second messengers are generated (see Figure 5). It is also clear that in the bulk of the experiments showing this, a relatively long time (> 60 s) was usually allowed to elapse between additions of glucagon/dibutyryl cyclic AMP/forskolin and the Ca²⁺-mobilizing agonist. It is possible that the rapid responses to glucagon action (i.e. those occurring within say 5 s) may result not from the generation of cyclic AMP, but from the generation of Ins(1,4,5)P_3 or some other, as yet unidentified, factor. Thus not only could the glucagon concentration used be important in determining the second messenger produced, but also the time at which measurements are made may have an effect: longer times of glucagon action may provide information different to that produced at shorter times, especially considering the rapid turnover of Ins(1,4,5)P_3 generated by low glucagon concentrations (Wakelam et al., 1986). Whether or not this turns out to be a crucial issue remains to be determined.

In principle, cross-talk between the second messenger-transducing systems activated by glucagon and by Ca²⁺-mobilizing agonists could occur before agonist-induced generation of Ins(1,4,5)P_3, or following the production of Ins(1,4,5)P_3, or at both sets of loci. It seems reasonable to suggest that, in those

experimental situations where intracellular cyclic AMP was elevated (by forskolin or via the membrane-permeant dibutyryl cyclic AMP) and synergism resulted, the interaction is one that is downstream of cyclic AMP and of Ins(1,4,5)P_3 formation.

A number of authors (Altin and Bygrave, 1986; Poggioli et al., 1986; Mauger and Claret, 1988; Kass et al., 1990) have expressed the view that the synergistic action of glucagon and Ca²⁺-mobilizing agonists occurs close to Ca²⁺ channels or on the carriers themselves located at the plasma membrane. The suggestions are that control might be effected through cyclic AMP perhaps phosphorylating Ca²⁺ channels and/or (Poggioli et al., 1986) through some product of Ins(1,4,5)P_3 metabolism. One other possible area where control by cyclic AMP could be exerted on Ins(1,4,5)P_3-mediated Ca²⁺ mobilization is at the level of the Ins(1,4,5)P_3 receptor. In this respect Mauger et al. (1989) have described an Ins(1,4,5)P_3 receptor from liver membranes with high and low affinity binding sites. In their view, vasopressin increased the number of high affinity binding sites, and phosphorylation (by dibutyryl cyclic AMP) through cyclic AMP-dependent protein kinase inhibited the effects of vasopressin.

Burgess et al. (1991) showed that in permeabilized guinea pig hepatocytes, cyclic AMP-dependent protein kinase can phosphorylate the Ins(1,4,5)P_3 receptor/channel on the endoplasmic reticulum, although, as the authors point out, the outcome of this phosphorylation is not yet clear. They found the action of cyclic AMP-dependent protein kinase to have at least two consequences on Ins(1,4,5)P_3-induced Ca²⁺ release; one to increase the total amount of Ca²⁺ released and the other to reduce the value of the EC₅₀ for Ins(1,4,5)P_3-induced Ca²⁺ release from approx. 0.7 mM to approx. 0.14 mM. Each of these groups (Mauger et al., 1989; Burgess et al., 1991) thus have suggested that the cross-talk between glucagon and Ca^{2+}-mobilizing agonists that results in potentiated Ca^{2+} mobilization occurs at the endoplasmic reticulum.

The other possibility, that cross-talk occurs at or before hormone-induced phosphatidylinositol metabolism, has been proposed by Blackmore and Exton (1986). The results of their experiments, in which the effects of 4β-phorbol 12-myristate 13-acetate and aluminium fluoride on Ca^{2+} mobilization in rat hepatocytes were examined, led them to suggest that the mobilization of Ca^{2+}, induced by glucagon, results from cyclic AMP-dependent phosphorylation of Np, phosphatidylinositol bisphosphate, or the event leading to $Ins(1,4,5)P_3$ formation.

In contrast to some of the reports mentioned above, Pittner and Fain (1989) observed a synergistic increase of $Ins(1,4,5)P_3$ formation when bromocyclic AMP was administered to rat hepatocytes with vasopressin. They suggested, not dissimilar to Blackmore and Exton (1986), sites of action of elevated cyclic AMP as possibly being one or other of the vasopressin receptor, the Gp (Np) protein or the phospholipase C enzyme itself.

While not directly related to the issue of the mechanism of synergism between glucagon and agonist-induced Ca^{2+} mobilization, it is relevant to point out that Hughes et al. (1987) reported that a GTP-binding protein is involved in the process by which vasopressin and EGF stimulate receptor-operated Ca^{2+} inflow transporters (see also Barritt and Hughes, 1991). EGF does not increase $Ins(1,4,5)P_3$ formation, and pertussis toxin treatment (which blocks the action of G proteins) of rat hepatocytes abolished EGF- and vasopressin-induced Ca^{2+} inflow. Also relevant are the reports showing that 4β-phorbol 12-myristate 13-acetate treatment of hepatocytes, which leads to activation of protein kinase C, attenuates the Ca^{2+}-mobilizing action of glucagon (Staddon and Hansford, 1986; see also Garcia-Sainz et al., 1985; Staddon and Hansford, 1989; Pittner and Fain, 1991). Whether some of the signals being generated here result from hormone-induced phosphatidylcholine breakdown (Exton, 1990) remains an interesting possibility.

CONCLUDING REMARKS AND FUTURE DIRECTIONS

This review has considered a number of observations, obtained from studies both with the whole organ (perfused rat liver) and with hepatocytes isolated therefrom, concerning the interaction of the second messenger-mediated signalling systems originating at the plasma membrane and for which glucagon and Ca^{2+}-mobilizing agonists are the stimuli. A number of experimental approaches and techniques, combined with the deployment of a range of pharmacological agents, have already been brought into play to examine the signalling cross-talk phenomenon. Apart from the perfused rat liver system, these include the increasingly widespread use of single-cell systems to examine the influence of various hormone regimes on Ca^{2+} mobilization in oscillatory phenomena (e.g. Schofl et al., 1991), in computer-assisted mathematical modelling (e.g. Somogyi and Stucki, 1991) and following the application of microinjection techniques (e.g. Glennon et al., 1992). All of these approaches will continue to provide relevant information and, in combination with the cloning techniques now available, results of some of which were mentioned above, exciting new insights into signalling cross-talk are assured. Two other recent developments with great potential for the present study are noteworthy. One is the ability to express a Ca^{2+}-mobilizing receptor, such as the V_1-vasopressin receptor cloned from rat liver, in *Xenopus* oocytes (Nathanson et al., 1992) and to then examine the subcellular distribution of Ca^{2+} signals by confocal microscopy following activation of the receptor. A second is the ability to target recombinant aequorin into mito-

chondria and observe the changes specifically in mitochondrial $[Ca^{2+}]$ in response to various stimuli (Rizzuto et al., 1992). No doubt all of this will be further aided by the development of even more diverse indicators including those designed to measure rapid changes in intracellular cyclic AMP (Tsien, 1992; Sammak et al., 1992).

While each of the experimental systems (whole organ and single cells) offers particular attributes to investigations of cross-talk, there is a remarkable similarity and confluence among the data already obtained. The integration of the information gained from each is providing considerable information in turn about the nature of the synergistic response. The view previously expressed in a number of reports, i.e. that enhanced mobilization and especially Ca^{2+} inflow is a major target for cross-talk between these signalling systems, has been reinforced. An interesting facet of hormone-induced Ca^{2+} inflow, not often considered, is that irrespective of what combination of hormones is present, Ca^{2+} inflow will not take place unless a substantial Ca^{2+} gradient exists across the plasma membrane.

The major features of the synergism so far revealed (see Figure 5) are (1) that it can occur at very low (physiological) concentrations of each of the two sets of hormones, (2) that responses are able to be induced which are very rapid in onset, and (3) that the earliest response, which is practically immediate, appears to be Ca^{2+} inflow; only after some 30 s is glucagon able to mobilize Ca^{2+} apparently from the same internal Ca^{2+} store as that mobilized by vasopressin; this is non-mitochondrial and most probably endoplasmic reticular in origin. Just as there appear to be $Ins(1,4,5)P_3$-sensitive and -insensitive Ca^{2+} pools in the endoplasmic reticulum, so one might also envisage glucagon-sensitive and -insensitive pools in the same organelle, all of which are in some way functionally interactive.

Despite these revelations, the mechanism by which cross-talk between the two second messenger-mediated signalling systems occurs to bring about the potentiation in Ca^{2+} mobilization is still largely unknown. Data from a number of separate reports, all examining different aspects of glucagon action as well the molecular biology of the glucagon and other closely related receptors, provide strong evidence for Ca^{2+} mobilization induced by glucagon alone. This apparently occurs not through the generation of cyclic AMP, but rather through the generation of $Ins(1,4,5)P_3$, albeit at lower concentrations than those generated by 'classical' Ca^{2+}-mobilizing agonists such as α_1-adrenergic agonists, vasopressin and angiotensin. The extent to which this action of glucagon (at low concentrations) contributes to the cross-talk phenomenon remains to be addressed, as does the question of whether this in turn is a component of the mechanism by which glucagon mobilizes Ca^{2+} from intracellular Ca^{2+} stores. Recent findings that activation of the cloned glucagon receptor results in Ca^{2+} mobilization independent of cyclic AMP production but with generation of $Ins(1,4,5)P_3$ is particularly intriguing and raises the question of the extent to which G proteins might play a part in cross-talk.

Suggested intracellular sites of action of glucagon are the Ca^{2+} transporters in mitochondria (Morgan et al., 1985), in the endoplasmic reticulum (Burgess et al., 1986; Mauger et al., 1989) and in the plasma membrane (Poggioli et al., 1986; Altin and Bygrave, 1986; Kass et al., 1990). It is possible that all sites are modified but perhaps to different extents and at different times following glucagon administration. Whilst glucagon action is known to be able to lead to the phosphorylation of cytoplasmic proteins (Garrison and Borland, 1977) including the multi-functional Ca^{2+}/calmodulin-dependent protein kinase (Connelly et al., 1987) as well as membrane proteins (Vargas et al., 1982), it is not yet established if any of these are components of, or are in any way associated with, the above-mentioned Ca^{2+} transport systems. Presumably details about possible phosphorylation events will become clearer once the molecular architecture of these Ca^{2+} transporters/channels becomes known.

Another area of recent study in which the outcome is of potential interest is the hormonal regulation of the $Ins(1,4,5)P_3$ receptor. It is known that the receptor can be phosphorylated by a cyclic AMP-dependent protein kinase resulting in a decrease in the ability of $Ins(1,4,5)P_3$ to release Ca^{2+} from microsomal stores (Ferris et al., 1991). It has been suggested (Mauger et al., 1989) that in liver such a mechanism could participate in the synergistic action of cyclic AMP and $Ins(1,4,5)P_3$-dependent hormones on Ca^{2+} mobilization. The extent to which this occurs remains to be assessed.

Finally, one might well question the physiological significance of the cross-talk phenomenon in liver. Clearly much of the current research in liver is focused on mechanisms at the cellular and subcellular level. A role for cross-talk between second messenger-generated signalling systems leading to modulated Ca^{2+} fluxes is seen to be crucial, for instance in excitation–contraction coupling in cardiac (see, e.g., Cohen, 1992) and smooth muscle (see, e.g., Abdel-Latif, 1991) cells, and in the rapid secretion of insulin (e.g. Holz and Habener, 1992) and saliva (e.g. McKinney et al., 1989), all of which are well established Ca^{2+}-dependent phenomena. We envisage that one of the roles of cross-talk between the two signalling systems in liver, described in this review, is to amplify and/or sensitize signals of which Ca^{2+} is a vital component. One physiological function in liver recognized only very recently as being subject to rapid, short-term modulation by glucagon and Ca^{2+}-mobilizing hormones is bile secretion (Bygrave et al., 1993b). This group has found that the ability of vasopressin to rapidly (in a time-frame of seconds) alter bile flow in the perfused rat liver is greatly modified when glucagon is preadministered also for only a brief time (Hamada et al., 1992a). This synergistic response, which is induced by physiological concentrations of the agonists, can be modulated in turn by the action of cholestatic and choleretic bile salts administered *in vitro* (Hamada et al., 1992b) and by agents that induce cholestasis *in vivo* (Bygrave et al., 1993b). All of these features should provide a useful set of probes with which one can gain further insights in liver, into a genuine physiological system modulated by cross-talk between the actions of glucagon and Ca^{2+}-mobilizing agonists.

F. L. B. is grateful to Dr. Alan Dawson (University of East Anglia), Drs Juan Llopis and Georges Kass (Karolinska Institutet) and to the editorial referees of the Biochemical Journal for helpful comments and suggestions.

REFERENCES

Abdel-Latif, A. A. (1991) Cell. Signalling **3**, 371–385
Abou-Samra, A.-B., Juppner, H., Force, T., Freeman, M. W., Kong, X.-F., Schipani, E., Urena, P., Richards, J., Bonventre, J. V., Potts, J. T., Jr., Kronenberg, H. M. and Segre, G. V. (1992) Proc. Natl. Acad. Sci. U.S.A. **89**, 2732–2736
Åkerman, K. E. O. and Nicholls, D. G. (1983) Rev. Physiol. Biochem. Pharmacol. **95**, 149–201
Althaus-Salzmann, M., Carafoli, E. and Jacob, A. (1980) Eur. J. Biochem. **106**, 241–248
Altin, J. G. and Bygrave, F. L. (1985) Biochem. J. **232**, 911–917
Altin, J. G. and Bygrave, F. L. (1986) Biochem. J. **238**, 653–661
Altin, J. G. and Bygrave, F. L. (1987a) Biochem. J. **242**, 43–50
Altin, J. G. and Bygrave, F. L. (1987b) Biochem. J. **247**, 613–619
Altin, J. G. and Bygrave, F. L. (1987c) Biochem. Biophys. Res. Commun. **142**, 745–753
Altin, J. G. and Bygrave, F. L. (1988) Biol. Rev. **63**, 551–611
Andia Waltenbaugh, A.-M., Tate, C. and Friedmann, N. (1981) Mol. Cell Biochem. **36**, 177–184
Assimacoupolos-Jeannet, F. D., Blackmore, P. F. and Exton, J. H. (1977) J. Biol. Chem. **252**, 2662–2669

Assimacoupolos-Jeannet, F. D., Blackmore, P. F. and Exton, J. H. (1982) J. Biol. Chem. **257**, 3759–3765
Assimacoupolos-Jeannet, F. D., McCormack, J. G. and Jeanrenaud, B. (1986) J. Biol. Chem. **261**, 8799–8804
Banhegyi, G., Fulceri, R., Bellomo, G., Romani, A., Pompella, A. and Benedetti, A. (1991) Arch. Biochem. Biophys. **287**, 320–328
Barritt, G. J. (1992) Communication Within Animal Cells, Oxford University Press, Oxford
Barritt, G. J. and Hughes, B. P. (1991) Cell. Signalling **3**, 283–292
Barritt, G. J., Parker, J. C. and Wadsworth, J. C. (1981) J. Physiol. (London) **312**, 29–55
Benedetti, A., Fulceri, R. and Comporti, M. (1985) Biochim. Biophys. Acta **816**, 267–277
Benedetti, A., Fulceri, R., Ferro, M. and Comporti, M. (1986) Trends Biochem. Sci. **11**, 284–285
Benedetti, A., Fulceri, R., Romani, A. and Comporti, M. (1987) Biochim. Biophys. Acta **928**, 282–286
Benedetti, A., Fulceri, R., Romani, A. and Comporti, M. (1988) J. Biol. Chem. **263**, 3466–3473
Benedetti, A., Graf, P., Fulceri, R., Romani, A. and Sies, H. (1989) Biochem. Pharmacol. **38**, 1799–1805
Berridge, M. J. (1987) Annu. Rev. Biochem. **56**, 159–193
Berry, M. N., Edwards, A. M. and Barritt, G. J. (1991) in Isolated Hepatocytes: Preparation, Properties and Applications, pp. 201–236, Elsevier, Amsterdam
Blackmore, P. F. and Exton, J. H. (1986) J. Biol. Chem. **261**, 11056–11063
Blackmore, P. F., Brumley, F. T., Marks, J. L. and Exton, J. H. (1978) J. Biol. Chem. **253**, 4851–4858
Blackmore, P. F., Waynick, L. E., Blackman, G. E., Graham, C. W. and Sherry, R. S. (1984) J. Biol. Chem. **259**, 12322–12325
Bond, M., Vadasz, G., Somlyo, A. V. and Somlyo, A. P. (1987) J. Biol. Chem. **262**, 15630–15636
Borle, A. B. (1972) J. Membr. Biol. **10**, 45–66
Borle, A. B. (1981) Rev. Physiol. Biochem. Pharmacol. **90**, 13–153
Borle, A. B. and Snowdowne, K. W. (1986) Methods Enzymol. **124**, 90–116
Brown, A. M. (1991) FASEB J. **5**, 2175–2179
Buhler, H. U., DaPrada, M., Haefely, W. and Picotti, G. B. (1978) J. Physiol. (London) **276**, 311–320
Burgess, G. M., Dooley, R. K., McKinney, J. S., Nanberg, E. and Putney, J. W., Jr. (1986) Mol. Pharmacol. **30**, 315–320
Burgess, G. M., Bird, G. S. J., Obie, J. F. and Putney, J. W., Jr. (1991) J. Biol. Chem. **266**, 4772–4781
Bygrave, F. L. (1966) Biochem. J. **101**, 488–494
Bygrave, F. L. (1967) Nature (London) **214**, 667–671
Bygrave, F. L. (1978a) Biol. Rev. **53**, 43–79
Bygrave, F. L. (1978b) Trends Biochem. Sci. **3**, 175–178
Bygrave, F. L. and Tranter, C. J. (1978) Biochem. J. **174**, 1021–1030
Bygrave, F. L., Lenton, L., Altin, J. G., Setchell, B. and Karjalainen, A. (1990) Biochem. J. **267**, 69–73
Bygrave, F. L., Gamberucci, A., Fulceri, R. and Benedetti, A. (1993a) Biochem. J. **292**, 19–22
Bygrave, F. L., Karjalainen, A. and Hamada, Y. (1993b) Cell. Signalling, in the press
Capiod, T., Noel, J., Combettes, L. and Claret, M. (1991) Biochem. J. **275**, 277–280
Chabre, O., Conklin, B. R., Lin, H. Y., Lodish, H. F., Wilson, E., Ives, H. E., Catanzariti, L., Hemmings, B. A. and Bourne, H. R. (1992) Mol. Endocrinol. **6**, 551–556
Charest, R., Blackmore, P. F., Berthon, B. and Exton, J. H. (1983) J. Biol. Chem. **258**, 8769–8773
Charest, R., Blackmore, P. F. and Exton, J. H. (1985a) J. Biol. Chem. **260**, 15789–15794
Charest, R., Prpic, V., Blackmore, P. F. and Exton, J. H. (1985b) Biochem. J. **227**, 79–90
Claret-Berthon, B., Claret, M. and Mazet, J. L. (1977) J. Physiol. (London) **272**, 529–552
Cockcroft, S. and Thomas, G. M. H. (1992) Biochem. J. **288**, 1–14
Cocks, T. M., Jenkinson, D. H. and Koller, K. (1984) Br. J. Pharmacol. **83**, 281–291
Cohen, P. (1992) Trends Biochem. Sci. **17**, 408–413
Combettes, L., Berthon, B., Binet, A. and Claret, M. (1986) Biochem. J. **237**, 675–683
Comerford, J. G. and Dawson, A. P. (1993) Biochem. J. **289**, 561–567
Connelly, P. A., Sisk, R. B., Schulman, H. and Garrison, J. C. (1987) J. Biol. Chem. **262**, 10154–10163
Crane, J. K., Campanile, C. P. and Garrison, J. C. (1982) J. Biol. Chem. **257**, 4959–4965
Creba, J. A., Downes, C. P., Hawkins, P. T., Brewster, G., Michell, R. H. and Kirk, C. J. (1983) Biochem. J. **212**, 733–747
Crofts, J. N. and Barritt, G. J. (1989) Biochem. J. **264**, 61–70
Crofts, J. N. and Barritt, G. J. (1990) Biochem. J. **269**, 579–587
Dasso, L. L. T. and Taylor, C. W. (1992) Mol. Pharmacol. **42**, 453–457
Denton, R. M. and McCormack, J. G. (1985) Am. J. Physiol. **249**, E543–E554
Eigler, W., Sacca, L. and Sherwin, R. (1979) J. Clin. Invest. **63**, 114–123
Epping, R. J. and Bygrave, F. L. (1984) Biochem. J. **223**, 735–745
Exton, J. H. (1975) Methods Enzymol. **39**, 25–36
Exton, J. H. (1981) Mol. Cell. Endocrinol. **23**, 233–264
Exton, J. H. (1985) Am. J. Physiol. **248**, E633–E647

Exton, J. H. (1988) Rev. Physiol. Biochem. Pharmacol. **111**, 117–224

Exton, J. H. (1990) J. Biol. Chem. **265**, 1–4

Ferris, C. D., Cameron, A. M., Bredt, D. S., Huganir, R. L. and Snyder, S. H. (1991) Biochem. Biophys. Res. Commun. **175**, 192–198

Freissmuth, M., Casey, P. J. and Gilman A. G. (1989) FASEB J. **3**, 2125–2131

Friedmann, N. (1972) Biochim. Biophys. Acta **274**, 215–225

Friedmann, N. and Dambach, G. (1973) Biochim. Biophys. Acta **307**, 339–403

Friedmann, N. and Park, C. R. (1968) Proc. Nat. Acad. Sci. U.S.A. **61**, 504–508

Fulceri, R., Bellomo, G., Gamberucci, A. and Benedetti, A. (1990) Biochem. J. **272**, 549–552

Fulceri, R., Bellomo, G., Mirabelli, F., Gamberucci, A. and Benedetti, A. (1991) Cell Calcium **12**, 431–439

Garcia-Sainz, J. A., Mendlovic, F. and Martinez-Olmedo, M. A. (1985) Biochem. J. **228**, 277–280

Garrison, J. C. and Borland, M. K. (1977) J. Biol. Chem. **253**, 7091–7100

Gevers, W. and Krebs, H. A. (1966) Biochem. J. **98**, 720–735

Glennon, M. C., Bird, G. S. J., Kwan, C.-Y. and Putney, J. W., Jr. (1992) J. Biol. Chem. **267**, 8230–8233

Graf, P., vom Dahl, S. and Sies, H. (1987) Biochem. J. **241**, 933–936

Hamada, Y., Karjalainen, A., Setchell, B. A., Millard, J. E. and Bygrave, F. L. (1992a) Biochem. J. **281**, 387–392

Hamada, Y., Karjalainen, A., Setchell, B. A., Millard, J. E. and Bygrave, F. L. (1992b) Biochem. J. **283**, 575–581

Hansen, C. A., Yang, L. and Williamson, J. R. (1991) J. Biol. Chem. **266**, 18573–18579

Heilbrunn, L. V. (1952) An Outline of General Physiology, 3rd edn., Saunders, Philadelphia

Holz, G. G. and Habener, J. F. (1992) Trends Biochem. Sci. **17**, 388–393

Houslay, M. D., Griffiths, S. L., Horton, Y. M., Livingstone, C., Lobban, M., Macdonald, F., Morris, N., Pryde, J., Scotland, G., Shakur, Y., Sweeney G. and Tang, E. K. Y. (1992) Biochem. Soc. Trans. **20**, 140–146

Hughes, B. P., Milton, S. E. and Barritt, G. J. (1986) Biochem. J. **238**, 793–800

Hughes, B. P., Crofts, J. N., Auld, A. M., Read, L. C. and Barritt, G. J. (1987) Biochem. J. **248**, 911–918

Irvine, R. F. (1992) FASEB J. **6**, 3085–3091

Jelinek, L. J., Lok, S., Rosenberg, G. B., Smith, R. A., Grant, F. J., Biggs, S., Bensch, P. A., Kuijper, J. L., Sheppard, P. O., Sprecher, C. A., O'Hara, P. J., Foster, D., Walker, K. M., Chen, L. H. J., McKernan, P. A. and Kindsvogel, W. (1993) Science **259**, 1614–1616

Jenkinson, D. H. and Koller, K. (1977) Br. J. Pharmacol. **59**, 163–175

Joseph, S. K., Coll, K. E., Thomas, A. P., Rubin, R. and Williamson, J. R. (1985) J. Biol. Chem. **260**, 12508–12515

Jouneaux, C., Audiger, Y., Goldsmith, P., Pecker, F. and Lotersztajn, S. (1993) J. Biol. Chem. **268**, 2368–2372

Kaibuchi, K., Takai, Y., Ogawa, Y., Kimura, S., Nishizuka, Y., Nakamura, T., Tonomaru, A. and Ichihara, A. (1982) Biochem. Biophys. Res. Commun. **104**, 105–112

Kass, G. E. N., Llopis, J., Chow, S. C., Duddy, S. K. and Orrenius, S. (1990) J. Biol. Chem. **265**, 17486–17492

Keppens, S. and de Wulf, H. (1977) Biochim. Biophys. Acta **588**, 63–69

Keppens, S. and de Wulf, H. (1984) Biochem. J. **222**, 277–280

Keppens, S., Vandenheede, J. R. and de Wulf, H. (1977) Biochim. Biophys. Acta **496**, 448–457

Kimura, S., Kugai, N., Tada, R., Kojima, I., Abe, K. and Ogata, E. (1982) Horm. Metab. Res. **14**, 133–138

Kleineke, J. and Soling, H.-D. (1985) J. Biol. Chem. **260**, 1040–1045

Kraus-Friedmann, N. (1986) Proc. Natl. Acad. Sci. U.S.A. **83**, 8943–8946

Lehninger, A. L. (1950) Physiol. Rev. **30**, 393–429

Lehninger, A. L., Carafoli, E. and Rossi, C. S. (1967) Adv. Enzymol. **29**, 259–320

Levine, M. J., Meuse, A. J., Watanabe, J., Bentivegna, L. A., Johnson, R. G., Grossman, W. and Morgan, J. P. (1990) Biophys. J. **57**, 173

Lin, S.-H., Wallace, M. A. and Fain, J. N. (1983) Endocrinology (Baltimore) **113**, 2268–2275

Llopis, J., Kass, G. E. N., Gahm, A. and Orrenius, S. (1992) Biochem. J. **284**, 243–247

Lotersztajn, S., Hanoune, J. and Pecker, F. (1981) J. Biol. Chem. **256**, 11209–11215

Lotersztajn, S., Epand, R., Mallat, A., Pavoine, C. and Pecker, F. (1985) Biochimie **67**, 1169–1176

Lotersztajn, S., Pavoine, C., Deterre, P., Capeau, J., Mallat, A., Le Nguyen, D., Dufour, M., Roust, B., Bataille, D. and Pecker, F. (1992) J. Biol. Chem. **267**, 2375–2379

Mauger, J.-P. and Claret, M. (1986) FEBS Lett. **195**, 106–110

Mauger, J.-P. and Claret, M. (1988) J. Hepatol. **7**, 278–282

Mauger, J.-P., Poggioli, J., Guesdon, F. and Claret, M. (1984) Biochem. J. **221**, 121–127

Mauger, J.-P. and Claret, M. (1985) J. Biol. Chem. **260**, 11635–11642

Mauger, J.-P., Claret, M., Pietri, F. and Hilly, M. (1989) J. Biol. Chem. **264**, 8821–8826

McKinney, J. S., Desole, M. S. and Rubin, R. P. (1989) Am. J. Physiol. **257**, C651–C657

Michell, R. W. (1975) Biochim. Biophys. Acta **415**, 81–147

Michell, R. W. (1992) Trends Biochem. Sci. **17**, 274–276

Milligan, G. (1992) Biochem. Soc. Trans. **20**, 135–140

Mine, T., Kojima, I. and Ogata, E. (1988a) Acta Endocrinol. **119**, 301–306

Mine, T., Kojima, I. and Ogata, E. (1988b) Biochim. Biophys. Acta **970**, 166–171

Morgan, N. G., Blackmore, P. F. and Exton, J. H. (1983) J. Biol. Chem. **258**, 5110–5116

Morgan, N. G., Charest, R., Blackmore, P. F. and Exton, J. H. (1984) Proc. Natl. Acad. Sci. U.S.A. **81**, 4208–4212

Morgan, N. G., Charest, R., Blackmore, P. F. and Exton, J. H. (1985) Biochem. Soc. Trans. **13**, 217–218

Nathanson, M. H., Moyer, M. S., Burgstahler, A. D., O'Carroll, A.-M., Brownstein, M. J. and Lolait, S. J. (1992) J. Biol. Chem. **267**, 23282–23289

Neher, E. (1992) Nature (London) **355**, 298–299

Newsholme, E. A. and Crabtree, B. (1976) Biochem. Soc. Symp. **41**, 61–109

Pittner, R. A. and Fain, J. N. (1989) Biochem. J. **257**, 455–460

Pittner, R. A. and Fain, J. N. (1991) Biochem. J. **277**, 371–378

Poggioli, J., Berthon, B. and Claret, M. (1980) FEBS Lett. **115**, 243–246

Poggioli, J., Mauger, J.-P. and Claret, M. (1986) Biochem. J. **235**, 663–669

Prpic, V. Spencer, T. L. and Bygrave, F. L. (1978) Biochem. J. **176**, 705–714

Prpic, V., Blackmore, P. F. and Exton, J. H. (1982) J. Biol. Chem. **257**, 11323–11331

Putney, J. W., Jr. (1990) Cell Calcium **11**, 611–624

Rashed, H. M. and Patel, T. B. (1987) J. Biol. Chem. **262**, 15953–15958

Rasmussen, H. (1970) Science **170**, 404–412

Rasmussen, H. (1981) Calcium and cAMP as Synarchic Messengers, Wiley, New York

Rasmussen, H. and Barrett, P. Q. (1984) Physiol. Rev. **64**, 938–984

Reinhart, P. H. and Bygrave, F. L. (1981) Biochem. J. **194**, 541–549

Reinhart, P. H., Taylor, W. M. and Bygrave, F. L. (1982) Biochem. J. **208**, 619–630

Reinhart, P. H., van de Pol, E., Taylor, W. M. and Bygrave, F. L. (1984a) Biochem. J. **218**, 415–420

Reinhart, P. H., Taylor, W. M. and Bygrave, F. L. (1984b) Biochem. J. **220**, 35–42

Reinhart, P. H., Taylor, W. M. and Bygrave, F. L. (1984c) Biochem. J. **220**, 43–50

Reinhart, P. H., Taylor, W. M. and Bygrave, F. L. (1984d) Biochem. J. **223**, 1–13

Ringer, S. (1883) J. Physiol. (London) **4**, 29–42

Rizzuto, R., Simpson, A. W. M., Brini, M. and Pozzan T. (1992) Nature (London) **358**, 325–327

Robison, G. A., Butcher, R. W. and Sutherland, E. W. (1971) Cyclic AMP, Academic Press, New York

Rusinko, N. and Lee, H. C. (1989) J. Biol. Chem. **264**, 11725–11731

Sammak, P. J., Adams, S. R., Harootunian, A. T., Schliwa, M. and Tsien, R. Y. (1992) J. Cell Biol. **117**, 57–72

Sanchez-Bueno, A. and Cobbold, P. H. (1993) Biochem. J. **291**, 169–172

Sanchez-Bueno, A., Marrero, I. and Cobbold, P. H. (1993) Biochem. J. **291**, 163–168

Sawanobori, T., Takanishi, H., Hiraoka, M., Iida, Y., Kamisaka, K. and Maezawa, H. (1989) J. Cell. Physiol. **139**, 580–585

Schofl, C., Sanchez-Bueno, A., Brabant, G., Cobbold, P. H. and Cuthbertson, K. S. R. (1991) Biochem. J. **273**, 799–802

Sies, H., Graf, P. and Estrela, J. M. (1981) Proc. Natl. Acad. Sci. U.S.A. **78**, 3358–3362

Sies, H., Graf, P. and Crane, D. (1983) Biochem. J. **212**, 271–278

Simon, M. I., Strathmann, M. P. and Gautam, N. (1991) Science **252**, 802–808

Sistare, F. D., Picking, R. A. and Haynes, R. C., Jr. (1985) J. Biol. Chem. **260**, 12744–12747

Soboll, S. and Sies, H. (1989) Methods Enzymol. **174**, 118–130

Somogyi, R. and Stucki, J. W. (1991) J. Biol. Chem. **266**, 11068–11077

Somogyi, R., Zhao, M. and Stucki, J. W. (1992) Biochem. J. **286**, 869–877

Staddon, J. M. and Hansford, R. G. (1986) Biochem. J. **238**, 737–743

Staddon, J. M. and Hansford, R. G. (1989) Eur. J. Biochem. **179**, 47–52

Studer, R. K., Snowdowne, K. W. and Borle, A. B. (1984) J. Biol. Chem. **259**, 3596–3604

Taylor, C. W. (1990) Biochem. J. **272**, 1–13

Taylor, W. M., Reinhart, P. H. and Bygrave, F. L. (1983) Biochem. J. **212**, 555–565

Tennes, K. A., McKinney, J. S. and Putney, J. W., Jr. (1987) Biochem. J. **242**, 797–802

Tsien, R. Y. (1992) Am. J. Physiol. **263**, C723–C728

Tsien, R. Y., Pozzan, T. and Rink, T. J. (1982) J. Cell Biol. **94**, 325–334

Unson, C. G., Gurzenda, E. M. and Merrifield, R. B. (1989) Peptides **10**, 1171–1177

Van de Werve, G., Hue, L. and Hers, H.-G. (1977) Biochem. J. **162**, 135–142

Vargas, A. M., Halestrap, A. P. and Denton, R. M. (1982) Biochem. J. **208**, 221–229

vom Dahl, S., Graf, P. and Sies, H. (1988) Biochem. J. **251**, 843–848

Wakelam, M. J. O., Murphy, G. J., Hruby, V. J. and Houslay, M. D. (1986) Nature (London) **323**, 68–71

Williamson, J. R. and Hansen, C. A. (1987) Biochem. Actions Horm. **14**, 29–71

Williamson, J. R., Cooper, R. H. and Hoek, J. B. (1981) Biochim. Biophys. Acta **639**, 243–295

Woods, N. M., Cuthbertson, K. S. R. and Cobbold, P. H. (1986) Nature (London) **319**, 600–602

Woods, N. M., Cuthbertson, K. S. R. and Cobbold, P. H. (1987) Cell Calcium **8**, 79–100

Yamaguchi, M. (1991) Mol. Cell. Endocrinol. **75**, 65–70

Biochem. J. (1993) **294**, 1–14 (Printed in Great Britain)

REVIEW ARTICLE
Phospholipids in animal eukaryotic membranes: transverse asymmetry and movement

Alain ZACHOWSKI

Institut de Biologie Physico-Chimique, 13 rue Pierre et Marie Curie, 75005 Paris, France

INTRODUCTION

The function attributed to the lipid bilayer, found in all membranes, has varied over the years. It is a barrier between the extracellular space and the cytoplasm (plasma membrane) or between the cytoplasm and the intra-organelle medium (internal membranes). It is a barrier primarily for strict hydrophilic solutes since the bilayer core is essentially hydrophobic. On the other hand, the lipid phase is a solvent for the proteins required to promote the exchange of hydrophilic solutes between the two compartments adjoining the membrane (carriers and transporters), or to recognize external signals (receptors). These are rather passive roles and it seems that almost any lipid molecule could fulfil this function. More recently, membrane lipids have been seen to play an active role, as a specificity of certain molecular species for given cellular requirements has been established. A typical example of this is the utilization by specific phospholipases of phosphoinositides, but also of phosphatidylcholine and phosphatidylethanolamine, to produce second messengers (such as diacylglycerol and arachidonic acid) following hormonal stimulus. An abundant literature exists concerning this topic and it will not be developed below. Some phospholipids are also crucial for the activity of certain enzymes, such as protein kinase C in the cytoplasm or prothrombinase at the outer surface of circulating cells. Recent studies have shown that, at the macrophage surface, phosphatidylserine receptors exist which are important for the recognition of cells to be cleared from the bloodstream. For phospholipid interaction with enzymes or receptors occuring at one side of a membrane, the distribution of the lipids between the two membrane leaflets is a key parameter. I will describe this distribution, showing that its symmetrical or asymmetrical nature results from transmembrane movements and that it can be perturbed (transiently or permanently) by a variety of events.

LIPID TRANSMEMBRANE DISTRIBUTION

Methods used to determine lipid asymmetry

None of the methods utilized is 'perfect'. Each one has its own drawbacks and advantages. One must also keep in mind that the membrane which is studied may be perturbed with respect to its cellular state: for instance, evaluation of lipid distribution in an intracellular membrane requires cell lysis and isolation of the organelles. We will see that, in some cases, ATP plays a pivotal role in membrane asymmetry. Since during the isolation procedure, the membrane is separated from cytoplasmic ATP, a new lipid repartition may occur, different from that *in situ*. If one is interested in the plasma membrane of nucleated cells, one should add that membrane rupture during cell lysis induces an important perturbation of lipid transverse distribution (Connor et al., 1990; Schrier et al., 1992b).

To determine the transmembrane distribution of endogenous lipids, several methods have been used (Etemadi, 1980; Op den Kamp, 1979; van Deenen, 1981). The principle of all these

techniques is to modify the exposed lipids of the outer leaflet and to quantify them in comparison with the unmodified molecules from the cryptic inner leaflet. This assumption that only the outer molecules are modified will be fulfilled if no appreciable transmembrane movement of lipids occurs during the assay. If this is the case, the kinetics of appearance of modified lipids can be fitted by a single exponential whose plateau corresponds to the outer leaflet composition. If it is not the case, the kinetics can be fitted by the sum of two exponentials, indicating that the membrane lipids are distributed in two pools (Figure 1, and Bloj and Zilversmit, 1976). When the limiting step is the transposition of lipid molecules from the inner to the outer monolayer ($k_a \gg k_{io}$, Figure 1), pool sizes and exchange constants can be derived from a model analysis. When the rate constant of transmembrane relocalization is close to or higher than the rate constant of the assay reaction ($k_{io} \geqslant k_a$) it is not possible to deduce the lipid disposition within the membrane.

A common assay employs hydrolysis of the membrane phospholipids by phospholipases (A_2 or C) and sphingomyelinase. Reaction time with these enzymes is rather long (30–60 min), and when a fast transmembrane redistribution of phospholipids occurs, such as after diamide treatment of human erythrocytes, the result leads to an erroneous description of the asymmetry (Franck et al., 1986; Haest et al., 1978). This drawback can be avoided if one uses palmitoylated phospholipases, which are able to hydrolyse the exposed phospholipids within 5 min. However, the reaction products are lysophospholipids and free fatty acids (phospholipase A_2) or diacylglycerol (phospholipase C) or ceramide (sphingomyelinase), which may also perturb membrane organization. Perturbations

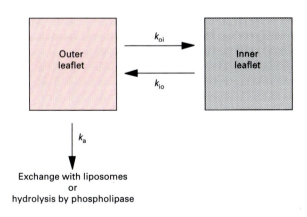

Figure 1   Principle of the two-pool model which has to be considered in the assay of phospholipid asymmetry (adapted from Bloj and Zilversmit, 1976)

When $k_a \gg k_{io}$ (k_{oi}), the kinetics of product formation (exchanged phospholipid or hydrolysed compounds) can be fitted by a single exponential. If not, the experimental points will be fitted by a sum of two exponentials, whose characteristics are a function of the three kinetic parameters.

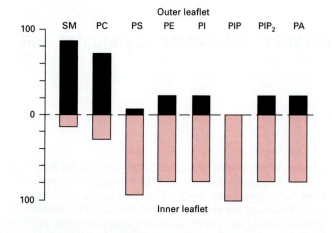

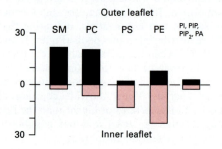

Figure 2   Asymmetrical distribution of phospholipids in the human erythrocyte membrane

Top: distribution is given as a percentage of each phospholipid; bottom: distribution is given as a percentage of total phospholipids. Abbreviations: SM, sphingomyelin; PC, phosphatidylcholine; PE, phosphatidylethanolamine; PS, phosphatidylserine; PI, phosphatidylinositol; PIP, phosphatidylinositol 4-phosphate; PIP$_2$, phosphatidylinositol 4,5-bisphosphate; PA, phosphatidic acid. Data are collected from: Bretscher (1971a,b), Butikofer et al. (1990), Gascard et al. (1991), Gordesky et al. (1972, 1975), van Meer et al. (1981) and Verkleij et al. (1973).

can be minimized by shortening the reaction time, thus degrading only a small fraction of the lipids, but one can ask whether results are truly representative of the whole outer leaflet. Cholesterol distribution is often studied using cholesterol oxidase; here again, reaction products could destabilize the bilayer. Moreover, the membrane has to be modified (addition of lysoderivatives or detergent) before the reaction can proceed.

Chemical labelling of phospholipids by specific group-reagents is often a fast reaction; however, it cannot be applied to all lipids, and the modification concerns only a fraction of the potentially reactive molecules. Moreover, one has to find experimental conditions where the reagent does not cross the bilayer and label the inner layer phospholipids (Gordesky et al., 1975).

The use of exogenous phospholipid exchange proteins may appear to be minimally perturbing for the bilayer, but the exchange of membrane phospholipids with exogenous counterparts is slow and can interfere with transmembrane movement (Bloj and Zilversmit, 1976). Exchange efficiency also depends on the lipid species (Geldwerth et al., 1991).

Use of antibodies specific to given phospholipids might be a promising tool, but, so far, this approach has been used qualitatively (to prove the presence of a lipid) rather than quantitatively (Gascard et al., 1991; Maneta-Peyret et al., 1989).

Transmembrane distribution can also be determined with exogenous lipid probes. The molecule must carry a reporter

group (e.g. fluorescent or paramagnetic) and the assumption is made that it is a 'good' analogue of the endogenous counterpart (which in fact remains to be demonstrated, especially when a bulky reporter group is involved). The probe is introduced (addition or exchange) in trace amounts into the membrane (to minimally perturb the bilayer), and its final distribution is determined, for instance, by selective quenching or extraction of the molecules on one side of the membrane only. As this last step can be performed at low temperature, the rate constant of transbilayer diffusion will be slower than the time required for the assay and there is no contribution from the other monolayer components. A drawback of the method is the chemical nature of the analogues used: in order to allow easy incorporation into the membrane from the external aqueous phase by a simple phase partition (without requiring a phospholipid exchange protein), these probes either are lysoderivatives, which may not be representative of the diacyl molecules, or possess a β-acyl chain short enough to provide some aqueous solubility. The latter probes ressemble nicked, peroxidized lipids and the cell repair system (de-acylase, re-acylase?) can recognize and metabolize them (Bishop and Bell, 1985; Seigneuret et al., 1984; Sune et al., 1987).

Asymmetry in various membranes

Plasma membranes

The plasma membrane of the human erythrocyte is a system in which all the transmembrane asymmetry determinations have given broadly similar results: 65–75 % of the phosphatidylcholine and > 85 % of the sphingomyelin are in the outer leaflet, while the inner one contains 80–85 % of the phosphatidylethanolamine and > 96 % of the phosphatidylserine (Figure 2). More recently, the distribution of the phosphoinositides was established (Büti-kofer et al., 1990; Gascard et al., 1991): 100 % of the phosphatidylinositol 4-phosphate, and 80 % of the phosphatidylinositol and of its 4,5-bisphosphate derivative are located in the cytoplasmic leaflet, as is 80 % of the phosphatidic acid. In fact, this distribution, with an outer surface composed principally of the choline-containing phospholipids phosphatidylcholine and sphingomyelin, seems to be a general trend for the plasma membrane of animal cells (see Table 1 and reviews by Devaux, 1992; Op den Kamp, 1979; Zachowski and Devaux, 1990). This outer monolayer is often poor in aminophospholipids, phosphatidylethanolamine and phosphatidylserine, but some plasma membranes can contain large amounts of them. For example, intestinal brush border or hepatocyte contiguous surface membranes have at least 20 % of their phosphatidylserine located in the extracellular leaflet. This distribution has to be considered in the light of the fact that these membranes were purified from a cell homogenate: cell membrane rupture occurring during lysis can induce important redistributions (up to 20–25 %) of all the phospholipids in the bilayer, as tested in the case of human erythrocyte membranes (Schrier et al., 1992b; Shukla et al., 1978). The values of exposed phosphatidylserine might thus not be entirely representative of the distribution in vivo. In the case of the myoblast plasma membrane, the large quantity of both phosphatidylserine and phosphatidylethanolamine located in the extracellular leaflet was supposed to reflect the ability of these cells to fuse into a myotube. Indeed, a surface rich in aminophospholipids is fusion-competent (see below).

Besides phospholipids, plasma membranes contain a high amount of cholesterol, but the distribution of this lipid is still a subject of debate: some reports are in favour of an almost symmetrical repartition in the two leaflets (Blau and Bittmann, 1978; Lange and Slayton, 1982), while others have determined

Table 1   Percentage of each main phospholipid class present in the outer leaflet of various animal plasma membranes

Abbreviations: PC, phosphatidylcholine; PE, phosphatidylethanolamine; PS, phosphatidylserine.

Cell	Sphingomyelin	PC	PE	PS	Reference
Human erythrocyte	80	77	20	< 4	See Figure 2
Mouse erythrocyte	85	50	20	0	Rawyler et al., 1985
Rat erythrocyte	100	62	20	6	Renooij et al., 1976
	100	63	0		Crain and Zilversmit, 1980
Monkey erythrocyte	82	67	13	0	van der Schaft et al., 1987
Human platelet	93	45	20	9	Perret et al., 1979
Pig platelet	91	40	34	6	Chap et al., 1977
Mouse erythroleukaemic cell	80	45	50	15	Rawyler et al., 1985
LM cell		48	24		Sandra and Pagano, 1978
Mouse synaptosome			10–15	20	Fontaine et al., 1980
Rabbit intestinal brush border	63	32	34	44	Lipka et al., 1991
		30	28		Barsukov et al., 1986
Rabbit kidney brush border	80	35	23	15	Venien and Le Grimellec, 1988
Trout intestinal brush border					
Middle			46	32	Pelletier et al., 1987
Posterior			36	31	
Rat cardiac sarcolemna	93	43	25	0	Post et al., 1988
Krebs ascites cell	47	52	45	19	Record et al., 1984
Rat hepatocyte					
Bile canalicular surface	65	85	50	0	Higgins and Evans, 1978
Contiguous surface		80	0	20	
Sinusoidal surface	65	85	55	0	
Chick embryo fibroblast			35	20	Sessions and Horwitz, 1983
Chick embryo myoblast			65	45	Sessions and Horwitz, 1983
Quail embryo myoblast			73	44	Sessions and Horwitz, 1983

Table 2   Percentage of each main phospholipid class present in the outer (cytoplasmic) leaflet of various animal intracellular membranes

Abbreviations: PC, phosphatidylcholine; PE, phosphatidylethanolamine; PS, phosphatidylserine.

Cell	Sphingomyelin	Cardiolipin	PC	PE	PS	Reference
Bovine rod outer segment disk			55	50	40	Drenthe et al., 1980
Rabbit muscle sarcoplasmic reticulum	25		48	69	15	Herbette et al., 1984
Pig gastric mucosa vesicle	45		80	85	60	Olaisson et al., 1985
Ray synaptic vesicle				60	40	Deutsch and Kelly, 1981
Rat liver endoplasmic reticulum						
Rough	58		68	40	26	Bollen and Higgins, 1980
Smooth	63		76	40	12	
Chick brain microsomes	60		90	32	0	Dominski et al., 1983
Bovine chromaffin granule	20		65	67	50	Buckland et al., 1978
Inner mitochondrial membrane						
Rat liver				35		Marinetti et al., 1976
		18	54	91		Nilsson and Dallner, 1977
				55		Crain and Marinetti, 1979
Beef heart		25	70	38		Krebs et al., 1979
Pig heart		10	50	80		Harb et al., 1981

an uneven distribution of cholesterol (Brasaemle et al., 1988; Fisher, 1976; Schroeder et al., 1991).

Intracellular organelles

The architecture of some intracellular membranes is given in Table 2. Not all the reported determinations are listed and I have chosen examples of the various organelles. Differences between these membranes can be explained by the specificity of their cellular function. In certain cases, groups have reported different results with the same membrane, such as the rat liver inner mitochondrial membrane. Possible sources of discrepancies are the isolation procedure (degree of contamination by other membranes, impermeability of the membrane during the assay) and, as cited above, the competition between the kinetics of the asymmetry assay and of transmembrane diffusion of the phospholipids. For inner mitochondrial membranes, the transbilayer motion of phosphatidylcholine and phosphatidylethanolamine can be rapid ($t_{\frac{1}{2}}$ of the order of 10 min; A. Maftah, personal communication) and even applying the two-pool model, it may be difficult to differentiate clearly between the two layers.

Asymmetric fluidity

An asymmetric distribution of the phospholipids may have important consequences for the physical properties of the mem-

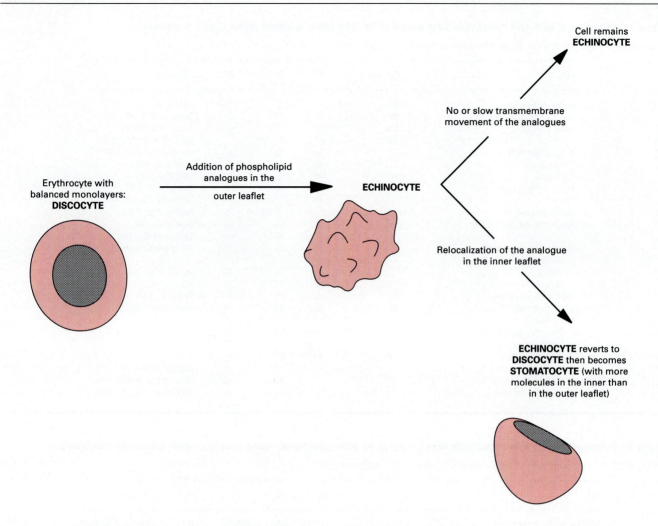

Figure 3   How the erythrocyte shape can help in studying the transmembrane redistribution of phospholipids

The requirement is that shape change depends only on the analogue used, and not on a membrane rearrangement of endogenous molecules induced by the analogue, such as scrambling.

brane. In fact, the nature of the fatty acyl chains varies with the head group as has been well documented for the human erythrocyte membrane (Myher et al., 1989): phosphatidylserine molecules have longer and more unsaturated chains than phosphatidylcholine molecules. As a consequence, the asymmetry of the polar head groups is accompanied by an asymmetry of the fatty acids and one can anticipate different chemicophysical properties for each leaflet. Thus lower local microviscosity (Morrot et al., 1989; Seigneuret et al., 1984; Tanaka and Ohnishi, 1976) and a faster lateral diffusion (Morrot et al., 1986; Rimon et al., 1984) have been ascribed to the lipids in the cytoplasmic erythrocyte leaflet. Similar properties exist also in other cells, such as fibroblasts (El-Hage Chahine et al., 1993). Studies with artificial membranes reconstituted from the characteristic phospholipids of each monolayer demonstrated that differences in microviscosity and diffusibility were primarily due to the phospholipid composition of the membrane (Cribier et al., 1990).

TRANSMEMBRANE MOTION

Assays

Kinetic parameters of lipid movements between the two hemi-leaflets can sometimes be inferred from the assay of the lipid distribution. Thus, data analysis using the two-pool model allows one to determine the characteristic time of diffusion from the inner to the outer leaflet. However, it is not possible to get information on movement in the opposite direction. A complete, detailed analysis requires the use of lipid analogues which are radioactive, fluorescent or paramagnetic. In all cases, the amount of analogue incorporated into the membrane will be low (< 1 % of the endogenous lipids) in order to minimize any perturbation due to the probe. At time zero, these analogues have to be localized in one leaflet, and their redistribution into the other leaflet must be determined as a function of the incubation time. If the analogues are strictly water-insoluble, their insertion into the membrane requires either the use of an exchange protein or the fusion of a liposome (Rousselet et al., 1976a,b; Schrier et al., 1992b; Tilley et al., 1986), and determination of the labelled fraction still exposed in the outer membrane half is made by selective modification of the molecules in this leaflet (degradation by phospholipases, reduction of the paramagnetic group by ascorbate). If the analogues are partially water-soluble, their spontaneous insertion in the membrane corresponds to a partition between the aqueous buffer and the hydrophobic bilayer. Their distribution between the two monolayers with time is determined either by selective destruction of the probe in the outer leaflet (Seigneuret and Devaux, 1984) or by selective extraction of the analogue from the external monolayer using

Table 3   Characteristics of passive diffusion of various lipids in some membrane bilayers

Abbreviations: FA, fatty acid; PL, phospholipid; PA, phosphatidic acid; PC, phosphatidylcholine; PS, phosphatidylserine; PE, phosphatidylethanolamine.

Membrane	Lipid	$t_{\frac{1}{2}}$	References
Egg PC	Spin label-PC	6.5 h	Kornberg and McConnell, 1971
	Spin label-FA	5–20 s	Yuann and Morse, 1991
	Bile acids	5–140 ms	Cabral et al., 1987
1-Palmitoyl-2-oleoyl-PC	Pyrene-PL	6–350 h	Homan and Pownall, 1988
Dioleoyl-PC	Dioleoyl-PC	> 11 h	Rothman and Dawidowicz, 1975
Dioleoyl-PC + glycophorin	Dioleoyl-PC	< 1 h	de Kruijff et al., 1978
Dioleoyl-PC/dioleoyl-PA	Neutral PA	25 s	Eastman et al., 1991
	Charged PA	12.6 days	
Influenza virus	PC	> 10 days	Rothman et al., 1976
Human erythrocyte	Diacyl-PC	3–26 h	Daleke and Huestis, 1985; Middelkoop et al., 1986
	Lyso-PC	10–30 h	Bergmann et al., 1984
	Spin label-PC	8–12 h	Bitbol and Devaux, 1988; Morrot et al., 1989
Rat erythrocyte	PC	5–11 h	Crain and Zilversmit, 1980
LM cell membrane	PC	88 h	Sandra and Pagano, 1978
Intestinal brush border	Endogenous PL	8–80 h	Lipka et al., 1991
		15–20 min	Barsukov et al., 1986

back-exchange into acceptor structures such as liposomes or bovine serum albumin (Bergmann et al., 1984; Bütikofer et al., 1990; Calvez et al., 1988; Connor et al., 1990; Daleke and Huestis, 1989). In the case of erythrocytes, whose shape is sensitive to the relative amount of lipid in each of the two membrane leaflets according to the bilayer-couple hypothesis (Deuticke, 1968; Sheetz and Singer, 1974), these changes (Figure 3) have also been used to follow the transmembrane motion of lipids (Daleke and Huestis, 1985; Seigneuret and Devaux, 1984). All the assays have given comparable results, with the exception of those performed with certain fluorescent short-chain phospholipids which proved to be poor analogues of their endogenous counterparts (Colleau et al., 1991). These measurements have allowed three types of transmembrane movements to be described.

Mechanism of transbilayer migration

Passive diffusion

The diffusion rate of a molecule is a function of its concentration gradient between the membrane halves and is not saturable with increasing amounts of substrate. It is typically the movement occurring for lipids in liposomes. However, it can also be encountered for some lipids in biological membranes, the best example being phosphatidylcholine in plasma membranes.

The characteristic time of this diffusion depends both on the lipid molecule and on the host membrane. Table 3 lists some of the systems in which such a movement was found. One can expect that neutral species (i.e. with no ionized residues) will diffuse very rapidly through the membrane, as no polar groups are present. This has been shown for diacylglycerol, which diffuses with a characteristic time of less than 1 min in artificial and biological membranes (Allan et al., 1978; Ganong and Bell, 1984). A priori, this should be true for cholesterol since related molecules such as bile salts equilibrate in a bilayer with characteristic times well below 1 s (Cabral et al., 1987). However, in the literature, characteristic times for cholesterol flip-flop varied from seconds to days, even in similar membranes. It is possible that this difference is due to the assay technique: with phosphatidylcholine/cholesterol liposomes, transmembrane movement was found to be slower when assayed by exchange ($t_{\frac{1}{2}} > 6$ days; Poznansky and Lange, 1978) than when assayed by

cholesterol oxidase ($t_{\frac{1}{2}} < 1$ min; Backer and Davidowicz, 1981). Some discrepancies may be due to the probes used to mimic membrane cholesterol: spin-labelled and fluorescent analogues are more polar than cholesterol and some important groups may be lacking, such as the 3-hydroxy group. Experimental conditions, such as the addition of detergent sometimes used in the cholesterol oxidase assay, or buffer composition (Brasaemle et al., 1988), may perturb the integrity of the membrane. One also has to keep in mind that, in membranes, cholesterol has preferential interactions with sphingomyelin (Demel et al., 1979), which may explain the different rates found with artificial and natural membranes.

Considering lipids which are weakly acidic or basic, the fraction which is not ionized should also experience easy transverse diffusion. This is true for fatty acids, phosphatidic acid and phosphatidylglycerol (Table 3). In contrast, even if they can exist as uncharged molecules, phosphatidylinositol and cardiolipin have a slow diffusion across bilayers (Bütikofer et al., 1990; Eastman et al., 1991), probably because they maintain an important hydration shell around their polar head groups, which prevents them from crossing the hydrophobic core of the bilayer.

The permanently charged lipids generally exhibit a slow passive transmembrane diffusion. In artificial membranes, the movement has a characteristic time of hours to days as a function of the membrane composition and its physical state (as illustrated by data on phosphatidylcholine in Table 3). The addition of proteins to artificial membranes may sometimes accelerate lipid diffusion (de Kruijff et al., 1978; van der Steen et al., 1982), but this effect remains unspecific, contrary to other effects which will be described below. In the human erythrocyte membrane, all the reported data agree that phosphatidylcholine undergoes a passive diffusion, with a half-time of several hours. This time depends strongly on the molecular phosphatidylcholine species (Fujii et al., 1986; Middlekoop et al., 1986; van Meer et al., 1982) and on the membrane sphingomyelin and cholesterol contents (Bergmann et al., 1984; Morrot et al., 1989). Similar data have been reported in the plasma membranes of erythrocytes belonging to other species (Connor and Schroit, 1989; Crain and Zilversmit, 1980; Devaux et al., 1988) and with cultured cells (Sandra and Pagano, 1978). A slightly faster phosphatidylcholine diffusion rate is characteristic of lymphocytes (Zachowski et al., 1987), platelets (Sune et al., 1987), renal brush border (Venien and Le

Table 4   Characteristics of transmembrane movements in rat liver endoplasmic reticulum

Abbreviations: PC, phosphatidylcholine; PS, phosphatidylserine; PE, phosphatidylethanolamine; PI, phosphatidylinositol.

Phospholipid	$t_{\frac{1}{2}}$ of reorientation	References
Endogenous PE	< 10 min	Hutson and Higgins, 1982
Endogenous PC	< 5 min	van den Besselaar et al., 1978
Endogenous PC, PE, PI, PS	< 40 min	Zilversmit and Hughes, 1977
Dibutyroyl-PC	5 min	Bishop and Bell, 1985
Lyso-butyroyl-PC	< 10 min	Kawashima and Bell, 1987
Spin-labelled PS, PC, PE, sphingomyelin	20 min	Herrmann et al., 1990

Table 5   Active transport of phosphatidylserine (PS) from the outer to the inner leaflet of various plasma membranes

Cell	PS analogue	$t_{\frac{1}{2}}$ of transport	References
Human erythrocyte	Spin-label	< 5 min	Bitbol and Devaux, 1988; Calvez et al., 1988; Morrot et al., 1989
	Long-chain, diacyl	8–10 min	Daleke and Huestis, 1985, 1989; Tilley et al., 1986
	Fluorescent (NBD)	10–30 min	Colleau et al., 1991; Connor and Schroit, 1987, 1989
	Lyso-derivative	3 h	Bergmann et al., 1984
K 562 erythroblast	Spin-label	< 2 min	Cribier et al., 1993
Erythroleukaemic cells	Long-chain, diacyl	< 15 min	Middelkoop et al., 1989
Platelets	Endogenous PS	3–5 min	Comfurius et al., 1990
	Spin-label	7 min	Sune et al., 1987
	Fluorescent (NBD)	4–8 min	Tilly et al., 1990
Lymphocyte	Spin-label	7 min	Sune et al., 1988; Zachowski et al., 1987
Macrophage	Spin-label	< 3 min	J. Rosso, personal communication
Spermatozoa	Spin-label	10 min	K. Müller, personal communication
Nerve endings	Spin-label	30–40 min	Zachowski and Morot-Gaudry Talarmain, 1990
Hamster fibroblast	Fluorescent (NBD)	5–15 min	El-Hage Chahine et al., 1993; Martin and Pagano, 1987

Grimellec, 1988) and nerve endings (Zachowski and Morot-Gaudry Talarmain, 1990). In rabbit small intestinal brush border membranes, contradictory results have been reported concerning the diffusion of the endogenous phospholipids, probed by phospholipase attack: Lipka et al. (1991) found a low re-orientation rate, while Barsukov et al. (1986) reported a very fast movement. In fact, these authors did not use enzymes from the same source and it has been shown that some phospholipases can induce an accelerated reorientation of the lipids (Venien and Le Grimellec, 1988). Passive diffusion also exists in some intracellular membranes, such as bovine chromaffin granules (Zachowski et al., 1989) or inner mitochondrial membranes (Rousselet et al., 1976a).

Facilitated diffusion

As in the case of simple diffusion, lipids will diffuse according to their gradient within the bilayer. Proteins act as channels or carriers, forming a pathway for the transmembrane movement of the lipids. The diffusion rate is diminished after modification of the membrane proteins by group-specific reagents or by proteolytic cleavage and is saturable with increasing substrate concentration. The typical example of such a system is found in the rat liver endoplasmic reticulum. Using a short-chain (dibutyroyl), highly water-soluble phosphatidylcholine analogue, Bishop and Bell (1985) described a movement fulfilling all the requirements of a facilitated diffusion. All the determinations of phospholipid reorientation rates in this membrane have shown that the movement is very fast (Table 4). The involvement of component(s) specific to the endoplasmic reticulum membrane was supported by reconstitution experiments (Backer and

Davidowicz, 1987): the fast diffusion appeared only in a mixture of microsomal lipids and proteins and not in microsomal lipids alone or a mixture of erythrocyte lipids and proteins. Spin-labelled phospholipid probes showed that the transfer rates were equivalent for different molecular species and that these molecules were in competition with one another (Herrmann et al., 1990). This argues in favour of a unique system of 'pores' used by all the phospholipids. Nothing is known concerning the mechanism of action although it has been proposed that the role of proteins would be to induce non-bilayer phases. These have been detected only in endoplasmic reticulum membranes and not in liposomes made from purified lipids (van Duijn et al., 1986). A similar diffusion system may exist in rod outer segments, in which a very fast transmembrane movement of the phospholipids is inhibited by chemical and enzymic modifications of the membrane proteins; this fast reorientation is absent from artificial membranes composed of purified lipids and rhodopsin (Wu and Hubbell, 1992). Nevertheless, it was possible to detect some asymmetrical distribution of the lipids in these membranes.

Active transport by aminophospholipid translocase

In this case, phospholipids are moved between the two membrane halves against a concentration gradient and sometimes against an electrical gradient The reorientation rate will tend to a maximum velocity at high amounts of substrate. Energy, provided by cytoplasmic ATP, is required to overcome the gradient(s). Such a system was first discovered in human erythrocytes by spin-label experiments (Seigneuret and Devaux, 1984) and later confirmed by a variety of other approaches (Table 5): the two aminophospholipids, phosphatidylserine and phosphatidylethanolamine, were rapidly moved from the outer membrane layer into the inner one, on condition that the red cells (or their ghosts) contained Mg-ATP. The same specific movement was later described in plasma membranes from other cells (Table 5). It was also detected in organelles, namely bovine adrenal chromaffin granules (Zachowski et al., 1989). We hypothesized that this transport was dependent on a protein that we named 'aminophospholipid translocase' (Zachowski et al., 1986) and whose properties are summarized in the following paragraph.

The energy required to transport the molecules is provided by the complex Mg-ATP, which is hydrolysed at the cytoplasmic membrane surface with an apparent K_m in the millimolar range (Seigneuret and Devaux, 1984). The aminophospholipids are recognized in the outer leaflet, phosphatidylserine being approximately a 10-fold better substrate than phosphatidylethanolamine (Zachowski et al., 1986). Binding of ATP and of the lipid are random, independent events (Devaux et al., 1988). The stoichiometry of the transport is one lipid moved per ATP hydrolysed (Beleznai et al., 1993). The reaction is inhibited by vanadate (Seigneuret and Devaux, 1984) and fluoride, which argues in favour of the involvement of a P-type ATPase. Transport is sensitive to chemical reagents such as N-ethylmaleimide (Daleke and Huestis, 1985; Zachowski et al., 1986) or pyridyldithioethylamine (Connor and Schroit, 1988) and to proteolytic enzymes (Bevers et al., 1989), and is inhibited by cytoplasmic free calcium in the micromolar range (Bitbol et al., 1987). The phospholipid pattern recognized by the enzyme is rather extended (Morrot et al., 1989). (a) The head group has to carry a dissociable amino group, separated from the phosphate group by two methylenes; a carboxyl group reinforces the association, but is unable to create it. (b) If the glyceride backbone is substituted by a ceramide backbone, the transport efficiency is reduced approximately 1000-fold, even though a serine head-group is present. (c) Part of the chain in the sn-2 position of the glycerol is important: lysophosphatidylserine is a poor substrate, but an esterified acetyl group already restores some specificity; saturated and unsaturated species are equally well recognized (Daleke and Huestis, 1985; Tilley et al., 1986). Long-chain lipids (Tilley et al., 1986) and lipids with a short 4-(doxyl)-pentanoyl chain (Morrot et al., 1989) exhibit similar transport kinetics, indicating that the nature of the chain is not very important. However, the presence in sn-2 of a (6-NBD)-caproyl chain decreased the affinity for phosphatidylserine by one order of magnitude and prevented phosphatidylethanolamine from being recognized (Colleau et al., 1991). This abnormal behaviour is no longer found when the fluorescent fatty acid is a dodecanoyl one (Connor et al., 1992). (d) The bond of the chain in sn-1 is not crucial since the transport of diacylphosphatidylethanolamine and of plasmalogen phosphatidylethanolamine are not distinguishable (P. Fellmann, personal communication).

Aminophospholipid translocase is not yet fully characterized at the molecular level. Morrot et al. (1990) and Daleke et al. (1991) have proposed that an Mg-ATPase of M_r 115000–120000 is the major component of the transport system. A partially purified protein fraction from the erythrocyte membrane possesses an ATPase activity which requires the presence of phosphatidylserine. No other phospholipid can substitute for phosphatidylserine, but phosphatidylethanolamine results in reduced activity. Moreover, the ATPase is inhibited by vanadate, vanadyl, fluoride and calcium ions, as is aminophospholipid translocase. Electrophoresis of the fraction, after incubation with radioactive ATP, shows a major labelled band at 115–120 kDa. Moreover, an ATPase of similar molecular mass is found in chromaffin granule membranes. Connor and Schroit (1987, 1988), on the basis of membrane labelling by inhibitors of

phosphatidylserine translocation, concluded that a polypeptide of 32 kDa is an important component of the transport system. This peptide would belong to the Rh blood group system (Schroit et al., 1990); as Rh protein does not contain ATP-binding sites, this peptide has to be associated with an ATP-utilizing protein (Schroit and Zwaal, 1991). Thus, these two proposals are not mutually exclusive.

Other mechanism(s)

It is also possible that phospholipids are moved through the membrane as they are synthesized. In rat (Renooij et al., 1976) and in human (Andrick et al., 1991) erythrocytes, phosphatidylcholine newly synthesized from lysophosphatidylcholine and fatty acid in the cytoplasmic leaflet is relocated in the outer leaflet with a halftime of translocation of approx. 20 min, i.e. ten times more rapidly than could be expected on the basis of a simple diffusion (see Table 3). Note that no outward translocation of newly synthesized phosphatidylethanolamine can be demonstrated, indicating some specificity for phosphatidylcholine (Andrick et al., 1991). Similarly, methylation of phosphatidylethanolamine into phosphatidylcholine at the inner face is accompanied by transmembrane reorientation of the latter lipid (Hirata and Axelrod, 1978).

RELATION BETWEEN MOVEMENT AND ASYMMETRY

A simple diffusion process alone is unable to create an asymmetrical distribution of lipids in a bilayer. To reach such a situation, there has to be a local environment able to trap a lipid in the leaflet where it is to be accumulated. Cullis's group has published several reports (Eastman et al., 1991; Hope and Cullis, 1987; Redelmeier et al., 1990) showing how to generate an asymmetric distribution of lipids which are weak acids or weak bases. It was noted above that such molecules (phosphatidic acid, for instance) diffuse very quickly through a bilayer when in their neutral form, which leads to a symmetrical distribution of the neutral molecules. As Table 3 shows, charged molecules have a slow diffusion through the bilayer. On each side of the membrane, the neutral species are in equilibrium with the charged ones as a function of the pK of the dissociable group and the local pH. Thus, if a pH gradient exists over the membrane, one can achieve an asymmetrical distribution of these lipids. However, this mechanism is applicable solely to minor components of an animal cell (free fatty acids, phosphatidic acid and phosphatidylglycerol) and a high asymmetry requires a strong pH gradient which exists only in few organelles (lysosomes, gastric mucosal vesicles). Since such a large proton gradient reflects the activity of an H^+-ATPase, the asymmetry of these lipids can be considered as energy-dependent. Hubbell (1990) has proposed a model in the disk membranes, where the electrostatic force trapping the lipids on one side of the membrane would result from an asymmetrical charge distribution on proteins, namely rhodopsin.

Based on the model of the human erythrocyte membrane, it was proposed that phosphatidylserine is maintained in the inner layer by interactions with cytoskeletal proteins, such as spectrin, which would prevent the lipid from diffusing towards the extracellular layer (Haest et al., 1978; Middelkoop et al., 1988). Recent experiments have shown that, if this phenomenon exists, it can only play a minor role in maintaining the asymmetry. The energy of the phosphatidylserine–spectrin interaction, under physiological conditions, is small and in the range of thermal agitation (Maksymiw et al., 1987). If an interaction exists, phosphatidylserine would be distributed in domains detectable by spectroscopic techniques, such as electron spin resonance; no signs of such domains were ever detected in spectra (Seigneuret et al., 1984). Moreover, it was shown that any lipid which is in the inner leaflet is able to move into the outer one (Bitbol and Devaux, 1988; Connor et al., 1992). In addition, heat-denaturation of spectrin did not modify the membrane asymmetry (Gudi et al., 1990) and spectrin-free erythrocyte micro-vesicles were claimed to retain a close-to-normal asymmetry (Raval and Allan, 1984). Finally, experiments with erythrocyte vesicles poor in or free from spectrin have clearly demonstrated that it is possible to obtain an asymmetrical distribution of the aminophospholipids as a result of aminophospholipid trans-locase activity (Beleznai et al., 1993; Calvez et al., 1988). In fact, phosphatidylserine, phosphatidylethanolamine and phospha-tidylcholine are subject to diffusion in both directions (Bitbol and Devaux, 1988; Connor et al., 1992); their steady-state distribution is the result of these movements. Phosphatidylserine (and to a lesser extent phosphatidylethanolamine) accumulates in the cytoplasmic layer because its inward transport is much more efficient than its outward motion. As for phosphatidyl-choline, the outward diffusion rate being 2–3 times higher than the inward one, it is more abundant in the outer layer. Sphingo-myelin distribution is more puzzling, as no inward motion can be seen in the erythrocyte (Morrot et al., 1989; Zachowski et al., 1985), which is in accordance with the fact that the sphingomyelin populations found in each leaflet differ in their species (Boegheim et al., 1983). Apparently, this absence of diffusion might be explained by sphingomyelin–cholesterol interactions, as depleting the erythrocyte membrane in cholesterol stimulated sphingo-myelin transmembrane movement (our unpublished data). This absence of sphingomyelin diffusion was also detected in other cells, namely hamster fibroblasts (El-Hage Chahine et al., 1993).

The inner-to-outer lipid movement is reported to be stimulated by cytoplasmic ATP and is dependent on protein integrity (Bitbol and Devaux, 1988; Connor et al., 1992). This can explain why in systems without ATP such as viruses (Allan and Quinn, 1989) or erythrocyte vesicles (Raval and Allan, 1984), membrane asymmetry is very slowly quenched.

PERTURBING THE TRANSMEMBRANE ASYMMETRY

During cell life, events can occur where asymmetry is partially or completely destroyed after sudden scrambling of the two mem-brane halves. Some of these events are 'laboratory perturbations' and are unlikely to happen physiologically. On the other hand, some are physiologically relevant, and correspond to a stimulated state.

Scrambling by membrane rupture

Lysis of a cell is obviously something which normally implies cell death. However, red cells can be lysed under controlled condi-tions, then tightly resealed, entrapping defined media (ghosts). Lipid asymmetry in ghosts is less marked than in intact cells (Connor et al., 1990; Schrier et al., 1992b; Shukla et al., 1978). Restoration of the aminophospholipid asymmetry (but not that of phosphatidylcholine or sphingomyelin, which diffuse too slowly) can occur rapidly, provided the ghosts contain Mg-ATP, to allow the aminophospholipid translocase to function. One can hypothesize that the asymmetry perturbation due to cell lysis is not peculiar to erythrocytes and may occur with any other cell. This implies that any determination of lipid disposition in isolated plasma membranes may not reflect the *in situ* asymmetry.

Scrambling by amphiphilic drugs

Cationic amphiphilic compounds, such as chlorpromazine or vinblastine, are known to induce a shape change of erythrocytes,

from discocytes to stomatocytes. According to the bilayer couple hypothesis (Deuticke, 1968; Sheetz and Singer, 1974), this change is due to a greater relative expansion of the cytoplasmic leaflet by accumulation of the compound in question. Surprisingly, this shape change depends on the hydrolysis of cytoplasmic ATP (Schrier et al., 1986), at least when moderate amounts of amphipathic species were used. In fact, the insertion of such molecules into the membrane induced, by an unknown mech-anism, the scrambling of a small fraction (10–20 %) of the membrane lipids (Schrier et al., 1983, 1992a). When ATP is available, the aminophospholipid translocase will remove phosphatidylserine and phosphatidylethanolamine from the outer leaflet and relocate them in the inner one. This movement is not compensated by an opposite and equal movement of phosphatidylcholine and sphingomyelin, and thus there will be an excess of phospholipids in the cytoplasmic monolayer, resulting in the appearance of stomatocytes. At very high amounts of cationic amphipath, the overpopulation of the internal surface is also due to the molecule itself, which is slightly asymmetrically distributed in the bilayer due to the pH gradient.

Scrambling by cellular activation

Platelets are activated by stimuli which induce an increase of the free cytoplasmic calcium concentration. Among the various cellular modifications which follow activation (for a review, see Siess, 1989), one is directly related to the orientation of phospholipids in the plasma membrane. Activated platelets offer a procoagulant surface which is due to the appearance of phosphatidylserine in the outer layer. The degree of phospha-tidylserine exposure strongly depends on the agonist used to activate the platelets (Bevers et al., 1983; Zwaal et al., 1989). Parallel to this outward and sudden movement of phosphatidyl-serine, and also phosphatidylethanolamine (originally present at > 90 % and 80 % respectively in the inner leaflet), there is an opposite movement of sphingomyelin (initially present at more than 90 % in the outer layer). Since phosphatidylcholine is almost symmetrically exposed, it is not easy to detect its redistribution. Strong agonists, such as a mixture of collagen and thrombin, or calcium ionophore, induce an almost complete randomization of the phospholipids. Other membrane events occur during the activation: weak agonists induce the fusion of intracellular granules with the plasma membrane and the release of their internal contents into the external medium; strong agonists induce, in addition, the budding of the plasma membrane followed by the shedding of vesicles. The liberation of these vesicles, as well as the exocytosis of the cell granules, involve the fusion (or fission) of opposing zones of membrane. At the point of fusion/fission, transient non-bilayer structures will appear (Burger and Verkleij, 1990; Sims et al., 1989), allowing a fast mixing of the components within the bilayers. Once these events have ceased, the aminophospholipid translocase should restore the internal localization of phosphatidylserine and phosphatidyl-ethanolamine. In fact, stimulation by weak agonists causes only a slight increase (from 0.1 to 1 μM) in cytosolic calcium, which remains below the inhibitory concentration for translocase activity, and transport can correct the phosphatidylserine and phosphatidylethanolamine distribution (in fact, aminophospho-lipid translocase activity is even stimulated under these condi-tions; Tilly et al., 1990). Strong agonists induce a much higher cytosolic calcium concentration, which completely inhibits the aminophospholipid translocase and thus the scrambled repar-tition of the phospholipids is maintained Moreover, in these cells, two other mechanisms also inhibit the transport system: oxidation of thiol groups and protein degradion by intracellular calpain (Comfurius et al., 1990).

Scrambling of the phospholipids between the two membrane halves can be induced in erythrocytes by increasing the cytosolic free calcium using an ionophore (Comfurius et al., 1990; Verhoven et al., 1992; Williamson et al., 1992). All the phospholipids are affected by this rearrangement, and its extent depends on the calcium concentration (Chandra et al., 1987). Aminophospholipid translocase is inhibited by intracellular calcium (Bitbol et al., 1987) and the normal asymmetry of phosphatidylserine can be restored after removal of calcium and regeneration of the usual ATP concentration (Comfurius et al., 1990). At high calcium load, the shedding of microvesicles from the erythrocyte (Allan and Michell, 1975) could explain the scrambling. Actually, the fusion between opposing faces of membranes, necessary for shedding, induces the formation of temporary non-bilayer structures allowing a rapid redistribution of molecules between the monolayers (Sims et al., 1989). At lower calcium concentrations, there is no such budding and the origin of the fast exchange of phospholipids between the membrane halves may involve phosphatidylinositol 4,5-bisphosphate. Recent experiments (Gascard et al., 1993) have shown that incorporation of 1% phosphatidylinositol 4,5-bisphosphate into the outer leaflet (i.e. an amount equivalent to that in the inner leaflet), followed by the addition of calcium, induces a membrane lipid scrambling whose intensity is a function of the amount of cation.

If we assume that any fusion or fission of membranes will induce a (partial) scrambling of their components, this lipid mixing should occur during the following events: endocytosis, exocytosis, virus fusion with or budding from a cell membrane, budding of vesicles from the endoplasmic reticulum or the Golgi apparatus, and cell division. At least during virus fusion, such a lipid redistribution has been observed (Hoekstra, 1983). There may still be other cellular events where non-bilayer phases occur in membranes, such as the insertion of signal peptides in a bilayer (Killian et al., 1990). It is important to point out that such physiological events would induce moderate mixing of membrane lipids: if a large fraction of phospholipid is exchanged and if the aminophospholipid translocase is still active and restores phosphatidylserine and phosphatidylethanolamine to their original location, slow-diffusing phosphatidylcholine and sphingomyelin will be unable to compensate for this movement, and an important imbalance will exist between the two layers. This may not be compensated by cholesterol. To escape from this mechanical stress, some parts of the membrane have to be extruded, as seen through the appearance of pseudo-endocytic vesicles in erythrocytes (Schrier et al., 1992a). It is understandable that a cell cannot survive such repetitive shocks.

Consequences of scrambling on membrane fluidity

It has been noted above that the fatty acyl composition of phospholipids differs with the polar head group, and that this has consequences for the 'fluidity' of each leaflet in an asymmetric membrane. Thus, any important redistribution of phospholipids within the membrane will affect this parameter. This assumption has been verified experimentally. Thus, when comparing intact red cells and resealed ghosts, where a partial scrambling has occurred (see above), the ghost outer leaflet is more fluid than the erythrocyte outer leaflet, and the converse is true for the inner leaflet, as probed with spin-labelled phospholipids (Schrier et al., 1992b; Tanaka and Ohnishi, 1976). Increasing the free cytoplasmic Ca^{2+} concentration, which induces a (partial) randomization of the membrane phospholipids, minimized the difference

in microviscosity between the membrane halves (Yamada and Ohnishi, 1983). An increase of 'fluidity' in the outer membrane layer of cells can also be detected with merocyanine 540 (MC 540); this fluorescent dye is a probe of lipid packing (Williamson et al., 1983), and it does not stain a normal erythrocyte or an ATP-containing resealed ghost. However, when the asymmetry was perturbed by elevating the internal calcium, the cell became fluorescent (Verhoven et al., 1992; Williamson et al., 1985). When the loss of asymmetry has been counteracted by the action of the aminophospholipid translocase, upon providing ATP to a ghost, the amount of dye bound decreased, as phosphatidylserine and phosphatidylethanolamine were imported back into the internal leaflet (Verhoven et al., 1992). Quiescent platelets did not take up MC 540, but activated ones were stained by the dye (Lupu et al., 1986). The destabilization of the cell membrane during cell fusion, that we previously postulated, is obvious if one looks at the lateral diffusibility of phospholipids: the fusion of Sendai virus with human red cells provoked not only a higher transmembrane diffusion rate (Bashford, 1988) but also a higher diffusion rate in the plane of the outer membrane leaflet (Aroeti et al., 1992) which we can attribute to the appearance of phosphatidylserine and phosphatidylethanolamine in this leaflet. If the cell was deprived of ATP, this change was permanent. In the presence of ATP (i.e. when the aminophospholipid translocase was able to re-establish the concentration of phosphatidylserine and phosphatidylethanolamine in the inner leaflet), the change was transient.

POSSIBLE CELLULAR ROLE FOR LIPID ASYMMETRY

The presence of an asymmetric distribution of phospholipids in certain membranes should be important for cellular life. This section is a partial survey of cellular activities which are dependent on the phosphatidylserine and phosphatidylethanolamine content of the lipid phase. Figure 4 illustrates these membranous events.

Phosphatidylserine and protein kinase C

Protein kinase C plays an important role in the transduction of extracellular signals which induce phospholipid modification. Hydrolysis of phosphatidylinositol 4,5-bisphosphate or phosphatidylcholine by phospholipase C generates diacylglycerol, which activates protein kinase C bound in a Ca^{2+}-dependent manner to phosphatidylserine-containing membranes (Nishizuka, 1986). In fact, protein kinase C can bind to any membrane containing acidic phospholipids, but its activity depends on an interaction with phosphatidylserine (Hannun et al., 1986) and Ca^{2+} and diacylglycerol increase the specific affinity of the protein towards phosphatidylserine (Orr and Newton, 1992a,b). Phosphatidylethanolamine can substitute for phosphatidylserine, but at least 10 μM Ca^{2+} is required to observe some phosphorylation, compared with 0.1 μM or less required in the presence of phosphatidylserine (Bazzi et al., 1992). Binding of the protein and kinase activity showed an identical sigmoidal dependence on the mole fraction of phosphatidylserine in the membrane (Newton and Koshland, 1989; Orr and Newton, 1992a,b), favouring a clustering of phosphatidylserine induced by the protein (Bazzi and Nelsuesten, 1991). Substrate specificity is affected in vitro by the phosphatidylserine concentration in membranes (Newton and Koshland, 1990): autophosphorylation is favoured at low and intermediate phosphatidylserine concentrations, while phosphorylation of exogenous substrates (histones) is more important at high phosphatidylserine concentrations. However, histone phosphorylation reaches a maximum

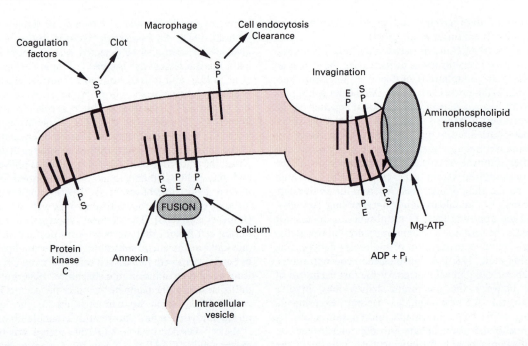

Figure 4   Schematic representation of some aminophospholipid-dependent reactions occurring at the extracellular or cytoplasmic surfaces of a plasma membrane

The polar head group of each phospholipid is represented by its usual abbreviation (see the legend to Figure 2).

at phosphatidylserine concentrations equal to 20 mol%, precluding modulation of the protein kinase C activity by the phosphatidylserine membrane content. Other substrates exhibit a greater possibility of regulation by membrane phosphatidylserine; the microtubule-associated protein τ, for instance, is poorly phosphorylated at phosphatidylserine concentrations lower than 20 mol%, and after that its phosphorylation increases linearly with phosphatidylserine content (Newton and Koshland, 1990). If such behaviour is true for other protein kinase C substrates (e.g. the glucose transporter and the insulin receptor), it may suggest how phosphatidylserine could modulate cellular activities through its regulation of protein kinase C activity.

Phosphatidylserine, phosphatidylethanolamine and membrane fusion

Fusion of two membranes, which occurs during cellular events such as exocytosis, is a very complex phenomenon, where proteins, lipids and ions are involved. It is beyond the scope of this Review to cover all the aspects of membrane fusion; however, I will consider two reactions, among many others, where lipid composition plays a critical role.

Binding of annexins

Annexins belong to a family of eight proteins which bind to membranes in a Ca^{2+}-dependent manner (Burgoyne and Geisow, 1989; Crumpton and Dedman, 1990). Their complete physiological role is as yet unknown, but they can aggregate and mediate the fusion of vesicles (Creutz et al., 1978; Drust and Creutz, 1988) and thus could modulate intracellular fusions. Annexin binding is supported by the presence of phosphatidylserine in the membrane, but other anionic phospholipids (phosphatidic acid or phosphatidylinositol) or even phosphatidylethanolamine can be sufficient (reviewed by Klee, 1988), i.e. the major phospholipids of a cytoplasmic membrane leaflet. The

amount of calcium required for annexin binding to membranes depends on the phospholipid composition and is related to the size and hydrophilicity of the lipid head group, considering that Ca^{2+} has to interact with the phosphoryl residue (Huber et al., 1990). Binding of annexins to the membrane modifies the physical state of the bilayer, as it reduces the lateral diffusion of phospholipids (Meers et al., 1991). However, even if annexins promote the aggregation of membranes, their fusion will require a mixing of the two bilayers and lipid composition modulates this phenomenon.

Creation of fusion-competent surfaces

To fuse, bilayers have first to appose; however, when these two hydrated surfaces are at 2 nm, or less, hydration pressure generates a repulsive barrier (LeNeveu et al., 1977). The fact that the fusion of two membranes requires the presence of anionic phospholipids (such as phosphatidylserine) and of Ca^{2+} (Papahadjopoulos et al., 1976) was interpreted, among other mechanisms, as a Ca^{2+}-induced close apposition of membranes without intervening water (Portis et al., 1979). This explains why lipids with a low hydration level of their head group, such as phosphatidylethanolamine, enhance liposome fusion. Note that the Ca^{2+} doses required for a fusion of two liposomes are in the millimolar range, indicating that, in the cell, the ion will be unable to promote fusion by acting directly at the lipid level. Membrane contact has to be followed by bilayer destabilization in order to fuse the membranes. It has been proposed that, in phosphatidylserine-containing membranes, another role of Ca^{2+} is to provoke an isothermic phase change from a fluid to a crystalline state (Papahadjopoulos et al., 1977), fusion occurring at the domain boundaries. Later, it was shown that fusion proceeds approximately ten times more rapidly than phase separation and that Mg^{2+}, which can induce fusion at high concentrations, does not induce phase separation (Hoekstra,

1982). A more recent hypothesis is that destabilization is generated by non-bilayer structures (Ellens et al., 1989) and that fusion between the two membranes occurs at a local point (Burger and Verkleij, 1990). Amongst lipids which easily form non-bilayer phases is phosphatidylethanolamine, abundant in cytoplasmic membrane leaflets. The nature of the phase responsible for fusion is not well defined. Formation of hexagonal H_{II} structures seems to be excluded as they would generate leaky fusion (Ellens et al., 1985). Some theoretical models favour the involvement of inverted micelles, precursors of H_{II} phases appearing in the physiological temperature range, whose lifetime agrees with the membrane fusion rate (Siegel, 1986). These inverted phases are stabilized by diacylglycerol (Kirk and Gruner, 1985) and the temperature at which they appear is noticeably lowered by the same molecules (Siegel et al., 1989). This would explain how phospholipase C induces the fusion of membranes (Burger et al., 1991).

Endocytic invagination as a consequence of exocytic events

Fusion of an exocytic vesicle with the plasma membrane will produce a redistribution of lipids in the bilayer. Phosphatidylserine sited in the outer leaflet will be relocated to the inner leaflet by aminophospholipid translocase. Movement in the opposite direction of phosphatidylcholine and sphingomyelin is much slower, leading to an excess of lipid present in the inner monolayer. This imbalance in the bilayer couple creates an invagination of the membrane and may even cause the formation of a vesicle. This has been obtained in human erythrocytes as a consequence of bilayer scrambling induced by amphipathic compounds (Schrier et al., 1992a). It is clear that endocytosis requires the action of other cellular components (Brodsky, 1988; Evans and Enrich, 1989), but lipid rearrangement is certainly one of the early steps of this phenomenon. That aminophospholipid translocase is involved in membrane curvature and invagination is supported by experiments with erythrocytes having a high ATP content. Under these conditions, an energy-dependent endocytosis is observed (Ben-Bassat et al., 1972; Penniston, 1972; Penniston and Green, 1968), attributed to the activity of a 'shape change Mg^{2+}-ATPase' (Morris et al., 1992; Patel and Fairbanks, 1986), probably the same protein as the aminophospholipid translocase. The shape change and the invagination are the consequence of an increased accumulation of phosphatidylethanolamine in the cytoplasmic monolayer as the ATP level increases (Calvez et al., 1988).

Consequences of the presence of phosphatidylserine in the outer leaflet of blood cells

Under normal conditions, almost no phosphatidylserine is present in the outer leaflet of circulating blood cells (erythrocytes, platelets, and probably leucocytes). However, this phospholipid is exposed in some pathologies, or after platelet stimulation. I will briefly list the consequences of such a change.

Formation of a procoagulant surface

Following platelet activation, there is a release into the extracellular space of coagulation proteins (such as fibrinogen and factor V), stored in the cellular α-granules, which are then activated by proteolytic degradations. This chain of reactions requires both Ca^{2+} and anionic phospholipids (phosphatidyl-

serine) exposed at the outer surface. Phosphatidylserine acts at different stages of the activation cascade, such as specific binding of activated factor VIIIa to membranes (Gilbert et al., 1992) and lowering the apparent K_m for factor X (van Dieijen et al., 1981) and for prothrombin (Rosing et al., 1980) well below their concentrations in plasma.

Phosphatidylserine and cellular recognition

It is possible to detect an abnormal amount of phosphatidylserine in the membrane outer leaflet in some pathological red cells, such as sickle cells. The phospholipid asymmetry is reduced in the old (dense) cell fraction in any conditions and in the young (light) fraction in a de-oxygenated atmosphere (Lubin et al., 1981; Blumenfeld et al., 1991). The high adherence of the sickled cells to the vascular endothelium (Hebbel et al., 1980; Hoover et al., 1979) is a direct consequence of the appearance of a significant amount of exposed phosphatidylserine: the same behaviour is obtained with ghosts from normal red cells where the asymmetry has been quenched (Schlegel et al., 1985). Erythrocytes with exposed phosphatidylserine, due to either natural (sickled cells) or artificial (lipid-symmetric ghosts and phosphatidylserine-enriched erythrocytes) causes, exhibit an increased adherence to monocytes and macrophages (McEvoy et al., 1986; Schwartz et al., 1985; Tanaka and Schroit, 1983) followed by an accelerated clearance (Schroit et al., 1985). This mechanism of recognition might be responsible for elimination of aged erythrocytes from the bloodstream, provided that the loss of phosphatidylethanolamine asymmetry found in the denser cells (Shukla and Hanahan, 1982; Herrmann and Devaux, 1990) precedes phosphatidylserine appearance on the outer leaflet. Note that artificial lipid vesicles which contain phosphatidylserine are also phagocytosed by macrophages (Schroit and Fidler, 1982), as are apoptotic lymphocytes which expose phosphatidylserine at their surface (Fadok et al., 1992). In a similar way, tumorigenic cells are bound and lysed by activated monocytes or macrophages, as they express much more phosphatidylserine in their membrane outer leaflet than their non-tumorigenic counterparts (Connor et al., 1989; Utsugi et al., 1991).

CONCLUDING REMARKS

Amazingly, research on phospholipid transmembrane asymmetry and movement is now focusing on proteins. So far, two 'flippase' proteins have been functionally characterized. The aminophospholipid translocase is responsible for the relocalization into the cytoplasmic leaflet of the plasma membrane of phosphatidylserine and phosphatidylethanolamine molecules which appear in the outer leaflet. This carrier, which is an integral Mg-ATPase, seems to be present in the plasma membrane of all animal cells. The characteristics of the transport reaction are now well established, in particular substrate specificities and stoichiometry. But nothing is known about its possible regulation by cellular components, with the exception of the inhibitory role of cell calcium. Since the enzyme is responsible for establishing and maintaining the asymmetry, it will be interesting to study if it is affected during cellular events, such as stimulation, in which the lipid phase is modified. To determine where the asymmetry is created, attempts have been made to localize the aminophospholipid translocase in cell subfractions. It is present in chromaffin granule membranes and absent from endoplasmic reticulum membranes. In this latter organelle, there exists another protein, functioning like a pore, which allows the fast equilibration of any phospholipid between the two membrane halves without requiring ATP. Transverse lipid diffusion has to be studied in other subcellular membranes in order to ascertain if one of the two 'flippases', or eventually a third one, is present. Unfortunately, all the currently described quantitative assays require a large amount of membranes which limits this search to particular cells.

More progress will be made when the various proteins have been purified and cloned. Antibodies raised against the 'flippases' will allow their localization within any cell and permit their presumably ubiquitous character to be confirmed. Structural studies of the purified proteins will help to answer the puzzling question of the architecture of a peptide which moves simultaneously the hydrophilic and hydrophobic moieties of a molecule through a lipid bilayer. Molecular biology, irrespective of information on protein sequence, will contribute to the understanding of the real involvement of the 'flippase(s)' in cell life, either by overexpression of the proteins or by selection of defective mutants. We can thus expect that the next few years will bring us a lot of interesting new information on this aspect of cell biology.

I am indebted to Dr. P. F. Devaux and Dr. R. Lavery for their helpful discussions and for their critical reading of the manuscript. This work was supported by grants from the Centre National de la Recherche Scientifique (UA 526), the Institut National de la Santé et de la Recherche Médicale and the Université Denis Diderot (Paris 7).

REFERENCES

Allan, D. and Michell, R. H. (1975) Nature (London) **258**, 348–349

Allan, D. and Quinn, P. (1989) Biochim. Biophys. Acta **987**, 199–204

Allan, D., Thomas, P. and Michell, R. H. (1978) Nature (London) **276**, 289–290

Andrick, C., Bröring, K., Deuticke, B. and Haest, C. W. M. (1991) Biochim. Biophys. Acta **1064**, 235–241

Aroeti, B., Gutman, O. and Hennis, Y. I. (1992) J. Biol. Chem. **267**, 13272–13277

Backer, J. M. and Dawidowicz, E. A. (1981) J. Biol. Chem. **256**, 586–588

Backer, J. M. and Dawidowicz, E. A. (1987) Nature (London) **327**, 341–343

Barsukov, L. I., Bergelson, L. D., Spiess, M., Hauser, H. and Semenza, G. (1986) Biochim. Biophys. Acta **862**, 87–99

Bashford, C. L. (1988) Studia Biophysica **127**, 155–162

Bazzi, M. D. and Nelsuesten, G. L. (1991) Biochemistry **30**, 7961–7970

Bazzi, M. D., Yonakim, M. A. and Nelsuesten, G. L. (1992) Biochemistry **31**, 1125–1134

Beleznai, Z., Zachowski, A., Devaux, P. F., Puente Navazo, M. and Ott, P. (1993) Biochemistry, **32**, 3146–3152

Ben-Bassat, I., Bensch, K. G. and Schrier, S. L. (1972) J. Clin. Invest. **51**, 1883–1844

Bergmann, W. L., Dressler, V., Haest, C. W. M. and Deuticke, B. (1984) Biochim. Biophys. Acta **772**, 328–336

Bevers, E. M., Comfurius, P. and Zwaal, R. F. A. (1983) Biochim. Biophys. Acta **736**, 57–66

Bevers, E. M., Tilly, R. H. J., Senden, J. M. G., Comfurius, P. and Zwaal, R. F. A. (1989) Biochemistry **28**, 2382–2387

Bishop, W. R. and Bell, R. M. (1985) Cell **42**, 51–60

Bitbol, M. and Devaux, P. F. (1988) Proc. Natl. Acad. Sci. U.S.A. **85**, 6783–6787

Bitbol, M., Fellmann, P., Zachowski, A. and Devaux, P. F. (1987) Biochim. Biophys. Acta **904**, 268–282

Blau, L. and Bittmann, R. (1978) J. Biol. Chem. **253**, 8366–8368

Bloj, B. and Zilversmit, D. B. (1976) Biochemistry **15**, 1277–1283

Bloj, B. and Zilversmit, D. B. (1977a) Biochemistry **16**, 3943–3948

Bloj, B. and Zilversmit, D. B. (1977b) Proc. Soc. Exp. Biol. Med. **156**, 539–543

Blumenfeld, N., Zachowski, A., Galacteros, F., Beuzard, Y. and Devaux, P. F. (1991) Blood **77**, 849–854

Boegheim, J. P. J., Jr., van Linde, M., Op den Kamp, J. A. F. and Roelofsen, B. (1983) Biochim. Biophys. Acta **735**, 438–442

Bollen, I. C. and Higgins, J. A. (1980) Biochem. J. **189**, 475–480

Brasaemle, D. L., Robertson, R. D. and Attie, A. D. (1988) J. Lipid Res. **29**, 481–489

Bretscher, M. S. (1972a) Nature (London) New Biol. **236**, 11–12

Bretscher, M. S. (1972b) J. Mol. Biol. **71**, 523–528

Brodsky, F. M. (1988) Science **242**, 1396–1402

Bröring, K., Haest, C. W. M. and Deuticke, B. (1989) Biochim. Biophys. Acta **986**, 321–331

Buckland, R. M., Radda, G. K. and Shennan, C. D. (1978) Biochim. Biophys. Acta **513**, 321–337

Burger, K. N. J. and Verkleij, A. J. (1990) Experientia **46**, 631–644

Burger, K. N. J., Nieva, J.-L., Alonso, A. and Verkleij, A. J. (1991) Biochim. Biophys. Acta **1068**, 249–253

Burgoyne, R. D. and Geisow, M. J. (1989) Cell Calcium **10**, 1–10

Bütikofer, P., Lin, Z. W., Chiu, D. T.-Y., Lubin, B. and Kuypers, F. A. (1990) J. Biol. Chem. **265**, 16035–16038

Cabral, D. J., Small, D. M., Lilly, H. S. and Hamilton, J. A. (1987) Biochemistry **26**, 1801–1804

Calvez, J.-Y., Zachowski, A., Herrmann, A., Morrot, G. and Devaux, P. F. (1988) Biochemistry **27**, 5666–5670

Chandra, R., Joshi, P. C., Bajpai, V. K. and Gupta, C. M. (1987) Biochim. Biophys. Acta **902**, 253–262

Chap, H. J., Zwaal, R. F. A. and van Deenen, L. L. M. (1977) Biochim. Biophys. Acta **467**, 146–164

Colleau, M., Hervé, P., Fellmann, P. and Devaux, P. F. (1991) Chem. Phys. Lipids **57**, 29–37

Comfurius, P., Senden, J. M. G., Tilly, R. H. J., Schroit, A. J., Bevers, E. M. and Zwaal, R. F. A. (1990) Biochim. Biophys. Acta **1026**, 153–160

Connor, J. and Schroit, A. J. (1987) Biochemistry **26**, 5099–5105

Connor, J. and Schroit, A. J. (1988) Biochemistry **27**, 848–851

Connor, J. and Schroit, A. J. (1989) Biochemistry **28**, 9680–9685

Connor, J., Bucana, C., Fidler, I. J. and Schroit, A. J. (1989) Proc. Natl. Acad. Sci. U.S.A. **86**, 3184–3188

Connor, J., Gillum, K. and Schroit, A. J. (1990) Biochim. Biophys. Acta **1025**, 82–86

Connor, J., Pak, C. H., Zwaal, R. F. A. and Schroit, A. J. (1992) J. Biol. Chem. **267**, 19412–19417

Crain, R. C. and Marinetti, G. V. (1979) Biochemistry **18**, 2407–2414

Crain, R. C. and Zilversmit, D. B. (1980) Biochemistry **19**, 1440–1447

Creutz, C. E., Pazoles, C. J. and Pollard, H. B. (1978) J. Biol. Chem. **253**, 2858–2866

Cribier, S., Morrot, G., Neumann, J.-M. and Devaux, P. F. (1990) Eur. Biophys. J. **18**, 33–41

Cribier, S., Sainte-Marie, J. and Devaux, P. F. (1993) Biochim. Biophys. Acta **1148**, 85–90

Crumpton, M. J. and Dedman, J. R. (1990) Nature (London) **345**, 212

Daleke, D. L. and Huestis, W. H. (1985) Biochemistry **24**, 5406–5416

Daleke, D. L. and Huestis, W. H. (1989) J. Cell Biol. **108**, 1375–1385

Daleke, D. L., Cornely-Moss, K. A. and Smith, C. M. (1991) Biophys. J. **59**, 381a

de Kruijff, B., van Zoelen, E. J. J. and van Deenen, L. L. M. (1978) Biochim. Biophys. Acta **509**, 537–542

Demel, R. A., Jansen, J. W. C. M., van Dijck, P. W. M. and van Deenen, L. L. M. (1977) Biochim. Biophys. Acta **465**, 1–10

Deuticke, B. (1968) Biochim. Biophys. Acta **163**, 494–500

Deutsch, J. W. and Kelly, R. B. (1981) Biochemistry **20**, 378–385

Devaux, P. F. (1992) Annu. Rev. Biophys. Biomol. Struct. **21**, 417–439

Devaux, P, F, Morrot, G., Herrmann, A. and Zachowski, A. (1988) Studia Biophysica **127**, 183–191

Dominski, J., Binaglia, L., Dreyfus, H., Massarelli, R., Mersel, M. and Freysz, L. (1983) Biochim. Biophys. Acta **734**, 257–266

Drenthe, E. H. S., Klompmakers, A. A., Bonting, S. L. and Daemen, F. J. M. (1980) Biochim. Biophys. Acta **603**, 130–141

Drust, D. S. and Creutz, C. E. (1988) Nature (London) **331**, 88–91

Eastman, S. J., Hope, M. J. and Cullis, P. R. (1991) Biochemistry **30**, 1740–1745

El-Hage Chahine, J.-M., Cribier, S. and Devaux, P. F. (1993) Proc. Natl. Acad. Sci. U.S.A. **90**, 447–451

Ellens, H., Bentz, J. and Szoka, F. C. (1985) Biochemistry **25**, 4141–4147

Ellens, H., Siegel, D. P., Alford, D., Yeagle, P. L., Boni, L., Lis, L. J., Quinn, P. J. and Bentz, J. (1989) Biochemistry **28**, 3692–3703

Etemadi, A.-H. (1980) Biochim. Biophys. Acta **604**, 423–475

Evans, W. H. and Enrich, C. (1989) Biochem. Soc. Trans. **17**, 619–622

Fadok, V. A., Voelker, D. R., Campbell, P. A., Cohen, J. J., Bratton, D. L. and Henson, P. M. (1992) J. Immunol. **148**, 2207–2216

Fisher, K. A. (1976) Proc. Natl. Acad. Sci. U.S.A. **73**, 173–177

Fontaine, R. N., Harris, R. A. and Schroeder, F. (1980) J. Neurochem. **34**, 269–277

Franck, P. F. H., Op den Kamp, J. A. F., Roelofsen, B. and van Deenen, L. L. M. (1986) Biochim. Biophys. Acta **857**, 127–130

Fujii, T., Tamura, A. and Yamane, T. (1986) J. Biochem. (Tokyo) **98**, 1221–1227

Ganong, B. R. and Bell, R. M. (1984) Biochemistry **23**, 4977–4983

Gascard, P., Tran, D., Sauvage, M., Sulpice, J.-C., Fukami, T., Claret, M. and Giraud, F. (1991) Biochim. Biophys. Acta **1069**, 27–36

Gascard, P., Sulpice, J.-C., Tran, D., Sauvage, M., Claret, M., Zachowski, A., Devaux, P. F. and Giraud, F. (1993) Biochem. Soc. Trans. **21**, 253–257

Geldwerth, D., de Kermel, A., Zachowski, A., Guerbette, F., Kader, J.-C., Henry, J.-P. and Devaux, P. F. (1991) Biochim. Biophys. Acta **1082**, 255–264

Gilbert, G. E., Drinkwater, D., Barter, S. and Clouse, S. B. (1992) J. Biol. Chem. **267**, 15861–15868

Gordesky, S. E., Marinetti, G. V. and Segel, G. B. (1972) Biochem. Biophys. Res. Commun. **47**, 1004–1009

Gordesky, S. E., Marinetti, G. V. and Love, R. (1975) J. Membr. Biol. **20**, 111–132

Gudi, S. R. P., Kumar, A., Bhakuni, V., Ghodale, S. M. and Gupta, C. M. (1990) Biochim. Biophys. Acta **1023**, 63–72

Haest, C. W. M., Plasa, G., Kamp, D. and Deuticke, B. (1978) Biochim. Biophys. Acta **509**, 21–32

Hannun, Y. A., Loomis, C. R. and Bell, R. M. (1986) J. Biol. Chem. **261**, 7184–7190

Harb, J. S., Comte, J. and Gautheron, D. C. (1981) Arch. Biochem. Biophys. **208**, 305–318

Hebbel, R. P., Yamada, O., Moldow, C. F., Jacob, H. S., White, J. G. and Eaton, J. W. (1980) J. Clin. Invest **65**, 154–160

Herbette, L., Blasie, J. K., Defoor, P., Fleischer, S., Bick, R. J., van Winkle, W. B., Tate, C. A. and Entman, M. L. (1984) Arch. Biochem. Biophys. **234**, 235–242

Herrmann, A. and Devaux, P. F. (1990) Biochim. Biophys. Acta **1027**, 41–46

Herrmann, A., Zachowski, A. and Devaux, P. F. (1990) Biochemistry **29**, 2023–2027

Higgins, J. A. and Evans, W. H. (1978) Biochem. J. **174**, 563–567

Hirata, F. and Axelrod, J. (1978) Proc. Natl. Acad. Sci. U.S.A. **75**, 2348–2352

Hoekstra, D. (1982) Biochemistry **21**, 2833–2840

Hoekstra, D. (1983) Exp. Cell Res. **144**, 482–488

Homan, R. and Pownall, H. J. (1987) J. Am. Chem. Soc. **109**, 4759–4760

Hoover, R., Rubin, R., Wise, G. and Warren, R. (1979) Blood **54**, 872–876

Hope, M. J. and Cullis, P. R. (1987) J. Biol. Chem. **262**, 4360–4366

Hubbell, W. L. (1990) Biophys. J. **57**, 99–108

Huber, R., Römisch, J. and Pâques, E.-P. (1990) EMBO J. **9**, 3867–3874

Hutson, J. L. and Higgins, J. A. (1982) Biochim. Biophys. Acta **687**, 247–256

Kawashima, Y. and Bell, R. M. (1987) J. Biol. Chem. **262**, 16495–16502

Killian, J. A., de Jong, A. M.Ph., Bijvelt, J., Verkleij, A. J. and de Kruijff, B. (1990) EMBO J. **9**, 815–819

Kirby, C. J. and Green, C. (1977) Biochem. J. **168**, 575–577

Kirk, G. L. and Gruner, S. M. (1985) J. Phys. **46**, 761–769

Klee, C. B. (1988) Biochemistry **27**, 6645–6653

Kornberg, R. D. and McConnell, H. M. (1971) Biochemistry **10**, 1111–1120

Krebs, J. J. R., Hauser, H. and Carafoli, E. (1979) J. Biol. Chem. **254**, 5308–5316

Lange, Y. and Slayton, J. M. (1982) J. Lipid Res. **23**, 1121–1127

Lange, Y., Dolde, J. and Steck, T. L. (1981) J. Biol. Chem. **256**, 5321–5323

Lenard, J. and Rothman, J. E. (1976) Proc. Natl. Acad. Sci. U.S.A. **73**, 391–395

LeNeveu, D. M., Rand, R. P., Parsegian, V. A. and Gingell, D. (1977) Biophys. J. **18**, 209–230

Lipka, G., Op den Kamp, J. A. F. and Hauser, H. (1991) Biochemistry **30**, 11828–11836

Lubin, B., Chiu, D., Bastacky, J., Roelofsen, B. and van Deenen, L. L. M. (1981) J. Clin. Invest. **67**, 1643–1649

Lupu, F., Caib, M., Scurei, C. and Simionescu, N. (1986) Lab. Invest. **54**, 1643–1649

Maksymiw, R., Sui, S., Gaub, H. and Sackmann, E. (1987) Biochemistry **26**, 2983–2990

Maneta-Peyret, L., Freyburger, G., Bessoule J.-J. and Cassagne, C. (1989) J. Immunol. Methods **122**, 155–159

Marinetti, G. V., Senior, A. E., Love, R. and Broadhurst, C. I. (1976) Chem. Phys. Lipids **17**, 353–362

Martin, O. C. and Pagano, R. E. (1987) J. Biol. Chem. **262**, 5890–5898

McEvoy, L., Williamson, P. and Schlegel, R. A. (1986) Proc. Natl. Acad. Sci. U.S.A. **83**, 3311–3315

Meers, P., Daleke, D., Hong, K. and Papahadjopoulos, D. (1991) Biochemistry **30**, 2903–2908

Middelkoop, E., Lubin, B. H., Op den Kamp, J. A. F. and Roelofsen, B. (1986) Biochim. Biophys. Acta **855**, 421–424

Middelkoop, E., Lubin, B. H., Bevers, E. M., Op den Kamp, J. A. F., Comfurius, P., Chiu, D. T.-Y., Zwaal, R. F. A., van Deenen, L. L. M. and Roelofsen, B. (1988) Biochim. Biophys. Acta **937**, 281–288

Middelkoop, E., Coppens, A., Llanillo, M., van der Hoek, E. E., Slotboom, A. J., Lubin, B. H., Op den Kamp, J. A. F., van Deenen, L. L. M. and Roelofsen, B. (1989) Biochim. Biophys. Acta **978**, 241–248

Morris, M. B., Monteith, G. and Roufogalis, B. D. (1992) J. Cell. Biochem. **48**, 356–366

Morrot, G., Cribier, S., Devaux, P. F., Geldwerth, D., Davoust, J., Bureau, J.-F., Fellmann, P., Hervé, P. and Frilley, B. (1986) Proc. Natl. Acad. Sci. U.S.A. **83**, 6863–6867

Morrot, G., Hervé, P., Zachowski, A., Fellmann, P. and Devaux, P. F. (1989) Biochemistry **28**, 3456–3462

Morrot, G., Zachowski, A. and Devaux, P. F. (1990) FEBS Lett. **266**, 29–32

Myher, J. J., Kuksis, A. and Pind, S. (1989) Lipids **24**, 396–407

Newton, A. C. and Koshland, D. E., Jr. (1989) J. Biol. Chem. **264**, 14909–14915

Newton, A. C. and Koshland, D. E., Jr. (1990) Biochemistry **29**, 6656–6661

Nilsson, O. S. and Dallner, G. (1977) Biochim. Biophys. Acta **464**, 453–458

Nishizuka, Y. (1986) Science **233**, 305–312

Olaisson, H., Mardh, S. and Arvidson, G. (1985) J. Biol. Chem. **260**, 11262–11267

Op den Kamp, J. A. F. (1979) Annu. Rev. Biochem. **48**, 47–71

Orr, J. W. and Newton, A. C. (1992a) Biochemistry **31**, 4661–4677

Orr, J. W. and Newton, A. C. (1992b) Biochemistry **31**, 4667–4673

Pagano, R. E. and Longmuir, K. J. (1985) J. Biol. Chem. **260**, 1909–1916

Papahadjopoulos, D., Vail, W. J., Pangborn, W. A. and Poste, G. (1976) Biochim. Biophys. Acta **448**, 265–283

Papahadjopoulos, D., Vail, W. J., Newton, C., Nir, S., Jacobson, K., Poste, G. and Lazo, R. (1977) Biochim. Biophys. Acta **465**, 579–598

Patel, V. P. and Fairbanks, G. (1986) J. Biol. Chem. **261**, 3170–3177

Patzer, E. J., Shaw, M., Moore, N. F., Thompson, T. E. and Wagner, R. R. (1978) Biochemistry **17**, 4192–4200

Pelletier, X., Mersel, M., Freysz, L. and Leray, C. (1987) Biochim. Biophys. Acta **902**, 223–228

Penniston, J. T. (1972) Arch. Biochem. Biophys. **153**, 410–412

Penniston, J. T. and Green, D. E. (1968) Arch. Biochem. Biophys. **128**, 339–350

Perret, B., Chap, H. J. and Douste-Blazy, L. (1979) Biochim. Biophys. Acta **556**, 434–446

Portis, A., Newton, C., Pangborn, W. and Papahadjopoulos, D. (1979) Biochemistry **18**, 780–790

Post, J. A., Langer, G. A., Op den Kamp, J. A. F. and Verkleij, A. J. (1988) Biochim. Biophys. Acta **943**, 256–266

Poznansky, M. and Lange, Y. (1976) Nature (London) **259**, 420–421

Raval, P. J. and Allan, D. (1984) Biochim. Biophys. Acta **772**, 192–196

Rawyler, A., van der Shaft, P. H., Roelofsen, B. and Op den Kamp, J. A. F. (1985) Biochemistry **24**, 1777–1783

Record, M., El Tamer, A., Chap, H. and Douste-Blazy, L. (1984) Biochim. Biophys. Acta **778**, 449–456

Redelmeier, T. E., Hope, M. J. and Cullis, P. R. (1990) Biochemistry **29**, 3046–3053

Renooij, W., van Golde, L. M. G., Zwaal, R. F. A. and van Deenen, L. L. M. (1976) Eur. J. Biochem. **61**, 53–58

Rimon, G., Meyerstein, N. and Henis, Y. I. (1984) Biochim. Biophys. Acta **775**, 283–290

Rosing, J., Tans, G., Govers-Riemslag, J. W. P., Zwaal, R. F. A. and Hemker, H. C. (1980) J. Biol. Chem. **255**, 274–283

Rothman, J. E. and Davidowicz, E. A. (1975) Biochemistry **14**, 2809–2816

Rousselet, A., Colbeau, A., Vignais, P. M. and Devaux, P. F. (1976a) Biochim. Biophys. Acta **426**, 372–384

Rousselet, A., Guthman, C., Matricon, J., Bienvenüe, A. and Devaux, P. F. (1976b) Biochim. Biophys. Acta **426**, 357–371

Sandra, A. and Pagano, R. E. (1978) Biochemistry **17**, 332–338

Schlegel, R. A., Prendergast, T. W. and Williamson, P. (1985) J. Cell. Physiol. **123**, 215–218

Schrier, S. L., Chiu, D. T.-Y., Yee, M., Sizer, K. and Lubin, B. (1983) J. Clin. Invest. **72**, 1698–1705

Schrier, S. L., Junga, I. and Ma, L. (1986) Blood **68**, 1008–1014

Schrier, S. L., Zachowski, A. and Devaux, P. F. (1992a) Blood **79**, 782–786

Schrier, S. L., Zachowski, A., Hervé, P., Kader, J.-C. and Devaux, P. F. (1992b) Biochim. Biophys. Acta **1105**, 170–176

Schroeder, F., Nemecz, G., Wood, W. G., Joiner, C., Morrot, G., Ayraut-Jarrier, M. and Devaux, P. F. (1991) Biochim. Biophys. Acta **1066**, 183–192

Schroit, A. J. and Fidler, I. J. (1982) Cancer Res. **42**, 161–167

Schroit, A. J. and Zwaal, R. F. A. (1991) Biochim. Biophys. Acta **1071**, 313–329

Schroit, A. J., Madsen, J. M. and Tanaka, Y. (1985) J. Biol. Chem. **260**, 5131–5138

Schroit, A. J., Bloy, C., Connor, J. and Cartron, J. P. (1990) Biochemistry **29**, 10303–10306

Schwartz, R. S., Tanaka, Y., Fidler, I. J., Chiu, D. T.-Y., Lubin, B. and Schroit, A. J. (1985) J. Clin. Invest. **75**, 1965–1972

Seigneuret, M. and Devaux, P. F. (1984) Proc. Natl. Acad. Sci. U.S.A. **81**, 3751–3755

Seigneuret, M., Zachowski, A., Herrmann, A. and Devaux, P. F. (1984) Biochemistry **23**, 4271–4275

Sessions, A. and Horwitz, A. F. (1983) Biochim. Biophys. Acta **728**, 103–111

Sheetz, M. P. and Singer, S. J. (1974) Proc. Natl. Acad. Sci. U.S.A. **71**, 4457–4461

Shukla, S. D. and Hanahan, D. J. (1982) Arch. Biochem. Biophys. **214**, 335–341

Shukla, S. D., Billah, M. M., Coleman, R., Finean, J. B. and Michell, R. H. (1978) Biochim. Biophys. Acta **509**, 48–57

Siegel, D. P. (1986) Biophys. J. **49**, 1171–1183

Siegel, D. P., Banschbach, J. and Yeagle, P. L. (1989) Biochemistry **28**, 5010–5019

Siess, W. (1989) Physiol. Rev. **69**, 58–178

Sims, P. J., Wiedmer, T., Esmon, C. T., Weiss, H. J. and Shattil, S. J. (1989) J. Biol. Chem. **264**, 17049–17057

Smith, R. J. M. and Green, C. (1974) FEBS Lett. **42**, 108–111

Sune, A., Bette-Bobillo, P., Bienvenüe, A., Fellmann, P. and Devaux, P. F. (1987) Biochemistry **26**, 2972–2978

Sune, A., Vidal, M., Morin, P., Sainte-Marie, J. and Bienvenüe, A. (1988) Biochim. Biophys. Acta **946**, 315–327

Tanaka, K. I. and Ohnishi, S.-I. (1976) Biochim. Biophys. Acta **426**, 218–231

Tanaka, Y. and Schroit, A. J. (1983) J. Biol. Chem. **258**, 11335–11343

Tilley, L., Cribier, S., Roelofsen, B., Op den Kamp, J. A. F. and van Deenen, L. L. M. (1986) FEBS Lett. **194**, 21–27

Tilly, R. H. J., Senden, J. M. G., Comfurius, P., Bevers, E. M. and Zwaal, R. F. A. (1990) Biochim. Biophys. Acta **1029**, 188–190

Utsugi, T., Schroit, A. J., Connor, J., Bucana, C. and Fiedler, I. J. (1991) Cancer Res. **51**, 3062–3066

van Deenen, L. L. M. (1981) FEBS Lett. **123**, 3–15

van den Besselaar, A. M. H. P., de Kruijff, B., van den Bosch, H. and van Deenen, L. L. M. (1978) Biochim. Biophys. Acta **510**, 242–255

van der Schaft, P. H., Beaumelle, B., Vial, H., Roelofsen, B., Op den Kamp, J. A. F. and van Deenen, L. L. M. (1987) Biochim. Biophys. Acta **901**, 1–14

van der Steen, A. T. M., de Kruijff, B. and de Gier, J. (1982) Biochim. Biophys. Acta **691**, 13–23

van Dieijen, G., Tans, G., Rosing, J. and Hemker, H. C. (1971) J. Biol. Chem. **256**, 3433–3442

van Duijn, G., Luiken, J., Verkleij, A. J. and de Kruijff, B. (1986) Biochim. Biophys. Acta **863**, 193–204

van Meer, G. and Op den Kamp, J. A. F. (1982) J. Cell Biochem. **19**, 193–204

van Meer, G., Gahmberg, C. G., Op den Kamp, J. A. F. and van Deenen, L. L. M. (1981) FEBS Lett. **135**, 53–55

Venien, C. and Le Grimellec, C. (1988) Biochim. Biophys. Acta **942**, 159–168

Verhoven, B., Schlegel, R. A. and Williamson, P. (1992) Biochim. Biophys. Acta **1104**, 15–23

Verkleij, A. J., Zwaal, R. F. A., Roelofsen, B., Comfurius, P., Kastelijn, D. and van Deenen, L. L. M. (1973) Biochim. Biophys. Acta **323**, 178–193

Williamson, P., Mattocks, K. and Schlegel, R. A. (1983) Biochim. Biophys. Acta **732**, 387–393

Williamson, P., Algarin, L., Bateman, J., Choe, H.-R. and Schlegel, R. A. (1985) J. Cell. Physiol. **123**, 209–214

Williamson, P., Kulick, A., Zachowski, A., Schlegel, R. A. and Devaux, P. F. (1992) Biochemistry **31**, 6355–6360

Wu, G. and Hubbell, W. L. (1992) FASEB J. **6**, A81

Yamada, S. and Ohnishi, S.-I. (1983) Acta Haematol. Jpn. **46**, 1406–1413

Yuann, J. M. and Morse, P. D., III (1991) Biophys. J. **59**, 627a

Zachowski, A. and Devaux, P. F. (1990) Experientia **46**, 644–656

Zachowski, A. and Morot-Gaudry Talarmain, Y. (1990) J. Neurochem. **55**, 1352–1356

Zachowski, A., Fellmann, P. and Devaux, P. F. (1985) Biochim. Biophys. Acta **815**, 510–514

Zachowski, A., Favre, E., Cribier, S., Hervé, P. and Devaux, P. F. (1986) Biochemistry **25**, 2585–2590

Zachowski, A., Herrmann, A., Paraf, A. and Devaux, P. F. (1987) Biochim. Biophys. Acta **897**, 197–200

Zachowski, A., Henry, J.-P. and Devaux, P. F. (1989) Nature (London) **340**, 75–76

Zilversmit, D. B. and Hughes, M. E. (1977) Biochim. Biophys. Acta **469**, 99–110

Zwaal, R. F. A., Bevers, E. M., Comfurius, P., Rosing, J., Tilly, R. H. J. and Verhallen, P. F. J. (1989) Mol. Cell. Biochem. **91**, 23–31

Biochem. J. (1993) **295**, 329–341 (Printed in Great Britain)

REVIEW ARTICLE

The glucose transporter family: structure, function and tissue-specific expression

Gwyn W. GOULD[*] and Geoffrey D. HOLMAN[†]

[*]Department of Biochemistry, University of Glasgow, Glasgow G12 8QQ, Scotland, and [†]Department of Biochemistry, University of Bath, Claverton Down, Bath BA2 7AY, U.K.

INTRODUCTION

The transport of glucose across the plasma membrane of mammalian cells represents one of the most important cellular nutrient transport events, since glucose plays a central role in cellular homeostasis and metabolism. It has long been established that the plasma membranes of virtually all mammalian cells possess a transport system for glucose of the facilitative diffusion type; these transporters allow the movement of glucose across the plasma membrane down its chemical gradient either into or out of cells. These transporters are specific for the D-enantiomer of glucose and are not coupled to any energy-requiring components, such as ATP hydrolysis or a H^+ gradient [1]. The facilitative glucose transporters are distinct from the Na^+-dependent transporters, which actively accumulate glucose [2,3].

The importance of glucose as a cellular metabolite has led to a great deal of research into the mechanism of this transport event. However, the realization that glucose transport into certain tissues of higher mammals is under both acute and chronic control by circulating hormones, and that defects in this transport system may underlie diseases such as diabetes mellitus, has led to an almost exponential growth in research effort in the transporter field over the past 5–10 years. Perhaps the most significant observation to arise during this time is the realization that, rather than being mediated by a single transporter expressed in all tissues, glucose transport is mediated by a family of highly related transporters which are the products of distinct genes and are expressed in a highly controlled tissue-specific fashion [2] (Table 1). The development and maintenance of this genetic diversity clearly implies a teleological requirement for multiple glucose transport proteins expressed in different tissues, with each being likely to play a distinct role in the regulation of whole-body glucose homeostasis. Some clues as to the relationship between the tissue-specific patterns of expression and the different kinetic characteristics of each of these transporters have recently been provided from an examination of the properties of the isolated transporters in expression systems such as the *Xenopus* oocyte system.

In this review we summarize the present state of knowledge of the currently identified members of the glucose transporter family, propose a basis for their diversity highlighting differences in kinetic properties, substrate specificity and hormonal regulation, and discuss aberrant expression and/or dysfunction of these transporters in disease states.

THE TRANSPORTER FAMILY

GLUT 1: the erythrocyte-type glucose transporter

Perhaps the best-studied glucose transporter is that present in

human red blood cell membranes. Erythrocytes provide a rich source of this transporter, with it comprising about 3–5 % of the membrane protein. The isolation of this protein by Lienhard and his co-workers in the early 1980s represented a major advance in the study of glucose transport [4]. The purified protein enabled a study of the kinetics of the transport system in defined lipid environments and also led to the generation of antibody probes [5–8]. These antibodies, together with partial sequence information from the protein, resulted in the isolation of a cDNA clone for the transporter in 1985 [9,10]. The gene encoding this transporter has also been isolated [11,12].

Utilizing both cDNA and antibody probes, many subsequent studies have demonstrated that both the GLUT 1 protein and its mRNA are present in many tissues and cells [13,14]. It is expressed at highest levels in brain but is also enriched in the cells of the blood–tissue barriers such as the blood–brain/nerve barrier, the placenta, the retina, etc. [15]. In addition, the GLUT 1 protein has been identified in muscle and fat, tissues which exhibit acute insulin-stimulated glucose transport, but only at very low levels in the liver, the other major tissue involved in whole-body glucose homeostasis [13].

It is well established that transformation of cell culture lines results in a pronounced elevation of GLUT 1 protein and mRNA levels, and that this general phenomenon is observed for all cell culture lines [16–21]. Moreover, it is clear that many, if not all, mitogens stimulate GLUT 1 transcription, and that glucose starvation can also stimulate GLUT 1 expression [22–27]. One potential advantage to the cell of increasing GLUT 1 may be related to the kinetic asymmetry property of this isoform [28,29]. The net influx K_m for glucose by GLUT 1 is 1.6 mM, significantly lower than either the equilibrium-exchange or net efflux K_m values (see Table 2 for a description of differences between net and exchange fluxes). The kinetic asymmetry of GLUT 1 appears to be allosterically regulated by binding of intracellular metabolites and is inhibited by intracellular ATP [28]. We would propose that this asymmetry would allow this transporter to function effectively as a unidirectional transporter under conditions where extracellular glucose is low and the intracellular demand for glucose is high, such as would occur during glucose starvation of cells in culture.

GLUT 2: the liver-type glucose transporter

The ability to detect only very low levels of GLUT 1 in hepatocyte membranes, coupled to the observation that the kinetics of glucose transport in hepatocytes were radically different from those in erythrocytes, led to the proposal that a distinct transporter may be expressed in hepatocytes [30,31]. To identify this transporter two laboratories independently developed a tech-

Abbreviations used: 3-O-MG, 3-O-methyl-D-glucose; IAPS-forskolin, 3-iodo-4-azidophenethylamido-7-O-succinyl-forskolin; ATB-BMPA, 2-N-4-(1-azi-2,2,2-trifluoroethyl)-benzoyl-1,3-bis-(D-mannos-4-yloxy)-2-propylamine.

Table 1   Major sites of expression of the different glucose transporters

Isoform	Tissue	References
GLUT 1	Placenta; brain; blood–tissue barrier; adipose and muscle tissue (low levels); tissue culture cells; transformed cells	9,10,13,15,17,18
GLUT 2	Liver; pancreatic β-cell; kidney proximal tubule and small intestine (basolateral membranes)	32–35
GLUT 3	Brain and nerve cells in rodents; brain, nerve; low levels in placenta, kidney, liver and heart (humans)	37–40
GLUT 4	Muscle, heart and adipose tissue	42–46,96–98
GLUT 5	Small intestine (apical membranes); brain, muscle and adipose tissue (muscle and brain at low levels)	60,61,64
GLUT 7	Microsomal glucose transporter; liver	65

Table 2   Kinetic parameters of the glucose transporter family expressed in *Xenopus* oocytes

A variety of different steady-state approaches can be used to determine kinetic constants for glucose transporters. These assays all measure the rate of glucose transport across the membrane, but under different conditions. It should be noted that GLUT 1, but not GLUTs 2 or 4, are asymmetrical with regard to the interactions of glucose at the two sides of the membrane. Note also that K_m and $V_{max.}$ need not be the same when measured under these different conditions because the re-orientation of the binding site may be faster in the presence of unlabelled sugar (the equilibrium-exchange experiment). (i) Equilibrium-exchange transport: the same concentration of sugar is present on both sides of the membrane, but the radioactive label is present only on one side. In exchange-influx experiments, the transporter can return to the outward-facing conformation with unlabelled sugar bound. (ii) Zero-trans transport: sugar is present only on one side of the membrane. In net influx experiments, the transporter will return to the outward-facing conformation unoccupied at initial time points when the intracellular sugar concentration is low. Values are from references [2], [36], [41], [57], [63] and [160].

Isoform	K_m (mM)		Asymmetrical?	Other transported substrates
	3-O-MG (equilibrium exchange)	2-Deoxyglucose (net influx)		
GLUT 1	20.9 ± 2.9	6.9 ± 1.5	Yes	Galactose ($K_m \sim 17$ mM)
GLUT 2	42.3 ± 4.1	11.2 ± 1.1	No	Fructose ($K_m \sim 66$ mM)
GLUT 3	10.6 ± 1.3	1.4 ± 0.06	Not known	Galactose ($K_m \sim 8.5$ mM)
GLUT 4	1.8	4.6 ± 0.3	No	Not studied in oocytes
GLUT 5	n.d.	n.d.	Not known	Fructose ($K_m \sim 6$ mM)

nique for isolating transporter-like cDNAs from additional tissues/cells, including liver. The approach developed involved the use of the GLUT 1 cDNA to probe libraries from hepatocytes under conditions of low stringency, with the rationale that only cDNAs which were similar to GLUT 1 would be identified. Using this approach, Thorens et al. [32] and Fukumoto et al. [33] were able to isolate a cDNA from hepatocytes which, upon analysis of the predicted amino acid sequence, proved to exhibit a high degree of homology to GLUT 1. Furthermore, hydropathy plots of the GLUT 1 and GLUT 2 proteins are virtually superimposable, suggesting that the two transporters are likely to adopt similar global shapes within the membrane.

Subsequent analysis of the sites of expression of GLUT 2 demonstrated that this isoform is expressed at highest levels in the liver, pancreatic β-cell (but not the α- or δ-cells), and on the basolateral surface of kidney and small-intestine epithelia [34,35]. Analysis of the equilibrium-exchange K_m of this isoform for a glucose analogue, 3-O-methyl-D-glucose (3-O-MG), when expressed in *Xenopus* oocytes revealed that this transporter exhibited a supraphysiological K_m for 3-O-MG for GLUT 2 of 42 mM [36]. This high K_m value for GLUT 2 is in agreement with data published for intact hepatocytes, where a K_m for glucose of approx. 66 mM has been reported [30]. The presence of a high-capacity high-K_m transporter in hepatocytes is therefore adventitious for rapid glucose efflux following gluconeogenesis.

This high K_m value may provide a rationale for GLUT 2 localization to those tissues that are involved in the net release of glucose during fasting (liver), glucose sensing (β-cells) and transepithelial transport of glucose (kidney and small intestine),

since glucose flux through this transporter at physiological glucose concentrations would be predicted to change in a virtually linear fashion with extracellular/intracellular glucose concentration. This would result in the highly favourable condition that transporter saturation by glucose would not be rate limiting.

Perhaps the more important functional consequence of the presence of GLUT 2 in kidney and intestinal epithelial cells is its high transport capacity compared with the other transporters. Glucose transport in both the intestine and the kidney is a two-step process, with the active accumulation of glucose via a Na^+-dependent transporter on the apical membrane of the small intestine transporting glucose against its concentration gradient [3]. The accumulated glucose is subsequently released into the capillaries via the high-capacity GLUT 2 which is present at the basolateral borders.

GLUT 3: the brain-type glucose transporter

The development of the low-stringency hybridization approach to cloning glucose transporter cDNAs was subsequently applied to other tissues. In an effort to identify the transporter species present in skeletal muscle, Bell and his co-workers screened a human fetal muscle library for the expression of transporter-like proteins. A novel transporter-like cDNA, GLUT 3, was isolated using this approach [37,38]. Surprisingly, Northern blot analysis revealed that this transporter was barely detectable in adult skeletal muscle, its predominant site of expression being the brain, with lower levels in fat, kidney, liver and muscle tissue. Anti-peptide antibodies specific for either the mouse or human isoforms of GLUT 3 have been used in an effort to further evaluate the role of GLUT 3 in these tissues [39,40]. Using the anti-(mouse GLUT 3) antibodies it has been demonstrated that the expression of GLUT 3 is restricted to brain and neural cell lines and is not immunologically detectable in highly purified mouse muscle, liver or fat membranes. Immunological analysis of human tissues revealed the presence of GLUT 3 at high levels in the brain, with lower amounts present in the placenta, liver, heart and kidney, but not in three different muscle groups, i.e. soleus, vastus lateralis and psoas major [40]. These latter results appear to be in discordance with the relative abundance of GLUT 3 mRNAs in these tissues: for example, the mRNA levels of GLUT 3 in kidney and placenta appear to be roughly 50% of that recorded in brain, but in contrast the level of GLUT 3 protein in these tissues is much lower. One explanation for the disparity between Northern and immunoblot levels of GLUT 3 could be the presence of significant neural contamination of the tissue sections used to prepare the mRNA for the Northern analysis. Alternatively, these tissues may exhibit a negative post-transcriptional regulation of this species of transporter.

Thus it appears that high GLUT 3 protein expression levels are confined generally only to tissues which exhibit a high glucose demand (brain, nerve). Therefore, this isoform may be specialized to act in tandem with GLUT 1 to meet the high energy demands of such tissues. The low level of apparent expression of GLUT 3 protein in liver and kidney may be the result of the localization of this isoform to a specific subset of cells within these tissues.

GLUT 3 exhibits a K_m for 3-O-MG exchange transport of about 10 mM [36,41]. It is well established that the major glucose transporter expressed at the blood–nerve and blood–brain barrier is GLUT 1, which has a higher equilibrium-exchange K_m than GLUT 3. In brain, under normal conditions the capacity of hexokinase for glucose (the preferred energy source) is considerably greater than the capacity of the glucose transport systems in this tissue. However, under conditions of either high glucose demand or hypoglycaemia, the expression of GLUT 3 in the brain with a low K_m for hexoses may be required to successfully utilize low concentrations of blood glucose.

GLUT 4: the insulin-responsive glucose transporter

Following the success in utilizing a GLUT 1 cDNA probe to obtain the homologous sequences of GLUT 2 and GLUT 3, there was enormous excitement in many laboratories as it was realized that the unique glucose transport regulation found in insulin-responsive fat and muscle tissue could be due to a fourth isoform. This was followed by feverish activity by several independent groups and there appeared in 1989 five separate reports of the cloning and sequencing of the GLUT 4 isoform [42–46]. This isoform was shown to occur only in muscle and adipose tissue.

In rat adipose cells, insulin produces an approximate 20–30-fold increase in glucose transport [47–51]. In human adipose cells the response to insulin is much smaller, approximately 2–4-fold [52]. Insulin has been shown to increase glucose transport activity in rat muscle by 7-fold [53] but only by 2-fold in human muscle [54]. Kinetic studies [47–50] have shown that the major effect of insulin is to increase the $V_{max.}$ of glucose uptake. Small changes in the K_m have been reported in rat adipose cells [55] and 3T3-L1 cells [56] but these differences are probably related to the greater proportion of GLUT 1 in the plasma membrane of non-insulin-treated cells. Several potential mechanisms for the increase in $V_{max.}$ can be considered. The increase in $V_{max.}$ for glucose uptake

could occur if insulin increased the intrinsic activity of the transporter (i.e. the catalytic rate constant of each transporter present in the membrane). Another related, and potentially plausible, mechanism would be an insulin-induced conformational redistribution of transport sites between the outside and inside surfaces of the plasma membrane in an asymmetrical transporter. Such a transporter would exhibit differences in K_m at the inner and outer surfaces and a difference between net and exchange flux as occurs in GLUT 1. However, studies investigating this mechanism have shown that the adipocyte transporter, now known to be GLUT 4, has kinetically symmetrical affinities for 3-O-MG influx and efflux and does not show accelerated exchange [47,49]. Another potential mechanism, and the one most supported by recent evidence, suggests that the majority of the acute insulin-stimulated increase in glucose transport measured in adipocytes and muscle is mediated by the appearance of additional GLUT 4 in the plasma membrane. The insulin regulation of GLUT 4 translocation is discussed later. There has been much debate concerning whether translocation can account for the full extent of glucose transport stimulation by insulin. Czech's group in particular have suggested that insulin induces changes in the intrinsic catalytic activity of transporters, possibly mediated by conformational redistribution of sites locked in an inwardly directed conformation due to the binding of an allosteric transport regulator [112]. While it is difficult to reconcile this hypothesis with the observation of kinetic symmetry of GLUT 4, there are precedents for such a mechanism, as GLUT 1 can exhibit allosteric regulation of asymmetry (see above) and mutations of GLUT 1 cause conformational locking into an inwardly directed conformation (see below). However, we consider that intrinsic activation is unlikely to be a major mechanism for GLUT 4 regulation by insulin and that the apparent discrepancies between the observed extent of translocation and the level of transport stimulation may be partly due to the presence of precursor intermediate states in the GLUT 4 trafficking pathway. In addition, many of the discrepancies between the level of transport stimulation induced by insulin and the fold change in GLUT 4 as detected by Western blotting of plasma membrane fractions are due to the difficulties inherent in obtaining highly purified plasma membrane fractions.

The most important property of GLUT 4, which distinguishes it from other isoforms, is its propensity to remain localized in intracellular vesicles in the absence of insulin. Insulin can then specifically recruit this transporter to the surface under metabolically appropriate conditions. The relatively low K_m value of this transporter (2–5 mM) [47–49,57,58] would ensure that it operates close to its $V_{max.}$ over the normal range of blood glucose concentrations, and this ensures the rapid removal of blood glucose into the body's energy stores of glycogen and triacylglycerol. In the absence of insulin (the basal state), glucose transport is rate limiting for metabolism, but insulin stimulates an increase in the plasma membrane abundance of GLUT 4 transporters so that insulin-stimulated transport does not limit metabolism [59].

GLUT 5: the small-intestine sugar transporter

Hexose transport/absorption in the small intestine is clearly an important aspect of whole-body glucose homeostasis. Bell and colleagues in his laboratory isolated another putative glucose transporter cDNA from human small intestine [60]. Northern blot analysis has suggested that this isoform is present at high levels in small intestine. Similar results have been obtained using specific anti-peptide antibodies; moreover, the protein appears to be localized exclusively to the apical brush border on the luminal side of the epithelial cells [61]. Since the transport of glucose from the lumen into the epithelial cells is, under normal circumstances, mediated predominantly by the unrelated Na$^+$-dependent glucose transporter [62], the presence of a putative facilitative glucose transporter in the brush border is not easily explained. The explanation for the presence of GLUT 5 in the brush border has been provided by the recent demonstration that GLUT 5 is a high-affinity fructose transporter, with an apparently poor ability to transport glucose [63]. Thus, on the luminal surface of the small intestine, the primary role of GLUT 5 would be the uptake of dietary fructose.

Northern and immuno-blot analyses have recently demonstrated that this protein is expressed in a range of tissues, including muscle (soleus, rectus abdominus, psoas major and vastus lateralis), brain and adipose tissue. This can be rationalized if GLUT 5 functions to supply these tissues with fructose. However, it is not clear if other fructose transporters also exist. It appears that, unlike GLUT 4, this transporter does not undergo insulin-stimulated translocation in adipocytes, consistent with an apparent lack of insulin-stimulated fructose transport in human adipocytes [64].

GLUT 6: a pseudogene-like sequence

The homology screening approach used by Bell and his colleagues has identified a further transporter-like transcript, with an apparently ubiquitous tissue distribution [60]. Sequence analysis of a cDNA clone for this transcript revealed a high level of base identity (79.6%) with GLUT 3. However, the cDNA was found to contain multiple stop codons and frame shifts, and is unlikely to encode a functional glucose transporter [60]. The extensive identity of the GLUT 6 cDNA with the GLUT 3 cDNA sequence suggests that the glucose transporter-like region of the GLUT 6 transcript may have arisen by the insertion of a reverse-transcribed copy of GLUT 3 into the non-coding region of a ubiquitously expressed gene [60].

GLUT 7: the hepatic microsomal glucose transporter

In the liver, glucose is produced from gluconeogenesis and glycogenolysis for export into the blood. The terminal step of both these processes is the removal of phosphate from glucose 6-phosphate by a specific phosphatase. Glucose-6-phosphatase is a multicomponent enzyme, and it is well established that the glucose produced as a result of the action of this phosphatase is initially confined to the lumen of the endoplasmic reticulum. Thus, in order for the glucose produced to be exported from the liver, it must first cross the endoplasmic reticulum membrane. The work of Burchell and her colleagues has recently revealed that the mechanism by which glucose crosses the endoplasmic reticulum membrane is via a unique member of the facilitated-diffusion-type transporters, now called GLUT 7 [65].

This latest member of the transporter family has been demonstrated to exhibit a close relationship to GLUT 2, there being 68% identity at the amino acid level. One important difference between GLUTs 2 and 7 is the presence of a unique sequence of six amino acids at the C-terminus of GLUT 7. These six amino acids contain a consensus motif for the retention of membrane-spanning proteins in the endoplasmic reticulum (KKMKND). Interestingly, the GLUT 7 protein is virtually identical with GLUT 2 throughout the first four membrane-spanning domains, and also in the regions of transmembrane helices 9 and 10. Moreover, the cDNA sequence is 100% identical with that of GLUT 2 at three locations. Surprisingly, these regions of identity

do not coincide with the intron–exon boundaries [2], suggesting that GLUT 7 is unlikely to be a simple splice variant. However, the lack of base-drift in the third position of the codons over significant stretches of the cDNA raises the intriguing possibility of a unique and complex splicing mechanism generating GLUTs 2 and 7 [65].

GLUCOSE TRANSPORTER STRUCTURE

Analysis of the predicted amino acid sequences of the mammalian glucose transporters shows that these are highly homologous with one another. The mammalian transporters possess high levels of sequence identity with transporters found in many species including cyanobacteria [66], *Escherichia coli* [67], *Zymomonas mobilis* [68], yeast [69,70], algae [71], protozoa [72,73] and plants [74]. This high level of sequence similarity is probably related both to a common mechanism of transport catalysis and also to the transport of a common type of substrate. There are, however, extremes to this generalization and the family includes transporters which differ in some aspects of mechanism, from those which are purely facilitative diffusion types in mammals to the H$^+$-coupled symporters that occur in bacteria [67]. Similarly, the range of preferred substrates includes hexoses, pentoses [67] and disaccharides [70]. Interestingly, the family of homologous proteins includes two transporters that transport the non-sugar substrate quinnate (a hydroxylated six-membered ring substrate) [75].

The common features revealed by sequence alignment and analysis of all the above-mentioned transporters include 12 predicted amphipathic helices arranged so that both the N- and C-termini are at the cytoplasmic surface (Figure 1). There are large loops between helices 1 and 2 and between helices 6 and 7. The large loop between helices 6 and 7 divides the structure into two halves, the N-terminal domain and the C-terminal domain. The loops between the remainder of the helices at the cytoplasmic surface are very short and the length of these loops (about eight residues) is a conserved feature of the whole family. These short loops place severe constraints on the possible tertiary structures and suggest very close packing of the helices at the inner surface of the membrane in each half of the protein. The length and sequence identity of the loops at the extracellular surface of these proteins are very varied but are generally longer than the loops at the cytoplasmic surface. This may potentially result in a less compact helical packing at the external surface. The two-dimensional topography with N- and C-termini on the cytoplasmic surface (Figure 1) has been confirmed using anti-peptide antibodies which react only when the inner surface of the transporter is exposed, as in inverted vesicles containing human erythrocyte GLUT 1. Infra-red spectroscopy has suggested a high (over 80%) helical content for the GLUT 1 protein [76,77].

Conserved motifs in the glucose transporters include GRR(K) between helices 1 and 2 in the N-terminal half, and correspondingly between helices 7 and 8 in the C-terminal half. Similarly, EXXXXXXR occurs between helices 4 and 5 in the N-terminal half and correspondingly between helices 10 and 11 in the C-terminal half. These motifs may be conserved to maintain conformational stability of the protein and may be involved in salt-bridging between helices. The repetition of these motifs between the two halves of the protein suggests that duplication of a gene encoding an ancestral six-membrane-spanning helical protein may have produced the two-domain 12-membrane-spanning helical structure that is so highly conserved in the sugar transporter family. The constraints imposed by the short cytoplasmic loops suggest that a single group of 12 helices is unlikely, but instead the six helices in each of the N- and C-terminal

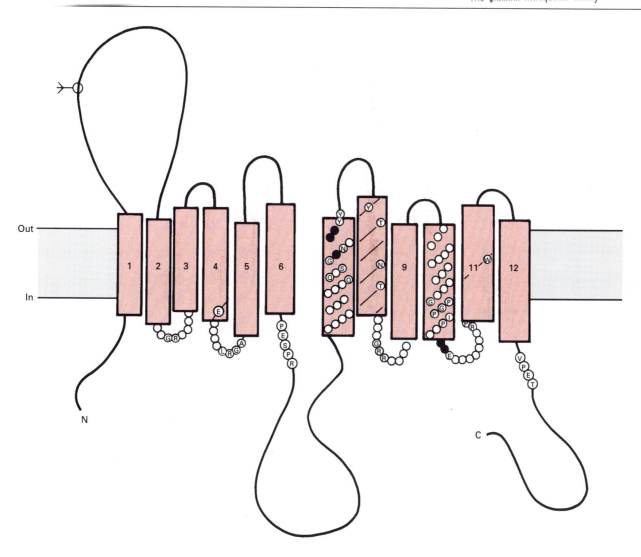

Figure 1   Hypothetical model for the structure of the glucose transporters

The protein is predicted to contain 12 transmembrane helices (1–12), with both the N- and C-termini intracellularly disposed. N-linked glycosylation can occur in the extracellular loop between helices 1 and 2 as shown. Conserved amino acids are indicated by the appropriate single-letter code; filled circles indicate conservative substitutions. Note that not all conserved amino acids are shown (see the text).

domains may be separately closely packed to produce a bilobular structure similar to that which has been observed in low-resolution electron microscopic images of the *E. coli* lactose permease [78]. This packing arrangement has been incorporated into a molecular model of the hexose transporter GLUT 1 [81] (Figure 2).

Molecular modelling suggests that most of the highly conserved residues in helical regions occur on the faces of helices that are directed to the centre of the protein and away from the membrane lipid. Conserved regions of particular interest occur in the C-terminal half of the protein and may be involved in ligand recognition. The motif QQXSGXNXXXYY in helix 7 is present in all the mammalian transporters and is highly conserved in all members of the wider glucose transporter superfamily. The first glutamine (Gln-282) has been implicated in recognition of the exofacial ligand ATB-BMPA [79] and the whole motif is likely to constitute an important part of the exofacial binding site. Immediately preceding this sequence are residues QLS that are highly conserved in the transporters (GLUT 1, GLUT 3 and GLUT 4) which accept D-glucose with high affinity, but not in

the transporters (GLUT 2 and GLUT 5) or the *Zymomonas molilis* [68] or trypanosome [73,80] transporters which accept D-fructose. The main difference between D-glucopyranose and D-fructofuranose is in the anomeric position at C-1 and C-2 respectively. The QLS residues may therefore be involved in docking the C-1 position of D-glucopyranose. Adjacent to the conserved regions in helix 7 are a series of conserved threonine and asparagine residues in helix 8. These may also constitute part of a hydrogen bonding channel allowing hexoses which are accepted at the exofacial site access to the inner binding site of the transporter. Release of sugars at the inner site may be controlled by conformational changes occurring in helices 10 and 11, where highly conserved tryptophan and proline residues are present. Molecular modelling and molecular dynamics studies suggest that prolines 383 and 385 are particularly important in facilitating an alternate opening and closing of the external site of these transporters [81] (Figure 3).

Ligand binding and labelling studies suggest some structural separation of external and internal binding sites. The bis-mannose labelling site has been mapped to helix 8 [83] and helix 9 [82].

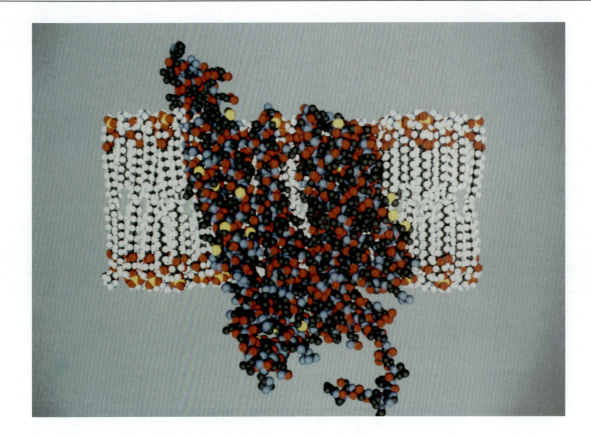

Figure 2   A hypothetical helix-packing arrangement for GLUT 1 (from [81])

The packing shown is that of a bilobular structure with six helices in each half of the protein. The shortness of the intracellular loops between helices restricts the allowable packing arrangements. Most of the protein is embedded in the membrane lipid but the N- and C-termini and the central loop connecting the two lobular domains are predicted to project into the cytoplasm.

However, the inside-specific ligand cytochalasin B labels a region between helices 10 and 11 [82–84], while the diterpine compound IAPS-forskolin labels helix 10 [85] and/or helix 9 [86].

Site-directed mutagenesis has already revealed some important features required for transport by GLUT 1. Truncation of the C-terminal region results in a mutated transporter that is locked in an inward-facing conformation that has low affinity for exofacial ligands such as the photolabel ATB-BMPA and results in a large reduction in sugar transport activity [87]. Mutation of Gln-282 in transmembrane helix 7 also results in the complete loss of exofacial binding of ATB-BMPA, but in this case the mutation only results in a 50 % reduction in sugar transport activity and the binding of the inside-specific binding ligand cytochalasin B [79]. Mutation of the conserved Trp-412 in helix 11 of GLUT 1 and GLUT 4 results in reduced transport activity but no loss of binding of cytochalasin B [88] or IAPS-forskolin [86]. Mutation of Trp-388 of GLUT 1 expressed in CHO cells results in reduced transport and labelling with forskolin [86]. However, when this mutant was expressed in oocytes, it failed to insert correctly in the oocyte plasma membrane [89]. Consistent with the proposed important role of Pro-385 in the mode of operation of the transporter, it has been observed that mutating this Pro-385 to isoleucine in GLUT 1 markedly reduces glucose transport activity and ATB-BMPA labelling, but not cytochalasin B labelling [90].

Recently, Oka and colleagues have demonstrated that the replacement of the C-terminal domain of GLUT 1 with that of GLUT 2 renders the mutated transporter with transport kinetics more like those of GLUT 2 than GLUT 1, but cytochalasin B binding, which is normally lower for GLUT 2 than for GLUT 1, remained unaffected [91].

There is some debate as to whether GLUT 1 is oligomeric in its native state within membranes. Carruthers has suggested [92] that in the absence of reducing agents the GLUT 1 protein is tetrameric. It is suggested that oligomerization produces a form of the transporter in which the substrate, during transport in one subunit, induces a conformational coupling between subunits so that an external site in another subunit is re-exposed more rapidly than would occur in a non-coupled (monomeric) transporter. It remains to be determined definitively whether this oligomerization would confer any biological advantage to the transporter. However, Carruthers has speculated that the co-operative interaction between GLUT 1 monomers may result in a 2–8-fold increase in the catalytic activity of the transporter. The possibility that the catalytic activity of the transporter may be modulated *in vivo* by such a mechanism awaits exploration. Further evidence for oligomerization of GLUT 1 has been obtained from co-immunoprecipitation studies in 3T3-L1 adipocytes; while GLUT 1 oligomers were identified, no co-oligomerization of GLUT 1 and GLUT 4 proteins was identified in these cells [93].

INSULIN REGULATION OF GLUT 4 TRANSLOCATION

In 1980, Cushman and Wardzala [94], and independently Suzuki and Kono [95], first showed that in unstimulated (basal) adipose cells, glucose transporters (now known to be GLUT 4) were

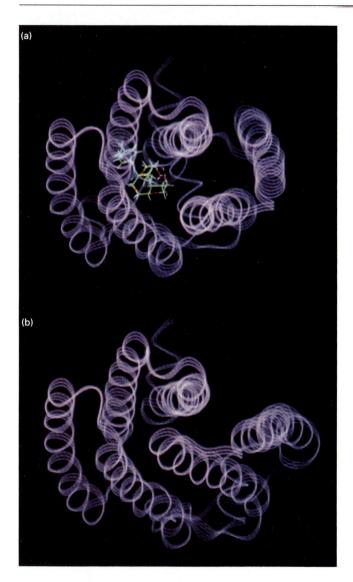

Figure 3   A putative mode of operation of the glucose transporters based on molecular dynamics simulations of GLUT 1 [81]

Highly conserved proline residues in GLUT 1 (residues 383 and 385 in helix 10) are predicted to act as a flexible region. Because of this flexibility, helices 11 and 12 can move relative to helices 7, 8 and 9, and open the outside glucose binding site and close the inner binding site (**a**). The C-terminal region of helix 12 is partly responsible for closing the inner site. Reversal of this helix flexing produces a closed site outside and an open site inside (**b**).

predominantly associated with a light microsome fraction of the cells, and that upon insulin stimulation, these transporters were recruited or translocated to the plasma membrane. An intracellular sequestration and an insulin-induced redistribution of these transporters to the plasma membrane was subsequently shown in other insulin-responsive tissues including brown adipose tissue [96], heart muscle [97,98], diaphragm muscle [99] and skeletal muscle [100–102].

A major obstacle that has hindered a resolution of the extent of the insulin-dependent subcellular redistribution of GLUT 4 has been the difficulty in obtaining pure membrane fractions. If plasma membranes from basal cells are cross-contaminated by light microsome membranes, which have high levels of GLUT 4, then this will lead to an underestimation of the extent of insulin-stimulated transporter redistribution.

Recent immunochemical techniques have circumvented the need to obtain subcellular membrane fractions [96–98]. These studies, involving the use of immunogold-tagged anti-GLUT 4 antibodies, have very clearly shown an intracellular location associated with tubulo-vesicular structures in the basal state, but a shift of GLUT 4 to the plasma membrane and early-endosome locations following insulin treatment. An additional approach to studying glucose transporter translocation (which also circumvents the requirement for obtaining membrane fractions to study insulin action) utilizes the cell-impermeant photolabel ATB-BMPA to selectively label the plasma membrane pool of transporters [51,103–106]. Because this label does not have access to the light-microsome-located transporters, it can be used to estimate both the extent and the rate of glucose transporter appearance in the plasma membrane following insulin-stimulation. Using this method, insulin was shown to increase the availability of GLUT 4 in the plasma membrane of rat adipocytes by 15–20-fold. In contrast, GLUT 1 labelling only increased by 3–5-fold.

The ATB-BMPA photolabel has been used to show that GLUT 4 is re-cycled to the light microsomes and back again to the plasma membrane even in the continuous presence of insulin [106]. GLUT 4 transporters (in insulin-stimulated rat adipose cells) were tracer-tagged with ATB-BMPA, and cells were then maintained either in the absence or in the continuous presence of insulin while subcellular trafficking was monitored. Under these conditions, it was found that the rate constant for endocytosis of the labelled transporters was similar in the presence and absence of insulin, but that re-exocytosis was markedly stimulated by insulin. Re-stimulation of cells in which the photolabelled transporter was internalized also showed that insulin increased the rate at which these transporters were transferred back (exocytosed) to the plasma membrane. Using the photolabel B3GL and an approach similar to that described for ATB-BMPA, Jhun et al. [107] suggested that insulin, in addition to increasing exocytosis, also reduced GLUT 4 endocytosis by 2.8-fold. Whatever the reason for these differences in the estimated endocytosis rates using ATB-BMPA or B3GL, it is clear that the major effect of insulin is to increase exocytosis.

GLUT 1 and GLUT 4 appear rapidly on the cell surface after insulin treatment of adipose cells, with half-times of about 2 min as detected by Western blotting and photolabelling. These half-times are about 1 min shorter that the half-time for the stimulation of transport, which increases with a half-time of 3 min. This lag between transporter appearance and participation in transport has been observed in both rat adipocytes [105,106,108] and 3T3-L1 adipocytes [109,110], and may occur because transporters, during the lag phase, are associated with trafficking proteins or may be present in occluded precursor states which do not fully expose transporters at the cell surface (Figure 4). The presence of these precursor states in the glucose transporter trafficking pathway may account for the observed disparities between the extent of translocation, as detected by Western blotting and photolabelling, and glucose transport activity under conditions of treatment with isoprenaline [111] and protein synthesis inhibitors [112]. Details of the trafficking intermediates involved in exocytosis have yet to be elucidated.

.Some details are emerging of the endocytosis intermediates involved in removing cell-surface transporters. Endocytosis of transporters may occur via clathrin-coated pits and involve similar mechanisms to those which have been demonstrated for removal of cell-surface receptors. Slot et al. [96] in their immunocytochemical study of GLUT 4 in brown adipose tissue observed a GLUT 4–clathrin association in the plasma membrane and early endosomes. Similarly, Robinson et al. [113] observed that

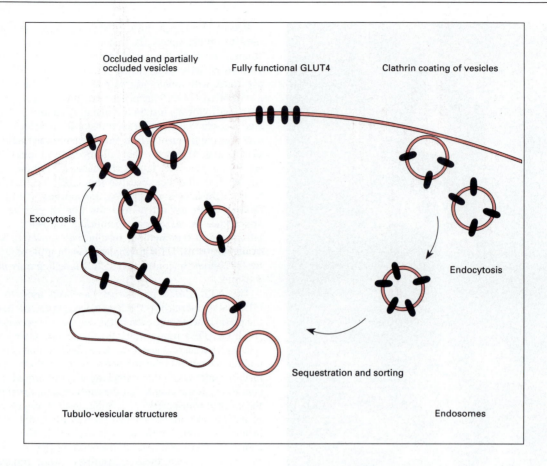

Figure 4   Insulin regulation of GLUT 4 translocation

GLUT 4 is predominantly present in intracellular tubulo-vesicular structures in the absence of insulin. Upon insulin stimulation, exocytosis is increased and GLUT 4 vesicles dock and fuse with the plasma membrane. Occluded and partially occluded structures in the plasma membrane may be responsible for slight discrepancies between the level of stimulation of glucose transport activity and GLUT 4 as detected by photolabelling (active and partially occluded forms) and by Western blotting (active, partially occluded and occluded forms). GLUT 4 is recycled in both the absence and the presence of insulin through clathrin-coated vesicles. Multiple GLUT 4 amino acid sequences responsible for targeting and sorting may be recognized by mechanisms present in the plasma membrane, the endosome recycling system and/or the tubulo-vesicular structures.

GLUT 4 was closely associated with flat clathrin lattices at the cell surface of 3T3-L1 cells.

The larger insulin stimulations of cell-surface availability of GLUT 4 (15–20-fold) compared with GLUT 1 (3–5-fold) are due to a lower rate of exocytosis of GLUT 4 in the basal state [114]. Although an intracellular sequestration of GLUT 1 has been demonstrated in several insulin-responsive cells, the proportion of the total cellular GLUT 1 which is maintained at the cell surface in the absence of insulin is much greater than for the GLUT 4 isoform. There is clearly some structurally distinct and unique property of GLUT 4 that results in its virtual absence from the plasma membrane in the basal state and this in turn results in this isoform responding acutely to insulin to produce the very large stimulations of glucose transport. There may be a unique targeting and sorting of the GLUT 4 protein, but not of GLUT 1, to a unique intracellular population of vesicles. These vesicles may have associated proteins that target the vesicles to a specific location in the tubulo-vesicular systems associated with the *trans* Golgi network. Evidence that GLUT 4 is located in different vesicles to the GLUT 1 isoform has been obtained in rat adipose cells [115] and skeletal muscle [116].

The GLUT 4 isoform has been shown to be sequestered to an intracellular pool when expressed in heterologous systems, including 3T3-L1 and NIH 3T3 fibroblasts [117,118], oocytes [119],

CHO cells [120] and COS cells [121]. The implication from these studies is that GLUT 4 has a unique amino acid sequence or sequences within its primary structure which direct its targeting to an intracellular location. Piper et al. [122] have proposed that the N-terminal region of GLUT 4 is both necessary and sufficient for targeting of this isoform to intracellular pools. The N-terminal region of GLUT 4 is slightly longer than in the other mammalian isoforms, and this extension (MPSGFQQIGSED-GEPPQQ) may comprise an amino acid sequence that is recognized by intracellular targeting processes. However, other investigators have suggested that the N-terminal region of GLUT 4 is not necessary for intracellular targeting and have suggested that more central regions are required [120]. The problem of identifying targeting sequences is complex as, in addition to N-terminal and central regions of the transporter, the C-terminal 30 amino acids of GLUT 4 (ASSFRRTPSLLEQEVKPSTELE-YLGPDEND) have similarities to regions of the cation-sensitive mannose 6-phosphate receptor that have been implicated in intracellular targeting. At present, it is not clear whether multiple regions within the three-dimensional structure of GLUT 4 are folded to form a single unique targeting region that is recognized by a single chaperone protein that sorts GLUT 4 into an appropriate compartment. A perhaps more likely possibility is that several targeting regions within GLUT 4 are recognized by separate chaperone proteins which are necessary at several steps in intracellular sorting (Figure 4). In their study of the targeting of the mannose 6-phosphate receptor, Johnson and Kornfeld [123] identified two separate motifs within the C-terminal tail that were required for sorting at the plasma membrane (YKYSKV) and to the *trans* Golgi network (LLHV). The C-terminal sequence of GLUT 4 therefore contains elements that can be considered as plasma membrane and *trans* Golgi network sorting signals. The investigation of the targeting of GLUT 4 to unique intracellular vesicles and the re-direction of these vesicles to the plasma membrane in response to insulin is currently a research area that is receiving intensive further study.

GLUCOSE TRANSPORTERS IN DISEASED STATES

Is GLUT 2 a component of the glucose-sensing apparatus of the β-cell?

Higher mammals can sense and respond to elevated blood sugar levels by secreting insulin within minutes. The glucose transporter expressed in β-cells has the same primary sequence as that expressed in the liver, i.e. it is GLUT 2 [124]. Immunolocalization has demonstrated that the protein is expressed predominantly in the microvillar portion of the plasma membrane, facing the adjacent endocrine cells, and that the protein is not expressed in the α- or δ-cells [35]. Some circumstantial evidence supports the proposal of a potential role for GLUT 2 in glucose sensing. Unger's laboratory showed that GLUT 2 expression in rat β-cells could be down-regulated by chronic hyperinsulinaemia. The islets from these animals were essentially devoid of GLUT 2 mRNA, and the glucose transport characteristics of the β-cells showed that the transport K_m was some 7-fold lower, indicative of a switch in the isoform of transporter expressed. The new K_m value, about 2.5 mM, is roughly half the typical fasting blood glucose concentration [125]. This result implies that the loss of GLUT 2 function would render β-cells unable to sense and respond to changes in circulating blood glucose levels above about 5 mM, and hence postprandial hyperglycaemia would not be corrected. It is notable that, as well as a reduction in the K_m, a significant reduction in the V_{max} for transport is also observed in islets from hyperinsulinaemic rats, and thus the overall capacity of the islets to transport glucose is markedly reduced [126]. Evidence for a possible role of GLUT 2 in glucose sensing has been suggested by investigators studying patterns of GLUT 2 expression in diabetes (see below).

However, it should be pointed out that two recent studies have seriously questioned the importance of GLUT 2 in the β-cell glucose-sensing apparatus [127,128]. Most significantly, it has recently been demonstrated that the glucose utilization rate of freshly isolated islets is 100-fold lower than the extent of glucose transport [127]. In freshly isolated islets, it would seem clear that the rate of glucose transport would have little consequence on the rate of glycolysis, and thus a role for GLUT 2 in glucose sensing would appear unlikely. Moreover, in a parallel study, it has been demonstrated that first-phase glucose-stimulated insulin release is unchanged or even enhanced in islets cells cultured in glucose, culture conditions which markedly reduce GLUT 2 levels in the β-cell. Further evidence against a role for GLUT 2 in glucose sensing has emerged from studies of a transgenic mouse engineered to express a transforming ras protein in the β-cells. Surprisingly, these animals have been shown to be completely normal with respect to the time course and extent of insulin secretion in response to a glucose bolus, but interestingly, the β-cells of these animals do not express GLUT 2 [128]. These observations are difficult to reconcile with the results alluded to

above, which suggest that GLUT 2 expression appears to modulate glucose-induced insulin secretion, and this dichotomy awaits resolution.

GLUT 2 expression changes in type I and type II diabetes

Insulin-dependent, or type I, diabetes mellitus is an autoimmune disease of β-cells. It affects predominantly children and younger adults, and is correlated with an inherited susceptibility linked to a class II major histocompatibility molecule [129]. The onset of type I diabetes occurs gradually, but a clinical manifestation of the metabolic abnormality does not occur until about 80% of the β-cells have been destroyed. During the pre-diabetic phase, the only identified symptom is the blunting of the first-phase insulin response to intravenous glucose. During this development period, several antibodies to β-cell proteins will be present in the serum of patients. These antibodies include antibodies to glutamic acid decarboxylase and heat-shock protein 65, as well as to other unidentified β-cell proteins, insulin auto-antibodies and β-cell surface antibodies [130–132].

One interesting and potentially important observation has come from the demonstration by Johnson et al. that immunoglobulins from newly diagnosed type I diabetics can affect GLUT 2-mediated glucose transport in normal islets [133]. Both the K_m and V_{max} of GLUT 2-mediated glucose transport in rat islets were reduced by IgG from diabetic patients compared to control patients. These and other data from Unger's laboratory suggest that GLUT 2, or a protein which modulates GLUT 2 activity, is a target for islet cell antibodies, but definitive evidence of a direct immunological reaction with GLUT 2 has yet to be demonstrated, and this inhibition of GLUT 2 activity by serum IgG from diabetics could not be demonstrated in oocytes expressing GLUT 2 [134].

It is of note that the amount of GLUT 2 protein is also reduced in the β-cells of animals undergoing autoimmune destruction. Thus, studies of the BB rat model of autoimmune diabetes have shown that less than half of the surviving β-cells express GLUT 2. This reduction of GLUT 2 levels is further magnified since the total number of β-cells is only 20% of that from a control animal, resulting in a 90% reduction in the number of GLUT 2-positive β-cells [135].

Type II (non-insulin-dependent) diabetes occurs in mature adults and is associated with abnormal insulin secretion and severe peripheral insulin resistance (insulin-resistant glucose transport is described in the next section). Evidence for a defect in β-cell function related to changes in GLUT 2 in type II diabetes has been provided by an analysis of the partially inbred glucose-intolerant Zucker fatty (*fa/fa*) rat [136]. All male rats become obese and develop overt type II diabetes between 7 and 9 weeks of age, whereas neither the *fa/fa* females, which are as obese as the *fa/fa* males, nor the lean male and female heterozygotes develop hyperglycaemia. Insulin secretion in the perfused pancreas from diabetic male rats responds to 10 mM arginine, but not to 20 mM glucose. In contrast, the age-matched female littermates respond to both [136]. Using immunofluorescence, Orci et al. showed that in male rats at the pre-diabetic stage GLUT 2 expression was normal in the β-cells, but upon development of overt diabetes GLUT 2 expression was essentially undetectable [137]. Similar decreases in GLUT 2 mRNA levels were recorded. This decrease in GLUT 2 expression was paralleled by a profound reduction in high-K_m glucose transport into isolated islets [137]. Two subsequent studies have demonstrated that the loss of immunoreactive GLUT 2 is not secondary to the onset of hyperglycaemia, thus establishing a potential link between the reduction of GLUT 2 expression, the loss of glucose-stimulated insulin secretion and the resulting steady-state hyper-glycaemia.

GLUT 1 and GLUT 4 in type II diabetes and insulin resistance

Type II diabetes is associated with severe peripheral insulin resistance. Insulin resistance is also linked with other syndromes and it is considered to be a major contributing factor to hypertension, atherosclerosis and coronary heart failure. It has been suggested by Dowse and Zimmet [138] that, with modern life-style changes, insulin resistance and its consequences can be considered as a major health epidemic.

Insulin resistance is characterized by a failure of insulin to result in efficient glucose disposal, and in particular by a failure of insulin to produce its normal increase in glucose transport in target tissues. The main site of glucose disposal is muscle and this tissue is therefore considered to be most important in terms of the site of insulin resistance. Adipose tissue accounts for only 5–20% of glucose disposal. However, much of the review of the experimental work on insulin-resistant glucose transport that is described here concerns studies on adipose tissue, because many of the mechanistic studies on this problem have been easier to address in this tissue. In muscle, mechanistic studies are rendered more difficult because of the inherent problems associated with preparing subcellular membrane fractions to assess the localization and translocation of the GLUT 4.

As described earlier, it is the propensity of GLUT 4 to become sequestered within the cells of non-stimulated adipose and muscle tissue that renders these tissues uniquely sensitive to insulin. GLUT 1 is not sequestered as efficiently as GLUT 4 and consequently only small increases in the recruitment of this isoform to the plasma membrane occur. A loss of cellular GLUT 4 could lead to insulin resistance, but a loss of the sequestration process for GLUT 4 and/or a decrease in its translocation to the plasma membrane may also contribute to impairment in insulin-responsiveness of glucose transport.

A depleted intracellular pool of glucose transporters in adipose tissue from obese and type II diabetes patients has been observed [139–141]. Similar changes can be induced by streptozotocin treatment of rats and in fasting rats, and these latter effects have been specifically attributed to GLUT 4 depletion. The mRNA and protein are decreased but, in the case of starved rats, the GLUT 4 level can be restored by re-feeding [142]. In rat adipocytes which are maintained in culture for 24 h the development of poor insulin-responsiveness of glucose transport is due to a decrease in GLUT 4 and to a shift in the ratio of GLUT 4 to GLUT 1 [143]. In freshly isolated cells this ratio is 9:1, but it is reduced to 3:1 in the cells maintained in culture for 24 h [143]. Thus there is a shift away from the acutely insulin-sensitive isoform GLUT 4 to the poorly sequestered and insulin-responsive isoform GLUT 1. A similar shift towards a greater contribution of GLUT 1 to the transport activity also occurs in adipocytes from obese rats. The shift in the GLUT 1/GLUT 4 ratio may be associated with a de-differentiation of the adipose cells. The insulin-resistance syndrome in general may be a consequence of de-differentiation of insulin target tissues. Consistent with this possibility, Block et al. [144] have shown that a shift in the GLUT 4/GLUT 1 ratio also occurs in muscle cells that are denervated, begin to de-differentiate and become insulin-resistant.

Cellular depletion of mRNA and protein cannot always account for the observed deficiency in glucose transport activity in obese and type II diabetes patients [145–149]. Similarly, in the *db/db* mouse model of insulin resistance there are no changes in

the total cellular content of GLUT 4 [147]. Pedersen et al. [147] have found that, in muscle from type II diabetes patients, there are no significant changes in GLUT 4 mRNA or protein. However, Dohm et al. [148] report that small (18%) decreases in GLUT 4 are observed in type II diabetes patients. Recent findings [150–152] have led to the suggestion that insulin resistance in glucose transport may be due to defective translocation of GLUT 4 in muscle.

An insulin resistance of glucose transport that is induced by chronic insulin treatment has been demonstrated to occur in human adipose cells. In the adipose cells from obese and type II diabetes patients, prolonged insulin treatment exacerbates the insulin resistance [153]. Several *in vitro* animal models of the insulin resistance that follows chronic insulin treatment have been developed. Garvey et al. [154] and Traxinger and Marshall [155] have shown that insulin resistance in glucose transport that is induced by chronic insulin treatment of primary cultured rat adipocytes is neither at the insulin receptor level nor due to a depleted intracellular pool of transporters. If rat adipose cells are maintained in the continuous presence of insulin during the culture period then GLUT 4 is down-regulated from the cell surface, but the total cellular level of this transporter does not fall below that found when cells are cultured without insulin [143]. The down-regulation of GLUT 4 from the cell surface is associated with a marked decrease in the ability of the cells to respond to a further challenge with insulin. In chronically insulin-treated adipose cells, re-challenging with insulin only increased transport to 30% of the normal response of cells cultured without insulin [143]. It is unclear whether insulin resistance in type II diabetes could be causally related to chronic insulin. Hyperinsulinaemia is always present in the early stages of type II diabetes but this could be a consequence of the resistance. Thus, more insulin may be secreted to compensate for the ineffectiveness of circulating insulin to produce its normal stimulation of glucose disposal.

Several drugs, including sulphonylureas and biguanides, which are used in the treatment of type II diabetes have been shown to potentiate insulin's action on glucose transport [156–159]. The biguanide metformin alleviates the insulin resistance found in cultured rat adipocytes which have been chronically treated with insulin. Using the photolabel ATB-BMPA it has been shown that, in this system, metformin treatment with chronic insulin treatment prevents the down-regulation of cell surface GLUT 4 [143]. Metformin has also been shown to enhance glucose transport activity in L6 muscle cell lines [158] and in skeletal muscle [159].

FUTURE DIRECTIONS

Transporter structure and kinetics

A major area of current and future investigation involves the use of molecular biology techniques to elucidate structural domains of the glucose transporters that are involved in substrate recognition and transport catalysis. It is likely that from further mutagenesis and labelling studies the relationship between the structure of the glucose transporters and their function will gradually emerge. Particular questions that can be addressed using molecular biology techniques are the identification of amino acids and the structural domains that confer the unique functional and kinetic properties to each of the glucose transporter isoforms, and the identification of the targeting signals in the protein sequence that are necessary for directing each of the isoforms to a specific subcellular location.

Diabetes

The investigation of the role of glucose transporters in diabetes is an area that is likely to produce considerable future advances. In particular, the investigation of the role of GLUT 2 in insulin secretion and the role of GLUT 4 in peripheral insulin action will be greatly facilitated by the detailed knowledge of the structure and function of these isoforms that will emerge from mutagenesis and chimera studies. Given the important role of GLUT 2 in the regulation of whole-body glucose homeostasis, the identification of the factor/factors which initiate the loss of islet cell GLUT 2 would represent a significant advance in our understanding of the development and control of the symptoms of diabetes. Future studies will also be aimed at determining the factors which regulate the expression of GLUT 4 in type II diabetes. Perhaps more importantly, the issue of GLUT 4 expression and regulation in muscle from type II diabetics needs to be resolved. Much emphasis will in future be placed on elucidating the signalling route between the insulin receptor and the GLUT 4 translocation pathway.

Research in the authors' laboratories is supported by The British Diabetic Association, The Science and Engineering Research Council, The Medical Research Council, The Juvenile Diabetes Foundation International, The Scottish Home and Health Department and The Scottish Hospitals Endowment Research Trust. In addition, the authors are grateful to Dr. Sam W. Cushman, Dr. E. Michael Gibbs, Dr. Graeme I. Bell and Dr. Barbara B. Kahn for helpful discussions, and to Dr. Paul Hodgson for providing the molecular graphics figures. G. W. G. is a Lister Institute of Preventive Medicine Research Fellow.

REFERENCES

1 Baldwin, S. A. and Lienhard, G. E. (1981) Trends Biochem. Sci. **6**, 208–211
2 Bell, G. I., Kayano, T., Buse, J. B., Burant, C. F., Takeda, J., Lin, D., Fukumoto, H. and Seino, S. (1990) Diabetes Care **13**, 198–208
3 Hediger, M. A., Coady, M. J., Ikeda, T. S. and Wright, E. M. (1987) Nature (London) **330**, 379–381
4 Baldwin, S. A., Baldwin, J. M. and Lienhard, G. E. (1982) Biochemistry **21**, 3837–3842
5 Baldwin, S. A., Gorga, J. C. and Lienhard, G. E. (1981) J. Biol. Chem. **256**, 3683–3689
6 Lowe, A. G. and Walmsley, A. R. (1986) Biochim. Biophys. Acta. **857**, 146–154
7 Walmsley, A. R. (1988) Trends Biochem. Sci. **13**, 226–231
8 Davies, A., Meeram, K., Cairns, M. T. and Baldwin, S. A. (1987) J. Biol. Chem. **262**, 9347–9352
9 Mueckler, M., Caruso, C., Baldwin, S. A., Panico, M., Blench, I., Morris, H. R., Allard, W. J., Lienhard, G. E. and Lodish, H. F. (1985) Science **229**, 941–945
10 Birnbaum, M. J., Haspel, H. C. and Rosen, O. M. (1986) Proc. Natl. Acad. Sci. U.S.A. **83**, 5784–5788
11 Fukumoto, H., Seino, S., Imura, H., Seino, Y. and Bell, G. I. (1988) Diabetes **37**, 647–661
12 Williams, S. A. and Birnbaum, M. J. (1988) J. Biol. Chem. **263**, 19513–19518
13 Flier, J., Mueckler, M., McCall, A. and Lodish, H. F. (1987) J. Clin. Invest. **79**, 657–661
14 Flier, J. S. and Kahn, B. B. (1990) Diabetes Care **13**, 548–564
15 Froehner, S. C., Davies, A., Baldwin, S. A. and Lienhard, G. E. (1988) J. Neurocytol. **17**, 173–178
16 Warburg, O. (1923) Biochem. Z. **142**, 317–321
17 Flier, J. S., Meuckler, M. M., Usher, P. and Lodish, H. F. (1987) Science **235**, 1492–1495
18 Hiraki, Y., de Herreros, A. G. and Birnbaum, M. J. (1989) Proc. Natl. Acad. Sci. U.S.A. **86**, 8252–8256
19 Birnbaum, M. J., Haspel, H. C. and Rosen, O. M. (1987) Science **235**, 1495–1498
20 Merrall, N. W., Plevin, R. J. and Gould, G. W. (1993) Cell. Signalling, in the press
21 Merrall, N. W., Plevin, R. J., Wakelam, M. J. O. and Gould, G. W. (1993) Biochim. Biophys. Acta **1177**, 191–198
22 Weber, M. J., Evans, P. K., Johnson, M. A., McNair, T. F., Nakamura, K. D. and Salter, D. W. (1984) Fed. Proc. Fed. Am. Soc. Exp. Biol. **43**, 107–112
23 Rollins, B. J., Morrison, E. D., Usher, P. and Flier, J. S. (1988) J. Biol. Chem. **263**, 16523–16526
24 Hiraki, Y., Rosen, O. M. and Birnbaum, M. J. (1988) J. Biol. Chem. **263**, 13655–13662
25 Kahn, B. B. and Flier, J. S. (1990) Diabetes Care **13**, 548–564
26 Haspel, H. C., Wilk, E. W., Birnbaum, M. J., Cushman, S. W. and Rosen, O. M. (1986) J. Biol. Chem. **261**, 6778–6789
27 Oriz, P. A., Honkanan, R. A., Klingman, D. E. and Haspel, H. C. (1992) Biochemistry **31**, 5386–5393
28 Diamond, D. L. and Carruthers, A. (1993) J. Biol. Chem. **268**, 6437–6444
29 Craik, J. D. and Elliott, K. R. F. (1979) Biochem. J. **182**, 503–508
30 Eliott, K. R. F. and Craik, J. D. (1983) Biochem. Soc. Trans. **10**, 12–13
31 Axelrod, J. D. and Pilch, P. F. (1983) Biochemistry **22**, 2222–2227
32 Thorens, B., Sarkar, H. K., Kaback, H. R. and Lodish, H. F., (1988) Cell **55**, 281–290
33 Fukumoto, H., Seino, S., Imura, H., Seino, Y., Eddy, R. L., Fukushima, Y., Byers, M. G., Shows, T. B. and Bell, G. I. (1988) Proc. Natl. Acad. Sci. U.S.A. **85**, 5434–5438
34 Thorens, B., Cheng, Z.-Q., Brown, D. and Lodish, H. F. (1990) Am. J. Physiol. **260**, C279–C285
35 Orci, L., Thorens, B., Ravazzola, M. and Lodish, H. F. (1989) Science **245**, 295–297
36 Gould, G. W., Thomas, H. M., Jess, T. J. and Bell, G. I. (1991) Biochemistry **30**, 5139–5145
37 Kayano, T., Fukumoto, H., Eddy, R. L., Fan, Y.-S., Byers, M. G., Shows, T. B. and Bell, G. I. (1988) J. Biol. Chem. **263**, 15245–15248
38 Nagamatsu, S., Kornhauser, J. M., Seino, S., Mayo, K. E., Steiner, D. F. and Bell, G. I. (1992) J. Biol. Chem. **267**, 467–472
39 Gould, G. W., Brant, A. M., Shepherd, P. R., Kahn, B. B., McCoid, S. and Gibbs, E. M. (1992) Diabetelogia **35**, 304–309
40 Shepherd, P. R., Gould, G. W., Colville, C. A., McCoid, S. C., Gibbs, E. M. and Kahn, B. B. (1992) Biochem. Biophys. Res. Commun. **188**, 149–154
41 Colville, C. A., Seatter, M. J., Jess, T. J., Gould, G. W. and Thomas, H. M. (1993) Biochem. J. **290**, 701–706
42 James, D. E., Strube, M. I. and Meuckler, M. (1989) Nature (London) **338**, 83–87
43 Birnbaum, M. J. (1989) Cell **57**, 305–315
44 Charron, M. J., Brosius, F. C., Alper, S. L. and Lodish, H. F. (1989) Proc. Natl. Acad. Sci. U.S.A. **86**, 2535–2539
45 Kaestner, K. H., Christry, R. J., McLenithan, J. C., Briterman, L. T., Cornelius, P., Pekala, P. H. and Lane, M. D. (1989) Proc. Natl. Acad. Sci. U.S.A. **86**, 3150–3154
46 Fukumoto, H., Kayano, T., Buse, J. B., Edwards, Y., Pilch, P. F., Bell, G. I. and Seino, S. (1989) J. Biol. Chem. **264**, 7776–7779
47 Taylor, L. P. and Holman, G. D. (1981) Biochim. Biophys. Acta **642**, 325–335
48 May, J. M. and Mikulecky, D. C. (1982) J. Biol. Chem. **257**, 11601–11608
49 Whitesell, R. R. and Gliemann, J. (1979) J. Biol. Chem. **254**, 5276–5283
50 Simpson, I. A. and Cushman, S. W. (1986) Annu. Rev. Biochem. **55**, 1059–1089
51 Holman, G. D., Kozka, I. J., Clark, A. E., Flower, C. J., Saltis, J., Habberfield, A. D., Simpson, I. A. and Cushman, S. W. (1990) J. Biol. Chem. **265**, 18172–18179
52 Pedersen, O. and Gliemann, J. (1981) Diabetologia **20**, 630–635
53 Ploug, T., Galbo, H., Vinten, J., Jorgensen, M. and Richter, E. A. (1987) Am. J. Physiol. **253**, E12–E20
54 Dohm, G. L., Tapscott, E. B., Pories, W. J., Flickinger, E. G., Meelheim, D., Fushiki, T., Atkinson, S. M., Elton, C. W. and Caro, J. F. (1988) J. Clin. Invest. **82**, 486–494
55 Whitesell, R. R. and Abumrad, N. A. (1986) J. Biol. Chem. **261**, 15090–15106
56 Palfreyman, R. W., Clark, A. E., Denton, R. M., Holman, G. D. and Kozka, I. J. (1992) Biochem. J. **284**, 275–281
57 Keller, K., Strube, M. and Mueckler, M. (1989) J. Biol. Chem. **264**, 18884–18889
58 Thomas, H. M., Takeda, J. and Gould, G. W. (1993) Biochem. J. **290**, 707–715
59 Gliemann, J. and Rees, W. D. (1983) Curr. Top. Membr. Transp. **18**, 339–379
60 Kayano, T., Burant, C. F., Fukumoto, H., Gould, G. W., Fan, Y.-S., Eddy, R. L., Byers, M. G., Shows, T. B., Seino, S. and Bell, G. I. (1990) J. Biol. Chem. **265**, 13276–13282
61 Davidson, N. O., Hausman, A. M. L., Ifkovits, C. A., Buse, J. B., Gould, G. W., Burant, C. F. and Bell, G. I. (1992) Am. J. Physiol. **262**, C795–C800
62 Hopfer, U. (1987) in Physiology of the Gastrointestinal Tract, 2nd edn. (Johnson, L. R., ed.), pp. 1499–1526, Raven Press, New York
63 Burant, C. F., Takeda, J., Brot-Laroche, E., Bell, G. I. and Davidson, N. O. (1992) J. Biol. Chem. **267**, 14523–14526
64 Shepherd, P. R., Gibbs, E. M., Weslau, C., Gould, G. W. and Kahn, B. B. (1992) Diabetes **41**, 1360–1365
65 Waddell, I. D., Zomerschoe, A. G., Voice, M. W. and Burchell, A. (1992) Biochem. J. **286**, 173–177
66 Zhang, C.-C., Durand, M.-C., Jeanjean, R. and Joset, F. (1989) Mol. Microbiol. **3**, 1221–1229
67 Maiden, M. C. J., Jones-Mortimer, M. C. and Henderson, P. J. H. (1988) J. Biol. Chem. **263**, 8003–8010
68 Barnell, W. O., Yi, K. C. and Conway, T. (1990) J. Bacteriol. **172**, 7227–7240
69 Celenza, J. L., Marshal-Carlson, L. and Carlson, M. (1988) Proc. Natl. Acad. Sci. U.S.A. **85**, 2130–2134

70 Szkutnicka, K., Tschopp, J. F., Andrews, L. and Cirrillo, V. P. (1989) J. Bacteriol. **171**, 4486–4493

71 Sauer, N. and Tanner, W. (1989) FEBS Lett. **259**, 43–46

72 Cairns, B. R., Collard, M. W. and Landfear, S. M. (1989) Proc. Natl. Acad. Sci. U.S.A. **86**, 7682–7686

73 Bringaud, F and Baltz, T. (1992) Mol. Biochem. Parasitol. **52**, 111–122

74 Sauer, N., Friedlander, K. and Graml-Wicke, U. (1990) EMBO J. **9**, 3045–3050

75 Geever, R. F., Hiuet, L., Baum, J. A., Tyler, B. M., Patel, V. B., Rutledge, B. J., Case, M. E. and Giles, N. H. (1989) J. Mol. Biol. **207**, 15–34

76 Chin, J. J., Jung, E. K. Y. and Jung, C. Y. (1986) J. Biol. Chem. **261**, 7101–7104

77 Alvarez, J., Lee, D. C., Baldwin, S. A. and Chapman, D. (1987) J. Biol. Chem. **262**, 3502–3509

78 Li, J. and Tooth, P. (1987) Biochemistry **26**, 4816–4823

79 Hashiramoto, M., Kadowaki, T., Clark, A. E., Muraoka, A., Momomura, K., Sakura, H., Tobe, K., Akamura, A., Yazaki, Y., Holman, G. D. and Kasuga, M. (1992) J. Biol. Chem. **267**, 17502–17507

80 Fry, A. J., Towner, P., Holman, G. D. and Eisenthal, R. (1993) Mol. Biochem. Parasitol. **60**, 9–18

81 Hodgson, P. A., Osguthorpe, D. J. and Holman, G. D. (1992) Molecular Modelling of the Human Erythrocyte Glucose Transporter, 11th Annual Molecular Graphics Society Meeting, Abstract

82 Holman, G. D. and Rees, W. D. (1987) Biochim. Biophys. Acta **987**, 395–405

83 Davies, A. F., Davies, A., Preston, R. A. J., Clark, A. E., Holman, G. D. and Baldwin, S. A. (1991) Molecular Basis of Biological Membrane Function meeting, abs S.E.R.C. (U.K.)

84 Cairns, M. T., Alvarez, J., Panico, M., Gibbs, A. F., Morris, H. R., Chapman, D. and Baldwin, S. A. (1987) Biochim. Biophys. Acta **905**, 295–310

85 Wadzinski, B. E., Shanahan, M. F., Clark, R. B. and Ruoho, A. E. (1988) Biochem. J. **255**, 983–990

86 Schurmann, A., Monden, I., Keller, K. and Joost, H. G. (1992) Biochim. Biophys. Acta **1131**, 245–252

87 Oka, Y., Asano, T., Shibasaki, Y., Lin, J.-L., Tsukuda, K., Akanuma, Y. and Takaku, F. (1990) Nature (London) **345**, 550–553

88 Katagiri, H., Asano, T., Shibasaki, Y., Lin, J.-L., Tsukuda, K., Isihara, H., Akanuma, Y., Takaku, F. and Oka, Y. (1991) J. Biol. Chem. **266**, 7769–7773

89 Garica, J. C., Strube, M., Leingang, K., Keller, K. and Mueckler, M. M. (1992) J. Biol. Chem. **267**, 7770–7776

90 Tamori, Y., Hashiramoto, M., Clarke, A. E., Mori, H., Muraoka, A., Kadowaki, T., Holman, G. D. and Kasuga, M. (1993) J. Biol. Chem., in the press

91 Katagiri, H., Asano, T., Ishihara, H., Tsukuda, K., Lin, J.-L., Inukai, K., Kikuchi, M., Yazaki, Y. and Oka, Y. (1992) J. Biol. Chem. **267**, 22550–22555

92 Hebert, D. M. and Carruthers, A. (1992) J. Biol. Chem. **267**, 23819–23838

93 Pessino, A., Hebert, D. N., Woon, C. W., Harrison, S. A., Clancey, B. M., Buxton, J. M., Carruthers, A. and Czech, M. P. (1991) J. Biol. Chem. **266**, 20213–20217

94 Cushman, S. W. and Wardzala, L. J. (1980) J. Biol. Chem. **255**, 4758–4762

95 Suzuki, K. and Kono, T. (1980) Proc. Natl. Acad. Sci. U. S. A. **77**, 2542–2545

96 Slot, J. W., Geuze, H. J., Gigengack, S., Lienhard, G. E. and James, D. E. (1991) J. Cell Biol. **113**, 123–135

97 Watanabe, T., Smith, M. M., Robinson, F. W. and Kono, T. (1984) J. Biol. Chem. **259**, 13117–13122

98 Slot, J. W., Geuze, H. J., Gigengack, S., Lienhard, G. E. and James, D. E. (1991) Proc. Natl. Acad. Sci. U. S. A. **88**, 7815–7819

99 Wardzala, L. J. and Jeanrenaud, B. (1983) Biochim. Biophys. Acta **730**, 49–56

100 Klip, A., Rampall, T., Young, D. and Holloszy, J. O. (1987) FEBS Lett. **224**, 224–230

101 Hirshman, M. F., Goodyear, L. J., Wardzala, L. J., Horton, E. D. and Horton, E. S. (1990) J. Biol. Chem. **265**, 987–991

102 Klip, A., Marette, A., Dimitrakoudis, D., Rampal, T., Gicca, A., Shi, Z. Q. and Vranic, M. (1992) Diabetes Care **15**, 1747–1776

103 Calderhead, D. M., Kitagawa, K., Tanner, L. I., Holman, G. D. and Lienhard, G. E. (1990) J. Biol. Chem. **265**, 13800–13808

104 Clark, A. E., Kozka, I. J. and Holman, G. D. (1991) J. Biol. Chem. **266**, 11726–11731

105 Clark, A. E., Holman, G. D. and Kozka, I. J. (1991) Biochem. J. **278**, 235–241

106 Satoh, S., Gonzalez-Mulero, O. M., Clark, A. E., Kozka, I. J., Quon, M. J., Cushman, S. W. and Holman, G. D. (1993) J. Biol. Chem. **268**, 17820–17829

107 Jhun, B. H., Rampal, A. L., Liu, H., Lachaal, M. and Jung, C. Y. (1992) J. Biol. Chem. **267**, 17710–17715

108 Karnielli, E., Karnowski, M. J., Hissin, P. J., Simpson, I. A., Salans, L. B. and Cushman, S. W. (1981) J. Biol. Chem. **256**, 4772–4777

109 Gibbs, E. M., Lienhard, G. E. and Gould, G. W. (1988) Biochemistry **27**, 6681–6685

110 Yang, J., Clark, A. E., Harrison, R., Kozka, I. J. and Holman, G. D. (1992) Biochem. J. **281**, 809–817

111 Vannucci, S. J., Nishimura, H., Satoh, S., Cushman, S. W., Holman, G. D. and Simpson, I. A. (1992) Biochem. J. **288**, 325–330

112 Czech, M. P., Clancy, B. M., Pessino, A., Woon, W. and Harrison, S. A. (1992) Trends Biochem. Sci. **17**, 197–201

113 Robinson, L. J., Pang, S., Hairns, D. A., Heuser, J. and James, D. E. (1992) J. Cell Biol. **117**, 1181–1196

114 Yang, J. and Holman, G. D. (1993) J. Biol. Chem. **268**, 4600–4603

115 Zorzano, A., Wilkinson, W., Kotlair, N., Thoidis, G., Wadzinski, B. E., Ruoho, A. E. and Pilch, P. F. (1989) J. Biol. Chem. **264**, 12358–12363

116 Piper, R. C. and James, D. E. (1991) J. Cell. Biochem. **15B**, 159

117 Haney, P. M., Slot, J. W., Piper, R. C., James, D. E. and Mueckler, M. (1991) J. Cell Biol. **114**, 689–699

118 Hudson, A. W., Uiz, M. and Birnbaum, M. J. (1992) J. Cell Biol. **116**, 785–797

119 Thomas, H. M., Takeda, J. and Gould, G. W. (1993) Biochem. J. **290**, 707–715

120 Asano, T., Takata, T., Katagiri, H., Tsukuda, K., Lin, J.-L., Ishihara, H., Inukai, K., Hirano, H., Yazaki, Y. and Oka, Y. (1992) J. Biol. Chem. **267**, 19636–19641

121 Schurmann, A., Monden, I., Joost, H. G. and Keller, K. (1992) Biochim. Biophys. Acta **1131**, 245–252

122 Piper, R. C., Tai, C., Slot, J. W., Hahn, C. S., Rice, C. M., Huang, H. and James, D. E. (1992) J. Cell Biol. **117**, 729–743

123 Johnson, K. F. and Kornfeld, S. (1992) J. Cell Biol. **119**, 249–257

124 Johnson, J. H., Newgard, C. B., Milburn, J. L., Lodish, H. F. and Thorens, B. (1990) J. Biol. Chem. **265**, 6548–6551

125 Chen., L., Alam, T., Johnson, J. H., Hughes, S., Newgard, C. B. and Unger, R. H. (1990) Proc. Natl. Acad. Sci. U. S. A. **87**, 4088–4092

126 Unger, R. H. (1991) Science **251**, 1200–1205

127 Tal, M., Liang, N., Najafi, H., Lodish, H. F. and Matschinsky, F. M. (1992) J. Biol. Chem. **267**, 17421–17427

128 Tal, M., Wu, Y.-J., Leiser, M., Surana, M., Lodish, H. F., Fleischer, N., Weir, G. and Efrat, S. (1992) Proc. Natl. Acad. Sci. U. S. A. **89**, 5744–5748

129 Todd, J. A., Bell, G. I. and McDervitt, H. O. (1987) Nature (London) **329**, 599–604

130 Baekkeskov, S., Nielsen, J. H., Marner, B., Bilde, T., Ludvigsson, J. and Lenmark, A. (1982) Nature (London) **298**, 167–169

131 Lernmark, A., Freedman, Z. R., Hofmann, C., Rubenstein, A. H., Steiner, D. F., Jackson, R. L., Winter, R. J. and Traisman, H. S. (1978) N. Engl. J. Med. **299**, 375–380

132 Srikanta, S., Ricker, A. T., McCullock, D. K., Soeldner, J. S., Eisenbarth, G. S. and Palmer, J. P. (1986) Diabetes **35**, 139–142

133 Johnson, J. H., Crider, B. P., McCorkle, K., Alford, M. and Unger, R. H. (1990) N. Engl. J. Med. **322**, 653–659

134 Marshall, M. O., Thomas, H. M., Seatter, M. J., Greer, K. R., Wood, P. J. and Gould, G. W. (1993) Biochem. Soc. Trans. **21**, 164–168

135 Tominaga, M., Komiya, I., Johnson, J. H., Inman, L., Alam, T., Holtz, J., Crider, B., Stefan, Y., Baetens, D., McCorkle, K., Orci, L. and Unger, R. H. (1986) Proc. Natl. Acad. Sci. U. S. A. **83**, 9749–9753

136 Johnson, J. H., Ogawa, A., Chen, L., Orci, L., Newgard, C. B., Alam, T. and Unger, R. H. (1990) Science **250**, 546–549

137 Orci, L., Unger, R. H., Ogawa, A., Koniya, I., Baetens, D., Lodish, H. F. and Thorens, B. (1990) J. Clin. Invest. **86**, 1615–1622

138 Dowse, G. K. and Zimmet, P. (1989) in Frontiers of Diabetes Research: Current Trends in Non-Insulin Dependent Diabetes Mellitus (Alberti, K. G. M. M. and Mazzo, R., eds.), pp. 370–389, Elsevier, Amsterdam

139 Sinha, M. K., Raineri-Maldonado, C., Buchanan C., Pories, W. J., Carter-Su, C., Pilch, P. F. and Caro, J. E. (1991) Diabetes **40**, 472–477

140 Garvey, W. T., Heucksteadt, T. P., Matthaei, S. and Olefsky, J. M. (1988) J. Clin. Invest. **81**, 1528–1536

141 Garvey, W. T., Maianu, L., Heucksteadt, T. P., Birnbaum, M. J., Molina, J. M. and Ciaraldi, T. P. (1990) J. Clin. Invest. **87**, 1072–1081

142 Berger, J., Biswas, B., Vicaro, P. P., Vincent, S. H., Spaerstein, R. and Pilch, P. F. (1989) Nature (London) **340**, 70–72

143 Kozka, I. J. and Holman, G. D. (1993) Diabetes **42**, 1159–1165

144 Block, N. E., Menick, D. R., Robinson, K. A. and Buse, M. G. (1991) J. Clin. Invest. **88**, 1546–1552

145 Kahn, B. B., Charron, M. J., Lodish, H. F., Cushman, S. W. and Flier, J. S. (1989) J. Clin. Invest. **84**, 404–411

146 Sivitz, W. I., DeSautel, S. L., Kayano, T., Bell, G. I. and Pessin, J. E. (1989) Nature (London) **340**, 72–74

147 Pedersen, O., Bak, J. F., Andersen, P. H., Lund, S., Moller, D. E., Flier, J. S. and Kahn, B. B. (1990) Diabetes **39**, 865–870

148 Dohm, G. L., Elton, G. W., Friedman, J. E., Pilch, P. F., Pories, W. J., Atkinson, S. M., Jr. and Caro, J. F. (1991) Am. J. Physiol. **260**, E459–E463

149 Koranyi, L., James, D. E., Mueckler, M. and Permutt, M. A. (1990) J. Clin. Invest. **85**, 962–967

150 King, P. A., Horton, E. D., Hirshman, M. F. and Horton, E. S. (1992) J. Clin. Invest. **90**, 1568–1575

151 Bader, S., Scholz, R., Kellerer, M., Rett, K., Freund, P. and Häring, H. U. (1992) Diabetologia **35**, 456–463

152 Garvey, W. T., Hardim, D., Juhaszova, M. and Dominguez, J. H. (1993) Am. J. Physiol. **264**, C837–C844

153 Sinha, M. K., Taylor, L. G., Poires, W. J., Flickinger, E. G., Meelheim, D., Atkinson, S., Sehgal, N. S. and Caro, J. F. (1987) J. Clin. Invest. **80**, 1073–1081

154 Garvey, W. T., Olefsky, J. M., Matthaei, S. and Marshall, S. (1987) J. Biol. Chem. **262**, 189–197

155 Traxinger, R. R. and Marshall, S. (1989) J. Biol. Chem. **264**, 8156–8163

156 Jacobs, D. B., Hayes, G. R. and Lockwood, D. H. (1989) Diabetes **38**, 205–211

157 Matthaei, S., Hamann, A., Klien, H. H., Benecke, H., Kreymann, G., Flier, J. S. and Greten, H. (1991) Diabetes **40**, 850–857

158 Hundal, H. S., Rampal, T., Reyes, R., Leiter, L. A. and Klip, A. (1992) Endocrinology (Baltimore) **131**, 1165–1173

159 Galuska, D., Zierath, J., Thorne, A., Sonnenfeld, T. and Wallberg-Henriksson, H. (1991) Diabetes Metab. **17**, 159–163

160 Burant, C. F. and Bell, G. I. (1992) Biochemistry **31**, 10414–10420

Biochem. J. (1993) **296**, 273–285 (Printed in Great Britain)

REVIEW ARTICLE

The Na+/H+ exchanger: an update on structure, regulation and cardiac physiology

Larry FLIEGEL*‡ and Otto FRÖHLICH†

*Departments of Biochemistry and Pediatrics, 417 Heritage Medical Research Building, University of Alberta, Edmonton, Alberta, Canada T6G 2S2, and †Department of Physiology, Emory University School of Medicine, Atlanta, GA 30322, U.S.A.

INTRODUCTION

Most mammalian cells maintain a physiological cytoplasmic pH of approximately 7.2 despite the prediction of a more acidic internal pH on the basis of thermodynamic and metabolic considerations. If the transmembrane gradient of hydrogen ions were determined solely by a hydrogen ion leak pathway in the presence of a membrane potential of -60 mV, then the cytoplasmic pH would be about 6.2, that is one pH unit more acidic than the extracellular pH. Besides membrane potential, metabolic conditions can contribute an intracellular acid load. This includes production of CO_2 in the cell and subsequent conversion to carbonic acid, and metabolic reactions producing other acids. Since there are many conditions, physiological and pathological, which tend to shift the metabolic acid balance, regulatory mechanisms need to be in place to maintain intracellular pH in the face of acidic or alkaline challenges. It is also necessary to change intracellular pH to support changes in the growth or functional state of the cell [1–3]. This is achieved by the interplay of several different mechanisms, including bicarbonate-transporting carriers and Na+/H+ exchange, the relative contributions of which vary among the different cell types [4]. The Na+/H+ exchanger is a universal pathway employed by essentially all eukaryotic cells to regulate intracellular pH [5]. Na+/H+ exchange activity is widely expressed in the animal and plant kingdoms in virtually all cell types. As implied by the name, the exchanger transports Na+ and H+ ions in opposite directions across the bilayer membrane. The direction of exchange is governed solely by the two ions' gradients and requires no additional metabolic energy. In higher organisms such as mammals, the free energy in the inward Na+ gradient is greater and therefore powers the H+ movement out of the cell against its own gradient. In mammalian cells the stoichiometry of Na+/H+ exchange is 1:1 [6]. In contrast, bacterial exchange is electrogenic [7] with a stoichiometry of 1:2 [8], reflecting the fact that the Na+/H+ exchangers in the different systems appear to be essentially unrelated to each other.

The best inhibitor of the Na+/H+ exchanger is amiloride and its analogues. Amiloride inhibits the mammalian Na+/H+ exchanger with a K_i of 1–100 μM, depending on the cell type. The more specific amiloride analogues also inhibit this activity with greater efficacy [9]. The exchanger is normally nearly quiescent when the cytoplasmic pH is at the physiological level. Activation occurs by various stimuli including hormones (insulin, vasopressin), growth factors (platelet-derived growth factor, epidermal growth factor), and other stimuli such as chemotactic factors and fertilization of eggs [5,6]. In higher organisms, Na+/H+ exchange fulfils different functions, depending on the cell type. The most common and important role of Na+/H+

exchange is to protect the cell from intracellular acidification, which is evident from mutant cell lines devoid of Na+/H+ exchange [10]. Na+/H+ exchange also participates in cell volume regulation after osmotic shrinkage [11].

Both pharmacological and kinetic criteria indicate that there are different isoforms of the Na+/H+ exchanger. Amiloride and its analogues distinguish two forms (or classes of isoforms): an epithelial luminal transporter with a K_i for amiloride near 0.1 mM, and the higher-affinity non-epithelial transporter ($K_i = 1$–10 μM). The latter form is now referred to as the ubiquitous or 'housekeeping' form, or NHE-1 (for Na+/H+ exchanger type 1). It appears to exist in the plasma membrane of most cells, including the basolateral membrane of epithelial cells [12]. Thus pharmacological criteria have established that different isoforms of antiporter can exist in the same cell albeit partitioned into different domains of the plasma membrane [12–14].

Some aspects of the Na+/H+ exchanger have recently been reviewed, such as its regulation by phosphorylation [15]. In this review we examine several newer and very interesting areas of research related to the Na+/H+ exchanger. We focus on three areas: (1) homology and diversity between the different members of the Na+/H+ exchanger family, (2) regulation of expression of the protein in response to external stimuli, and (3) the role of the Na+/H+ exchanger in the myocardium. Several different NHE isoforms have been isolated and their similarities and differences will be discussed. We will also survey studies on the regulation of the Na+/H+ exchanger gene which demonstrate how the level of the antiporter responds to chronic acidosis and other stimuli. Finally, we will examine the Na+/H+ exchanger in the myocardium. In this tissue in particular, pH regulation and the consequences of this regulation may play an important role in health and disease.

THE Na+/H+ EXCHANGER AND ITS ISOFORMS

Several different isoforms of the Na+/H+ exchanger have been identified and the number of species from which exchanger clones have been isolated is growing (Table 1). Sardet et al. [16,17] isolated the first known Na+/H+ exchanger cDNA clone through a series of elegant experiments involving complementation of exchanger-deficient cell lines. The cDNA coded for the human housekeeping (amiloride-sensitive) isoform that is now referred to as NHE-1 [16,17]. Portions of the cDNA of NHE-1 have been used for screening libraries for other isoforms and from other animals. To date, four isoforms from a number of species are known from mammals: NHE-1 to NHE-4 (from man, sheep, rabbit, rat, hamster and pig; Table 1) [16–28a]. In addition, cDNA clones have been isolated from trout [29], turtle [35] and the nematode, *Caenorhabditis elegans* [30]. As expected

---

Abbreviations used: TM segment, membrane-spanning segment; CaM kinase II, calmodulin-dependent protein kinase II; PKA, cyclic AMP-dependent protein kinase; PKC, protein kinase C.

‡ To whom correspondence and reprint requests should be addressed.

Table 1   Cloned Na⁺/H⁺ exchangers

NA, not applicable.

Species	Isoform	Tissue	Accession nos.	Reference
Mammalian				
Human	NHE-1	*	M81768, J03163	16
Human	NHE-1	Kidney, placenta, breast carcinoma	M81768, J03163	18
Human	NHE-1	Heart	M96066, M96067	19
Hamster	NHE-1	CCL39	CGNHE1	20
Pig	NHE-1	LLC-PK₁ (Kidney)	M89631, S71135	21
Rabbit	NHE-1	Heart	X56536	22
Rabbit	NHE-1	Ileal villus	X59935, S40845	23
Rabbit	NHE-2	Ileal villus	L13733	28a
Rabbit	NHE-3	Kidney (cortex)	M87007, S99441	24
Rabbit	NHE-1	Kidney	X61504, S73423	25
Rat	NHE-1	Heart	M85299	26
Rat	NHE-2	Intestine	L11004	27
Rat	NHE-2	Stomach	L11236	28
Rat	NHE-3	Kidney	M85300	26
Rat	NHE-4	Stomach	M85301	26
Non-mammalian eukaryotes				
Trout	β-NHE	Cephalic kidney	M94581	29
Caenorhabditis elegans		NA	M23064	30
Schizosaccharomyces pombe	sod2	NA	Z11736	31
Prokaryotes				
Bacillus firmus	nhaC	NA	M73530	32
Enterococcus hirae	napA	NA	M81961	33
Escherichia coli	nhaA	NA	J03879	7
Escherichia coli	nhaB	NA	M83655	34

* This cDNA clone was isolated by complementation of an exchanger-deficient cell line.

from kinetic and pharmacological studies of Na⁺/H⁺ exchange in different tissues, the various isoforms differ in their tissue distribution and in their kinetic and pharmacological properties. In mammals, NHE-1 appears to exist essentially in every tissue examined with a message size of 5.0–5.4 kb [16,17]. NHE-2 and NHE-3 are found in intestinal and renal epithelial tissues. NHE-4 occurs mainly in the stomach, which also contains large amounts of NHE-1 and some NHE-3 mRNA [26]. The two epithelial isoforms, NHE-2 and NHE-3, are believed to reside in the apical membrane [28a]. They participate in trans-epithelial NaCl transport. Expression of NHE-2 in polarized human epithelial cells results in a functional transporter in the apical membrane [28a]. NHE-1 resides in the basolateral membrane of epithelia [36], as well as in the plasma membrane of non-polarized cells, where it participates in pH regulation. It should be noted that in at least one epithelial tissue (the human placenta), NHE-1 resides in the apical membrane instead of the basolateral membrane [37].

A great deal of effort is currently focused on comparing the transport properties of the different isoforms in transfected cells with those found previously in the different tissues. Such expression experiments are performed in cells which have been mutated and selected for the absence of intrinsic Na⁺/H⁺ exchange activity [16]. After expression in host cells, NHE-1, NHE-2 and NHE-3 can be activated by growth factors [38,39], but they differ in their response to phorbol esters in that NHE-1 and NHE-2 are activated while NHE-3 is inhibited [39]. These phorbol ester responses agree with findings with the endogenous exchangers, where basolateral Na⁺/H⁺ exchange is stimulated whereas apical Na⁺/H⁺ exchange is inhibited by phorbol esters and agents that activate protein kinase [40]. In an analogous

manner, the trout Na⁺/H⁺ exchanger can be activated by isoprenaline after expression in a fibroblast cell line [29]. The amiloride sensitivity of the expressed isoforms is also close to that expected from previous studies: NHE-1 exhibits the same high affinity for amiloride and its higher-potency analogues while NHE-2 and NHE-3 have a lesser affinity for these drugs. When NHE-2 is expressed in exchanger-deficient cell lines, its sensitivity to amiloride is the same as that of NHE-1, but it is 25-fold more resistant to ethyl isopropyl amiloride than is NHE-1 [28a]. NHE-3 is the most resistant, consistent with the observed low inhibitory affinity of amiloride for apical epithelial transport [41]. The first identification of a site that influenced amiloride binding was made by Counillon et al. [42]. They identified Leu-167 as the amino acid that was altered in an amiloride-resistant mutant and whose mutation into a phenylalanine residue conferred the same resistance phenotype to NHE-1. Preliminary studies also examined the kinetic properties of the expressed proteins, in particular examining the activation of transport by intracellular hydrogen ions and the extent of apparent co-operativity in the activation curve. In this respect, however, the correlation with the *in situ* studies is not as clear: all reconstituted isoforms exhibited an apparent Hill coefficient of greater than 2 [39], whereas in the originating tissues the coefficient ranged from 1.3 in epithelia to about 2 in fibroblasts and to 3 and above in muscle cells. It remains to be seen to what extent the cellular environment might influence the activation state and protein modifier site of the Na⁺/H⁺ exchanger, and thus determine what appears to be an isoform-specific phenotype. This is particularly important since a recent observation shows that the same NHE-1 clone, transfected into different cell lines, is differentially regulated by different stimuli [43].

Table 2   Putative phosphorylation and glycosylation sites on Na$^+$/H$^+$ exchanger isoforms

| Species | Isoform | Tissue | Phosphorylation sites | | | Glycosylation sites | References |
			XRXXSX	XRRXSX	XPX(ST)P	NX(ST)	
Human	NHE-1	†	56,324,648,703,796		723,726	75,370,410	16
Hamster	NHE-1	CCL39	57,328,652,707,803		727,730,780	76,374,414	20
Pig	NHE-1	LLC-PK$_1$	56,324,648,703,799		723,726,774	75,370,410	21
Rabbit	NHE-1	Ileal villus	11, 56,648,703,797		723,726,772	75,370,410	23
Rat	NHE-1	Heart	57,328,652,801		727,730,776	41,76,374,414	26
Rat	NHE-2	Intestine	503,554,665	528*,546,683		235,508,577	27
Rat	NHE-2	Stomach	47,619,670,781	644*,662,799	52	351,624,693	28
Rabbit	NHE-2	Ileal villus	618,669,777,806	643*,661,795	51	350,623,685	28a
Rat	NHE-3	Kidney	560,661,691	552,605,690	791,825	323,588,689,705,805	26
Rabbit	NHE-3	Kidney (cortex)	515,562,663,694	554,607*,693	797	325,692,811	24
Rat	NHE-4	Stomach	10,400,609,660			31,297,342	26

† This cDNA clone was isolated by complementation of an exchanger-deficient cell line.

* This site also contains an R residue, two amino acids after the serine, forming consensus sequence XRXXSXRX. One example of each species was used for each isoform except for NHE-2, which varied between the tissues. Only full-length sequences were used for analysis.

Although Na$^+$/H$^+$ exchange has also been observed in bacteria and yeast, and in organelles such as mitochondria (Table 1 and [44]), it appears that these antiporters belong to different gene families. Several of these exchangers have been cloned. However, none show significant homology to NHE-1 or other members of the NHE family. The different Na$^+$/H$^+$ exchanger proteins may have diverged from each other very long ago, or they may have evolved independently from each other and are not related at all. The stoichiometry and physiological missions of these exchangers differ markedly from those of the NHE family members.

Are there yet other isoforms to be discovered? Despite the fact that four different forms have already been identified (not counting trout and *C. elegans*), this is still a possibility. Several studies have shown the presence of other possible isoforms or have provided evidence for the possible differential splicing of some messages. Northern analysis revealed a message in testis which is smaller than the NHE-1 message in other tissues [26]. Dyck et al. [45] recently reported a smaller (3.8 kb) message in ischaemic hearts. This message hybridized with different NHE-1 cDNA probes derived from the open reading frame but not from the untranslated regions. Finally, there are reports of an Na$^+$/H$^+$ exchange function that is completely resistant to amiloride [46–48]. Since the currently expressed isoforms only exhibit reduced affinities as opposed to no binding at all, the underlying transporter could be considered as a candidate Na$^+$/H$^+$ exchanger isoform as well.

STRUCTURE OF THE Na$^+$/H$^+$ EXCHANGER

The members of the NHE family have very closely related structures. The different Na$^+$/H$^+$ exchanger isoforms range between 717 and 835 amino acids in length. Human NHE-1, for example, consists of 815 residues. It may occur in the cell membrane as a dimer with an apparent molecular mass of over 200 kDa [49]. The monomeric protein has a higher apparent mass (110 kDa) than expected from the amino acid sequence because it is glycosylated [17] on at least two glycosylation sites [50]. The human NHE-1 isoform has three consensus sites for glycosylation, and NHE-3 and NHE-4 have between three and five depending on the species (Table 2). All isoforms share a hydropathy profile which clearly delineates two different domains

for the protein. The first 500–510 amino acids form alternating hydrophilic/hydrophobic stretches which suggest membrane-spanning (TM) segments. The remainder of the molecule, the C-terminal domain, is on average quite polar so that one is compelled to assign it to an aqueous environment (cytoplasm). In order to detect the exchanger protein immunologically with an antibody against the C-terminal region, one needs to permeabilize the cell, which provides evidence for the intracellular location of this region [17].

The structures of the hydropathy plots of the N-terminal membrane-intrinsic domains are nearly identical for the different isoforms [16,23,26]. These plots suggest that the peptide crosses the membrane up to 12 times. However, this suggestion of 12 TM segments can at best be considered tentative since different algorithms can lead to differing arrangements with a lower or higher TM number [51]. The near identity of the hydropathy profiles is based on the close sequence similarity among the different isoforms in the membrane-resident domain. In the hydrophilic domain, the hydropathy profile varies much more, suggesting that much of this domain confers isoform-specific properties.

In general, the NHE proteins are very well conserved within a given isoform and among different species, and even among different isoforms of the same species. On the peptide level, the different NHE-1 representatives from human, pig, hamster, rabbit and rat are 93–96% identical among themselves. The NHE-3 forms from rat and rabbit are also 87% identical in their peptide sequence. The catecholamine-responsive trout isoform, β-NHE, is 64% identical with human NHE-1 over its entire sequence length. When compared from the beginning of the second until the end of the twelfth TM segment (corresponding to residues 100–550 of human NHE-1), the trout sequence is 80% identical with human NHE-1. Even the exchanger of *C. elegans* is 44% identical with human NHE-1 over that membrane-intrinsic stretch, and 38% identical when all 699 residues of the partial-length clone are compared. Good conservation also exists among the different isoforms within one species, in particular for the stretch from TM-2 to TM-12. Comparing three published rat sequences (NHE-1, NHE-3 and NHE-4), for example, one finds for this stretch identity values of 47–53%.

The different isoforms do differ completely, however, in their

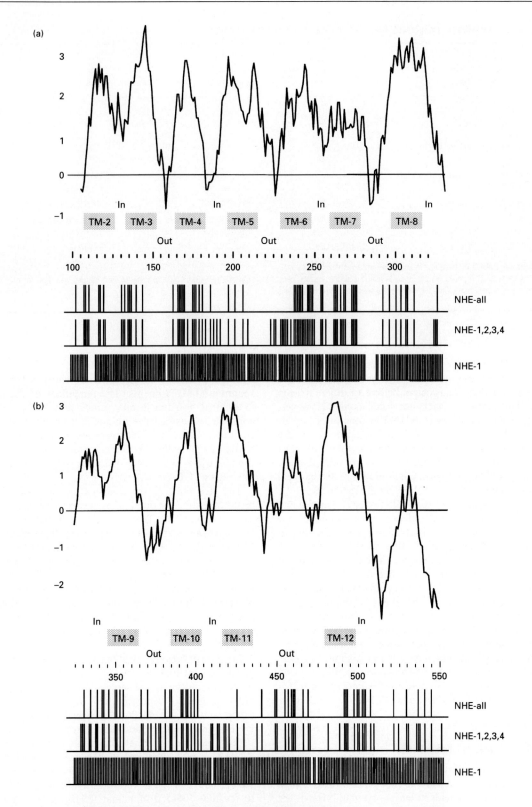

Figure 1   Analysis of the hydropathy profile and sequence identity of different NHE isoforms

The x-axis of the graph represents the amino acid chain of the exchanger protein, with the numbering corresponding to that of human NHE-1. Only the portion of the sequence in which there is identity between the different NHE isoforms is shown. The top line shows the hydropathy profile of human NHE-1, calculated as a moving average of the hydrophobicity values of Kyte and Doolittle [138] with a moving window size of 13. The boxes below indicate where one could expect the peptide to cross the membrane (TM-2–TM-12). TM-1 is not shown because it lies in a region where no identity exists among the different isoforms. The three bar graphs at the bottom give a measure of the identity among the different NHE members: among all published sequences (top), among NHE-1, NHE-2, NHE-3 and NHE-4 from rat (middle), and among all published mammalian NHE-1 members (bottom). The presence of a bar indicates identity at this position.

first putative TM segment (TM-1) and the following first extracellular loop. The strong identity begins only shortly before the peptide presumably enters the membrane for the second time. This curious phenomenon becomes more plausible in light of the suggestion that the first TM segment may comprise a leader sequence that is cleaved off during post-translational processing [15]. If this is the case, then the first TM segment would be lost after processing, leaving behind mature proteins that contain 11 TM segments of very similar structure. Hence, the only significant remaining isoform-specific difference within the membrane-resident domain would then be the first 20–60 (depending on the isoform) hydrophilic residues of the mature protein whose N-terminus would reside in the extracellular space. Perhaps the first transmembrane stretch contains the information that guides the protein into the proper region of the cell, such as the apical membrane in the case of the epithelial isoforms NHE-2 and NHE-3. However, this information could also reside in the C-terminal domain, which also varies between isoforms. Future studies are necessary to show whether this first transmembrane region can target the various exchangers to their proper location.

Figure 1 is a graphical summary of an analysis of the identity among the transmembrane regions of different transporter forms beginning after the first transmembrane passage. It shows vertical bars at the (equivalent) position of the human NHE-1 peptide sequence where the amino acid is conserved. The different NHE-1 representatives (human, pig, sheep, rabbit, rat and hamster) are closely related and most of the amino acids are conserved so that a bar is found in almost all positions (bottom row). In contrast, there is much less identity between NHE-1, NHE-2, NHE-3 and NHE-4 (from rat). When all sequences are included and compared (top row), the conservation is reduced further. It is apparent that most of the selection for the conserved sites on the exchanger protein has already occurred among the different isoforms within a species. Even including evolutionarily more distant representatives (trout and *C. elegans*) eliminated only about 30% of the identical sites among NHE-1, NHE-2, NHE-3 and NHE-4. However, it should be noted that this analysis deals with strict identity; if conservative replacements are included, the degree of similarity is considerably higher.

The strictly conserved residues belong almost exclusively to the membrane-associated domain. One can argue that these residues are somehow essential for basic transport function, or at least for structural integrity. Unfortunately, the number of conserved sites is fairly high (109 residues in the stretch corresponding to positions 100 and 550 of human NHE-1), and it is not possible to tell with any certainty which sites are crucial for transport function. This is aggravated by the lack of experimental verification of the putative topology of the TM segments. However, within the framework of the 12-TM model, there are several conserved charged sites in positions which one might suspect to reside in a TM segment, and which are therefore the best candidates for cation-binding sites. McDaniel et al. [52] have tentatively assigned a 'charge-relay' mechanism to the combination of Lys-116, His-120 and Glu-131 in TM-2, based on a molecular model which places them in close proximity to each other. Only one of these three sites is strictly conserved among all NHE members: in *C. elegans*, His-120 is replaced by Asn and Glu-131 is replaced by Asp. Since one could consider these substitutions to be conservative, this proposed mechanism is probably still a viable hypothesis. It is interesting to note that the most highly conserved region of the protein, corresponding to the stretch that includes putative segments TM-6 and TM-7, contains eight conserved negatively charged amino acids. These residues could be part of a cation-binding or transport site, or

they could be essential for the proper local environment around an access channel to the binding site. They would provide a high negative charge density which in turn would result in a higher surface concentration of cations (protons?) and a higher apparent affinity for these cations. There are several other positions along the NHE sequence which one could consider as potential candidates for essential transport sites. However, further structural information is necessary before their analysis.

GLYCOSYLATION AND PHOSPHORYLATION SITES

A comparison of the glycosylation sites of the various isoforms is shown in Table 2. In most species, including man, NHE-1 possesses three potential sites of glycosylation, located in the first and fifth exoplasmic loops, and near the cytoplasmic entry of TM-11 (all in terms of the 12-TM hypothetical model); an additional site can be found still closer to the beginning of the protein in the rat exchanger. According to this 12-TM model, the exchanger would only be glycosylated at the first two sites. Although NHE-1 is known to be glycosylated [17] and the glycosylation is of the complex biantennary type [50], the exact glycosylation pattern is not yet clear. For example, there is evidence from partial glycolytic digestion experiments that the human placental transporter contains at least two and possibly three glycosylation sites [50]. The glycosylation sites are well conserved throughout the different NHE-1 representatives, but the carbohydrates do not appear to be essential for transport function [50,53]. Perhaps the carbohydrates' role lies in providing post-translational processing or targeting information. Only the second site is conserved among the different isoforms. NHE-3 and NHE-4 (from rat) contain additional potential glycosylation sites without NHE-1 counterparts; however, with NHE-3 they are mainly in the cytoplasmic domain or otherwise in locations that the 12-TM model would consider intracellular and therefore not glycosylated. NHE-2 is similar to NHE-3 in this regard. Of the three glycosylation sites, two are probably within the cytoplasmic domain [27].

The greatest differences among the different isoforms are found in the C-terminal domain. NHE-3 and the Na$^+$/H$^+$ exchanger of *C. elegans* exchanger lose their resemblance to NHE-1 beyond positions 540–550 of NHE-1. NHE-4 [26] and β-NHE of trout [29] retain significant identity with NHE-1 for approximately another 100 residues, beyond which only small patches of potential identity are observed. One major functional difference among the different isoforms lies in the way they are hormonally regulated. One might suspect that the structural correlate of this difference is found in the cytoplasmic domain since it comprises the interface between cytoplasmic second-messenger signalling and the exchanger function. However, the list of physiological roles of the cytoplasmic domain quite likely does not end here. For example, this portion of the protein could also contain binding sites for cytoskeletal elements [15,38]. Of greatest current interest, however, is to test the hypothesis that the C-terminal domain contains the phosphorylation site(s) that are utilized during hormonal regulation of the exchanger's activity. This hypothesis has been confirmed with C-terminally truncated exchanger protein. When the cytoplasmic domain was removed, the protein could no longer be stimulated by growth factors and the cytoplasm did not become more alkaline [38]. Interestingly, the truncated protein also exhibited a strongly acid-shifted pH$_i$ dependence, as if the C-terminal domain also contained elements that controlled the hydrogen ion affinity of the presumed regulatory modifier site. To examine the functional role of the C-terminal domain in fibroblasts, Winkel et al. [54] microinjected polyclonal antibodies raised against the C-terminal

157 amino acids of NHE-1. The antibodies blocked activation by endothelin and α-thrombin, but did not block activation by phorbol esters, cell acidity and osmotic shrinkage. This suggests that the exchanger can be activated by phosphorylation or another form of regulation at different loci, with each locus specific for a different mechanism.

Both the C- and N-terminal domains of the exchanger proteins contain a number of putative sites of phosphorylation. Table 2 provides a summary of consensus phosphorylation sequences in the different isoforms of the Na^+/H^+ exchanger. These include the ideal recognition motifs for multifunctional calmodulin-dependent protein kinase II (CaM kinase II) (XRXXSX), for cyclic AMP-dependent protein kinase (PKA) (XRRXSX), and for protein kinase C (PKC) (XRXXSXRX) [55]. The presence of the sequence XPX(ST)P is also examined in Table 2. This represents the optimal consensus sequence for most of the isoforms of the mitogen-activated protein (MAP) kinase family and for p34^{cdc2}, the cyclin-dependent protein kinase [56]. The analysis of Table 2 shows that mammalian NHE-1 does not contain the consensus sequences XRRXSX and XRXXSXRX. The human, hamster and porcine NHE-1 isoforms all contain five conserved XRXXSX sites: two in the N-terminal and three in the C-terminal domain. One of the N-terminal and two of the C-terminal sites are also conserved in rabbit and rat NHE-1, and one even in rat NHE-4. NHE-2, NHE-3 and NHE-4 (rat) also contain additional sites for potential phosphorylation by CaM kinase II. Of the exchanger isoforms, only NHE-2, NHE-3 and β-NHE contain XRRXSX(RX) sites, which means that direct phosphorylation of the protein by PKA and PKC would be possible.

Mitogen-activated protein kinases (MAP kinases) and cyclin-dependent protein kinases have recently been implicated in the control of many cellular events. In particular, they are suggested to be involved in the cell cycle and meiosis [56]. For most isoforms of MAP kinase and for p34^{cdc2} the consensus sequence XPX(ST)P defines the preferred site of phosphorylation [56]. In most species the NHE-1 isoform of the Na^+/H^+ antiporter contains three such sites (Table 2). Two of these are interesting in that they overlap, while a third site further downstream occurs in all species, except in human NHE-1. NHE-3 contains one or two such sites depending on the species and the rat NHE-4 contains none. In all representatives of NHE-3 and NHE-4, these consensus sequences are located in the C-terminal domain of the proteins. The Na^+/H^+ exchanger is known to be important in regulation of cell proliferation and in control of the cell cycle [57,58]; it remains to be seen whether these sites are involved in these processes and in the hormonal stimulation of the Na^+/H^+ exchanger in general.

Specifically which sites on the C-terminal domain are phosphorylated in vivo is not yet known since they possess many serine and threonine residues. It is known that serine residues are phosphorylated [17] but their exact location is not known. The situation is rather complex in that most NHE-1 members do not possess a good consensus sequence for PKC even though this kinase is involved in many signalling chains following receptor activation [17]. The same set of serine residues appears to be phosphorylated in response to different agonists, whether these activate a tyrosine kinase or phosphoinositide hydrolysis [59]. Therefore one or more additional, yet to be identified, kinases must be involved. Also, dephosphorylation by ATP depletion and rephosphorylation by ATP repletion phosphorylates what appears to be another class of sites (probably serine residues as well [17,60]). Finally, under at least one condition, namely in response to cell shrinkage, transport activation can also occur without detectable phosphorylation [61,62]. It was suggested that

dual control of NHE-1 may occur by phosphorylation-dependent and phosphorylation-independent mechanisms [62].

There are some sites that would appear to be reasonable candidates for regulation through other signalling pathways. The NHE-1 sequence contains several consensus sites for CaM kinase II. Several of these sites from the C-terminal region have been successfully phosphorylated in vitro by CaM kinase II but not by PKA or PKC [63]. Also, β-NHE has two consensus sequences for phosphorylation by PKA which are probably the sites through which the trout red cell exchanger is stimulated by β-adrenergic pathways [29,64]. However, the existence of a consensus sequence does not necessarily mean that this site is phosphorylated. Despite indications that the cytoplasmic domain between residues 566 and 635 is required for mitogen activation of transport, mutating each of the eight serine residues in this stretch did not alter this activation [15]. The same negative result was obtained when Ser-648, a promising candidate in a consensus context for phosphorylation, was mutated [15]. This could mean that more than one residue needs to be phosphorylated for transport activation, and/or that one has to look elsewhere for additional phosphorylation sites, possibly even on the membrane-resident N-terminal domain. An alternative explanation is that phosphorylation of this domain is degenerate. With the cystic fibrosis transmembrane regulator it was found that although four sites were phosphorylated in vivo, no one specific site was necessary for responsiveness to cyclic AMP. One site alone was sufficient for regulation [65]. Future studies will be necessary to determine if this is the case for NHE-1.

It should be noted that while phosphorylation of NHE-1 is usually considered to be stimulatory, for the apical exchanger this is not necessarily the case. Stimulation of either PKC or CaM kinase II can inhibit the ileal brush border membrane exchanger [66,67]. Both NHE-2 and NHE-3 are expressed in this tissue [24,27,28a] and both contain consensus sequences for phosphorylation by either kinase. Which sites are used for phosphorylation is not yet known at this time.

REGULATION OF THE Na^+/H^+ EXCHANGER

Na^+/H^+ exchange is subject to regulation on several different levels: (1) through modification of exchanger turnover rate and (2) through mechanisms which modulate the number of exchanger units available for transport. On the first level, the enzymic activity can be modified by an intrinsic H^+ modifier site giving rise to the steep activation of exchange at a lowered intracellular pH. This gives the protein the means to adjust its activity sensitively to changes in intracellular pH. Extrinsically, hormones and mitogens modulate the exchanger's activity through phosphorylation, which can further depend on the cellular milieu in which the Na^+/H^+ exchanger is expressed [43]. These two aspects of regulation of activity have been reviewed most recently by Wakabayashi et al. [38] and will not be described here. The other important level of regulation of the Na^+/H^+ exchanger deals with the numbers of exchanger units in the plasma membrane. Regulation may occur during transcription of the Na^+/H^+ exchanger gene into message and during translation from the message to protein. In addition, recruitment of Na^+/H^+ exchanger from intracellular stores to the plasma membrane may occur. We focus on transcriptional control of Na^+/H^+ exchanger numbers and to a lesser extent on translational control and recruitment. We will concentrate on the NHE-1 isoform (or generic Na^+/H^+ exchange), since little information is available in this area on the other isoforms.

Although the Na^+/H^+ exchanger exists in all tissues, it is generally present in only small amounts and is not easily detectable by immunological methods [49]. One possible explanation for the low protein levels would be an intrinsically low rate of translation of the message into protein. Translational control of the exchanger has not yet been studied in detail, but at least for NHE-1 there are suggestions that the message contains elements that serve to keep translational activity low. Wakabayashi et al. [38] showed that the 5′ untranslated region of the NHE-1 message inhibits translation of NHE-1. Removal of most of the 5′ untranslated region of the cDNA resulted in higher cellular Na$^+$/H$^+$ exchange transport activity after transfection and expression in an exchanger-deficient cell line. Depending on the species, the 5′ untranslated region of NHE-1 contains up to three minicistrons. These are short, open reading frames which contain an initiation codon, code for only a few amino acids and end in a termination codon. In NHE-1 these minicistrons are all in-frame with each other, but are out-of-frame with the initiating ATG of the main open reading frame. Such out-of-frame minicistrons can be inhibitory to translational efficiency. They cause the translational machinery to fall off the mRNA before reaching the major open reading frame or can cause a failure to recognize the relevant major ATG codon [68]. Whether or not this is the reason for the effects noted earlier on translational efficiency [38] is not yet known. Interestingly, rat NHE-3 possesses one in-frame minicistron a few codons upstream of the major open reading frame. On the other hand, rat NHE-4 possesses several minicistrons in all three reading frames, with one bracketing the main initiation site. There are no studies yet that have tested the potential role(s) of these minicistrons in more detail. In addition, it is not known whether these hypothetical minicistron effects are constitutive or are subject to additional control by modulation of the translational machinery.

Probably a greater regulatory influence lies in the transcription step from Na$^+$/H$^+$ exchanger gene to mRNA. All other factors being equal, increased mRNA levels by transcriptional up-regulation will result in increased levels of the gene product. That the Na$^+$/H$^+$ exchanger may be transcriptionally regulated has been suspected for some time. A number of external environmental stimuli have been shown to affect the maximal rate of Na$^+$/H$^+$ exchanger activity, suggesting a possible increase in Na$^+$/H$^+$ exchanger message and protein levels. Earlier studies examined the maximal rate of Na$^+$/H$^+$ antiport activity, since specific antibodies or probes for mRNA were not yet available. For example, brush border membrane vesicles from rat proximal tubules showed increased Na$^+$/H$^+$ exchange after treatment of the animals with glucocorticoids [69]. The activation was specific in that glucose uptake was not affected and Na$^+$ gradient-dependent phosphate uptake was decreased. Also, the mineral-corticoid aldosterone caused no such increase [69]. A similar effect was observed in isolated proximal cells from kidney where glucocorticoid treatment increased the V_{max} of Na$^+$/H$^+$ exchange. The stimulation was blocked by actinomycin D or cycloheximide, suggesting that both RNA and protein synthesis were required [70]. The mechanism by which glucocorticoids activate the exchanger is not known but is generally thought to be by classical steroid hormone action [71]. A recent study suggests that, at least under some circumstances, it is the NHE-3 isoform of the exchanger that is regulated by glucocorticoids and not the NHE-1 or NHE-2 isoforms [72]. Glucocorticoids were shown to elevate the ileal brush border exchanger NHE-3 message levels and the activity of the protein in rabbits treated with methylprednisolone [72]. Thyroid hormone also affects the rate of Na$^+$/H$^+$ exchange in kidney brush border membranes from rats. Hypothyroid rats show decreased Na$^+$/H$^+$ exchange and hyperthyroid rats show increased maximal exchange rates [71,73,74]. Mineralocorticoids such as aldosterone are not thought to affect the protein levels; however, in some special cell types they may act similarly to the glucocorticoids [75]. These findings, combined with the other studies, show that glucocorticoids and other hormones regulate the level of Na$^+$/H$^+$ exchange activity and that the variations occur in a physiologically adaptive way.

Regulatory effects have also been observed in different renal models. In renal hypertrophy, increased binding of the amiloride analogue, [^{3}H]ethyl isopropyl amiloride, also supports the theory that Na$^+$/H$^+$ exchanger numbers can increase in response to external stimuli [69]. Unilateral nephrectomy causes an increase in the Na$^+$/H$^+$ exchange activity of renal cortical brush border membrane vesicles. The increase in Na$^+$/H$^+$ exchange after 48 h is abolished by prior administration of actinomycin D, indicating that protein synthesis is involved [76]. Several other conditions such as renal disease and diabetic nephropathy may also affect Na$^+$/H$^+$ exchange activity, possibly acting through other mechanisms such as systemic acidosis (reviewed in Fine et al. [77]).

Perhaps the most interesting physiological response is the increase in level of the Na$^+$/H$^+$ exchanger activity in response to long-term acidosis. There appears to be a physiologically adaptive mechanism by which some eukaryotic cell types are able to upregulate this acid-removing transporter in response to chronic acid load. In an early study, chronic metabolic acidosis, induced by addition of NH$_4$Cl to drinking water, increased the V_{max} of Na$^+$/H$^+$ exchange in rat renal cortical brush border membrane vesicles [73]. Similarly, chronic acid feeding increased the V_{max} of apical Na$^+$/H$^+$ exchange in the rat proximal tubule [78]. At that time, little was known about the different isoforms of the exchanger and it is likely that these workers were not examining the activity of the NHE-1 isoform in the apical membranes. Probably their results originated from a more amiloride-resistant isoform of the Na$^+$/H$^+$ antiporter [12–14], possibly the NHE-3 isoform [26]. However, Akiba et al. [79] found similar effects in both brush border and basolateral membranes of rabbit renal cortex. Others [80,81] have also shown that chronic metabolic acidosis induced by NH$_4$Cl feeding results in increased renal cortical mRNA levels (NHE-1). Similar results were obtained with whole renal proximal tubule cells incubated in acidic media. This treatment resulted in increased Na$^+$/H$^+$ antiporter activity, and this increase depended on protein synthesis. There was no stimulation in fibroblasts treated in the same way [82], suggesting that the effects of acidosis on the Na$^+$/H$^+$ antiporter are tissue-specific. Similar effects occur in mouse renal cortical tubule cells and in opossum kidney cells [80]. Incubation in acid media increases Na$^+$/H$^+$ exchanger mRNA by up to 90% as detected by an NHE-1 probe. The same treatment decreased the mRNA abundance of the Na$^+$/H$^+$ exchanger of 3T3 fibroblasts, again demonstrating the tissue specificity of the effect [80]. LLC-PK$_1$ renal epithelial cells also show increased expression and activity after 48 h of treatment at pH 6.9 [83]. When rat hearts are subjected to relatively short periods of ischaemia, intracellular acidosis occurs. This results in a small increase in NHE-1 mRNA levels and a larger increase in a related isoform (3.8 kb) of the message [45]. Primary cultures of isolated myocytes also increase Na$^+$/H$^+$ exchanger activity in response to exposure to external media of low pH (L. Fliegel, unpublished work). The stimulatory effects of acidosis are not unique to the Na$^+$/H$^+$ exchanger. Chronic metabolic acidosis also causes increases in activities of the basolateral membrane Na$^+$/HCO$_3^-$ cotransporter and phosphoenolpyruvate carboxykinase [78,79,84].

The exact mechanism by which low external pH increases Na$^+$/H$^+$ exchanger mRNA levels and transport activity is still in question. Some of the increases in message observed *in vivo* may be accounted for by acidosis-induced increases in circulating adrenal corticosteroids [71]. However, the effect has been shown in cultured cell lines, suggesting that it is intrinsic to the cells and does not depend on additional external factors. In the case of phosphoenolpyruvate carboxykinase, both an increased transcription rate [85] and increased mRNA stability may be responsible for some of the acidosis-induced increases in message [86]. A potential mediator of the acidosis effect is the transcription factor AP-1, the activity of which is increased during acid stimulation [87]. The human NHE-1 gene contains three AP-1 binding sites [88]. In comparison, the mouse NHE-1 promoter region contains only one AP-1-like site (L. Fliegel and J. Dyck, unpublished work), as does the rabbit gene [89]. However, it has not yet been conclusively demonstrated that AP-1 directly activates the antiporter gene or that the acid-induced increases in NHE-1 message are solely transcriptionally mediated. AP-1 could also activate other genes whose products activate the exchanger, or other protein kinase pathways could mediate the effect [87,90]. Experiments with reporter genes coupled to a string of six AP-1 sites demonstrated the involvement of cellular AP-1 protein in acidosis-induced upregulation in renal epithelial cells [87]. The next step will be to demonstrate the AP-1 effect with an NHE-1 promoter-coupled reporter gene, and to show that AP-1 actually binds to the expected sites on the promoter. In the case of phosphoenolpyruvate carboxykinase, the 5' flanking region of the gene has been shown to mediate increased transcription of a reporter gene in response to pH [91]. In addition, other mechanisms such as changes in RNA stability could also be involved. It is, however, clear that a mechanism of compensation has evolved by which chronic increases in acid load in the cell result in a compensatory increase in the Na^+/H^+ exchanger. Figure 2 shows several hypothetical mechanisms that would result in increased NHE-1 expression. Future studies may confirm, disprove or replace these mechanisms.

In addition to transcription, recruitment of protein from intracellular stores may participate in the acidosis-mediated upregulation of Na^+/H^+ exchanger activity. Soleimani et al. [92] showed that acute treatment of rabbit renal proximal tubule cells with acidotic media can affect Na^+/H^+ exchanger activity. Treatment for 2 h with an acidic solution resulted in an increased $V_{max.}$ in both brush border and basolateral membrane vesicles from these cells. This upregulation might reflect activities of two different isoforms of the exchanger. The presence of cycloheximide did not block the increased activities, suggesting that protein biosynthesis was not required for this effect. In this case the increased activity could be due to changes in regulation of the protein or in recruitment from intracellular stores. Regulation of the protein by phosphorylation is a viable candidate since it is known that the exchanger can be phosphorylated [17]. In addition, the Na^+/H^+ exchanger is present in endosomal vesicles [93], and hormonally induced translocation has been reported previously in proximal tubules [94]. This short-term stimulation of Na^+/H^+ exchange appears different from that observed after longer periods of acidosis, which involves an increased message and a requirement for protein biosynthesis [80-82]. Thus acid-induced upregulation of Na^+/H^+ exchange appears to be the result of several different regulatory pathways, in parallel or sequentially, by which cells protect themselves from acute or chronic acidosis.

Several other factors have also been shown to affect or be involved in regulation of NHE-1 mRNA levels in a variety of cell types. In vascular smooth muscle, serum and platelet-derived growth factor increase Na^+/H^+ exchanger mRNA levels by up to 25-fold [95], but serum only causes a minor increase in the $V_{max.}$ of the protein [96]. The human NHE-1 gene [88] and the mouse

gene (L. Fliegel et al., unpublished work) contain only portions of the serum response element, leaving the mechanism of the effect of serum in question [97]. Other mitogens such as phorbol esters, fibroblast growth factor and platelet-derived growth factor also cause 5–15-fold increases in mRNA levels, while angiotensin II has only marginal effects. In smooth muscle, the response can be characterized as an increase in Na^+/H^+ exchanger mRNA levels in response to mitogenic but not hypertrophic stimuli [96]. Rao et al. [98] examined the regulation of NHE-1 gene expression during monocytic differentiation of HL60 cells. During phorbol ester-induced differentiation, mRNA levels increased 50-fold and protein levels also increased 30-fold. Nuclear run-on assays showed that increased transcription accompanied the increased mRNA levels [98]. Activation of PKC alone with a synthetic diacylglycerol which does not induce differentiation did not cause the large increase in transcription. This study showed that increased transcription can, at least with some types of stimulation, be responsible for the increases in Na^+/H^+ exchanger mRNA levels. Similar results were observed with HL60 cells during differentiation to granulocytes induced by retinoic acid. In this case the level of the Na^+/H^+ exchanger message increased 7–18-fold. The increase was also due to an increased rate of gene transcription, and the level of protein also increased 7-fold [99].

To investigate further the role of PKC, Horie et al. [90] examined the effect of phorbol esters on exchanger levels in proximal tubule cells. Long-term stimulation of the cells with phorbol esters resulted in increased activity of the exchanger that could be blocked by cycloheximide and actinomycin D. However, Northern blot analysis showed only a 2-fold increase in the message levels upon treatment with phorbol ester. Preliminary experiments by Horie et al. [90] suggest that inhibition of PKC can prevent the increase in antiporter activity induced by acid media. Future experiments are necessary to confirm this suggestion. Overall, it is apparent that besides acid-induced increases in Na^+/H^+ exchanger message and activity, stimuli such as activation of PKC and differentiation of some cell types can have profound effects on Na^+/H^+ exchanger levels.

THE MYOCARDIAL Na^+/H^+ EXCHANGER IN HEALTH AND DISEASE

The regulation of internal myocardial pH is of especial importance to the function of the heart. The resting intracellular pH is typically near 7.2. It can drop dramatically during ischaemia, in the process depressing contractility of the myocardium. The negative inotropic effect of acidosis has been demonstrated in a variety of cardiac preparations ranging from cardiac muscle fibres [100] to the isolated perfused heart [101] and the rabbit heart in vivo [102]. The lowered pH_i depresses the contractility by affecting a number of steps of excitation–contraction coupling (reviewed in [103]). In order to protect the physiological role of the heart muscle as a pump, one would therefore expect that the heart muscle cells possess mechanisms to maintain the cellular pH_i within fairly narrow limits. One important mechanism is Na^+/H^+ exchange.

The Na^+/H^+ exchanger shares the physiological mission of regulating the myocardial pH with at least two other transporters: a Cl^-/HCO_3^- exchanger and a $Na^+–HCO_3^-$ cotransporter [104,105]; in some cardiac tissues possibly also a Na^+-dependent Cl^-/HCO_3^- exchanger [106]. In this collective mission the different transporters appear to have specialized roles with relatively little overlap. The Cl^-/HCO_3^- exchanger, which at least in some cells is activated by a more alkaline pH_i [107,108], serves to acidify the cell when needed. The Na^+/H^+ exchanger, which is activated by acidic pH [109], helps the cell to recover

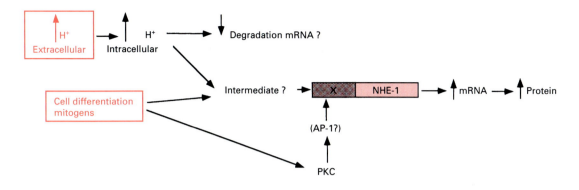

Figure 2   Possible mechanisms of induction of expression of the Na$^+$/H$^+$ exchanger

The temporal sequence is shown from left to right. A number of stimuli including acidosis and mitogens are known to stimulate Na$^+$/H$^+$ exchanger activity. In some cases increased transcription of the NHE-1 gene has been demonstrated. Whether all stimuli act through the same mechanisms, such as through AP-1 or another intermediate, is not yet known. Increased levels of message, protein and activity have been demonstrated as a result of some types of stimuli. For some other proteins, decreased mRNA degradation has been shown to be responsible for increased message. Whether this occurs with the Na$^+$/H$^+$ exchanger is not yet known. X represents the promoter–enhancer region of the gene.

from a strong acid challenge. The Na$^+$–HCO$_3^-$ cotransporter also exports acid equivalents by importing bicarbonate together with Na$^+$, and supports the Na$^+$/H$^+$ exchanger during recovery from acidosis. However, its contribution relative to Na$^+$/H$^+$ exchange varies with the cardiac preparation between 20 and 40% [104,105]. Because of the steep activation curve of Na$^+$/H$^+$ exchange with lowered pH$_i$ [109], it is reasonable to assume that at resting pH$_i$ the Na$^+$/H$^+$ exchanger is only marginally active and that the job of regulation of resting pH is physiologically assigned mainly to Na$^+$–HCO$_3^-$ cotransport.

In other words, steady-state pH$_i$ can be seen as the balance between acidifying mechanisms (residual Cl$^-$/HCO$_3^-$ exchange, a background hydrogen ion leak through the membrane or cation channels, metabolic acid production, etc.) and the alkalinizing effects of residual Na$^+$/H$^+$ exchange and the Na$^+$–HCO$_3^-$ cotransporter. The notion of a relatively minor role for Na$^+$/H$^+$ exchange near resting pH$_i$ is supported by the observation that amiloride has only a small acidifying effect on steady-state pH in heart cells. One should keep in mind, however, that it takes only a small alkaline shift (by 0.1 to 0.2 pH units) in the pH-dependence curve of Na$^+$/H$^+$ exchange to raise the relative importance of this exchange and cause cytoplasmic alkalinization, such as is observed after exposure to α-adrenergic agonists [110,111], especially if this is accompanied by inhibition of Na$^+$–HCO$_3^-$ cotransport [104,105].

Myocardial Na$^+$/H$^+$ exchange exhibits a transport kinetic feature which seems to distinguish it from Na$^+$/H$^+$ exchange in other cells, namely its dependence on the intracellular H$^+$ concentration. Intracellular hydrogen ions do not activate Na$^+$/H$^+$ exchange in a Michaelis–Menten-type manner. Rather, the pH$_i$-dependence exhibits positive co-operativity, as if more than one hydrogen ion is involved in the activation. When fitted to a Hill-type expression, Na$^+$/H$^+$ exchange in renal brush border vesicles could be characterized by a Hill coefficient of $h = 1.2$–1.5 [112,113]. This value of h appears to be typical for epithelial cells (and by implication probably for the epithelial isoforms), but it is in general higher (1.5–2.0) [5] in tissues where NHE-1 is expected to exist. In heart and skeletal muscle, however, the fitted Hill coefficient is still higher, with $h = 2.5$–3 [109,114,115].

Such a difference in transport kinetics would suggest that the heart expresses its own cardiac isoform. However, this notion is not borne out by experiments and the fact that NHE-1 is the predominant isoform in the heart. Screening of rabbit and rat

heart cDNA libraries led only to clones of NHE-1 [22,26]. In addition, Northern blot analysis of heart muscle mRNA with NHE-1 probes reveals mainly the message size of 5 kb that is consistent with NHE-1 [26,45]. However, a smaller message of 3.8 kb also hybridizes to NHE-1 probes under low stringency, which becomes more pronounced after ischaemic exposure [45]. In conclusion, it is not clear whether the kinetic difference is due to the contribution by an additional, as yet unidentified, exchanger isoform or whether it represents a difference in the modulation state of the exchanger in different cell types.

The high value of the Hill coefficient observed in cardiac myocytes is a quantitative description of the fact that the Na$^+$/H$^+$ exchanger is activated by intracellular hydrogen ions over a quite narrow pH range. The Na$^+$/H$^+$ exchanger goes from being nearly inactive to nearly maximally activated within one pH unit [109]. This sensitive response to intracellular acidification is needed if the Na$^+$/H$^+$ exchanger's main role is to protect the cell from an acidic challenge. Na$^+$/H$^+$ exchange can handle even a strong acidification because of the large cellular transport capacity (of all Na$^+$/H$^+$ exchangers combined). At $V_{max.}$, Na$^+$/H$^+$ exchange has probably the highest transport capacity among the different carrier systems in the sarcolemma. Unfortunately, this also means that during recovery from strong acidosis there is a high influx rate of Na$^+$ ions in exchange for extruded H$^+$ ions. This rate is higher than the transport capacity of the Na$^+$/K$^+$-ATPase whose role is to maintain a low intracellular Na$^+$ concentration. As a consequence, the intracellular Na$^+$ concentration rises and with it, because of the tight coupling of the Ca^{2+} gradient to the Na$^+$ gradient, the intracellular free Ca^{2+} concentration rises. A high intracellular Ca^{2+} concentration in turn can cause cellular damage in many different ways: biochemically through uncoupling mitochondrial function, mechanically through local cellular hypercontraction, and by triggering arrhythmias that impede the pumping function of the heart and further aggravate the ischaemic cause (Figure 3).

Under normal physiological conditions this scenario does not occur, and even a mild acidosis is handled by the combination of Na$^+$/H$^+$ exchange and Na$^+$ extrusion. However, the advantage of a responsive, powerful acid extrusion mechanism becomes detrimental to the heart when the acid challenge is too large, such as during and after an ischaemic episode. This is not to say that the sequence of events described above is the only mechanism that contributes to reperfusion injury after ischaemia. However, it represents a new viable hypothesis for one process which has

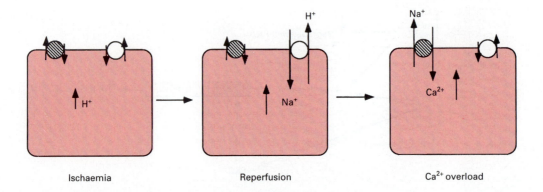

Figure 3   Putative series of events illustrating how intracellular acidosis can lead to Ca²⁺ overload

Excess intracellular protons as a result of ischaemia lead to decreased intracellular pH. During reperfusion the resulting acid load activates the Na⁺/H⁺ exchanger, resulting in increased intracellular Na⁺. Subsequently, the increased intracellular Na⁺ results in increased intracellular Ca²⁺ through the actions of the Na⁺/Ca²⁺ exchanger. Excess intracellular Ca²⁺ is known to have a variety of detrimental effects resulting in cell damage and possibly in generation of arrhythmias.

recently received strong support from a number of different laboratories.

When hearts are rendered ischaemic for long periods of time, they do not completely recover their contractile strength after perfusion is resumed. Myocardial damage associated with these episodes occurs with reperfusion of the myocardium and not with the initial ischaemic episode [116,117]. Therefore it is not the low intracellular pH or the change in oxidation state of the cell that elicits injury, but an event that starts with reperfusion. This is very interesting in terms of the observation that Na⁺/H⁺ exchange is inhibited by a low extracellular pH [109]. During ischaemic acid build-up of mainly lactate, both intracellular and extracellular pH levels drop, but the low intracellular pH cannot stimulate Na⁺/H⁺ exchange since the low extracellular pH has shut down the exchanger. Once reperfusion is started, the first change is a return to normal extracellular pH, which removes the block on the Na⁺/H⁺ exchanger. The rising intracellular Na⁺ concentration then activates Ca²⁺ influx via the Na⁺/Ca²⁺ exchanger, resulting in Ca²⁺ overload and cell death or arrhythmias [118–120]. Figure 3 shows this proposed mechanism of damage to the myocardium.

The individual steps of the reaction sequence described above have been subject to experimental scrutiny by examining the effect of the inhibitor amiloride on reperfusion injury. First, in n.m.r. experiments to examine intracellular pH in rat hearts, it was shown that the recovery of myocardial pH from ischaemia is slowed down in the presence of amiloride [121]. Several studies also showed that intracellular Na⁺ levels increase significantly during reperfusion and that inhibiting the antiporter with amiloride can reduce Na⁺ accumulation during this period [119,122]. In the same manner, amiloride analogues prevent reperfusion-induced increases in Na⁺ and Ca²⁺ concentrations [119,122–124] and lead to an improved return of mitochondrial function [125]. They also reverse lactate-induced depression in post-ischaemic ventricular recovery [126]. The amiloride analogues ethyl isopropyl amiloride, dimethyl amiloride and hexamethyl amiloride are much more potent inhibitors of the exchanger than is amiloride [9]. They are also more potent in preventing the detrimental effects of reperfusion. A similar mechanism may occur when amiloride and its analogues improve recovery from hypoxia and subsequent reoxygenation [127,128].

Amiloride and its analogues also improve the recovery of cardiac contractility and reduce the severity of the arrhythmias after ischaemia and reperfusion in isolated myocardial tissue

preparations [118,119,122,123,129,130]. The beneficial effect of amiloride and ethyl isopropyl amiloride on reperfusion-triggered arrhythmias was also observed *in vivo* with rats whose hearts were made ischaemic by coronary artery ligation [131]. Again, this is consistent with the idea of a central role of the Na⁺/H⁺ exchanger in the mechanism of cellular damage and necrosis [131,132]. Significantly, amiloride needs to be present during the reperfusion phase and there is no noticeable improvement in the protection when amiloride is also present at the time perfusion is stopped and the ischaemic period begins [123]. This suggests that protons accumulate during ischaemia through the formation of metabolic acid, but their exchange with intracellular Na⁺ through the antiporter occurs only during reperfusion of the myocardium. In support of this notion, when the heart is reperfused after an ischaemic period with media of a more acidic pH, it is protected relative to the extent of damage observed with reperfusion media of normal pH (Figure 3 and [22]).

These observations demonstrate the importance of Na⁺/H⁺ exchange under physiological and pathological conditions. A surprising and seemingly paradoxical conclusion is that inhibition of the normal function of this transporter actually improves the health of the myocardium in certain pathological disease states. A reasonable explanation would be that it is more beneficial for the heart muscle as a pump to respond efficiently to a minor acid challenge and rapidly restore its contractility than to be ready for a catastrophic event that usually occurs at a later stage in the life of the individual. As a logical extension, one could surmise that if the protein were overexpressed in certain pathological states *in vivo*, it could have an even more deleterious effect on the myocardium. Does chronic ischaemia or acidosis cause increased expression of the antiporter in the heart, similar to what is observed in some other tissues? Could increased expression of the antiporter further aggravate some pathological conditions? The answers to these question lie in future experiments and may lead to a better understanding of the mechanism of myocardial dysfunction in disease.

FUTURE PROSPECTS

The Na⁺/H⁺ exchanger with its several isoforms represents a major pathway for the removal of acid equivalents. It plays important roles in cellular homeostasis and development, as well as under certain pathological conditions. While a fundamental understanding of its functioning is emerging, many questions remain unanswered about this family of transport proteins. For example, little is known about the mechanism by which the exchanger is regulated by extrinsic factors, a topic which is not addressed in detail here but which has received extensive coverage elsewhere [15,133]. A number of other important questions remain. How many different phosphorylation sites (or classes of sites) are there? Does their phosphorylation lead to the same kinetic modulation of the exchanger? Which are the kinases that directly phosphorylate and regulate the exchangers? To what extent does the regulation involve interactions among different exchanger molecules in a multimeric complex? Or could it involve interactions between the membrane and the protein's membrane anchoring site, as could potentially happen during volume-dependent modulation?

Targeting and localization of the Na$^+$/H$^+$ exchanger is another largely unexplored issue. In epithelial tissues, the epithelial exchanger isoforms (NHE-2 and NHE-3, and probably also NHE-4) are presumed to be localized in the apical membrane and NHE-1 on the basolateral side, but in the human placenta the localization of the different isoforms is reversed [37]. How this differential localization is programmed and how the targeting to the correct membrane is executed is not yet known. It could be the result of specific signals present in the sequence of the protein itself that are recognized in a tissue-specific fashion. This question may receive an answer in the near future with the expression of the different isoforms and the development of isoform-specific antibodies.

The physical mechanism of transport through the Na$^+$/H$^+$ exchanger is a complete unknown and has previously only been accessible through transport-kinetic experiments. Now that the protein can be expressed in cells, as the wild type or containing specific mutations, its kinetics can be studied again with the goal of correlating structure with function. Recent studies have provided evidence for sites whose mutation drastically lowers the affinity for amiloride [134]. Other experiments suggest that the cytoplasmic domain contains elements of the internal H$^+$ modifier site [15]. However, a rational search for the structural correlates of the kinetic properties will only be possible once the topology of the protein is better understood.

Beyond the division of the exchanger protein into two domains, little is certain about the structure of the protein. An analysis of the hydrophobicity profile of the amino acid sequence suggests that the hydrophobic N-terminal domain can contain as many as 12 membrane-spanning segments. However, predicting the transmembrane orientation by hydropathy analysis is not overly reliable [51]. Other algorithms of predicting structural properties of the protein lead to differing estimates of how many TM segments one can expect (down to 10 or even less). For example, Orlowski et al. [26] proposed that TM-1 and TM-2 of the 12-TM model do not span the plasma membrane but constitute a cytoplasmic membrane-associated domain. Only specifically designed experiments will reveal details of the protein's topography. A first handle could be a careful analysis of the glycosylation pattern since extensive glycosylation is found essentially only on the exoplasmic face of the protein. NHE-1 is indeed glycosylated [17], and the placental form possesses at least two, perhaps even three, glycosylation trees [50]. If indeed three such sites were found by additional experimentation, this would challenge the 12-TM model that is currently favoured (see Figure 1), since this model would place the third site on the cytoplasmic face. A comparative study with the other NHE isoforms could be helpful since not all sites are conserved within the NHE family (Table 2). Further tests on cloned exchangers await the development of an expression system that provides sufficient quantities of fully functional protein. Initial attempts to over-produce the protein in insect cells using the baculovirus system have met with mixed success, with much of the protein apparently existing in a non-functional form [53]. We have noted that attempts to produce the protein *in vitro* have not been as successful as with other proteins (L. Fliegel and O. Fröhlich, unpublished work). It is not clear to what extent possible low-usage codons, secondary structures in the message or other mechanisms can account for these results.

A newly emerging and interesting aspect of the Na$^+$/H$^+$ exchanger is the regulation of the gene. As discussed above, cells possess mechanisms which respond to acid loads by increasing the levels of Na$^+$/H$^+$ exchanger mRNA and protein. It is up to future experiments to determine the details of how this response is mediated. Do similar mechanisms exist for the other isoforms? How widespread is this phenomenon among the different tissues? Apparently it is not observed in all cells [82].

In this context, future studies will in addition deal with the different transcription factors which may also be involved in regulation of the exchanger. There are consensus sequences for several DNA-binding proteins present on the human and mouse NHE-1 genes, including Sp-1, AP-1, AP-2 glucocorticoid response element, cyclic AMP response element and others ([88]; L. Fliegel, unpublished work). Finding how this gene is regulated will not only lead to the mechanism behind the response to chronic acid load, but it may also reveal information on the possible involvement of the exchanger in the cell cycle and in some pathological states.

There is at least one experimental pathological model in which the normally beneficial action of the Na$^+$/H$^+$ exchanger can trigger events that lead to tissue damage, namely during the reperfusion injury of the isolated animal heart after an ischaemic period. However, whether these findings are applicable in the clinical setting has yet to be determined. Of great value in demonstrating this role have been amiloride and its more potent and specific analogues such as methyl isobutyl amiloride and ethyl isopropyl amiloride. Unfortunately, the action of amiloride and these analogues is not confined to inhibiting the Na$^+$/H$^+$ exchanger [9,135]. These compounds are moderately membrane-permeant and can enter the tissue in significant quantities, and at the necessary dosage they affect other cellular functions in the cytosol [136], in addition to amiloride's action as a diuretic. It would be of great interest to develop either more specific amiloride analogues or other unrelated inhibitors of the exchanger. Such drugs would be valuable during the treatment of myocardial infarctions. They would also have tremendous potential in the treatment of cancer tissue, since they could promote cellular acidification of the transformed cells to the point of their death (reviewed in Harguindey [137]). The challenges therefore are great and may require multidisciplinary approaches for their solution.

We are grateful for the support of our appropriate granting agencies. Research by L. F. is supported by the Medical Research Council of Canada and the Heart and Stroke Foundation of Canada. L. F. is a scholar of the Heart and Stroke Foundation of Canada and of the Alberta Heritage Foundation for Medical Research. O. F. was supported by a Grant-in-Aid from the American Heart Association/Georgia Affiliate. We are particularly grateful to the members of our laboratory who have helped with their contributions to several relevant publications; these include J. R. B. Dyck and R. S. Haworth. We thank M.-J. Boeglin and S. Fliegel for their assistance in preparation of the manuscript. We are also grateful to Drs. D. Warnock, G. Pierce and D. Dawson for making some of their groups' work available to us prior to publication.

REFERENCES

1 Nucitelli, R., Webb, D. J., Lagier, S. T. and Matson, G. B. (1981) Proc. Natl. Acad. Sci. U.S.A. **78**, 4421–4425

2   Hesketh, R. T., Moore, J. P., Morris, J. D. H., Taylor, M. V., Rogers, J., Smith, G. A. and Metcalfe, J. C. (1985) Nature (London) **313**, 481–484

3   Berk, B. C., Canessa, M., Vallega, G. and Alexander, R. (1988) J. Cardiovasc. Pharmacol. **5**, S104–S114

4   Hoffmann, E. and Simonsen, L. O. (1989) Physiol. Rev. **69**, 315–382

5   Grinstein, S. and Rothstein, A. (1986) J. Membr. Biol. **90**, 1–12

6   Seifter, J. L. and Aronson, P. S. (1986) J. Clin. Invest. **78**, 859–864

7   Karpel, R., Olami, Y., Taglicht, D., Schuldiner, S. and Padan, E. (1988) J. Biol. Chem. **263**, 10408–10414

8   Taglicht, D., Padan, E. and Schuldiner, S. (1993) J. Biol. Chem. **268**, 5382–5387

9   Kleyman, T. R. and Cragoe, E. J., Jr. (1988) J. Membr. Biol. **105**, 1–21

10   Pouysségur, J., Sardet, C., Franchi, A., L'Allemain, G. and Paris, S. (1984) Proc. Natl. Acad. Sci. U.S.A. **81**, 4833–4837

11   Rothstein, A. (1989) Rev. Physiol. Biochem. Pharmacol. **112**, 235–257

12   Haggerty, J. G., Agarwal, N., Reilli, R. F., Adelberg, E. A. and Slayman, C. W. (1988) Proc. Natl. Acad. Sci. U.S.A. **85**, 6797–6801

13   Casavola, V., Helmle-Kolb, C. and Murer, H. (1989) Biochem. Biophys. Res. Commun. **165**, 833–837

14   Kulanthaivel, P., Leibach, F. H., Mahesh, V. B., Cragoe, E. J., Jr. and Ganapathy, V. (1990) J. Biol. Chem. **265**, 1249–1252

15   Wakabayashi, S., Sardet, C., Fafournoux, P., Counillon, L., Meloche, S., Pages, G. and Pouysségur, J. (1992) Rev. Physiol. Biochem. Pharmacol. **119**, 157–186

16   Sardet, C., Franchi, A. and Pouysségur, J. (1989) Cell **56**, 271–280

17   Sardet, C., Counillon, L., Franchi, A. and Pouysségur, J. (1990) Science **247**, 723–726

18   Takaichi, K., Wang, D., Balkovetz, D. F. and Warnock, D. G. (1992) Am. J. Physiol. **262**, C1069–C1076

19   Fliegel, L., Dyck, J. R. B., Wang, H. and Haworth, R. S. (1993) Mol. Cell. Biochem., in the press

20   Counillon, L. and Pouysségur, J. (1993) Biochim. Biophys. Acta **1172**, 343–345

21   Reilly, R., Hildebrandt, F., Biemesderfer, D., Sardet,C., Pouysségur, J., Aronson, P. S., Slayman, C. W. and Igarashi, P. (1991) Am. J. Physiol. **261**, F1088–F1094

22   Fliegel, L., Sardet, C., Pouysségur, J. and Barr, A. (1991) FEBS Lett. **279**, 25–29

23   Tse, C.-M., Ma, A. I., Yang, V. W., Watson, A. J. M., Levine, S., Montrose, J. M., Potter, J., Sardet, C., Pouysségur, J. and Donowitz, M. (1991) EMBO J. **10**, 1957–1967

24   Tse, C.-M., Brant, S. R., Walker, M. S., Pouysségur, J. and Donowitz, M. (1992) J. Biol. Chem. **267**, 9340–9346

25   Hildebrandt, F., Pizzonia, J. H., Reilly, R. F., Reboucas, N. A., Sardet, C., Pouysségur, J., Slayman, C. W., Aronson, P. S. and Igarashi, P. (1991) Biochim. Biophys. Acta **1129**, 105–108

26   Orlowski, J., Kandasamy, R. A. and Shull, G. E. (1992) J. Biol. Chem. **267**, 9331–9339

27   Collins, J. F., Honda, T., Knobel, S., Bulus, N. M., Conary, J., Dubois, R. and Ghishan, F. K. (1993) Proc. Natl. Acad. Sci. U.S.A. **90**, 3938–3942

28   Wang, Z., Orlowski, J. and Schull, G. E. (1993) J. Biol. Chem. **268**, 11925–11928

28a   Tse, C.-M., Levine, S. A., Yun, C. H. C., Montrose, M. H., Little, P. J., Pousségur, J. and Donowitz, M. (1993) J. Biol. Chem. **268**, 11917–11924

29   Borgese, F., Sardet, C., Cappadoro, M., Pouysségur, J. and Motais, R. (1992) Proc. Natl. Acad Sci U.S.A. **89**, 6765–6769

30   Marra, M. A., Prasad, S. S. and Baillie, D. L. (1992) Genomics **5**, 185–198

31   Jia, Z.-P., McCullough, N., Martel, R., Hemmingsen, S. and Young, P. G. (1992) EMBO J. **11**, 1631–1640

32   Ivey, D. M., Guffanti, A. A., Bossewitch, J. S., Padan, E. and Krulwich, T. A. (1991) J. Biol. Chem. **266**, 23483–23489

33   Waser, M., Hess-Bienz, D., Davies, K. and Solioz, M. (1992) J. Biol. Chem. **267**, 5396–5400

34   Pinner, E., Padan, E. and Schuldiner, S. (1992) J. Biol. Chem. **267**, 11064–11068

35   Harris, S. P., Richards, N. W., Logsdon, C. D., Pouysségur, J. and Dawson, D. C. (1992) J. Gen. Physiol. **100**, 34a (Abstr.)

36   Biemesderfer, D., Reilly, R. F., Exner, M., Igarashi P. and Aronson P. S. (1992) Am. J. Physiol. **263**, F833–F840

37   Kulanthaivel, P. L., Furesz, T. C., Moe, A. J., Smith, C. H., Mahesh, V. B., Leibach, F. H. and Ganapathy, V. (1992) Biochem. J. **284**, 33–38

38   Wakabayashi, S., Fafournoux, P., Sardet, C. and Pouysségur, J. (1992) Proc. Natl. Acad. Sci. U.S.A. **89**, 2424–2428

39   Levine, S., Montrose, M., Pouysségur, J., Tse, M. and Donowitz, M. (1992) J. Gen. Physiol. **100**, 64a (Abstr.)

40   Casavola, V., Reshkin, S. J., Murer, H. and Helmle-Kolb, C. (1992) Pflugers Arch. **420**, 282–289

41   Tse, C.-M., Levine, S., Montrose, M., Brant, S., Walker, S., Pouysségur, J. and Donowitz, M. (1992) J. Gen. Physiol. **100**, 60a (Abstr.)

42   Counillon, L., Franchi, A. and Pouysségur, J. (1992) J. Gen. Physiol. **100**, 41 (Abstr)

43   Takaichi, K., Balkovetz, D. F., Meir, E. V. and Warnock, D. G. (1993) Am. J. Physiol. **264**, C944–C950

44   Garlid, K. D., Shariat-Mada, Z., Nath, S. and Jezek, P. (1991) J. Biol. Chem. **266**, 6518–6523

45   Dyck, J. R. B., Lopaschuk, G. D. and Fliegel, L. (1992) FEBS Lett. **310**, 255–259

46   Grinstein, S. and Cohen, S. (1984) J. Gen. Physiol. **83**, 341–369

47   Raley-Susman, K. M., Cragoe, E. J., Jr., Sapolsky, R.M. and Kopito, R. R. (1991) J. Biol. Chem. **266**, 2739–2745

48   Poronnik, P., Young, J. A. and Cook, D. I. (1993) FEBS Lett. **315**, 307–312

49   Fliegel, L., Haworth, R. S. and Dyck, J. R. B. (1993) Biochem. J. **289**, 101–107

50   Haworth, R. S., Fröhlich, O. and Fliegel, L. (1993) Biochem. J. **289**, 637–640

51   Fasman, G. D. and Gilbert, W. A. (1990) Trends Biochem. Sci. **15**, 89–92

52   McDaniel, H. B., Huang, Z. Q., Cook, W. J. and Warnock, D. G. (1990) Kidney Int. **37**, 232 (Abstr.)

53   Fafournoux, P., Ghysdael, J., Sardet, C. and Pouysségur, J. (1991) Biochemistry **30**, 9510–9515

54   Winkel, G. K., Sardet, S. Pouysségur, J. and Ives, H. E. (1993) J. Biol. Chem. **268**, 3396–3400

55   Kemp, B. E. and Pearson, R. B. (1990) Trends Biochem. Sci. **15**, 342–346

56   Pelech, S. L. and Sanghera, J. S. (1992) Trends Biochem. Sci. **17**, 233–238

57   Grinstein, S., Rotin, D. and Mason, M. J. (1989) Biochim. Biophys. Acta **988**, 73–97

58   Vairo, G., Cock, B. G., Cragoe, E. J., Jr. and Hamilton, J. A. (1992) J. Biol. Chem. **267**, 19043–19046

59   Chambard, J. C., Paris, S. L'Allemain, G. and Pouysségur, J. (1987) Nature (London) **326**, 900–903

60   Sardet, C., Fafournoux, P., and Pouysségur, J. (1991) J. Biol. Chem. **266**, 19166–19171

61   Bianchini, L., Woodside, M., Sardet, C., Pouysségur, J., Takai A. and Grinstein, S. (1991) J. Biol. Chem. **266**, 15406–15413

62   Grinstein, S., Woodside, M., Sardet, C., Pouysségur, J. and Rotin, D. (1992) J. Biol. Chem. **267**, 23823–23828

63   Fliegel, L., Walsh, M. P., Singh, D., Wong, C. and Barr, A. (1992) Biochem. J. **282**, 139–145

64   Guizouarn, H., Borgese, F., Pellisier, B., Garcia-Romeu, F. and Motais, R. (1993) J. Biol. Chem. **267**, 8632–8639

65   Cheng, S. H., Rich, D. P., Marshall, J., Gregory, R. J., Welsh, M. J. and Smith, A. E. (1991) Cell **66**, 1027–1036

66   Rood, R. P., Emmer, E., Wesolek, J., McCullen, J., Husain, Z., Cohen, M. E., Braithwaite, R. S., Murer, H., Sharp, G. W. G. and Donowitz, M. (1988) J. Clin. Invest. **82**, 1091–1097

67   Cohen, M. E., Wesolek, J., McCullen, J., Rys-Sikora, K., Pandol, S., Rood, R. P., Sharp, G. W. G. and Donowitz, M. (1991) J. Clin. Invest. **88**, 855–863

68   Kozak, M. (1991) J. Cell Biol. **115**, 887–903

69   Freiberg, J. M., Kinsella, J. and Sacktor, B. (1982) Proc. Natl. Acad. Sci. U.S.A. **79**, 4932–4936

70   Bidet, M., Merot, J., Tauc, M. and Poujeol, P. (1987) Am. J. Physiol. **253**, F945–F951

71   Sacktor, B. and Kinsella, J. (1988) in Na⁺/H⁺ Exchange (Grinstein, S., ed.), pp. 307–324, CRC Press, Boca Raton, Florida

72   Yun, C. H. C., Gurughagavatula, S., Levine, S. A., Montgomery, J. L. M., Brant, S. R., Coehn, M. E., Cragoe, E. J., Jr., Pouysségur, J., Tse, C.-M. and Donowitz, M. (1993) J. Biol. Chem. **268**, 206–211

73   Kinsella, J. L., Cujdik, T. and Sacktor, B. (1984) J. Biol. Chem. **259**, 13224–13227

74   Kinsella, J. L., Cujdik, T. and Sacktor, B. (1986) J. Membr. Biol. **91**, 183–191

75   Mishina, T., Scholer, D. W. and Edelman, I. S. (1981) Am. J. Physiol. **240**, F38–F45

76   Salihagic, A., Mackovic, M., Bangic, H. and Sabolic., I. (1988) Pflugers Arch. **413**, 190–196

77   Fine, L. G., Nord, E. P., Gunther, R. and Kurtz, I. (1988) in Na⁺/H⁺ Exchange (Grinstein, S., ed.), pp. 325–334, CRC Press, Boca Raton, Florida

78   Presig, P. A. and Alpern, R. J. (1988) J. Clin. Invest. **82**, 1445–1453

79   Akiba, T., Rocco, V. K. and Warnock, D. G. (1987) J. Clin. Invest. **80**, 308–315

80   Moe, O. W., Miller, R. T., Horie, S., Cano, A., Presif, P. A. and Alpern, R. J. (1991) J. Clin. Invest. **88**, 1703–1708

81   Krapf, R., Pearce, D., Lynch, C., Xi, X.-P., Reudelhuber, T. L., Pouysségur, J. and Rector, F. C., Jr. (1991) J Clin Invest. **87**, 747–751

82   Horie, S., Moe, O., Tejedor, A. and Alpern, R. J. (1990) Proc. Natl. Acad. Sci. U.S.A. **87**, 4742–4745

83   Igarashi, P., Freed, M. I., Ganz, B. and Reilly, R. F. (1992) Am. J. Physiol. **263**, F83–F88

84   Jenkins, A. D., Dousa, T. P. and Smith, L. H. (1985) Am. J. Physiol. **249**, F590–F595

85   Hwang, J.-J. and Curthoys, N. P. (1991) J. Biol. Chem. **266**, 9392–9396

86   Kaiser, S. and Curthoys, N. P. (1991) J. Biol. Chem. **266**, 9397–9402

87   Horie, S., Moe, O., Yamaji, Y., Cano, A. and Miller, T. (1992) Proc. Natl. Acad. Sci. U.S.A. **89**, 5236–5240

88   Miller, R. T., Counillon, L., Pages, G., Lifton, R. P., Sardet, C. and Pouysségur, J. (1991) J. Biol. Chem. **266**, 10813–10819

89   Reboucas, N. A, Blaurock, M. and Igarashi, P. (1991) J. Am. Soc. Nephrol. **2**, 710 (Abstr.)

90   Horie, S., Moe, O., Miller, R. T. and Alpern, R. J. (1992) J. Clin. Invest. **89**, 365–372

91   Pollock, A. S. and Long, J. A. (1989) Biochem. Biophys. Res. Commun. **164**, 81–87

92   Soleimani, M., Bizal, G. L., McKinney, D. and Hattabaugh, Y. J. (1992) J. Clin. Invest. **90**, 211–218

93   Sabolic, I. and Brown, D. (1990) Am. J. Physiol. **258**, F1245–F1253

94   Hensley, C. B., Bradley, M. E. and Mircheff, A. K. (1989) Am. J. Physiol. **257**, C637–C645

95   Rao, G. N., Sardet, C., Pouysségur, J. and Berk, B. C. (1990) J. Biol. Chem. **265**, 19393–19396

96   Berk, B. C., Elder, E. and Mitsuka, M. (1990) J. Biol. Chem. **265**, 19632–19637

97   Boxer, L. M., Prywes, R., Roeder, R. G. and Kedes, L. (1989) Mol. Cell. Biol. **9**, 515–522

98   Rao, G. N., Roux, N., Sardet, C., Pouysségur, J. and Berk, B. C. (1991) J. Biol. Chem. **266**, 13485–13488

99   Rao, G. N., Sardet, C., Pouysségur, J. and Berk, B. C. (1992) J. Cell. Physiol. **151**, 361–366

100   Godt, R. E. and Nosek, T. M. (1989) J. Physiol. (London) **412**, 155–180

101   Jeffrey, F. M. H., Mallow, C. R. and Radda, G. K. (1987) Am. J. Physiol. **253**, H1499–H1505

102   Malloy, C., Matthews, P., Smith, M. and Radda, G. K. (1985) Adv. Myocardiol. **6**, 461–464

103   Orchard, C. H. and Kentish, J. C. (1990) Am. J. Physiol. **258**, C967–C981

104   Lagadic-Gossmann, D., Buckler, K. J. and Vaughan-Jones, R. D. (1992) J. Physiol. (London) **458**, 361–384

105   Lagadic-Gossmann, D., Vaughan-Jones, R. D. and Buckler, K. J. (1992) J. Physiol. (London) **458**, 385–407

106   Liu, S., Piwnica-Worms, D. and Lieberman, M. (1990) J. Gen. Physiol. **96**, 1247–1269

107   Vigne, P., Breittmayer, J. P., Frelin, C. and Lazdunski, M. (1988) J. Biol. Chem. **263**, 18023–18029

108   Mason, M. J., Smith, J. D., Garcia-Soto, J. J. and Grinstein. S. (1989) Am. J. Physiol. **256**, C428–C433

109   Wallert, M. A. and Fröhlich, O. (1989) Am. J. Physiol. **257**, C207–C213

110   Terzic, A., Puceat, M., Clement, O., Scamp, F. and Vassort, G. (1992) J. Physiol. (London) **447**, 275–292

111   Wallert, M. A. and Fröhlich, O. (1992) Am. J. Physiol. **263**, C1096–C1102

112   Aronson, P. S., Nee J. and Suhm, M. A. (1982) Nature (London) **299**, 161–163

113   Aronson, P. S. (1985) Annu Rev. Physiol. **47**, 545–560

114   Vigne, P., Frelin, C. and Lazdunski, M. (1984) EMBO J. **3**, 1865–1870

115   Frelin, C., Vigne, P. and Lazdunski, M. (1985) Eur. J. Biochem. **149**, 1–4

116   Hearse, D. J. (1977) J. Mol. Cell. Cardiol. **9**, 605–616

117   Becker, L. C. and Ambrosio, G. (1987) Progr. Cardiovasc. Dis. **30**, 23–44

118   Karmazyn, M. (1988) Am. J. Physiol. **255**, H608–H615

119   Tani, M. and Neely, J. R. (1989) Circ. Res. **65**, 1045–1056

120   Avkiran, M. and Ibuki, C. (1992) Circ. Res. **71**, 1429–1440

121   Khandoudi, N., Bernard, M., Cozzone, P. and Feuvray, D. (1990) Cardiovasc. Res. **24**, 873–878

122   Meng, H.-P. and Pierce, G. N. (1990) Am. J. Physiol. **258**, H1615–H1619

123   Meng, H.-P. and Pierce, G. N. (1991) J. Pharmacol. Exp. Ther. **256**, 1094–1100

124   Murphy, E., Perlman, M., London, R. E. and Steenbergen, C. (1991) Circ. Res. **68**, 1250–1258

125   Duan, J. and Karmazyn, M. (1992) Eur. J. Pharmacol. **210**, 149–157

126   Karmazyn, M. (1993) Br. J. Pharmacol. **108**, 50–56

127   Anderson, S. E., Murphy, E., Steengergen, C., London, R. E. and Cala, P. M. (1990) Am. J. Physiol. **259**, C940–C948

128   Karmazyn, M., Ray, M. and Haist, J. V. (1993) J.Cardiovasc. Pharmacol. **21**, 172–178

129   Dennis, S. C., Coetzee, W. A., Cragoe, E. J., Jr. and Opie, L. H. (1990) Circ. Res. **66**, 1156–1159

130   Meng, H. P., Lonsberry, B. and Pierce, G. N. (1991) J. Pharmacol. Exp. Ther. **258**, 772–777

131   Scholz, W., Albus, U., Linz, W., Martorana, P., Lang, H. J. and Scholkens, B. A. (1992) J. Mol. Cell. Cardiol. **24**, 731–740

132   Meng, H. P., Maddaford, T. G. and Pierce, G. N. (1993) Am. J. Physiol. **264**, H1831–H1835

133   Sardet, C., Wakabayashi, S., Fafournoux, P. and Pouysségur, J. (1991) NATO ASI Ser. Ser. B **52**, 253–269

134   Pouysségur, J. (1992) Physiologist **35**, 7.0 (Abstr.)

135   Reilly, R. F. and Aronson, P. S. (1992) in Amiloride and Its Analogs (Cragoe, E. G., Jr. and Kleyman, T. R., eds.), pp. 57–77, VCH Publishers, New York

136   Benos, D. J., Reyes, J. and Shoemaker, D. G. (1983) Biochim. Biophys. Acta **734**, 99–104

137   Harguindey, S. (1992) in Amiloride and Its Analogs (Cragoe, E. J., Jr. and Kleyman, T. R., eds.), pp. 317–334, VCH Publishers, New York

138   Kyte, J. and Doolittle, R. F. (1982) J. Mol. Biol. **157**, 105–132

Biochem. J. (1993) **291**, 1–10 (Printed in Great Britain)

REVIEW ARTICLE
Proteasomes: multicatalytic proteinase complexes

A. Jennifer RIVETT

Department of Biochemistry, University of Leicester, Leicester LE1 7RH, U.K.

INTRODUCTION

The multicatalytic proteinase complex (proteasome) is a high-molecular-mass (approximately 700 kDa) intracellular proteinase which has been isolated under a variety of different names from a wide variety of eukaryotic cells and tissues (reviewed, Rivett, 1989a; Orlowski, 1990). The proteinase complex is composed of at least 24 subunits which include many different polypeptides arranged in a cylindrical structure. Other multi-subunit complexes with cylindrical structures include 'prosomes', that are widely distributed 19 S ribonucleoprotein particles which were thought to be involved in the control of translation (Schmid et al., 1984; Martins de Sa et al., 1986), erythrocyte cylindrin (Harris, 1988), and a number of other partially characterized 16–22 S cylindrical particles. 'Prosomes' and certain related particles have been shown to have proteolytic activity and to share antigenic cross-reactivity with the multicatalytic proteinase (Falkenburg et al., 1988) and are therefore believed to be the same. The particles are most often referred to either as multicatalytic proteinase complexes (Dahlmann et al., 1988; Orlowski and Wilk, 1988) or, more recently, as proteasomes (Arrigo et al., 1988) and their properties are broadly similar irrespective of the source (Martins de Sa et al., 1986; Tanaka et al., 1988b; Rivett, 1989a; Orlowski, 1990). A related proteinase with a simpler subunit composition has been isolated from the archaebacterium *Thermoplasma acidophilum* (Dahlmann et al., 1989).

The abundance (up to 1% of the soluble protein in cell extracts; e.g. Tanaka et al., 1986; Hendil, 1988) and ubiquitous distribution of proteasomes in eukaryotic cells, as well as observations that some proteasome subunit genes in yeast are essential for cell growth and viability, suggest that they have important functions within the cell. There is evidence to suggest that they play a major role in nonlysosomal pathways of intracellular protein turnover and they have recently been implicated in antigen processing. Proteasomes are not closely related to any other known proteases and it is likely that they represent a novel family of proteolytic enzymes.

POLYPEPTIDE AND RNA COMPONENTS OF PROTEASOMES

Analysis of the subunit composition of highly purified proteasomes by two-dimensional polyacrylamide gel electrophoresis reveals a characteristic pattern of polypeptide components of varying molecular masses (usually in the range of 22–35 kDa) and pI values in the range of pH 4 to 10 (Figure 1). Some changes in subunit pattern have been observed during development in *Drosophila melanogaster* (Haass and Kloetzel, 1989) and chicken (Ahn et al., 1991). The different polypeptides are not always present in equal amounts and the number appears to vary depending upon the species. For example, yeast proteasomes have 14 different types of subunits (Heinemeyer et al., 1991), plant proteasomes 12 to 15 (Schliephacke et al., 1991), and other eukaryotic complexes have been reported to contain up to 25 distinct polypeptides (e.g. Martins de Sa et al., 1986; Rivett and Sweeney, 1991). The proteasome isolated from thermoacidophilic archaebacteria, on the other hand, is a much simpler molecule, being composed of only two different types of

subunit (α, 27kDa; β, 25kDa; Dahlmann et al., 1989) and has therefore been particularly useful for structural studies. Although a few of the spots observed on two-dimensional polyacrylamide gels of eukaryotic proteasomes may be related to each other by **proteolysis** (Lee et al., 1990; Rivett and Sweeney, 1991; Weitman and Etlinger, 1992), **phosphorylation** (Haass and Kloetzel, 1989) or **glycosylation** (Tomek et al., 1988; Schliephacke et al., 1991), many are antigenically distinct (Rivett and Sweeney, 1991; Kaltoft et al., 1992). However, it is not yet possible to put a precise value to the number of proteasomal subunits which are the products of different, but related (see below), genes. Even the archaebacterial proteasome can give rise to a more complex pattern on two-dimensional PAGE gels than the two spots expected (Zwickl et al., 1992). Proteolysis of some subunits of eukaryotic proteasomes may occur during purification, storage or activation of the enzyme (Lee et al., 1990; Rivett and Sweeney, 1991; Weitman and Etlinger, 1992) and some specific processing events may also occur *in vivo*. For example, a 41 kDa cross-reacting protein has been detected by using a monoclonal antibody raised against the 32 kDa subunit of human erythrocyte proteasomes (Weitman and Etlinger, 1992), a 38kDa protein cross-reacts with a monoclonal antibody for a 27kDa subunit of "prosomes" (Kreutzer-Schmid and Schmid, 1990), and the archaebacterial β subunit is known to be post-translationally processed (Zwickl et al., 1992).

There are discrepancies in the literature with regard to the presence and amount of RNA associated with proteasomes. Recent studies have confirmed the presence of one or more small RNA species (approximately 80 nucleotides) in highly purified multicatalytic proteinase and prosome preparations, but the RNA is not present in stoichiometric amounts (Skilton et al., 1990; Coux et al., 1992). Since the RNA is resistant to nuclease attack in the intact particles, the latter observation suggests the possibility of a subpopulation of particles containing RNA. The

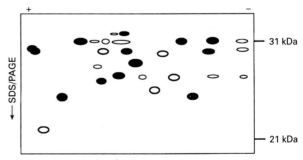

Figure 1   Two-dimensional PAGE of proteasomes

Schematic diagram showing the characteristic two-dimensional PAGE pattern of proteasomes purified from rat liver. Closed spots represent subunits which are usually present. Open spots are variable in different preparations and/or present in lower amounts than the dark polypeptides. Small differences are observed in the ratios of the different subunits in different proteinase preparations and differences are apparent between the particles isolated from different eukaryotic sources. All show a complex subunit composition with some major and some minor components. The precise stoichiometry of eukaryotic proteasomes is not yet clear.

Table 1   Size and shape of proteasomes

Particle Shape/subunit arrangement	Diameter (nm)	Height (nm)	Symmetry	References
Rat muscle MCP	Hollow cylinder	11	16	6	Kopp et al., 1986
	Stack of four hexagonal rings				
	Reel-shaped			Not true 6-fold	Baumeister et al., 1988
Rat liver MCP	Prolate ellipsoid	16	11		Tanaka et al., 1988a
				8	Arrigo et al., 1988
	Cylindrical pseudo-helical arrangement	11	17	6–7	Djaballah et al., 1993
Duck erythroblast prosomes	Hollow cylinder	12	17		Coux et al., 1992
Related particles	Cylindrical	10–13	16–18	6–8	Reviewed in Harris, 1988
Archaebacterial proteasome	Cylindrical barrel	11	15	7	Dahlmann et al., 1989; Hegerl et al., 1991
				7	Pühler et al., 1992

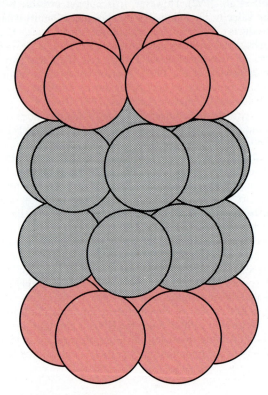

Figure 2   Model of the proteasome

The model shows the cylindrical structure which appears side-on as a stack of four rings. There are six or seven subunits around each ring. The arrangement of subunits within the mammalian complex is pseudo-helical (Djaballah et al., 1993). Based on analogy with the archaebacterial enzyme (Grziwa et al., 1991), A-type subunits (red) may be located in the outer rings while B-type subunits probably form the inner rings.

number of RNA species appears to depend on the source of the particles and a major proteasomal RNA from HeLa cells and from duck erythroblasts has recently been identified as tRNA[Lys3] (Nothwang et al., 1992b). The functional significance of these observations remains to be established.

CYLINDRICAL STRUCTURE

The subunits of proteasomes are arranged in a stack of four rings to form a hollow cylindrical structure. Recent detailed structural investigations of the archaebacterial proteasome by image analy-

sis of negatively stained particles (Dahlmann et al., 1989; Hegerl et al., 1991) and of small three-dimensional crystals (Pühler et al., 1992) show clear 7-fold symmetry similar to that of molecular chaperones of the GroEL family (Zwickl et al., 1990). Each ring contains seven subunits and the stoichiometry of the molecule is $\alpha_{14}\beta_{14}$. An immunoelectronmicroscopic investigation has demonstrated that the α subunits form the outer rings while the β subunits form the inner ones (Grziwa et al., 1991).

There are variations in the literature with respect to the size and shape of eukaryotic proteasomes (Table 1) but these now appear to be resolved. Electron microscopy of negatively stained rat muscle and liver proteasome preparations suggest a cylindrical structure (Baumeister et al., 1988; Djaballah et al., 1993). The particle viewed end-on appears to be ring-shaped and side-on a rectangular shape with four bands, suggesting a stack of four rings. The molecule has two-fold rotational symmetry. The rings do not show the true 6-fold symmetry nor the direct stacking of the model previously proposed for the rat muscle enzyme (Kopp et al., 1986). The arrangement of subunits in the rat liver enzyme appears to be pseudo-helical (Djaballah et al., 1993; Figure 2). This model resembles that for the archaebacterial enzyme (Pühler et al., 1992), but differs somewhat from the prolate ellipsoid structure suggested earlier for the rat liver proteinase based on X-ray scattering data (Tanaka et al., 1988a). However, it has already been pointed out that the structure of the archaebacterial enzyme is consistent with the X-ray scattering data of Tanaka et al., (1988a) and that the differences between models lies largely in the interpretation of data (Pühler et al., 1992). The dimensions of the archaebacterial proteasome and various eukaryotic proteasomes are all similar, bearing in mind differences in measurements obtained using different negative stains (diameter 11–12nm, height 16–20nm; Table 1).

PRIMARY STRUCTURES OF PROTEASOMAL PROTEINS

During the last few years the amino acid sequences of several subunits of human, rat, yeast, and *Drosophila* proteasomes have been deduced from the sequence of cloned cDNA (see Table 2). The sequences of these proteasome subunits show significant identities to each other but not to any other known proteins. Similarity scores obtained by pairwise alignment of different human, rat or yeast proteasomal subunits of the same species (often 20–40 % identity) imply that many of the genes belong to the same gene family. Moreover, the significant sequence similarity of the α subunit of the archaebacterial particle (Zwickl et al., 1991) to various subunits of eukaryotic proteasomes suggests that these proteasomal proteins are encoded by a gene family of

Table 2   **Sequence information for proteasomal subunits**

Source	Subunit	Reference
Archaebacteria (*Thermoplasma acidophilum*)	α	Zwickl et al., 1991
	β	Zwickl et al., 1992
Yeast (*Saccharomyces cerevisiae*)	YC1, YC7α	Fujiwara et al., 1990
	Y7, Y8, Y13	Emori et al., 1991
	PRE1	Heinemeyer et al., 1991
	PRS3	Lee et al., 1992
	scl1$^+$ (YC7α)	Balzi et al., 1990
	PUP1	Haffter and Fox, 1991
	PUP2	Georgatsou et al., 1992
Drosophila melanogaster	PROS35	Haass et al., 1989
	PROS28.1	Haass et al., 1990a
	PROS29	Haass et al., 1990b
Xenopus laevis	XC3	Fujii et al., 1991
	β	van Riel and Martens, 1992
Rat	C2	Fujiwara et al., 1989
	C3	Tanaka et al., 1990a
	C5	Tamura et al., 1990
	C8	Tanaka et al., 1990b
	C9	Kumatori et al., 1990b; Sorimachi et al., 1990
	N-terminal seq. 1–7	Lilley et al., 1990
	C1	Aki et al., 1992
Mouse	LMP2	Martinez and Monaco, 1991
	MC13	Frentzel et al., 1992a
Human	C2, C3, C5, C8, C9	Tamura et al., 1991
	$\nu, \iota, \zeta, \delta$	De Martino et al., 1991
	RING10	Glynne et al., 1991
	RING12	Kelly et al., 1991
	N-terminal seq. $\alpha, \beta, \gamma, \delta, \epsilon$	Lee et al., 1990

ancient origin. Several gene duplication events may have contributed to the complex subunit composition of eukaryotic proteasomes, probably via selection for a variety of functional requirements, although there are clearly structural constraints on the extent of subunit divergence within the cylindrical molecule.

The possibility of at least two **groups of closely-related proteasome subunits**, which seemed likely from comparison of directly determined N-terminal amino acid sequence data (Lilley et al., 1990; Lee et al., 1990) with deduced sequences of N-terminally blocked subunits (e.g. Fujiwara et al., 1989; Tanaka et al., 1990a; Tamura et al., 1991) of the rat and human proteinases, has been confirmed by the recent sequence analysis of the second (β) subunit of the archaebacterial enzyme (Zwickl et al., 1992). Although the two amino acid sequences of the *Thermoplasma acidophilum* proteasome can be aligned to show 24 % identity, there are significant differences, particularly in the N-terminal regions. Close examination of sequences of proteasome subunits from other species shows that they can be divided into two groups, A and B (Figure 3), based upon whether they are more closely related to the archaebacterial α or β subunit, respectively. However, not all eukaryotic proteasome subunits fit neatly into group A or B. For example, the amino acid sequence deduced for C5, although related to the A and B groups, is not characteristic of either (Tamura et al., 1990; Lee et al., 1992). Many of the cloned subunits fall into the A group. The *Thermoplasma acidophilum* α subunit is most similar to the eukaryotic C3, which is very highly conserved between rat (Tanaka et al., 1990a), human (98 % identity to rat C3; Tamura et al., 1991) and *Xenopus* (95 % identity to rat C3; Fujii et al., 1991). Other A-type subunits are also very highly conserved between different animal species (Tamura et al., 1991).

The archaebacterial β subunit has an unblocked N-terminus (Zwickl et al., 1992) as do seven of the related rat liver proteasomal polypeptides (Lilley et al., 1990), while A subunits RC2, RC3, and RC8 have been shown to be N-acetylated (Tokunaga et al., 1990). Sequences of N-terminal regions of different proteasomal A-type subunits are particularly highly conserved, whereas the C-terminal regions are often not. A-type subunits each contain three highly conserved regions (termed PROS-Box I, II, and III by Haass et al., 1990b; Zwickl et al., 1991). Sequences of B subunits share some highly conserved regions but have variable lengths at the N-terminus (Figure 3b). Members of the proteasome gene family are found on several different chromosomes (Lee et al., 1992; Frentzel et al., 1992b).

In general there are no sequence motifs in proteasome subunits which are characteristic of other proteases. Sequences which are related to the conserved regions around catalytic residues in subtilisin-like serine proteases which were identified in the primary structure deduced from the cDNA sequence of RING10 (Glynne et al., 1991) are not conserved in other B-type proteasomal subunits, and in RC1, which is the rat homologue of RING10, the putative catalytic histidine residue is replaced by an asparagine residue (Aki et al., 1992). The fact that proteasome subunits show no obvious relationship to other known proteases suggests that the proteasome represents a **novel family of proteolytic enzymes** (multicatalytic endopeptidase complex; EC 3.4.99.46; and see Rawlings and Barrett, 1993, for a listing of proteinase families).

KINETIC PROPERTIES OF PROTEASOMES

The term multicatalytic was applied to this proteinase complex because it has an unusual enzymological property when compared to other proteinases; it can catalyse peptide bond cleavage on the carboxyl side of basic, hydrophobic and acidic amino acid residues (Wilk and Orlowski, 1983). These three types of pro-

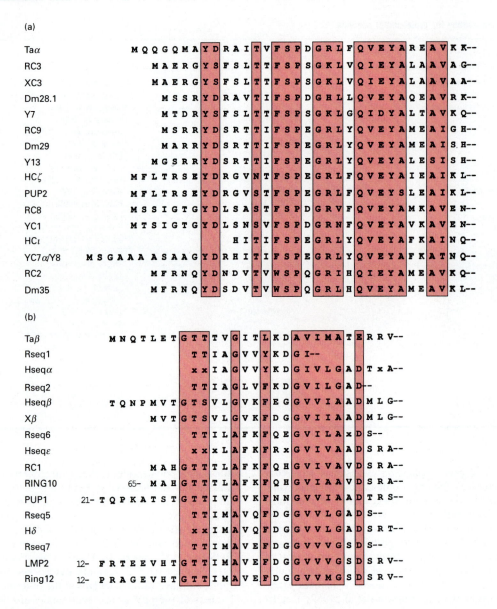

(a)

Taα		M Q Q G Q M A **Y D** R A I **T** V **F S P D** G R L **F Q V E Y A** R E **A V** K K --
RC3		M A E R G **Y** S **F** S L **T T F S P** S G K L **V Q I E Y A** L A **A V** A G --
XC3		M A E R G **Y** S **F** S L **T T F S P** S G K L **V Q I E Y A** L A **A V** A A --
Dm28.1		M S S R **Y D** R A V **T** I **F S P D** G H L **L Q V E Y A** Q E **A V** R K --
Y7		M T D R **Y** S **F** S L **T T F S P** S G K L G **Q I D Y A** L T **A V** K Q --
RC9		M S R R **Y D** S R **T** T I **F S P E** G R L **Y Q V E Y A** M E **A I** G H --
Dm29		M A R R **Y D** S R **T** T I **F S P E** G R L **Y Q V E Y A** M E **A I** S H --
Y13		M G S R **Y D** S R **T** T I **F S P E** G R L **Y Q V E Y A** L E **S I** S H --
HCζ		M F L T R S E **Y D** R G V N **T F S P E** G R L **F Q V E Y A** I E **A I** K L --
PUP2		M F L T R S E **Y D** R G V S **T F S P E** G R L **F Q V E Y** S L E **A I** K L --
RC8		M S S I G T G **Y D** L S A S **T F S P D** G R V **F Q V E Y A** M K **A V** E N --
YC1		M T S I G T G **Y D** L S N S V **F S P D** G R N **F Q V E Y A** V K **A V** E N --
HCι		H I **T** I **F S P E** G R L **Y Q V E Y A** F K **A I** N Q --
YC7α/Y8	M S G A A A A S A A G **Y D** R H I **T** I **F S P E** G R L **Y Q V E Y A** F K **A T** N Q --
RC2		M F R N Q **Y D** N D V **T** V **W S P** Q G R I H **Q I E Y A** M E **A V** K Q --
Dm35		M F R N Q **Y D** S D V **T** V **W S P** Q G R L H **Q V E Y A** M E **A V** K L --

(b)

Taβ		M N Q T L E T **G T T** T V **G I T L** K D **A** V I M **A T E** R R V --
Rseq1		**T T** I **A G V V** Y K D **G I** --
Hseqα		x x I **A G V V** Y K D **G I** V L G **A D** T x A --
Rseq2		**T T** I **A G L V F** K D **G** V I L G **A D** --
Hseqβ		T Q N P M V **T G T** S V L **G V K F E G G** V V I **A A D** M L G --
Xβ		M V **T G T** S V L **G V K F D G G** V I I **A A D** M L G --
Rseq6		**T T** I L **A F K F** Q **E G** V I L **A** x **D** S --
Hseqε		x x x L **A F K F** R x **G** V I V **A A D** S R A --
RC1		M A H **G T T** T L **A F K F** Q H **G** V I V **A** V **D** S R A --
RING10	65-	M A H **G T T** T L **A F K F** Q H **G** V I A **A** V **D** S R A --
PUP1	21-	T Q P K A T S **T G T** T I V **G V K F** N N **G** V V I **A A D** T R S --
Rseq5		**T T** I M **A V** Q **F D G G** V V L G **A D** S --
Hδ		x x I M **A V** Q **F D G G** V V L G **A D** S R T --
Rseq7		**T T** I M **A V E F D G G** V V V **G** S **D** S --
LMP2	12-	F R T E E V H **T G T** T I M **A V E F D G G** V V V **G** S **D** S R V --
Ring12	12-	P R A G E V H **T G T** T I M **A V E F D G G** V V M **G** S **D** S R V --

Figure 3   Sequence relationships in proteasome subunits

Proteasomal subunits can usually be assigned to one of two groups based on their similarity to the *Thermoplasma acidophilum* α (A-type) or β (B-type) subunit. Sequence references are given in Table 1. (**a**) A-type subunits are closely related to the *Thermoplasma acidophilum* α subunit. Sequences have been deduced from the nucleotide sequence of cloned cDNA. The N-terminal sequences are shown with highly conserved residues highlighted. There are two other regions of highly conserved residues within the full-length sequences which are characteristic of A-type subunits (PROS-BOXII and III, Haass et al., 1990a). At least some A-type subunits are N-acetylated (Tokunaga et al., 1990). (**b**) B-type subunits have sequences which show some similarities to the *Thermoplasma acidophilum* β subunit. Several B-type subunits are not N-terminally blocked and the sequences shown are aligned to maximize identities with directly determined N-terminal amino acid sequences (Lee et al., 1990; Lilley et al., 1990). Highly conserved residues have been highlighted. Archaebacterial α and β subunits are related to each other and can be aligned to give 24% identical residues (Zwickl et al., 1992).

teolytic activity have usually been referred to as "trypsin-like", "chymotrypsin-like" and "peptidylglutamyl-peptide hydrolase" activities, respectively. Information concerning the kinetic characteristics of the proteinase has been derived from many studies carried out with the enzyme from a variety of different sources (reviewed in Rivett, 1989a; Orlowski, 1990).

The **multiple proteolytic functions** are believed to be catalysed at independent sites within the complex because the different proteolytic activities respond differently to various activators and inhibitors. From detailed studies with the rat liver and bovine pituitary proteinases (Wilk and Orlowski, 1983; Rivett, 1989b; Orlowski et al., 1991; Yu et al., 1991; Djaballah and

Rivett, 1992; Pereira et al., 1992; Djaballah et al., 1992), it now seems likely that there may be at least five distinct catalytic components within the mammalian proteasome, while the much simpler archaebacterial enzyme possesses primarily chymotrypsin-like activity (Dahlmann et al., 1989). The early names for the multiple catalytic activities were clearly an over-simplification. For example, peptidylglutamyl-peptide hydrolase activity of the mammalian proteasome, which is usually assayed with Z-Leu-Leu-Glu-β-naphthylamide, is catalysed by at least two distinct components, LLE1 and LLE2 (Djaballah and Rivett, 1992) and there may be as many as three distinct components catalysing chymotrypsin-like activity (Cardozo et al., 1992;

Djaballah et al., 1992). Rather little is known in detail about the specificities of the different catalytic centres of eukaryotic proteasomes and a better system of naming them must therefore await further characterization of catalytic components.

It is difficult to predict sites of cleavage in peptide substrates, but many different bonds can be cleaved (Wilk and Orlowski, 1980; Rivett, 1985a; McDermott et al., 1991). The rat liver and human erythrocyte proteasomes show the same major cleavage sites in insulin B chain (Rivett, 1985a; Dick et al., 1991). Some differences in kinetic properties of the proteinases isolated from bovine pituitary (Orlowski and Michaud, 1989; Orlowski et al., 1991; Pereira et al., 1992) and rat liver (Rivett, 1989b; Arribas and Castaño, 1991; Djaballah and Rivett, 1992; Djaballah et al., 1992) have been reported, but in view of the complexity of the particles and differences in purification protocols it is not easy to assess the significance of such differences.

The **effects of protease inhibitors** on proteasome activities depends on the substrate used. Dramatic differences in reactivity with different inhibitors have been observed. For example, the peptidyl arginine aldehyde leupeptin inhibits only the trypsin-like activity (Wilk and Orlowski, 1983; Rivett, 1989b). Although results of early studies with protease inhibitors led to some confusion about whether the proteinase should be classified as a serine or cysteine endopeptidase (reviewed by Rivett, 1989a), recent results obtained with the serine protease inhibitors, 3,4-dichloroisocoumarin (Orlowski and Michaud, 1989; Mason, 1990; Djaballah et al., 1992) and 4-(2-aminoethyl)-benzenesulphonyl fluoride (Djaballah et al., 1992), which, at least for some proteasomes, are more potent inhibitors than those used previously (e.g. di-isopropyl fluorophosphate and phenylmethanesulphonyl fluoride), support the view that the enzyme is possibly an **unusual type of serine endopeptidase** although several of the activities are sensitive to inhibition by thiol-reactive reagents. The apparent rate constants for inactivation by inhibitors of serine proteases are mostly very low when compared to those for other proteinases (Djaballah et al., 1992). The precise determination of catalytic mechanism will only become possible with the identification of catalytic residues. Surprising differences in reactivity have been observed between the different catalytic centres of the mammalian proteinase complex. For example, 3,4-dichloroisocoumarin reacts rapidly with some, but does not inhibit all, catalytic components of the rat liver or bovine pituitary enzyme (Djaballah et al., 1992; Pereira et al., 1992) and 4-(2-aminoethyl)-benzenesulphonyl fluoride can selectively inhibit trypsin-like activity (Djaballah et al., 1992). Therefore, although it seems likely that the proteolytic sites are mechanistically related, they obviously differ in important aspects which determine specificity.

PROTEIN DEGRADATION BY PROTEASOMES

The proteinase can degrade a variety of protein and peptide substrates, and activity at all sites is optimal at neutral to weakly alkaline pH values. The multiple proteolytic activities of the proteasome should be advantageous for rapid degradation of the variety of intracellular proteins. Protein substrates, which include certain oxidized proteins (Rivett, 1985b), crystallins (Ray and Harris, 1985), myofibrillar proteins (Mykles and Haire, 1991) and casein, can be degraded to small acid soluble peptides without accumulation of intermediates detectable by SDS/PAGE (Rivett, 1985b). Although it seems likely that the various peptidase activities contribute to protein degradation, an additional distinct catalytic centre has been implicated in the degradation of casein (Pereira et al., 1992). However, this may

simply be a catalytic centre for which a convenient synthetic peptide substrate has not yet been identified.

REGULATION OF PROTEINASE ACTIVITIES

In vitro experiments have shown that sodium dodecyl sulphate, at low concentrations, as well as a variety of other treatments (including polylysine, low concentrations of guanidine hydrochloride, dialysis against distilled water and heat treatment) can cause a dramatic stimulation of one or more activities of proteasomes isolated from different sources (see e.g. McGuire et al., 1989; Tanaka et al., 1990b; Arribas and Castaño, 1990; Mylkes and Haire, 1991; Djaballah et al., 1992). Such observations have sometimes led to use of the term "**latent form**" for the non-activated complex, especially for human erythrocyte proteasomes (Lee et al., 1990; Weitman and Etlinger, 1992) but their significance to the regulation of activity *in vivo* is not clear. Sedimentation velocity studies have shown significant changes in conformation associated with activation and inhibition of proteinase activities (Djaballah et al., 1993).

Such stimulatory or inhibitory effects may be mediated *in vivo* by specific **endogenous activator and inhibitor proteins.** Several different endogenous inhibitors have been identified: a 200 kDa (subunit molecular mass 50kDa) inhibitor from human erythrocytes (Li et al., 1991), a 60 kDa (subunit molecular mass 31 kDa) protein from bovine erythrocytes (Chu-Ping et al., 1992a) and a 250 kDa (subunit molecular mass 40 kDa) proteasome inhibitor (Murakami and Etlinger, 1986; Li et al., 1991; Driscoll et al., 1992). Activator proteins have also been purified (Yukawa et al., 1991; Goldberg and Rock, 1992; Chu-Ping et al., 1992b) but their physiological function and mechanism of action have not been established.

It is difficult to measure the activity of proteasomes *in vivo* and the significance of observed differences in proteolytic activity attributed to proteasomes during differentiation of murine erythroleukaemia cells (Tsukahara et al., 1990), during development (Ahn et al., 1991) and during maturation of red blood cells (DiCola et al., 1991) cannot be evaluated because assays of activities in crude extracts do not provide an accurate measure of levels of proteasome activity due to possible interference by other proteins.

In addition to modulation of proteasome activities and function by association with other proteins, the mammalian proteinase complex may also be activated by certain substrates. For example, it shows **positive co-operativity** with a substrate of the peptidylglutamyl-peptide hydrolase activity (Orlowski et al., 1991; Djaballah and Rivett, 1992) and high concentrations of this substrate can induce a conformational change (Djaballah et al., 1993). Proteolysis may also be regulated by a requirement for **modification of some protein substrates** (e.g. ubiquitination, oxidation or other modification) prior to their recognition by proteasomes (Rivett, 1985b, 1989a).

FUNCTIONS OF INDIVIDUAL COMPONENTS OF THE COMPLEX

Dissociation of mammalian proteasomes by denaturing agents causes rapid loss of the activities associated with the complex and it has not proved possible so far to demonstrate functions of the isolated subunits nor to reconstitute active molecules from the dissociated subunits. It is not yet clear exactly how many subunits are responsible for the proteolytic activities and, although it seems likely that individual subunits are responsible for each of the proteolytic activities, the possibility that active sites are each shared between two subunits cannot be ruled out. Two approaches have been taken to identify catalytic components of the complex. In yeast, mutants defective in an individual proteasome gene (PRE1) show reduced chymotrypsin-like activity which suggests the possibility of a catalytic role for this subunit (Heinemeyer et al., 1991). The alternative method for identification of catalytic subunits involves active site labelling experiments which are also essential to learn something about the catalytic residues and mechanism. The problem with the latter approach has, until recently, been the identification of suitable selective inhibitors for the distinct catalytic centres. Leupeptin is a specific reversible inhibitor of the trypsin-like activity (Wilk and Orlowski, 1983) but subunits identified by leupeptin protection agaist modification by radiolabelled N-ethylmaleimide (Dick et al., 1992; Savory and Rivett, 1993) may not be the catalytic components. However, some selective affinity labels have now been identified (P. J. Savory, H. Angliker, E. Shaw and A. J. Rivett, unpublished work).

Although computer analysis has revealed no overall homology to other known proteins, **predicted tyrosine phosphorylation sites** have been identified in Taa, C3, C9, Dm29 and YC7a subunits (Tanaka et al., 1990a; Haass et al., 1990b; Zwickl et al., 1991), as have **putative cyclic AMP/cyclic GMP-dependent phosphorylation sites** in Dm28.1 and Taa (Haass et al., 1990a; Zwickl et al., 1991). These observations, as well as the fact that the archaebacterial α subunit does not have a histidine residue and would therefore be unable to function independently as a classical serine protease, have led to the suggestion that that A-type subunits of the proteasome have a regulatory and targeting function, while the B-type subunits may be catalytic (Zwickl et al., 1992). However, this may be an oversimplification, especially for eukaryotic proteasomes.

SUBCELLULAR LOCALIZATION

Immunocytochemical studies have shown proteasomes to be present both in the nucleus and cytoplasm of a variety of cells and tissues (see Tanaka et al., 1990c for references; Beyette and Mykles, 1992). More detailed localization studies using electron microscopic immunogold labelling techniques with anti-proteasome antibodies have shown that a proportion of proteasomes are also close to or actually on the rough endoplasmic reticulum and in polysomes in rat hepatocytes and human L-132 cells (Rivett et al., 1992; Table 3). "Prosomes" have been suggested to be associated with intermediate filaments (Grossi de Sa et al., 1988; Briane et al., 1992) and a significant amount of the complex has been purified from erythrocyte membranes (Kinoshita et al., 1990).

The functional significance of the distribution of proteasomes and the relationship between particles in the different cellular compartments are not yet well understood (Rivett and Knecht, 1993). Proteasomes isolated from rat liver nuclei and cytoplasm show the same basic properties (Tanaka et al., 1989). Several proteasome components do contain **putative nuclear localization signals** and it has been suggested that tyrosine phosphorylation could affect exposure of nuclear localization signals and thereby regulate movement of proteasomes between the nucleus and cytoplasm (Tanaka et al., 1990c). The similarity between sequences in some A-type proteasome subunits and the nuclear targeting sequence of SV40 large T antigen has already been noted (Tanaka et al., 1990c; Zwickl et al., 1991; Figure 4). However, it has recently been suggested (Dingwall and Laskey, 1991) that many nuclear targeting sequences may be more complex than the SV40 large T antigen sequence, and that a bipartite motif comprising two basic amino acids, a spacer usually of ten amino acids, and a basic cluster in which three out

Table 3 Subcellular localization determined by immunogold electron microscopy

The data are from Rivett et al. (1992) obtained using polyclonal antibodies for rat liver proteasomes. Percentage labelling is calculated from the number of gold particles/μm^2. V_v is the volume density of the compartment. There was no significant labelling in mitochondria, lysosomes, peroxisomes or Golgi.

	Significant labelling (%)	
Cell compartment	Rat hepatocytes	Human L-132 cells
Nucleoli	3.2	3.3
Nuclear matrix	13.5 (V_v 8.4%)	47.3 (V_v 37.0%)
Mitochondria	—	—
Lysosomes	—	—
Endoplasmic reticulum	14.3	3.5
Peroxisomes	—	—
Golgi	—	—
Cytoplasmic matrix	69.0	45.9

of the next five amino acids are basic residues, is required. Of the mammalian proteasome subunits cloned so far, only the C9 component contains a putative bipartite motif (Figure 4), but this is incomplete in the closely related *Drosophila* Dm29 and yeast Y13 subunits.

The distribution of proteasomes between different cell compartments seems to vary with cell type (Haass et al., 1989) and changes in localization of proteasomes have been found to occur during oogenesis, embryogenesis and development in lower eukaryotic organisms (e.g. Grainger and Winkler, 1989; Klein et al., 1990; Kawahara and Yokosawa, 1992). There are also changes with the cell cycle (Knecht et al., 1991; Kawahara and Yokosawa, 1992; Amsterdam et al., 1992). In transformed cells and proliferating tissues, proteasomes have been found to occur predominantly in the nuclei (Grossi de Sa et al., 1988; Kumatori et al., 1990a; Kanayama et al., 1991).

PROTEASOMES AND CELL PROLIFERATION

Gene disruption experiments in yeast have demonstrated that several proteasomal proteins are each encoded by a single copy gene and that some of these genes (YC1, YC7α or Y8, Y7, PRE1, PRS3 and PUP2; Fujiwara et al., 1990; Emori et al., 1991; Heinemeyer et al., 1991; Lee et al., 1992; Georgatsou et al., 1992) are essential for cell proliferation, whereas others are not (e.g. Y13; Emori et ai., 1991). An increase in the level of expression of the gene for the one subunit of proteasomes which was investigated has been found in malignant human haematopoietic cell lines and growth-stimulated mononuclear cells, in leukaemic cells in bone marrow, and in cells during blastogenic transformation induced by phytohaemagglutinin and interleukin-2 (Kumatori et al., 1990a). Significantly increased levels of several proteasomal mRNAs have also been observed in malignant tumour cells and in transformed cells in culture (Kanayama et al., 1991; Balson et al., 1992; Shimbara et al., 1992) but there is no change in the level of proteasomal proteins detected by Western blotting. There appears to be an increase in the turnover of proteasomes under some conditions (Shimbara et al., 1992).

Although changes in proteasomal gene expression appear to be important during growth and differentiation of cells (Shimbara et al., 1992) the reason for this is presently unclear. Proteasomes may participate in the specific breakdown of certain cellular

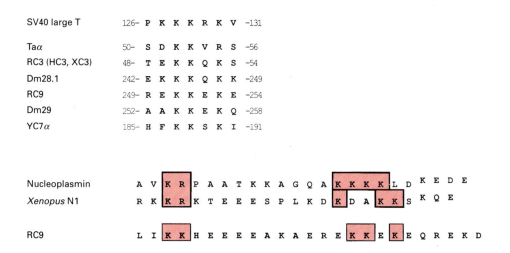

Figure 4   Putative nuclear localization signals in proteasomal subunits

Sequences related to the SV40 nuclear localization signal have been identified in several subunits (Tanaka et al., 1990c; Haass et al., 1990b) as shown. A sequence in the RC9 subunit is similar to the bipartite motif described by Dingwall and Laskey (1991) and illustrated by nucleoplasmin and the *Xenopus* N1 protein.

proteins, such as cyclins, for which rapid turnover is important in regulating the cell cycle. Short-lived proteins, including many key regulatory proteins, are primarily degraded by non-lysosomal pathways of intracellular protein degradation (Rivett, 1990; Rechsteiner, 1991; Goldberg and Rock, 1992; see below). Proteasomes are believed to play a major role in these pathways (Orlowski, 1990; Rivett, 1990) and decreased availability of functional proteasomes could allow increases in the levels of short-lived proteins. Alternatively, although such regulatory proteolytic properties of proteasomes may be important, the possibility that other functions of proteasomes predominate in cell proliferation cannot be ruled out. Their localization in the nucleus as well as in the cytoplasm and the increase in nuclear localization in rapidly proliferating cells argues for a more direct involvement in characteristic nuclear functions.

FUNCTIONS IN INTRACELLULAR PROTEIN DEGRADATION

The abundance and widespread distribution of proteasomes, as well as their ability to degrade protein substrates, suggest that they may play an **important role in non-lysosomal pathways of protein turnover** (Orlowski, 1990; Rivett and Knecht, 1993). There are both ubiquitin-dependent and ubiquitin-independent pathways. The **ubiquitin system**, a multicomponent, ATP-dependent pathway, which has been well-characterized in reticulocytes and yeast (see Hershko and Ciechanover, 1992; Jentsch, 1992, for reviews), is believed to be responsible for the removal of abnormal proteins and for the rapid degradation of normal short-lived proteins which are important for cell cycle progression such as cyclin and nuclear oncoproteins (Rechsteiner, 1991). Not all ubiquitinated proteins are rapidly degraded and the signal for ubiquitin-mediated proteolysis appears to be multiple ubiquitination. Other short-lived proteins and some long-lived proteins may be degraded by ubiquitin-independent non-lysosomal pathways, which are less clearly defined but may also be ATP-dependent (Rivett, 1990; Goldberg and Rock, 1992).

Although purified proteasomes are not activated by ATP (reviewed in Rivett, 1989a) and are apparently unable to degrade ubiquitin–protein conjugates (Hough et al., 1987), proteasomes have been implicated in both ATP- and ubiquitin-dependent

pathways of protein breakdown. Immunodepletion of proteasomes from crude cell extracts reduces both ATP-dependent and ubiquitin-dependent proteolysis (e.g. Matthews et al., 1989). Direct evidence for the involvement of proteasomes in the ubiquitin pathway is that ubiquitin-mediated proteolysis is decreased in the yeast PRE1 mutants which are defective in proteasomal chymotrypsin-like activity. Disruption of the PRE1 gene results in a decreased ability of cells to cope with stress conditions (Heinemeyer et al., 1991) and to degrade known substrates of the ubiquitin pathway (Richter-Ruoff et al., 1992; Seufert and Jentsch, 1992). The relative contribution of proteasomes to ubiquitin-dependent and ubiquitin-independent pathways may be determined by association with other proteins. They appear to form part of a larger (26 S) proteinase complex (see below) which is believed to be responsible for degradation of ubiquitin–protein conjugates (Hershko and Ciechanover, 1992) but which also appears to be involved in the ubiquitin-independent degradation of ornithine decarboxylase (Murakami et al., 1992).

In addition to their general role in intracellular protein turnover, proteasomes may have **specific proteolytic functions**. For example, a proteinase which is believed to contribute to cell-mediated cytotoxicity in interleukin-2-activated natural killer cells shows similar characteristics and immunochemical relatedness to proteasomes isolated from rat liver (K. Wasserman, R. P. Kitson, A. J. Rivett, S. T. Sweeney, M. K. Gabauer, R. B. Herberman, S. Watkins and R. H. Goldfarb, unpublished work) and proteasomes have also been implicated in antigen processing (see below). The balance of functions may vary in different cell types.

PROTEASOMES AND THE 26 S PROTEINASE

Based on results from a number of different laboratories it seems likely that the proteasome forms part of the 26 S proteinase (Hough et al., 1987; Eytan et al., 1989; Driscoll and Goldberg, 1990; Kanayama et al., 1992) but this view is not unanimous (Seelig et al., 1991) and the precise nature of the other components of the 26 S proteinase is not entirely clear. On SDS/PAGE gels a range of polypeptides of 35 kDa to 110 kDa, in addition to lower-molecular-mass bands corresponding to proteasome subunits, has been reported for the 26 S proteinase isolated from reticulocytes (Hough et al., 1987), promyelocytic leukaemic HL60 cells (Orino et al., 1991) and *Xenopus* (Peters et al., 1991). The reticulocyte 26 S proteinase, which also has ATPase activity (Armon et al., 1990), is formed by the ATP-dependent association of proteasomes with two other components, CF1 and CF2 (Eytan et al., 1989; Driscoll and Goldberg, 1990). One of these has been suggested to be a 250 kDa inhibitor of the proteasome (Driscoll et al., 1992) while the other component may be an activator (Goldberg and Rock, 1992). However, in muscle, the 26 S proteinase has been reported to be composed of the proteasome and a 600 kDa ATP-dependent proteinase called multipain (Goldberg, 1992). There are some differences in the structural models which have been proposed from electron microscopy studies with negatively stained preparations (Peters et al., 1991; Ikai et al., 1991) and further work is required to clearly establish the components of the 26 S proteinase and its precise relationship to the proteasome.

PROTEASOMES AND ANTIGEN PROCESSING

Peptides which bind to newly synthesized major histocompatibility complex class I molecules in the endoplasmic reticulum are presumed to be generated in the cytoplasm prior to transport across the endoplasmic reticulum which is facilitated by peptide transporters (DeMars and Spies, 1992). A complex of low-molecular-mass proteins ("LMP"; Monaco and McDevitt, 1984), which was suggested to be involved in the generation of peptides from cellular proteins or viral proteins produced within infected cells for presentation in association with the class I major histocompatibility complex, has properties in common with proteasomes, suggesting that the two complexes might be related. The demonstration that immunoprecipitates of mouse macrophage or spleen cell extracts obtained with anti-LMP antibodies and anti-proteasome antibodies give very similar patterns on two-dimensional polyacrylamide gels (Brown et al., 1991; Ortiz-Navarette et al., 1991) supports this view.

A possible role of proteasomes in antigen processing was also suggested by the identification of two genes within the MHC class II region (LMP7/RING10 and LMP2/RING12; Glynne et al., 1991; Martinez and Monaco, 1991) which are up-regulated by γ-interferon (Kelly et al., 1991; Ortiz-Navarette et al., 1991; Yang et al., 1992) and which encode proteins containing deduced amino acid sequences similar to those of known proteasomal subunits (Lee et al., 1990; Lilley et al., 1990). Neither of the two proteins is essential for cell viability. Although the possibility that proteasomes contribute to antigen processing has received a lot of attention recently (see Goldberg and Rock, 1992, for a review) there is recent evidence that the two polymorphic MHC-encoded subunits are not essential for the processing of peptides bound by MHC class I molecules (Momburg et al., 1992; Arnold et al., 1992). However, such observations do not preclude some role for proteasomes in antigen processing.

CONCLUSIONS AND FUTURE DIRECTIONS

Proteasomes are high-molecular-mass cylindrical particles which contain at least five distinct types of catalytic components and are capable of degrading proteins to small peptides. They are found in all eukaryotic cells and tissues and are usually present in the nucleus as well as in the cytoplasm. They are composed of many different subunits which are encoded by members of the same gene family. Individual subunits can be grouped into A or B types depending on whether they are most similar to the α or β subunit of the archaebacterial proteasome. The latter enzyme provides a simple model for structural studies.

The diversity of eukaryotic proteasome localization and subunit composition, as well as the presence of RNA in only a small proportion of particles, suggests the possible existence of subpopulations of proteasomes. The precise subunit stoichiometry remains unclear as does the mechanism for the assembly or exchange of selected subunits to produce particles having different collections of polypeptide components. The fact that some but not all components of proteasomes are essential for cell viability presumably reflects essential functions of these subunits, which may either be structural or functional. Further knowledge of the function and distribution of individual components of the complex will help to elucidate some of these issues.

Since the proteinase does not fall into one of the recognized families of proteases, the identification of proteolytic components as well as the determination of catalytic residues and mechanism are of particular interest. The assembly of different proteolytic components into a single complex allows them to share regulatory or targeting mechanisms and provides an efficient means by which proteins recognized as substrates can be rapidly degraded to small peptides. The broad specificity of the proteinase would allow degradation of many different substrates by a single proteinase complex.

It is widely believed that proteasomes play a major part in both ubiquitin-dependent and ubiquitin-independent non-lysosomal pathways of protein breakdown. There is direct evidence for a role in ubiquitin-dependent proteolysis which may involve association of the proteasome with other components to form the 26 S proteinase which has recently also been implicated in ubiquitin-independent protein turnover (Murakami et al., 1992). Proposed mechanisms for the modulation of proteasome function and localization are speculative, and although it seems likely that the function of the proteasome can be modulated by association with other proteins such as endogenous inhibitors, activators, or other components of the 26 S proteinase, further studies are required to establish which of these factors are important *in vivo*. The functional significance of the proteasomal RNA remains to be established, as does the proposed role of proteasomes in the control of translation (Nothwang et al., 1992a).

A. J. R. is a Lister Institute–Jenner Research Fellow. Research in the author's laboratory was also supported by the Medical Research Council and the Wellcome Trust.

REFERENCES

Ahn, J. Y., Hong, S. O., Kwak, K. B., Kang, S. S., Tanaka, K., Ichihara, A., Ha, D. B. and Chung, C. H. (1991) J. Biol. Chem. **266**, 15746–15749

Aki, M., Tamura, T., Tokunaga, F., Iwanaga, S., Kawamura, Y., Shimbara, N., Kagawa, S., Tanaka, K. and Ichihara A. (1992) FEBS Lett. **301**, 65–68

Amsterdam, A., Pitzar, F. and Baumeister, W. (1992) Proc. Natl. Acad. Sci. U.S.A., in the press

Armon, T., Ganoth, D. and Hershko, A. (1990) J. Biol. Chem. **265**, 20723–20726

Arnold, D., Driscoll, J., Androlewicz, M., Hughes, E., Cresswell, P. and Spies, T. (1992) Nature (London) **360**, 171–174

Arribas, J. and Castaño, G. (1990) J. Biol. Chem. **265**, 13969–13973

Arrigo, A.-P., Tanaka, K., Goldberg, A. L. and Welch, W. J. (1988) Nature (London) **331**, 192–194.

Balson, D. F., Skilton, H. E., Sweeney, S. T., Thomson, S. and Rivett A. J. (1992) Biol. Chem. Hoppe-Seyler **373**, 623–628

Balzi, E., Chen, W., Capieaux, E., McCusker, J. H., Haber, J. E. and Goffeau, A. (1990) Gene **89**, 151

Baumeister, W., Dahlmann, B., Hegerl, R., Kopp, F., Kuehn, L. and Pfeifer, G. (1988) FEBS Lett. **241**, 239–245

Beyette, J. R. and Mykles, D. L. (1992) Muscle Nerve **15**, 1023–1035

Briane, D., Olink-Coux, M., Vassy, J., Oudar, O., Huesca, M., Scherrer, K. and Foucrier, J. (1992) Eur. J. Cell Biol. **57**, 30–39

Brown, M.G, Driscoll, J. and Monaco, J. J. (1991) Nature (London) **353**, 355–357

Cardozo, C., Vinitsky, A., Hidalgo, M. C., Michaud, C. and Orlowski, M. (1992) Biochemistry **31**, 7373–7380

Chu-Ping, M., Slaughter, C. A. and DeMartino, G. N. (1992a) Biochim. Biophys. Acta **1119**, 303–311

Chu-Ping, M., Slaughter, C. A. and DeMartino, G. N. (1992b) J. Biol. Chem. **267**, 10515–10523

Coux, O., Nothwang, H. G., Scherrer, K., Bergsma-Schutter, W., Arnberg, A. C., Timmins, P. A., Langowski, J. and Cohen-Addad, C. (1992) FEBS Lett. **300**, 49–55

Dahlmann, B., Kopp, F., Kuehn, L., Niedel, B., Pfeifer, G., Hegerl, R. and Baumeister, W. (1989) FEBS Lett. **251**, 125–131

Dahlmann, B., Kuehn, L., Ishiura, S., Tsukahara, T., Sugita, H., Tanaka, K., Rivett, J., Hough, R. F., Rechsteiner, M., Mykles, D. L., Fagan, J. M., Waxman, L., Ishii, S., Sasaki, M., Kloetzel, P. M., Harris, H., Ray, K., Behal, F. J., DeMartino, G. N. and McGuire, M. J. (1988) Biochem. J. **255**, 750–751

DeMars, R. and Spies, T. (1992) Trends Cell Biol. **2**, 81–86

DeMartino, G. N., Orth, K., McCullough, M. L., Lee, L. W., Munn, T. Z., Moomaw, C. R., Dawson, P. A. and Slaughter, C. A. (1991) Biochim. Biophys. Acta **1079**, 29–38

Dick, L. R., Moomaw, C. R., DeMartino, G. N. and Slaughter, C. A. (1991) Biochemistry **30**, 2725–2734

Dick, L. R., Moomaw, C. R., Pramanik, B. C., DeMartino, G. N. and Slaughter, C. A. (1992) Biochemistry **31**, 7347–7355

DiCola D., Pratt, G. and Rechsteiner, M. (1991) FEBS Lett. **280**, 137–140

Dingwall, C. and Laskey, R. A. (1991) Trends Biochem Sci. **16**, 478–481

Djaballah, H. and Rivett, A. J. (1992) Biochemistry **31**, 4133–4141

Djaballah, H., Harness, J. A., Savory, P. J. and Rivett, A. J. (1992) Eur. J. Biochem. **209**, 629–634

Djaballah, H., Rowe, A. J., Harding, S. E. and Rivett, A. J. (1993) Biochem. J., in the press

Driscoll, J. and Goldberg, A. L. (1990) J. Biol. Chem. **265**, 4789–4792

Driscoll, J., Frydman, J., Goldberg, A. L. (1992) Proc. Natl. Acad. Sci. U.S.A. **89**, 4986–4990

Emori, Y., Tsukahara, T., Kawasaki, H., Ishiura, S., Sugita, H. and Suzuki, K. (1991) Mol. Cell. Biol. **11**, 344–353

Eytan, E., Ganoth, D., Armon, T. and Hershko, A. (1989) Proc. Natl. Acad. Sci. U.S.A. **86**, 7751–7755

Falkenberg, P. E., Haass, C., Kloetzel, P. M., Niedel, B., Kopp, F., Kuehn, L. and Dahlmann, B. (1988) Nature (London) **331**, 190–192

Frentzel, S., Gräf, U., Hämmerling, G. J. and Kloetzel, P.-M. (1992a) FEBS Lett. **302**, 121–125

Frentzel, S., Troxell, M., Haass, C., Pesold-Hurt, B., Glätzer, K. H. and Kloetzel, P.-M. (1992b) Eur. J. Biochem. **205**, 1043–1051

Fujii, G., Tashiro, K., Emori, Y., Saigo, K., Tanaka, K. and Shiokawa, K. (1991) Biochem. Biophys. Research Commun. **178**, 1233–1239

Fujiwara, T., Tanaka, K., Kumatori, A., Shin, S., Yoshimura, T., Ichihara, A., Tokunaga, F., Aruga, R., Iwanaga, S., Kakizuka, A. and Nakanishi, S. (1989) Biochemistry **28**, 7332–7340

Fujiwara, T., Tanaka, K., Orino, E., Yoshimura, T., Kumatori, A., Tamura, T., Chung C. H., Nakai, T., Yamaguchi, K., Shin, S., Kakizuka, A., Nakanishi, S. and Ichihara, A. (1990) J. Biol. Chem. **265**, 16604–16613

Georgatsou, E., Georgakopoulos, T. and Thireos, G. (1992) FEBS Lett. **299**, 39–43

Glynne, R., Powis, S. H., Beck, S., Kelly, A., Kerr, L. A. and Trowsdale, J. (1991) Nature (London) **353**, 357–360

Goldberg, A. L. (1992) Eur. J. Biochem. **203**, 9–23

Goldberg, A. L. and Rock, K. L. (1992) Nature (London) **357**, 375–379.

Grainger, J. L. and Winkler, M. M. (1989) J. Cell Biol. **109**, 675–683

Grossi de Sa, M. F., Martins de Sa, C., Harper, F., Coux, O., Akhayat, O., Pal, J. K., Florentin, Y. and Scherrer, K. (1988) J. Cell Sci. **89**, 151–165

Grziwa, A., Baumeister, W., Dahlmann, B. and Kopp, F. (1991) FEBS Lett. **290**, 186–190

Haass, C. and Kloetzel, P. M. (1989) Exp. Cell. Res. **180**, 243–252

Haass, C., Pesold-Hurt, B., Multhaup, G., Beyreuther, K. and Kloetzel, P. M. (1989) EMBO J. **8**, 2373–2379

Haass, C., Pesold-Hurt, B., Multhaup, G., Beyreuther, K. and Kloetzel, P. M. (1990a) Gene **90**, 235–241

Haass, C., Pesold-Hurt, B. and Kloetzel, P. M. (1990b) Nucleic Acids Res. **18**, 4018

Haffter, P. and Fox, T. D. (1991) Nucleic Acids Res. **19**, 5075

Harris, J. R. (1988) Ind. J. Biochem. **25**, 459–466

Hegerl, R., Pfeifer, G., Pühler, G., Dahlmann, B. and Baumeister, W. (1991) FEBS Lett. **283**, 117–121

Heinemeyer, W., Kleinschmidt, J. A., Saidowsky, J., Escher, C. and Wolf, D. H. (1991) EMBO J. **10**, 555–562

Hendil, K. B. (1988) Biochem. Int. **17**, 471–477

Hershko, A. and Ciechanover, A. (1992) Annu. Rev. Biochem. **61**, 761–807

Hough, R., Pratt, G. and Rechsteiner, M. (1987) J. Biol. Chem. **262**, 8303–8313

Ikai, A., Nishigai, M., Tanaka, K. and Ichihara, A. (1991) FEBS Lett. **292**, 21–24

Jentsch, S. (1992) Trends Cell Biol. **2**, 98–103

Kaltoft, M. B., Koch, C., Uerkvitz, W. and Hendil, K. B. (1992) Hybridoma **11**, 507–517

Kanayama, H., Tanaka, K., Aki, M., Kagawa, S., Miyaji, H., Satoh, M., Okada, F., Sato, S., Shimbara, N. and Ichihara, A. (1991) Cancer Res. **51**, 6677–6685

Kanayama, H., Tamura, T., Ugai, S., Kagawa, S., Tanahashi, N., Yoshimura, T., Tanaka, K. and Ichihara, A. (1992) Eur. J. Biochem. **206**, 567–578

Kawahara, H. and Yokosawa, H. (1992) Dev. Biol. **151**, 27–33

Kelly, A., Powis, S. H., Glynne, R., Radley, E., Beck, S. and Trowsdale, J. (1991) Nature (London) **353**, 667–668

Klein, U., Gernold, M. and Kloetzel, P. M. (1990) J. Cell. Biol. **111**, 2275–2282

Kinoshita, M., Hamakubo, T., Fukui, I., Murachi, T. and Toyohara, H. (1990) J. Biochem. (Tokyo) **107**, 440–444

Knecht, E., Palmer, A., Sweeney, S. T. and Rivett, A. J. (1991) Biochem. Soc. Trans. **19**, 293S

Kopp, F., Steiner, R., Dahlmann, B., Kuehn, L. and Reinauer, H. (1986) Biochim. Biophys. Acta **872**, 253–260

Kreutzer-Schmid, C. and Schmid, H. P. (1990) FEBS Lett. **267**, 142–146

Kumatori, A., Tanaka, K., Inamura, N., Sone, S., Ogura, T., Matsumoto, T., Tachikawa, T., Shin, S. and Ichihara, A. (1990a) Proc. Natl. Acad. Sci. U.S.A. **87**, 7071–7075

Kumatori, A., Tanaka, K., Tamura, T., Fujiwara, T., Ichihara, A., Tokunaga, F., Onikura, A. and Iwanaga, S. (1990b) FEBS Lett **264**, 279–282

Lee, L. W., Moomaw, C. R., Orth, K., McGuire, M. J., DeMartino, G. N. and Slaughter, C. A. (1990) Biochim. Biophys. Acta **1037**, 178–185

Lee, D. H., Tanaka, K., Tamura, T., Chung, C. H. and Ichihara, A. (1992) Biochem. Biophys. Res. Commun. **182**, 452–460

Li, X., Gu, M. and Etlinger, J. D. (1991) Biochemistry **30**, 9709–9715

Lilley, K. S., Davison, M. D. and Rivett, A. J. (1990) FEBS Lett. **262**, 327–329

Martinez, C. K. and Monaco, J. J. (1991) Nature (London) **353**, 664–667

Martins de Sa, C., Grossi de Sa, F., Akhayat, O., Broders, F., Scherrer, K., Horsch, A. and Schmid, H.-P. (1986) J. Mol. Biol. **187**, 479–493

Mason, R. W. (1990) Biochem. J. **265**, 479–484

Matthews, W., Tanaka, K., Driscoll, J., Ichihara, A. and Goldberg, A. L. (1989) Proc. Natl. Acad. Sci. U.S.A. **86**, 2597–2601

McDermott, J. R., Gibson, A. M., Oakley, A. E. and Biggins, J. A. (1991) J. Neurochem. **56**, 1509–1517

McGuire, M. J., McCullough, M. L., Croall, D. E. and DeMartino, G. N. (1989) Biochim. Biophys. Acta **994**, 181–186

Momburg, F., Ortiz-Navarette, V., Neefjes, J., Goulmy, E., van de Wal, Y., Spits, H., Powis, S. J., Butcher, G. W., Howard, J. C., Walden, P. and Hämmerling, G. J. (1992) Nature (London) **360**, 174–177

Monaco, J. J. and McDevitt, H. O. (1984) Nature (London) **309**, 797–799

Murakami, K. and Etlinger, J. D. (1986) Proc. Natl. Acad. Sci. U.S.A. **83**, 7588–7592

Murakami, Y., Matsufuji, S., Kameji, T., Hayashi, S., Igarashi, K., Tamura, T., Tanaka, K. and Ichihara, A. (1992) Nature (London) **360**, 597–599

Mykles, D. L. and Haire, M. F. (1991) Arch. Biochem. Biophys. **288**, 543–551

Nothwang, H. G., Coux, O., Bey, F. and Scherrer, K. (1992a) Biochem. J. **207**, 621–630

Nothwang, H. G., Coux, O., Keith, G. Silva-Pereira, I. and Scherrer, K. (1992b) Nucleic Acids Res. **20**, 1959–1965

Orino, E., Tanaka, K., Tamura, T., Sone, S., Ogura, T. and Ichihara, A. (1991) FEBS Lett. **284**, 206–210

Orlowski, M. (1990) Biochemistry **29**, 10289–10297

Orlowski, M. and Michaud, C. (1989) Biochemistry **28**, 9270–9278

Orlowski, M. and Wilk, S. (1988) Biochem. J. **255**, 751

Orlowski, M., Cardozo, C., Hidalgo, M. C. and Michaud, C. (1991) Biochemistry **30**, 5999–6005

Ortiz-Navarrete, V., Seelig, A., Gernold, M., Frentzel, S., Kloetzel, P. M. and Hämmerling, G. J. (1991) Nature (London) **353**, 662–664

Pereira, M. E., Nguyen, T., Wagner, B. J., Margolis, J. W., Yu, B. and Wilk, S. (1992) J. Biol. Chem. **267**, 7949–7955

Peters, J. M., Harris, J. R. and Kleinschmidt, J. A. (1991) Eur. J. Cell. Biol. **56**, 422–432

Pühler, G., Weinkauf, S., Bachmann, L., Müller, S., Engel, A., Hegerl, R. and Wolfgang Baumeister, W. (1992) EMBO J. **11**, 1607–1616

Rawlings, N. and Barrett, A. J. (1993) Biochem. J. **290**, 205–218

Ray, K. and Harris, H. (1985) Proc. Natl. Acad. Sci. U.S.A. **82**, 7545–7549

Rechsteiner, M. (1991) Cell **66**, 615–618

Richter-Ruoff, B., Heinemeyer, W. and Wolf, D. H. (1992) FEBS Lett. **302**, 192–196

Rivett, A. J. (1985a) J. Biol. Chem. **260**, 12600–12606

Rivett, A. J. (1985b) Arch. Biochem. Biophys. **243**, 624–632

Rivett, A. J. (1989a) Arch. Biochem. Biophys. **268**, 1–8

Rivett, A. J. (1989b) J. Biol. Chem. **264**, 12215–12219

Rivett, A. J. (1990) Curr. Opin. Cell Biol. **2**, 1143–1149

Rivett, A. J. and Knecht, E. (1993) Curr. Biol. **3**, 127–129

Rivett, A. J. and Sweeney, S. T. (1991) Biochem. J. **278**, 171–177

Rivett, A. J., Palmer, A. and Knecht, E. (1992) J. Histochem. Cytochem. **40**, 1165–1172

Savory, P. J. and Rivett, A. J. (1993) Biochem. J. **289**, 45–48

Schliephacke, M., Kremp, A., Schmid, H. P., Kohler, K. and Kull, U. (1991) Eur. J. Cell Biol. **55**, 114–121

Schmid, H. P., Akhayat, O., Martins de Sa, C., Puvion, F., Koehler, K. and Scherrer, K. (1984) EMBO J. **3**, 29–34

Seelig, A., Kloetzel, P.-M., Kuehn, L. and Dahlmann, B. (1991) Biochem. J. **280**, 225–232

Seufert, S. and Jentsch, S. (1992) EMBO J. **11**, 3077–3080

Shimbara, N., Orino, E., Sone, S., Ogura, T., Takashina, M., Shono, M., Tamura, T., Yasuda, H., Tanaka, K. and Ichihara, A. (1992) J. Biol. Chem. **267**, 18100–18109

Skilton, H. S., Eperon, I. C. and Rivett, A. J. (1991) FEBS Lett. **279**, 351–355

Sorimachi, H., Tsukahara, T., Kawasaki, H., Ishiura, S., Emori, Y., Sugita, H. and Suzuki, K. (1990) Eur. J. Biochem. **193**, 775–781

Tamura, T., Tanaka, K., Kumatori, A., Yamada, F., Tsurumi, C., Fujiwara, T., Ichihara, A., Tokunaga, F., Aruga, R. and Iwanaga, S. (1990) FEBS Lett. **264**, 91–94

Tamura, T., Lee, D. H., Osaka, F., Shin, S., Chung, C. H., Fujiwara, T., Tanaka, K. and Ichihara, A. (1991) Biochim. Biophys. Acta **1089**, 95–102

Tanaka, K., Ii, K., Ichihara, A., Waxman, L. and Goldberg, A. L. (1986) J. Biol. Chem. **261**, 15197–15203

Tanaka, K., Yoshimura, T., Ichihara, A., Ikai, A., Nishigai, M., Morimoto, Y., Sato, M., Tanaka, N., Katsube, Y., Kameyama, K. and Takagi, T. (1988a) J. Mol. Biol. **203**, 985–996

Tanaka, K., Yoshimura, T., Kumatori, A., Ichihara, A., Ikai, A., Nishigai, M., Kameyama, K. and Takagi., T. (1988b) J. Biol.Chem. **263**, 16209–16217

Tanaka, K., Kumatori, A., Ii, K. and Ichihara, A. (1989) J. Cell. Physiol. **139**, 34–41

Tanaka, K., Fujiwara, T., Kumatori, A., Shin, S., Yoshimura, T., Ichihara, A., Tokunaga, F., Aruga, R., Iwanaga, S., Kakizuka, A. and Nakanishi, S. (1990a) Biochemistry **29**, 3777–3785

Tanaka, K., Kanayana, H., Tamura, T., Lee, D. H., Kumatori, A., Fujiwara, T., Ichihara, A., Tokunaga, F., Aruga, R. and Iwanaga, S. (1990b) Biochem. Biophys. Res. Commun. **171**, 676–683

Tanaka, K., Yoshimura, T., Tamura, T., Fujiwara, T., Kumatori, A. and Ichihara, A. (1990c) FEBS Lett. **271**, 41–46

Tokunaga, F., Aruga, R., Iwanaga, S., Tanaka, K., Ichihara, A., Takao, T. and Shimonishi, Y. (1990) FEBS Lett. **262**, 373–375

Tomek, W., Adam, G. and Schmid, H. P. (1988) FEBS Lett, **239**, 155–158

Tsukahara, T., Ishihura, S., Kominami, E. and Sugita, H. (1990) Exp. Cell Res. **188**, 111–116

Van Riel, M. C. H. M. and Martens, G. J. M. (1991) FEBS Lett. **291**, 37–40

Weitman, D. and Etlinger, J. D. (1992) J.Biol. Chem. **267**, 6977–6982

Wilk, S. and Orlowski, M. (1980) J. Neurochem. **35**, 1172–1182

Wilk, S. and Orlowski, M. (1983) J. Neurochem. **40**, 842–849

Yang, Y., Waters, J. B., Früh, K. and Peterson, P. A. (1992) Proc. Natl. Acad. Sci. U.S.A. **89**, 4928–4932

Yu, B., Pereira, M. and Wilk, S. (1991) J. Biol. Chem. **266**, 17396–17400

Yukawa, M., Sakon, M., Kambayashi, J., Shiba, E., Kawasaki, T., Ariyoshi, H. and Mori, T. (1991) Biochem. Biophys. Res. Commun. **178**, 256–262

Zwickl, P., Pfeifer, G., Lottspeich, F., Kopp, F., Dahlmann, B. and Baumeister, W. (1990) J. Struct. Biol. **103**, 197–203

Zwickl, P., Lottspeich, F., Dahlmann, B. and Baumeister, W. (1991) FEBS Lett. **278**, 217–221

Zwickl, P., Grziwa, A., Pühler, G., Dahlmann, B., Lottspeich, F. and Baumeister, W. (1992) Biochemistry **31**, 964–972

Biochem. J. (1993) **293**, 1–13 (Printed in Great Britain)

REVIEW ARTICLE
Mammalian glucokinase and its gene

Patrick B. IYNEDJIAN

Division of Clinical Biochemistry, University of Geneva School of Medicine, 1 rue Michel Servet, CH-1211 Geneva 4, Switzerland

INTRODUCTION

Mammalian glucokinase was identified 30 years ago as a distinct form of hexokinase in rat liver. The hexokinases (ATP:hexose 6-phosphotransferases, EC 2.7.1.1) constitute a family of evolutionarily and stucturally related enzymes present in eukaryotic cells from yeast to mammals. In the cells of higher organisms, the physiologically significant substrate for these enzymes is D-glucose. The reaction catalysed by the hexokinases, $ATP + D$-glucose $\rightarrow ADP + D$-glucose 6-phosphate, is the first and obligatory step for glucose utilization after transport of the sugar into the cell. Mammalian tissues contain four different hexokinases which can be isolated by conventional protein separation techniques and for which cDNAs have been cloned. The isoenzymes of the rat have been designated hexokinases type I–IV or A–D in order of increasing negative net charge. The subject of this Review is hexokinase type IV or D, usually called glucokinase.

Glucokinase stands apart from all the other hexokinases by a number of criteria. The first and most striking is its low affinity for glucose. The enzyme is half-saturated with glucose at 6 mM, compared with K_m values in the micromolar range for the three other mammalian hexokinases. This feature led to the discovery of the enzyme and underlies its key role in the physiology of glucose homeostasis. The second hallmark of mammalian glucokinase is its highly typical tissue distribution. The glucokinase gene is transcribed and the mRNA translated into active enzyme only in hepatocytes and insulin-secreting β-cells of the pancreatic islets of Langerhans, reflecting the great functional specialization of this isoenzyme. A third outstanding feature is the developmental and multihormonal regulation of the enzyme, illustrated most dramatically by the transcriptional induction of the glucokinase gene by insulin in the liver.

The distinctive kinetics of glucokinase, its tissue-specific expression and its hormonal regulation were recognized within a few years of the discovery of the enzyme. However, our understanding of these particular characteristics has remained superficial until recently. The main reason for limited progress was the difficulty of purifying the enzyme, hence of raising specific antibodies or obtaining peptide sequence for the isolation of cDNA clones. Once this obstacle was surmounted, studies on glucokinase became a very fertile field of research. The most recent and medically rewarding outcome of this research has been the discovery of mutations of the glucokinase gene as the cause of one subtype of non-insulin-dependent diabetes mellitus (NIDDM). The purpose of this article is to review the recent developments on the structure and function of the glucokinase gene and its gene products, as they relate to our understanding of blood glucose homeostasis. The reader interested in historical perspectives and a complete background on the biochemistry of glucokinase should refer to the classical reviews of Walker [1], Weinhouse [2] and Colowick [3]. Recent commentaries on topical aspects are also available [4–8].

ONE GENE, TWO ENZYMES

The glucokinase gene was first cloned from the rat and the structure of the gene in this species can serve as the standard of reference (Figure 1a). The most remarkable feature is the presence of alternative promoters, responsible for the initiation of transcription at different sites on the DNA in hepatic and endocrine cells. The first clue to the existence of cell-type-specific promoters came from the sequences of two quasi-full-length cDNAs isolated from rat liver and insulinoma libraries [9,10]. The sequences were essentially identical for more than 2000 nucleotides starting from the 3′ ends of the cDNAs, but segments of approximately 100 nucleotides at the 5′ ends were found to differ entirely. The 5′ specific sequences of the cDNAs were mapped to widely separated sites in genomic DNA by Magnuson and co-workers [10,11]. These investigators further identified nine exons, numbered 2–10 in the transcription unit, whose assembly gives rise to the common sequence found in the liver and insulinoma-derived cDNAs. The leader exon encoding the 5′ end of the hepatic mRNA, termed exon 1L, was contained in a phage λ clone which also carried the common exons 2–4. The leader exon for the 5′ end of the insulinoma mRNA, termed exon 1β in reference to the β-cells of the islets of Langerhans, was localized in a different phage clone with non-overlapping genomic DNA. It was therefore concluded that the liver-specific exon 1L was contiguous to the body of the structural gene, whereas the islet-type exon 1β was located at an unspecified distance further upstream. The intervening sequence between the two leader exons has yet to be mapped accurately. Several attempts to isolate rat genomic DNA clones for the entire region have remained unsuccessful in my laboratory, perhaps suggesting unusual features of this DNA. In any event, more than 22 kb of DNA separate the two leader exons in the rat gene (Figure 1a).

The fact that the upstream exon 1β is used exclusively in islet-derived cells, and the downstream exon 1L exclusively in liver, was established by primer extension and nuclease protection experiments and further confirmed by reverse transcription and PCR [10]. It should be noted that the two tissue-specific exons 1 of the glucokinase gene specify not only the 5′ untranslated regions of the islet and liver mRNAs, but also their initial 45 nucleotides of protein coding sequence. It follows that the rat islet and liver glucokinase enzymes will differ in primary structure by 15 amino acids (including initiator Met) at the N-terminal ends of the molecules, for a total sequence of 465 amino acid residues.

In addition to the differential splicing of leader exons associated with the cell-specific control of transcription initiation, other modes of alternative splicing are known to affect glucokinase transcripts. An optional cassette exon has been identified in the rat gene between the originally described exons 1L and 2 (Figure 1a). This cassette exon, termed exon 2A, is retained in a minor fraction of glucokinase mRNA in rat liver [12]. Alternative

---

(a)

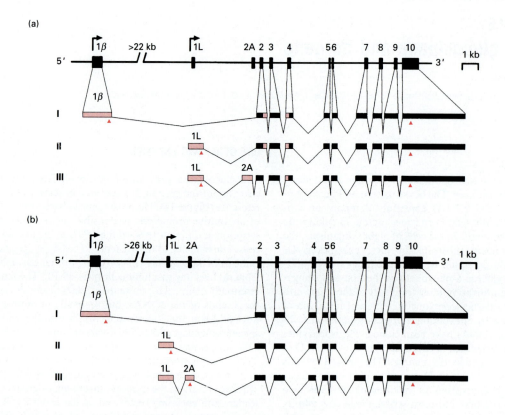

(b)

Figure 1   Glucokinase gene in the rat (a) and man (b)

The **top lines** show the exon–intron organization of the genes. Filled boxes represent the exons, curved arrows the sites of transcription initiation. Lines **I–III** show tissue-specific initiation of transcription and alternative splicing. Alternatively spliced exons or parts of exons are coloured red; arrowheads denote the initiation and termination codons. **I**, transcription and pre-mRNA splicing in β-cells of the islets of Langerhans; **II** and **III**, transcription and pre-mRNA splicing in hepatocytes. In the rat, the mRNA in line III is quantitatively minor and encodes an enzymically inactive protein. The relative abundance of mRNAs shown in lines II and III in human liver has not been determined.

donor sites in exon 2 are utilized with low frequency and lead to deletions of either 52 or 25 nucleotides from mature mRNA. Finally, an acceptor site in exon 4 located 51 nucleotides downstream of the major splice site has been described [13]. These different modes of alternative splicing are responsible for the formation of minor mRNA species encoding enzymically inactive polypeptides, as further discussed in the section on the structure of glucokinase.

Recent reports on the cloning of the human glucokinase gene demonstrate that the major structural features of the rat gene have been conserved in the human species [14–16]. Most importantly, the dual promoter arrangement with upstream islet-specific exon 1β and downstream liver-specific exon 1L is very similar in the human and rat transcription units (Figure 1b). As in the rat, the mature glucokinase mRNAs of human islets and liver are assembled from cell-type specific exons 1 and common exons 2–10 [14,15]. The cell-specific exons harbor the initiator ATG triplets, which are located 45 and 48 nucleotides upstream to the donor splice sites in exons 1β and 1L respectively. Reminiscent of the situation in the rat, an alternative transcript in human liver is produced by splicing of an optional cassette exon 2A between exons 1L and 2. In this case, the open reading frame for the protein initiates at an ATG codon situated 42 nucleotides upstream of the cassette exon's donor splice site, predicting a variant N-terminal domain of 14 amino acid residues in the corresponding protein [15]. The gene has been mapped to human chromosome 7p [17].

TISSUE-SPECIFIC EXPRESSION OF THE GLUCOKINASE GENE

The hexokinases taken as a group can be viewed as ubiquitous housekeeping enzymes since glucose phosphorylation is an absolute requirement for further metabolism of this key energetic substrate. Glucokinase, however, subserves highly specialized physiological functions (see below). Accordingly, its gene is expressed with stringent tissue specificity. Early studies using enzymic assays and chromatographic analyses have documented the presence of glucokinase activity only in the liver and islets of Langerhans [18–20] This narrow tissue distribution was confirmed by a sensitive immunoblotting method, which revealed a typical glucokinase polypeptide band with apparent molecular mass of 56 kDa in the cytosolic fraction of liver and islet homogenates, but not in brain, spleen, intestinal mucosa, pancreas, kidney and white adipose tissue [21]. Similarly, by Northern blotting of poly(A)-containing RNA, glucokinase mRNA was initially detected only in the livers and islets of Langerhans of glucose-fed rats [22]. The glucokinase mRNAs of rat liver and islets migrate in electrophoretic gels as 2.5 and 2.8 kb mRNA species respectively. The size difference is explained entirely by the longer 5′ untranslated region of the islet mRNA, encoded by the cell-specific exon 1β. Recently, Hughes et al. [23] have reported the presence of the long version of glucokinase mRNA, reflecting transcription from the β-cell type promoter, in the rat pituitary gland and the corticotroph pituitary cell line AtT-20. However, most if not all of the pituitary mRNA represents alternatively spliced transcripts which do not code for enzymically active glucokinase[13,23].

As already mentioned, transcription in islet β-cells (and pituitary cells) is initiated exclusively at the upstream promoter and adjacent exon 1β. Conversely, only the downstream promoter and associated exon 1L are active in liver cells. For the rat [10,11] as well as the human [15] genes, the nucleotide sequences of both promoters and flanking regions have been published. Transcription was shown to initiate at well-defined cap sites for the rat and human liver promoters and the human β-cell promoter. In contrast, multiple start sites scattered over a region spanning 62 nucleotides were noted at the β-cell promoter of the rat gene, probably in relation to the absence of a TATA box in this promoter. Elements of DNA sequence with similarity to binding sites for known *trans*-acting factors, cell-specific or ubiquitous, have been highlighted in the promoters and putative regulatory regions of the rat and human genes [10,11,15].

Cis-acting elements important for the transcriptional activity of the β-cell promoter of the rat gene have recently been delineated by Shelton et al. [24]. In short-term transfection experiments, a β-cell promoter fragment, including nucleotides -280 to $+14$ (with respect to the most downstream of the multiple start sites) was able to drive the expression of the luciferase reporter gene in insulin-producing insulinoma cells of the hamster HIT line, but not in NIH 3T3 fibroblasts. The glucokinase β-cell promoter was indeed more efficient than a rat proinsulin II promoter fragment of comparable length. Block mutations of 10 bp introduced throughout the 280 bp glucokinase promoter led to the identification of two types of *cis*-acting elements of functional significance. One important element is a perfect palindrome TGGTCACCA found at positions -169 to -161 and again at positions -90 to -82. This element was investigated by electrophoretic mobility shift assay for its ability to form specific complexes with nuclear proteins. Several complexes with nuclear proteins from a variety of cell types were observed, suggesting that the palindromic element is a binding site for *trans*-acting factors of the ubiquitous class. A second interesting element, which contributed substantially to promoter activity is found at three locations in the promoter (-215 to -210, -135 to -126 and -102 to -99). The consensus core sequence for this motif, termed the upstream promoter element (UPE) by Shelton et al. [24], is CAT(T/C)A(G/C). Two specific DNA–protein complexes were formed in the electrophoretic mobility shift assay between this type of oligonucleotide and nuclear proteins from β-cell lines. Nuclear extracts from other cell types did not give rise to these complexes, with the possible exception of a pancreatic α-cell line. Interestingly, the consensus UPE motif is identical to *cis*-acting elements of the human proinsulin gene called CT-boxes [25]. In addition, the promoter of the rat proinsulin I gene presents a *cis*-acting element with 5/6 identity with the consensus glucokinase UPE motif [26]. As pointed out by Shelton et al., the β-cell enriched protein factor(s) binding to the glucokinase UPE may be related to or identical with previously described factors which bind to the CT boxes of the human proinsulin promoter or the related element of the rat proinsulin I gene [26,27]. If all these factors prove to be one and the same, they may well represent a key determinant for the transcription of genes typically expressed in β-cells of the islets of Langerhans. It should be pointed out that the pituitary AtT-20 cells, although transcribing the glucokinase gene from the upstream promoter, do not contain the same UPE-binding protein factor(s).

The search for *cis*-acting elements involved in the control of transcription at the hepatic promoter is lagging behind. To my knowledge, liver-specific activity of this promoter in a transient transfection system has not been demonstrated. Hepatoma cell lines in general use do not express the endogenous glucokinase gene [28] and therefore appear an unlikely model for uncovering critical regulatory elements of the promoter. Primary rat hepatocytes can be efficiently transfected by electroporation or lipofection, and high activity of marker genes can be elicited with a variety of viral or cellular promoters [29–31]. However, efforts in our laboratory and elsewhere to express reporter enzymes at the direction of the hepatic glucokinase promoter have remained inconclusive, with plasmid constructs containing as much as 7 kb of 5′ flanking sequence or as little as 110 bp of proximal promoter sequence [31,32]. Inspection of the DNA sequence suggests possible recognition elements for both ubiquitous and liver-enriched *trans*-acting factors. Footprinting by the DNAase I protection assay reveals putative sites of DNA–protein interaction, but their relevance has yet to be tested functionally.

HORMONES AND GENE REGULATION

Alternative promoters allow for versatility in gene control, inasmuch as the initiation of transcription at each promoter in the transcription unit can be regulated by a distinct combination of *trans*-acting factors. Several genes that are expressed at different levels during ontogenic development or in various cell types are known to be transcribed from alternative promoters [33]. The rat glucokinase gene provides an interesting example of transcription unit with tissue-specific promoters which are differentially affected by nutritional and hormonal stimuli.

The level of glucokinase activity in the liver of the rat and other mammalian species has long been known to vary with the nutritional status of the animal. Hepatic glucokinase activity falls during fasting and is restored by glucose refeeding [34,35]. Marked changes in the amount of glucokinase mRNA occur in rat liver under these conditions [36,37]. The message for the enzyme is undetectable by Northern blot analysis in total liver RNA from rats fasted for 24–72 h. Oral glucose administration to such animals causes a rapid, massive and transient accumulation of the hepatic 2.5 kb mRNA for glucokinase. Induction of the mRNA culminates 6–10 h after refeeding, at a level many times as high as seen in the livers of normal animals fed *ad libitum* [36]. In contrast, in the islets of Langerhans, the typical 2.8 kb mRNA for glucokinase is maintained at a constant level during prolonged starvation. Moreover, there is no increase in islet mRNA after an oral glucose load [22]. In line with the mRNA data, Western blotting failed to show any significant change in glucokinase protein amount in pancreatic islets during the fasting–refeeding transition, whereas hepatic glucokinase increased 3-fold within 18 h of glucose refeeding [22]. These results suggest that transcription of the glucokinase gene in liver is turned on by a nutritional signal. On the contrary, the islet β-cell promoter is unresponsive to this stimulus. As a consequence, the enzyme in the islets of Langerhans is expressed constitutively, regardless of the nutritional status of the animal, whereas hepatic glucokinase behaves as a typical adaptive enzyme. This notion has been challenged by Tiedge and Lenzen on the basis of experiments with rats fed a copper-free diet for 3 months [38]. The diet was used to induce atrophy of the exocrine pancreas and a relative enrichment of endocrine tissue, making it possible to assay glucokinase mRNA by Northern blot of poly(A)-containing RNA from total pancreas without prior isolation of islets. In copper-deficient animals, glucose refeeding after a fast caused the rapid appearance of a 2.5 kb mRNA (liver-type) in the pancreas, leading Tiedge and Lenzen to conclude that glucokinase mRNA in islet β-cells was indeed responsive to dietary glucose. In my view, a more likely interpretation is that the glucose-induced 2.5 kb mRNA originated not in islets, but rather in ectopic foci of hepatocytes that are known to differentiate from putative stem cells in the exocrine pancreas of rodents fed copper-deficient diets, a process called transdifferentiation [39,40].

The response of hepatic glucokinase to glucose refeeding is abolished in animals simultaneously treated with anti-insulin serum [41]. A role for insulin as positive effector of hepatic glucokinase expression in the whole animal is further illustrated in diabetes mellitus. Both enzyme protein and enzyme mRNA are absent from the livers of streptozotocin-diabetic rats. Insulin treatment is accompanied by a prompt build-up of glucokinase mRNA, with a marked overshoot above the normal level for 10 h after the first injection of insulin [42]. At later times of treatment, the mRNA falls towards and below the reference level. The glucokinase protein rises with some lag with respect to the mRNA and reaches normal levels in 16–24 h. The delay in time-course of enzyme accumulation compared to mRNA can be explained by the fairly long half-life of approx. 30 h of gluco-kinase in rat liver. The mechanism responsible in the first place for the build-up of mRNA and enzyme is a transient burst in the transcriptional activity of the glucokinase gene, as evidenced by run-on assays with isolated liver nuclei [42].

Rat hepatocytes in primary culture have been used extensively to study the role of individual hormones in the regulation of the glucokinase gene. Liver cells isolated from fasted animals and maintained in basal medium are devoid of glucokinase mRNA. Addition of insulin to the medium elicits a time-dependent increase in specific mRNA, with physiological concentrations of hormone. Insulin acts at the transcriptional level, as demonstrated by run-on assays with hepatocyte nuclei [43]. The effect of insulin occurs with or without glucose in the culture medium [43,44]. In this respect, the glucokinase gene stands apart from a group of other genes, such as the L-type pyruvate kinase and S14 genes, which require both insulin and high glucose concentration for transcriptional activation [44–47]. The phorbol ester phorbol myristate acetate, which elicits insulin-like effects in a number of systems [48], does not mimic the effect of insulin on the glucokinase gene (T. Nouspikel and P. B. Iynedjian, unpublished work).

The activation of the glucokinase gene by insulin might be a primary effect, or it might be mediated by an insulin-inducible 'early gene', whose newly synthesized protein product would in turn stimulate glucokinase gene transcription. Insulin induction of glucokinase mRNA was largely or totally suppressed in hepatocytes cultured in the presence of cycloheximide, anisomycin or pactamycin [44,49]. However, a slight stimulation of specific gene transcription was still detectable in these cells by run-on assay. The interpretation of the data was further complicated by a non-specific negative effect of the inhibitors on general transcription [49]. Moreover, protein synthesis inhibitors have been shown to activate protein kinases in cultured cells and to interfere with signal transduction pathways [50,51]. Taken together, the available data suggest that the effect of insulin on the glucokinase gene is at least in part independent of concomitant protein synthesis. Specific activation of the glucokinase gene as early as 30 min after hormone addition (P. B. Iynedjian, unpublished work) supports the idea that the gene is a primary target for insulin action.

Another central aspect of the regulation of hepatic glucokinase is the acute repressor effect of cyclic AMP on the gene. Insulin induction of the message is inhibited by simultaneous addition to the culture medium of glucagon or derivatives of cyclic AMP. At maximal doses of glucagon and insulin, the negative effect of glucagon is dominant and induction is completely abolished [43]. Glucagon or cyclic AMP are also the dominant effectors in cells

fully induced by prior incubation with insulin alone. Addition of these effectors in the continued presence of insulin causes an almost immediate cessation of glucokinase gene transcription. Under these circumstances, the mRNA decays with an apparent half-life of 40 min [43]. Interestingly, the negative effect of glucagon is not mimicked by amylin, a polypeptide produced by the pancreatic β-cell and able to counteract the effects of insulin in some target cells [52].

Since cyclic AMP exerts dominant negative control over glucokinase gene transcription, it is legitimate to ask whether insulin might induce the gene by relieving it from basal level cyclic AMP repression. Insulin can antagonize cyclic AMP in liver and adipose tissue by activation of a low-K_m cyclic nucleotide phosphodiesterase termed type III phosphodiesterase [53,54]. At a more distal level in the signal transduction pathway, insulin can oppose cyclic AMP-dependent protein phosphorylation by the stimulation of serine/threonine protein phosphatases [55,56]. We have shown that insulin activation of the glucokinase gene in cultured hepatocytes is prevented by several inhibitors of the cyclic nucleotide phosphodiesterases, in particular by a preferential inhibitor of type III phosphodiesterase [49]. In addition, the inductive effect of insulin was suppressed in presence of low concentrations of okadaic acid, a specific inhibitor of protein phosphatases PP1 and PP2A [49]. These observations underline the importance of the interaction between insulin and the cyclic AMP signalling system for the control of the hepatic glucokinase gene. A highly simplified scheme of the hormonal interactions in the control of specific gene transcription is presented in Figure 2. Glucagon acting via a cyclic AMP-dependent protein kinase is shown to phosphorylate and thereby inactivate a trans-acting factor essential for glucokinase gene transcription. Conversely, the putative factor is de-phosphorylated and converted to the active form by an insulin-activated phosphatase. The insulin-dependent kinase cascade resulting in phosphatase activation is derived from the one thought to be active in the control of glycogen synthase activity in muscle ([56–60]; see legend to Figure 2 for details). The putative regulatory factor of transcription is modelled after ADR1, a trans-acting factor involved in the control of the alcohol dehydrogenase (ADH II) gene in yeast. In that system, ADR1 is phosphorylated by a cyclic AMP-dependent kinase responsive to glucose and phosphorylation of the factor results in its inactivation and repression of the alcohol dehydrogenase gene [61,62]. The scheme shown in Figure 2 should be regarded as a minimal model. At the gene level, it implies the presence of a single hormone-response element as the target for both the positive effect of insulin and the negative effect of cyclic AMP. It is also possible that insulin and cyclic AMP regulate the gene via separate regulatory proteins and hormone-response elements. Identifying one or several regulatory DNA elements will be the first step to distinguish between these possibilities.

In cultured hepatocytes from newborn rats, the thyroid hormone tri-iodothyronine was shown to be an effective inducer of glucokinase mRNA [64]. The hormone promoted de novo appearance of the message, suggesting an effect at the transcriptional level. A direct proof of this point remains to be provided by run-on assays. The effect was additive with that of insulin. In the same study, the synthetic glucocorticoid dexamethasone was devoid of effect by itself, but augmented the response to insulin 2-fold. In the whole animal, the thyroid hormones appear to play a permissive role for the induction of hepatic glucokinase mRNA during the fasting–refeeding transition [65]. It has also been reported that biotin administration to starved rats results in glucokinase gene induction, but this effect may well be indirect and reflect a stimulation of insulin secretion [66].

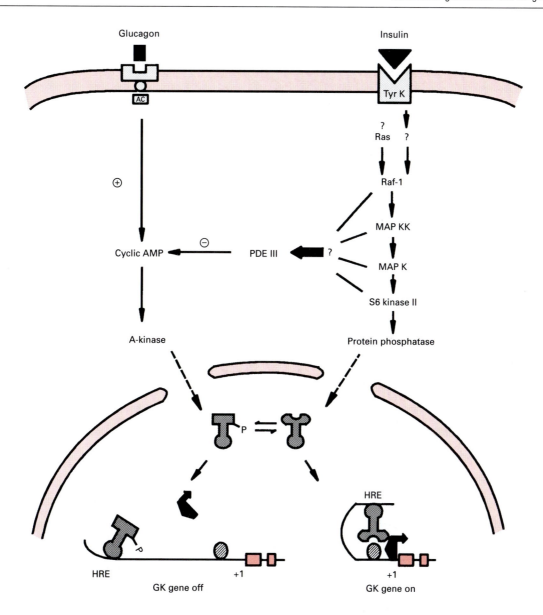

Figure 2   Hypothetical model of glucokinase gene regulation in rat hepatocytes

Two conformations of a putative *trans*-acting factor interconvertible by phosphorylation/dephosphorylation are shown in the cell nucleus. Glucagon binding to its plasma membrane receptor triggers the activation of adenylate cyclase (AC), followed by a stimulation of cyclic AMP-dependent protein kinase (A-kinase) and the migration of the active catalytic subunit of A-kinase to the nucleus. The subunit phosphorylates the *trans*-acting factor, altering the conformation of the transcription activating domain and making it inactive. Insulin activates a protein phosphatase via a protein kinase cascade whose initial steps following the activation of the receptor tyrosine kinase (Tyr K) may or may not involve the cellular proto-oncogenes Ras [60] and Raf-1. The serine/threonine protein kinase Raf-1 phosphorylates and activates a kinase (MAP KK) [59], which in turn phosphorylates and activates a mitogen-activated protein kinase (MAP K, also termed extracellular signal regulated kinase, ERK) [58]. MAP K phosphorylates and activates an insulin-sensitive protein kinase (S6 kinase II) [57] which can specifically phosphorylate a subunit of a protein phosphatase and activate it [56]. The activated phosphatase is shown to migrate to the nucleus and dephosphorylate the *trans*-acting factor, making it competent for the activation of transcription initiation by RNA polymerase II (solid symbol with arrow) at the liver promoter of the glucokinase (GK) gene. Insulin activation of the protein kinase cascade is also shown to result in phosphorylation and activation of a hormone-sensitive cyclic nucleotide phosphodiesterase (PDE III) [54], increasing the turnover of cyclic AMP. The particular kinase for this effect is unknown. The scheme should be taken as a minimal model. The *trans*-acting factor is shown to be bound to the DNA hormone-response element (HRE) even when inactive. It is also possible that the binding affinity of the putative factor is reduced by phosphorylation. Phosphorylation may affect more than one factor, or regulatory subunit(s) of factor(s). Finally, phosphorylation/dephosphorylation of the regulatory factor(s) may take place in the cytoplasm with subsequent migration of the protein to the nucleus (for a review on regulation of transcription by protein phosphorylation, see [63]).

Glucokinase gene transcription in islet β-cells does not appear to be regulated by hormones or other effectors. Down-regulation of islet glucokinase mRNA after prolonged exercise training has been reported, but the decrease in enzyme mRNA was accompanied by a similar decrease in total RNA content of the islets [67]. In RIN insulinoma cells, long-term culture with dexamethasone resulted in a modest increase of glucokinase mRNA relative to γ-actin mRNA. No evidence for a transcriptional effect of the glucocorticoid was produced [68]. In islets of Langerhans maintained in organ culture, glucokinase activity and to a lesser extent glucokinase protein are increased by incubation in the presence of high glucose. However, the level of glucokinase mRNA was unchanged under these conditions [69,70]. The effect on the enzyme concentration might therefore

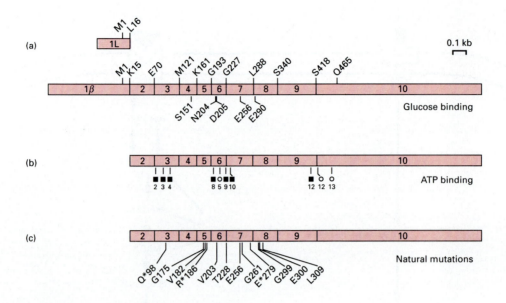

Figure 3   Functional assignments of amino acid residues in glucokinase

The exon structure of islet (1β) and liver (1L) glucokinase mRNAs is represented by the shaded bars. (**a**) **Residues involved in glucose binding.** Above the bar, the first and last amino acids of the enzyme sequence are shown, as well as the last amino acids encoded by each of the exons in the human mRNAs. The one-letter code of abbreviations for amino acids is used. Below the bar, the positions of residues thought to participate in the formation of hydrogen bonds with the hydroxyls of glucose are shown. (**b**) **Sequence elements lining the ATP-binding pocket**: (■) β-strand, (○) α-helix. Thirteen β-strands (numbered 1–13) and 13 α-helices (1–13) have been tentatively assigned along the glucokinase sequence by analogy with the structure of yeast hexokinase [79]. Numbers below the symbols correspond to the numbering in [79]. Residues directly involved in binding ATP are: β-phosphate of ATP, D78–T82 at the loop between β-strands 2 and 3; Mg of Mg-ATP, D205; γ-phosphate of ATP, T228 at the loop between β-strands 9 and 10; adenine of ATP, D409–Y413 between β-strand 12 and α-helix 12. (**c**) **Mutations linked to early-onset non-insulin-dependent diabetes mellitus (MODY).** Residues of the normal sequence affected by nonsense (*) or missense mutations are shown. The amino acid substitutions are given below. For mutations that have been analysed by *in vitro* mutagenesis and bacterial expression, figures in parentheses indicate the -fold decrease in $V_{max.}$ and the -fold increase (+) or decrease (−) in apparent K_m for glucose: G175R (2; +5), V182M (2; +9), V203A (200; +12), T228M (200; 0), E256K (400; −3), G261R (200; −3), E279Q (2; +5), G299R (300; −3), E300K (3; +3), E300Q (0; +3), L309P (100; −4). Data were compiled from [79].

reflect a stimulation of enzyme synthesis at the translational level and/or a stabilization of the enzyme protein.

ENZYME STRUCTURE

The amino acid sequence of rat liver glucokinase has been deduced from the nucleotide sequence of a full-length cDNA by Andreone et al. [9] and independently confirmed by Hayzer and Iynedjian [12]. The sequence is 465 residues in length, with a calculated molecular mass of 51919 daltons and a isoelectric point of 4.85. The sequence is highly conserved in man, with 98 % identical residues in the human and rat enzymes [14]. There is high sequence similarity between glucokinase and the three other mammalian hexokinases. The rat hexokinases I–III have amino acid chain lengths of 918, 917 and 924 residues respectively [71–74]. Their genes arose in evolution by a duplication and fusion process from an ancestral gene encoding a primordial enzyme with molecular mass of 50 kDa, similar to the present day yeast hexokinases A and B and mammalian glucokinase [75]. As a reflection of the duplication event, the large-size hexokinases present homologous N-terminal and C-terminal halves, with approximately 50 % sequence identity between them [76]. Each half-molecule bears strong similarity to glucokinase. The gluco-kinase sequence is 49 % identical to the C-terminal half of hexokinase I and displays conservative replacements at an additional 15 % of the residues; identity with the N-terminal half of the hexokinase is 46 % and conservative substitution occurs at 18 % of positions [74].

Rat liver glucokinase exhibits 28 % identity with yeast hexo-kinase A and conservative amino acid substitution at 14 % of

the residues [77]. Extensive similarity in primary structure suggests the conservation of secondary and tertiary structures as well. The crystallographic studies of yeast hexokinase constitute therefore a valuable resource for modelling the mammalian hexokinases and predicting the location of functionally important amino acid residues [78]. In glucokinase, residues directly involved in the binding of glucose, by the formation of hydrogen bonds with the hexose hydroxyls, are thought to be Ser-151, Asn-204, Asp-205, Glu-256 and Glu-290 [9,79]. The positions of these residues in relation to the exon organization of the human glucokinase mRNA are shown in Figure 3(a). Tests of these assignments by site-directed mutagenesis of cDNA and bacterial expression have been initiated by Pilkis and collaborators [80]. The replacement of Asp-205 by Ala has been shown to result in a 500-fold reduction of the enzyme specific activity, without significant change in affinity for either glucose or ATP. This observation is consistent with the putative role of Asp-205 as a base catalyst, promoting the nucleophilic attack of the 6-hydroxyl group of glucose on the γ-phosphate of ATP [80].

A previously unsuspected structural similarity between the hexokinase family and a larger group of ATP-binding proteins has recently been uncovered [81]. Besides the hexokinases, the newly defined superfamily includes a number of bacterial sugar kinases, as well as actin and the heat shock protein Hsp70 and cognate proteins. The three-dimensional structure typical of all these proteins comprises two domains enclosing a large cleft at the bottom of which ATP binds [82]. The ATP-binding pocket is formed by elements of secondary structure which appear to be conserved in similar relative positions in all members of the superfamily. These elements include seven β-strands and three α-helices [81]. Their tentative locations in the human glucokinase sequence are depicted in Figure 3(b). Further details on residues which participate directly in ATP binding are given in the Figure legend. In yeast hexokinase, binding of a glucose molecule to the enzyme leads to the closing about a hinge of the two enzyme domains mentioned above, resulting in the formation of the ATP-binding site [78]. This mechanism, classically referred to as 'induced fit', is probably also operative in mammalian gluco-kinase, given the structural similarity between these enzymes.

The liver and islet β-cell forms of glucokinase differ by short N-terminal chains of 14–16 amino acids encoded by tissue-specific leader exons, whereas the remaining 450 residues in the polypeptide sequence are common to both molecules (Figure 1). Is there any particular role for the distinct N-terminal domains ? Earlier biochemical studies have shown that the kinetic properties of glucokinase partially purified from rat liver and insulinoma cells are indistinguishable [83]. More recently, forced expression of cDNAs encoding both types of sequences in transfected NIH 3T3 cells produced glucokinase with similar affinity for glucose in both cases, although there was a suggestion for higher specific enzyme activity in cells transfected with the islet-type cDNA [13]. In yeast hexokinase, the first 20 amino acids at the N-terminus appear to form a flexible chain without any specific assigned function [72]. However, a surprising finding in the glucokinase cDNA expression study with NIH 3T3 cells was that a protein with an incomplete islet-type N-terminal domain, due to deletion of the first seven amino acids, was enzymically inactive [13]. Thus, the function of the tissue-specific N-terminal domains of mammalian glucokinase remains elusive. It has been speculated that this domain might be engaged in protein–protein interaction, perhaps with the glucose transporter GLUT 2, but there is no experimental support for this hypothesis [84].

Some alternatively spliced forms of glucokinase mRNA code for proteins with insertion or deletion of polypeptide segments. The implication of these structural alterations for enzymic activity has been tested by expression of cDNAs in bacteria. One cDNA with insertion of the 151-nucleotide cassette exon 2A and deletion of 52 nucleotides at the end of exon 2 has been isolated from a rat liver cDNA library (Figure 1a, line III). The corresponding message was shown to represent a small proportion of gluco-kinase mRNA in the livers of glucose-refed rats [12]. At the protein level, a novel domain of 87 amino acids, encoded by exon 2A and the retained part of exon 2, is inserted between Leu-15 and Glu-70 of the conventional glucokinase sequence, in place of the 54 amino acids normally encoded by exon 2. In bacteria, the above cDNA directed the synthesis of a protein with the predicted slightly larger size than authentic glucokinase, but entirely devoid of enzyme activity [85]. Another variant cDNA with deletion of 51 nucleotides at the junction of exons 3 and 4, due to the use of an alternative acceptor site in exon 4, was originally cloned from a rat insulinoma library [10]. Polymerase chain reaction with first-strand cDNA has subsequently shown that mRNAs with the 51-nucleotide deletion were present as minor forms in both islets of Langerhans and liver. The insertion of this cDNA in a vector for bacterial expression of a glutathione S-transferase–glucokinase fusion protein resulted in the synthesis of a hybrid protein without any detectable glucokinase activity, whereas the non-deleted cDNA form produced an active gluco-kinase fusion protein [13]. Were pre-mRNA splicing regulated, it could represent a means for the control of cellular glucokinase activity, by varying the ratio of messages coding for active enzyme or inactive protein. This hypothesis was tested to explain the decrease in glucokinase activity reported in islets of Lan-gerhans cultured in medium with low glucose concentration.

However, a semi-quantitative mRNA assay by reverse tran-scription–PCR did not reveal a glucose-dependent shift in the relative abundance of the two mRNA species [13].

KINETIC PROPERTIES AND SHORT-TERM REGULATION OF ENZYME ACTIVITY

Three enzymological properties distinguish glucokinase from the other mammalian hexokinases: (i) low affinity for glucose; (ii) lack of inhibition by glucose 6-phosphate and (iii) kinetic co-operativity with glucose. Low affinity for glucose is the diagnostic feature of glucokinase. Half-saturation of the enzyme at 6 mM glucose is put in perspective by considering the K_m of 50, 150 and 7 μM respectively for hexokinases I–III [86]. The physiological advantage derived from half-saturation with glucose at the normal blood glucose concentration is discussed in a separate section. The lack of inhibition by glucose 6-phosphate at low concentration is also unique to glucokinase and relates to the absence of an allosteric binding site for the reaction product [87,88]. Biochemical studies with rat brain hexokinase I have suggested that the glucose 6-phosphate binding site had evolved after gene duplication from the original catalytic site in the N-terminal half of the bipartite enzyme molecule, but this view has recently been shown to be incorrect [89,90].

The third property of glucokinase, its co-operative kinetics with glucose, is of both physiological and theoretical interest. The sigmoid curve of saturation with glucose allows for sharper changes of the reaction rate in response to shifts in the glucose concentration below or above the half-saturation value [91–93]. Glucokinase has a single glucose-binding site and functions in the monomeric state [94]. The classical models of co-operativity for multimeric enzymes are therefore not applicable. A mech-anism called 'ligand-induced slow transition' is currently the favoured explanation for the co-operative behaviour of gluco-kinase [95,96]. The basic tenets of the model are the existence of two kinetically distinct conformational states of the enzyme, and the possibility of slow interconversion between them in function of the ambient substrate concentration. The slowness of the conformational transition relative to the catalytic rate confers to the enzyme a 'memory' of its interaction with the substrate. This property is also the central aspect of the 'mnemonical' model of co-operativity for monomeric enzymes [97,98]. In the latter model, the conformational transition is possible only for the free enzyme, whereas in the ligand-induced slow transition model, the interconversion is also possible for the enzyme–substrate and enzyme–product reaction intermediaries [99]. Kinetic and physicochemical evidence for glucose-induced int-erconversion between two conformations of rat liver glucokinase has recently been reported by Neet and co-workers. A lag in the reaction velocity could be observed during assay of the enzyme in presence of glycerol, if the enzyme was previously stored with glucose at lower than the assay concentration; conversely, a 'burst' transient was observed when storage of the enzyme was at a higher glucose concentration than during the assay [100]. The physicochemical assay for conformational change relied on spectroscopic measurement of the intrinsic tryptophan fluor-escence of glucokinase. A slow enhancement of fluorescence was recorded upon glucose addition and, subsequently, the decay of fluorescence could be followed upon glucose dilution [101]. Given the assumed similarity in tertiary structure between glucokinase and yeast hexokinase, one can speculate that the glucose-induced slow conformational transition of glucokinase is mechanistically related to the classical induced fit occurring in yeast hexokinase upon glucose binding ([78], and see above).

Glucokinase is inhibited by long-chain fatty acyl-CoAs [102].

The inhibitory effect of these compounds is immediate, instantly reversible and occurs with concentrations of fatty acyl-CoAs lower than the critical micelle concentration [103]. On this basis, Tippet and Neet [103] have argued that the inhibition reflects specific binding of the fatty acid derivatives to the enzyme, rather than a trivial detergent effect or unspecific lipid-protein interaction. Kinetic studies have suggested that palmitoyl-CoA and oleoyl-CoA bind to an allosteric site on the glucokinase molecule and elicit a structural change in the enzyme which decreases the binding affinity for glucose and Mg-ATP without impairing $V_{max.}$. Inhibition constants were calculated to be 1.3 μM and 0.75 μM for palmitoyl-CoA and oleoyl-CoA, which appears to be within the range of intrahepatic concentrations [104]. Inhibition of glucokinase activity by fatty acyl-CoAs may be significant *in vivo* in situations of increased lipolysis such as fasting or diabetes mellitus.

Highly purified glucokinase from rat liver can serve as a substrate for cyclic AMP-dependent protein kinase *in vitro* [105]. Phosphorylation was shown to be exclusively on serine and amounted to 1 mol of phosphate/mol of enzyme after prolonged incubation with the kinase. Under these conditions, the glucose affinity and $V_{max.}$ of glucokinase were decreased. The minimal motif Arg-Xaa-Ser, which can serve as substrate site for *in vitro* phosphorylation by cyclic AMP-dependent protein kinase, occurs twice at positions 358–360 and 394–396 in the amino acid sequence of rat liver glucokinase. However, neither of the target motifs Arg-Arg-Xaa-Ser or Arg-Arg-Ser found in proteins phosphorylated *in vivo* by the kinase [106] are present in glucokinase. Indeed, before a prediction based on sequence analysis could be made, attempts to demonstrate a charge shift in hepatic glucokinase according to the nutritional or hormonal condition of the animal had been unsuccessful (P. B. Iynedjian, unpublished work). The possibility of physiological control of glucokinase activity by phosphorylation–dephosphorylation appears therefore unlikely.

An interesting mechanism for the short-term control of glucokinase activity has recently been described by Van Schaftingen and collaborators [107]. These investigators have identified a novel regulatory protein capable of binding to and inhibiting glucokinase in presence of fructose 6-phosphate. Inhibition is relieved in presence of fructose 1-phosphate. The regulatory protein has been purified to near homogeneity. It has a molecular mass of 62 kDa. The formation of a one-to-one complex between this protein and glucokinase in the presence of fructose 6-phosphate has been demonstrated by sedimentation in sucrose gradients. The assembly of the complex was prevented by excess fructose 1-phosphate [108]. A binding assay based on protein precipitation by poly(ethylene glycol) provided direct evidence for the binding to the purified regulatory protein of sorbitol 6-phosphate, an analogue of fructose 6-phosphate, and of fructose 1-phosphate [109]. The model deduced from these studies suggests the existence in rat liver (and in lower amounts in islets of Langerhans [110]) of a glucokinase regulatory protein capable of binding fructose 6-phosphate or fructose 1-phosphate in a reversible and mutually exclusive manner. The fructose 6-phosphate bound form makes contact with glucokinase and inhibits it competitively with respect to glucose. In contrast, the unliganded or fructose 1-phosphate bound form of the regulatory protein does not associate with glucokinase and consequently does not interfere with its activity. These new observations may provide an explanation for the stimulatory effect of fructose on glucose phosphorylation in intact hepatocytes or in the liver of the anaesthetized animal [111,112]. They may also have important implications in normal physiology. In the post-absorptive state, the intra-hepatic concentration of fructose 6-phosphate appears

sufficient for substantial inhibition of glucokinase by the regulatory protein. Following a meal, ingested fructose will cause a rapid rise in the hepatic level of fructose 1-phosphate (via fructokinase), with concomitant de-inhibition of glucokinase and thereby increase in the rate of glucose phosphorylation and disposal.

ASSAY AND PURIFICATION OF GLUCOKINASE

The standard method for measuring glucokinase activity in tissue extracts is to assay the rate of formation of the reaction product glucose-6-phosphate with the help of an accessory or 'coupling' enzyme, glucose-6-phosphate dehydrogenase [18,113–115]. The coupling enzyme of choice is the glucose-6-phosphate dehydrogenase from *Leuconostoc mesenteroides*, which is commercially available. This bacterial enzyme uses NAD^+ as coenzyme, the reduction of which is followed spectrophotometrically or, for enhanced sensitivity, fluorometrically. The assay is run in duplicate tubes with 100 mM and 0.5 mM glucose, to score total hexokinase activity and the activity of the three low-K_m hexokinases respectively. The difference between total and low-K_m activity is taken as a measure of glucokinase activity. The above procedure is well suited to measurements in rat liver extracts, in which glucokinase represents 90% of the total hexokinase activity. However, in tissues or tumour cell lines that are rich in low-K_m hexokinases, such as insulinoma cells, the difference in reaction rate at the two glucose concentrations can be small and become experimentally uncertain. Several types of radiometric assays using isotopically labelled glucose have been designed to circumvent this problem [116,117]. In one variant, the synthesis of radioactive glucose 6-phosphate is measured directly by binding the product onto DEAE-cellulose filter disks. The reaction can be performed at high glucose concentration, in the presence of added unlabelled glucose 6-phosphate to inhibit the activity of the low-K_m hexokinases, so as to register only glucokinase activity [116]. Whatever the assay procedure, care should be taken to avoid inactivation of glucokinase *in vitro* (see [118]).

Glucokinase is a low abundance cytosolic protein in liver and is present at even lower levels in islets of Langerhans. From the enrichment factor during purification and quantitative immunoblotting data, it can be estimated that the enzyme represents 0.01–0.1% of total cytosolic protein in rat liver and approximately one-tenth to one-twentieth of this amount in islets of Langerhans. Owing to its scarcity and instability *in vitro*, glucokinase is difficult to purify. An efficient purification scheme has been designed by Holroyde et al. [119], relying essentially on affinity chromatography on Sepharose-*N*-(6-aminohexanoyl)-2-amino-2-deoxy-D-glucopyranose. This procedure has been instrumental in allowing Seitz and co-workers to purify rat liver glucokinase to homogeneity and raise specific antibodies to the enzyme in sheep. These antibodies were used initially for the quantification of hepatic glucokinase mRNA by translational assay [37] and subsequently for the immunological screening of a rat liver cDNA library which led to the isolation of the first glucokinase cDNA clone [36].

GLUCOKINASE AND INTEGRATION OF GLUCOSE METABOLISM

The liver can alternatively take up glucose from the blood for the synthesis of glycogen and fatty acids, or release glucose formed via glycogenolysis and gluconeogenesis into the circulation. Net glucose uptake occurs in the postprandial state, when the portal plasma glucose concentration rises above 8 mM, whereas net glucose output occurs at lower glucose levels, namely in the

HEPATOCYTES

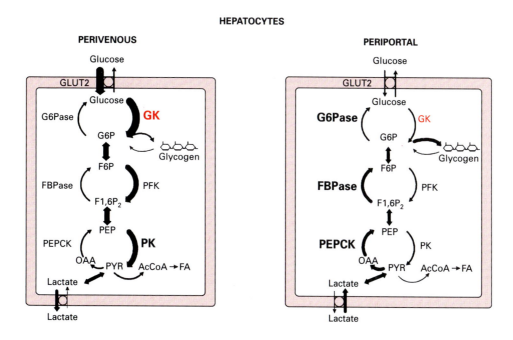

Figure 4   Role of glucokinase in hepatic glucose metabolism according to the concept of functional zonation

Typical hepatocytes of the perivenous and periportal zones of the liver acinus are illustrated. Enzymes enriched in a particular zone are shown in large-size bold letters. The relative thickness of the arrows indicates the importance of carbon flux in a reaction or series of reactions. The metabolic pathways are illustrated in the postprandial state in which net uptake of glucose occurs in the whole liver. In spite of a high glucose-6-phosphatase amount in the periportal cell, most of the carbon flux is channelled toward glycogen synthesis and no net glucose output occurs. See the text for details. Abbreviations: G6P, glucose 6-phosphate. F6P, fructose 6-phosphate; F1,6P$_2$, fructose 1,6-bisphosphate; PEP, phosphoenolpyruvate; PYR, pyruvate; OAA, oxaloacetate; AcCoA, acetyl-CoA; FA, fatty acids; GK, glucokinase; G6Pase, glucose-6-phosphatase; PFK, 6-phosphofructose-1-kinase; FBPase, fructose-1,6-bisphosphatase; PK, pyruvate kinase; PEPCK, phosphoenolpyruvate carboxykinase; GLUT, glucose transporter.

postabsorptive and fasting states [120,121]. By virtue of its particular affinity for glucose, glucokinase is the molecular device which allows the liver to 'assay' the glucose concentration and to shift between net glucose uptake or release. In contrast to the skeletal muscle cell and adipocyte, the hepatocyte is endowed with a high-capacity glucose transport system which is independent of insulin [122–124]. Transport across the cell membrane occurs by facilitated diffusion via the specific glucose transporter isotype called GLUT 2 [125]. The transporter can move glucose in and out the cells, is half-saturated at around 20 mM glucose and is expressed at high level, such that the transport step is never rate-limiting for glucose metabolism. This system allows free equilibration of glucose inside the cell and glucose in the extracellular fluid. Consequently, fluctuations of the blood glucose concentration are instantly followed by parallel changes in the intracellular glucose concentration. These variations will be monitored by glucokinase in the cytosol of the cell and translated into an increase or decrease in the rate of synthesis of glucose 6-phosphate destined for glycogen synthesis or glycolysis.

A second enzyme directly involved in hepatic glucose uptake or release is glucose-6-phosphatase. In the postabsorptive and fasting states, glycogenolysis and gluconeogenesis provide a steady supply of glucose 6-phosphate. Glucose-6-phosphatase converts this metabolite into free glucose for release into the circulation. Glucose-6-phosphatase and glucokinase catalyse opposing reactions, giving rise to a substrate cycle between glucose and glucose 6-phosphate [126]. The net carbon flux between these substrates, hence the net movement of glucose in or out of the liver, is determined by the difference in rates (if any) of the two reactions. This in turn will depend on the amounts of the two enzymes and on the balance of all factors involved in the

acute control of the reactions rates. Hyperglycaemia and dietary fructose (acting via fructose-1-phosphate and release of the glucokinase inhibitory protein) are factors which enhance the rate of the glucokinase reaction and promote glucose uptake in the postprandial period. Conversely, hypoglycaemia causes a decrease in the rate of glucose phosphorylation and favors hepatic glucose output in the fasting state. Unrestrained glucose output will ensue from a reduction of the glucokinase enzyme concentration in prolonged fasting or diabetes, due to the turning off of specific gene transcription.

Although substrate cycling is a valid notion for the liver taken as a whole, it appears to be limited in extent at the single cell level owing to the functional specialization of individual hepatocytes. The unidirectional enzymes of gluconeogenesis and glucose-6-phosphatase are enriched in the periportal hepatocytes of the liver acinus, whereas glucokinase and the unidirectional enzymes of glycolysis predominate in the perivenous hepatocytes. This topological separation of metabolic pathways is referred to as metabolic zonation [127]. Concentration gradients for oxygen, substrates and hormones in the liver microcirculation are thought to be responsible for region-specific gene expression. Both immunohistochemistry and microdissection data have shown that the concentration of glucokinase is approximately two times higher in perivenous than in periportal hepatocytes, and inversely for glucose-6-phosphatase [128–131]. Glucokinase mRNA has been localized to the perivenous areas by *in situ* hybridization [132]. In the perfused rat liver, glycogen synthesis from glucose was shown to take place primarily in the perivenous zone, whereas the periportal cells synthesized glycogen preferentially from lactate and pyruvate [133,134]. Thus, metabolic zonation may account for the capacity of the liver to synthesize glycogen after a glucose load by a direct route from glucose itself as well

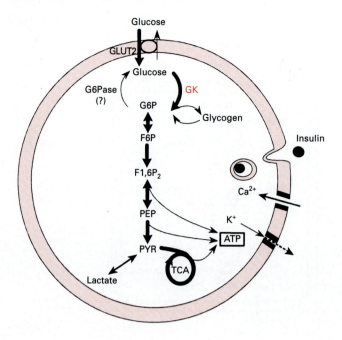

**Figure 5   Role of glucokinase in the physiology of insulin secretion in the
β-cells of the islets of Langerhans**

Glucose metabolism is necessary for regulated insulin release. The glycolytic pathway and
tricarboxylic acid cycle (TCA) are schematically illustrated. The metabolic flux is controlled at
the glucokinase step. The production of ATP results in a rise in the ATP/ADP ratio, which effects
the closure of ATP-sensitive potassium channels. The ensuing membrane depolarization causes
the opening of voltage-sensitive calcium channels and an influx of Ca^{2+} ions necessary for
insulin release. See the text for details, and the legend to Figure 4 for explanation of
abbreviations.

as by an indirect route from lactate [135]. The direct route would
be active mostly in perivenous hepatocytes, which would take up
glucose from the blood and phosphorylate it via glucokinase.
Part of the glucose 6-phosphate would be converted directly into
glycogen and the remainder metabolized to lactate by glycolysis.
After release into the general circulation, lactate would be taken
up by periportal hepatocytes and converted into glycogen via the
gluconeogenic pathway. It is noteworthy in this scheme (illustrated in Figure 4) that hepatic glucokinase handles all the
glucose eventually utilized for glycogen synthesis, even though
some of the carbon flux after glucose phosphorylation is diverted
via glycolysis and gluconeogenesis prior to the synthesis of
glycogen [136].

The physiological impact of glucokinase gene expression in β-
cells of the islets of Langerhans is well understood. The single
most important stimulus for insulin secretion by the β-cells is a
rise in the blood glucose concentration. Cellular uptake of
glucose and metabolism via the glycolytic pathway and tricarboxylic acid cycle are requisites for the insulinotropic effect
[137]. Three biochemical properties of the β-cell are essential for
a physiological insulin secretory response: (i) rapid equilibration
of the glucose concentrations in the extra-and intra-cellular
compartments; (ii) rate-limiting step of glucose metabolism
catalysed by glucokinase and (iii) coupling link between rate of
glucose metabolism and the insulin release process.

The glucose transporter isotype normally expressed in islet β-
cells is GLUT 2 [125,138,139]. Maximal rates of glucose transport
measured in isolated islets or dispersed β-cells exceed the rates of
glucose metabolism approximately 100-fold, making it clear that

metabolism is not controlled at the plasma membrane transport
step [140–142]. Several lines of evidence suggest that the phosphorylation of glucose is the rate-limiting step in whole islets,
and by inference in β-cells, which constitute approx. 80 % of the
islet mass [143]. Enzyme assays under $V_{max.}$ conditions indicate
that total hexokinase activity, including both low-K_m hexo-
kinase(s) and glucokinase, is by far lower than the activities of
the two regulatory enzymes of glycolysis, 6-phosphofructo-1-
kinase and pyruvate kinase [144]. Although glucokinase accounts
for only 25 % of total hexokinase activity in islet extracts, it
appears to contribute approx. 90 % of the glucose 6-phosphate
synthesized at normal glucose concentration in intact islet cells,
owing to effective inhibition of the low-K_m hexokinase(s) by the
ambient glucose 6-phosphate concentration [145]. It is therefore
glucokinase which assumes control over the carbon flux in the
glycolytic pathway, allowing physiological fluctuations of glucose
concentration to translate into alterations of the overall glycolytic
rate. The coupling link between glycolytic metabolism and insulin
secretory activity is thought to reside in ATP-sensitive potassium
channels in the β-cell plasma membrane [146–148]. These channels close following a rise in the intracellular ATP/ADP ratio.
The ensuing membrane depolarization and electrical activity
cause the opening of voltage-gated calcium channel, an influx of
calcium ions from the extracellular fluid into the cell and
ultimately, via steps that remain to be elucidated in detail, the
release of insulin from secretory granules (Figure 5). In the final
analysis, the glucose responsiveness of the entire system is vested
in glucokinase. This notion is encapsulated in the term 'glucose
sensor' used by Meglasson and Matschinsky to designate the
β-cell glucokinase [149].

A recent immunohistochemical study by Jetton and Magnuson
[150] has suggested large differences in glucokinase amount
between individual β-cells of single islets. Whether β-cells enriched in or devoid of glucokinase also vary in the levels of other
enzymes is unknown, but it is plausible that metabolic zonation
occurs in the β-cell population of the islets. This may represent
an underlying factor for the reported functional heterogeneity
among these cells [151]. Glucose 6-phosphatase immunoreactive
protein and enzyme activity have been detected in whole islet
extracts [152]. Moreover, substrate cycling between glucose and
glucose 6-phosphate has been demonstrated [153,154]. Whether
glucokinase and glucose-6-phosphatase coexist in the same cells
and whether glucose cycling plays a role in the regulation of
insulin secretion remains to be determined. It is also interesting
to note that tumoural β-cell lines can present marked alterations
in enzyme expression as compared to the parent cells. An
example is provided by an insulin-producing rat insulinoma cell
line (RINm5F) studied in our laboratory. The cells have near-
normal levels of glucokinase, but dramatically overexpress hexokinase II [83]. We have proposed that the imbalance between
glucokinase and low-K_m hexokinase activities in this and probably other β-cell lines is a major determinant for the lack of
glucose-stimulated insulin release within the physiological range.

A MOLECULAR DISEASE OF GLUCOKINASE

NIDDM is a common metabolic disorder characterized by
impaired insulin secretion in response to glucose and resistance
of the target tissues to the action of insulin, including exaggerated
hepatic glucose output in the postabsorptive state. Both defects
contribute to the development of fasting hyperglycaemia at full-
blown stages of the disease. The aetiology of NIDDM is not
understood in detail, but is known to involve both genetic and
environmental factors [155]. As a general rule, the disease is
classified among the multifactorial–polygenic syndromes. How-

ever, there is a rare and mild form of NIDDM with simple autosomal dominant mode of inheritance, the early-onset NIDDM or maturity-onset diabetes of the young (MODY) [156]. The clear-cut inheritance of MODY makes this syndrome attractive for the search of disease genes linked to NIDDM [157]. Candidate genes can be targeted for study, based on our extensive biochemical knowledge of glucoregulatory mechanisms. Because of the major impact of glucokinase on blood glucose homeostasis, in the control of hepatic glucose disposal as well as in the regulation of pancreatic insulin secretion, the glucokinase gene has been placed high on the list of potential diabetes susceptibility genes.

Two separate studies have recently reported linkage of the MODY phenotype to the glucokinase locus in a number of multigenerational pedigrees with the disease [158,159]. The transmission of glucokinase alleles was followed using two polymorphic microsatellite DNA markers in the immediate vicinity of the gene [17,158]. Specific mutations have been identified in MODY patients in several of the above families [160,161]. The mutations were not found in unaffected relatives nor in control subjects in the general population, suggesting their critical role in the disease. The reported mutations include nonsense, missense and splicing site mutations. The positions of the amino acid residues affected by nonsense or missense mutations are shown in Figure 3(c). In order to assess the effect of the missense mutations on the enzymic activity of glucokinase, Gidh-Jain et al. [79] have reproduced these mutations by *in vitro* mutagenesis of a human islet cDNA and analysed the kinetic properties of the resulting proteins after bacterial expression. The amino acid substitutions were shown to have more or less pronounced deleterious effects on the $V_{max.}$ and/or glucose affinity of the enzyme, as detailed in the legend to Figure 3.

What are the clinical consequences of mutations in the glucokinase gene? Patients from four MODY kindreds with four distinct glucokinase mutations were evaluated for islet β-cell function by Velho et al. [162]. The patients presented a normal first-phase elevation of plasma insulin in response to a priming injection of glucose, but failed to sustain adequate insulin levels during a continuous glucose infusion set to maintain the plasma glucose concentration at 10 mM. In the basal state, these subjects also displayed inappropriately low plasma levels of insulin in relation to the slightly elevated blood glucose levels. Together, data from these patients are compatible with an increase in the glucose threshold for insulin secretion. Such a defect is a predicted consequence of a reduction in β-cell glucokinase activity, as shown in a theoretical model of glucose-stimulated insulin secretion presented several years ago by Meglasson and Matschinsky [149]. Investigations of hepatic glucose metabolism in subjects with MODY have not yet been reported. From our current knowledge, the MODY syndrome associated with mutations of glucokinase emerges as a disorder characterized by mild hyperglycaemia appearing during childhood, resulting primarily from a discrete defect in glucose-induced insulin secretion [161,162]. In that, it differs from more prevalent forms of NIDDM in which the primary abnormality appears to be the resistance of the liver and peripheral tissues to the action of insulin [163]. Mutations of the glucokinase gene could also be a factor in some forms of NIDDM with late onset, although their occurrence in these forms appears to be uncommon [164,165].

CONCLUSIONS AND PROSPECTS

The discovery of glucokinase mutations as a cause or contributing factor of MODY and other forms of NIDDM will come to be regarded as a paradigm in human genetics. Given its essential

role in blood glucose homeostasis, glucokinase was near the top of the list of candidate genes potentially involved in NIDDM. Conversely, the strong evidence now produced for the association of glucokinase mutations with NIDDM reinforces the physiological importance of the enzyme. Classically, genetic enzyme deficiencies with clinical consequences are recessive diseases, since a single normal allele can in principle sustain sufficient enzyme synthesis for unimpaired cellular function. Glucokinase, however, is a rate-limiting enzyme of metabolism in liver as well as islet β-cells and it is to be expected that seemingly small deficits in enzyme activity would lead to both exaggerated hepatic glucose output and reduced insulin release at a given blood glucose level. To put things in perspective, it should be reminded that hepatic glucokinase activity is reduced by 'only' 50% in rats after 3 days of fasting, a long fast for rodent species. An intriguing question that remains to be answered is whether mutations in the regulatory regions of the tissue-specific promoters or in the leader exons 1β or 1L occur in human disease, affecting either islet or liver glucokinase separately.

The technique of targeted gene disruption in embryonic stem cells [166] makes it feasible to produce mouse strains with partial (heterozygous) or total (homozygous) glucokinase deficiencies. Such strains would provide invaluable animal models for studying: (i) the metabolic consequences of liver and/or islet glucokinase deficiencies; (ii) the interaction of the genetic background and environmental (e.g. nutritional) factors in the development of glucose intolerance or diabetes and (iii) the effects of drugs and other preventive or therapeutic manoeuvres.

Glucokinase will continue to attract great interest as a model system for studies of differential gene regulation in various cell types. The β-cell specific promoter has already been dissected into discrete binding sites for transcription regulatory *trans*-acting factors. The characterization of these proteins, the cloning of their genes and their comparison with factors involved in the transcription of other β-cell expressed genes (e.g. the insulin gene) will provide important clues on the differentiation of this endocrine cell type. At present, it appears that the β-cell promoter is not acutely regulated by hormones or other signals, but this remains to be definitively established. The possibility of specific regulation of enzyme synthesis at the translational level should also be pursued. Although the distribution of glucokinase is stringently restricted in the organism, it cannot be ruled out that cell-types other than hepatocytes and endocrine β-cells express the enzyme.

Transcription of the glucokinase gene in hepatocytes is absolutely dependent on the presence of insulin, at least in the rat, and indeed the activation of the glucokinase gene has emerged as a major hepatic effect of insulin. It will be important to determine whether this holds true in human hepatocytes. The effect of insulin is long-term in that it will affect the synthesis of a fairly stable enzyme; it is nevertheless a very rapid effect when monitored by transcriptional assay. Just as impressive is the instant turning off of gene transcription under the influence of cyclic AMP. Rapid repression of gene transcription contrasts with the delayed and blunted response at the enzyme level. Stringent and acute control of gene transcription for an enzyme with relatively long half-life seems somewhat paradoxical; its value may reside in the sparing of high energy nucleoside triphosphates when carbohydrate food becomes scarce. As regards molecular mechanisms, the regulation of the glucokinase gene in the hepatocyte remains much of a mystery. A decisive step would be accomplished if one could understand why the expression of the gene is extinguished in hepatoma cell lines. The developmental regulation of the gene in liver, with the initial appearance of the mRNA at weaning time, may also provide interesting cues when studied at the molecular level. Finally, working backward from the gene to regulatory *trans*-acting factors and signal transduction cascades is an exciting prospect that is sure to add important new insight into the mechanism of insulin action. In the course of such studies, more may be learned about insulin resistance and non-insulin-dependent diabetes mellitus.

I thank Dr Milena Girotti for designing the figures. Thanks are also expressed to Dr Philippe Froguel for making preprints available before publication. Research in my laboratory was supported by grants from the SNSF and the Wolfermann–Nägeli Foundation.

REFERENCES

1 Walker, D. G. (1966) Essays Biochem. **2**, 33–67
2 Weinhouse, S. (1976) Curr. Top. Cell Regul. **11**, 1–50
3 Colowick, S. P. (1973) in The Enzymes (Boyer, P. D. ed.), pp. 1–48, Academic Press, New York
4 Watford, M. (1990) Trends Biochem. Sci. **15**, 1–2
5 Magnuson, M. A. (1990) Diabetes **39**, 523–527
6 Middleton, R. J. (1990) Biochem. Soc. Trans. **18**, 180–183
7 Magnuson, M. A. (1992) J. Cell. Biochem. **48**, 115–121
8 Permutt, M. A., Chiu, K. C. and Tanizawa, Y. (1992) Diabetes **41**, 1367–1372
9 Andreone, T. L., Printz, R. L., Pilkis, S. J., Magnuson, M. A. and Granner, D. K. (1989) J. Biol. Chem. **264**, 363–369
10 Magnuson, M. A. and Shelton, K. D. (1989) J. Biol. Chem. **264**, 15936–15942
11 Magnuson, M. A., Andreone, T. L., Printz, R. L., Koch, S. and Granner, D. K. (1989) Proc. Natl. Acad. Sci. U.S.A. **86**, 4838–4842
12 Hayzer, D. J. and Iynedjian, P. B. (1990) Biochem. J. **270**, 261–263
13 Liang, Y., Jetton, T. L., Zimmerman, E. C., Najafi, H., Matschinsky, F. M. and Magnuson, M. A. (1991) J. Biol. Chem. **266**, 6999–7007
14 Tanizawa, Y., Koranyi, L. I., Welling, C. M. and Permutt, M. A. (1991) Proc. Natl. Acad. Sci. U.S.A. **88**, 7294–7297
15 Tanizawa, Y., Matsutani, A., Chiu, K. C. and Permutt, M. A. (1992) Mol. Endocrinol. **6**, 1070–1081
16 Stoffel, M., Froguel, Ph., Takeda, J., Zouali, H., Vionnet, N., Nishi, S., Weber, I. T., Harrison, R. W., Pilkis, S. J., Lesage, S., Vaxillaire, M., Velho, G., Sun, F., Iris, F., Passa, Ph., Cohen, D. and Bell, G. I. (1992) Proc. Natl. Acad. Sci. U.S.A. **89**, 7698–7702
17 Matsutani, A., Janssen, R., Donis-Keller, H. and Permutt, M. A. (1992) Genomics **12**, 319–325
18 DiPietro, D. L., Sharma, C. and Weinhouse, S. (1962) Biochemistry **1**, 455–462
19 Matschinsky, F. M. and Ellermann, J. E. (1968) J. Biol. Chem. **243**, 2730–2736
20 Ashcroft, S. J. H. and Randle, P. J. (1970) Biochem. J. **119**, 5–15
21 Iynedjian, P. B., Mobius, G., Seitz, H. J., Wollheim, C. B. and Renold, A. E. (1986) Proc. Natl. Acad. Sci. U.S.A. **83**, 1998–2001
22 Iynedjian, P. B., Pilot, P.-R., Nouspikel, T., Milburn, J. L., Quaade, C., Hughes, S., Ucla, C. and Newgard, C. B. (1989) Proc. Natl. Acad. Sci. U.S.A. **86**, 7838–7842
23 Hughes, S. D., Quaade, C., Milburn, J. L., Cassidy, L. and Newgard, C. B. (1991) J. Biol. Chem. **266**, 4521–4530
24 Shelton, K. D., Franklin, A. J., Khoor, A., Beechem, J. and Magnuson, M. A. (1992) Mol. Cell. Biol. **12**, 4578–4589
25 Boam, D. S. W. and Docherty, K. (1989) Biochem. J. **264**, 233–239
26 Ohlsson, H., Thor, S. and Edlund, T. (1991) Mol. Endocrinol. **5**, 897–904
27 Scott, V., Clark, A. R., Hutton, J. C. and Docherty, K. (1991) FEBS Lett. **290**, 27–30
28 Shatton, J. B., Morris, H. P. and Weinhouse, S. (1969) Cancer Res. **29**, 1161–1172
29 Jacoby, D. B., Zilz, N. D. and Towle, H. C. (1989) J. Biol. Chem. **264**, 17623–17626
30 Jarnagin, W. R., Debs, R. J., Wang, S.-S. and Bissel, D. M. (1992) Nucleic Acids Res. **20**, 4205–4211
31 Nouspikel, T., Marie, S. and Iynedjian, P. B. (1993) in Progress in Endocrinology: Proceedings of the Ninth International Congress on Endocrinology, Nice, 1992 (Mornex, R., Jaffiol, C. and Leclerc, J., eds.), Parthenon Publishing, London
32 Noguchi, T., Takenaka, M., Yamada, K., Matsuda, T., Hashimoto, M. and Tanaka, T. (1989) Biochem. Biophys. Res. Commun. **164**, 1247–1252
33 Schibler, U. and Sierra, F. (1987) Annu. Rev. Genet. **21**, 237–257
34 DiPietro, D. L. and Weinhouse, S. (1960) J. Biol. Chem. **235**, 2542–2545
35 Sharma, C., Manjeshwar, R. and Weinhouse, S. (1963) J. Biol. Chem. **238**, 3840–3845
36 Iynedjian, P. B., Ucla, C. and Mach, B. (1987) J. Biol. Chem. **262**, 6032–6038
37 Minderop, R. H., Hoeppner, W. and Seitz, H. J. (1987) Eur. J. Biochem. **164**, 181–187
38 Tiedge, M. and Lenzen, S. (1991) Biochem. J. **279**, 899–901
39 Rao, M. S., Dwivedi, R. S., Yeldandi, A. V., Subbarao, V., Tan, X., Usman, M. I., Thangada, S., Nemali, M. R., Kumar, S., Scarpelli, D. G. and Reddy, J. K. (1989) Am. J. Pathol. **134**, 1069–1086
40 Yeldandi, A. V., Tan, X., Dwivedi, R. S., Subbarao, V., Smith, D. D., Scarpelli, D. G., Rao, S. M. and Reddy, J. K. (1992) Proc. Natl. Acad. Sci. U.S.A. **87**, 881–885
41 Niemeyer, H., Perez, N. and Codoceo, R. (1967) J. Biol. Chem. **242**, 860–864
42 Iynedjian, P. B., Gjinovci, A. and Renold, A. E. (1988) J. Biol. Chem. **263**, 740–744
43 Iynedjian, P. B., Jotterand, D., Nouspikel, T., Asfari, M. and Pilot, P.-R. (1989) J. Biol. Chem. **264**, 21824–21829
44 Matsuda, T., Noguchi, T., Yamada, K., Takenaka, M. and Tanaka, T. (1990) J. Biochem. (Tokyo) **108**, 778–784
45 Decaux, J. F., Antoine, B. and Kahn, A. (1989) J. Biol. Chem. **264**, 11584–11590
46 Thompson, K. S. and Towle, H. C. (1991) J. Biol. Chem. **266**, 8679–8682
47 Clarke, S. D. and Abraham, S. (1992) FASEB J. **6**, 3146–3152
48 Blackshear, P. J., McNeill Haupt, D. and Stumpo, D. J. (1991) J. Biol. Chem. **266**, 10946–10952
49 Nouspikel, T. and Iynedjian, P. B. (1992) Eur. J. Biochem. **210**, 365–373
50 Mahadevan, L. C., Willis, A. C. and Barratt, M. J. (1991) Cell **65**, 775–783
51 Edwards, D. R. and Mahadevan, L. C. (1992) EMBO J. **11**, 2415–2424
52 Nouspikel, T., Gjinovci, A., Li, S. and Iynedjian, P. B. (1992) FEBS Lett. **301**, 115–118
53 Flawn, P. and Loten, E. G. (1990) Int. J. Biochem. **22**, 983–988
54 Smith, C. J., Vasta, V., Degerman, E., Belfrage, P. and Manganiello, V. C. (1991) J. Biol. Chem. **266**, 13385–13390
55 Toth, B., Bollen, M. and Stalmans, W. (1988) J. Biol. Chem. **263**, 14061–14066
56 Dent, P., Lavoinne, A., Nakielny, S., Caudwell, F. B., Watt, P. and Cohen, P. (1990) Nature (London) **348**, 302–308
57 Lavoinne, A., Erikson, E., Maller, J. L., Price, D. J., Avruch, J. and Cohen, P. (1991) Eur. J. Biochem. **199**, 723–728
58 Nakielny, S., Cohen, P., Wu, J. and Sturgill, T. (1992) EMBO J. **11**, 2123–2129
59 Kyriakis, J. M., App, H., Zhang, X.-F., Banerjee, P., Brautigan, D. L., Rapp, U. R. and Avruch, J. (1992) Nature (London) **358**, 417–421
60 Satoh, T., Nakafuku, M. and Kaziro, Y. (1992) J. Biol. Chem. **267**, 24149–24152
61 Cherry, J. R., Johnson, T. R., Dollard, C. A., Shuster, J. R. and Denis, C. L. (1989) Cell **56**, 409–419
62 Taylor, W. E. and Young, E. T. (1990) Proc. Natl. Acad. Sci. U.S.A. **87**, 4098–4102
63 Hunter, T. and Karin, M. (1992) Cell **70**, 375–387
64 Narkewicz, M. R., Iynedjian, P. B., Ferre, P. and Girard, J. (1990) Biochem. J. **271**, 585–589
65 Hoppner, W. and Seitz, H. J. (1989) J. Biol. Chem. **264**, 20643–20647
66 Chauhan, J. and Dakshinamurti, K. (1991) J. Biol. Chem. **266**, 10035–10038
67 Koranyi, L. I., Bourey, R. E., Slentz, C. A., Holloszy, J. O. and Permutt, M. A. (1992) Diabetes **40**, 401–404
68 Fernandez-Mejia, C. and Davidson, M. B. (1992) Endocrinology **130**, 1660–1668
69 Liang, Y., Najafi, H. and Matschinsky, F. M. (1990) J. Biol. Chem. **265**, 16863–16866
70 Liang, Y., Najafi, H., Smith, R. M., Zimmerman, E. C., Magnuson, M. A., Tal, M. and Matschinsky, F. M. (1992) Diabetes **41**, 792–806
71 Schwab, D. A. and Wilson, J. E. (1988) J. Biol. Chem. **263**, 3220–3224
72 Schwab, D. A. and Wilson, J. E. (1989) Proc. Natl. Acad. Sci. U.S.A. **86**, 2563–2567
73 Thelen, A. P. and Wilson, J. E. (1991) Arch. Biochem. Biophys. **286**, 645–651
74 Schwab, D. A. and Wilson, J. E. (1991) Arch. Biochem. Biophys. **285**, 365–370
75 Ureta, T. (1982) Comp. Biochem. Physiol. **71B**, 549–555
76 Nishi, S., Seino, S. and Bell, G. I. (1988) Biochem. Biophys. Res. Commun. **157**, 937–943
77 Stachelek, C., Stachelek, J., Swan, J., Botstein, D. and Konigsberg, W. (1986) Nucleic Acids Res. **14**, 945–963
78 Bennett, W. S., Jr. and Steitz, T. A. (1980) J. Mol. Biol. **140**, 211–230
79 Gidh-Jain, M., Takeda, J., Xu, L. Z., Lange, A. J., Vionnet, N., Stoffel, M., Froguel, P., Velho, G., Sun, F., Cohen, D., Patel, P., Lo, Y. M. D., Hattersley, A. T., Luthman, H., Wedell, A., Charles, R. St., Harrison, R. W., Weber, I. T., Bell, G. I. and Pilkis, S. J. (1993) Proc. Natl. Acad. Sci. U.S.A. **90**, 1932–1936
80 Lange, A. J., Xu, L. Z., Van Poelwijk, F., Lin, K., Granner, D. K. and Pilkis, S. J. (1991) Biochem. J. **277**, 159–163
81 Bork, P., Sander, C. and Valencia, A. (1992) Proc. Natl. Acad. Sci. U.S.A. **89**, 7290–7294
82 Holmes, K. C., Sander, C. and Valencia, A. (1993) Trends Cell Biol. **3**, 53–59
83 Vischer, U., Blondel, B., Wollheim, C. B., Höppner, W., Seitz, H. J. and Iynedjian, P. B. (1987) Biochem. J. **241**, 249–255
84 Newgard, C. B., Quaade, C., Hughes, S. D. and Milburn, J. L. (1990) Biochem. Soc. Trans. **18**, 851–853

85 Quaade, C., Hughes, S. D., Coats, W. S., Sestak, A. L., Iynedjian, P. B. and Newgard, C. B. (1991) FEBS Lett. **280**, 47–52

86 Grossbard, L. and Schimke, R. T. (1966) J. Biol. Chem. **241**, 3546–3560

87 Vinuela, E., Salas, M. and Sols, A. (1963) J. Biol. Chem. **238**, 1175–1177

88 Parry, M. J. and Walker, D. G. (1966) Biochem. J. **99**, 266–274

89 White, T. K. and Wilson, J. E. (1989) Arch. Biochem. Biophys. **274**, 375–393

90 Magnani, M., Bianchi, M., Casabianca, A., Stocchi, V., Daniele, A., Altruda, F., Ferrone, M. and Silengo, L. (1992) Biochem. J. **285**, 193–199

91 Parry, M. J. and Walker, D. G. (1967) Biochem. J. **105**, 473–482

92 Niemeyer, H., Cardenas, M. L., Rabajille, E., Ureta, T., Clark-Turri, L. and Penaranda, J. (1975) Enzyme **20**, 321–333

93 Storer, A. C. and Cornish-Bowden, A. (1976) Biochem. J. **159**, 7–14

94 Connolly, B. A. and Trayer, I. P. (1979) Eur. J. Biochem. **99**, 299–308

95 Neet, K. E. and Ainslie, G. R., Jr. (1980) Methods Enzymol. **64**, 192–226

96 Cardenas, M. L., Rabajille, E. and Niemeyer, H. (1984) Eur. J. Biochem. **145**, 163–171

97 Storer, A. C. and Cornish-Bowden, A. (1977) Biochem. J. **165**, 61–69

98 Cornish-Bowden, A. and Storer, A. C. (1986) Biochem. J. **240**, 293–296

99 Ricard, J. and Cornish-Bowden, A. (1987) Eur. J. Biochem. **166**, 255–272

100 Neet, K. E., Keenan, R. P. and Tippett, P. S. (1990) Biochemistry **29**, 770–777

101 Lin, S. X. and Neet, K. E. (1990) J. Biol. Chem. **265**, 9670–9675

102 Dawson, C. M. and Hales, C. N. (1969) Biochim. Biophys. Acta **176**, 657–659

103 Tippett, P. S. and Neet, K. E. (1982) J. Biol. Chem. **257**, 12839–12845

104 Tippett, P. S. and Neet, K. E. (1982) J. Biol. Chem. **257**, 12846–12852

105 Ekman, P. and Nilsson, E. (1988) Arch. Biochem. Biophys. **261**, 275–282

106 Kemp, B. E. and Pearson, R. B. (1990) Trends Biochem. Sci. **15**, 342–346

107 Van Schaftingen, E., Vandercammen, A., Detheux, M. and Davies, D. R. (1992) Adv. Enzyme Regul. **32**, 133–148

108 Vandercammen, A. and Van Schaftingen, E. (1990) Eur. J. Biochem. **191**, 483–489

109 Vandercammen, A., Detheux, M. and Van Schaftingen, E. (1992) Biochem. J. **286**, 253–256

110 Malaisse, W. J., Malaisse Lagae, F., Davies, D. R., Vandercammen, A. and Van Schaftingen, E. (1990) Eur. J. Biochem. **190**, 539–545

111 Davies, D. R., Detheux, M. and Van Schaftingen, E. (1990) Eur. J. Biochem. **192**, 283–289

112 Van Schaftingen, E. and Davies, D. R. (1991) FASEB J. **5**, 326–330

113 Walker, D. G. and Parry, M. J. (1966) Methods Enzymol. **9**, 381–392

114 Storer, A. C. and Cornish-Bowden, A. (1974) Biochem. J. **141**, 205–209

115 Pilkis, S. J. (1975) Methods Enzymol. **42**, 31–39

116 Stanley, J. C., Dohm, G. L., McManus, B. S. and Newsholme, E. A. (1984) Biochem. J. **224**, 667–671

117 Bedoya, F., Meglasson, M. D., Wilson, J. and Matschinsky, F. M. (1985) Anal. Biochem. **144**, 504–513

118 Davidson, A. L. and Arion, W. J. (1987) Arch. Biochem. Biophys. **253**, 156–167

119 Holroyde, M. J., Allen, M. B., Storer, A. C., Warsy, A. S., Chesher, J. M. E., Trayer, I. P., Cornish-Bowden, A. and Walker, D. G. (1976) Biochem. J. **153**, 363–373

120 Cahill, G. F., Jr., Ashmore, J., Renold, A. E. and Hastings, A. B. (1959) Am. J. Med. **26**, 264–282

121 Huang, M.-T. and Veech, R. L. (1988) J. Clin. Invest. **81**, 872–878

122 Cahill, G. F., Jr., Ashmore, J., Earle, A. S. and Zottu, S. (1958) Am. J. Physiol. **192**, 491–496

123 Williams, T. F., Exton, J. H., Park, C. R. and Regen, D. M. (1968) Am. J. Physiol. **215**, 1200–1209

124 Ciaraldi, T. P., Horuk, R. and Matthaei, S. (1986) Biochem. J. **240**, 115–123

125 Thorens, B., Sarkar, H. K., Kaback, H. R. and Lodish, H. F. (1988) Cell **55**, 281–290

126 Hue, L. (1981) Adv. Enzymol. **52**, 247–331

127 Jungermann, K. and Katz, N. (1989) Physiol. Rev. **69**, 708–764

128 Lawrence, G. M., Trayer, I. P. and Walker, D. G. (1984) Histochem. J. **16**, 1099–1111

129 Katz, N., Teutsch, H. F., Jungermann, K. and Sasse, D. (1977) FEBS Lett. **83**, 272–276

130 Teutsch, H. F. (1978) Histochemistry **58**, 281–288

131 Katz, N., Teutsch, H. F., Sasse, D. and Jungermann, K. (1977) FEBS Lett. **76**, 226–230

132 Moorman, A. F. M., De Boer, P. A. J., Charles, R. and Lamers, W. H. (1991) FEBS Lett. **287**, 47–52

133 Bartels, H., Vogt, B. and Jungermann, K. (1987) FEBS Lett. **221**, 277–283

134 Agius, L., Peak, M. and Alberti, K. G. M. M. (1990) Biochem. J. **266**, 91–102

135 Schulman, G. I. and Landau, B. R. (1992) Physiol. Rev. **72**, 1019–1035

136 Pilkis, S. J., Regen, D. M., Claus, T. H. and Cherrington, A. D. (1985) BioEssays **2**, 273–276

137 Ashcroft, S. J. H. (1980) Diabetologia **18**, 5–15

138 Johnson, J. H., Newgard, C. B., Milburn, J. L., Lodish, H. F. and Thorens, B. (1990) J. Biol. Chem. **265**, 6548–6551

139 Pessin, J. E. and Bell, G. I. (1992) Annu. Rev. Physiol. **54**, 911–930

140 Hellman, B., Sehlin, J. and Täljedal, I.-B. (1971) Biochim. Biophys. Acta **241**, 147–154

141 Gorus, F. K., Malaisse, W. J. and Pipeleers, D. G. (1984) J. Biol. Chem. **259**, 1196–2000

142 Tal, M., Liang, Y., Najafi, H., Lodish, H. F. and Matschinsky, F. M. (1992) J. Biol. Chem. **267**, 17241–17247

143 Meglasson, M. D. and Matschinsky, F. M. (1986) Diabetes Metab. Rev. **2**, 163–214

144 Trus, M. D., Zawalich, W. S., Burch, P. T., Berner, D. K., Weill, V. A. and Matschinsky, F. M. (1981) Diabetes **30**, 911–922

145 Giroix, M.-H., Sener, A., Pipeleers, D. G. and Malaisse, W. J. (1984) Biochem. J. **223**, 447–453

146 Ashcroft, F. M., Harrison, D. E. and Ashcroft, S. J. H. (1984) Nature (London) **312**, 446–448

147 Cook, D. L. and Hales, C. N. (1984) Nature (London) **311**, 271–273

148 Ashcroft, F. M. and Rorsman, P. (1989) Prog. Biophys. Mol. Biol. **54**, 87–143

149 Meglasson, M. D. and Matschinsky, F. M. (1984) Am. J. Physiol. **246**, E1–E13

150 Jetton, T. L. and Magnuson, M. A. (1992) Proc. Natl. Acad. Sci. U.S.A. **89**, 2619–2623

151 Pipeleers, D. G. (1992) Diabetes **41**, 777–781

152 Waddell, I. D. and Burchell, A. (1988) Biochem. J. **255**, 471–476

153 Khan, A., Chandramouli, V., Östenson, C.-G., Ahren, B., Schumann, W. C., Löw, H., Landau, B. R. and Efendic, S. (1989) J. Biol. Chem. **264**, 9732–9733

154 Khan, A., Chandramouli, V., Östensson, C.-G., Löw, H., Landau, B. R. and Efendic, S. (1990) Diabetes **39**, 456–459

155 Bennett, P. H., Bogardus, C., Tuomilehto, J. and Zimmet, P. (1992) in International Textbook of Diabetes Mellitus, Volume I (Alberti, K. G. M. M., de Fronzo, R. A., Keen, H. and Zimmet, P., eds.), pp. 147–176, John Wiley and Sons, Chichester

156 Fajans, S. S. (1990) Diabetes Care **13**, 49–64

157 Cox, N. J., Xiang, K.-S., Fajans, S. S. and Bell, G. I. (1992) Diabetes **41**, 401–407

158 Froguel, Ph., Vaxillaire, M., Sun, F., Velho, G., Zouali, H., Butel, M. O., Lesage, S., Vionnet, N., Clément, K., Fougerousse, F., Tanizawa, Y., Weissenbach, J., Beckmann, J. S., Lathrop, G. M., Passa, Ph., Permutt, M. A. and Cohen, D. (1992) Nature (London) **356**, 162–164

159 Hattersley, A. T., Turner, R. C., Permutt, M. A., Patel, P., Tanizawa, Y., Chiu, K. C., O'Rahilly, S., Watkins, P. J. and Wainscoat, J. S. (1992) Lancet **339**, 1307–1310

160 Vionnet, N., Stoffel, M., Takeda, J., Yasuda, K., Bell, G. I., Zouali, H., Lesage, S., Velho, G., Iris, F., Passa, Ph., Froguel, Ph. and Cohen, D. (1992) Nature (London) **356**, 721–722

161 Froguel, P., Zouali, H., Vionnet, N., Velho, G., Vaxillaire, M., Sun, F., Lesage, S., Stoffel, M., Takeda, J., Passa, P., Permutt, A., Beckmann, J. S., Bell, G. I. and Cohen, D. (1993) N. Engl. J. Med. **328**, 697–702

162 Velho, G., Froguel, P., Clement, K., Pueyo, M. E., Rakotoambinina, B., Zouali, H., Passa, P., Cohen, D. and Robert, J.-J. (1992) Lancet **340**, 444–448

163 Martin, B. C., Warram, J. H., Krolewski, A. S., Bergman, R. N., Soeldner, J. S. and Kahn, C. R. (1992) Lancet **340**, 925–929

164 Katagiri, H., Asano, T., Ishihara, H., Inukai, K., Anai, M., Miyazaki, J.-I., Tsukuda, K., Kikuchi, M., Yazaki, Y. and Oka, Y. (1992) Lancet **340**, 1316–1317

165 Cook, J. T. E., Hattersley, A. T., Christopher, P., Bown, E., Barrow, B., Patel, P., Shaw, J. A. G., Cookson, W. O. C. M., Permutt, M. A. and Turner, R. C. (1992) Diabetes **41**, 1496–1500

166 Capecchi, M. R. (1989) Science **244**, 1288–1292

Biochem. J. (1993) **289**, 313–330 (Printed in Great Britain)

REVIEW ARTICLE
Glycosaminoglycans and the regulation of blood coagulation

Marie-Claude BOURIN* and Ulf LINDAHL†

Laboratoire de Biotechnologie des Cellules Eucaryotes, Université Paris XII, 94010 Créteil, France, and †Department of Medical and Physiological Chemistry, University of Uppsala, S-751 23 Uppsala, Sweden

INTRODUCTION

Blood coagulation involves the sequential activation of a series of serine proteinases, which culminates in the generation of thrombin and subsequent thrombin-catalysed conversion of fibrinogen into insoluble fibrin (Furie and Furie, 1988). Inhibitory modulation of this process, of paramount physiological importance, is primarily achieved by two principally different mechanisms (Figure 1). The enzymes may be inactivated by serine proteinase inhibitors (known as 'serpins'), which act by formation of stable 1:1 molar complexes with their target enzymes (Travis and Salvesen, 1983). Alternatively, the so-called protein C pathway leads to inactivation of auxiliary coagulation proteins (factors V_a and VIII_a) by cleavage at distinct sites (Esmon, 1989; Dahlbäck, 1991). The prime site of regulation is the surface of vascular endothelial cells, which have been known to possess anticoagulant properties (Colburn and Buonassisi, 1982). These properties are particularly conspicuous in the microcirculation, with its high wall surface to blood volume ratio (Busch, 1984).

The ability of certain sulphated polysaccharides, glycosaminoglycans, to interfere with blood coagulation has a long-standing record, as illustrated by the extensive clinical use of heparin as an antithrombotic agent (see Rodén, 1989). The main effect of heparin (and of its relative, heparan sulphate) is to accelerate the inactivation of coagulation enzymes by the serpin antithrombin (Rosenberg, 1977; Björk and Lindahl, 1982). A more complex picture emerged with the finding of an additional serpin, heparin cofactor II, which is 'activated' not only by heparin, but also by another glycosaminoglycan, dermatan sulphate, and which selectively inactivates thrombin (Tollefsen et al., 1982; Tollefsen, 1989). Remarkably, also the other major regulatory mechanism, the protein C pathway, involves a glycosaminoglycan-containing molecular species, since the protein C activation cofactor, thrombomodulin, turned out to be a proteoglycan with a functionally important, covalently bound glycosaminoglycan chain (Bourin and Lindahl, 1990; Bourin et al., 1990). In this Review we attempt to summarize our current understanding of glycosaminoglycan involvement in the regulation of blood coagulation.

THE GLYCOSAMINOGLYCANS

The proteoglycans comprise a heterogeneous group of macromolecular glycoconjugates that are composed of sulphated glycosaminoglycan chains covalently linked to a protein core. They are widely distributed in animal tissues and appear to be synthesized by virtually all types of cells. All glycosaminoglycans identified, except the nonsulphated polysaccharide hyaluronan, which occurs as free glycosaminoglycan chains, are synthesized in proteoglycan form. The large (and growing) number of core proteins identified, the variable extent of substitution with glycosaminoglycan chains and the variability in glycosaminoglycan structure contribute to the overall structural diversity of the proteoglycans. A detailed consideration of these features is beyond the scope of this Review, in which the structural aspects will be restricted to a brief presentation of relevant glycosaminoglycan sequences. For more comprehensive overviews on proteoglycan biochemistry the reader is referred to reviews by Fransson (1985, 1987), Hassell et al. (1986), Poole (1986), Ruoslahti (1988, 1989), Gallagher (1989) and Kjellén and Lindahl (1991).

Classification of glycosaminoglycans takes note of the basic structure of the glycan backbone, which may be composed of (1) $(HexA-GalN)_n$, (2) $(HexA-GlcN)_n$, or (3) $(Gal-GlcN)_n$ type disaccharide units. The type-3 disaccharide unit occurs in keratan sulphate only, which has so far not been implicated with blood coagulation and will not be considered further in this Review. Types 1 and 2 may be further subdivided, type 1 into chondroitin sulphate and dermatan sulphate, type 2 into heparan sulphate and heparin (Figure 2). The type-2 saccharides include also hyaluronan, which differs from the heparin/heparan sulphate family with regard to position of glycosidic linkages and by lacking sulphate substituents; again, this glycosaminoglycan species does not seem to be directly involved in blood coagulation. The definition of subspecies within each class of glucosamino- or galactosamino-glycans is complicated by extensive microheterogeneity of the glycan structures, which is best understood through a short discourse of glycosaminoglycan biosynthesis.

The biosynthesis of heparin/heparan sulphate (Lindahl, 1989; Lindahl and Kjellén, 1991) is initiated by formation of a polysaccharide chain with the structure $(-GlcA\beta1,4-GlcNAc\alpha1,4-)_n$. This polymer is N-deacetylated/N-sulphated and subsequently undergoes, in the order mentioned, C-5 epimerization of GlcA to IdoA units, 2-O-sulphation of IdoA, and 6-O-sulphation of GlcN units. Additional O-sulphate substituents may be incorporated at C-3 of GlcN units (Kusche et al., 1988) and at C-2 (or C-3) of GlcA units (Bienkowski and Conrad, 1985; Kusche and Lindahl, 1990). Due to the stepwise nature of the process, and the substrate specificities of the enzymes involved, the product of any given reaction will be the substrate for the subsequent reaction. Polymer modification is incomplete in the sense that the enzymes generally act upon only a fraction of the potential substrate residues; hence, the structural complexity and heterogeneity of the polysaccharide chain under modification increase throughout the process. Following the initial N-deacetylation/N-sulphation reaction all subsequent modifications will depend on N-sulphate groups for substrate recognition. Therefore, the IdoA units and O-sulphate groups of the final products are accumulated in the N-sulphated regions of the glycosaminoglycan chains, while the N-acetylated sequences retain GlcA units and remain largely nonsulphated. Heparin is extensively N-sulphated, and therefore rich in IdoA and O-sulphate groups,

---

Abbreviations used: EGF, epidermal growth factor; HexA, unspecified hexuronic acid; GlcN, glucosamine; GalN, galactosamine; Gal, galactose; GlcA, D-glucuronic acid; IdoA, L-iduronic acid.

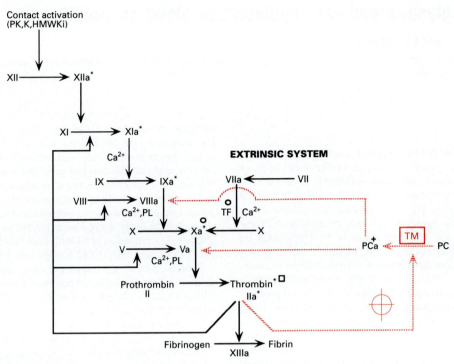

Figure 1   Overview of blood coagulation

The coagulation factors are designated with Roman numerals and the suffix a indicates a proteolytically activated factor. The solid arrows indicate pathways resulting in acceleration of blood clotting, whereas broken arrows (red) represent inhibitory mechanisms. The symbols associated with certain coagulation factors or reactions indicate target sites for various proteinase inhibitors (*, antithrombin; □, heparin cofactor II; ○, tissue factor pathway inhibitor; +, protein C inhibitor). Other abbreviations: PK, prekallikrein, K, kallikrein, HMWKi, high-molecular-weight kininogen; PL, phospholipid; TF, tissue factor. Note the protein C activation cofactor function of thrombomodulin. The factor XI-activating role of thrombin was recently postulated by Gailani and Broze (1991).

whereas heparan sulphate contains more N-acetylated, unmodified, regions (Höök et al., 1974; Gallagher and Walker, 1985; Lindahl and Kjellén, 1991). However, mixed-type, 'irregular' regions may occur in both heparin and heparan sulphate and, furthermore, may be of functional importance, as illustrated by the antithrombin-binding region (Figure 3). This pentasaccharide sequence is composed of three GlcN units, one of which is preferentially N-acetylated, one GlcA unit, and one IdoA unit, with O-sulphate groups in various positions.

The galactosaminoglycans, chondroitin sulphate and dermatan sulphate, are generated by principally similar modifications of an initial polymerization product, which has the structure $(GlcA\beta1,3\text{-}GalNAc\beta1,4\text{-})_n$ (Rodén, 1980; Fransson, 1985). The structural diversity is less pronounced than for the heparin-related glycosaminoglycans, since the GalNAc residues remain exclusively N-acetylated. Nevertheless, owing to the variable location of (O-)sulphate groups and the presence of GlcA as well as IdoA units, as many as nine different HexA-GalNAc disaccharide units have been identified (Seldin et al., 1984). By definition, chondroitin sulphate contains GlcA as the only HexA component, whereas any galactosaminoglycan with detectable amounts of IdoA will be referred to as a dermatan sulphate. The sulphate content is usually ∼1/disaccharide unit, thus less variable than in the heparin/heparan sulphate family, although 'oversulphated' species have been described.

Most of the biological activities known to be associated with proteoglycans are due to interactions between the negatively charged glycosaminoglycan chains and various proteins (Jackson et al., 1991; Kjellén and Lindahl, 1991). In general, IdoA-

containing glycosaminoglycans interact more avidly than do those containing GlcA only (Casu et al., 1988). With regard to the proteins involved in blood coagulation and its regulation the interactions with glycosaminoglycans vary from specific, 'lock-and-key' type binding to relatively nonspecific, co-operative electrostatic association.

EFFECTS OF GLYCOSAMINOGLYCANS ON PROTEINASE INHIBITORS

The major inhibitors of serine proteinases involved in blood coagulation are antithrombin and heparin cofactor II. Their mechanisms of action are profoundly influenced by glycosaminoglycans, which accelerate the rates of inhibition by binding to the inhibitors and, albeit with some exceptions, to their target enzymes. Other inhibitors, such as protein C inhibitor, protease nexin-1, and tissue factor pathway inhibitor are also affected by glycosaminoglycans; however, these interactions have not been elucidated to the same extent and will therefore be considered in less detail.

Antithrombin

Antithrombin is an α_2-glycoprotein of M_r ∼58000 that is synthesized in the liver and occurs in human blood at ∼2.7 μM concentration (for review, see Björk and Danielsson, 1986). Antithrombin is the major inhibitor of thrombin in plasma, but also inactivates the other serine proteinases of the intrinsic pathway, factors IXa, Xa, XIa and XIIa (Figure 1).

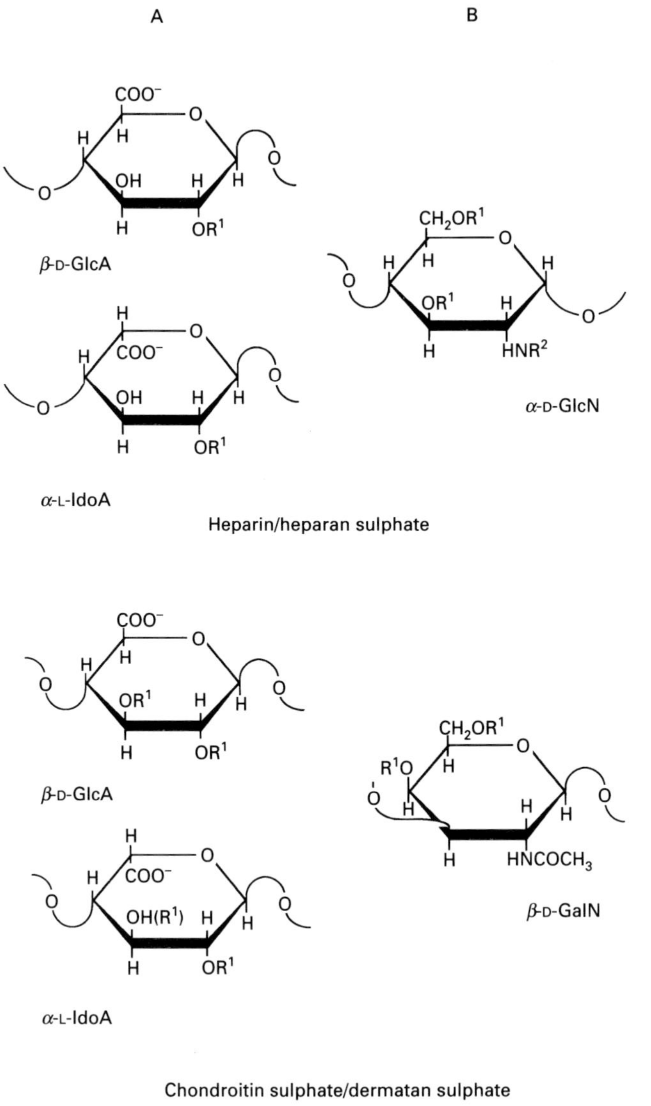

A   B

β-D-GlcA

α-L-IdoA

α-D-GlcN

Heparin/heparan sulphate

β-D-GlcA

α-L-IdoA

β-D-GalN

Chondroitin sulphate/dermatan sulphate

Figure 2   Structures of glycosaminoglycans

The sulphated glycosaminoglycans are linear polymers of alternating A and B units, where A is a HexA residue (GlcA or its C-5 epimer, IdoA), and B is either GlcN (in the glucosaminoglycans, heparin and heparan sulphate) or GalN (in the galactosaminoglycans, chondroitin sulphate and dermatan sulphate). Additional species such as hyaluronan and keratan sulphate, without any apparent relation to blood coagulation, are not included. R^1 = -H or -SO_3^-; R^2 = -SO_3^- or -$COCH_3$. The assignment of positions for the sulphate groups on the GlcA units remains somewhat tentative. For additional information see the text.

Antithrombin inhibits serine proteinases by forming tight, equimolar complexes through interaction between a specific reactive bond of the inhibitor and the active site of the enzyme (Rosenberg and Damus, 1973; Björk et al., 1989a; Olson and Björk, 1992). The stability of these complexes has suggested that they represent acyl-intermediates formed during cleavage of the reactive bond as in reaction with a normal substrate. That the proteinase may cleave the reactive bond in the inhibitor instead of forming a stable complex is indicated by the observation that small amounts of free inhibitor, cleaved at the reactive site, are produced during the reaction of antithrombin with proteinases (Björk and Fish, 1982; Olson, 1985). The reactive bond of antithrombin has been identified as the Arg-393–Ser-394 bond

near the C-terminus of the inhibitor (see Björk et al., 1989a; Olson and Björk, 1991a). A peptide sequence 8–12 amino-acid residues N-terminal to the reactive bond (designated residues P8–P12) appears to be of critical importance for antithrombin to function as an inhibitor of proteinases. Natural antithrombin variants in which amino acids within this region are mutated (Devraj-Kizuk et al., 1988; Molho-Sabatier et al., 1989), were shown to be inactive as inhibitors, but are instead excellent substrates of their target enzymes, which efficiently cleave antithrombin at the reactive bond (Caso et al., 1991; Ireland et al., 1991). Moreover, Asakura et al. (1990) showed that a monoclonal antibody that binds to the P8–P12 region also transforms antithrombin from an inhibitor to a substrate of thrombin. A more detailed understanding of this inhibitor–substrate transition has emerged from X-ray crystallographic studies of homologous serpins (Huber and Carrell, 1989; Mourey et al., 1990), including the noninhibitory serpin, ovalbumin (Stein et al., 1990), and from molecular dynamics simulations (Engh et al., 1990). Cleavage of the reactive bond induces a drastic conformational change of the P1–P16 region, an exposed peptide loop in the native state, such that it becomes inserted into the major β-sheet of the protein. These findings suggest that a partial insertion of the P1–P16 loop might be involved in trapping a proteinase in a stable complex (Skriver et al., 1991). Mutations in the P8–P12 region of the loop would interfere with such insertion and thereby allow the exposed reactive bond to be cleaved as a normal substrate. In support of this hypothesis, addition of a synthetic, competing P1–P14 peptide, blocking the insertion site for the reactive loop, resulted in a loss of the ability of antithrombin to inhibit thrombin, and in concomitant cleavage by the enzyme of the reactive bond of the inhibitor (Björk et al., 1992). Recent X-ray diffraction studies on human plasminogen activator inhibitor-1 support the general concept of a conformationally flexible reactive loop adjacent to the scissile bond in serpins (Mottonen et al., 1992).

While it had been known since the work of Howell in the 1920s that heparin requires a plasma cofactor for its anticoagulant action, it was not until 1968 that Abildgaard (1968) established the identity of this cofactor with antithrombin. The purified protein was found to account for both the 'progressive antithrombin activity' (slow inhibition of thrombin in the absence of heparin) and 'heparin cofactor activity' (rapid inhibition of thrombin in the presence of heparin). Rosenberg and Damus (1973) then suggested that heparin binds to antithrombin and effects a conformational change which results in a greatly accelerated reaction with thrombin. Following complex formation with thrombin, antithrombin loses its high affinity for heparin, which will be released and ready to 'activate' another antithrombin molecule (Björk and Nordenman, 1976; Olson and Shore, 1986; Peterson and Blackburn, 1987a). Heparin thus acts as a catalyst.

The accelerating effect of heparin on antithrombin–proteinase reactions depends on the presence of a unique antithrombin-binding pentasaccharide sequence in the glycosaminoglycan chain (Lindahl et al., 1980, 1984; Casu et al., 1981; Thunberg et al., 1982; Atha et al., 1984, 1985). This region is composed of one GlcA unit, one IdoA unit and three GlcN units, two of which are invariably N-sulphated whereas the remaining one may be either N-acetylated or N-sulphated (Figure 3). The structure/function relationships pertaining to this sequence have been elucidated in detail and have been confirmed by chemical synthesis (Choay et al., 1983; Petitou et al., 1988a,b; Petitou, 1989; Grootenhuis and van Boeckel, 1991). Both N-sulphate groups (Riesenfeld et al., 1981), the nonreducing-terminal 6-O-sulphate group (Lindahl et al., 1983) and the 3-O-sulphate group on the internal GlcN unit

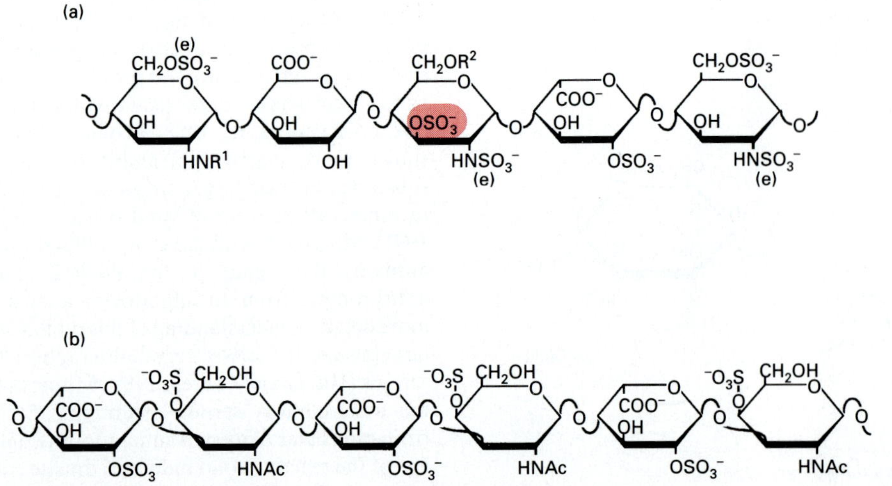

Figure 3   Serpin-binding regions in glycosaminoglycans

The structures shown represent (**a**) the antithrombin-binding region in heparin (heparan sulphate), and (**b**) the heparin cofactor II-binding region in dermatan sulphate. (**a**) Structural variants in the antithrombin-binding region are indicated by R^1 (-SO_3^- or -$COCH_3$) and R^2 (-H or -SO_3^-). The 3-O-sulphate group highlighted in red is a marker group for the antithrombin-binding region, and is essential for the high-affinity binding of the polysaccharide to antithrombin. The three sulphate groups marked (e) are also highly important to this interaction. (**b**) The heparin cofactor II-binding sequence (of highest affinity) in dermatan sulphate is composed of three consecutive -IdoA(2-OSO_3)-GalNAc(4-OSO_3)- disaccharide units that create a region of high negative charge density. For further information see the text.

(Atha et al., 1985; Petitou et al., 1988a) are essential for the biological activity. The latter residue is a distinguishing structural feature of the antithrombin-binding sequence, and thus, by and large, serves to indicate anticoagulant activity, although it has also been detected in other regions of heparin (Kusche et al., 1990) and heparan sulphate (Pejler et al., 1987a; Edge and Spiro, 1990; Kojima et al., 1992a) chains. The occurrence of 3-O-sulphate groups in only 30–40 % of the molecules in commercially available heparin preparations (Kusche et al., 1990) explains the previous, at the time highly unexpected, finding that only a fraction of such preparations exhibited high affinity for anti-thrombin (Andersson et al., 1976; Höök et al., 1976; Lam et al., 1976). Heparin (Horner, 1986; Horner et al., 1988) as well as heparan sulphate (Hovingh et al., 1986; Lane et al., 1986; Marcum et al., 1986; Pejler and David, 1987; Pejler et al., 1987b; Horner, 1990; Lindblom et al., 1991; Kojima et al., 1992a) preparations from various sources vary with regard to the proportion of molecules having high affinity for antithrombin. Of particular interest is the demonstration of proteoglycans with antithrombin-binding heparan sulphate chains, synthesized by vascular endothelial cells (Marcum et al., 1986, Kojima et al., 1992a). Such proteoglycans showed no apparent correlation between the proportion of antithrombin-binding heparan sulphate chains and core protein structure (Kojima et al., 1992a,b).

The heparin-binding region in antithrombin appears to be a composite site involving peptide sequences from different parts, largely in the N-terminal domain, of the protein. Identification of individual amino-acid residues contributing to binding has been based on studies of antithrombin variants or of chemically modified derivatives with decreased or abolished heparin binding (Brennan et al., 1988; Chang, 1989; Björk et al., 1989a; Borg et al., 1992; Gandrille et al., 1990; Sun and Chang, 1990; Olson and Björk, 1992), but also on n.m.r. spectroscopy (Gettins and Wooten, 1987). Two separate regions are implicated, one involving His-1, Ile-7, Arg-24, Pro-41 (Chang and Tran, 1986), Arg-47 (Koide et al., 1984) and Trp-49 (Blackburn et al., 1984),

and the other, further upstream in the sequence, Leu-99 (Olds et al., 1992), Lys-107, Lys-114, Lys-125, Arg-129, Asn-135, Lys-136, and Arg-145. A disulphide bond connecting these two regions also seems to be essential for heparin binding (Sun and Chang, 1989). These amino-acid residues may either directly contribute to heparin binding by participating in ionic inter-actions with the polysaccharide, or they may be essential by maintaining the structural integrity of the binding site (Shah et al., 1990). The two peptide regions implicated in heparin binding map to the A and D α-helices on the surface of a three-dimensional model of antithrombin, such that the basic residues were noted to form a band of positive charge, of appropriate size for interaction with the antithrombin-binding pentasaccharide sequence in heparin (Huber and Carrell, 1989). Interestingly, a monoclonal antibody that recognizes one of these peptide blocks (residues 104–251) not only inhibited binding of heparin to antithrombin but actually induced an increase in the rate of antithrombin–thrombin complex formation (Smith et al., 1990). More recent data suggest that also mutants having single amino-acid substitutions within the C-terminal sequence 402–407 show defective heparin binding (D. A. Lane, personal communication). By contrast, an adjacent mutation (Arg-393 His/Pro), involving the reactive site of the inhibitor, was found to actually increase heparin affinity (Owen et al., 1991).

There is now ample evidence that the interaction between antithrombin and heparin chains that contain the specific anti-thrombin-binding pentasaccharide sequence is accompanied by a conformational change in the inhibitor (see Björk et al., 1989a; Olson and Björk, 1992). In a comparative study Shore et al. (1989) found that full-length, high-affinity heparin and synthetic, antithrombin-binding pentasaccharide induced highly similar conformational changes in antithrombin, as evidenced by spec-troscopic methods. Rapid kinetic experiments showed that full-length heparin and the pentasaccharide both bind antithrombin in a two-step process, comprised by an initial weak interaction that is essentially identical for the two saccharides, followed by

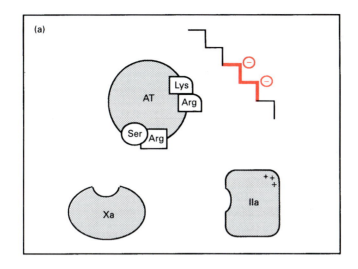

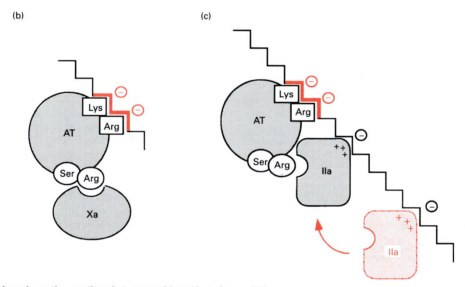

Figure 4   Effects of heparin on the reactions between antithrombin and coagulation enzymes

The scheme is specially designed to illustrate the importance of glycosaminoglycan chain length in the various antithrombin–proteinase reactions. (**a**) Schematic display of the interacting species. The adjacent Ser and Arg residues in antithrombin (AT) represent the reactive site of the inhibitor, normally in a conformation not conducive to interaction with proteinases. The Lys/Arg designation illustrates the heparin-binding site of antithrombin. (**b**) Binding to heparin induces a conformational change in the antithrombin that will facilitate its reaction with factor Xa. This mechanism works also with short heparin oligosaccharides, provided that they contain the specific antithrombin-binding pentasaccharide sequence (indicated by red segments). (**c**) The effect of heparin on the reaction between antithrombin and thrombin (IIa) involves binding of both the enzyme and the inhibitor to the heparin chain, which thus needs to contain a saccharide segment of certain length in addition to the antithrombin-binding region. The thrombin molecule will bind in a nonspecific fashion, through positive surface charges (Heuck et al., 1985; Beresford et al., 1990), to any site along the glycosaminoglycan chain, and will then move along the chain until it encounters the bound antithrombin ('template' or 'surface approximation' mechanism).

a conformational change that is responsible for generating the high-affinity binding (see also Olson et al., 1981; Peterson and Blackburn, 1987b). While this conformational change appears to be stabilized to a somewhat greater extent by the full-length heparin than by the pentasaccharide, binding studies indicated that the major portion (>90 %) of the binding energy of the full-length heparin interaction is due to the pentasaccharide.

The mechanism behind the heparin-induced acceleration of antithrombin-proteinase reactions has been an issue of controversy. The conformational change of the antithrombin molecule, largely induced by the specific antithrombin-binding penta-saccharide sequence was assumed to be an important contributor to the rate enhancement (Rosenberg and Damus, 1973; Jordan et al., 1980; Carrell et al., 1987), presumably by causing the

reactive bond of the inhibitor to be more accessible to the active-site of the proteinases (Figure 4b). This conclusion appeared to be supported by the findings that the reactions of antithrombin with proteinases such as factor Xa, factor XIIa and plasma kallikrein were potentiated in approximately similar fashion by full-length heparin and by the antithrombin-binding pentasac-charide, or small oligosaccharides containing this specific se-quence. On the other hand, the reactions of antithrombin with other proteinases such as thrombin, factor IXa and factor XIa were negligibly influenced by small-sized, high-affinity oligo-saccharides; instead, a minimum chain length of ~18 saccharides was found to be required to significantly enhance the rates of inhibition (Laurent et al., 1978; Holmer et al., 1980, 1981; Oosta et al., 1981; Choay et al., 1983; Lane et al., 1984; Shore et al.,

1989). These findings were better explained by an alternative mechanism, which predicted that heparin was acting as a surface or bridge to approximate antithrombin and the enzyme by the binding of both proteins to the same heparin chain (Figure 4c; Laurent et al., 1978; Machovich and Arányi, 1978; Pomeranz and Owen, 1978; Holmer et al., 1979). This proposal was supported by chemical modification studies in which the rate-enhancing effect of heparin could be selectively abolished by modification of basic residues of the proteinase assumed to be involved in heparin binding (Pomeranz and Owen, 1978; Machovich et al., 1978). Further, kinetic experiments showed inhibition of the rate enhancement at high heparin concentrations which correlated with the binding of inhibitor and proteinase to separate heparin chains (Jordan et al., 1979, 1980; Oosta et al., 1981; Griffith, 1982; Nesheim, 1983; Hoylaerts et al., 1984; Olson, 1988). Finally, binding studies indicated that the smallest heparin fragment capable of significantly accelerating the antithrombin–thrombin reaction corresponded to the smallest saccharide that could bind both antithrombin and the active-site blocked proteinase (Danielsson et al., 1986).

While a substantial body of evidence thus supports the importance of the surface approximation or bridging mechanism for the acceleration of the antithrombin–thrombin reaction by heparin, some investigators maintain that this mechanism plays only a secondary role and that activation of antithrombin through the conformational change would be the primary basis for the effects of heparin on all antithrombin–proteinase reactions (Huber and Carrell, 1989; Beresford and Owen, 1990; see also Björk et al., 1989a; Olson and Björk, 1992). Additional information was obtained through detailed studies of the ionic-strength dependence of the various macromolecular interactions involved in heparin-mediated antithrombin–proteinase complex formation (Shore et al., 1989; Olson and Björk, 1991). Thrombin binding to heparin essentially involves a nonspecific electrostatic association of the proteinase with any three contiguous disaccharide units of the polysaccharide chain (Olson et al., 1991). Quantitatively similar electrostatic contributions were found for the (specific) antithrombin–heparin and (nonspecific) thrombin–heparin interactions, contrary to the predominantly non-ionic mode of antithrombin–thrombin interaction. Moreover, the salt-dependence of the accelerating effect of heparin on the thrombin–antithrombin complex formation was found to be indistinguishable from the salt dependence of thrombin binding to heparin. Heparin-dependent approximation of antithrombin and thrombin bound to the polysaccharide thus quantitatively accounts for the rate-enhancing effect on the antithrombin–thrombin reaction, whereas the antithrombin conformational change contributes to a minor degree only. It therefore seems reasonable to conclude that whereas both the antithrombin conformational change and surface approximation mechanisms contribute to the accelerating effect of heparin on antithrombin–proteinase reactions, the relative contribution of each mechanism varies with the target proteinase. Additional complexity, and further support for the bridging mechanism, emerged from studies of the rate-enhancing effect of heparin on the reactions of antithrombin with the proteinases plasma kallikrein and factor XIa (Olson, 1989; Shore et al., 1989; Björk et al., 1989b). This (rather weak) effect is reinforced by a protein cofactor, high-molecular-weight kininogen, which apparently acts by promoting the interactions between the heparin chain and the target proteinases.

Heparin cofactor II

Heparin cofactor II, another serpin that occurs in plasma at micromolar concentration (for review, see Tollefsen, 1989), is identical (or closely related) to human Leuserpin 2 (hLS2; Ragg, 1986; Ragg and Preibisch, 1988). It consists of a single polypeptide chain composed of 480 amino acid residues, as deduced from cDNA analysis, and the gene has been localized to chromosome 22 (Ragg, 1986; Inhorn and Tollefsen, 1986; Blinder et al., 1988). An acidic domain near the N-terminus contains two sulphated tyrosine residues (Hortin et al., 1986), and peptides cleaved from this portion of heparin cofactor II by neutrophil proteinases have potent chemotactic activity (Pratt et al., 1990). Heparin cofactor II inactivates thrombin (Figure 1) by formation of a stable 1:1 complex, but does not react with factor Xa (Tollefsen et al., 1982; Parker and Tollefsen, 1985; Griffith et al., 1985a). It functions as a pseudosubstrate for thrombin, the Leu-444–Ser-445 reactive site peptide bond, located near the C-terminus, containing a leucine rather than the more typical arginine residue in the P1 position (Griffith et al., 1985b). While hereditary heparin cofactor II deficiency has been documented in a few patients suffering from thrombosis (Tran et al., 1985; Sié et al., 1985; Weisdorf and Edson, 1991), the role of heparin cofactor II in the prevention of thrombosis remains unclear (Bertina et al., 1987; Tollefsen, 1990; Toulon et al., 1991).

Because of the P1 leucine, the inhibition of thrombin is very slow in the absence of a glycosaminoglycan. Both dermatan sulphate and heparin increase the rate of inhibition of thrombin by heparin cofactor II more than 1000-fold, whereas chondroitin 4- or 6-sulphate have no effect (Tollefsen et al., 1983). According to Scully et al. (1986) chondroitin 4,6-disulphate (also known as CS-E) has some ability to promote the heparin cofactor II–thrombin reaction. An important step toward the characterization of the glycosaminoglycan-binding domain of heparin cofactor II was the discovery of an inhibitor variant ('HCII Oslo', isolated from a healthy Norwegian blood donor), capable of binding heparin but unable to interact with dermatan sulphate (Andersson et al., 1987). The aberrant properties of HCII Oslo could be traced to a nucleotide substitution in the codon for Arg-189, resulting in a histidine residue at this position (Blinder et al., 1989). This finding suggested that the positive charge on Arg-189 may be involved in the binding of dermatan sulphate but not of heparin, and provided the first evidence that the two glycosaminoglycans are bound to different sites. The further characterization of these sites relied essentially on site-directed mutagenesis in the normal heparin cofactor II cDNA, followed by analysis of the resulting expressed recombinant proteins. Mutations of Arg-189, Arg-192, and Arg-193, with loss of the corresponding positive charges, were found to specifically interfere with dermatan sulphate binding but did not affect the interaction with heparin. Conversely, mutation of Lys-173 decreased binding to heparin but not to dermatan sulphate. Finally, mutations of Arg-184 or Lys-185 affected the interactions of heparin cofactor II with both glycosaminoglycans. The binding sites for heparin and dermatan sulphate in heparin cofactor II thus appear to be overlapping but not identical (Blinder et al., 1989; Church et al., 1989; Blinder and Tollefsen, 1990; Ragg et al., 1990a,b; Whinna et al., 1991).

The effect of heparin on the inhibition of thrombin by heparin cofactor II appears to be governed mainly by overall charge (Hurst et al., 1983), without requirement for any specific oligosaccharide sequence akin to that involved in antithrombin binding (Griffith, 1983; Petitou et al., 1988c; Sié et al., 1988; Kim and Linhardt, 1989; Tollefsen et al., 1990). Moreover, the catalytic efficiency of heparin saccharides increased continuously with the molecular size of the chain, up to > 20 monosaccharide residues (Maimone and Tollefsen, 1988; Sié et al., 1988; Bray et al., 1989; Tollefsen et al., 1990). The heparin cofactor II-mediated anticoagulant activities of dermatan sulphate preparations cor-

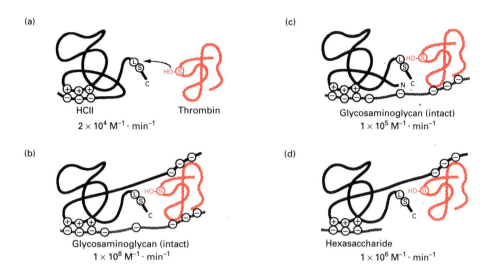

Figure 5   Model for the effects of glycosaminoglycans on the inhibition of thrombin by heparin cofactor II

The figure shows interactions of native or truncated heparin cofactor II with thrombin (red), in the presence or absence of glycosaminoglycan chains/oligosaccharides, and the corresponding approximative second-order rate constants for thrombin inhibition. (**a**) Native heparin cofactor II and thrombin in the absence of glycosaminoglycan; (**b**) native heparin cofactor II and thrombin in the presence of full-sized glycosaminoglycan chain; (**c**) heparin cofactor II, lacking the N-terminal domain, and thrombin in the presence of full-sized glycosaminoglycan chain; (**d**) native heparin cofactor II and thrombin in the presence of dermatan sulphate hexasaccharide. For further information see the text. (From Van Deerlin and Tollefsen, 1991, with permission from the authors.)

related with the degrees of sulphation (Akiyama et al., 1982; Munakata et al., 1987; Scully et al., 1988) and appeared to be promoted by the occurrence of disulphated -IdoA(2-OSO$_3$)-GalNAc(4-OSO$_3$)- disaccharide units (Scully et al., 1988). Among the oligosaccharides produced by partial chemical depolymerization of dermatan sulphate, the smallest fragment displaying both appreciable anticoagulant activity and affinity for heparin cofactor II was a dodecasaccharide (Tollefsen et al., 1986), suggesting that binding of dermatan sulphate to both heparin cofactor II and the thrombin target molecule was important for efficient proteinase inhibition. The smallest heparin cofactor II-binding high-affinity oligosaccharide, a hexasaccharide, generated in a similar fashion and isolated by affinity chromatography, was found to consist of three consecutive disulphated -IdoA(2-OSO$_3$)-GalNAc(4-OSO$_3$)- units (Maimone and Tollefsen, 1990; Figure 3b). Such clusters of disulphated disaccharide units are generally rare in dermatan sulphate, which thus seems to bind heparin cofactor II in a more selective manner than does heparin.

More detailed information regarding the role of glycosaminoglycans in heparin cofactor II-mediated thrombin inhibition was obtained through analysis of recombinant inhibitor mutants (Ragg et al., 1990a,b; VanDeerlin and Tollefsen, 1991). The N-terminal region contains two repeated acidic domains that are homologous to thrombin-binding domains in the C-terminal portion of hirudin and have been implicated as a site of interaction between heparin cofactor II and thrombin (Hortin et al., 1989). Deletion of this region did not affect the rate of inactivation of thrombin by heparin cofactor II in the absence of glycosaminoglycan but dramatically (by about three orders of magnitude) impeded the increase in reaction rate normally obtained in the presence of heparin or dermatan sulphate. This and other observations have been interpreted in terms of the model illustrated in Figure 5, which shows the effects of glycosaminoglycans on the interactions of native or truncated heparin cofactor II with thrombin. In the absence of glycosaminoglycan, the N-terminal, hirudin-like acidic domain of heparin cofactor II binds

intramolecularly to a glycosaminoglycan-binding site, and covalent complex formation with thrombin occurs at the basal rate (Figure 5a). Glycosaminoglycans displace the acidic region from the internal binding site, thereby enabling it to interact with the hirudin-binding site of thrombin (referred to as the 'anion-binding exosite' by Fenton, 1986); simultaneous binding of the glycosaminoglycan chain also to a glycosaminoglycan-binding site of thrombin results in maximal acceleration of the rate of thrombin inhibition (Figure 5b). Heparin cofactor II lacking the N-terminal region is unable to interact with the hirudin-binding site of thrombin, even in the presence of glycosaminoglycan, the resulting modest increase in reaction rate presumably being due to approximation of the two proteins bound to the glycosaminoglycan chain (Figure 5c). Oligosaccharides too short to simultaneously accommodate both proteins may nevertheless promote heparin cofactor II–thrombin complex formation (van Deerlin and Tollefsen, 1991; see also Bray et al., 1989) by displacing the acidic N-terminal region, which may then interact with thrombin (Figure 5d). Accordingly, deletion of the acidic region reduces thrombin inhibition to the basal rate in the presence of small oligosaccharides. Results in accord with these conclusions were obtained in studies on the inhibition of proteolytically modified thrombin by heparin cofactor II in the presence of heparin (Rogers et al., 1992). Ultimate confirmation of the model will require crystallization of heparin cofactor II and its complexes with saccharides and thrombin.

Protein C inhibitor

Activated protein C is a serine proteinase with anticoagulant properties (see below and Figures 1 and 6), not inhibitable by antithrombin (Suzuki et al., 1983). A 55 kDa inhibitor of protein Ca has been isolated from plasma (Marlar and Griffin, 1980; Canfield and Kisiel, 1982; Suzuki et al., 1983, 1984; Laurell and Stenflo, 1989). The rather low rate of protein C inactivation by this inhibitor was found to be increased ~30-fold in the presence of heparin (Suzuki et al., 1984). A non-heparin dependent inhibitor of protein C was identified as α1-antitrypsin (Heeb and Griffin, 1988). Further, proteins with protein C inhibitory activity, unaffected by heparin, may be released from platelets (Fay and Owen, 1989; Jane et al., 1989). A novel heparin-dependent inhibitor of protein C, $M_r \sim 50000$, was isolated from human urine (Geiger et al., 1988) and found to be immuno-logically distinct from other known proteinase inhibitors, in-cluding antithrombin and plasminogen activator inhibitor-1. This inhibitor was considered to be the urinary counterpart of the previously described plasma protein C inhibitor (Marlar and Griffin, 1980; Marlar et al., 1982; Suzuki et al., 1983, 1984) and to be functionally related to urinary urokinase inhibitor (Stump et al., 1986) which also shows heparin-dependence. In the absence of heparin, the urinary inhibitor inhibits both protein C and urokinase at similar rates, whereas in the presence of heparin the inhibitory activity is preferentially increased toward protein C (Geiger et al., 1988). The heparin-dependent plasma and urinary protein C inhibitors were immunologically identical and, more-over, closely related to the plasminogen activator inhibitor-3 (Heeb et al., 1987).

The heparin-dependent protein C inhibitor/plasminogen ac-tivator inhibitor-3 belongs to the serpin family and shows some sequence similarity to other members of the family (Suzuki et al., 1987a). The effect of heparin is strongly dependent on charge interactions and can be mimicked by dextran sulphate, the inhibitory activity increasing with both the molecular size and the degree of sulphation of the polymer (Suzuki, 1985; Kazama et al., 1987). These observations are in accord with a 'template' model by which binding of inhibitor and target enzyme to the same glycan chain are required for efficient protein C inhibition.

A three-dimensional model for protein C inhibitor, based on the structural homology with α1-antitrypsin, implicated a two-helix motif in glycosaminoglycan binding (Kuhn et al., 1990). The N-terminal A+ helix along with the internal H helix expose altogether 12 positive charges, sufficient to accommodate an oligosaccharide of 8–10 residues [a corresponding saccharide sequence in heparin will contain 12–16 negative charges (Casu et al., 1988)]. Monoclonal antibodies directed against the positively charged N-terminal peptide sequence (amino acid residues 5–19) prevented the interaction between protein C inhibitor and heparin as well as the stimulatory effect of heparin on protein C inactivation by the inhibitor, in agreement with the postulated role of the A+ helix (Meijers et al., 1988). The model showed no positive surface associated with the D helix, which is implicated in heparin binding to antithrombin (Carrell et al., 1987). Recent findings by Pratt and Church (1992) suggest that the ability of glycosaminoglycans to accelerate proteinase inhibition by protein C inhibitor depends on the formation of ternary inhibitor-glycosaminoglycan–enzyme complexes.

Other proteinase inhibitors

Infusion of heparin *in vivo* releases an anticoagulant protein into the blood from binding sites, presumably involving glycos-aminoglycan structures, at the endothelial cell surface (Sandset et al., 1988). This protein, currently known as tissue factor pathway inhibitor (formerly extrinsic pathway inhibitor or lipoprotein-associated coagulation inhibitor) is a multivalent proteinase inhibitor that directly inhibits factor Xa and, in a factor Xa-dependent fashion, the factor VIIa/tissue factor catalytic com-plex (Broze et al., 1988; see Figure 1). The different activities have been ascribed to two separate Kunitz-type proteinase inhibitor domains that are arranged in tandem along with a

third, similar domain with unknown function (Wun et al., 1988; Girard et al., 1989). The inhibitor binds to heparin–agarose and heparin reportedly promotes the inhibition of factor Xa by tissue factor pathway inhibitor (Broze et al., 1990). The interaction between heparin and this inhibitor is not entirely elucidated, but appears to involve a basic sequence close to the C-terminus of the molecule. Studies of full-length and truncated forms of recombi-nant tissue factor pathway inhibitor indicate that the basic N-terminal sequence is essential for the inhibitory potency (Broze et al., 1992).

Protease nexin-1 is a 43 kDa proteinase inhibitor which shares $\sim 30\%$ sequence identity with antithrombin (MacGrogan et al., 1988). It inactivates certain serine proteinases, such as thrombin, plasmin and urokinase, by forming a complex with their catalytic site serine residues. The complexes bind back to the cells, via a receptor for the protease nexin-1 moiety, and are rapidly internal-ized and degraded. This process provides a mechanism for inhibiting and clearing proteinases from the extracellular en-vironment (Low et al., 1981). Protease nexin-1 is synthesized and released by a variety of cultured cells including fibroblasts, smooth muscle cells, astrocytes and to a lesser extent, neurons (Baker et al., 1980; Laug et al., 1989; Rosenblatt et al., 1987; Wagner et al., 1991). Heparin binds to protease nexin-1 and increases the rate of thrombin inhibition by protease nexin-1 about 200-fold (Baker et al., 1980; Scott et al., 1985). Binding of protease nexin-1 to the extracellular matrix of fibroblasts, in particular to heparan sulphate chains, accelerates its reaction with thrombin, and at the same time modulates its target proteinase specificity such that it no longer inhibits urokinase or plasmin (Farrell and Cunningham, 1986, 1987; Farrell et al., 1988; Wagner et al., 1989; Cunningham et al., 1992). These observations suggest that thrombin is a likely physiological target of protease nexin-1 in the extracellular environment. Interestingly, protease nexin-1 shows neurite outgrowth-pro-moting activity, which has been ascribed to blocking of the thrombin-induced retraction of neurites (Gurwitz and Cunning-ham, 1988, 1990).

Physiological aspects

Although heparin displays potent anticoagulant properties, the extravascular location of the connective-tissue type mast cells that harbour this polysaccharide argues against a role as a natural anticoagulant/antithrombotic agent. Alternative func-tions proposed for anticoagulant mast-cell heparin include modu-lation of inflammatory reactions involving macrophage proco-agulant activities (Lindahl et al., 1989). However, vascular glycosaminoglycans, primarily heparan sulphate, are believed to bind thrombin and antithrombin and to catalyse the anti-thrombin–thrombin reaction (Lollar and Owen, 1980; Busch and Owen, 1982; Marcum et al., 1984; Marcum and Rosenberg, 1985; Stern et al., 1985). Binding of thrombin and antithrombin to endothelial heparan sulphate would permit the control of haemostasis at the blood-cell interface where the coagulation enzymes are generated. According to Stern et al. (1985) heparan sulphate provides $\sim 50 \times 10^3$ antithrombin-binding sites per endothelial cell. As noted in the section on antithrombin, heparan sulphate derived from vascular endothelial cells has been shown to contain the specific antithrombin-binding region. On the other hand, the alleged catalytic effect of heparan sulphate on the antithrombin–thrombin reaction could not be confirmed in recirculating rabbit Langendorff heart preparations (Lollar et al., 1984). Moreover, attempts to localize the antithrombin-binding sites at the cell surface indicated that the high-affinity heparan sulphate proteoglycans expressed in cell culture, or occurring in aorta, were preferentially facing the extracellular matrix compartment, thus not in direct contact with the blood (de Agostini et al., 1990). The anticoagulant/antithrombotic functions ascribed to heparan sulphate at the vascular cell surface remain to be conclusively established.

Despite the uncertainty concerning the details of the inhibitory process it seems clear that antithrombin has a major functional role in the normal control of haemostasis. Individuals with decreased levels of antithrombin in plasma are at risk of developing thrombosis (Egeberg, 1965). Other thrombin-inhibitory serpins, such as heparin cofactor II (Tollefsen et al., 1983; Church et al., 1991) or protease nexin-1 (Cunningham et al., 1992), are believed to have primarily extravascular functions, modulating the various activities (mitogenic, inflammatory, chemotactic, anti-neurite forming etc.) ascribed to thrombin in the extracellular matrix.

Interactions of intravascular glycosaminoglycans may influence the coagulation process in a number of ways. Formation of a ternary complex with fibrin and heparin (regardless of affinity for antithrombin) modulates the susceptibility of thrombin to antithrombin inhibition (Hogg and Jackson, 1990a,b). Fibrin monomers thus will protect thrombin from inhibition by antithrombin–heparin (Hogg and Jackson, 1989) but not from inhibition by heparin cofactor II in the presence of dermatan sulphate (Okwusidi et al., 1991). A number of protein ligands, including platelet factor 4, histidine-rich glycoprotein, vitronectin (S-protein) (Lane, 1989; Lane et al., 1984, 1986, 1987; Preissner and Jenne, 1991) and the enzymes lipoprotein lipase (Olivecrona and Bengtsson-Olivecrona, 1989) and extracellular superoxide dismutase C (Karlsson et al., 1988) may potentially sequester endothelial heparan sulphate from interaction with antithrombin. In fact, such blocking might explain the unexpected promoting effect of low-affinity heparin on the antithrombotic activity of heparin oligosaccharides having high affinity for antithrombin (Barrowcliffe et al., 1984); the low-affinity heparin would bind the endogenous protein ligands (except antithrombin) and thus clear the heparan sulphate chains. Heparin is also believed to release the tissue factor pathway inhibitor from endothelial cells (Sandset et al., 1988) and to promote its inhibition of the factor VII/tissue factor complex (see above). It has been proposed that tissue factor pathway inhibitor may account for a significant proportion of the anticoagulant/antithrombotic action of heparin in vivo, in spite of the fact that its concentration in plasma is only approximately one-thousandth that of antithrombin (Abildgaard, 1992). Finally, interactions of glycosaminoglycans that do not directly affect the actual coagulation process may instead modulate important associated systems. For instance, Ehrlich et al. (1991) found that heparin promotes the reaction between thrombin and plasminogen activator inhibitor 1, a serpin with regulatory function in the fibrinolytic pathway, thus leading to depletion of this inhibitor and increased fibrinolysis.

THROMBOMODULIN: A NOVEL PROTEOGLYCAN

Thrombomodulin is an integral membrane protein that provides high-affinity binding sites for thrombin at the luminal surface of the vascular endothelium (Esmon and Owen, 1981; Esmon et al., 1982a). Thrombomodulin forms a 1:1 molar complex with thrombin (Esmon et al., 1982b) which thus loses its procoagulant properties and instead acquires specific and complex anticoagulant activities (Figure 6). Thrombomodulin has a widespread distribution in the mammalian organism. It has been isolated from rabbit lung (Esmon et al., 1982b), bovine lung (Jakubowski et al., 1986; Suzuki et al., 1986), human placenta

(Salem et al., 1984b; Kurosawa and Aoki, 1985) and human platelets (Suzuki et al., 1988). Immunochemical staining demonstrated the presence of thrombomodulin antigen on the vascular endothelium of both large and small vessels, including lymphatics, on syncytiotrophoblasts of human placenta (Maruyama et al., 1985a; DeBault et al., 1986), and in certain extravascular compartments (Boffa et al., 1987). Smooth muscle cells in culture were shown to express thrombomodulin (Soff et al., 1991). While initial reports (Maruyama et al., 1985a; Ishii et al., 1986; Wen et al., 1987) suggested that thrombomodulin is absent from brain, a more recent study provided immunochemical evidence for the occurrence of thrombomodulin in cryosections of brain capillaries (Wong et al., 1991).

Thrombomodulin is an acidic protein with an isoelectric point around 4 (human species; Kurosawa and Aoki, 1985). It consists of a single polypeptide chain, with an apparent M_r of $(68–78) \times 10^3$ before reduction and $(74–105) \times 10^3$ after reduction. It has a marked tendency to form multimers (Winnard et al., 1989), and is heavily glycosylated with both N- and O-linked sugar substituents. The structure of thrombomodulin (Figure 7) was elucidated by cloning of the (intron-less) genes for the human, bovine, and murine proteins (Jackman et al., 1986, 1987; Wen et al., 1987; Suzuki et al., 1987a,b; Dittman et al., 1988). An N-terminal domain with structural homology to animal lectins (Patthy, 1988) is followed by six tandem-organized regions homologous to epidermal growth factor (EGF repeats), a serine/threonine-rich domain, a transmembrane hydrophobic domain and a cytoplasmic tail. The M_r calculated from thrombomodulin cDNA sequences is $\sim 60 \times 10^3$. The domain containing the EGF-like repeats appears to be the most conserved ($\sim 75\%$ identity) between species. Indeed, this region was implicated as the 'template' for the activation of protein C (see below) (Kurosawa et al., 1987, 1988; Stearns et al., 1989; Suzuki et al., 1989; Zushi et al., 1989). The thrombin-binding site, located within the fifth and sixth EGF-like repeats, involves in particular the amino acid sequence (Glu-408–Glu-426) of the fifth repeat, whereas protein C binds to thrombomodulin through the third and fourth EGF-like repeats, with a specific Ca^{2+}-dependent binding site within the latter region (Hayashi et al., 1990). The minimal unit of thrombomodulin capable of binding thrombin and accelerating protein C activation appears to be comprised by the fourth, fifth and sixth EGF-like repeats.

Effects of thrombomodulin on specific functions of thrombin

Binding of thrombin to (rabbit) thrombomodulin results in profound changes in the mode of action of the enzyme (Figure 6), especially in relation to macromolecular substrates (Esmon et al., 1982a; Hofsteenge et al., 1986; Bourin et al., 1988). Thrombin bound to thrombomodulin no longer cleaves fibrinogen (this inhibitory action will in the following be referred to as the direct anticoagulant activity of thrombomodulin; reaction 2 in Figure 6), nor is it able to activate factor V or platelets (Esmon et al., 1982a, 1983; Murata et al., 1988). Instead, thrombomodulin dramatically (~ 20000-fold) accelerates the rate by which thrombin activates protein C (protein C activation cofactor activity; reaction 1 in Figure 6). Moreover, thrombomodulin accelerates the rate by which thrombin is inhibited by antithrombin (antithrombin-dependent anticoagulant activity; reaction 3 in Figure 6) (Bourin et al., 1986; Hofsteenge et al., 1986; Preissner et al., 1987).

Bourin et al. (1986) noted that a freshly isolated preparation of rabbit thrombomodulin, characterized by its protein C activation cofactor activity, bound to an anion-exchange resin and thus displayed acidic properties. During storage this preparation

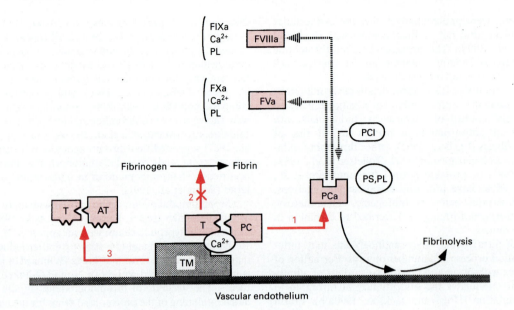

Figure 6   Anticoagulant pathways associated with thrombomodulin

Binding of thrombin (T) to thrombomodulin (TM) will influence the biological activities of thrombin, as indicated by arrows 1–3. (1) Acceleration of the activation of protein C by thrombin (referred to in the text as protein C activation cofactor activity of thrombomodulin). Activated protein C, a serine proteinase, specifically inactivates (broken arrows) the coagulation cofactors Va (involved in factor Xa-catalysed activation of prothrombin) and VIIIa (involved in factor IXa-catalysed activation of factor X). Inactivation of factors Va and VIIIa by activated protein C requires protein S (PS) and phospholipids (PL) as cofactors. According to Sakata et al. (1985) activated protein C may also promote fibrinolysis. (2) Inhibition of thrombin-catalysed cleavage of fibrinogen (direct anticoagulant activity of thrombomodulin). (3) Accelerated inactivation of thrombin by antithrombin (antithrombin-dependent anticoagulant activity of thrombomodulin).

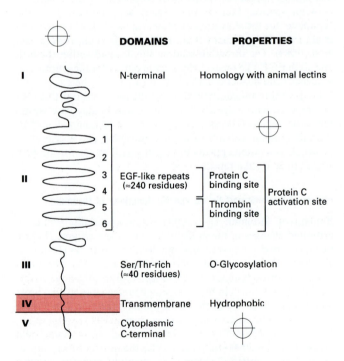

Figure 7   The molecular organization of thrombomodulin

The schematic representation is based on data relating to human thrombomodulin (Wen et al., 1987; Suzuki et al., 1987b). The molecule is composed of (I) an N-terminal domain, (II) a domain comprised of six EGF-like repeats, (III) a domain rich in serine and threonine residues, (IV) a trans-membrane hydrophobic region, and (V) a cytoplasmic C-terminal domain. The EGF-like regions 3–4 and 5–6 provide the major binding sites for protein C and thrombin, respectively.

gradually lost its acidic properties, apparently due to proteolytic cleavage. While the acidic form of thrombomodulin expressed all the three anticoagulant activities indicated above, the resulting non-acidic form showed protein C activation cofactor activity but no direct nor antithrombin-dependent anticoagulant activity. The two latter activities were tentatively ascribed to the presence of an acidic domain, containing a 'heparin-like' component, separated from the protein C activation site of thrombomodulin by a proteinase-sensitive region (see Figure 8). Before discussing the properties of this acidic domain the characteristics of the various biological activities 1–3 will be outlined.

Protein C activation cofactor activity

Protein C is a plasma protein with vitamin K-dependent anticoagulant properties that are expressed following activation by limited proteolytic cleavage, catalysed by thrombin. Activated protein C, a serine proteinase, inactivates the auxiliary co-agulation proteins, factors Va and VIIIa, hence preventing the generation of factor Xa and thrombin (Stenflo, 1976; Kisiel, 1979; Kisiel et al., 1977; Walker et al., 1979; Dahlbäck and Stenflo, 1980; Owen and Esmon, 1981; Stenflo and Fernlund, 1982; Marlar et al., 1982; Esmon, 1987). Activated protein C requires protein S, another vitamin K-dependent protein, as a cofactor in order to inactivate factors Va and VIIIa (Walker, 1980; Walker et al., 1987). The anticoagulant protein C pathway is outlined in Figure 6. A crucial physiological role of this pathway is indicated by the strong correlation between congenital deficiencies of protein C or protein S and recurring thrombo-embolic episodes (High, 1988). Thrombomodulin has a key role in promoting the activation of protein C by thrombin. Accordingly, injection of recombinant human thrombomodulin into mice was found to prevent thrombin-induced thrombo-

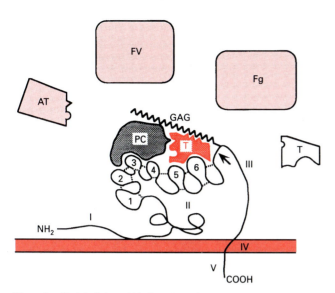

Figure 8   Model of the rabbit thrombomodulin proteoglycan

The depicted mode of interaction of thrombomodulin with the endothelial cell surface shows the thrombin molecule to be bound 'on top of' thrombomodulin (Lu et al., 1989). The EGF-like structures are stabilized by intrachain disulphide bridges that appear to be essential for expression of all the various anticoagulant activities of thrombomodulin. The glycosaminoglycan (GAG) chain is shown to be inserted in domain III which provides potential O-glycosylation sites (Figure 7; see also Figure 9 for identification of potential glycosaminoglycan-attachment sites). The arrow indicates a site of proteolytic cleavage that will generate a nonacidic form of thrombomodulin (Bourin et al., 1986). The model is intended to illustrate that the polysaccharide component of thrombomodulin is not essential for the activation of protein C (PC) by thrombin, whereas it is required to promote the inhibition of thrombin (T) by antithrombin (AT) and to prevent the cleavage of fibrinogen (Fg) as well as the activation of factor V (FV) by thrombin. Binding of thrombin to thrombomodulin will induce a conformational change in the catalytic centre of the enzyme (Musci et al., 1988) and unmask a site capable of interaction with the glycosaminoglycan chain. This secondary interaction with the glycosaminoglycan chain in turn induces a further conformational change of the thrombin molecule, which facilitates its reaction with antithrombin. For further information see the text. (Modified from Bourin and Lindahl, 1990.)

embolism (Gomi et al., 1990). No clinical condition associated with thrombomodulin deficiency has yet been described.

Direct anticoagulant activity

The major, direct procoagulant function of thrombin is to catalyse the cleavage of fibrinogen to form the fibrin clot. Addition of rabbit thrombomodulin to thrombin, in amounts equal to or slightly exceeding molar equivalency, was found to eliminate the clotting ability of thrombin (Esmon et al., 1982a). Thrombomodulin isolated from other species varied with regard to direct anticoagulant activity (Jakubowski et al., 1986; Maruyama et al., 1985). As mentioned above, this activity was associated with the occurrence of an acidic domain in the thrombomodulin molecule. Lack of activity thus could be due to the absence of such a component, possibly a result of proteolytic cleavage (see above), or to the presence of a 'neutralizing' basic protein (see below).

Antithrombin-dependent anticoagulant activity

Rabbit thrombomodulin induced a 4–8-fold acceleration of the rate of inhibition of thrombin by antithrombin (Bourin et al.,

1986; Hofsteenge et al., 1986; Preissner et al., 1987). The process was saturable with regard to thrombomodulin (Hofsteenge et al., 1986; Bourin and Lindahl, 1990), irrespective of antithrombin concentration, the maximal rate of thrombin inhibition being achieved at a molar ratio of thrombomodulin/thrombin ≈ 1. Similar conditions were required to completely inhibit the thrombin-catalysed cleavage of fibrinogen (Esmon et al., 1982a) and to maximally stimulate protein C activation (Esmon et al., 1982b). Once inactivated, the bound thrombin is released from thrombomodulin, which is then ready to accommodate another thrombin molecule (Bourin et al., 1988). Albeit less efficient than heparin, thrombomodulin thus acts as a catalyst for the inhibition of thrombin by antithrombin. Hirahara et al. (1990) described the formation of complexes between either rabbit or human thrombomodulin and antithrombin. On the other hand, thrombomodulin failed to bind to an antithrombin–Sepharose matrix (Hofsteenge et al., 1986; Preissner et al., 1987).

Contrary to rabbit thrombomodulin, neither bovine lung thrombomodulin (Jakubowski et al., 1986; Suzuki et al., 1986) nor human placenta thrombomodulin (Hirahara et al., 1990) were found to accelerate the inhibition of thrombin by antithrombin. However, Preissner et al. (1990) observed that preparations of human thrombomodulin were contaminated by vitronectin (S-protein), one of the so-called heparin-neutralizing proteins that would be likely to interfere with any glycosaminoglycan-dependent phenomenon such as the antithrombin-dependent anticoagulant activity (see below). Another seemingly contradictory feature was noted in studies on the effects of exogenous heparin on the thrombomodulin–antithrombin–thrombin system (Bourin, 1989). Whereas thrombomodulin promoted the inactivation of thrombin by antithrombin, it protected the bound thrombin against the rapid reaction with antithrombin normally induced by exogenous heparin.

The effects of thrombomodulin on the inactivation of thrombin by heparin cofactor II were again inconsistent. Rabbit thrombomodulin did not accelerate the reaction but, in fact, protected the thrombin from inactivation by heparin cofactor II (Koyama et al., 1991). Human recombinant thrombomodulin, on the other hand, promoted both the antithrombin–thrombin and the heparin cofactor II–thrombin reactions (Koyama et al., 1991). Moreover, the recombinant thrombomodulin protected thrombin against fast inhibition by heparin–antithrombin as well as by dermatan sulphate–heparin cofactor II complexes (Koyama et al., 1991).

The thrombomodulin proteoglycan

Functional domains of thrombomodulin

Experiments to be described below indicate that the acidic properties associated with some of the modulatory activities of thrombomodulin are due to a (presumably single) chondroitin sulphate (or dermatan sulphate) chain. The mode of interaction of this glycosaminoglycan component with thrombomodulin-bound thrombin is depicted by the model shown in Figure 8, which suggests that the diverse activities of thrombomodulin all relate to the same thrombin molecule, bound at the protein C activation site (Bourin et al., 1988). This proposal was based in particular on the effects of monoclonal antibodies that were found to abrogate the protein C activation cofactor activity, and at the same time reverse the antithrombin-dependent as well as the direct anticoagulant activities. Binding of the glycosaminoglycan chain to the thrombin appeared to be prerequisite to the two latter activities, since these were eliminated in the presence of a synthetic polyamine, Polybrene (Bourin et al., 1986), or of proteins such as platelet factor 4 and S-protein (vitronectin)

```
          481                               *    ↓              499
Human     - - D S G K V         D G G      D S G S G E P P P S P
Bovine    - - D P T Q V N E E R G T P E D    Y G G S G E P P V S P
Mouse     - - D P I P V         R E D T K E E E G S G E P P V S P

          500                    516
Human     T P G S T L T P   P A   V G L V H S G - -
Bovine    T P G A T A R P S P A P A G P L H S G - -
Mouse     T P G S P T G P   P S   A R P V H S G - -
```

Figure 9   Potential glycosaminoglycan-attachment sites in thrombomodulin from various species

Alignment of amino acid (represented by the one-letter code) sequences from domain III of thrombomodulin from different species, as deduced from the corresponding cDNA structures. The -Ser-Gly- sequences representing potential glycosaminoglycan-attachment sites are shown in red. Ser-490 (marked by an asterisk) in human thrombomodulin is part of the 'conventional' -Ser-Gly-Xaa-Gly- sequence implicated as recognition structure for the xylosyltransferase that initiates formation of glycosaminoglycan chains (Bourdon et al., 1987). Alternatively, Ser-492 (arrow) which occurs within the conserved -Gly-Ser-Gly-Glu- sequence might provide a glycosaminoglycan-attachment site common to the three different thrombomodulin species. The numbering of the amino acid residues 481–516 of human thrombomodulin is according to Wen et al. (1987). The sequence alignments are according to Suzuki et al. (1987b) for human and bovine thrombomodulin and according to Dittman and Majerus (1989) for human and murine thrombomodulin.

which are known to interact with glycosaminoglycan chains (Bourin et al., 1988; see also Preissner et al., 1990; Koyama et al., 1991). None of these compounds had any apparent effect on the protein C activation cofactor activity. These findings are in accord with the initial observation (see above) of an acidic form of thrombomodulin which displayed all three different anticoagulant activities and a nonacidic form that promoted protein C activation but lacked the antithrombin-dependent and direct anticoagulant activities (Bourin et al., 1986). Moreover, non-acidic fragments of rabbit thrombomodulin generated by proteolysis or cyanogen bromide treatment showed protein C activation cofactor activity but only minimal direct anticoagulant activity (Kurosawa et al., 1987; Stearns et al., 1989). These fragments contained region II, composed of the EGF-like repeats, but lacked region III which contains potential O-glycosylation sites (see Figures 7 and 8).

Characterization of the glycosaminoglycan component

Conclusive evidence as to the nature of the acidic domain of rabbit thrombomodulin was obtained by digestion with the bacterial eliminase, chondroitinase ABC (Bourin, 1989; Bourin et al., 1988; Bourin and Lindahl, 1990). Such treatment reduced the apparent M_r of unreduced thrombomodulin on SDS/PAGE from ~90000 to ~74000, along with loss of the direct and antithrombin-dependent anticoagulant activities (see below). A total of ~180 μg of polysaccharide was recovered by ion-exchange chromatography, following release by alkaline β-elimination from 6 mg of rabbit thrombomodulin (Bourin et al., 1990). The isolated product was subjected to partial N-deacetylation by hydrazinolysis and was then re-N-[³H]acetylated by treatment with [³H]acetic anhydride. The resulting ³H-labelled product showed an M_r of $(10–12) \times 10^3$ on gel chromatography and was susceptible to digestion by chondroitinase ABC or AC and by testicular hyaluronidase (Bourin et al., 1990). Analysis of digestion products by use of the radiolabel enabled a fairly detailed structural characterization of the molecule. Most of the internal region of the glycosaminoglycan chain consisted of monosulphated disaccharide units, the sulphate groups being located at C-4 or C-6 of the N-acetylgalactosamine residues. By

contrast, the nonreducing end of the chain showed an unusual structure, with a terminal tetrasulphated [GalNAc(4,6-di-OSO₃)-GlcA-GalNAc(4,6-di-OSO₃)-] trisaccharide sequence (Bourin et al., 1990). This local accumulation of negative charges may contribute to the functional properties of the thrombomodulin-bound glycosaminoglycan component.

The proteoglycan nature of rabbit thrombomodulin was confirmed by the isolation of ³⁵S-labelled thrombomodulin from rabbit heart endothelial cells that had been cultured in the presence of Na₂³⁵SO₄ (Bourin et al., 1990). Chondroitinase digestion indicated that virtually all of the label had been incorporated into a galactosaminoglycan chain, composed of essentially 4/6-monosulphated disaccharide units and about 8 % disulphated disaccharide units. Analysis of the glycosaminoglycan component of human recombinant thrombomodulin indicated mainly 4-mono-O-sulphated disaccharide units (Nawa et al., 1990).

Amino-acid sequences within region III of thrombomodulin from different species were compared with regard to potential glycosaminoglycan attachment sites (Figure 9). The tetrapeptide structure -Ser-Gly-Xaa-Gly- (where Xaa may be any amino acid), previously postulated to constitute a consensus sequence for glycosaminoglycan attachment (Bourdon et al., 1987), was identified within this domain (residues 490–493), Ser-490 providing a likely potential acceptor site for the xylosyltransferase. However, this sequence was not present in bovine or murine thrombomodulin. Alternatively, -Ser-Gly- structures adjacent to acidic amino acid residues may be substituted by glycosaminoglycan chains (Rodén et al., 1985; see also Fransson, 1989; Gallagher, 1989). This requirement is fulfilled by the sequence -Gly-Ser-Gly-Glu-, residues 491–494 in human thrombomodulin, which is conserved in all three thrombomodulin species. The constituent serine unit (residue 492 in human thrombomodulin) thus may provide a glycosaminoglycan attachment site that is common to all three thrombomodulin species (see also Parkinson et al., 1992a).

Functional role of the glycosaminoglycan component

Digestion of rabbit thrombomodulin with chondroitinase ABC had no apparent effect on the protein C activation cofactor activity (Figure 10a; however, see recombinant thrombomodulin below) but virtually eliminated both the direct (Figure 10b) and the antithrombin-dependent (Figure 10c) anticoagulant activities (Bourin, 1989; Bourin et al., 1988; Bourin and Lindahl, 1990; Preissner et al., 1990). Moreover, chondroitinase-digested thrombomodulin had lost the ability of the native compound (Esmon et al., 1982a) to prevent the activation of factor V by thrombin (Bourin and Lindahl, 1990). Similarly, the glycosaminoglycan component of thrombomodulin was implicated in the inhibition of other thrombin-dependent phenomena, including activation of platelets and endothelial cells (Parkinson et al., 1991a,b). Finally, the presence or absence of the glycosaminoglycan component was found to profoundly influence the effects of added polysaccharides (Bourin, 1989; Koyama et al., 1991). In contrast to accelerating the slow thrombin inhibition by antithrombin and heparin cofactor II in the absence of exogenous glycosaminoglycans, the chondroitin (dermatan) sulphate component of thrombomodulin was found to prevent the rapid thrombin inhibition by antithrombin in the presence of added heparin and by heparin cofactor II in the presence of dermatan sulphate. Presumably, the interaction between the thrombomodulin-bound glycosaminoglycan chain and thrombin will preclude formation of the ternary serpin–glycosaminoglycan–thrombin complexes that are required to mediate the effects of

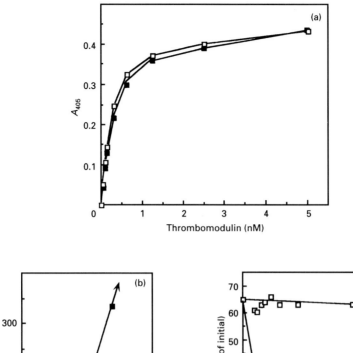

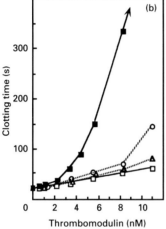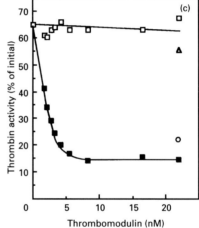

Figure 10   Effects of glycosaminoglycan-degrading enzymes on the various anticoagulant activities of rabbit thrombomodulin

Thrombomodulin was analysed for (**a**) protein C activation cofactor activity (determination of activated protein C formed, as measured using a chromogenic substrate), (**b**) direct anticoagulant activity (ability to prolong the clotting of fibrinogen; the arrow indicates that samples remained non-coagulable for >10 min); (**c**) antithrombin-dependent anticoagulant activity (determination of residual thrombin activity following incubation with antithrombin). The thrombomodulin was tested at the concentrations indicated, either in native form (■; thrombomodulin incubated with chondroitinase buffer but without enzyme), or after digestion with chondroitinase ABC (□), chondroitinase AC (○) or testicular hyaluronidase (△). The less pronounced effect of chondroitinase AC, as compared with chondroitinase ABC, might suggest the occurrence of some IdoA (in addition to GlcA) units, but could also reflect the more pronounced exo-enzyme character of the former enzyme (Fransson, 1985). (From Bourin and Lindahl, 1990.)

the exogenous glycosaminoglycans (Bourin, 1989; Parkinson et al., 1992b).

More detailed information regarding the influence of the thrombomodulin-bound chondroitin sulphate chain on thrombin function was obtained through studies of recombinant thrombomodulin (Parkinson et al., 1990a, 1992a,b; Nawa et al., 1990). Distinct soluble 'glycoforms' of deletion mutants of recombinant human thrombomodulin were found to differ with regard to the presence or absence of the single glycosaminoglycan chain. The major anticoagulant activities of the two forms differed as predicted from previous studies on native and chondroitinase-digested rabbit thrombomodulin; moreover, the functional properties of the glycosaminoglycan-substituted form could be converted into those of the unsubstituted form by such digestion. In addition, the two forms differed significantly with regard to affinity for thrombin, maximal rates of protein C activation by the thrombin–thrombomodulin complex and optimal Ca^{2+} ion concentration for protein C activation (Parkinson et al., 1990a,

1992a,b). The glycosylation state of thrombomodulin *in vivo*, particularly with regard to the occurrence of chondroitin sulphate, is thus likely to have profound functional effects. In accord with this view, recent studies with cultured human and bovine endothelial cells showed altered thrombin affinity and kinetics of protein C activation following treatment of the cells with a β-D-xyloside, known to inhibit glycosaminoglycan initiation on proteoglycan core proteins (Parkinson et al., 1990b). Further studies are required to define the mechanisms that control the posttranslational modifications in thrombomodulin biosynthesis, in particular the formation and elaboration of glycosaminoglycan structures.

The studies outlined above suggest that the properties of thrombomodulin as a modulator of thrombin function rely heavily on its ability to efficiently compete with various ligands, such as fibrinogen and certain thrombin receptors, for binding to thrombin (Figure 8). A glycosaminoglycan chain in the O-glycosylation region appears to be critically involved in the

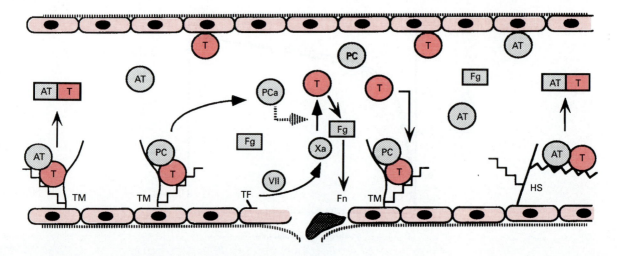

Figure 11   Regulation of blood coagulation at the vascular endothelial surface

Damage to the endothelial surface (exposure of subendothelial structures) leads to expression of tissue factor (TF), activation of the coagulation cascade and finally to fibrin (Fn) deposition. Mechanisms are activated to ensure that clot formation be restricted to the area of vascular damage. Surplus thrombin (T) is either directly inactivated by antithrombin (AT) bound to heparan sulphate (HS) proteoglycans or bound to thrombomodulin (TM), another endothelial proteoglycan. Due to the shielding effect of the constituent glycosaminoglycan chain, the thrombomodulin-bound thrombin is unable to clot fibrinogen (Fg) and to amplify coagulation by positive feed-back mechanisms. Instead, it activates the anticoagulant protein C (PC) pathway, thus leading to down-regulation of the coagulation cascade. The glycosaminoglycan component of thrombomodulin also promotes the continuous inactivation of bound thrombin, which is subsequently released in complex with antithrombin, and replaced by new thrombin molecules. This mechanism will ensure that activation of protein C is readily terminated once the excess thrombin has been cleared.

inhibition of the interactions of thrombin with procoagulant substrates and cellular receptors. The precise mechanism behind these effects is still unclear. A number of thrombin ligands, including thrombomodulin, fibrinogen (Wu et al., 1991), hirudin (Hofsteenge and Stone, 1987; Rydel et al., 1990) and the platelet/endothelial cell receptor for thrombin (Vu et al., 1991) seem to interact with a positively charged surface region located some distance from the thrombin active site, the so-called anion-binding exosite of the proteinase (see Fenton, 1986). Parkinson et al. (1992b) proposed that the acidic glycosaminoglycan component of thrombomodulin interferes with these interactions by binding to and thus blocking the exosite. While this possibility appears attractive, there are reports suggesting that other (EGF-like peptide) regions of the thrombomodulin molecule may bind to the thrombin exosite (see e.g. Hortin and Benutto, 1990). On the other hand, thrombin apparently contains more than one such site (Rogers et al., 1992). The precise nature of the glycosaminoglycan–thrombin interaction remains to be defined. If, in fact, such interaction serves to increase the overall affinity of thrombin for thrombomodulin then some of the anticoagulant effects ascribed to the glycosaminoglycan component could simply be due to shifting the equilibrium of free versus thrombomodulin-bound thrombin.

The complex interactions between thrombomodulin and thrombin become particularly intriguing in relation to the antithrombin-dependent anticoagulant activity. This activity depends on the presence of the thrombomodulin-bound glycosaminoglycan chain. The isolated labelled polysaccharide showed no appreciable affinity for either immobilized thrombin nor antithrombin (Bourin et al., 1990). Although it expressed weak antithrombin-dependent thrombin inhibition, approximately equal to that observed with the 'oversulphated' chondroitin sulphate-E (see Scully et al., 1986), this activity was about 20-fold lower, on a molar basis, than that of the corresponding thrombomodulin-bound glycosaminoglycan chain (Bourin et al., 1990). Possible reasons for this discrepancy have been considered in constructing the model shown in Figure 8. It is proposed that

thrombin first binds to the core protein of the thrombomodulin proteoglycan. The resulting conformational change of the thrombin provides for efficient protein C activation, but also promotes binding of the glycosaminoglycan chain to the thrombin molecule (possibly at an anion-binding exosite; see above). The dual effects of binding to both the protein and the glycosaminoglycan components induces a further conformational change of the thrombin which thereby becomes more amenable to inhibition by antithrombin (see Figure 8). The antithrombin-dependent anticoagulant activity of thrombomodulin seems to present an unprecedented case of functional co-operation between the carbohydrate and protein moieties of a glycoconjugate.

AN INTEGRATED VIEW

An integrated view on the role(s) of glycosaminoglycans in the regulation of blood coagulation is presented in Figure 11. By necessity, the concepts are simplified and a number of potentially significant aspects have been deliberately ignored. The scheme is focused on the major roles of endogenous, intravascular proteoglycans and does not account for effects of exogenous glycosaminoglycans such as heparin or dermatan sulphate, used as drugs. Moreover, heparin cofactor II is considered to have mainly an extravascular function and thus will not be considered.

Efficient inhibition of a 'cascade' mechanism, such as that involved in blood coagulation, should theoretically aim for 'upstream' targets early in the sequence. In fact, however, inhibition of thrombin, the last proteinase to be activated in the process, appears to be most critical for control of haemostasis (Ofosu et al., 1985, 1987, 1990; Béguin et al., 1988). The probable reason for this seeming anomaly is that thrombin, in addition to clotting fibrinogen, also catalyses two reactions that lead to amplification of the coagulation cascade, i.e. activation of factors V and VIII (Figure 1). In addition, Gailani and Broze (1991) recently proposed that thrombin is a major activator of factor XI (Figure 1), thus initiating the intrinsic pathway of coagulation. Suppression of these thrombin-dependent amplification reactions delays prothrombin activation and is believed to explain, in large part, the antithrombotic effect of heparin (Ofosu et al., 1989).

The occurrence at the endothelial cell surface of proteoglycans containing heparan sulphate chains with high affinity for antithrombin (see Kojima et al., 1992a,b) provides a means of inactivating not only thrombin, but also the other serine proteinases of the coagulation cascade, by complex formation with antithrombin (Figure 11). The potential of this system depends in particular on the regulation, still poorly understood (Kusche et al., 1990; Rosenberg and de Agostini, 1992), of the biosynthetic machinery involved in generating the specific antithrombin-binding pentasaccharide sequence. Furthermore, preferential inactivation of certain proteinase species may conceivably be achieved by modulating the size and charge properties of the individual heparan sulphate chains (see above).

A more complex mechanism, also dependent on glycosaminoglycan involvement, for inhibition of coagulation is utilized by the thrombomodulin system (Figure 11), which is specifically designed to modulate thrombin function. Once bound to thrombomodulin, virtually all known procoagulant functions of thrombin are switched off, the catalytic power of the enzyme instead being channelled into generation of the anticoagulant protein Ca. Most of the thrombin formed, at least in the microcirculation, is believed to be bound to thrombomodulin (Busch and Owen, 1982) and McIntosh and Owen (1987) proposed that complex formation of thrombin with thrombomodulin represents the major anticoagulant mechanism normally in operation at the intact endothelial surface (see Figure 11). The chondroitin sulphate component of thrombomodulin contributes to the anticoagulant function by shielding thrombin from procoagulant interactions with various macromolecular substrates (see detailed discussion above). In addition, given the large proportion of thrombomodulin-bound thrombin, inactivation of such thrombin molecules by antithrombin, promoted by the chondroitin sulphate chain (Figure 11), may well constitute the functionally predominant pathway for elimination of thrombin from the circulation. On the other hand, this effect of the glycosaminoglycan chain may seem paradoxical, since it will also terminate the anticoagulant action (protein C activation) of the target thrombin. However, once inactivated, the thrombin, complexed with antithrombin, will be released from the thrombomodulin and give way for another thrombin molecule (Bourin et al., 1988). The functional purpose of this process conceivably relates to the overall regulation of thrombomodulin action. Activation of protein C should continue only as long as excess thrombin is being generated. The antithrombin-dependent anticoagulant action thus may be perceived as a mechanism to ascertain that protein C activation is terminated without undue delay.

REFERENCES

Abildgaard, U. (1968) Scand. J. Clin. Lab. Invest. **21**, 89–91

Abildgaard, U. (1992) in Heparin and Related Polysaccharides (Lane, D. A., Björk, I. and Lindahl, U., eds.), pp. 199–204, Plenum, New York

Akiyama, F., Seno, N. and Yoshida, K. (1982) Tohoku J. Exp. Med. **136**, 359–365

Andersson, L.-O., Barrowcliffe, T. W., Holmer, E., Johnson, E. A. and Sims, G. E. C. (1976) Thromb. Res. **9**, 575–583

Andersson, T. R., Larsen, M. L. and Abildgaard, U, (1987) Thromb. Res. **47**, 243–248

Asakura, S., Hirata, H., Okazaki, H., Hashimoto-Gotoh, T. and Matsuda, M. (1990) J. Biol. Chem. **265**, 5135–5138

Atha, D. H., Lormeau, J.-C., Petitou, M., Rosenberg, R. D. and Choay, J. (1985) Biochemistry **24**, 6723–6729

Atha, D. H., Stephens, A. W. and Rosenberg, R. D. (1984) Proc. Natl. Acad. Sci. U. S. A. **81**, 1030–1034

Baker, J. B., Low, D. A., Simmer, R. L. and Cunningham, D. D. (1980) Cell **21**, 37–45

Barrowcliffe, T. W., Merton, R. E., Havercroft, S. J., Thunberg, L., Lindahl, U. and Thomas, D. P. (1984) Thromb. Res. **34**, 125–133

Béguin, S., Lindhout, T. and Hemker, H. C. (1988) Thromb. Haemostasis **60**, 457–462

Beresford, C. H. and Owen, M. C. (1990) Int. J. Biochem. **22**, 121–128

Bertina, R. M., Van Der Linden, I. K., Engeser, L., Muller, H. P. and Brommer, E. J. P. (1987) Thromb. Haemostasis **57**, 196–200

Bienkowski, M. J. and Conrad, H. E. (1985) J. Biol. Chem. **260**, 356–365

Björk, I. and Danielsson, Å. (1986) in Proteinase Inhibitors (Barrett, A. J. and Salvesen, G., eds.), pp. 489–513, Elsevier, Amsterdam

Björk, I. and Fish, W. W. (1982) J. Biol. Chem. **257**, 9487–9493

Björk, I. and Lindahl, U. (1982) Mol. Cell. Biochem. **48**, 161–182

Björk, I. and Nordenman, B. (1976) Eur. J. Biochem. **68**, 507–511

Björk, I., Olson, S. T. and Shore, J. D. (1989a) in Heparin; Chemical and Biological Properties, Clinical Applications (Lane, D. A. and Lindahl, U., eds.), pp. 229–255, Edward Arnold, London

Björk, I., Olson, S. T., Sheffer, R. G. and Shore, J. D. (1989b) Biochemistry **28**, 1213–1221

Björk, I., Ylinenjärvi, K. Olson, S. T. and Bock, P. E. (1992) J. Biol. Chem. **267**, 1976–1982

Blackburn, M. N., Smith, R. L., Carson, J. and Sibley, C. C. (1984) J. Biol. Chem. **259**, 939–941

Blinder, M. A., Marasa, J. C., Reynolds, C. H., Deaven, L. L. and Tollefsen, D. M. (1988) Biochemistry **27**, 752–759

Blinder, M. A., Andersson, T. R., Abildgaard, U. and Tollefsen, D. M. (1989) J. Biol. Chem. **264**, 5128–5133

Blinder, M. A. and Tollefsen, D. M. (1990) J. Biol. Chem. **265**, 286–291

Boffa, M.-C., Burke, K., and Haudenschild, C. C. (1987) J. Histochem. Cytochem. **35**, 1267–1276

Borg, J. Y., Brennan, S. O., Carrell, R. W., George, P., Perry, D. J. and Shaw, J. (1990) FEBS Lett. **266**, 163–166

Bourdon, M. A. Krusius, T., Campbell, S., Schwartz, N. B. and Ruoslahti, E. (1987) Proc. Natl. Acad. Sci. U. S. A. **84**, 3194–3198

Bourin, M.-C. (1989) Thromb. Res. **54**, 27–39

Bourin, M.-C., Boffa, M.-C., Björk, I. and Lindahl, U. (1986) Proc. Natl. Acad. Sci. U. S. A. **83**, 5924–5928

Bourin, M.-C., Öhlin, A.-K., Lane, D. A., Stenflo, J. and Lindahl, U. (1988) J. Biol. Chem. **263**, 8044–8052

Bourin, M.-C. and Lindahl, U. (1990) Biochem. J. **270**, 419–425

Bourin, M.-C., Lundgren-Åkerlund, E. and Lindahl, U. (1990) J. Biol. Chem. **265**, 15424–15431

Bray, B., Lane, D. A., Freyssinet, J. M., Pejler, G. and Lindahl, U. (1989) Biochem. J. **262**, 225–232

Brennan, S. O., Borg, J. Y., George, P. M., Soria, C., Soria, J., Caen, J. and Carrell, R. W. (1988) FEBS Lett. **237**, 118–122

Broze, G. J., Jr., Warren, L. A., Novotny, W. F., Higuchi, D. A., Girard, J. J. and Miletich, J. P. (1988) Blood **71**, 335–343

Broze, G. J., Jr., Girard, T. J. and Novotny, W. F. (1990) Biochemistry **29**, 7539–7546

Broze, G. J., Jr., Wesselschmidt, R., Higuchi, D., Girard, T., Likert, K., MacPhail, L. and Wun, T.-C. (1992) in Heparin and Related Polysaccharides (Lane, D. A., Björk, I. and Lindahl, U., eds.), pp. 189–197, Plenum, New York

Busch, C. (1984) in Biology of Endothelial Cells (Jaffe, E. A., ed.), pp. 178–188, Martinus Nijhoff, The Hague

Busch, C. and Owen, W. G. (1982) J. Clin. Invest. **69**, 726–729

Canfield, W. M. and Kisiel, W. (1982) J. Clin. Invest. **70**, 1260–1272

Carrell, R. W., Christey, P. B. and Boswell, D. R. (1987) in Thrombosis and Haemostasis (Verstraete, M., Vermylen, J., Lijnen, H. R. and Arnout, J., eds.) pp. 1–15, Leuven University Press, Leuven

Caso, R., Lane, D. A., Thompson, E. A., Olds, R. J., Thein, S. L., Pagnico, M., Blench, I., Morris, H. R., Freyssinet, J. M., Aiach, M., Rodeghiero, F. and Finazzi, G. (1991) Br. J. Haematol. **77**, 87–92

Casu, B., Oreste, P., Torri, G., Zoppetti, G., Choay, J., Lormeau, J.-C., Petitou, M. and Sinaÿ, P. (1981) Biochem. J. **197**, 599–609

Casu, B., Petitou, M., Provascoli, M. and Sinaÿ, P. (1988) Trends Biochem. Sci. **13**, 221–225

Chang, J. Y. (1989) J. Biol. Chem. **264**, 3111–3115

Chang, J. Y. and Tran, T. H. (1986) J. Biol. Chem. **261**, 1174–1176

Choay, J., Petitou, M., Lormeau, J.-C., Sinaÿ, P., Casu, B. and Gatti, G. (1983) Biochem. Biophys. Res. Commun. **116**, 492–499

Church, F. C., Meade, J. B., Treanor, R. T. and Whinna, H. C. (1989) J. Biol. Chem. **264**, 3618–3623

Church, F. C., Pratt C. W. and Hoffman M. (1991) J. Biol. Chem. **266**, 704–709

Colburn, P. and Buonassisi, V. (1982) Biochem. Biophys. Res. Commun. **104**, 220–227

Cunningham, D. D., Wagner, S. L. and Farrell, D. H. (1992) in Heparin and Related Polysaccharides (Lane, D. A., Björk, I. and Lindahl, U., eds.), pp. 297–306, Plenum, New York

Dahlbäck, B. (1991) Thromb. Haemostasis **66**, 49–61

Dahlbäck, B. and Stenflo, J. (1980) Eur. J. Biochem. **107**, 331–335

Danielsson, Å., Raub, E., Lindahl, U., and Björk, I. (1986) J. Biol. Chem. **261**, 15467–15473

de Agostini, A. I., Watkins, S. C., Slayter, H. S., Youssoufian, H. and Rosenberg, R. D. (1990) J. Cell Biol. **111**, 1293–1304

DeBault, L. E., Esmon, N. L., Olson, J. R. and Esmon, C. T. (1986) Lab. Invest. **54**, 172–178

Devraj-Kizuk, R., Chui, D. H. K., Prochownik, E. V., Carter, C. J., Ofosu, F. A. and Blajchman, M. A. (1988) Blood **72**, 1518–1523

Dittman, W. A., Kumada, T., Sadler, J. E. and Majerus, P. W. (1988) J. Biol. Chem. **263**, 15815–15822

Dittman, W. A. and Majerus, P. W. (1989) Nucleic Acids Res. **17**, 802

Edge, A. S. B. and Spiro, R. G. (1990) J. Biol. Chem. **265**, 15874–15881

Egeberg, O. (1965) Thromb. Diath. Haemorrh. **13**, 516–530

Ehrlich, H. J., Keijer, J., Preissner, K. T., Gebbink, R. K. and Pannekoek (1991) Biochemistry **30**, 1021–1028

Engh, R. A., Wright, H. T. and Huber, R. (1990) Protein Eng. **3**, 469–477

Esmon, C. T. (1987) Science **235**, 1348–1352

Esmon, C. T. (1989) J. Biol. Chem. **264**, 4743–4746

Esmon, C. T. and Owen, W. G. (1981) Proc. Natl. Acad. Sci. U. S. A. **78**, 2249–2252

Esmon, C. T., Esmon, N. L. and Harris, K. W. (1982a) J. Biol. Chem. **257**, 7944–7947

Esmon, N. L., Owen, W. G. and Esmon, C. T. (1982b) J. Biol. Chem. **257**, 859–864

Esmon, N. L., Carroll, R. C. and Esmon, C. T. (1983) J. Biol. Chem. **258**, 12238–12242

Farrell, D. H. and Cunningham, D. D. (1986) Proc. Natl. Acad. Sci. U. S. A. **83**, 6858–6852

Farrell, D. H. and Cunningham, D. D. (1987) Biochem. J. **245**, 543–550

Farrell, D. H., Wagner, S. L., Yuan, R. H. and Cunningham, D. D. (1988) J. Cell. Physiol. **134**, 179–188

Fay, W. P. and Owen, W. G. (1989) Biochemistry **28**, 5773–5778

Fenton, II, J. W. (1986) Ann. N. Y. Acad. Sci. **485**, 5–15

Fransson, L.-Å. (1985) in The Polysaccharides, Vol. III (Aspinall, G. O., ed.), pp. 337–415, Academic Press, New York

Fransson, L.-Å. (1987) Trends Biochem. Sci. **12**, 406–411

Fransson, L.-Å. (1989) in Heparin, Chemical and Biological Properties, Clinical Applications (Lane D. A. and Lindahl, U., eds.), pp. 115–133, Edward Arnold, London

Furie, B. and Furie, B. C. (1988) Cell **33**, 505–518

Gailani, D. and Broze, G. J., Jr. (1991) Science **253**, 909–912

Gallagher, J. T. (1989) Curr. Opin. Cell Biol. **1**, 1201–1218

Gallagher, J. T. and Walker, A. (1985) Biochem. J. **230**, 665–674

Gandrille, S., Aiach, M., Lane, D. A., Vidaud, D., Molho-Sabatier, P., Caso, R., deMoerloose, P., Fiessinger, J. N. and Clauser, E. (1990) J. Biol. Chem. **265**, 18997–19001

Geiger, M., Heeb, M. J., Binder, B. R. and Griffin, J. H. (1988) FASEB J. **2**, 2263–2267

Gettins, P. and Wooten, E. W. (1987) Biochemistry **26**, 4403–4408

Girard, T. J., Warren, L. A., Novotny, W. F., Likert, K. M., Brown, S. G., Miletich, J. P. and Broze, G. J., Jr. (1989) Nature (London) **338**, 518–520

Gomi, K., Zushi, M., Honda, G., Kawahara, S., Matsuzaki, O., Kanabayashi, T., Yamamoto, S., Maruyama, I. and Suzuki, K. (1990) Blood **75**, 1396–1399

Griffith, M. J. (1982) J. Biol. Chem. **257**, 7360–7365

Griffith, M. J. (1983) Proc. Natl. Acad. Sci. U. S. A. **80**, 5460–5464

Griffith, M. J., Noyes, C. M. and Church, F. C. (1985a) J. Biol. Chem. **260**, 2218–2225

Griffith, M. J., Noyes, C. M., Tyndall, J. A. and Church, F. C. (1985b) Biochemistry **24**, 6777–6782

Grootenhuis, P. D. J. and van Boeckel, C. A. A. (1991) J. Am. Chem. Soc. **113**, 2743–2747

Gurwitz, D. and Cunningham, D. D. (1988) Proc. Natl. Acad. Sci. U. S. A. **85**, 3440–3444

Gurwitz, D. and Cunningham, D. D. (1990) J. Cell. Physiol. **142**, 155–162

Hassell, J. R., Kimura, J. H. and Hascall, V. C. (1986) Annu. Rev. Biochem. **55**, 539–567

Hayashi, T., Zushi, M., Yamamoto, S. and Suzuki, K. (1990) J. Biol. Chem. **265**, 20156–20159

Heeb, M. J., Espana, F., Geiger, M., Collen, D., Stump, D. C. and Griffin, J. H. (1987) J. Biol Chem. **262**, 15813–15816

Heeb, M. J. and Griffin, J. H. (1988) J. Biol. Chem. **263**, 11613–11616

Heuck, C. C., Schiele, U., Horn, D., Fronda, D. and Tirz, E. (1985) J. Biol. Chem. **260**, 4598–4603

High, K. A. (1988) Arch. Pathol. Lab. Med. **112**, 28–36

Hirahara, K., Koyama, M., Matsuishi, T. and Kurata, M. (1990) Thromb. Res. **57**, 117–126

Hofsteenge, J., Taguchi, H. and Stone, R. S. (1986) Biochem. J. **237**, 243–251

Hofsteenge, J. and Stone, S. R. (1987) Eur. J. Biochem. **168**, 49–56

Hogg, P. J. and Jackson, C. M. (1989) Proc. Natl. Acad. Sci. U. S. A. **86**, 3619–3623

Hogg, P. J. and Jackson, C. M. (1990a) J. Biol. Chem. **265**, 241–247

Hogg, P. J. and Jackson, C. M. (1990b) J. Biol. Chem. **265**, 248–255

Holmer, E., Söderström, G. and Andersson, L.-O. (1979) Eur. J. Biochem. **93**, 1–5

Holmer, E., Lindahl, U., Bäckström, G., Thunberg, L., Sandberg, H., Söderström, G. and Andersson, L.-O. (1980) Thromb. Res. **18**, 861–869

Holmer, E., Kurachi, K. and Söderström, G. (1981) Biochem. J. **193**, 395–400

Höök, M., Björk, I., Hopwood, J. and Lindahl, U. (1976) FEBS Lett. **66**, 90–93

Höök, M., Lindahl, U. and Iverius, P.-H. (1974) Biochem. J. **137**, 33–43

Horner, A. A. (1986) Biochem. J. **240**, 171–179

Horner, A. A. (1990) Biochem. J. **266**, 553–559

Horner, A. A., Kusche, M., Lindahl, U. and Peterson, C. B. (1988) Biochem. J. **251**, 141–145

Hortin, G., Tollefsen, D. M. and Strauss, A. W. (1986) J. Biol. Chem. **261**, 15827–15830

Hortin, G. L., Tollefsen, D. M. and Benutto, B. M. (1989) J. Biol. Chem. **264**, 13979–13982

Hortin, G. L. and Benutto, B. M. (1990) Biochem. Biophys. Res. Commun. **169**, 437–442

Hovingh, P., Piepkorn, M. and Linker, A. (1986) Biochem. J. **237**, 573–581

Hoylaerts, M., Owen, W. G. and Collen, D. (1984) J. Biol. Chem. **259**, 5670–5677

Huber, R. and Carrell, R. W. (1989) Biochemistry **28**, 8951–8966

Hurst, R. E., Poon, M. C. and Griffith, M. J. (1983) J. Clin. Invest. **72**, 1042–1045

Inhorn, R. C. and Tollefsen, D. M. (1986) Biochem. Biophys. Res. Commun. **137**, 431–436

Ireland, H., Lane D. A., Thompson, E., Walker, I. D., Blench, I., Morris, H. R., Freyssinet, J. M., Grunebaun, L., Olds, R. and Thein, S. L. (1991) Br. J. Haematol. **79**, 70–74

Ishii, H., Salem, H. H., Bell, C. E., Laposata, E. A. and Majerus, P. W. (1986) Blood **67**, 362–365

Jackman, R. W., Beeler, D. L., VanDeWalter, L. and Rosenberg, R. D. (1986) Proc. Natl. Acad. Sci. U. S. A. **83**, 8834–8838

Jackman, R. W., Beeler, D. L., Fritze, L., Soff, G. and Rosenberg, R. D. (1987) Proc. Natl. Acad. Sci. U. S. A. **84**, 6425–6429

Jackson, R. L., Busch, S. J. and Cardin, A. D. (1991) Physiol. Rev. **71**, 481–539

Jakubowski, H. V., Kline, M. D. and Owen, W. G. (1986) J. Biol. Chem. **261**, 3876–3882

Jane, S. M., Mitchell, C. A., Hau, L. and Salem, H. H. (1989) J. Clin. Invest. **83**, 222–226

Jordan, R. E., Beeler, D. and Rosenberg, R. D. (1979) J. Biol. Chem. **254**, 2902–2913

Jordan, R. E., Oosta, G. M., Gardner, W. T. and Rosenberg, R. D. (1980) J. Biol. Chem. **255**, 10081–10090

Karlsson, K., Lindahl, U. and Marklund S. L. (1988) Biochem. J. **256**, 24–33

Kazama, Y., Niwa, M., Yamagishi, R., Takahashi, K., Sakuragawa, N. and Koide, T. (1987) Thromb. Res. **48**, 179–185

Kim, Y. S. and Linhardt, R. J. (1989) Thromb. Res. **53**, 55–71

Kisiel, W. (1979) J. Clin. Invest. **64**, 761–769

Kisiel, W., Canfield, W. M., Ericson, L. H. and Davie, E. W. (1977) Biochemistry **16**, 5824–5831

Kjellén, L. and Lindahl, U. (1991) Annu. Rev. Biochem. **60**, 443–475

Koide, T., Odani, S., Takahashi, K., Ono, T. and Sakuragawa, N. (1984) Proc. Natl. Acad. Sci. U. S. A. **81**, 289–293

Kojima, T., Leone, C. W., Marchildon, G. A., Marcum J. A. and Rosenberg, R. D. (1992a) J. Biol. Chem. **267**, 4859–4869

Kojima, T., Shworaki, N. W. and Rosenberg, R. D. (1992b) J. Biol. Chem. **267**, 4870–4877

Koyama, T., Parkinson, J. F., Sié, P., Bang, N. U., Müller-Berghaus, G. and Preissner, K. T. (1991) Eur. J. Biochem. **198**, 563–570

Kuhn, L. A., Griffin, J. H., Fisher, C. L., Greengard, J. S., Bouma, B. N., España, F. and Tainer, J. A. (1990) Proc. Natl. Acad. Sci. U. S. A. **87**, 8506–8510

Kurosawa, S. and Aoki, N. (1985) Thromb. Res. **37**, 353–364

Kurosawa, S., Galvin, J. B., Esmon, N. L. and Esmon, C. T. (1987) J. Biol. Chem. **262**, 2206–2212

Kurosawa, S., Stearns, D., Jackson, K. W. and Esmon, C. T. (1988) J. Biol. Chem. **263**, 5993–5996

Kusche, M. and Lindahl, U. (1990) J. Biol. Chem. **265**, 15403–15409

Kusche, M., Bäckström, G., Riesenfeld, J., Petitou, M., Choay, J. and Lindahl, U. (1988) J. Biol. Chem. **263**, 15474–15484

Kusche, M., Torri, G., Casu, B. and Lindahl, U. (1990) J. Biol. Chem. **265**, 7292–7300

Lam, L. H., Silbert, J. E. and Rosenberg, R. D. (1976) Biochem. Biophys. Res. Commun. **69**, 570–577

Lane, D. A. (1989) in Heparin, Chemical and Biological Properties, Clinical Applications, (Lane D. A. and Lindahl, U., eds.), pp. 363–390, Edward Arnold, London

Lane, D. A., Denton, J., Flynn, A. M., Thunberg, L. and Lindahl, U. (1984) Biochem. J. **218**, 725–732

Lane, D. A., Pejler, G., Flynn, A. M., Thompson, E. A. and Lindahl, U. (1986) J. Biol. Chem. **261**, 3980–3986

Lane, D. A., Flynn, A. A., Pejler, G., Lindahl, U., Choay, J. and Preissner, K. T. (1987) J. Biol. Chem. **262**, 16343–16348

Laug, W. E., Aebersold, R., Jong, A., Rideout, W., Bergman, B. L. and Baker, J. B. (1989) Thromb. Haemostasis **61**, 517–521

Laurell, M. and Stenflo, J. (1989) Thromb. Haemostasis **62**, 885–891

Laurent, T. C., Tengblad, A., Thunberg, L., Höök, M. and Lindahl, U. (1978) Biochem. J. **175**, 691–701

Lindahl, U. (1989) in Heparin, Chemical and Biological Properties, Clinical Applications, (Lane, D. A. and Lindahl, U., eds.), pp. 159–189, Edward Arnold, London

Lindahl, U., Bäckström, G., Thunberg, L. and Leder, I. G. (1980) Proc. Natl. Acad. Sci. U.S.A. **77**, 6551–6555

Lindahl, U., Bäckström, G. and Thunberg, L. (1983) J. Biol. Chem. **258**, 9826–9830

Lindahl, U., Thunberg, L., Bäckström, G., Riesenfeld, J., Nordling, K. and Björk, I. (1984) J. Biol. Chem. **259**, 12368–12376

Lindahl, U., Pejler, G., Bögwald, J. and Seljelid, R. (1989) Arch. Biochem. Biophys. **273**, 180–188

Lindahl, U. and Kjellén, L. (1991) Thromb. Haemostasis **66**, 44–48

Lindblom, A., Bengtsson-Olivecrona, G. and Fransson, L.-Å. (1991) Biochem. J. **279**, 821–829

Lollar, P. and Owen, W. G. (1980) J. Clin. Invest. **66**, 1222–1230

Lollar, P., MacIntosh, S. and Owen, W. G. (1984) J. Biol. Chem. **259**, 4335–4338

Low, D. A., Baker, J. B., Koonce, W. C. and Cunningham, D. D. (1981) Proc. Natl. Acad. Sci. U. S. A. **78**, 2340–2344

Lu, R., Esmon, N. L., Esmon, C. T. and Johnson, A. E. (1989) J. Biol. Chem. **264**, 12956–12962

MacGrogan, M., Kennedy, J., Li, M. P., Hsu, C., Scott, R., Simonsen, C. C. and Baker, J. B. (1988) Bio/Technology **6**, 172–177

Machovich, R. and Arányi, P. (1978) Biochem. J. **173**, 869–875

Machovich, R., Staub, M. and Patthy, L. (1978) Eur. J. Biochem. **83**, 473–477

Maimone, M. M. and Tollefsen, D. M. (1988) Biochem. Biophys. Res. Commun. **152**, 1052–1061

Maimone, M. M. and Tollefsen, D. M. (1990) J. Biol. Chem. **265**, 18263–18271

Marcum, J. A., McKenney, J. B. and Rosenberg, R. D. (1984) J. Clin. Invest. **74**, 341–350

Marcum, J. A. and Rosenberg, R. D. (1985) Biochem. Biophys. Res. Commun. **126**, 365–372

Marcum, J. A., Atha, D. H., Fritze, L. M. S., Nawroth, P., Stern, D. and Rosenberg, R. D. (1986) J. Biol. Chem. **261**, 7507–7515

Marlar, R. A. and Griffin, J. H. (1980) J. Clin. Invest. **66**, 1186–1189

Marlar, R. A., Kleiss, A. J. and Griffin, J. H. (1982) Blood, **59**, 1067–1072

Maruyama, I., Bell, C. E., and Majerus, P. W. (1985a) J. Cell. Biol. **101**, 363–371

Maruyama, I., Salem, H. H., Ishii, H. and Majerus, P. W. (1985b) J. Clin. Invest. **75**, 987–991

McIntosh, S. and Owen, W. G. (1987) Sanofi Thromb. Res. Found. **1**, 8–18

Meijers, J. C. M., Vlooswijk, R. A. A., Kanters, D. H A. J., Hessing, M. and Bouma, B. N. (1988) Blood **72**, 1401–1403

Molho-Sabatier, P., Aiach, M., Gaillaird, I., Fiessinger, J. N., Fischer, A. M., Chadeuf, G. and Clause, E. (1989) J. Clin. Invest. **84**, 1236–1242

Mottonen, J., Strand, A., Symersky, J., Sweet, R. M., Danley, D. E., Geoghegan, K. F., Gerard, R. D. and Goldsmith, E. J. (1992) Nature (London) **355**, 270–273

Mourey, L., Samama, J. P., Delarue, M., Choay, J., Lormeau, J. C., Petitou, M. and Moras, D. (1990) Biochimie **72**, 599–608

Munakata, H., Hsu, C. C., Kodama, C., Aikawa, J., Sakurada, M., Ototani, N., Isemura, M., Yosizawa, Z. and Hayashi, N. (1987) Biochim. Biophys. Acta **925**, 325–331

Murata, M., Ikeda, Y., Araki, Y., Murakami, H., Sato, K., Yamamoto, M., Watanabe, Y., Ando, Y., Igawa, T. and Maruyama, I. (1988) Thromb. Res. **50**, 647–656

Musci, G., Berliner, L. J. and Esmon, C. T. (1988) Biochemistry **27**, 769–773

Nawa, K., Sakano, K., Fujiwara, H., Sato, Y., Sugiyama, N., Teruuchi, T., Iwamoto, M. and Marumoto, Y. (1990) Biochem. Biophys. Res. Commun. **171**, 729–737

Nesheim, M. E. (1983) J. Biol. Chem. **258**, 14708–14717

Ofosu, F. A., Blajchman, M. A., Modi, G. J., Smith, L. M., Buchanan, M. R. and Hirsh, J. (1985) Br. J. Haematol. **60**, 695–705

Ofosu, F. A., Sié, P., Modi, G. J., Fernandez, F., Buchanan, M. R., Boneu, B. and Hirsh, J. (1987) Biochem. J. **243**, 579–588

Ofosu, F. A., Hirsh, J., Esmon, C. T., Modi, G. J. Smith, L. M., Anvari, N., Buchanan, M. R., Fenton, J. W., II and Blajchman, M. A. (1989) Biochem. J. **257**, 143–150

Ofosu, F. A., Choay, J., Anvari, N., Smith, L. M. and Blajchman, M. A. (1990) Eur. J. Biochem. **193**, 485–493

Okwusidi, J. I., Anvari, N., Kulczycky, M., Blajchman, M. A., Buchanan, M. R. and Ofosu, F. A. (1991) J. Lab. Clin. Med. **117**, 359–364

Olds, R. J., Lane, D. A., Boisclair, M., Sas, G., Bock, S. C. and Thein, S. L. (1992) FEBS Lett. **300**, 241–246

Olivecrona, T. and Bengtsson-Olivecrona, G. (1989) in Heparin, Chemical and Biological Properties, Clinical Applications (Lane, D. A. and Lindahl, U., eds.) pp. 335–361, Edward Arnold, London

Olson, S. T. (1985) J. Biol. Chem. **260**, 10153–10160

Olson, S. T. (1988a) J. Biol. Chem. **263**, 1698–1708

Olson, S. T. (1988b) J. Cell Biol. **107**, 827a

Olson, S T., Srinivasan, K. R., Björk, I. and Shore, J. D. (1981) J. Biol. Chem. **256**, 11073–11079

Olson, S. T. and Shore, J. D. (1986) J. Biol. Chem. **261**, 13151–13159

Olson, S. T. and Björk, I. (1992) in Thrombin: Structure and Function (Berliner, L. J., ed.), Plenum, New York, in the press

Olson, S. T. and Björk, I. (1991b) J. Biol. Chem. **266**, 6353–6364

Olson, S. T., Halvorson, H. R. and Björk, I. (1991) J. Biol. Chem. **266**, 6342–6352

Oosta, G. M., Gardner, W. T., Beeler, D. L. and Rosenberg, R. D. (1981) Proc. Natl. Acad. Sci. U. S. A. **78**, 829–833

Owen, W. G. and Esmon, C. T. (1981) J. Biol. Chem. **256**, 5532–5535

Owen, M. C., George, P. M., Lane, D. A. and Boswell, D. R. (1991) FEBS Lett. **280**, 216–220

Parker, K. A. and Tollefsen, D. W. (1985) J. Biol. Chem. **260**, 3501–3505

Parkinson, J. F., Grinnell, B. W., Moore, R. E., Hoskins, J., Vlahos, C. J. and Bang, N. U. (1990a) J. Biol. Chem. **265**, 12602–12610

Parkinson, J. F., Garcia, J. G. N. and Bang, N. U. (1990b) Biochem. Biophys. Res. Commun. **169**, 177–183

Parkinson, J. F., Bang, N. U. and Garcia, J. G. N. (1991a) Thromb. Haemostasis **65**, 825 (abstr.)

Parkinson, J. F., Vlahos, C. J., Yan, S. C. B. and Bang, N. U. (1991b) Thromb. Haemostasis **65**, 874 (abstr.)

Parkinson, J. F., Vlahos, C. J., Yan, S. C. B. and Bang, N. U. (1992a) Biochem. J. **283**, 151–157

Parkinson, J. F., Koyama, T., Bang, N. U. and Preissner, K. T. (1992b) in Heparin and Related Polysaccharides (Lane, D. A., Björk, I. and Lindahl, U., eds.), pp. 177–188. Plenum, New York

Patthy, L. (1988) J. Mol. Biol. **202**, 689–696

Pejler, G. and David, G. (1987) Biochem. J. **248**, 69–77

Pejler, G., Bäckström, G., Lindahl, U., Paulsson, M., Dziadek, M., Fujiwara, S. and Timpl, R. (1987b) J. Biol. Chem. **262**, 5036–5043

Pejler, G., Danielsson, Å., Björk, I., Lindahl, U., Nader, H. B. and Dietrich, C. P. (1987a) J. Biol. Chem. **262**, 11413–11421

Peterson, C. B. and Blackburn, M. N. (1987a) J. Biol. Chem. **262**, 7552–7558

Peterson, C. B. and Blackburn, M. N. (1987b) J. Biol. Chem. **262**, 7559–7566

Petitou, M. (1989) in Heparin: Chemical and Biological Properties, Clinical Applications (Lane, D. A. and Lindahl, U., eds.) pp. 65–79, Edward Arnold, London

Petitou, M., Duchaussoy, P., Lederman, I., Choay, J. and Sinaÿ, P. (1988a) Carbohydr. Res. **179**, 163–172

Petitou, M., Lormeau, J.-C. and Choay, J. (1988b) Eur. J. Biochem. **176**, 637–640

Petitou, M., Lormeau, J. C., Perly, B., Berthault, P., Bossennec, V., Sié, P. and Choay, J. (1988c) J. Biol. Chem. **263**, 8685–8690

Pomerantz, M. W. and Owen, W. G. (1978) Biochim. Biophys. Acta **535**, 66–77

Poole, A. R. (1986) Biochem. J. **236**, 1–14

Pratt, C. W., Tobin, R. B. and Church, F. C. (1990) J. Biol. Chem. **265**, 6092–6097

Pratt, C. W. and Church, F. C. (1992) J. Biol. Chem. **267**, 8789–8794

Preissner, K. T., Delvos, U. and Müller-Berghaus, G. (1987) Biochemistry **26**, 2521–2528

Preissner, K. T., Koyama, T., Müller, D., Tschopp, J. and Müller-Berghaus, G. (1990) J. Biol. Chem. **265**, 4915–4922

Preissner, K. T. and Jenne, D. (1991) Thromb. Haemostasis **66**, 123–132

Ragg, H. (1986) Nucleic Acids Res. **14**, 1073–1088

Ragg, H. and Preibisch, G. (1988) J. Biol. Chem. **263**, 12129–12134

Ragg, H., Ulshöfer, T. and Gerewitz, J. (1990a) J. Biol. Chem. **265**, 5211–5218

Ragg, H., Ulshöfer, T. and Gerewitz, J. (1990b) J. Biol. Chem. **265**, 22386–22391

Riesenfeld, J., Thunberg, L., Höök, M. and Lindahl, U. (1981) J. Biol. Chem. **256**, 2389–2394

Rodén, L. (1980) in The Biochemistry of Glycoproteins and Proteoglycans, (Lennarz, W. J., ed.), pp. 267–371, Plenum, New York

Rodén, L. (1989) in Heparin: Chemical and Biological Properties, Clinical Applications, (Lane, D. A. and Lindahl, U., eds.), pp. 1–23, Edward Arnold, London

Rodén, L., Koerner, T., Olson, C. and Schwartz, N. B. (1985) Fed. Proc. Fed. Am. Soc. Exp. Biol. **44**, 373–380

Rogers, S. J., Pratt, C. W., Whinna, H. C. and Church, F. C. (1992) J. Biol. Chem. **267**, 3613–3617

Rosenberg, R. D. (1977) Fed. Proc. Fed. Am. Soc. Exp. Biol. **36**, 10–18

Rosenberg, R. D. and Damus, P. S. (1973) J. Biol. Chem. **248**, 6490–6505

Rosenberg, R. D. and de Agostini, A. I. (1992) in Heparin and Related Polysaccharides (Lane, D. A., Björk, I. and Lindahl, U., eds.), pp. 307–316, Plenum, New York

Rosenblatt, D. E., Cotman, C. W., Nieto-Sampedro, M., Rowe, J. W. and Knauer, D. J. (1987) Brain Res. **415**, 40–48

Ruoslahti, E. (1988) Annu. Rev. Cell Biol. **4**, 229–255

Ruoslahti, E. (1989) J. Biol. Chem. **264**, 13369–13372

Rydel, T. J., Ravichandran, K. G., Tulinsky, A., Bode, W., Huber, R., Roitsch, C and Fenton, J. W. (1990) Science **249**, 277–280

Sakata, Y., Curriden, S., Lawrence, D., Griffin, J. M. and Loskutoff, D. J. (1985) Proc. Natl. Acad. Sci. U.S.A. **82**, 1121–1125

Salem, H. H., Maruyama, I., Ishii, H. and Majerus, P. W. (1984) J. Biol. Chem. **259**, 12246–12251

Sandset, P. M., Abildgaard, U. and Larsen, M. L. (1988) Thromb. Res. **50**, 803–813

Scott, R. W., Bergman, B. L., Bajpai, A., Hersh, R. T., Rodriguez, H., Jones, B. N., Barreda, C., Watts, S. and Baker, J. B. (1985) J. Biol. Chem. **260**, 7029–7034

Scully, M. F., Ellis, V., Seno, N. and Kakkar, V. V. (1986) Biochem. Biophys. Res. Commun. **137**, 15–22

Scully, M. F., Ellis, V., Seno, N. and Kakkar, V. V. (1988) Biochem. J. **254**, 547–551

Seldin, D. C., Seno, N., Austen, K. F., Stevens, R. L. (1984) Anal. Biochem. **141**, 291–300

Shah, N., Scully, M. F., Ellis, V. and Kakkar, V. V. (1990) Thromb. Res. **57**, 343–352

Shore, J. D., Olson, S. T., Craig, P. A., Choay, J. and Björk, I. (1989) Ann. N. Y. Acad. Sci. **556**, 75–80

Sié, P., Dupouy, D., Pichon, J. and Boneu, B (1985) Lancet **ii**, 414–416

Sié, P., Petitou, M., Lormeau, J. C., Dupouy, D., Boneu, B. and Choay, J. (1988) Biochim. Biophys. Acta **966**, 188–195

Skriver, K., Wikoff, W. R., Patston, P. A., Tausk, F., Schapira, M., Kaplan A. P. and Bock, S. C. (1991) J. Biol. Chem. **266**, 9216–9221

Smith, J. W., Dey, N. and Knauer, D. J. (1990) Biochemistry **29**, 8950–8957

Soff, G. A., Jackman, R. W. and Rosenberg, R. D. (1991) Blood **77**, 515–518

Stearns, D. J., Kurosawa, S. and Esmon, C. T. (1989) J. Biol. Chem. **264**, 3352–3356

Stein, P. E., Leslie, A. G. W., Finch, J. T., Turnell, W. G., McLaughlin, P. J. and Carrell, R. W. (1990) Nature (London) **347**, 99–102

Stern, D., Nawroth, P., Marcum, J. A., Handley, D., Kisiel, W., Rosenberg, R. D. and Stern, K. (1985) J. Clin. Invest. **75**, 272–279

Stenflo, J. (1976) J. Biol. Chem. **251**, 355–363

Stenflo, J. and Fernlund, P. (1982) J. Biol. Chem. **257**, 12180–12190

Stump, D. C., Thienpont, M. and Collen, D. (1986) J. Biol. Chem. **261**, 12759–12766

Sun, X. J. and Chang, J. Y. (1989) J. Biol. Chem. **264**, 11288–11293

Sun, X. J. and Chang, J. Y. (1990) Biochemistry **29**, 8957–8962

Suzuki, K. (1985) in Protein C, Biological and Medical aspects (Witt, I., ed.), pp. 43–58, Walter de Gruyter, Berlin, New York

Suzuki, K., Nishioka, J. and Hashimoto, S. (1983) J. Biol. Chem. **258**, 163–168

Suzuki, K., Nishioka, J., Kusumoto, H. and Hashimoto, S. (1984) J. Biochem. (Tokyo) **95**, 187–195

Suzuki, K., Kusumoto, H. and Hashimoto, S. (1986) Biochim. Biophys. Acta **882**, 343–352

Suzuki, K., Deyashiki, Y., Nishioka, J., Kurachi, K., Akira, M., Yamamoto, S. and Hashimoto, S. (1987a) J. Biol. Chem. **262**, 611–616

Suzuki, K., Kusumoto, H., Deyashiki, Y., Nishioka, J., Maruyama, I., Zushi, M., Kawahara, S., Honda, G., Yamamoto, S. and Horiguchi, S. (1987b) EMBO J. **6**, 1891–1897

Suzuki, K., Nishioka, J., Hayashi, T. and Kosaka, Y. (1988) J. Biochem. (Tokyo) **104**, 628–632

Suzuki, K., Hayashi, T., Nishioka, S., Kosaka, Y., Zushi, M., Honda, G. and Yamamoto, S. (1989) J. Biol. Chem. **264**, 4872–4876

Thunberg, L., Bäckström, G. and Lindahl, U. (1982) Carbohydr. Res. **100**, 393–410

Tollefsen, D. M. (1989) in Heparin: Chemical and Biological Properties, Clinical Applications, (Lane, D. A. and Lindahl, U., eds.) pp. 256–273, Edward Arnold, London

Tollefsen, D. M. (1990) Semin. Thromb. Haemostasis **16**, 162–168

Tollefsen, D. M., Majerus, D. W. and Blank, M. K. (1982) J. Biol. Chem. **257**, 2162–2169

Tollefsen, D. M., Petska, C. A. and Monafo, W. J. (1983) J. Biol. Chem. **258**, 6713–6716

Tollefsen, D. M., Peacock, M. E. and Monafo, W. J. (1986) J. Biol. Chem. **261**, 8854–8858

Tollefsen, D. M., Sugimori, T. and Maimone, M. M. (1990) Semin. Thromb. Haemostasis **16**, 66–70

Toulon, P., Moulonguet-Doloris, L., Costa, J. M. and Aiach, M. (1991) Thromb. Haemostasis **65**, 20–24

Tran, T. H., Marbet, G. A. and Duckert, F. (1985) Lancet **ii**, 413–414

Travis, J. and Salvesen, G. S. (1983) Annu. Rev. Biochem. **52**, 655–709

Van Deerlin, V. M. D. and Tollefsen, D. M. (1991) J. Biol. Chem. **266**, 20223–20231

Vu, T.-K. V., Hung, D. T., Wheaton, V. I. and Coughlin, S. R. (1991) Cell **64**, 1057–1068

Wagner, S. L., Lau, A. L. and Cunningham, D. D. (1989) J. Biol. Chem. **264**, 611–615

Wagner, S. L., Lau, A. L., Nguyen, A., Mimuro, J., Loskutoff, D. J., Isackson, P. J. and Cunningham, D. D. (1991) J. Neurochem. **56**, 234–242

Walker, F. J. (1980) J. Biol. Chem. **255**, 5521–5524

Walker, F. J., Sexton, P. W. and Esmon, C. T. (1979) Biochim. Biophys. Acta **571**, 333–342

Walker, F. J., Chavin, S. I. and Fay, P. J. (1987) Arch. Biochem. Biophys. **252**, 322–328

Weisdorf, D. J. and Edson, J. R. (1991) Br. J. Haematol. **77**, 125–126

Wen, D., Dittman, W. A., Ye, R. D., Deaven, L. L., Majerus, P. W. and Sadler, J. E. (1987) Biochemistry **26**, 4350–4357

Whinna, H. C., Blinder, M. A., Szewczyk, M., Tollefsen, D. M. and Church, F. C. (1991) J. Biol. Chem. **266**, 8129–8135

Winnard, P. T., Esmon, C. T. and Laue, T. M. (1989) Arch. Biochem. Biophys. **239**, 339–344

Wong, V. L. Y., Bready, J., Berliner, J., Cancilla, P. A. and Fisher, M. (1991) Thromb. Haemostasis **65**, 947 (abstr.)

Wu, Q., Sheehan, J. P., Tsiang, M., Lentz, W. R., Birktoft, J. J. and Sadler, J. E. (1991) Proc. Natl. Acad. Sci. U. S. A. **88**, 6775–6779

Wun, T.-C., Kretzmer, K. K., Girard, T. J., Miletich, J. P. and Broze, G. J., Jr. (1988) J. Biol. Chem. **263**, 6001–6004

Zushi, M., Gomi, K., Yamamoto, S., Maruyama, I., Hayashi, T. and Suzuki, K. (1989) J. Biol. Chem. **264**, 10351–10353

Reviews published in the *Biochemical Journal* 1990–1992

Enzymes

5′-Nucleotidase: molecular structure and functional aspects
H. Zimmermann (1992) **285**, 345–365

Gene structure and expression

On the biological role of histone acetylation
A. Csordas (1990) **265**, 23–38

DNA methylation. The effect of minor bases on DNA–protein interactions
R.L.P. Adams (1990) **265**, 309–320

Eukaryotic transcription factors
D.S. Latchman (1990) **270**, 281–289

Function and regulation of expression of pulmonary surfactant-associated proteins
T.E. Weaver and J.A. Whitsett (1991) **273**, 249–264

Expression and interactions of human adenovirus oncoproteins
P.A. Boulanger and G.E. Blair (1991) **275**, 281–299

Structural aspects of protein–DNA recognition
P.S. Freemont, A.N. Lane and M.R. Sanderson (1991) **278**, 1–23

Xenopus transcription factors: key molecules in the developmental regulation of differential gene expression
A.P. Wolffe (1991) **278**, 313–324

Regulation of gene expression by insulin
R.M. O'Brien and D.K. Granner (1991) **278**, 609–619

Phenobarbital induction of cytochrome *P*-450 gene expression
D.J. Waxman and L. Azaroff (1992) **281**, 577–592

Factors controlling the expression of mouse mammary tumour virus
W.H. Günzburg and B. Salmons (1992) **283**, 625–632

Nuclear protein phosphorylation and growth control
D.W. Meek and A.J. Street (1992) **287**, 1–15

Cell biology and development

Interleukin-6 and the acute phase response
P.C. Heinrich, J.V. Castell and T. Andus (1990) **265**, 621–636

Prion liposomes
R. Gabizon and S.B. Prusiner (1990) **266**, 1–14

Scavenger functions of the liver endothelial cell
B. Smedsrød, H. Pertoft, S. Gustafson and T.C. Laurent (1990) **266**, 313–327

The role of transferrin in the mechanism of cellular iron uptake
K. Thorstensen and I. Romslo (1990) **271**, 1–10

Mammalian glycosylation mutants as tools for the analysis and reconstitution of protein transport
A.W. Brändli (1991) **276**, 1–12

The *Ah* receptor and the mechanism of dioxin toxicity
J.P. Landers and N.J. Bunce (1991) **276**, 273–287

Interaction between mRNA, ribosomes and the cytoskeleton
J.E. Hesketh and I.F. Pryme (1991) **277**, 1–10

Targeting proteins to mitochondria: a current overview
L.A. Glover and J.G. Lindsay (1992) **284**, 609–620

Import of proteins into peroxisomes and other microbodies
M.J. de Hoop and G. Ab (1992) **286**, 657–669

Proteins

γ-Carboxyglutamate-containing proteins and the vitamin K-dependent carboxylase
C. Vermeer (1990) **266**, 625–636

The prolamin storage proteins of cereal seeds: structure and evolution
P.R. Shewry and A.S. Tatham (1990), **267**, 1–12

Protein folding
T.E. Creighton (1990) **270**, 1–16

Receptors and signal transduction

Membranes and bioenergetics

Regulation of metabolism

Carbohydrates and lipids